Ethnobotany and Ethnopharmacology of Medicinal and Aromatic Plants

Medicinal and aromatic plants are beneficial to human health. Plant-derived molecules possess biological activities that can be used to prevent many infectious diseases and metabolic disorders. *Ethnobotany and Ethnopharmacology of Medicinal and Aromatic Plants* summarizes techniques and methods used to study the biological activities of plant-derived extracts and compounds to study ethnobotanical and ethnopharmacological features of medicinal and aromatic plants.

This book:

- Includes computational approaches to study the pharmacological properties of biomolecules in medicinal and aromatic plants.
- Details methods in ethnopharmacology including chromatographical and analytical techniques.
- Demonstrates trends in sustainable use and management of medicinal and aromatic plants.
- Features information on databases and tools used in computational phytochemistry for drug designing and discovery.
- Elucidates the importance of phytochemicals as immunomodulators in herbal drug development including their nanoformulations.

A volume in the Exploring Medicinal Plants series, *Ethnobotany and Ethnopharmacology of Medicinal and Aromatic Plants* will be of interest to those working with plant extracts, including botanists and ethnobotanists, pharmacologists and ethnopharmacologists, as well as scientists and researchers interested in natural compounds and their potential applications.

Exploring Medicinal Plants

Series Editor:
Azamal Husen
Wolaita Sodo University, Ethiopia

Medicinal plants render a rich source of bioactive compounds used in drug formulation and development; they play a key role in traditional or indigenous health systems. As the demand for herbal medicines increases worldwide, supply is declining as most of the harvest is derived from naturally growing vegetation. Considering global interests and covering several important aspects associated with medicinal plants, the Exploring Medicinal Plants series comprises volumes valuable to academia, practitioners, and researchers interested in medicinal plants. Topics provide information on a range of subjects including diversity, conservation, propagation, cultivation, physiology, molecular biology, growth response under extreme environment, handling, storage, bioactive compounds, secondary metabolites, extraction, therapeutics, mode of action, and healthcare practices.

Led by Azamal Husen, Ph.D., this series is directed to a broad range of researchers and professionals consisting of topical books exploring information related to medicinal plants. It includes edited volumes, references, and textbooks available for individual print and electronic purchases.

Sustainable Uses of Medicinal Plants
Learnmore Kambizi and Callistus Bvenura

Medicinal Plant Responses to Stressful Conditions
Arafat Abdel Hamed Abdel Latef

Aromatic and Medicinal Plants of Drylands and Deserts: Ecology, Ethnobiology and Potential Uses
David Ramiro Aguillón Gutiérrez, Cristian Torres León, and Jorge Alejandro Aguirre Joya

Secondary Metabolites from Medicinal Plants: Nanoparticles Synthesis and their Applications
Rakesh Kumar Bachheti, Archana Bachheti

Aquatic Medicinal Plants
Archana Bachheti, Rakesh Kumar Bachheti, and Azamal Husen

Antidiabetic Medicinal Plants and Herbal Treatments
Azamal Husen

Ethnobotany and Ethnopharmacology of Medicinal and Aromatic Plants: Steps Towards Drugs Discovery
Mohd Adnan, Mitesh Patel and Mejdi Snoussi

Wild Mushrooms and Health Diversity, Phytochemistry, Medicinal Benefits, and Cultivation
Kamal Ch. Semwal, Steve L. Stephenson, and Azamal Husen

Ethnobotany and Ethnopharmacology of Medicinal and Aromatic Plants

Steps Towards Drug Discovery

Edited by
Mohd Adnan, Mitesh Patel, and Mejdi Snoussi

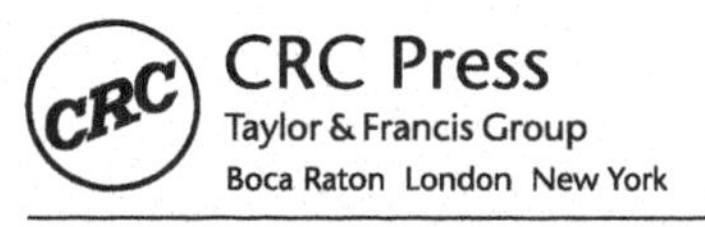

CRC Press is an imprint of the
Taylor & Francis Group, an **informa** business

Designed cover image: © Shutterstock

First edition published 2024
by CRC Press
6000 Broken Sound Parkway NW, Suite 300, Boca Raton, FL 33487-2742

and by CRC Press
4 Park Square, Milton Park, Abingdon, Oxon, OX14 4RN

CRC Press is an imprint of Taylor & Francis Group, LLC

ISBN: 9781032256085 (hbk)
ISBN: 9781032256092 (pbk)
ISBN: 9781003284215 (ebk)

DOI: 10.1201/b22842

Typeset in Times
by codeMantra

Contents

About the Editors

Dr. Mohd Adnan is an Associate Professor at Department of Biology, College of Science, University of Ha'il, Saudi Arabia. He did his Ph.D. from University of Central Lancashire, UK; Master's and Bachelor's degree from Bangalore University, India; Post-Graduate Diploma in Bioinformatics from SJAIT, India. He has more than 10 years of research, teaching, and administrative experience. In his professional capacity, he has received various travel, observership, and research grants as a Principal and Co-investigator from various prestigious organizations. He has successfully published 140+ publications in internationally recognized peer-reviewed reputed journals, several book chapters for internationally renowned publishers, and presented many papers and posters in various conferences/workshops globally. He has published widely in the field of phytomedicine, biofilms, drug discovery, natural products, nutraceuticals, and functional foods with specialization in plant-based antibiofilm and anticancer agents, microbial biosurfactants, biofilms in food industry and medical settings, probiotics and cancer biology, and novel biomolecules for health and as antimicrobial agents.

He has acted as a referee/reviewer for 70+ internationally recognized peer-reviewed journals and grant reviewer for many prestigious universities and organizations. In addition, he is a member of SFAM, UK and ESCMID, Switzerland, and an Elected Member of Royal Society of Biology, UK. He currently holds an Associate, Academic, Review, and Guest Editor positions in various reputed journals. Dr. Adnan is recently (September 2022) listed among the Top 2% Scientists in the World.

Dr. Mitesh Patel has completed his Bachelor's from Veer Narmad South Gujarat University, Master's from Uka Tarsadia University, and Doctorate from Veer Narmad South Gujarat University, India. Dr. Patel has worked on different aspects of biology with a multidisciplinary thinking specifically on natural product research, pharmacological applications of natural products, plant growth promotion, and systematic studies of fungi and pteridophytes. Within a short span of time and during his doctoral studies, Dr. Patel has described five new species and one new variety of plant. Dr. Patel discovered the "World's Smallest Terrestrial Pteridophyte". He has successfully published 60+ publications in internationally recognized peer-reviewed journals and several book chapters for internationally renowned publishers. Dr. Patel has received BSR and NF-OBC fellowships from India and conducted several research projects as a consultant, principal, and co-investigator sponsored by various eminent organizations. He has been on the panel of academic editor, guest editor, and reviewers of several reputed journals and a member of ESCMID, Switzerland, and Indian Fern Society, India.

Dr. Mejdi Snoussi is an Associate Professor graduated from the Faculty of Science of Sfax, University of Sfax, Tunisia, with a Bachelor's degree in Natural Sciences in June 2001. He completed his Master's degree in Biology and Health in October 2005 and his Ph.D. in Biological Sciences and Biotechnology in December 2009 from the University of Monastir, Tunisia. Previously, he held several posts including Assistant Professor in Water and Technology Center (Tunisia), Associate Professor in High Institute of Biotechnology in Monastir (Tunisia), Post-Doctoral Fellow in Valencia (Spain), CICOPS-Research Fellow in the Department of Drug Sciences, University of Pavia (Italy), and Research Fellow in Department of Pharmacy, University of Salerno (Italy). Currently, he is working as an Associate Professor in the Department of Biology, College of Science, University of Ha'il, Saudi Arabia. Over the past 16 years of his career, Dr. Snoussi has published more than 180 scientific papers in peer-reviewed reputed journals in the field of Microbiology/Plant Science and

Phytochemistry, one national patent, and one book chapter. Dr. Snoussi conducted several research projects as a consultant and co-investigator in the field of Microbiology, Phytochemistry, and Plant Science. He is a referee in several international journals, member of the Editorial Board and Guest Editor in *Frontiers in Microbiology*, *Environmental Health and Exposome*, *Canadian Journal of Infectious Diseases and Medical Microbiology*, *BioMed Research International*, and *Molecules*.

Contributors

Sohail Ahmad
Gomal Center of Biochemistry and Biotechnology, Gomal University
D.I. Khan, Pakistan

Salman Akhtar
Department of Bioengineering, Integral University
Lucknow, Uttar Pradesh, India

Manzar Alam
Centre for Interdisciplinary Research in Basic Sciences, Jamia Millia Islamia
New Delhi, India

Alexios Alexopoulos
Department of Agriculture, University of the Peloponnese
Antikalamos, Kalamata, Messinia, Greece

Shaheen Ali
Department of Biotechnology, School of Chemical and Life Sciences
Jamia Hamdard, New Delhi, India

Sami G. Almalki
Department of Medical Laboratory Sciences, College of Applied Medical Sciences, Majmaah University
Al-Majmaah, Saudi Arabia

Mousa Alreshidi
Department of Biology, College of Sciences, University of Ha'il
Ha'il, Saudi Arabia

Suliman A. Alsagaby
Department of Medical Laboratory Sciences, College of Applied Medical Sciences, Majmaah University
Al-Majmaah, Saudi Arabia

Irfan Ahmad Ansari
IIRC1, Department of Biosciences, Integral University
Lucknow, India

Uzair Ahmad Ansari
System Toxicology & Health Risk Assessment Group, CSIR-Indian Institute of Toxicology Research (CSIR-IITR), Vishvigyan Bhavan, 31, Mahatma Gandhi Marg
Lucknow, Uttar Pradesh, India
and
CSIR-Human Resource Development Centre (CSIR-HRDC) Campus, Academy of Scientific and Innovative Research (ACSIR), Postal Staff College Area
Kamla Nehru Nagar, Ghaziabad, Uttar Pradesh, India

Syed Amir Ashraf
Department of Clinical Nutrition, College of Applied Medical Sciences, University of Ha'il
Ha'il, Saudi Arabia

Abdelhamid Azeroual
Agri-Food and Health Laboratory (AFHL), Faculty of Sciences and Techniques of Settat, Hassan First University
Settat, Morocco

Riadh Badraoui
Department of Biology, University of Ha'il
Ha'il, Saudi Arabia
and
Section of Histology-Cytology, Faculty of Medicine of Tunis, University of Tunis El Manar
La Rabta-Tunis, Tunisia

Fevzi Bardakci
Department of Biology, University of Ha'il
Ha'il, Saudi Arabia

Hmed Ben-Nasr
Department of Biology, Faculty of Science of Gafsa, University of Gafsa
Gafsa, Tunisia

Urvashi Bhardwaj
Department of Biochemistry, School of Chemical and Life Sciences
Jamia Hamdard, New Delhi, India

Valeria Cavalloro
Department of Drug Sciences, University of Pavia
Pavia, Italy
and
Department of Earth and Environmental Sciences, University of Pavia
Pavia, Italy

Noureddine Chaachouay
Interdisciplinary Research Laboratory in the Sciences, Education, and Training (IRLSET), Hassan First University
Settat, Morocco

Simona Collina
Department of Drug Sciences, University of Pavia
Pavia, Italy

Raffaella Colombo
Drug Sciences Department, University of Pavia
Pavia, Italy

Abd Elmoneim O. Elkhalifa
Department of Clinical Nutrition, College of Applied Medical Sciences, University of Ha'il
Ha'il, Saudi Arabia

Maria Faraz
Gomal Center of Biochemistry and Biotechnology, Gomal University
D.I. Khan, Pakistan

Arshad Farid
Gomal Center of Biochemistry and Biotechnology, Gomal University
D.I. Khan, Pakistan

Shireen Fatima
IIRC1, Department of Biosciences, Integral University
Lucknow, India

Ana I. Faustino-Rocha
School of Sciences and Technology, University of Évora
Évora, Portugal
and
Center for the Research and Technology of Agro-Environmental and Biological Sciences (CITAB), Inov4Agro
Vila Real, Portugal
and
Comprehensive Health Research Center (CHRC)
Évora, Portugal

Ilaria Frosi
Drug Sciences Department, University of Pavia
Pavia, Italy

Shakira Ghazanfar
National Institute for Genomics Advanced Biotechnology (NIGAB) National Agricultural Research Centre
Islamabad, Pakistan

Kyriakos Giannoulis
Department of Agriculture, Crop Production and Rural Environment, University of Thessaly
Volos, Greece

Walid Sabri Hamadou
Department of Biology, College of Sciences, University of Ha'il
Ha'il, Saudi Arabia
and
Department of Molecular Biology and Genetics, Faculty of Science and Literature, Kilis 7 Aralik University
Kilis, Turkey

Stephano Hanolo Mlozi
Department of Chemistry, University of Dar es Salaam, Mkwawa University College of Education
Iringa, Tanzania

Ashanul Haque
Department of Chemistry, College of Science, University of Ha'il
Ha'il, Saudi Arabia

Syed Imran Hassan
Department of Chemistry, College of Science, Sultan Qaboos University
Seeb, Oman

Md Imtaiyaz Hassan
Centre for Interdisciplinary Research in Basic Sciences, Jamia Millia Islamia
New Delhi, India

Danish Iqbal
Department of Medical Laboratory Sciences, College of Applied Medical Sciences, Majmaah University
Al-Majmaah, Saudi Arabia

Sadaf Jahan
Department of Medical Laboratory Sciences, College of Applied Medical Sciences, Majmaah University
Al Majma'ah, Saudi Arabia

Imran Khan
Department of Chemistry, College of Science, Sultan Qaboos University
Seeb, Oman

Iqra Khan
Department of Bioengineering, Integral University
Lucknow, Uttar Pradesh, India

Jahanarah Khatoon
Department of Biosciences, Integral University
Lucknow, Uttar Pradesh, India

Mohammad Kalim Ahmad Khan
Department of Bioengineering, Integral University
Lucknow, Uttar Pradesh, India

Sanjeev Kumar
Department of Environmental Sciences, School of Natural Resource Management, Central University of Jharkhand
Brambe, Ranchi, India

Pasquale Linciano
Department of Drug Sciences, University of Pavia
Pavia, Italy

Danish Mahmood
Department of Pharmacology and Toxicology, Unaizah College of Pharmacy, Qassim University
Qassim, Saudi Arabia

Emanuela Martino
Department of Earth and Environmental Sciences, University of Pavia
Pavia, Italy

Beatriz Medeiros-Fonseca
Center for the Research and Technology of Agro-Environmental and Biological Sciences (CITAB), Inov4Agro
Vila Real, Portugal

Chiara Milanese
Department of Chemistry, University of Pavia
Pavia, Italy

Prakriti Mishra
IIRC1, Department of Biosciences, Integral University
Lucknow, India

Taj Mohammad
Centre for Interdisciplinary Research in Basic Sciences, Jamia Millia Islamia
Jamia Nagar, New Delhi, India

Shouvik Mukherjee
Department of Biotechnology, School of Chemical and Life Sciences
Jamia Hamdard, New Delhi, India

Md. Mushtaque
Department of Chemistry, Samastipur College (L.N. Mithila University)
Samastipur, Bihar, India

Mohd Nehal
Department of Bioscience, Integral University
Lucknow, Uttar Pradesh, India

Paula A. Oliveira
Center for the Research and Technology of Agro-Environmental and Biological Sciences (CITAB), Inov4Agro
Vila Real, Portugal
and
Department of Veterinary Sciences, University of Trás-os-Montes and Alto Douro
Vila Real, Portugal

Adele Papetti
Drug Sciences Department, University of Pavia
Pavia, Italy

Mirav Patel
Department of Biotechnology, Parul Institute of Applied Sciences and Centre of Research for Development, Parul University
Vadodara, India

Spyridon A. Petropoulos
Department of Agriculture, Crop Production and Rural Environment, University of Thessaly
Volos, Greece

Maria J. Pires
Center for the Research and Technology of Agro-Environmental and Biological Sciences (CITAB), Inov4Agro
Vila Real, Portugal
and
Department of Veterinary Sciences, University of Trás-os-Montes and Alto Douro
Vila Real, Portugal

Md. Ataur Rahman
Experimental Research Building, Department of Chemistry, New York University Abu Dhabi
Abu Dhabi, United Arab Emirates

Daniela Rossi
Department of Drug Sciences, University of Pavia
Pavia, Italy

Mohd Saeed
Department of Biology, College of Sciences, University of Ha'il
Ha'il, Saudi Arabia

Ilma Shakeel
Centre for Interdisciplinary Research in Basic Sciences, Jamia Millia Islamia
Jamia Nagar, New Delhi, India

Anas Shamsi
Centre for Interdisciplinary Research in Basic Sciences, Jamia Millia Islamia
New Delhi, India

Amit Shrivastav
Vasu Research Centre, Makarpura GIDC
Vadodara, Gujarat, India

Arif Jamal Siddiqui
Department of Biology, University of Ha'il
Ha'il, Saudi Arabia

Ritu Singh
Department of Environmental Science, School of Earth Sciences, Central University of Rajasthan
Ajmer, Rajasthan, India

Malvi Surti
Bapalal Vaidya Botanical Research Centre, Department of Biosciences, Veer Narmad South Gujarat University
Surat, India

Bektas Tepe
Department of Molecular Biology and Genetics, Faculty of Science and Literature, Kilis 7 Aralik University
Kilis, Turkey

Helena Vala
Center for the Research and Technology of Agro-Environmental and Biological Sciences (CITAB), Inov4Agro
Vila Real, Portugal
and
CERNAS-IPV Research Centre, Polytechnic Institute of Viseu
Viseu, Portugal
and
Agrarian School of Viseu, Polytechnic Institute of Viseu
Viseu, Portugal

Daniela Vallelonga
Drug Sciences Department, University of Pavia
Pavia, Italy

Akash Vanzara
Vasu Research Centre, Makarpura GIDC
Vadodara, Gujarat, India

Cármen Vasconcelos-Nóbrega
Center for the Research and Technology of Agro-Environmental and Biological Sciences (CITAB), Inov4Agro
Vila Real, Portugal
and
Department of Veterinary Sciences, University of Trás-os-Montes and Alto Douro
Vila Real, Portugal

Mahima Verma
IIRC1, Department of Biosciences, Integral University
Lucknow, India

Lahcen Zidane
Plant, Animal Productions and Agro-industry Laboratory, Department of Biology, Faculty of Sciences, Ibn Tofail University
Kenitra, Morocco

1 Introduction to Medicinal and Aromatic Plants

Diversity, Biogeographic Distribution and Conservation Status

Malvi Surti
Veer Narmad South Gujarat University

Mirav Patel
Parul University

CONTENTS

1.1 INTRODUCTION

The definition of "medicinal plants" can be broadened to include all plants and herbs possessed of therapeutic properties or that have been shown to exert a beneficial pharmacological effect on the body of animals or humans (Bajaj 2012), whereas the term "aromatic plant" refers to a special kind of plant that is used for its flavours and aromas. A number of these plants are also used as medicines for a variety of ailments. A wide variety of aromatic compounds can be found in different parts of a plant (Rao, Palada and Becker 2004). In general, the term "Medicinal and Aromatic Plants" (MAPs) refers to botanical raw materials, also called herbal drugs, used as components of

DOI: 10.1201/b22842-1

cosmetics and medicinal products and other natural health products. MAPs are primarily utilized as therapeutic, aromatic or culinary ingredients (Lubbe and Verpoorte 2011). There are a number of MAPs that represent a significant segment of the flora, providing raw materials that are used in the pharmaceutical, beauty, and for the production of other natural products. Medicinal plants, also known as medicinal herbs, are plants that possess therapeutic properties or exert pharmacological effects that are beneficial to the human or animal body in some way (Miguel 2010). Aromatic plants are plants whose aroma is pleasing to the human or animal senses. The aroma of aromatic plants is used extensively as spices, flavouring agents, perfumery and medicine, all of which have a wide range of uses. Additionally, these plants also serve as a major source of raw materials for the production of a number of important industrial chemicals (Costa et al. 2015).

It is generally accepted that plants that possess medicinal and aromatic properties, which are frequently used in pharmacy and/or perfumery, should be defined as medicinal and aromatic plants; however, as many medicinal and aromatic plants are also utilized in cosmetics, it might be better to call them medicinal, aromatic and cosmetic plants (Fierascu et al. 2021). In order to be classified as aromatic, a plant must contain aromatic compounds – basically volatile essential oils that are released, when the plant is in its natural state. It is well-known that essential oils are highly concentrated compounds with an odour, volatile properties and hydrophobic properties. Among the sources of these substances are roots, wood, bark, twigs, leaves, buds, flowers, fruits and seeds (Christaki et al. 2020). There are a number of complex mixtures of secondary metabolites found in essential oils, the majority of which consist of low-boiling-point phenylpropenes and terpenes. Aromatherapy and health care are two of the many industries where essential oils are widely used due to their characteristic flavour and fragrance properties, as well as their biological properties (Skendi et al. 2022). They have also been used extensively in spices, flavouring and fragrance, herbal beverages, pesticides, repellents and cosmetics. Some of the most popular examples of essential oils include eucalyptus, spearmint, ajwain, lemongrass, turpentine, grapefruit oil, mentha, patchouli, palmarosa and citronella (Aftab and Hakeem 2021) (Table 1.1).

There has been a long history of the use of MAPs for medicinal purposes, dating back well before the prehistoric period. The use of MAPs has been described in ancient Unani manuscripts as well as Egyptian papyrus and Chinese writings (Giannenas et al. 2020). Since over 4,000 years ago, the practice of MAPs medicine has been used as a form of treatment in Europe and throughout the Mediterranean region by different medicinal practitioners. MAPs have been traditionally used as part of the healing rituals of indigenous cultures throughout the world, such as Rome, Egypt, Iran, Africa and America, and there are a number of traditional systems of medicine that also incorporate herbal therapies into their traditional medicine systems (Faydaoğlu and Sürücüoğlu 2011). As far as traditional systems of medicine are concerned, they continue to be widely practiced on a variety of levels. In recent years, a number of factors have led to the increasing use of plant materials as a source of medicines for a wide variety of human ailments. These factors include an increase in population, insufficient supply of drugs, high prices for treatments, side effects of several synthetic drugs, and the emergence of resistance to current medicines used for infectious diseases (Máthé 2015).

Recent statistics by WHO estimates that 80% of the population of the world relies on herbal medicines for one or more aspects of their primary healthcare needs. There are approximately 21,000 species of plants that have the potential to be used as medicinal plants according to the WHO (Schippmann, Leaman and Cunningham 2002). There are more than 30% of all plant species on earth that have been used in some way or another for medicinal purposes at some point in time. In developed countries such as the United States, there has been an estimate that plant medicines contribute to as much as 25% of the total amount of medicine that is prescribed, while in fast developing countries such as India and China, this number could reach as high as 80% (Farnsworth and Soejarto 1991). Due to this fact, in countries like India, the economic importance of medicinal plants is much greater than in other countries in the world. Several of these countries are the source

TABLE 1.1
List of Several Important Medicinal and Aromatic Plants, Their Origin and Uses

Medicinal and Aromatic Plants	Place of Origin	Medicinal Uses	Drug Derived from MAPs
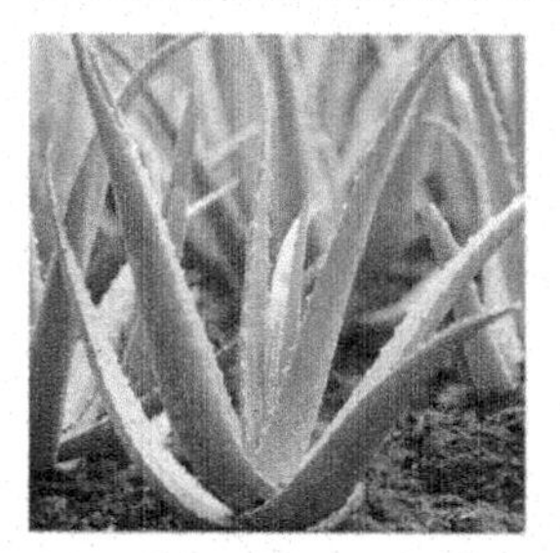 *Aloe vera*	South Africa, Arabian Peninsula	In wounds and burn treatment	Aloe
 *Salix babylonica*	Europe	Good heart health, thinning of blood, pain relief	Aspirin
Cimicifuga racemosa	Eastern United States and Canada	Hormonal disorder treatment, rheumatic disorder treatment	Black cohosh
 *Sanguinaria canadensis*	Southeastern United States	Cancer and skin disorder treatment	Bloodroot
 *Cinnamomum camphora*	Asia	Relief in rheumatic pain	Camphor

(*Continued*)

TABLE 1.1 (*Continued*)
List of Several Important Medicinal and Aromatic Plants, Their Origin and Uses

Medicinal and Aromatic Plants	Place of Origin	Medicinal Uses	Drug Derived from MAPs
 *Papaver somniferum*	Southeastern Europe, Western Asia	Cough suppression, pain relief	Codeine
 *Colchicum autumnale*	Eurasia	Gouty arthritis and cancer treatment	Colchicine
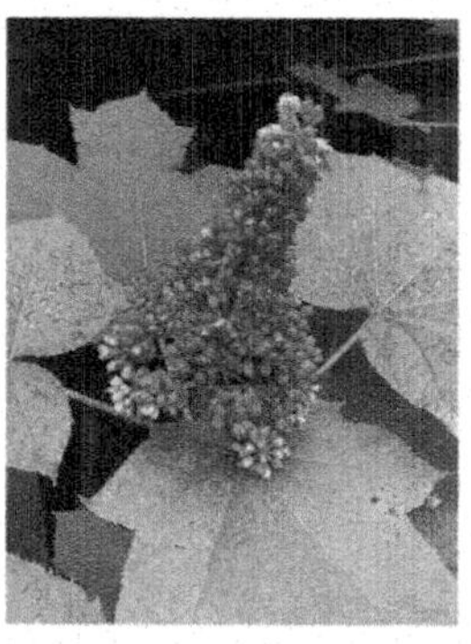 *Oplopanax horridus*	Western North America	Cure infection, tuberculosis and diabetes treatment	Devil's club shrub
 *Digitalis purpurea*	Europe	Cardiac arrest treatment	Digitalis

(*Continued*)

TABLE 1.1 (*Continued*)
List of Several Important Medicinal and Aromatic Plants, Their Origin and Uses

Medicinal and Aromatic Plants	Place of Origin	Medicinal Uses	Drug Derived from MAPs
 *Oenothera biennis*	North America	Atopic dermatitis, cancer and nerve damage treatment	Gamma-linolenic acid
 *Hoodia gordonii*	South Africa	Reduce weight	Hoodia
Camptotheca acuminata	China	Lung, ovarian and colon cancer treatment	Irinotecan, Topotecan
 *Catharanthus roseus*	Madagascar	Treatment of Hodgkin's lymphoma and leukaemia	Madagascar periwinkle

(*Continued*)

TABLE 1.1 (*Continued*)
List of Several Important Medicinal and Aromatic Plants, Their Origin and Uses

Medicinal and Aromatic Plants	Place of Origin	Medicinal Uses	Drug Derived from MAPs
 *Ephedra sinica*	China	Stuffy nose or nasal congestion, allergies, bronchodilation	Pseudoephedrine (Sudafed)
 *Pilocarpus pennatifolius*	Amazon Basin	Chronic glaucoma treatment	Pilocarpine
 *Cinchona pubescens*	South America	Heart and Malaria disease treatment	Quinine, quinidine
 *Datura stramonium*	Americas	Motion sickness, sedation	Scopolamine

(*Continued*)

TABLE 1.1 (*Continued*)
List of Several Important Medicinal and Aromatic Plants, Their Origin and Uses

Medicinal and Aromatic Plants	Place of Origin	Medicinal Uses	Drug Derived from MAPs
 *Illicium verum*	Asia	Viral diseases treatment (Influenza)	Shikimic acid (e.g. Tamiflu)
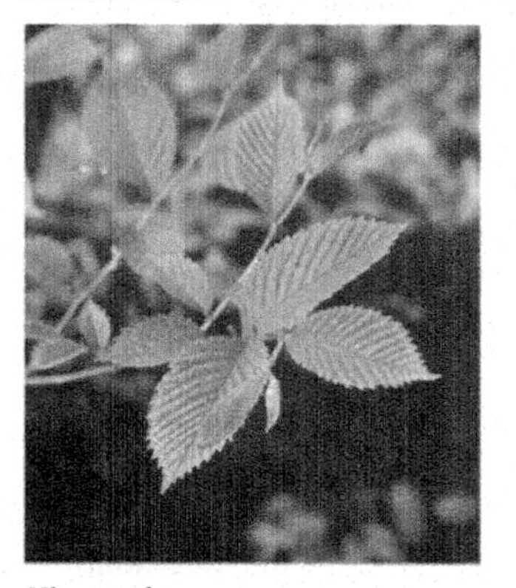 *Ulmus rubra*	Southeastern United States	Gastrointestinal ailments, relief in coughs, skin irritations	Slippery elm bark lining
 *Taxus baccata*	North America, Asia	Breast cancer treatment	Taxol
 *Strychnopsis thouarsii*	Madagascar	Treatment of Malaria (in development)	Tazopsine

(*Continued*)

TABLE 1.1 (*Continued*)
List of Several Important Medicinal and Aromatic Plants, Their Origin and Uses

Medicinal and Aromatic Plants	Place of Origin	Medicinal Uses	Drug Derived from MAPs
 *Chondrodendron tomentosum*	South America	Relaxation in surgical muscle	Tubocurarine

Roberson (2008).

of two thirds of the plants used in the modern system of medicine, and indigenous systems of medicine still provide the majority of health care to rural populations (Pan et al. 2014).

The use of medicinal plants for the treatment of various ailments is considered very safe since there are no or very few side effects (Adnan, Siddiqui, Arshad, et al. 2021; Siddiqui et al. 2021; Awadelkareem et al. 2022; Elasbali et al. 2022). One of the main advantages of these remedies is that they are in sync with nature. The golden truth is that the use of herbal treatments is not affected by age or gender (Reddy et al. 2020; Adnan, Siddiqui, Hamadou, et al. 2021). However, in the past few decades, certain groups of people have significantly benefited from the exploitation of biodiversity and the conversion of natural ecosystems to human-dominated ecosystems. The unfortunate thing is that such gains have often been achieved at the expense of biodiversity loss and degradation, which in turn has resulted in the exacerbation of poverty for other groups (Chandra 2016). In spite of our growing appreciation of the potential benefits that might accrue from systematic exploration of this vast storehouse of plants, we are also becoming increasingly aware that there is a simultaneous decline in the number of species available, which may lead to catastrophic results. Biodiversity is in decline, largely due to the activities of humans, such as the destruction of natural landscapes or the deforestation (Braddick 1994).

As a result of these occurrences, sustainable development is posed with a serious danger because our planet's species diversity is among the most valuable and irreplaceable resources we have. The loss of biodiversity is a grave matter of concern for human civilization, as it has become a matter of paramount concern (Shafi et al. 2021). As such, emergency measures need to be taken in order to prevent further diminution of the number of potential medicinal and biological agents (Heywood 2002). It is therefore the purpose of this chapter to introduce the readers to MAPs, their biodiversity, biogeographic distribution and discussion on their conservation.

1.2 DIVERSITY OF MAPs WORLDWIDE

In some cultures, there have been several plant species that have been used medicinally at one time or another in some form, but it is impossible to estimate the exact number. From an enumeration that was conducted by the WHO in the late 1970s, 21,000 medicinal species are listed (Penso 1978). According to Rafieian-Kopaei (2012), of 26,092 native species originating from China alone, 4,941 are used as medicines in Chinese traditional medicine, resembling 18.9% of the total species. If this calculation is considered for other well-known medicinal floras and then applied to the total number of species of flowering plants worldwide that exist (422,000), it has been estimated that in the world, there are more than 50,000 species of plants that can be used as medicinal plants. Worldwide,

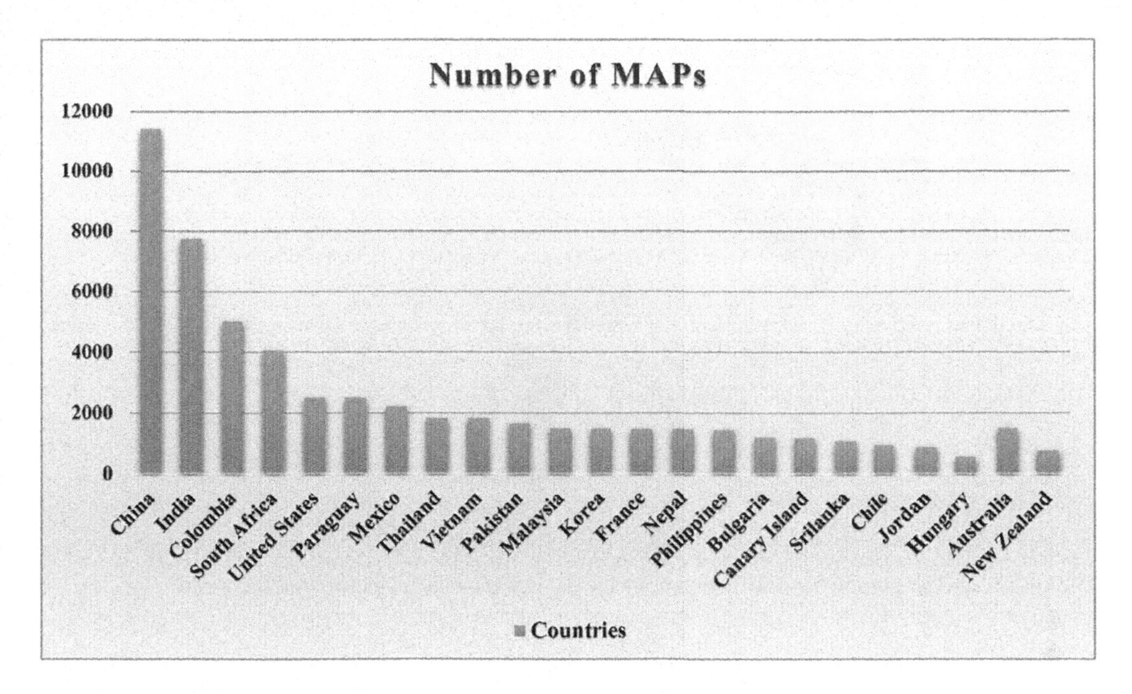

FIGURE 1.1 Number of medicinal and aromatic plants species in different countries of the world.

India and China are the two countries that have the highest utilization rate of medicinal and aromatics plants. Other countries, such as Colombia, South Africa, the United States, and 16 other countries, also utilize medicinal plants in great numbers (Figure 1.1) (A. Hamilton 2003; Balunas and Kinghorn 2005; Srujana, Konduri, and Rao 2012).

However, it is important to recognize that there are some plant families that are more likely to have medicinal values than other plant families. The following are some examples: Apocynaceae, Apiaceae, Araliaceae, Asclepiadaceae, Canellaceae, Guttiferae and Menispermaceae. Furthermore, it is imperative to note that these families are not distributed uniformly throughout the world. Thus, it is no surprise that some floras have higher proportions of medicinal plants than others and that certain plant families have a higher proportion of species that are threatened than others (Huang 2011).

1.3 BIOGEOGRAPHIC DISTRIBUTION OF MAPs

The science of biogeography studies how living things are distributed across the globe, and how abiotic and biotic factors influence the distribution of those living things. The abiotic factors like temperature and rainfall are mainly influenced by factors, such as latitude and elevation. As a consequence of these changes in abiotic factors, plant communities as well as animal communities are also likely to change in composition. It is important to note that some species are endemic, which means they can only be found in certain regions, while others are generalists, which means they can be found anywhere. The geographical distribution of important MAPs is presented in Figure 1.2.

1.4 MAPs OF AMERICA AND THEIR IMPORTANCE

The Amazon rainforest is home to a huge variety of medicinal plants that can be used to treat various ailments. There are many medicinal plants used in the United States that are listed in the "United States Pharmacopoeia" (Revision 2008). Similarly, the medicinal plants used in Brazil are listed in the "Brazilian Pharmacopoeia" (Zhou et al. 2021). There is a palm tree in the Amazon Basin

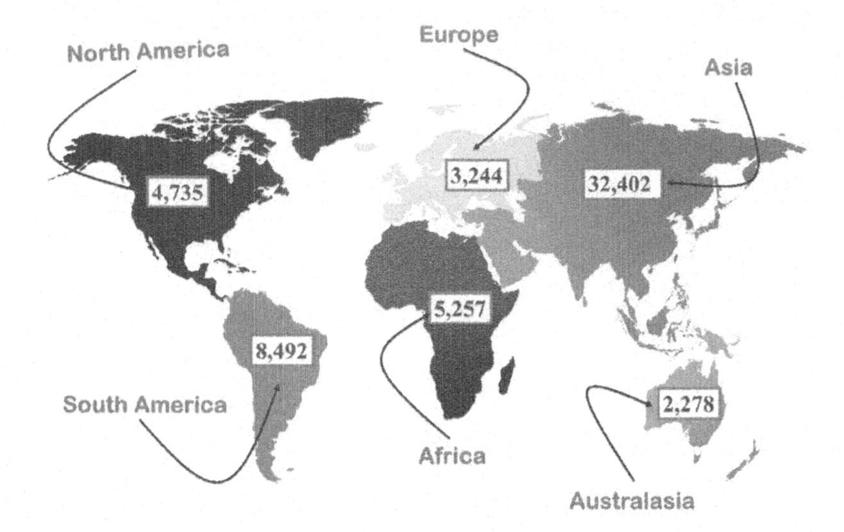

FIGURE 1.2 The geographic distribution of medicinal and aromatic plants around the world.

called *Euterpe oleracea* (Arecaceae). Due to the presence of anthocyanins in its fruits, it exerts medicinal properties (Desmarchelier 2010). In regions such as Peru, *Myrciaria dubia* (Myrtaceae) commonly known as Camu-Camu is another plant that grows in the Western Amazon Basin. There is a high level of vitamin C in the fruit that makes it an antioxidant (Desmarchelier 2010). *Croton lechleri* (Euphorbiaceae) is a small tree or shrub that produces red viscous latex that is referred to as Dragon's Blood. Traditionally, this latex has been used as a substance for wound-healing, anti-inflammatory, antiviral, and antitumour purposes in the Amazon Valley. This plant has several bioactive compounds, such as the alkaloid taspine, the lignan 3,4-O-dimethylcedrusin, a variety of polyphenols and anthocyanidins (Desmarchelier 2010). There is a large and woody vine called *Uncaria tomentosa* (Rubiaceae), commonly known as "Cat's claw," found in the rainforests of the western Amazon Basin of Peru, Bolivia and Ecuador. An anti-inflammatory action is provided by its quinovic acid glycosides and alkaloids, which contains quinovic acid. As high as 4,000 meters above sea level, there is a plant known as *Lepidium meyenii* (Brassicaceae) grows in the Andes. As a result of its isothiocyanate content, it has stimulant properties. There is also another Asteraceae plant that can be found in the Andes called *Smallanthus sonchifolius* (also known as Yacón). Due to its contents of oligofructans and phenolic compounds, it has been shown to be effective in treating hyperglycaemia and kidney problems. Moreover, fructo-oligosaccharides are also found in high levels in this tuber (Desmarchelier 2010).

In the Amazon rainforest, *Paullinia cupana* (Sapindaceae) is a plant that produces a fruit that contains high levels of caffeine and has stimulant effects (Desmarchelier 2010). It has been found that there are many medicinal plants in Brazil (Benko-Iseppon and Crovella 2010). The root decoction from *Aleurites moluccanus* (Euphorbiaceae) can be applied for the treatment of inflammations of the urinary tract and the ovaries. In particular, the stem bark of *Amburana cearensis* (Fabaceae) and also fresh leaves and seeds can be highly effective in the treatment of external ulcers as well as vaginal and throat infections caused by infected bacteria. There is a certain antimicrobial property in the rhizomes of *Costus spiralis* (Zingiberaceae). In order to treat hepatitis, other liver diseases, as well as diarrhoea, *Bromelia laciniosa* (Bromeliaceae) is used. In the north-eastern part of Brazil, *Hancornia speciosa* (Apocynaceae) is found. It has also been found that Mate (*Ilex paraguariensis*, Aquifoliaceae) can also be used as a remedy for stomach diseases and diabetes (Benko-Iseppon and Crovella 2010). Mate can be found in north-eastern Argentina, Southern Brazil and Paraguay. In addition to being used as a tea, it can also be used as an ingredient in dietary supplements and as a food. There are several types of xanthine alkaloids in this plant, which contribute to its stimulant and tonic effects. Paraguay's Stevia plant (*Stevia rebaudiana*, Asteraceae) contains a substance called stevioside in its leaves, which is a natural sweetener. One of the uses of stevia is as a sweetener in some foods that is low in calories (Desmarchelier 2010).

1.5 MAPs OF AFRICA AND THEIR IMPORTANCE

There is a wide variety of medicinal plants in Africa that are described in the "African Herbal Pharmacopoeia" (Brendler et al. 2010). The African continent is home to a rich diversity of MAPs. In the Dioncophyllaceae family, one species of such plant is *Triphyophyllum peltatum*. The drug has been shown to have antimalarial properties against *Plasmodium falciparum* (Khalid 2009). This activity can be attributed to the alkaloid 5-O-demethyl-8-O-methyl-7-epidioncophylline A, which has a unique structure. The antifungal activity of *Dolichos marginata* ssp. *erecta* (Fabaceae, contains sphenostylis in the bark of the roots) and *Chenopodium procerum* (Chenopodiaceae, contains isoflavonoids in the roots) has been observed recently. There are prenylated xanthones present in the root bark of South African *Garcinia gerrardii* (Clusiaceae) that are fungicidal (Khalid 2009). A plant called *Diospyros usambarensis* (Ebendaceae) is found in Malawi that contains quinones that have both molluscicidal and fungicidal effects, and *Hypericum revolutum* (Clusiaceae) contains chromenes, which are also fungicidal (Khalid 2009). Diterpene lactones present in *Parinari capensis* (Chrysobalanaceae) are antifungal compounds that are found in Zimbabwe. *Bauhinia rufescens* Lam (Fabaceae), growing in the Niger region of West Africa, has also been found to contain tetracycline, an antifungal compound. In Madagascar, *Lepidagathis alopecuroides* (Acanthaceae) has antifungal properties because of two diterpenes it contains: fluricoserpol A and dolabeserpenoic acid A, both of which are diterpenes. There is a crop called *Swartzia madagascariensis* (Fabaceae) that is grown in East Africa. There is a saponin in the seeds that is molluscicidal. It has been demonstrated that the saponins contained in the leaves of the endemic tree *Polyscias dichroostachya* (Araliaceae), growing in Mauritius, are capable of killing molluscs (Khalid 2009).

1.6 MAPs OF ASIA AND THEIR IMPORTANCE

"The Indian Pharmacopoeia" (Pharmacopoeia and Commission 2010) and the "Pharmacopoeia of the People's Republic of China" are two important pharmacopoeias in Asia (Xu et al. 2021). Asian MAPs are very diverse, when it comes to their medicinal properties. There are several plants included in this chapter that come from India, Vietnam, Laos and China. A few of these plants include: *Swertia angustifolia* (Gentianaceae), which can be used in the treatment of fever and malaria; *Stemona tuberosa,* which can be used to treat asthma and tuberculosis and *Dillenia indica* (Dilleniaceae), which can be used to treat diarrhoea and dysentery (Rai 2012). In the Lauraceae family, *Litsea verticillata* is an example of a medicinal plant from Vietnam. Plants such as this one contain an anti-HIV substance known as litesane sesquiterpene and litsea verticillol, which can be isolated from the leaves and twigs of the plant. There are various macrocyclic trichothecene sesquiterpenoids that can be found in the leaves and stem bark of *Ficus fistulosa* (Moraceae), but one of the most active ones is verrucasin L acetate. There is evidence that it has antimalarial activity (Soejarto et al. 2006). *Asparagus cochinchinensis* (Asparagaceae) has been found to have anti-HIV activity in Laos, while *Nauclea orientalis* (Rubiaceae) has been discovered to have antimalarial activity (Soejarto et al. 2006). Thailand and China both have species of *Crassocephalum crepidioides*, which are the members of the Asteraceae family. Jacoline and jacobine are the main components of the plant. A decoction of the whole plant is used as a treatment for fever, dysentery, gastroenteritis, urinary tract infections and mastitis (Roeder 2000). An Asteraceae plant called *Senecio integrifolius* var. *fauriri* (also known as *Senecio integrifolius*) is a traditional Chinese herb that is used as a remedy for dysentery, conjunctivitis and tumefactions. A number of alkaloids are present in it, including integrifoline, 7-angeloyltumeforcidine, 1,2-dihydrosenkirkine and 7-angeloylheliotridine.

1.7 MAPs OF AUSTRALIA AND THEIR IMPORTANCE

A comprehensive list of medicinal plants found in Australia and New Zealand is available in the "Australian and New Zealand Pharmaceutical Formulary" (Gill 1934). It has been shown that *Amyema quandang* leaves (Loranthaceae), *Eremophila duttonii* leaves (Myoporaceae) and *Lepidosperma viscidum* stem bases (Cyperaceae) have antibacterial activity against gram-positive bacteria such as *Bacillus cereus*, *Enterococcus faecalis*, *Staphylococcus aureus* and *Streptococcus pyogenes*. There are a number of traditional medicinal plants that are used to treat colds (*Lepidosperma viscidum*), fever (*Amyema quandang*), respiratory tract infections, sore throats, skin cuts, earaches or eye inflammation (*Eremophila duttonii*), as well as other conditions (Palombo and Semple 2001). Several gram-negative bacteria, such as *Escherichia coli*, *Klebsiella pneumoniae*, *Pseudomonas aeruginosa* and *Salmonella typhimurium*, are partially inhibited by *Euphorbia australis* (Euphorbiaceae). The traditional use of *Euphorbia australis* has been to treat skin sores and to make medicinal washes. Among the plants that have been found to have antiviral activity against poliovirus are *Dianella longifolia* var. *grandis* (Liliaceae, roots) and *Pterocaulon sphacelatum* (Asteraceae, green aerial parts). The extracts from *Euphorbia australis* and *Scaevola spinescens*, both of which belong to the Euphorbiaceae, have also shown activity against human cytomegalovirus. A variety of plant extracts have been found to be effective in the treatment of Ross River virus, such as *Eremophila latrobei* (Myoporaceae) and *Pittosporum phillyraeoides* var. *microcarpa* (Pittosporaceae) (Semple et al. 1998).

1.8 MAPs OF EUROPE AND THEIR IMPORTANCE

"The European Pharmacopoeia," "The German Pharmacopoeia" and "The British Pharmacopoeia" (Plitzko et al. 2009; Anderson 2010; Yacoub, Cibis, and Risch 2014) list a wide variety of medicinal plants that are used in Europe. There are also a number of medicinal plants from Italy and Spain. There is an Italian traditional medicine that uses the juice of *Chelidonium majus* (Papaveraceae) to treat warts. Topically, it is applied to the skin. To treat abdominal pain, the aerial parts of *Echium italicum* (Boraginaceae) are applied externally. In addition to its use as an antiseptic, the buds and flowers of *Crocus napolitanus* (Liliaceae) are also used as an external poultice in order to prevent lice from spreading. For the treatment of shingles (*Herpes zoster*), the bulbs of *Lilium candidum* (Liliaceae) can be useful. There is a use for the entire *Parmelia* sp. (Parmeliaceae) as a cholagogue (Pieroni 2000). A Spanish plant called *Phlomis lychnitis* (Lamioideae) can be found in the Mediterranean region. A decoction made from this plant proves to be effective at treating haemorrhoids. As a vulnerary and disinfectant, the bark of *Quercus suber* (Fagaceae) can be used to treat wounds and for disinfection of wounds. The use of *Marrubium vulgare* (Lamiaceae) has also been found to be effective in the treatment of asthma. *Leuzea conifera* (Asteraceae) should also be mentioned, as its decoction is used for treating gastritis and colitis (Vázquez, Suarez, and Pérez 1997).

1.9 CONSERVATION STATUS

It is becoming increasingly common to harvest medicinal plants in increasing quantities, mainly from wild populations of plants. Over the course of the last decade, Europe, North America and Asia have experienced an increase in the demand for wild resources of 8–15% per year (Simmonds 2006; Bentley et al. 2014). It has been shown that reproductive capacities of important MAPs can become irreversibly reduced below a certain threshold (Soulé et al. 2005; Semwal et al. 2007). Different sets of recommendations regarding the conservation of MAPs have been developed over the years, including the provision of *in situ* and *ex situ* conservation measures (Huang 2011; Liu, Chen, and Yu 2011). It is well established that natural reserves, as well as wild nurseries have an important role to play in preserving the medicinal efficacy of plants in their natural habitats, while botanic gardens and seed banks hold an important place for the conservation of *ex situ* plants and the replanting of future crops (Sheikh, Ahmad, and Khan 2002; Coley et al. 2003). In order to determine whether a species is best conserved in nature or in a nursery, the geographical distribution and biological characteristics of medicinal plants must be known in order to guide conservation activities.

1.10 IN SITU CONSERVATION

In most endemic MAPs, the medicinal properties are caused by secondary metabolites that are formed when plants are exposed to certain stimuli, and it seems that these secondary metabolites do not have the same effect in cultivation conditions (Coley et al. 2003; Figueiredo and Grelle 2009). It is only by preserving whole communities out of *in situ* that it is possible for us to preserve indigenous plants, maintain natural communities and also take into account their intricate network of relationships in order to conserve them (Gepts 2006). Furthermore, the practice of *in situ* conservation is known to increase the amount of biodiversity that can be protected (Forest et al. 2007), and it facilitates the connection between resource conservation and sustainable use (Chunlin et al. 2003). Globally, *in situ* conservation efforts focus on the establishment of protected areas and ecosystem-centric approaches, rather than species-specific approaches (Ma, Rong, and Cheng 2012). For MAPs to be successfully conserved in their natural habitats, rules, regulations and potential compliance are essential (Soulé et al. 2005; Volis and Blecher 2010).

Degradation of habitats and destruction of MAPs are major factors, resulting in the loss of important resources (Camm et al. 2002). The purpose of natural reserves is to protect and restore biodiversity by preserving and restoring important wild resources (Chiarucci, Maccherini, and De Dominicis 2001; Rodríguez et al. 2007). There are over 12,700 protected areas around the world, covering 13.2 million km^2, or 8.81% of the Earth's surface (Huang et al. 2002). It is important to assess the contributions and ecosystem functions of individual habitats when conserving MAPs by protecting natural habitats (Liu et al. 2001). Natural wild plant habitats cannot all be protected due to competing land uses and cost considerations (Soulé et al. 2005; Kramer and Havens 2009). In a natural habitat, protected area or a place that is close to where the plants naturally grow, a wild nursery is established to cultivate and domesticate endangered MAPs (Schipmann et al. 2005; A. C. Hamilton 2004; Strandby and Olsen 2008). Overexploitation, habitat degradation and invasive species are putting heavy strain on many wild species, but wild nurseries can provide an effective method for conserving endemic and endangered MAPs *in-situ* (Li and Chen 2007; Wani et al. 2021).

1.11 EX SITU CONSERVATION

There is no sharp line between *ex situ* conservation and *in situ* conservation; however, the latter is often an effective complement to the former, especially when it comes to those MAPs that have been overexploited and are endangered, because they have slow growth, low abundance and a high susceptibility to replanting (A. C. Hamilton 2004; Havens et al. 2006; Yu et al. 2010). As part of *ex situ* conservation, plants are cultivated and naturalized to ensure that they can survive for as long as possible, and they are also sometimes used to create large amounts of plant material for creating drugs, so it is a direct step towards saving medicinal plants (Pulliam 2000; Swarts and Dixon 2009). In addition to retaining their potency even when grown far from their natural habitat, many previously wild medicinal plants can store their reproductive materials for future replanting in seed banks (A. C. Hamilton 2004).

1.12 BOTANIC GARDENS

Ex situ conservation is important at botanic gardens (Havens et al. 2006), which keeps ecosystems healthy to ensure the survival of rare and endangered plants (Huang et al. 2002). In terms of genetic conservation, living collections generally contain only a few individuals per species (Yuan et al. 2010), botanic gardens offer multiple advantages. Usually, they feature diverse taxonomical and ecological flora grown together under common conditions (Primack and Miller-Rushing 2009). MAPs can be conserved by botanic gardens through propagation and cultivation protocols, as well as breeding and domestication programs (Maunder, Higgens, and Culham 2001). Genetic diversity of medicinal plants can be stored better *ex situ* in seed banks than in botanic gardens, and seed banks should be used to preserve biological diversity (Schoen and Brown 2001; Li and Pritchard 2009). Royal Botanic Gardens in Britain is the most noteworthy seed bank (Schoen and Brown 2001). By allowing quick access to samples, seed banks help conserve the remaining natural populations by providing information about their properties (Schoen and Brown 2001; Li and Pritchard 2009). To restore wild populations, seed banks must reintroduce plant species back into the wild and assist in their reintroduction (Li and Pritchard 2009).

1.13 PRACTICES RELATED TO CULTIVATION

Domestic cultivation of MAPs is widely used and generally accepted, despite wild-harvested plants being considered more efficacious (Gepts 2006; Leung and Wong 2010; Joshi and Joshi 2014). Through cultivation, new techniques can be used to resolve common problems encountered in the production of MAPs, such as toxic components, pesticide contamination, low active ingredient content and misidentification of herbal origin (Raina, Chand, and Sharma 2011). In order to improve yields of active compounds, which are almost invariably secondary metabolites, cultivation under controlled growth conditions can enhance yields and ensure stable production of active compounds. In order to obtain increased yields of the target products, cultivation practices are designed to provide the optimal levels of water, nutrients, optional additives and environmental factors, such as temperature, light and humidity, in order to provide the best conditions for cultivation (Liu, Chen, and Yu 2011; Wong et al. 2014). It has been shown that increased cultivation of MAPs results in a decrease in the harvest volume, assists in the recovery of their wild resources and lowers their prices to levels that are more reasonable (Schipmann et al. 2005; A. C. Hamilton 2004; Larsen and Olsen 2006).

1.14 GOOD AGRICULTURAL PRACTICES (GAP)

There have been a number of good agricultural practices (GAPs) developed for MAPs with the aim of regulating production, ensuring quality and assisting in the standardization of herbal drugs. An effective GAP approach ensures that herbal drugs (or crude drugs) of high quality are safe for use and are pollution free by making use of available knowledge to address a variety of problems (Muchugi et al. 2008). An environmental gap analysis includes germplasm, cultivation, collection and pesticide detection quality aspects, forensic authentication, identification of bioactive compounds and metal inspection (Makunga, Philander, and Smith 2008). GAP is actively promoted by many countries. It can be seen, for example, that Chinese authorities are promoting GAP as a means to cultivate plants that are commonly used medicinally in regions where they are traditionally grown (Ma, Rong, and Cheng 2012; Wani et al. 2021).

Moreover, the benefits of organic farming for MAPs are becoming increasingly well-known for their ability to create an integrated, humane, environmental and economic sustainably produced system (Rigby and Cáceres 2001; Macilwain 2004). The primary objectives of organic farming of MAPs include producing plants of higher quality and with greater productivity, ensuring their conservation and sustainable utilization and ensuring their quality and productivity. Among the most distinctive characteristics of organic farming is the lack of the use of synthetic nutrients, pesticides and herbicides. According to many current organic certification standards in both Europe and North America, synthetic fertilizers, pesticides and herbicides are not allowed in organic cultivation (Rigby and Cáceres 2001). As a natural method of farming, organic farming is benign to the environment, and using farm-derived renewable resources is one of the most effective means of maintaining the biological processes of medicinal plants and maintaining the ecological balance of their habitats (Rigby and Cáceres 2001; Chan et al. 2012). In addition to providing nutrients to soil continuously, organic fertilizers improve soil stability, thereby contributing to the growth of medicinal plants and the biosynthesis of essential compounds. For example, when organic fertilizers were applied, the biomass yield of *Chrysanthemum balsamita* was increased and its essential oil content was high relative to those free from organic fertilizers (Reddy 2010). It is becoming increasingly important to cultivate medicinal plants organically to ensure their longevity and sustainability (Macilwain 2004).

1.15 UTILIZATION WITH SUSTAINABILITY

Destructive harvesting is generally associated with resource exhaustion and even extinction of MAPs that are limited in abundance and grow slowly (Larsen and Olsen 2006; Baker et al. 2007). As a result, good harvesting practices should be established in order to ensure the sustainable use of MAPs. Harvesting roots and whole plants (e.g. herbs, shrubs and trees) cause more damage to medicinal plants than harvesting leaves and flowers. Leaves can be used as an alternative to whole plants or roots in herbal remedies.

1.16 CONCLUSION

Over the centuries, different cultures around the world have used MAPs for a variety of purposes, including the treatment of illness and the maintenance of good health. MAPs are readily available to rural populations who are much more dependent on them for their healthcare regimens than they are on modern medicines. Currently, the majority of drugs that are available on the market have been derived from MAPs and have been largely used for various purposes. Consequently, it has become increasingly important for scientists and businesses to pay attention to MAPs that can be found in natural environments. However, we have yet to discover the treasure trove of MAPs that inhabit around the world. It has been estimated that less than 1% of all tropical plants had been screened for the possibility of being used as a pharmaceutical. Moreover, the harvesting of MAPs from wild populations is becoming more difficult as scientific and commercial attention increases on them. Several medicinal species are thought to be at risk of extinction due to the overharvesting of their medicinal properties. Therefore, it is essential to study and conserve MAPs. A growing number of species and habitats are being lost across the globe as a result of global climate change, which intensifies this urgency. Around 15,000 medicinal plants may be on the verge of extinction within the next few decades around the world. It is estimated that approximately one major drug is lost every two years, according to experts. Therefore, there is an urgent requirement for the protection and conservation of MAPs. Consequently, the need to conserve and protect the MAPs is urgent, as they have a great deal of importance. Several *in situ* and *ex situ* conservation approaches, practices related to cultivation, good agricultural practices can be widely utilized for the conservation purposes of MAPs. There are several advanced technologies that can be used in order to boost the conservation approach. For example, genetic engineering can make it possible to synthesize natural products on a large scale, and advances in tissue culture of MAPs can make it possible to produce important bioactive compounds at scale and with high efficiency. By encapsulating propagules in tissue, micropropagation can allow both storage and transportation of propagules and also promote faster regeneration. Synthetic seeds can also offer an alternative to normal seeds that may be used for cultivation, when they are not sufficient for propagation. In addition, molecular markers are available that can be used in breeding improvements in order to make improvements to the genome, and the breeding process can be significantly shortened through the application of genetic marker.

REFERENCES

Adnan, Mohd, Arif Jamal Siddiqui, Jamal Arshad, Walid Sabri Hamadou, Amir Mahgoub Awadelkareem, Manojkumar Sachidanandan, and Mitesh Patel. 2021. "Evidence-Based Medicinal Potential and Possible Role of Selaginella in the Prevention of Modern Chronic Diseases: Ethnopharmacological and Ethnobotanical Perspective." *Records of Natural Products* 15(5): 355.

Adnan, Mohd, Arif Jamal Siddiqui, Walid Sabri Hamadou, Mitesh Patel, Syed Amir Ashraf, Arshad Jamal, Amir Mahgoub Awadelkareem, Manojkumar Sachidanandan, Mejdi Snoussi, and Vincenzo De Feo. 2021. "Phytochemistry, Bioactivities, Pharmacokinetics and Toxicity Prediction of Selaginella Repanda with Its Anticancer Potential against Human Lung, Breast and Colorectal Carcinoma Cell Lines." *Molecules* 26(3): 768.

Aftab, Tariq, and Khalid Rehman Hakeem. 2021. *Medicinal and Aromatic Plants: Healthcare and Industrial Applications*. Springer Nature.

Anderson, Stuart. 2010. "Pharmacy and Empire: The "British Pharmacopoeia" as an Instrument of Imperialism 1864 to 1932." *Pharmacy in History* 52(3/4): 112–21.

Awadelkareem, Amir Mahgoub, Eyad Al-Shammari, Abd Elmoneim O Elkhalifa, Mohd Adnan, Arif Jamal Siddiqui, Mejdi Snoussi, Mohammad Idreesh Khan, Z R Azaz Ahmad Azad, Mitesh Patel, and Syed Amir Ashraf. 2022. "Phytochemical and In Silico ADME/Tox Analysis of Eruca Sativa Extract with Antioxidant, Antibacterial and Anticancer Potential against Caco-2 and HCT-116 Colorectal Carcinoma Cell Lines." *Molecules* 27(4): 1409.

Bajaj, Yashpal P Singh. 2012. *Medicinal and Aromatic Plants I*. Vol. 4. Springer Science & Business Media.

Baker, Dwight D, Min Chu, Uma Oza, and Vineet Rajgarhia. 2007. "The Value of Natural Products to Future Pharmaceutical Discovery." *Natural Product Reports* 24(6): 1225–44.

Balunas, Marcy J, and A Douglas Kinghorn. 2005. "Drug Discovery from Medicinal Plants." *Life Sciences* 78(5): 431–41.

Benko-Iseppon, Ana Maria, and Sergio Crovella. 2010. "Ethnobotanical Bioprospection of Candidates for Potential Antimicrobial Drugs from Brazilian Plants: State of Art and Perspectives." *Current Protein and Peptide Science* 11(3): 189–94.

Bentley, R Alexander, Alberto Acerbi, Paul Ormerod, and Vasileios Lampos. 2014. "Books Average Previous Decade of Economic Misery." *PloS One* 9(1): e83147.

Braddick, M. 1994. "Critical Condition: Human Health and the Environment." *British Medical Journal* 309: 548.

Brendler, Thomas, Eloff, J N, A Gurib-Fakim, and L D Phillips. 2010. "African Herbal Pharmacopoeia; Association for African Medicinal Plants Standards (AAMPS)." Viva Publication New Delhi, India.

Camm, Jeffrey D, Susan K Norman, Stephen Polasky, and Andrew R Solow. 2002. "Nature Reserve Site Selection to Maximize Expected Species Covered." *Operations Research* 50(6): 946–55.

Chan, Kelvin, Debbie Shaw, Monique S J Simmonds, Christine J Leon, Qihe Xu, Aiping Lu, Ian Sutherland, Svetlana Ignatova, You-Ping Zhu, and Rob Verpoorte. 2012. "Good Practice in Reviewing and Publishing Studies on Herbal Medicine, with Special Emphasis on Traditional Chinese Medicine and Chinese Materia Medica." *Journal of Ethnopharmacology* 140(3): 469–75.

Chandra, L D. 2016. "Bio-Diversity and Conservation of Medicinal and Aromatic Plants." *Advances in Plants & Agriculture Research* 5(4): 186.

Chiarucci, Alessandro, Simona Maccherini, and Vincenzo De Dominicis. 2001. "Evaluation and Monitoring of the Flora in a Nature Reserve by Estimation Methods." *Biological Conservation* 101(3): 305–14.

Christaki, E, Ilias Giannenas, Eleftherios Bonos, and Panagiota Florou-Paneri. 2020. "Innovative Uses of Aromatic Plants as Natural Supplements in Nutrition." In Panagiota Florou-Paneri, Efterpi Christaki, and Ilias Giannenas (eds.), *Feed Additives*, 19–34. Elsevier.

Chunlin, Long, Li Heng, Ouyang Zhiqin, Yang Xiangyun, Li Qin, and Trangmar Bruce. 2003. "Strategies for Agrobiodiversity and Promotion: A Case Study from Yunnan, China." *Biodiversity and Conservation* 12(6): 1145–56.

Coley, Phyllis D, Maria V Heller, Rafael Aizprua, Blanca Araúz, Nayda Flores, Mireya Correa, Mahabir Gupta, Pablo N Solis, Eduardo Ortega-Barría, and Luz I Romero. 2003. "Using Ecological Criteria to Design Plant Collection Strategies for Drug Discovery." *Frontiers in The Ecology Environment* 1(8): 421–28.

Costa, D Carvalho, H S Costa, T Gonçalves Albuquerque, Fernando Ramos, Maria Conceição Castilho, and Ana Sanches-Silva. 2015. "Advances in Phenolic Compounds Analysis of Aromatic Plants and Their Potential Applications." *Trends in Food Science & Technology* 45(2): 336–54.

Desmarchelier, Cristian. 2010. "Neotropics and Natural Ingredients for Pharmaceuticals: Why Isn't South American Biodiversity on the Crest of the Wave?" *Phytotherapy Research* 24(6): 791–99.

Elasbali, Abdelbaset Mohamed, Waleed Abu Al-Soud, Ziad H Al-Oanzi, Husam Qanash, Bandar Alharbi, Naif K Binsaleh, Mousa Alreshidi, Mitesh Patel, and Mohd Adnan. 2022. "Cytotoxic Activity, Cell Cycle Inhibition, and Apoptosis-Inducing Potential of Athyrium Hohenackerianum (Lady Fern) with Its Phytochemical Profiling." *Evidence-Based Complementary and Alternative Medicine* 2022, Article ID 2055773, 13 pages. https://doi.org/10.1155/2022/2055773.

Farnsworth, Norman R, and Djaja D Soejarto. 1991. "Global Importance of Medicinal Plants." *The Conservation of Medicinal Plants* 26: 25–51.

Faydaoğlu, E, and M S Sürücüoğlu. 2011. "History of the Use of Medical and Aromatic Plants and Their Economic Importance." *Kastamonu Üniversitesi Orman Fakültesi Dergisi* 11(1): 52–67.

Fierascu, Radu Claudiu, Irina Fierascu, Anda Maria Baroi, and Alina Ortan. 2021. "Selected Aspects Related to Medicinal and Aromatic Plants as Alternative Sources of Bioactive Compounds." *International Journal of Molecular Sciences* 22(4): 1521.

Figueiredo, Marcos S L, and Carlos Eduardo V Grelle. 2009. "Predicting Global Abundance of a Threatened Species from Its Occurrence: Implications for Conservation Planning." *Diversity and Distributions* 15(1): 117–21.

Forest, Félix, Richard Grenyer, Mathieu Rouget, T Jonathan Davies, Richard M Cowling, Daniel P Faith, Andrew Balmford, John C Manning, Şerban Procheş, and Michelle van der Bank. 2007. "Preserving the Evolutionary Potential of Floras in Biodiversity Hotspots." *Nature* 445(7129): 757–60.

Gepts, Paul. 2006. "Plant Genetic Resources Conservation and Utilization: The Accomplishments and Future of a Societal Insurance Policy." *Crop Science* 46(5): 2278–92.

Giannenas, Ilias, E Sidiropoulou, Eleftherios Bonos, E Christaki, and P Florou-Paneri. 2020. "The History of Herbs, Medicinal and Aromatic Plants, and Their Extracts: Past, Current Situation and Future Perspectives." In Panagiota Florou-Paneri, Efterpi Christaki, and Ilias Giannenas (eds.), *Feed Additives*, 1–18. Elsevier.

Gill, D A. 1934. "Some Aspects of Entero-Toxaemia in New Zealand." *Australian Veterinary Journal* 10: 212–16.

Hamilton, A. C. 2003. "Medicinal Plants and Conservation: Issues and Approaches." *International Plants Conservation Unit, WWF-UK*, Pandahouse, Catteshall Lane. pp. 29–33.

Hamilton, Alan C. 2004. "Medicinal Plants, Conservation and Livelihoods." *Biodiversity Conservation* 13(8): 1477–1517.

Havens, Kayri, Pati Vitt, Mike Maunder, Edward O Guerrant, and Kingsley Dixon. 2006. "Ex Situ Plant Conservation and Beyond." *BioScience* 56(6): 525–31.

Heywood, Vernon H. 2002. "The Conservation of Genetic and Chemical Diversity in Medicinal and Aromatic Plants." In Bilge Şener (ed.), *Biodiversity*, 13–22. Springer.

Huang, H, X Han, L Kang, P Raven, P W Jackson, and Y Chen. 2002. "Conserving Native Plants in China." *Science* 297(5583): 935–36.

Huang, Hongwen. 2011. "Plant Diversity and Conservation in China: Planning a Strategic Bioresource for a Sustainable Future." *Botanical Journal of the Linnean Society* 166(3): 282–300.

Joshi, Bipin Chandra, and Rakesh K Joshi. 2014. "The Role of Medicinal Plants in Livelihood Improvement in Uttarakhand." *International Journal of Herbal Medicine* 1(6): 55–58.

Khalid, Sami A. 2009. "Decades of Phytochemical Research on African Biodiversity." *Natural Product Communications* 4(10): 1934578X0900401020.

Kramer, Andrea T, and Kayri Havens. 2009. "Plant Conservation Genetics in a Changing World." *Trends in Plant Science* 14 (11): 599–607.

Larsen, Helle Overgaard, and Carsten Smith Olsen. 2006. "Unsustainable Collection and Unfair Trade? Uncovering and Assessing Assumptions Regarding Central Himalayan Medicinal Plant Conservation." In David L. Hawksworth and Alan T. Bull (eds.), *Plant Conservation and Biodiversity*, 105–23. Springer.

Leung, Kar Wah, and Alice Sze-Tsai Wong. 2010. "Pharmacology of Ginsenosides: A Literature Review." *Chinese Medicine* 5(1): 1–7.

Li, De-Zhu, and Hugh W Pritchard. 2009. "The Science and Economics of Ex Situ Plant Conservation." *Trends in Plant Science* 14(11): 614–21.

Li, Xi-Wen, and Shi-Lin Chen. 2007. "Conspectus of Ecophysiological Study on Medicinal Plant in Wild Nursery." *Zhongguo Zhong Yao Za Zhi= Zhongguo Zhongyao Zazhi= China Journal of Chinese Materia Medica* 32(14): 1388–92.

Liu, Chang, Hua Yu, and Shi-Lin Chen. 2011. "Framework for Sustainable Use of Medicinal Plants in China." *Plant Diversity* 33(01): 65–68.

Liu, Jianguo, Marc Linderman, Zhiyun Ouyang, Li An, Jian Yang, and Hemin Zhang. 2001. "Ecological Degradation in Protected Areas: The Case of Wolong Nature Reserve for Giant Pandas." *Science* 292(5514): 98–101.

Lubbe, Andrea, and Robert Verpoorte. 2011. "Cultivation of Medicinal and Aromatic Plants for Specialty Industrial Materials." *Industrial Crops and Products* 34(1): 785–801. https://doi.org/10.1016/j.indcrop.2011.01.019.

Ma, Jianzhang, Ke Rong, and Kun Cheng. 2012. "Research and Practice on Biodiversity in Situ Conservation in China: Progress and Prospect." *Biodiversity Science* 20(5): 551–58.

Macilwain, Colin. 2004. "Organic: Is It the Future of Farming?" *Nature* 428(6985): 792–94.

Makunga, N P, L E Philander, and M Smith. 2008. "Current Perspectives on an Emerging Formal Natural Products Sector in South Africa." *Journal of Ethnopharmacology* 119(3): 365–75.

Máthé, Ákos (2015). "Introduction: Utilization/Significance of Medicinal and Aromatic Plants." In Máthé, Á. (ed), *Medicinal and Aromatic Plants of the World. Medicinal and Aromatic Plants of the World*, vol. 1, Springer. https://doi.org/10.1007/978-94-017-9810-5_1.

Maunder, Mike, Sarah Higgens, and Alastair Culham. 2001. "The Effectiveness of Botanic Garden Collections in Supporting Plant Conservation: A European Case Study." *Biodiversity & Conservation* 10(3): 383–401.

Miguel, Maria Graça. 2010. "Antioxidant Activity of Medicinal and Aromatic Plants. A Review." *Flavour and Fragrance Journal* 25(5): 291–312.

Muchugi, A, G M Muluvi, R Kindt, Caroline A C Kadu, A J Simons, and R H Jamnadass. 2008. "Genetic Structuring of Important Medicinal Species of Genus Warburgia as Revealed by AFLP Analysis." *Tree Genetics & Genomes* 4(4): 787–95.

Palombo, Enzo A, and Susan J Semple. 2001. "Antibacterial Activity of Traditional Australian Medicinal Plants." *Journal of Ethnopharmacology* 77(2–3): 151–57.

Pan, Si-Yuan, Gerhard Litscher, Si-Hua Gao, Shu-Feng Zhou, Zhi-Ling Yu, Hou-Qi Chen, Shuo-Feng Zhang, Min-Ke Tang, Jian-Ning Sun, and Kam-Ming Ko. 2014. "Historical Perspective of Traditional Indigenous Medical Practices: The Current Renaissance and Conservation of Herbal Resources." *Evidence-Based Complementary and Alternative Medicine* 2014, Article ID 525340. doi: 10.1155/2014/525340.

Penso, Giuseppe. 1978. *Inventory of Medicinal Plants Used in the Different Countries*. Organisation Mondiale de la Santé.

Pharmacopoeia, Indian, and Indian Pharmacopoeia Commission. 2010. "Ghaziabad." *Government of India Ministry of Health and Family Welfare* 1: 192–93.

Pieroni, Andrea. 2000. "Medicinal Plants and Food Medicines in the Folk Traditions of the Upper Lucca Province, Italy." *Journal of Ethnopharmacology* 70(3): 235–73.

Plitzko, Inken, Tobias Mohn, Natalie Sedlacek, and Matthias Hamburger. 2009. "Composition of Indigo Naturalis." *Planta Medica* 75(08): 860–63.

Primack, Richard B, and Abraham J Miller-Rushing. 2009. "The Role of Botanical Gardens in Climate Change Research." *New Phytologist* 182(2): 303–13.

Pulliam, H Ronald. 2000. "On the Relationship between Niche and Distribution." *Ecology Letters* 3(4): 349–61.

Rafieian-Kopaei, Mahmoud. 2012. "Medicinal Plants and the Human Needs." *Journal of HerbMed Pharmacology* 1(1): 1–2.

Rai, Prabhat Kumar. 2012. "Assessment of Multifaceted Environmental Issues and Model Development of an Indo-Burma Hotspot Region." *Environmental Monitoring and Assessment* 184(1): 113–31.

Raina, R, Romesh Chand, and Yash Pal Sharma. 2011. "Conservation Strategies of Some Important Medicinal Plants." *International Journal of Medicinal Plants, Aromatic* 1(3): 342–47.

Rao, M., Palada, M. & Becker, B. 2004. "Medicinal and Aromatic Plants in Agroforestry Systems." *Agroforestry Systems* 61: 107–122. https://doi.org/10.1023/B:AGFO.0000028993.83007.4b

Reddy, B Suresh. 2010. "Organic Farming: Status, Issues and Prospects–a Review." *Agricultural Economics Research Review* 23(347–2016–16927): 343–58.

Reddy, Mandadi N, Mohd Adnan, Mousa M Alreshidi, Mohd Saeed, and Mitesh Patel. 2020. "Evaluation of Anticancer, Antibacterial and Antioxidant Properties of a Medicinally Treasured Fern Tectaria Coadunata with Its Phytoconstituents Analysis by HR-LCMS." *Anti-Cancer Agents in Medicinal Chemistry* 20(15): 1845–56.

Revision, United States Pharmacopeial Convention. Committee of. 2008. "United States Pharmacopeia, the National Formulary," vol. 31, Parts 2–3. In United States Pharmacopeial Convention, Committee of Revision. United States Pharmacopeial Convention, Incorporated, 2008. University of Chicago. ISBN: 9781889788531

Rigby, Dan, and Daniel Cáceres. 2001. "Organic Farming and the Sustainability of Agricultural Systems." *Agricultural Systems* 68(1): 21–40.

Roberson, E. 2008. *Nature's Pharmacy, Our Treasure Chest: Why We Must Conserve Our Natural Heritage*. Center For Biological Diversity. www.biologicaldiversity.org.

Rodríguez, Jon Paul, Lluís Brotons, Javier Bustamante, and Javier Seoane. 2007. "The Application of Predictive Modelling of Species Distribution to Biodiversity Conservation." *Diversity and Distributions* 13: 243–51.

Roeder, Erhard. 2000. "Medicinal Plants in China Containing Pyrrolizidine Alkaloids." *Pharmazie* 55(10): 711–26.

Schipmann, U, D J Leaman, A B Cunningham, and S Walter. 2005. "Impact of Cultivation and Collection on the Conservation of Medicinal Plants: Global Trends and Issues." *Acta Horticulturae* 676: 31–44.

Schippmann, Uwe, A B Cunningham, and Danna J Leaman. 2002. "Impact of Cultivation and Gathering of Medicinal Plants on Biodiversity: Global Trends and Issues (Case Study No. 7)." In *Biodiversity and the Ecosystem Approach in Agriculture, Forestry and Fisheries,* Satellite Event Session on the Occasion of the 9th Regular Session of the Commission on "Genetic Resources for Food and Agriculture", Rome, 12–13 October 2002. FAO Document Repository of United Nations.

Schoen, Daniel J, and Anthony H D Brown. 2001. "The Conservation of Wild Plant Species in Seed Banks: Attention to Both Taxonomic Coverage and Population Biology Will Improve the Role of Seed Banks as Conservation Tools." *BioScience* 51(11): 960–66.

Semple, Susan J, G D Reynolds, M C O'leary, and R L P Flower. 1998. "Screening of Australian Medicinal Plants for Antiviral Activity." *Journal of Ethnopharmacology* 60(2): 163–72.

Semwal, D P, P Pardha Saradhi, B P Nautiyal, and A B Bhatt. 2007. "Current Status, Distribution and Conservation of Rare and Endangered Medicinal Plants of Kedarnath Wildlife Sanctuary, Central Himalayas, India." *Current Science* 92(12), 1733–38.

Shafi, Amrina, Farhana Hassan, Insha Zahoor, Umer Majeed, and Firdous A Khanday. 2021. "Biodiversity, Management and Sustainable Use of Medicinal and Aromatic Plant Resources." In T Aftab and K R Hakeem (eds), *Medicinal and Aromatic Plants*, 85–111. Springer.

Sheikh, Kashif, Tahira Ahmad, and Mir Ajab Khan. 2002. "Use, Exploitation and Prospects for Conservation: People and Plant Biodiversity of Naltar Valley, Northwestern Karakorums, Pakistan." *Biodiversity Conservation* 11(4): 715–42.

Siddiqui, Arif Jamal, Sadaf Jahan, Syed Amir Ashraf, Mousa Alreshidi, Mohammad Saquib Ashraf, Mitesh Patel, Mejdi Snoussi, Ritu Singh, and Mohd Adnan. 2021. "Current Status and Strategic Possibilities on Potential Use of Combinational Drug Therapy against COVID-19 Caused by SARS-CoV-2." *Journal of Biomolecular Structure and Dynamics* 39(17): 6828–41.

Simmonds, Monique S J. 2006. "Medicinal Plants of the World: Volume 3 Chemical Constituents, Traditional and Modern Medicinal Uses, Ivan A. Ross, Humana Press Inc., New Jersey (2005), pp. 623, ISBN: 1-58829-129-4." Pergamon.

Skendi, Adriana, Maria Irakli, Paschalina Chatzopoulou, Elisavet Bouloumpasi, and Costas G Biliaderis. 2022. "Phenolic Extracts from Solid Wastes of the Aromatic Plant Essential Oil Industry: Potential Uses in Food Applications." *Food Chemistry Advances*1: 100065.

Soejarto, Djaja Doel, Hong Jie Zhang, Harry H S Fong, Ghee T Tan, Cui Ying Ma, Charlotte Gyllenhaal, Mary C Riley, Marian R Kadushin, Scott G Franzblau, and Truong Quang Bich. 2006. "'Studies on Biodiversity of Vietnam and Laos' 1998– 2005: Examining the Impact." *Journal of Natural Products* 69 (3): 473–81.

Soulé, Michael E, James A Estes, Brian Miller, and Douglas L Honnold. 2005. "Strongly Interacting Species: Conservation Policy, Management, and Ethics." *BioScience* 55(2): 168–76.

Srujana, T Susan, Raveendra Babu Konduri, and Bodavula Samba Siva Rao. 2012. "Phytochemical Investigation and Biological Activity of Leaves Extract of Plant Boswellia Serrata." *The Pharma Innovation* 1(5, Part A): 22.

Strandby, Uffe, and Carsten Smith Olsen. 2008. "The Importance of Understanding Trade When Designing Effective Conservation Policy–The Case of the Vulnerable Abies Guatemalensis Rehder." *Biological Conservation* 141(12): 2959–68.

Swarts, Nigel D, and Kingsley W Dixon. 2009. "Terrestrial Orchid Conservation in the Age of Extinction." *Annals of Botany* 104(3): 543–56.

Vázquez, F M, M A Suarez, and A Pérez. 1997. "Medicinal Plants Used in the Barros Área, Badajoz Province (Spain)." *Journal of Ethnopharmacology* 55(2): 81–85.

Volis, Sergei, and Michael Blecher. 2010. "Quasi in Situ: A Bridge between Ex Situ and in Situ Conservation of Plants." *Biodiversity and Conservation* 19(9): 2441–54.

Wani, Naseema Aqbar, Younas Rasheed Tantray, Mohammad Saleem Wani, and Nazir Ahmad Malik. 2021. "The Conservation and Utilization of Medicinal Plant Resources." In T Aftab and K R Hakeem (eds), *Medicinal and Aromatic Plants*, 691–715. Springer.

Wong, Kam Lok, Ricky Ngok Shun Wong, Liang Zhang, Wing Keung Liu, Tzi Bun Ng, Pang Chui Shaw, Philip Chi Lip Kwok, Yau Ming Lai, Zhang Jin Zhang, and Yanbo Zhang. 2014. "Bioactive Proteins and Peptides Isolated from Chinese Medicines with Pharmaceutical Potential." *Chinese Medicine* 9(1): 1–14.

Xu, Xinyi, Huayu Xu, Yue Shang, Ran Zhu, Xiaoxu Hong, Zonghua Song, and Zhaopeng Yang. 2021. "Development of the General Chapters of the Chinese Pharmacopoeia 2020 Edition: A Review." *Journal of Pharmaceutical Analysis* 11(4): 398–404.

Yacoub, Kirsten, Katharina Cibis, and Corinna Risch. 2014. "Biodiversity of Medicinal Plants." In Victor Kuete and Thomas Efferth (eds.), *Biodiversity, Natural Products and Cancer Treatment*, 1–32. World Scientific.

Yu, Hua, Caixiang Xie, Jingyuan Song, Yingqun Zhou, and Shilin Chen. 2010. "TCMGIS-II Based Prediction of Medicinal Plant Distribution for Conservation Planning: A Case Study of Rheum Tanguticum." *Chinese Medicine* 5(1): 1–9.

Yuan, Qing-Jun, Zhi-Yong Zhang, Juan Hu, Lan-Ping Guo, Ai-Juan Shao, and Lu-Qi Huang. 2010. "Impacts of Recent Cultivation on Genetic Diversity Pattern of a Medicinal Plant, Scutellaria Baicalensis (Lamiaceae)." *BMC Genetics* 11(1): 1–13.

Zhou, Guan-Ru, Bao-Sheng Liao, Qiu-Shi Li, Jiang Xu, and Shi-Lin Chen. 2021. "Establishing a Genomic Database for the Medicinal Plants in the Brazilian Pharmacopoeia." *Chinese Medicine* 16(1): 1–10.

2 Botanical Bases of Medicinal and Aromatic Plants

Akash Vanzara and Amit Shrivastav
Vasu Research Centre

CONTENTS

DOI: 10.1201/b22842-2

2.1 INTRODUCTION

Aromatic plants, also known as medicinal herbs, have been used for centuries to treat a wide range of health conditions. These plants are characterized by their strong and distinct odors, which are often attributed to the presence of essential oils. Aromatic plants have been used in traditional medicine systems such as Ayurveda, Chinese medicine, and Western herbalism, and they continue to be popular in modern times due to their perceived health benefits and natural origins. An amazing treasure of various chemical compounds can be found in aromatic medicinal plants, and these compounds have diverse applications in human health care. According to World Health Organization (WHO), medicinal plants are plants that either as a whole or in one or more of their organs, contain substance capable of creating useful drugs, or contain substance with medicinal properties that can be used directly in accordance with the therapeutic potential, which can help to counter many diseases (Farnsworth and Soejarto 1991). Various parts of these plants, such as their roots, stems, barks, fruits, or seeds, can be useful to control or cure disease. There are many non-nutrient substances or compounds present in these plants that are useful for medicinal purposes. They are referred to be as phytochemicals, bioactive chemicals, or active components (Liu 2004).

2.2 AN OVERVIEW OF MAPs' HISTORY

In the plant system of economic botany, Medicinal and Aromatic Plants (MAPs) are considered as an independent unit (Simpson and Ogorzaly 1995). According to a study conducted by the WHO, there are approximately 3,000 aromatic medicinal plants that are used globally for their therapeutic properties (World Health Organization 2013). These plants are commonly found in various countries around the world, with some regions having a higher diversity of aromatic medicinal plants compared to others.

In India, there are over 1,000 aromatic medicinal plants that are used in traditional and modern medicine. These plants are widely available in the country and are commonly used in the preparation of Ayurvedic remedies. Similarly, in China, there are over 500 aromatic medicinal plants that are used in traditional medicine (Chan 1995). These plants are commonly found in the country's diverse climates and are used in the preparation of traditional Chinese medicine (TCM) remedies.

In the United States, there are around 300 aromatic medicinal plants that are used in traditional and modern medicine (Chen et al. 2016). These plants are commonly used in the preparation of herbal remedies and supplements. There is a wide variety of aromatic medicinal plants that are found in different countries around the world. These plants are commonly used in traditional and modern medicine due to their therapeutic properties and are widely available in different regions.

2.2.1 An Overview of Medicinal and Aromatic Plants

Medicinal herbs are plants that have medicinal properties and are used in the production of medicine. They can be classified based on their use (such as medicinal, culinary, aromatic, or ornamental), active constituents, life period, and botanical taxonomy. Aromatic herbs have pleasant odors and may be stimulants or nervines. Active constituents found in herbs include volatile oils, tannins, phenols, saponins, alkaloids, polysaccharide substances, and food stuffs. These can be used to classify herbs into five groups: aromatic, astringent, bitter, mucilaginous, and nutritive. Astringent herbs have tannins that can tighten and contract living tissue and have various medicinal properties. Bitter herbs contain bitter principles like phenols, phenol glycosides, alkaloids, and saponins and can be classified into laxative herbs, diuretic herbs, saponin-containing herbs, and alkaloid-containing herbs. Mucilaginous herbs contain polysaccharide substances that produce mucilage when mixed with water and can soothe and protect mucous membranes. Nutritive herbs contain food substances that nourish the body and can be used as tonics or to rebuild and strengthen the body.

2.2.2 Application of MAPs in Alternative Medicine

Aromatic plants are plants that contain essential oils, which are used in a variety of applications, including alternative medicine. These oils are extracted from the plants through a process called distillation and are used in a number of ways, including:

- Aromatherapy: Aromatherapy is the practice of using essential oils to improve physical and emotional well-being. Aromatherapy is often used to treat stress, anxiety, insomnia, and other conditions.
- Massage: Essential oils are often used in massage to improve circulation, reduce muscle tension, and promote relaxation.
- Inhalation: Inhaling the aroma of essential oils can help to improve respiratory function, reduce symptoms of colds and flu, and improve mood.
- Topical application: Essential oils can be applied topically to the skin to treat a variety of conditions, including acne, eczema, and other skin conditions.

2.3 TAXONOMICAL SYSTEMATICS IN MAPs

MAPs produce chemical compounds called secondary metabolites and pigments, such as alkaloids, phenolics, terpenes, anthocyanins, and carotenoids, which are used in the production of crude drugs. The botanical or taxonomic identity of these plants is important for their effective therapeutic use, and determining this identity is a fundamental step in the scientific study of medicinal plants. Plant classification, taxonomy, phylogeny, and systematics are interrelated and involve the arrangement of organisms into groups, the study of principles and procedures of classification, the evolutionary hierarchical structure of life forms, and the study of the comparative and evolutionary relationships of plants based on their anatomy, physiology, and biochemistry. Classification is important for studying plant diversity and involves the use of taxonomic ranks such as species, genus, family, order, class, subdivision, division, and kingdom (Cain, 2023).

Medicinal plants are diverse and belong to various families of the plant kingdom, which often have similar active principal components due to biosynthetic pathway similarities. For example, the Solanaceae family includes several alkaloid-containing species, while species in the Labiatae family are characterized by the presence of a large number of essential oil ingredients. The species is the basic taxonomic unit of medicinal plants, with the genus comprising related species. Subspecies, variety, and form are used to differentiate dissimilar populations among wild-growing species. In some cases, genetic variations can affect a species' chemical constituents, known as chemical races or chemodemes. Both natural and cultivated species can be divided into infraspecific varieties based on eco-botanical characteristics, while cultivars are usually differentiated based on their features valued by human societies. Correct identification is essential in the study and use of medicinal plants, and vouchered plants should be deposited in recognized herbaria. Botanical sciences, including plant systematics, plant morphology, physiology, molecular traits, and chemo-differentiation based on plant metabolism, can assist in the correct identification and description of medicinal plants. A multidisciplinary collaborative approach combining these approaches can not only aid in correct identification, but also open up promising perspectives for the breeding of chemo-cultivars of medicinal taxa (Bennett and Balick 2008; M. G. Simpson 2019; Singh 2019)

The binomial system of nomenclature, developed by Carl von Linnaeus, is used to name plants and consists of a genus and species name. The International Code of Nomenclature for algae, fungi, and plants is used to determine the correct scientific names of plants. Therefore, classification of MAPs is important for their identification, standardization, and quality control (Tables 2.1 and 2.2).

2.3.1 Lavandula angustifolia Mill

2.3.1.1 Medicinal Uses of Lavender

Lavandula angustifolia is a popular aromatic herb that has a long history of use in traditional medicine. It is a perennial flowering plant that is native to the Mediterranean region and has been

TABLE 2.1
List of Certain Aromatic Medicinal Plants with Therapeutic Activity

Plant Name	Botanical Name	Family	Medicinal Use
Lavender	*Lavandula angustifolia* Mill.	Lamiaceae	Calming, relaxing
Basil	*Ocimum basilicum* L.	Lamiaceae	Anti-inflammatory, antioxidant
Rosemary	*Rosmarinus officinalis* L.	Lamiaceae	Memory improvement, circulation improvement
Thyme	*Thymus vulgaris* L.	Lamiaceae	Respiratory health, immune system boost
Oregano	*Origanum vulgare* L.	Lamiaceae	Digestion improvement, inflammation reduction
Sage	*Salvia officinalis* L.	Lamiaceae	Memory improvement, cognitive function improvement
Mint	*Mentha spicata* L.	Lamiaceae	Digestion improvement, freshening breath
Chamomile	*Matricaria chamomilla* L.	Asteraceae	Calming, sleep aid
Geranium	*Pelargonium graveolens* L'Hér.	Geraniaceae	Stress relief, skin care
Eucalyptus	*Eucalyptus globulus* Labill.	Myrtaceae	Respiratory health, inflammation reduction
Lemon balm	*Melissa officinalis* L.	Lamiaceae	Calming, sleep aid
Marjoram	*Origanum majorana* L.	Lamiaceae	Digestion improvement, anxiety relief
Peppermint	*Mentha piperita* L.	Lamiaceae	Digestion improvement, freshening breath
Clary sage	*Salvia sclarea* L.	Lamiaceae	Menstrual cramp relief, stress relief
Cypress	*Cupressus sempervirens* L	Cupressaceae	Respiratory health, anxiety relief
Juniper	*Juniperus communis* L.	Cupressaceae	Digestion improvement, diuretic
Palmarosa	*Cymbopogon martini* (Roxb.) W. Watson	Poaceae	Skin care, stress relief

TABLE 2.2
List of Different Active Compounds with Chemical Structure and Molecular Formula

Sr. No.	Name of the Plant	Name of Actives	Chemical Structure	Molecular formula
1	*Lavandula angustifolia* Mill.	Linalool	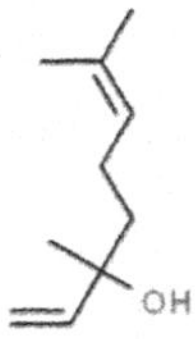	$C_{10}H_{18}O$
		Linalyl Acetate	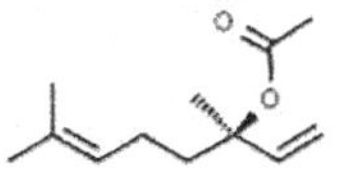	$C_{12}H_{20}O_2$
		Lavandulol	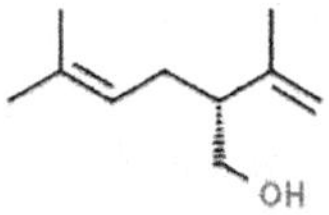	$C_{10}H_{18}O$
		Lavandulyl Acetate	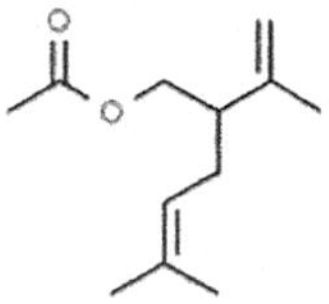	$C_{12}H_{20}O_2$
		Eugenol	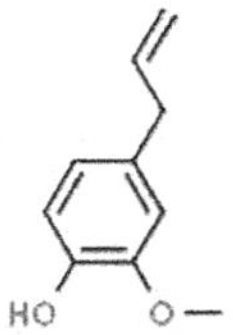	$C_{10}H_{12}O_2$
2	*Ocimum basilicum* L.	Carvacrol	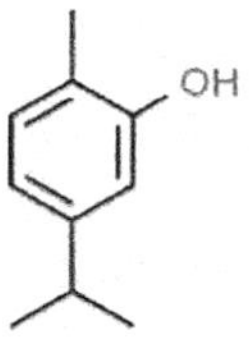	$C_{10}H_{14}O$
		Estragole	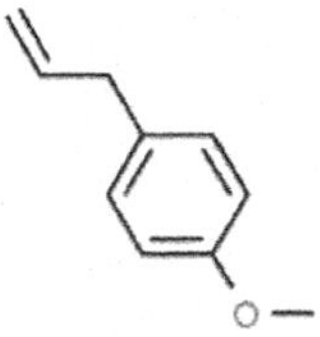	$C_{10}H_{12}O$
		Carnosol	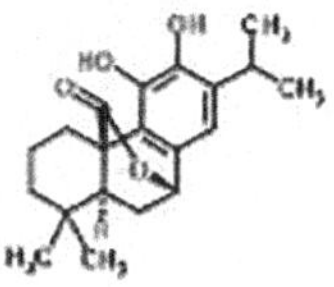	$C_{20}H_{26}O_4$
3	*Rosmarinus officinalis* L.	Ursolic Acid	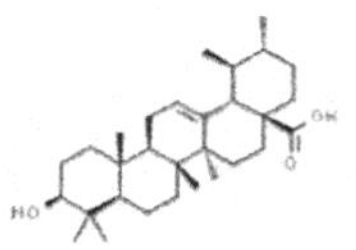	$C_{30}H_{48}O_3$

(*continued*)

TABLE 2.2 (*Continued*)
List of Different Active Compounds with Chemical Structure and Molecular Formula

Sr. No.	Name of the Plant	Name of Actives	Chemical Structure	Molecular formula
4	*Thymus vulgaris* L.	Thymol	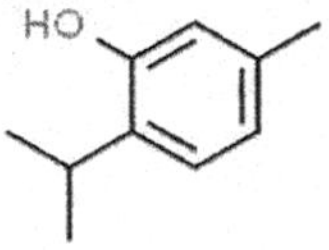 HO	$C_{10}H_{14}O$
		Thujone		$C_{10}H_{16}O$
5	*Salvia officinalis L*	Rosmarinic acid	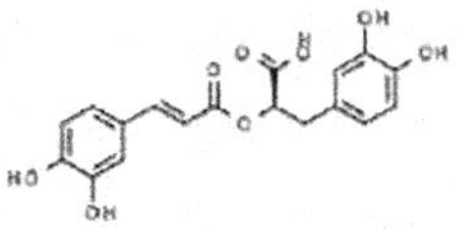	$C_{18}H_{16}O_8$
		Eucalyptol	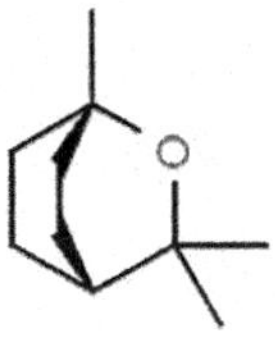 O	$C_{10}H_{18}O$
6	*Mentha spicata* L.	Limonene		$C_{10}H_{16}$
		Pulegone	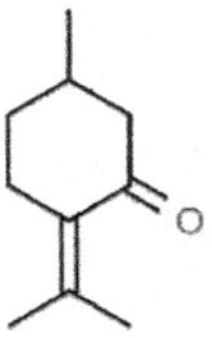 O	$C_{10}H_{16}O$
		Carvone	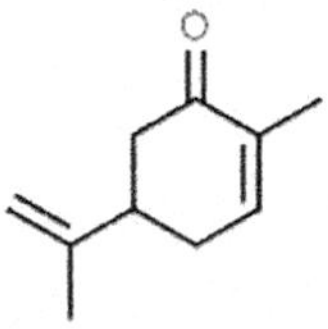 O	$C_{10}H_{14}O$
7	*Matricaria chamomilla* L.	Apigenin	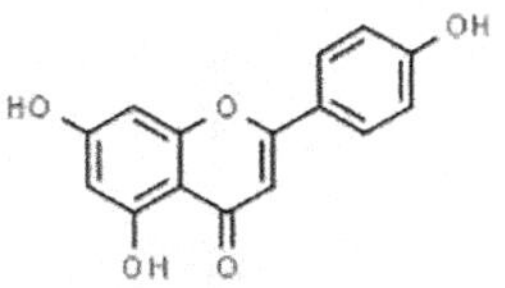 HO O OH OH O	$C_{15}H_{10}O_5$
		Chamazulene	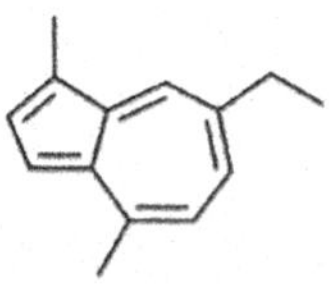	$C_{14}H_{16}$

widely cultivated for its fragrant and medicinal properties. It belongs to the Lamiaceae (mint) family and is known for its violet or blue flowers and sweet, floral aroma. It is most commonly used for its calming and relaxing effects and is often used to treat anxiety, insomnia, and stress-related conditions (Buchbauer et al., 1993). Lavender oil is also used topically for its antimicrobial and anti-inflammatory properties and is commonly used to treat skin irritations, burns, and wounds (Tisserand and Young 2013).

2.3.1.2 Chemical Constituents

Lavender essential oil is composed of over 100 chemical compounds, with the most abundant being linalool; it is a terpene alcohol with the chemical formula $C_{10}H_{18}O$, and is known for its sweet, floral aroma (Burt 2004). It also contains linalyl acetate, which is an ester of linalool and acetic acid with the chemical formula $C_{12}H_{20}O_2$, and is also responsible for the floral aroma of lavender essential oil, and lavandulyl acetate (Buchbauer et al. 1993). The plant is rich in monoterpenes, monoterpenoids, and phenylpropanoids (Burt 2004). It also contains a range of other terpenes, including cineole, camphor, and terpinen-4-ol.

2.3.1.3 Chemical Structure

Linalool is a monoterpene alcohol that is responsible for much of the characteristic aroma of lavender oil. It has the chemical formula $C_{10}H_{18}O$, and is composed of a linear chain of ten carbon atoms with an OH group attached to the third carbon atom. Linalyl acetate is a monoterpene ester that is also present in high concentrations in lavender oil. It has the chemical formula $C_{12}H_{2}0O_{2}$, and is composed of a linear chain of 12 carbon atoms with acetate.

2.3.1.4 Taxonomical Characteristics

Genus and Species: Lavender belongs to the genus *Lavandula* in the family Lamiaceae. It consists of about 39 species of flowering plants in the Lamiaceae family. There are several species of lavender, including *L. angustifolia* (also known as English lavender), *L. latifolia* (also known as spike lavender), and *L. stoechas* (also known as French lavender) (Nelson et al. 2007).

Description: Lavender plants are perennial herbs with distinctive square stems and fragrant purple or white flowers. The leaves are narrow and pointed, and the flowers are typically arranged in spikes (Nelson et al. 2007).

Habitat: Lavender is native to the Mediterranean region and can be found in countries such as Spain, France, and Italy. It is also grown in other parts of the world, including the United States, for its essential oils and as an ornamental plant (Nelson et al. 2007).

2.3.1.5 Pharmacological Action of Lavender

Lavender has a long history of use in traditional medicine as an anxiolytic, sedative, and anti-inflammatory agent. Lavender essential oil has been shown to have a range of pharmacological effects. It has been found to have anxiolytic and sedative effects in animals and humans (Shaw et al. 2007; Moss et al. 2003). Lavender essential oil has also been found to have anti-inflammatory effects, both in animal and in vitro studies (Ramos da Silva et al. 2021). It has been shown to inhibit the production of pro-inflammatory mediators, such as nitric oxide and prostaglandins, and to reduce inflammation in various models of inflammation, including carrageenan-induced paw edema and xylene-induced ear edema in mice (Ramos da Silva et al. 2021). In addition to these effects, lavender essential oil has been found to have antimicrobial activity against a range of microorganisms, including bacteria, fungi, and viruses (Ramos da Silva et al. 2021). It has also been found to have analgesic effects in animal studies (Ramos da Silva et al. 2021). Overall, lavender essential oil has a range of pharmacological effects, including anxiolytic activities (Figure 2.1).

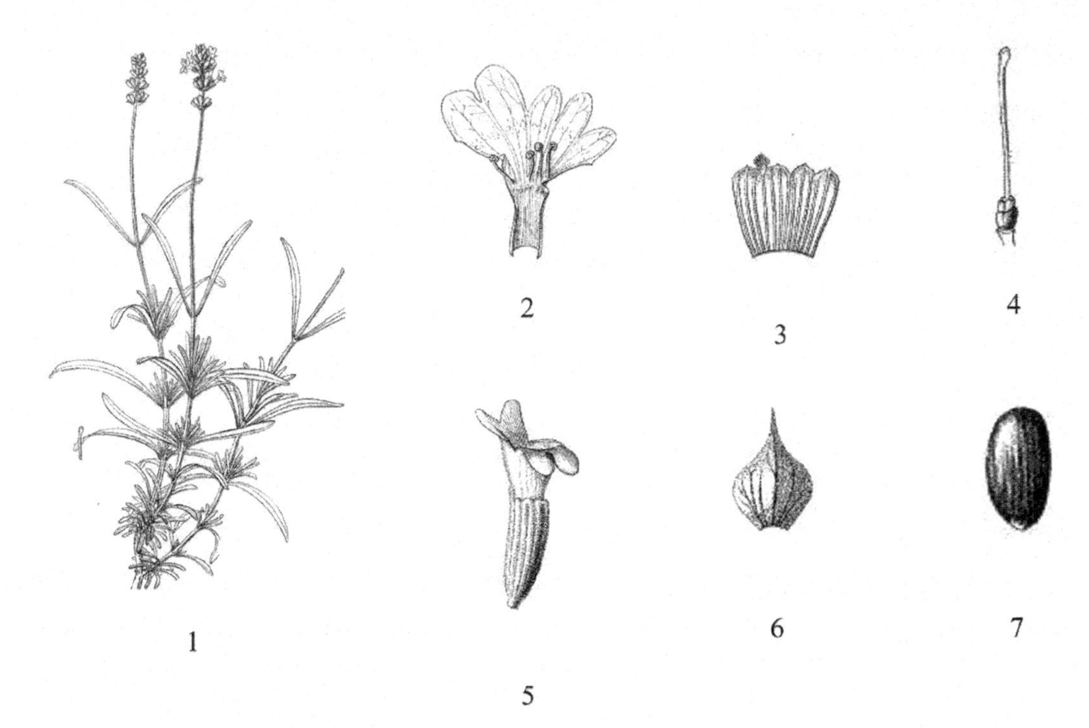

FIGURE 2.1 *Lavandula angustifolia* Mill. (1) Habit, (2) Open flower, (3) Open calyx, (4) Gynoecium, (5) Whole flower, (6) Bract abaxial view, (7) Nutlet.

2.4 *OCIMUM SANCTUM* L.

2.4.1 Medicinal Uses of *Ocimum sanctum* (Tulsi)

Ocimum sanctum (O. sanctum), also known as tulsi, is a medicinal plant native to the Indian subcontinent. It has been used in traditional Ayurvedic and Siddha medicine for centuries. In modern times, it is still widely used in India and other parts of the world for its numerous medicinal properties. *O. sanctum*, commonly known as holy basil or tulsi, is a popular medicinal herb in Ayurvedic and traditional medicine. It is believed to have a wide range of therapeutic properties, including anti-inflammatory, antioxidant, and stress-relieving effects (Kumar et al. 2013). In traditional medicine, *O. sanctum* is often used to treat respiratory conditions, such as asthma and bronchitis, as well as digestive disorders and skin conditions (Kumar et al. 2013). It is also believed to have immune-boosting properties and is commonly used to prevent and treat infections (S. K. Srivastava and Singh 2020).

2.4.2 Chemical Constituents

The chemical constituents of *O. sanctum* include a variety of compounds, such as flavonoids, tannins, saponins, and essential oils. The essential oil of tulsi is composed of a variety of monoterpenoids and sesquiterpenoids, including eugenol, caryophyllene, and linalool, ursolic acid, rosmarinic acid, and carnosic acid (Pradhan et al. 2022).

2.4.3 Chemical Structure

Eugenol, a major component of *O. sanctum*, has the chemical structure $C_{10}H_{12}O_2$ (Kim et al. 2021). Ursolic acid has the structure $C_{30}H_{48}O_3$, rosmarinic acid has the structure $C_{18}H_{16}O_8$, and carnosic acid has the structure $C_{22}H_{32}O_4$ (Kim et al. 2021).

2.4.4 Taxonomical Characteristics

The plant is an annual or perennial herb that grows to a height of 50–80 cm. It has square stems and opposite, green to purple leaves that are 4–6 cm long and 1–2 cm wide. The leaves are ovate to lanceolate in shape, and have a distinctive aroma when crushed. The plant produces purple or white flowers in terminal spikes that are 4–8 cm long. The flowers are 2-lipped and have four stamens.

2.4.5 Pharmacological Action

O. sanctum has various pharmacological actions including anti-inflammatory, antioxidant, and adaptogenic effects. These properties are attributed to the presence of various compounds such as eugenol, rosmarinic acid, flavonoids, and phenolic acids. The pharmacological actions of *O. sanctum* include anti-inflammatory, antioxidant, and adaptogenic effects (Triveni et al. 2013). It has been shown to have a potent anti-inflammatory effect, which is attributed to the presence of compounds such as eugenol and rosmarinic acid. These compounds have been found to inhibit the production of pro-inflammatory cytokines, such as tumor necrosis factor-alpha (TNF-α) and interleukin-6 (IL-6), which play a role in the development of inflammation (Edwin et al., 2016).

O. sanctum also has potent antioxidant activity, which is attributed to the presence of compounds such as flavonoids and phenolic acids (Triveni et al. 2013). These compounds have been found to scavenge reactive oxygen species (ROS) and neutralize the damaging effects of oxidative stress. In addition, *O. sanctum* has adaptogenic effects, which means it helps the body adapt to stress and improve overall physical and mental well-being (Triveni et al. 2013). This is attributed to the presence of compounds such as ursolic acid and oleanolic acid, which have been found to have a modulatory effect on the hypothalamic–pituitary–adrenal (HPA) axis (Edwin et al. 2016). The HPA axis plays a key role in the body's response to stress, and its modulation can help improve the body's overall stress-coping mechanisms (Figure 2.2).

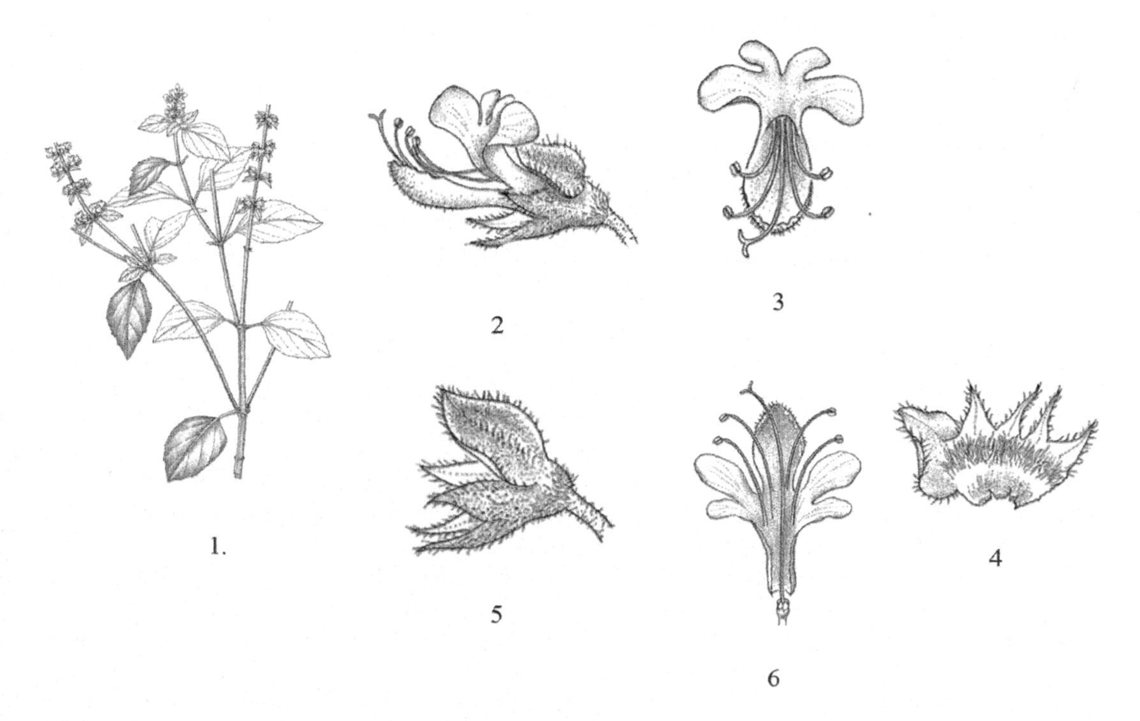

FIGURE 2.2 *Ocimum sanctum.* (1) Habit, (2) Whole flower, (3) Upper view of corolla, (4) Open calyx, (5) Whole calyx, (6) Open flower.

2.5 *ROSMARINUS OFFICINALIS* L.

It is also known as rosemary is an aromatic medicinal plant that is commonly used in traditional and modern medicine. It is native to the Mediterranean region and is widely cultivated in various parts of the world due to its various therapeutic properties.

2.5.1 Medicinal Uses

R. officinalis has been used to improve cognitive function, reduce stress and anxiety, and improve cardiovascular health (Rahbardar and Hosseinzadeh 2020).

2.5.2 Chemical Constituents

R. officinalis contains a wide range of chemical constituents, including flavonoids, phenolic acids, and terpenoids (Rahbardar and Hosseinzadeh 2020). The most abundant flavonoids in rosemary are quercetin, kaempferol, and apigenin, while the main phenolic acids are rosmarinic acid and caffeic acid. The main terpenoids found in rosemary include monoterpenes, such as α-pinene and camphene, and sesquiterpenes, such as caryophyllene and bisabolene (Rahbardar and Hosseinzadeh 2020).

2.5.3 Chemical Structure

The chemical structure of the various chemical constituents of *R. officinalis* is complex and diverse. The flavonoids present in rosemary have a basic flavone structure, which is characterized by the presence of a ring system consisting of two benzene rings joined by a pyrone ring. The phenolic acids present in rosemary have a basic phenol structure, which is characterized by the presence of a hydroxyl group (-OH) attached to a benzene ring (Rahbardar and Hosseinzadeh 2020). The terpenoids present in rosemary have a basic terpene structure, which is characterized by the presence of multiple isoprene units.

2.5.4 Morphological Characters

R. officinalis is an evergreen shrub that grows up to 1–2 meters in height. It has linear, needle-like leaves that are green on top and white on the bottom, and small, blue, pink, or white flowers that grow in clusters. The plant has a woody stem and a strong, distinctive aroma (Rahbardar and Hosseinzadeh 2020).

2.5.5 Pharmacological Activity

One of the main pharmacological actions of *R. officinalis* is its antioxidant activity. The plant contains a range of antioxidants, including flavonoids and phenolic acids, which have been found to scavenge reactive oxygen species (ROS) and neutralize the damaging effects of oxidative stress. The antioxidant activity of *R. officinalis* has been linked to its potential to prevent or reduce the risk of various diseases, such as cancer, cardiovascular disease, and neurodegenerative disorders (Rahbardar and Hosseinzadeh 2020).

In addition, *R. officinalis* has anti-inflammatory activity, which is attributed to the presence of compounds such as rosmarinic acid and ursolic acid. These compounds have been found to inhibit the production of pro-inflammatory cytokines, such as TNF-α and IL-6, which play a role in the development of inflammation (Rahbardar and Hosseinzadeh 2020). The anti-inflammatory activity of *R. officinalis* has been linked to its potential to reduce the severity of various inflammatory

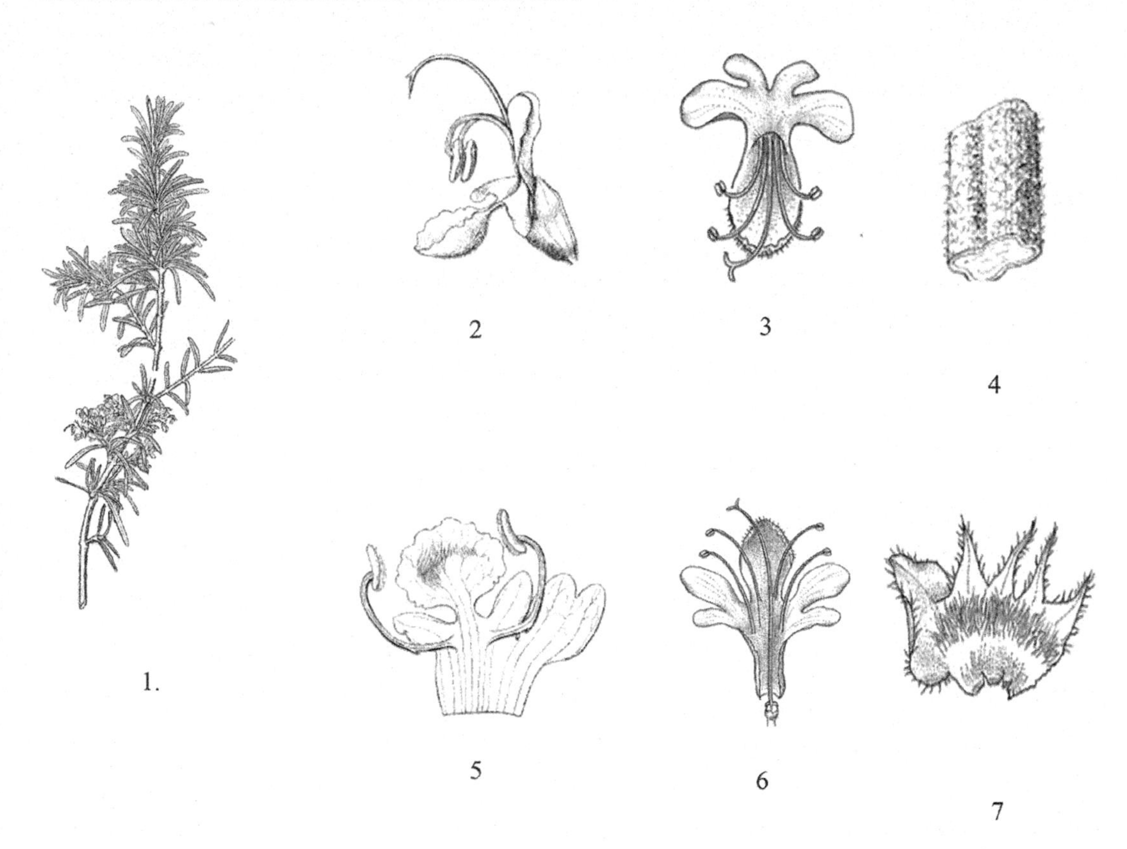

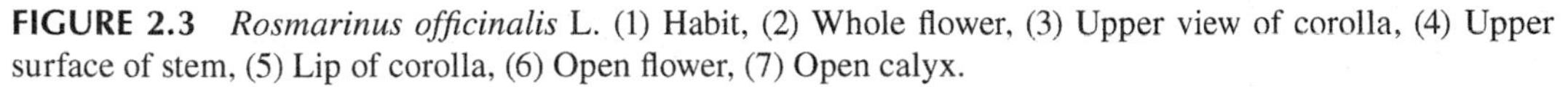

FIGURE 2.3 *Rosmarinus officinalis* L. (1) Habit, (2) Whole flower, (3) Upper view of corolla, (4) Upper surface of stem, (5) Lip of corolla, (6) Open flower, (7) Open calyx.

conditions, such as rheumatoid arthritis and asthma. It has neuroprotective effects, which are attributed to the presence of compounds such as rosmarinic acid and ursolic acid. These compounds have been found to have a protective effect on neurons and to improve cognitive function. The neuroprotective effects of *R. officinalis* have been linked to its potential to prevent or reduce the risk of various neurodegenerative disorders, such as Alzheimer's disease and Parkinson's disease (Rahbardar and Hosseinzadeh 2020). (Figure 2.3).

2.6 *THYMUS VULGARIS* L.

T. vulgaris L., also known as common thyme, is an aromatic medicinal plant that has been used for centuries in traditional and modern medicine. It is native to the Mediterranean region and is commonly used in the preparation of herbal remedies due to its therapeutic properties.

2.6.1 Medicinal Uses

T. vulgaris has various medicinal uses, including its use as an antimicrobial, expectorant, and carminative agent (Patil et al. 2021). It has been found to have a potent antimicrobial effect, which is attributed to the presence of compounds such as thymol and carvacrol (Imelouane et al. 2009). These compounds have been found to inhibit the growth of various bacterial and fungal strains (Borugă et al. 2014). *T. vulgaris* is also commonly used as an expectorant due to its ability to help clear mucus from the respiratory tract and ease congestion (Patil et al. 2021). In addition, it is used as a carminative agent due to its ability to relieve gas and bloating (Patil et al. 2021).

2.6.2 Chemical Constituents

T. vulgaris contains a variety of chemical constituents, including essential oils, flavonoids, tannins, saponins, and mucilage (Patil et al. 2021). The essential oils found in *T. vulgaris* are composed of compounds such as thymol, carvacrol, and p-cymene (Borugă et al. 2014). The flavonoids found in *T. vulgaris* include apigenin, luteolin, and thymonin. The tannins found in *T. vulgaris* include catechin, epicatechin, and procyanidin. The saponins found in *T. vulgaris* include oleanolic acid and ursolic acid. The mucilage found in *T. vulgaris* is composed of polysaccharides such as xylose and galactose (Borugă et al. 2014).

2.6.3 Chemical Structure

Thymol, a compound found in *T. vulgaris*, has the chemical structure $C_{10}H_{14}O$. Apigenin, a flavonoid found in *T. vulgaris*, has the chemical structure $C_{15}H_{10}O_5$ (Imtara et al. 2021; Kim et al. 2021).

2.6.4 Morphological Characters

Among the mint family (Lamiaceae), *T. vulgaris* L. is native to southern Europe from the western Mediterranean to southern Italy (Brickell, 2019). It is a bushy, woody based evergreen subshrub growing to 15-30 cm & 40 cm in wide with yellow or purple flowers and small, highly aromatic, gray-green leaves (Brickell, 2019).

2.6.5 Pharmacological Activities

Antimicrobial effect: *T. vulgaris* has a potent antimicrobial effect, which is attributed to the presence of compounds such as thymol and carvacrol (Imelouane et al. 2009). These compounds have been found to inhibit the growth of various bacterial and fungal strains (Borugă et al. 2014). Thymol has been found to have a broad-spectrum antimicrobial activity against various bacterial and fungal strains, including *Escherichia coli, Staphylococcus aureus*, and *Candida albicans*. Similarly, carvacrol has been found to have a strong antimicrobial activity against various bacterial strains, including *Pseudomonas aeruginosa* and *Bacillus subtilis*.

Expectorant effect: *T. vulgaris* is commonly used as an expectorant due to its ability to help clear mucus from the respiratory tract and ease congestion (Patil et al. 2021). The expectorant effect of *T. vulgaris* is attributed to the presence of compounds such as thymol and carvacrol, which have been found to stimulate the secretion of mucus and increase the flow of bronchial secretions (Borugă et al. 2014).

Carminative effect: *T. vulgaris* is used as a carminative agent due to its ability to relieve gas and bloating (Patil et al. 2021). The carminative effect of *T. vulgaris* is attributed to the presence of compounds such as thymol and carvacrol, which have been found to have a relaxing effect on the smooth muscles of the gastrointestinal tract and help relieve gas and bloating (Borugă et al. 2014) (Figure 2.4).

2.7 *ORIGANUM VULGARE* L.

O. vulgare, commonly known as oregano, is a perennial herb that is native to the Mediterranean region and is widely used in cooking and as a medicinal plant.

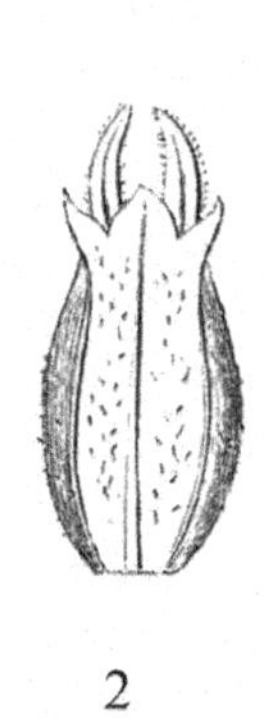

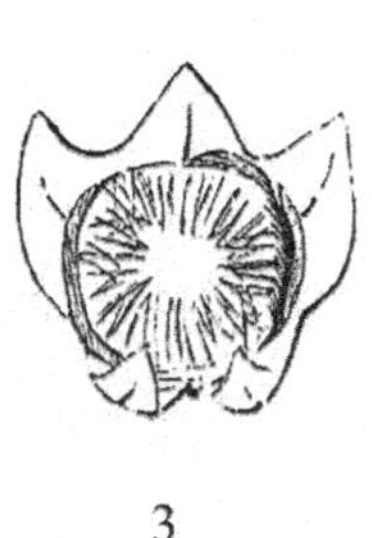

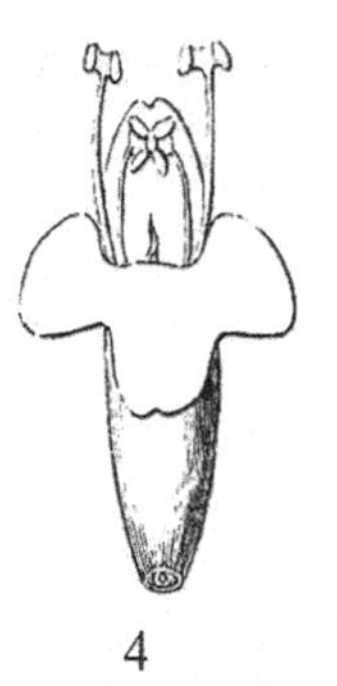

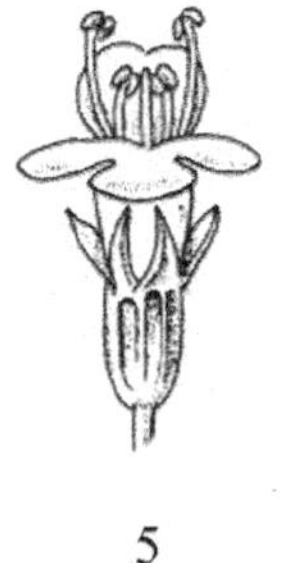

FIGURE 2.4 *Thymus vulgaris* L. (1) Habit, (2) Calyx, (3) Upper view of gynoecium, (4) Single flower, (5) Whole flower.

2.7.1 Medicinal Uses

It has been used for centuries in traditional medicine to treat a variety of ailments, including respiratory problems, gastrointestinal disorders, and skin conditions.

2.7.2 Chemical Constituents

The chemical constituents of oregano include a variety of volatile oils, such as carvacrol and thymol, as well as flavonoids and terpenes. These compounds are thought to be responsible for the plant's medicinal properties, and have been shown to have antimicrobial, antioxidant, and anti-inflammatory effects.

2.7.3 Chemical Structure

The chemical structure of oregano is complex, with a variety of compounds contributing to its unique properties. Carvacrol, for example, is a monoterpene that has been shown to have antibacterial and antifungal properties. Thymol, another monoterpene found in oregano, is also known for its antimicrobial properties and is commonly used in mouthwashes and toothpaste.

2.7.4 Morphological Characters

In terms of morphological characteristics, oregano is a perennial herb that grows to a height of about 20–60 cm. It has dark green, hairy leaves and small, purple flowers that bloom in the summer months. The plant is often grown in gardens or pots and is easy to cultivate, making it a popular choice for those interested in growing their own medicinal plants.

2.7.5 Pharmacological Activity

One of the main pharmacological activities of oregano is its antibacterial effect. A number of studies have shown that oregano and its essential oil have the ability to inhibit the growth of a range of bacterial strains, including *Escherichia coli, Staphylococcus aureus*, and *Salmonella* typhi. This activity is thought to be due to the presence of carvacrol in the herb, which has been shown to disrupt the bacterial cell wall and inhibit the synthesis of proteins (Soltani et al. 2021).

In addition to its antibacterial effects, oregano also has antiviral and anti-inflammatory properties. A study found that oregano essential oil was effective in inhibiting the replication of yellow fever virus. Similarly, the anti-microbial effects of oregano have been demonstrated in animal studies, with oregano extract reducing acne and wound healing of mice (Soltani et al. 2021).

(Figure 2.5).

2.8 *SALVIA OFFICINALIS* L.

S. officinalis, also known as sage, is a perennial herb that has been used for centuries as a natural remedy.

2.8.1 Medicinal Uses

The herb has a number of medicinal uses, including the treatment of digestive issues, menopausal symptoms, and respiratory problems (Culpeper, 2016).

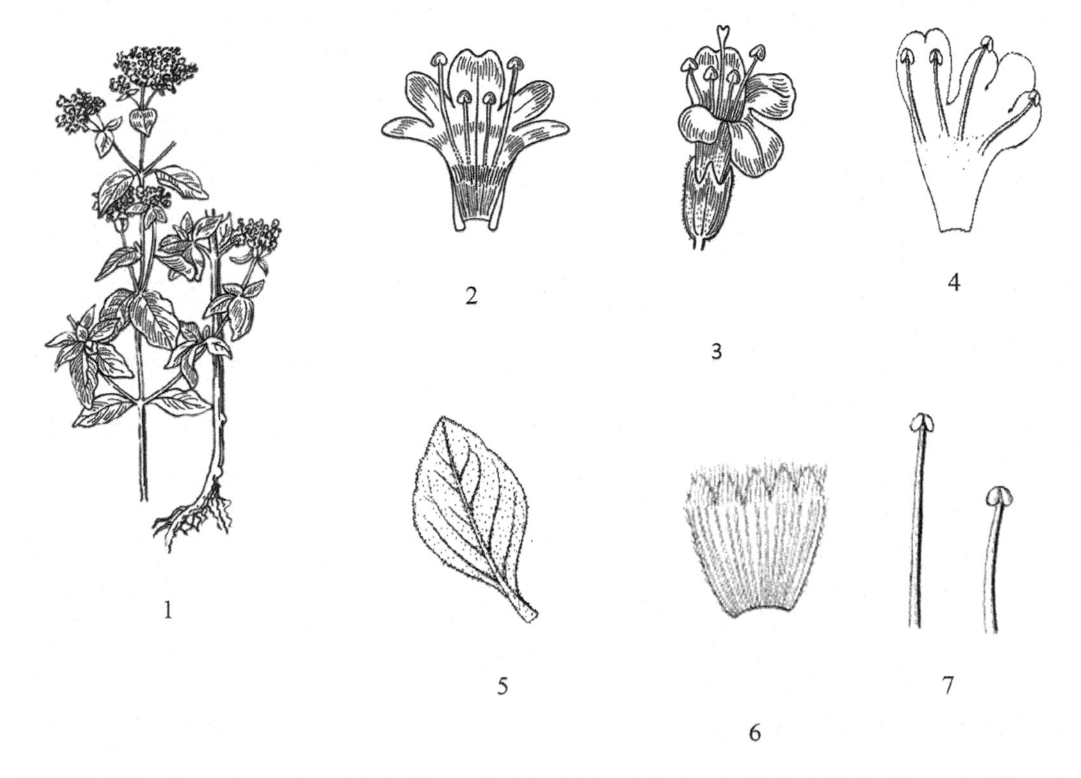

FIGURE 2.5 *Origanum vulgar* L. (1) Habit, (2) Open flower, (3) Single flower, (4) Position of anthers on corolla, (5) Single leaf, (6) Open calyx, (7) Androecium.

2.8.2 Chemical Constituents

One of the main chemical constituents of sage is salvinorin A, which is a potent psychoactive compound with hallucinogenic effects (Prisinzano et al., 2005). Other chemical constituents of sage include thujone, cineole, and rosmarinic acid, which have medicinal properties (Culpeper, 2016).

2.8.3 Chemical Structure

The chemical structure of salvinorin A is characterized by a number of structural features, including a diterpene skeleton and an acetate group (Prisinzano et al., 2005).

2.8.4 Morphological Character

In terms of morphological characteristics, sage is an aromatic herb that grows to a height of up to 1.5 m. The plant has oval leaves and small, blue or purple flowers (Culpeper, 2016).

2.8.5 Pharmacological Activities

One of the main pharmacological activities of sage is its psychoactive effects. Salvinorin A, a chemical constituent of sage, is a potent psychoactive compound that has been shown to produce hallucinations in humans (Prisinzano et al., 2005). This activity is thought to be due to the ability of salvinorin A to bind to specific receptors in the brain, leading to changes in perception and cognition (Prisinzano et al., 2005). In addition to its psychoactive effects, sage also has a range of other pharmacological activities. For example, the herb has been shown to have antioxidant, anti-inflammatory, and antimicrobial properties (Culpeper, 2016). A study by Culpeper (2016) found that sage tea was effective in reducing inflammation and swelling in the mouth, while other studies have shown that the herb has the ability to inhibit the growth of various bacterial strains (Culpeper, 2016) (Figure 2.6).

2.9 *MENTHA SPICATA* L.

Mentha spicata (M. spicata), commonly known as spearmint, is a perennial herb that is widely used in traditional medicine and as a natural remedy.

2.9.1 Medicinal Uses

The herb has a number of medicinal uses, including the treatment of digestive issues, respiratory problems, and skin conditions (El Menyiy et al. 2022).

2.9.2 Chemical Constituents

One of the main chemical constituents of spearmint is menthol, which has been shown to have analgesic, anti-inflammatory, and antimicrobial properties (El Menyiy et al., 2022). Other chemical constituents of spearmint include limonene, pulegone, and carvone, which also have medicinal properties (El Menyiy et al. 2022).

2.9.3 Chemical Structure

The chemical structure of menthol, one of the main chemical constituents of spearmint, is characterized by a methyl group attached to a cyclohexanol ring (El Menyiy et al. 2022).

FIGURE 2.6 *Salvia officinalis* L. (1) Habit, (2) Whole flower, (3) Open flower, (4) Anthers, (5) Calyx, (6) Ovary.

2.9.4 Morphological Characters

In terms of morphological characteristics, spearmint is an aromatic herb that grows to a height of up to 50 cm. The plant has lanceolate leaves and small, pink or white flowers (El Menyiy et al. 2022).

2.9.5 Pharmacological Characters

M. spicata L., commonly known as spearmint, is a perennial herb that has a number of pharmacological activities. These activities are due to the presence of various chemical constituents in the herb, including menthol, limonene, pulegone, and carvone (El Menyiy et al. 2022).

One of the main pharmacological activities of spearmint is its analgesic effect. A number of studies have shown that spearmint and its essential oil have the ability to reduce pain and discomfort. This activity is thought to be due to the presence of menthol in the herb, which has been shown to stimulate the production of pain-relieving compounds in the body (El Menyiy et al. 2022).

In addition to its analgesic effects, spearmint also has anti-inflammatory and antimicrobial properties. A study found that spearmint essential oil was effective in reducing inflammation and swelling in mice. Similarly, the antimicrobial properties of spearmint have been demonstrated in a number of studies, with spearmint extract inhibiting the growth of a range of bacterial and fungal strains (El Menyiy et al. 2022). (Figure 2.7).

2.10 *MATRICARIA CHAMOMILLA* L.

Matricaria chamomilla (M. chamomilla), also known as chamomile, is a plant that is commonly used in traditional medicine for its medicinal properties.

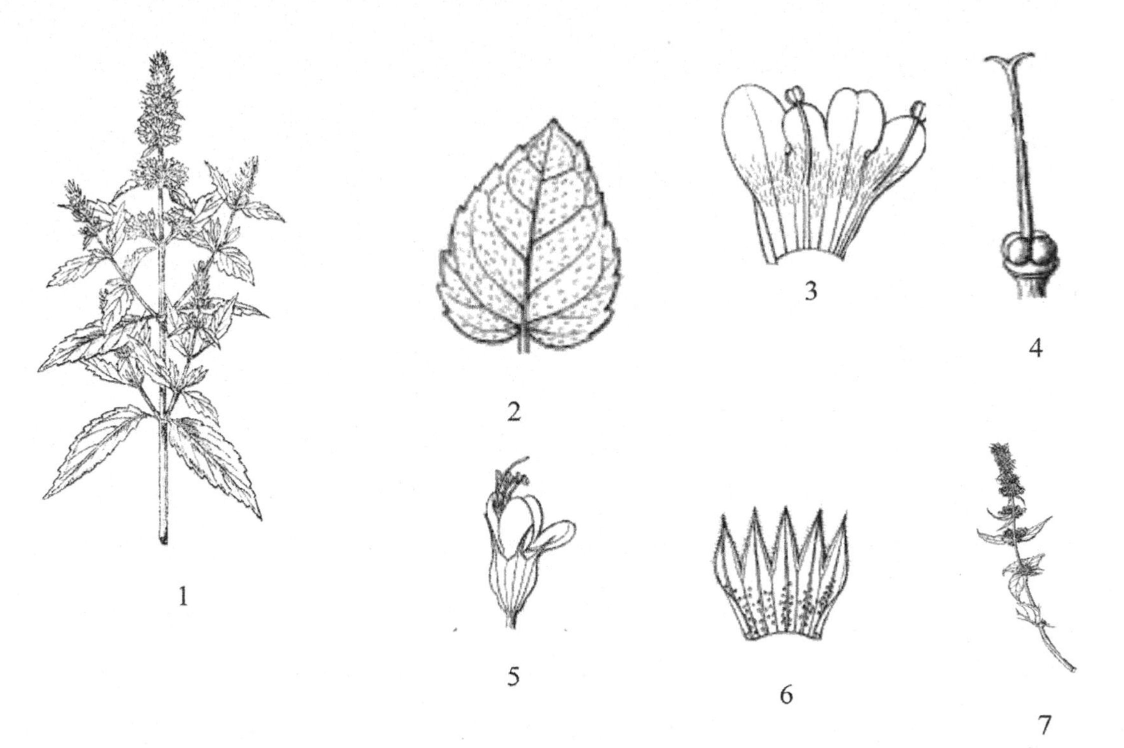

FIGURE 2.7 *Mentha spicata* L. (1) Habit, (2) Leaf, (3) Open corolla, (4) Gynobasic ovary, (5) Single flower, (6) Open calyx, (7) Inflorescence.

2.10.1 Medicinal Uses

The plant is often used to treat a range of conditions, including anxiety, insomnia, and digestive issues (El Mihyaoui et al. 2022).

2.10.2 Chemical Constituents

One of the main chemical constituents of chamomile is apigenin, which is a flavonoid that has been shown to have a number of medicinal properties (Salehi et al. 2019). Other chemical constituents of chamomile include chamazulene, quercetin, and rutin, which also have medicinal properties (El Mihyaoui et al. 2022).

2.10.3 Chemical Structure

The chemical structure of apigenin, one of the main chemical constituents of chamomile, is characterized by a ring structure with a pyran ring and a hydroxyl group attached to a six-carbon ring (Salehi et al. 2019).

2.10.4 Morphological Characters

In terms of morphological characteristics, chamomile is a small, annual herb that grows to a height of up to 60 cm. The plant has delicate, feathery leaves and small, daisy-like flowers that are white or yellow in color (El Mihyaoui et al. 2022).

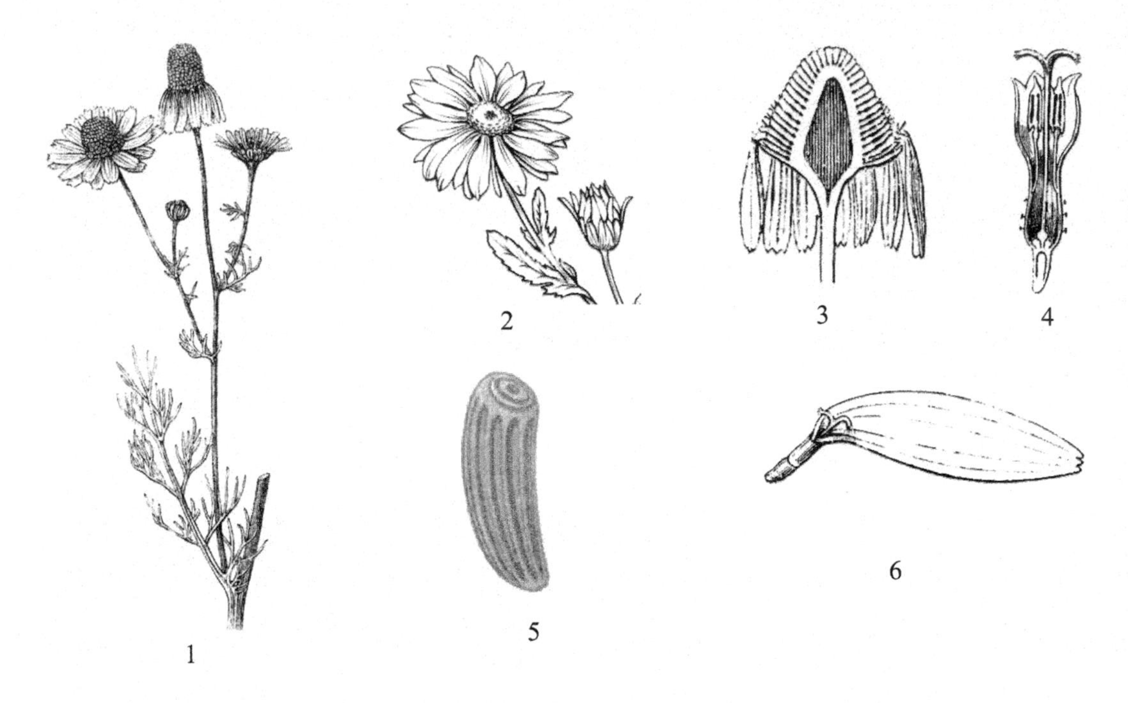

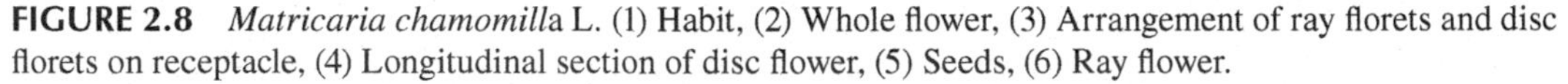

FIGURE 2.8 *Matricaria chamomilla* L. (1) Habit, (2) Whole flower, (3) Arrangement of ray florets and disc florets on receptacle, (4) Longitudinal section of disc flower, (5) Seeds, (6) Ray flower.

2.10.5 Pharmacological Activities

One of the main active constituents of chamomile is apigenin, which has been shown to have a number of medicinal properties. For example, apigenin has been shown to have anti-inflammatory effects in animal studies (Srivastava, Shankar, and Gupta 2010). It has also been shown to have antimicrobial properties, as it has been found to inhibit the growth of several different types of bacteria (Kamatou and Viljoen 2010). In addition to apigenin, chamomile also contains other active constituents, including chamazulene and bisabolol, which have been shown to have antioxidant and anti-inflammatory effects (Kamatou and Viljoen 2010) (Figure 2.8).

2.11 CONCLUSION

Aromatic medicinal plants are plants that have a distinctive, pleasant aroma and are used for medicinal purposes. These plants are often used in traditional and folk medicine, and have been used for centuries to treat a wide range of ailments. The morphology, or physical structure, of aromatic medicinal plants can vary widely, depending on the specific species. Some plants may have small, delicate flowers, while others may have large, bold leaves or seeds. Many aromatic medicinal plants are also characterized by their strong fragrance, which is produced by essential oils found within the plant. In terms of taxonomy, aromatic medicinal plants belong to a wide range of plant families, including the Lamiaceae, Apiaceae, and Rutaceae, Asteraceae, etc. families. These plants can be found all over the world, although they are more common in certain regions, such as the Mediterranean and Asia. Aromatic medicinal plants are used for a wide range of medicinal purposes, including the treatment of respiratory problems, digestive issues, and skin conditions. They are also used to help with relaxation and stress relief, and are often used in aromatherapy. The pharmacological action of aromatic medicinal plants is often due to the presence of active compounds, such as essential oils and flavonoids. These compounds can have a range of effects on the body, including anti-inflammatory, antibacterial, and antiviral effects. In terms of biogeography, aromatic medicinal plants can be found in a wide range of habitats, including forests, grasslands, and deserts. Some species may be native to a particular region, while others may have been introduced from other parts of the world. Overall, aromatic medicinal plants are an important resource for traditional and folk medicine, and have a wide range of medicinal uses and pharmacological actions. Further research is needed to fully understand the potential therapeutic benefits of these plants, and to develop new, effective treatments for a variety of health conditions.

REFERENCES

Bennett, Bradley C., and Michael J. Balick. 2008. "Phytomedicine 101: Plant Taxonomy for Preclinical and Clinical Medicinal Plant Researchers." *Journal of the Society for Integrative Oncology* 6 (4): 150.

Borugă, O., C. Jianu, C. Mişcă, I. Goleţ, A. T. Gruia, and F. G. Horhat. 2014. "Thymus Vulgaris Essential Oil: Chemical Composition and Antimicrobial Activity." *Journal of Medicine Life* 7 (Spec Iss 3): 56–60.

Buchbauer, Gerhard., Leopold Jirovetz, Walter Jager, Christine Plank, and Hermann Dietrich. 1993. Fragrance Compounds and Essential Oils with Sedative Effects upon Inhalation. *Journal of Pharmaceutical Sciences*, 82 (6): 660–664.

Burt, Sara. 2004. "Essential Oils: Their Antibacterial Properties and Potential Applications in Foods—a Review." *International Journal of Food Microbiology* 94 (3): 223–53.

Cain, A. J. "taxonomy". Encyclopedia Britannica, 27 Apr. 2023, https://www.britannica.com/science/taxonomy. Accessed 5 May 2023.

Chan, Kelvin. 1995. "Progress in Traditional Chinese Medicine." *Trends in Pharmacological Sciences* 16 (6): 182–187.

Chen, Shi-Lin, Hua Yu, Hong-Mei Luo, Qiong Wu, Chun-Fang Li, and André Steinmetz. 2016. "Conservation and Sustainable Use of Medicinal Plants: Problems, Progress, and Prospects." *Chinese Medicine* 11: 1–10.

Culpeper, Nicholas. 2016. *The complete herbal*. anboco.

Edwin, Jothie Richard, Illuri Ramanaiah, Bethapudi Bharathi, Senthilkumar Anandhakumar, Anirban Bhaskar, Chandrasekaran Chinampudur Velusami, Deepak Mundkinajeddu, and Amit Agarwal. 2016. "Anti-stress Activity of Ocimum Sanctum: Possible Effects on Hypothalamic–Pituitary–Adrenal Axis." *Phytotherapy Research* 30 (5): 805–814.

Farnsworth, Norman R, and Djaja D Soejarto. 1991. "Global Importance of Medicinal Plants." *The Conservation of Medicinal Plants* 26: 25–51.

Imelouane, B., Amhamdi Hassan, J. P. Wathelet, M. Ankit, Khadija Khedid, and Ali Ali El Bachiri. 2009. "Chemical composition and antimicrobial activity of essential oil of thyme (Thymus vulgaris) from Eastern Morocco." *International Journal of Agriculture and Biology* 11(2): 205–208.

Imtara, Hamada, Noori Al-Waili, Abderrazak Aboulghazi, Abdelfattah Abdellaoui, Thia Al-Waili, and Badiaa Lyoussi. 2021. "Chemical Composition and Antioxidant Content of Thymus Vulgaris Honey and Origanum Vulgare Essential Oil; Their Effect on Carbon Tetrachloride-Induced Toxicity." *Veterinary World* 14 (1): 292.

Kamatou, Guy P P, and Alvaro M Viljoen. 2010. "A Review of the Application and Pharmacological Properties of α-Bisabolol and α-Bisabolol-Rich Oils." *Journal of the American Oil Chemists' Society* 87: 1–7.

Kim, Sunghwan, Jie Chen, Tiejun Cheng, Asta Gindulyte, Jia He, Siqian He, Qingliang Li, Benjamin A Shoemaker, Paul A Thiessen, and Bo Yu. 2021. "PubChem in 2021: New Data Content and Improved Web Interfaces." *Nucleic Acids Research* 49 (D1): D1388–D1395.

Kumar, Amit, Anu Rahal, Sandip Chakraborty, Ruchi Tiwari, Shyma K Latheef, and Kuldeep Dhama. 2013. "Ocimum Sanctum (Tulsi): A Miracle Herb and Boon to Medical Science-A Review." *International Journal of Agronomy and Plant Production* 4 (7): 1580–1589.

Liu, Rui Hai. 2004. "Potential Synergy of Phytochemicals in Cancer Prevention: Mechanism of Action." *The Journal of Nutrition* 134 (12): 3479S–3485S.

Menyiy, Naoual El., Hanae Naceiri Mrabti., Nasreddine El Omari., Afaf EI Bakili., Saad Bakrim., Mouna Mekkaoui, Abdelaali Balahbib., Ehsan Amiri-Ardekani., Riaz Ullah., Ali S. Alqahtani., Abdelaaty A. Shahat., Abdelhakim Bouyahya. 2022. "Medicinial Uses, Phytochemistry, Pharmacology, and Toxicology of Mentha Spicata." *Evid Based Complement Alternat Med.* Apr 12; 2022: 7990508.

Mihyaoui, Amina El, Joaquim C. G. Esteves da Silva, Saoulajan Charfi, María Emilia Candela Castillo, Ahmed Lamarti, and Marino B. Arnao. 2022. "Chamomile (Matricaria Chamomilla L.): A Review of Ethnomedicinal Use, Phytochemistry and Pharmacological Uses." *Life* 12 (4): 479.

Moss, Mark, Jenny Cook, Keith Wesnes, and Paul Duckett. 2003. "Aromas of Rosemary and Lavender Essential Oils Differentially Affect Cognition and Mood in Healthy Adults." *International Journal of Neuroscience* 113 (1): 15–38.

Nelson, Lewis S., Richard D. Shih, Michael J. Balick, and Kenneth F. Lampe. 2007. *Handbook of poisonous and injurious plants.* Springer, New York: New York Botanical Garden.

Patil, Shashank M., Ramith Ramu, Prithvi S. Shirahatti, Chandan Shivamallu, and Raghavendra G. Amachawadi. 2021. "A Systematic Review on Ethnopharmacology, Phytochemistry and Pharmacological Aspects of Thymus Vulgaris Linn." *Heliyon* 7 (5): e07054.

Pradhan, Deepak, Prativa Biswasroy, Jitu Haldar, Priya Cheruvanachari, Debasmita Dubey, Vineet Kumar Rai, Biswakanth Kar, Durga Madhab Kar, Goutam Rath, and Goutam Ghosh. 2022. "A Comprehensive Review on Phytochemistry, Molecular Pharmacology, Clinical and Translational Outfit of Ocimum Sanctum L." *South African Journal of Botany* 150: 342–360.

Prisinzano, Thomas E. 2005. Psychopharmacology of the hallucinogenic sage Salvia divinorum. *Life sciences* 78(5): 527–531.

Rahbardar, Mahboobeh Ghasemzadeh, and Hossein Hosseinzadeh. 2020. "Therapeutic Effects of Rosemary (Rosmarinus Officinalis L.) and Its Active Constituents on Nervous System Disorders." *Iranian Journal of Basic Medical Sciences* 23 (9): 1100.

Ramos da Silva, Luiz Renan, Oberdan Oliveira Ferreira, Jorddy Nevez Cruz, Celeste de Jesus Pereira Franco, Taina Oliveira dos Anjos, Marcia Moraes Cascaes, Wanessa Almeida da Costa, Eloisa Helena de Aguiar Andrade, and Mozaniel Santana de Oliveira. 2021. "Lamiaceae Essential Oils, Phytochemical Profile, Antioxidant, and Biological Activities." *Evidence- Based Complementary Medicine.* 1–18.

Brickell, Christopher. 2019. RHS encyclopedia of plants and flowers. Dorling Kindersley Ltd.

Salehi, Bahare, Alessandro Venditti, Mehdi Sharifi-Rad, Dorota Kręgiel, Javad Sharifi-Rad, Alessandra Durazzo, Massimo Lucarini, Antonello Santini, Eliana B. Souto, and Ettore Novellino. 2019. "The Therapeutic Potential of Apigenin." *International Journal of Molecular Sciences* 20 (6): 1305.

Shaw, David, Judith Marion Annett, B. Doherty, and J. C. Leslie. 2007. "Anxiolytic Effects of Lavender Oil Inhalation on Open-Field Behaviour in Rats." *Phytomedicine* 14 (9): 613–620.

Simpson, Beryl Brintnall, and Molly Conner Ogorzaly. 1995. *Economic Botany: Plants in Our Worl.* McGraw-Hill Inc., New York, USA.

Simpson, Michael G. 2019. *Plant Systematics.* Academic press, Elsevier Science.

Singh, Gurcharan. 2019. *Plant Systematics: An Integrated Approach.* CRC Press, Boca Raton.

Soltani, Saba, Abolfazl Shakeri, Mehrdad Iranshahi, and Motahareh Boozari. 2021. "A Review of the Phytochemistry and Antimicrobial Properties of Origanum Vulgare L. and Subspecies." *Iranian Journal of Pharmaceutical Research: IJPR* 20 (2): 268.

Srivastava, Janmejai K., Eswar Shankar, and Sanjay Gupta. 2010. "Chamomile: A Herbal Medicine of the Past with a Bright Future." *Molecular Medicine Reports* 3 (6): 895–901.

Srivastava, Sunil Kumar, and Naveen Kumar Singh. 2020. "General Overview of Medicinal and Aromatic Plants: A." *Journal of Medicinal Plants* 8 (5): 91–93.

Tisserand, Robert, and Rodney Young. 2013. *Essential Oil Safety: A Guide for Health Care Professionals.* Elsevier Health Sciences, London.

Triveni, Kuldeep Kumar, Amit Kumar Singh, Rahul Kumar, Vaishnavee Gupta, and Kishu Tripathi. 2013. "Ocimum Sanctum Linn: A Review on Phytopharmacology and Therapeutic Potential of Tulsi." *International Journal of Pharmaceutical Research* 3 (2): 148–151.

World Health Organization. 2013. *WHO Traditional Medicine Strategy: 2014-2023.* World Health Organization.

3 Trends in Sustainable Use and Management of Medicinal and Aromatic Plants

Utilization and Development

Kyriakos Giannoulis and Spyridon A. Petropoulos
University of Thessaly

Alexios Alexopoulos
University of the Peloponnese

CONTENTS

3.1 INTRODUCTION: BACKGROUND AND DRIVING FORCES

Medicinal and aromatic plants (MAPs) and their herbal extracts have been widely used throughout the centuries for their bioactive properties (Alexieva, Popova, and Mihaylova 2020). The healing effects of these species were identified and exploited in the folk and traditional medicine in various forms, while they constituted the roots of modern pharmacology and drug science (Goldman 2001; Gurib-Fakim 2006). The growing awareness of consumers about the safety of food products that were consumed on a regular basis has resulted in the increasing availability of a wide range of herbal products obtained from MAPs in many points of retail sale (supermarket and local store shelves) around the world (Khan and Rauf 2014). Many of these herbal products are sourced from native populations in various nations, resulting in the (over)-exploitation and eventually, along with land use changes and anthropogenic activities, in the extinction and the genetic erosion of some species, as well as the loss of genetic diversity (Canter, Thomas, and Ernst 2005). Moreover, the increasing consumers' interest in natural compounds has rekindled bioprospecting, which combined with the limited availability of raw materials could increase the risk of biopiracy (McGaw et al. 2005). Therefore, domestication of wild MAPs and cultivation under commercial conditions is presented as a viable and sustainable alternative to harvesting plants from their wild habitats, although several issues have to be addressed for this purpose. For example, some doubts arise related to the impact of the growth in MAP-based products that demands may have on the ecosystems and whether farmers

DOI: 10.1201/b22842-3

will continue to pick native plants or will be forced to cultivate these species under commercial conditions (Schippmann, Leaman, and Cunningham 2006). Moreover, the consumer behavior toward wild picked and commercially produced plants will be an important driver for the development and sustainability of this sector (Williams, Victor, and Crouch 2013; Smith-Hall, Larsen, and Pouliot 2012). Finally, the propagation of several species has to be extensively studied due to several difficulties in sexual propagation (low seed germination percentage, dormancy of seeds), while alternative techniques which involve asexual propagation (*in vitro* propagation, micro-propagation) also need to be developed and standardized (Rout, Samantaray, and Das 2000; Mehalaine and Chenchouni 2021).

Medicinal plants are globally valuable sources of herbal products, but unfortunately they are disappearing at an alarmingly increasing speed due to anthropogenic activities (Bakoumé 2016). Furthermore, the research on MAPs is attracting a lot of national and international attention, especially because of its many applications in poverty alleviation and healthcare in undeveloped and developing countries (Mukherjee, Venkatesh, and Kumar 2007; Okigbo, Eme, and Ogbogu 2008), as well as the increasing usage in developed countries (Salmerón-Manzano, Garrido-Cardenas, and Manzano-Agugliaro 2020). Scientific data on the capacities of commercial plant production systems, and techniques of usage, on the other hand, are still in the early stages of development.

According to recent estimates, there are around 18,000 aromatic plant species, 60,000 medicinal plant species, and at least 50 families that include several MAPs (*Apiaceae*, *Asteraceae*, *Geraniaceae*, *Lamiaceae*, etc.) (Salmerón-Manzano, Garrido-Cardenas, and Manzano-Agugliaro 2020), while several of these species are indigenous in specific regions of the world with interesting and promising bioactive properties and health effects (Craker, Gardner, and Etter 2003; Axiotis, Halabalaki, and Skaltsounis 2018; Petrakou, Iatrou, and Lamari 2020; Okigbo, Eme, and Ogbogu 2008). MAPs are pest- and disease-resistant and can be grown "organically" or under sustainable cropping systems since their input requirements are minimal, while their essential oils are usually part of plant defense mechanisms that can be used instead of chemical compounds for crop protection (Chrysargyris, Petropoulos, and Tzortzakis 2022; Seidler-Łożykowska et al. 2015).

For better understanding of MAPs, it is important to note the distinction between medicinal and non-medicinal plants, which, according to the literature (Slikkerveer 2006; Schippmann, Leaman, and Cunningham 2006; Smith-Hall, Larsen, and Pouliot 2012), are those plants which are utilized for healthcare in both allopathic and traditional medical systems, and also include a wide range of species such as condiments, food, aromatics, and cosmetics (Kunwar et al. 2006; He 2015; Yu et al. 2006).

Herbal medicines have their origins in each country's traditional knowledge with a high rate of use (Gurib-Fakim 2006), while several of these plants and the bioactive compounds they contain are commonly used in modern medicine (Karunamoorthi et al. 2013; Jeelani et al. 2018). However, the lack of a comprehensive data base makes it difficult to conduct a comprehensive research program to advance the current state of knowledge. Due to the increased demand for natural products with healing properties today, the following global trends can be identified: (a) there is a gradual shift from cultivation of wild native plants to collection from their natural environment, (b) there is increasing research related to the composition of the ingredients in these plants and the association of the contained substances with healing properties, (c) there is still a lack of research on the impact of major factors (such as climate, irrigation, and nutrition, among others) on the yield and quality characteristics of several aromatic and medicinal plants, and (d) there are few studies on the proper and safe commercialization of medicinal plants.

The present chapter aims to provide information regarding the current trends in the cropping sector of MAPs, as well as the key drivers for the sustainable management and development of MAPs and MAP-based products. A special section is devoted to the propagation techniques and the research needed to optimize and standardize the sexual and asexual propagation protocols which are essential for the domestication and commercialization of wild and unexploited species and their protection against the threat of genetic erosion and extinction. Finally, future and conclusionary remarks will highlight the next steps and requirements needed to ensure the sustainable development of MAPs and MAP-related products.

3.2 CURRENT TRENDS IN MEDICINAL AND AROMATIC PLANTS

Recently, a shift in the management of various MAPs as cultivated plants in the last decades has been recorded following the increasing market demands and consumer awareness for healthy and functional foods (Chrysargyris et al. 2019; Gurib-Fakim 2006; Smith-Hall, Larsen, and Pouliot 2012; Jeelani et al. 2018; Canter, Thomas, and Ernst 2005). Recent studies have tested the possible effects of irrigation and nitrogen fertilization on various MAP crops growth characteristics (Geerts and Raes 2009; Giannoulis, Kamvoukou, Gougoulias, and Wogiatzi 2020a; Sifola and Barbieri 2006). Despite the fact that MAPs can be cultivated as non-irrigated crops, several studies have demonstrated that irrigation has a major impact on both quantitative and qualitative features depending on the growth stage of plants (Geerts and Raes 2009; Giannoulis, Kamvoukou, et al. 2020a; Giannoulis, Kamvoukou, Gougoulias, and Wogiatzi 2020b; Marques, Bernardi Filho, and Frizzone 2012). For example, the application of irrigation in chamomile (*Matricaria chamomilla*) plants increased significantly the yield of fresh and dried flowers, whereas it reduced essential oil content of plants (Giannoulis, Kamvoukou, et al. 2020b). On the other hand, *Tetragonia tetragonioides* (Pall) O. Kuntze plants grown under arid conditions recorded a higher yield than irrigated ones, suggesting different mechanisms of plant adaptation to water deficit (Bekmirzaev, Beltrao, and Ouddane 2019), while Abbaszadeh, Farahani, and Morteza (2009) suggested that moderate water stress may increase the essential oil yield of lemon balm plants. In this context, deficit irrigation is proposed as a sustainable practice that is suitable to MAPs cropping and may allow the exploitation of arid and semi-arid regions under low-input cropping systems (Okwany, Peters, Ringer, and Walsh 2012; Okwany, Peters, Ringer, Walsh, and Rubio 2012).

Another important cropping management practice is nitrogen fertilization, and several studies have investigated its effect on growth and development of MAPs and suggested a favorable impact on harvested yield (Baranauskiene et al. 2003; Omer 1999; Ozguven, Ayanoglu, and Ozel 2006; Karioti et al. 2003; Sifola and Barbieri 2006). According to Sotiropoulou and Karamanos (2010), nitrogen is a key ingredient which has a considerable impact on dry matter production and essential oil output without negative effects on essential oil content. The majority of studies have discovered that employing chemical fertilizers to increase nitrogen availability can improve not only quantitative but also qualitative traits of MAPs (Anwar et al. 2005; Giannoulis, Kamvoukou, et al. 2020a; Giannoulis et al. 2021; Karioti et al. 2003; Khalil, Kandil, and Swaefy Hend 2008). Apart from nitrogen, micronutrients may also have beneficial effects on the essential oil yield and quality. For example, the foliar application of Zn resulted in better essential oil yield in *Ocimum sanctum* plants, whereas a variable effect on the major essential oil components (eugenol, 1,8-cineole and methyl chavicol) was recorded (Moghimipour et al. 2017). Moreover, the foliar spraying of chamomile plants with Zn and Fe significantly improved essential oil content and yield (Nasiri et al. 2010).

Nowadays, there is an urgent need to achieve a more environmentally friendly transition from wild harvested MAPs to commercial production following best practice guides. This would be helpful in specific regions of the world where medical care is based mostly on wild MAPs; for example, in India about 7,500 species out 17,000 are commonly used for medicinal purposes, while in Africa, 80% of the population relies on traditional medicine for its healthcare (McGaw et al. 2005). According to WHO, the demand for MAPs is increasing with an annual growth of 15–25% and is expected to reach US $5 trillion by 2050 (Kala, Dhyani, and Sajwan 2006). For this purpose, the International Standard for Sustainable Wild Collection of Medicinal and Aromatic Plants has been established to provide protocols for harvesting of wild MAPs while ensuring the sustainability of ecosystems and natural habitats and addressing the social and economic needs (Leaman 2008). On the other hand, the Interactive European Network for Industrial Crops and their Applications (IENICAs) provides information regarding the potential industrial uses of MAPs, which except for food purposes include the production of essential, pharmaceuticals, herbal health products, natural dyes and colorants, cosmeceuticals, products for personal care, and products for plant protection (Lubbe and Verpoorte 2011).

Determining the overall quality of the final product (medicine) depends on the manufacturing techniques used; therefore, it makes perfect sense that the idea of sustainable production is crucial in the growth of MAPs. Aloe, chamomile, rose, geranium, and other MAPs are grown with sustainable practices; however, a standardized cultivation procedure needs to be devised for as many MAPs as feasible that are utilized solely for making medications but also for human consumption. However, it is important to highlight that cropping managements should conform to specific protocols, especially when the final products are indented for direct human use (e.g., herbal preparations or essential oils); therefore, crop production should ensure human safety, for example, conform to the designation of generally recognized as safe (GRAS) of the Food and Drug Administration (FDA) of the United States or to the European Union's food safety policy (Skendi et al. 2020; Steinhoff 2005; Lubbe and Verpoorte 2011). For this purpose, biological fertilizers, as well as the utilization of beneficial microorganisms in the soil and their association with plant roots, have become commonplace in organic crops and aromatic–medicinal plants. The research of mycorrhizae fungi is growing, and the obtained results indicate the positive impact on several MAPs cultivation, such as *Salvia officinalis*, *Origanum vulgare*, *Ocimum basilicum*, Thymus *vulgaris*, *Melissa officinalis*, *Lavandula angustifolia,* and *Achillea millefolium,* among others (Allison 2002; Copetta, Lingua, and Berta 2006; Eshaghi Gorgi et al. 2021; Geneva et al. 2010; Giannoulis, Evangelopoulos, Gougoulias, and Wogiatzi 2020a,b; Giannoulis et al. 2021; Helgason and Fitter 2005; Khaosaad et al. 2006; Tarraf et al. 2015; Zolfaghari et al. 2013).

Numerous abiotic factors today severely affect a wide range of crops. Salinity, heavy metals, loss of organic matter/desertification, and weather extremities (droughts, high temperatures, excessive and untimely rainfalls, etc.) are a few of them that are frequently highlighted. Moreover, the use of biostimulants in organic agriculture has been a trend in recent years that is highly appreciated in MAPs cultivation, especially under environmental constraints. Sustainable agriculture production is critical for natural resource preservation in addition to sustaining or even boosting yields. In a time of climate change, reducing dependency on agrochemicals and/or boosting the efficiency of their use is pivotal. Because of the possible detrimental impacts on the environment, using agrochemicals to improve yield has to be reconsidered and alternative means have to be implemented (Ruzzi and Aroca 2015). Biostimulants are natural products that contain physiologically active compounds with properties that improve plant growth attributes such as plant height and yield, nutrients and water-use efficiency, crop and product quality, and resistance to abiotic stress (El-Khateeb, El-Attar, and Nour 2017; Giannoulis, Evangelopoulos, Gougoulias, and Wogiatzi 2020; Posmyk and Szafrańska 2016; Yin et al. 2012; Zulfiqar et al. 2020). Furthermore, according to Kwiatkowski and Juszczak (2012), biostimulants may boost growth rates, while they also affect chlorophylls, micronutrients, and carotenoids content (Hanafy, Ashour, and Sedek 2018). Finally, the application of biostimulants may have an impact on oil yield and essential oil quality, where an increase in oil content of dried herb and herb yield could increase the essential oil yield/plant, thus increasing the overall yield per harvested area (Giannoulis, Evangelopoulos, et al. 2020; Mohammadipour et al. 2012; Said-Al Ahl, El Gendy, and Ea 2016). Other compounds with biostimulatory effects are salicylic and jasmonic acid and chitosan, whose foliar application on MAPs, such as lavender (*Lavandula angustifolia*), lemon balm (*Melissa officinalis*), lemongrass (*Cymbopogon flexuosus*), summer savory (*Satureja hortensis*), and Mediterranean thyme (*Thymbra spicata*), promoted plant growth and alleviated the negative effects of abiotic stressors (Miclea et al. 2020; Safari et al. 2019; Idrees et al. 2010; Momeni et al. 2020; Pirbalouti et al. 2014).

3.3 THERMOCHEMICAL ROUTES FOR BIOMASS CONVERSION TO FUELS

Currently, a variety of MAPs are consumed and used in order to support a healthy lifestyle following the consumer-driven market trends. Although more work has to be carried out to bring together knowledge and skills in plant science to investigate alternate uses of MAPs and to promote new bio-based products, education and training about the culinary and therapeutic characteristics of natural herbs and plants are also needed. The majority of the raw material supply available in the market originates from plants collected in the wild, with the exception of a few species that are grown commercially (Schippmann et al. 2005). Moreover, it is rather difficult to understand the current state of the MAPs sector since, unlike other agricultural commodities, organized data about the cultivated area, total production, and the amounts of traded/exported products are usually unavailable or partially available. Therefore, both buyers and growers are confused as a result of this broken link between supply and demand, which hampers the further exploitation and development of the MAPs sector. Without a reliable database and market information system, accurate targeting of both domestic- and foreign-specialized markets is impossible.

The development of such a database through collaboration among many stakeholders could increase the overall effectiveness of the production and marketing channels. Such a cooperation between stakeholders could lead to the creation of a reliable system for exchanging MAPs both locally and internationally. In the late 1970s, World Health Organization (WHO) published a list of 21,000 medicinal species. However, more natural species are used as medicines in China, which can lead to the misconception that there are many more plant species used for medicinal purposes. This is especially true when one considers that there are approximately 500,000 blooming plants in the world (Pimm and Joppa 2015).

One of the main issues for the promotion and development of MAPs commercial cultivation is the shortage of recognized grade propagation material. The plants are currently harvested using unethical methods, inhibiting natural reproduction in the wild. This is especially dangerous for plants whose roots, bark, flowers, and fruits are employed in traditional medicine (Smitha 2013). Plant propagating material is of great importance for the commercial cultivation of MAPs due to its significant effect on yield and the quality of the final products. The production of propagating material on a large scale allows the commercial cultivation of MAPs and at the same time limits the negative impacts on biodiversity and, in general, on the environment in which these plants grow naturally as a consequence of collecting plants or propagating material from the wild (Grigoriadou et al. 2019).

The cultivation of MAPs, as in other cultivated plant species, requires the use of propagating material with high-quality characteristics for the successful establishment of a new crop in the field. Such quality characteristics include the health status (plant material free from diseases), the absence of impurities from other plant species and the genetic purity, the high rate of seed germination, the high rate of seedling vitality, the high rate of rooting of organs/tissues used as propagating material (e.g., stem cuttings, bulbs), etc. (Mehalaine and Chenchouni 2021). To achieve the above objectives, a program to produce propagating material must be properly adapted to the way each MAP species reproduces.

3.3.1 Sexual Propagation

Many MAPs are easily propagated with seeds as a means of sexual reproduction, and the seeds of many annual and perennial MAPs are used as the main propagating material. This method is inexpensive and allows the reproduction of plant species on a large scale. In addition, the production of seeds free of pathogens is always easier than the production of asexual planting material (e.g., rhizomes, bulbs, cuttings, etc.). Sexual propagation also allows to maintain genetic variability and consequently plant biodiversity. This inherent variability results in having propagating material available with good adaptability and high resistance to possible abiotic and biotic pressure from climate change, enemies, and diseases (Alcántara-Flores et al. 2017).

For the establishment of new MAPs, crop seeds are sown either directly in the field or in a nursery and then seedlings transplanted in the field. Direct seed sowing is mainly preferred for seeds with large size and/or high germination rates. In contrast, plant species with low seed germination or small seed size (e.g., oregano, basil) or difficulty in finding large quantities of seeds are preferred to be sown either in open nurseries (e.g., a part of the field) or protected (under cover) nurseries.

Seeds are either directly sown in the field or in nurseries and then transplanted in the field, usually after two to four months. In some cases, plants are cultivated in hydroponic cultivation systems, for example, floating system or aeroponic cultivation system, which allows the fastest plant growth mainly due to the better plant nutrition, the constant, and appropriate temperature of the root growth environment and better root aeration (especially in aeroponic cultivation system) which allows faster root growth and consequently faster seedling growth (Traykova et al. 2019; Hayden 2006; Lakhiar et al. 2018; Maluin et al. 2021; DeKalb et al. 2014).

Seeds are used as the main propagating material for the establishment of many MAPs, such as several species of the genus *Origanum* (e.g., *Origanum compactum*, *Origanum vulgare*, *Origanum onites*, *Origanum dictamnus* (Thanos and Doussi 1995; Laghmouchi et al. 2017), various species of the genus *Sideritis* (e.g., *Sideritis syriaca* ssp. *Syriaca*) (Thanos and Doussi 1995), *Matricaria chamomilla* (Albrecht and Otto 2020), *Ocimum basilicum* (B. Kumar 2012), *Valeriana officinalis* (Wiśniewski et al. 2016), *Mentha spicata* (Vining et al. 2020), *Hypericum perforatum* (Alan, Murch, and Saxena 2015), *Pimpinella anisum* (Amer and Omar 2019), *Petroselinum crispum* (Marthe 2021), *Glycyrrhiza glabra* (Zimnitskaya 2009; Karkanis et al. 2016), *Laurus nobilis* (Konstantinidou, Takos, and Merou 2008), *Rosmarinus officinalis* (Novak and Blüthner 2020), *Echinacea* species (Abbasi et al. 2007), *Saturejia thymbra* (Papadatou et al. 2015), *Thymus vulgaris* (Vouillamoz and Christ 2020), *Capparis spinosa* (Mohammad, Kashani, and Azarbad 2012), *Coriandrum sativum* (Ali et al. 2017), *Cuminum cyminum* (Paatre Shashikanthalu, Ramireddy, and Radhakrishnan 2020), *Lavandula angustifolia* (Urwin, Horsnell, and Moon 2007), *Humulus lupulus* (Clapa and Hârţa 2021), *Origanum majorana* (Baatour et al. 2010), *Foeniculum vulgare* (Tahaei, Soleymani, and Shams 2016), *Melissa officinalis* (Saglam et al. 2004), *Achillea millefolium* (Nejad, Bistgani, and Barker 2022; Allison 2002), *Pelargonium roseum* (Narnoliya, Jadaun, and Singh 2019), *Salvia officinalis* (Dastanpoor et al. 2013), *Crithmum maritimum* (Atia et al. 2011), and *Pancratium maritimum* (Nikopoulos and Alexopoulos 2008; Nikopoulos, Nikopoulou, and Alexopoulos 2008).

However, the extended collection of MAPs seeds from habitats where they are naturally grown for their commercial cultivation negatively affects the ability to maintain the plant species in the wild. On the other hand, seed production by mother plants grown in areas with high differences in climate conditions may be a risk due to a possible failure to transfer genetic material to the progeny after many years of cultivation (Aguilar et al. 2019). In addition, in several occasions, the use of seeds as propagating material is not economically advantageous because of the low germination rates, the low seedling vigor, or the great phenotypic variability of seedlings which does not favor the production of plants with high and stable quality characteristics. For example, the propagation of *Pimpinella anisum* by seeds results in non-uniform growth of seedlings (Amer and Omar 2019).

3.3.2 Asexual Propagation

Asexual (vegetative) propagation of MAPs is based on the inherent ability of plant parts or organs, such as stem cuttings (usually green or woody shoots), underground shoots (e.g., offshoots, rhizomes, stolons), and bulbs or corms to reproduce the entire plant. In some perennial MAPs, the formation of rooted underground shoots in high numbers occurs when they become old (usually after three to four years of cultivation). Those rooted shoots can be separated from the mother plant (division) and being used as propagating material for the establishment of a new crop. The asexual propagation is applied for the commercial cultivation of many MAPs, despite their ability to produce seeds, due to various reasons related to cost production and the production of non-uniform seedlings. In addition, the vegetative propagation is applied for the establishment of new MAP crops when mother plants cannot be propagated by seed due to failure in seed production. The propagation of MAPs by cuttings (stem cuttings) is a highly effective method for mass production propagating material. It is also relatively cheap in comparison with other vegetative methods. The success of the method depends on the ability of the cuttings to produce roots, which in turn depends significantly on the plant species and the type of cutting. It is therefore necessary to have special knowledge about the ability to reproduce each plant species from the appropriate cutting.

The method of dividing existing plants (roots, rooted stems, rhizomes) can be carried out on many perennial MAPs but requires the existence of old plants from which a large number of propagating materials can be obtained. This method is also recommended, since in many perennial MAPs, the plants are thinned (e.g., from crowns) to maintain a healthy crop and to enhance plant growth and productivity. For example, *O. vulgare*, a species of great economic interest, is not propagated with seeds because of seed dormancy (both endogenous and exogenous) which results in low seed germination rates and lack of uniform germination and consequently in delays in plant growth and entry into production (Kuris, Altman, and Putievsky 1981; Vleeshouwers, Bouwmeester, and Karssen 1995). Thus, the vegetative propagation of the species is often preferred to sexual propagation and green stem cuttings (Pratap et al. 2017) or rooted stem cuttings derived from the division of older plants are used for this purpose (Dehghanpour Farashah et al. 2011; Kuris, Altman, and Putievsky 1981; Vleeshouwers, Bouwmeester, and Karssen 1995). Similarly, propagation of *O. majorana* by cuttings and by division is reported as an alternative way to avoid the phenomenon of limited essential oil production and degraded product quality recorded in seed-derived plants (EL-Keltawi, Abdou, and Bishay 1985).

In the case of *Valeriana officinalis* and *Valeriana jatamansi*, the seeds often have a very slow germination rate and remain dormant for a long time period (Kaur et al. 1999). For the rapid and large-scale production of propagating material of *Valeriana* species, the use of offshoots or the *in vitro* propagation is usually preferred to seed propagation (Wiśniewski et al. 2016). *Lavandula angustifolia* is propagated by stem cuttings (woody ones are not preferred) because plants derived from seeds have slow growth rates and great variability in plant growth and composition of the essential oil produced (Echeverrigaray, Basso, and Andrade 2005). *Rosmarinus officinalis* is propagated vegetatively by stem cuttings or division of roots (Hammer and Junghanns 2020), while *Laurus nobilis* is propagated by stem cuttings (Parlak and Semizer-Cuming 2012). *Thymus vulgaris* can be propagated by plant division or stem cuttings when clones are needed for plant reproduction; however, the method is more expensive than seed propagation (Vouillamoz and Christ 2020).

Low seed viability and low germination rate in species of the genus *Ocimum* have been reported consistently causing difficulties in the production of planting material on a large scale. Alternatively, Ahuja, Verma, and Grewal (1982) reported the ability of producing plants from nodal segments and axillary buds in *O. gratissimum* and *O. viride*. Similarly, Lim and Eom (2013) proposed the propagation of *Ocimum basilicum* by green stem cuttings on a small scale.

Pelargonium roseum is propagated by fresh (green) stem cuttings (Mahdieh, Yazdani, and Mahdieh 2013), while *Saturejia thymbra* is propagated by stem cuttings (Papadatou et al. 2015) and *Melissa officinalis* is propagated by stem cuttings with roots (Saglam et al. 2004). *Glycyrrhiza glabra* is propagated by underground stem cuttings with two or more meristems (buds) (Dagar et al. 2015), while propagation of *Humulus lupulus* and *Mentha spicata* is performed with stem cuttings and rhizome parts, respectively (Clapa and Hârţa 2021; Vining et al. 2020). For *Echinacea* species, root cuttings or rooted shoots derived from the crown divisions are the preferred plant parts for reproduction (Abbasi et al. 2007).

Apart from the above-mentioned MAPs, the asexual reproduction is the only means of producing propagating material for some other species. For example, *Mentha piperita* is a sterile hybrid (only some varieties exhibit a small degree of fertility), and for this reason, it is solely propagated by rooted stem cuttings or with rhizome cuttings (Kumar, Mishra, Malik, and Satya 2011; Tucker 2012; Vining et al. 2020).

Zingiber officinale is primarily propagated by rhizomes, because plants fail to produce flowers and seeds, mainly due to the environmental conditions effects (especially the photoperiod regime) and the high sterility rates of flowers (Kumari, Kumar, and Solankey 2020; Ravindran, Pillai, and Divakaran 2012). Rhizomes are usually divided into small segments that have two or more buds (Kumari, Kumar, and Solankey 2020; Ravindran, Pillai, and Divakaran 2012).

Crocus sativus is propagated by corms (underground stems with protective scaly leaves modified into skins or tunics) that grow around the mother corm of long-aged plants. Each corm must have one or more nodes on which there are buds/meristems from which new plants develop (Negbi et al. 1989). *Artemisia dracunculus* (also known as French tarragon) is propagated by rooted stem cuttings or by rhizome division (Mackay and Kitto 1988).

Origanum scabrum, an endemic plant of Greece (found only in the mountains of Parnonas and Taygetos in the Southern Peloponnese and in the mountain Dirfi in Cantral Greece, Evia), with medicinal properties does not produce any seeds and can be propagated only vegetatively by underground shoots – rhizomes (Alexopoulos et al. 2011).

3.3.3 Micropropagation

Plant propagation by tissue culture (micropropagation) is an up-to-date technique for the production of propagating material of MAPs and has attracted the interest of several researchers. Micropropagation is not only a basic research tool, but also allows the production of healthy propagating material in large quantities and in a short period of time, the faster and more efficient completion of research and breading programs, as well as the preservation of genetic material (George, Hall, and Klerk 2008). The latter parameter is of particular importance in cases where the plants are endemic with limited expansion and are classified as threatened, as is the case with several medicinal plants. Plant micropropagation is carried out through the *in vitro* growth of plant parts, that is, cultures of seeds, meristems, shoots or shoot tips, calluses, embryos, isolated roots, cell suspensions, or protoplasts. The selection of the most suitable method for *in vitro* propagation is strongly affected by the characteristics of each plant species (George, Hall, and Klerk 2008).

Micropropagation is now applied for the production of propagating material in a large number of MAPs, and there is a lot of literature concerning either commercially cultivated or endemic plants. Morone-Fortunato and Avato (2008) have reported that *Origanum vulgare* ssp. *hirtum* plantlets derived from micropropagation did not differ from the mother plants in terms of quality and quantity of the essential oils. In the case of *Mentha piperita*, the application of the in *vitro* technique is expected to produce new varieties with a high production of mint oil (Zheljazkov, Yankov, and Topalov 1996). *Hypericum perforatum* is also successfully propagated through tissue culture (Čellárová et al. 1994; Brutovská, Čellárová, and Doležel 1998), as well as several valerian species such as *Valeriana wallichii* (Mathur 1992), *V. jatamansi* (Kaur et al. 1999), and *V. officinalis* (Wiśniewski et al. 2016). Similarly, the micropropagation technique leads to higher multiplication rates in *Rosmarinus officinalis* (Hammer and Junghanns 2020), while Vouillamoz and Christ (2020) reported that micropropagation is suitable in the case of *Thymus vulgaris* since it allows the maintenance of mother plants obtained from clonal varieties, although it is more expensive for propagation in comparison with the stem cuttings or plant division.

Due to high cost, the implementation of micropropagation should be justified by economic parameters as well as by the particularities of each species. However, in the case of several MAPs, the contribution of micropropagation in the production of propagating material of superior genotypes, endemic or rare plant species, and the genotypic improvement via mutagenesis and/or genetic engineering is unquestionable.

3.4 FUTURE REMARKS

Since ancient times, aromatic plants, commonly known as "herbs and spices," have been used extensively to cure bacteria, fungus, viruses, and insects, while they are also highly appreciated for their important health benefits (Jamshidi-Kia, Lorigooini, and Amini-Khoei 2018; Tariq et al. 2019). Nowadays, they are increasingly used in food production as natural antioxidants, particularly as alternatives to synthetic ones (Christaki et al. 2012). Additionally, apart from improving the antioxidant potential of food products, the food sector has noted that using aromatic plant compounds as natural preservatives lengthens product shelf lives (Fernandes et al. 2016). The improvement of food's nutritional value, which is directly tied to their health-promoting qualities, is also a benefit associated with the addition of aromatic plant compounds to food products. However, it is regarded important to conduct further research in encapsulation which will allow the extended use of essential oils by the food sector for food preservation and food functionality without impairing the sensorial quality of the final product (Tajkarimi, Ibrahim, and Cliver 2010). The manner of food preservation, any side effects on flavor and aroma from the usage of aromatic plants, etc., should all be the subject of this study (Fernandes et al. 2016).

In addition, the training of producers should be focused on increasing their skills in the distillation and drying process, as well as any other post-harvest treatments that may improve the commercial raw material (essential oils, dried plants, etc.). Other crops have demonstrated that using high-quality raw materials and suitable equipment during the conversion process results in good final products. As a result, it is vital to encourage MAP certification, enhance integrated product traceability, take advantage of sustainable harvesting methods, and give consumers comprehensive information about the origin and the cultivation practices of the marketed product.

The limited information regarding the propagation requirements for the majority of species, as well as the lack of knowledge about the best practice guides for the majority of the species, especially the endemic ones, necessitates the development in the agro-technology sector that could address these issues and allow the further development of the MAP sector (Kala, Dhyani, and Sajwan 2006). Bioprospecting and biopiracy are two major issues that need to be considered, since they are associated with the prosperity of local rural communities and the *in situ* conservation of valuable natural resources such as the various endemic MAPs (Chen et al. 2016).

Furthermore, efforts should be made to mechanize the cultivation of MAPs, which will help to lower cultivation costs and further expand the industry, which is now lagging due to a labor shortage. Utilizing the traditional knowledge that rural people have is possible through the collection of cultivated plants, processing, and sale of MAPs, which will also help to apply co-management techniques in these places. The majority of the population in rural and forest areas, however, encounters numerous obstacles due to limited market access and insufficient capital investment.

Regarding the issue of boosting entrepreneurship, national authorities should encourage small- and medium-sized businesses, because they have a lot of potential to increase employment and income growth in the areas of cultivation and processing. So far, the marketing chain of MAPs is unregulated and inequitable, while illegal trading channels also exist, especially regarding the threat or species under the risk of extinction (Kala, Dhyani, and Sajwan 2006). There is a significant chance to address these issues at hand by increasing the public interest in MAP products. Therefore, policymakers ought to give the upstream base of the supply chain (management plans, legislation, etc.) and move forward in the subsequent phase priorities, when implementing measures to enhance the sector.

There are a number of deficiencies that could threaten the development of the MAPs sector, notwithstanding its expansion and progress. This has an impact on the private sector, and the policymakers and the private sector should collaborate to address these issues and any further challenges that may arise. MAPs could be a great tool for enhancing local residents' quality of life and generating new employment opportunities, particularly for women and young people. However, there are significant worries regarding the market's growth because it is thought that in order to overcome the main obstacles to market access, local producers, harvesters, and processors must form clusters and organizations.

Finally, innovations in the field of MAPs are uncommon, typically managed by a limited number of companies, and infrequently encouraged by the institutional framework. Numerous initiatives should be taken at the product level to encourage novel MAP applications and increase consumer awareness. For improved horizontal collaboration between small manufacturers and better vertical integration, it is crucial to establish joint marketing strategies.

3.5 CONCLUSION

Despite the fact that the market for MAPs appears to be very lucrative, numerous challenges are recorded, such as the irrational plant collection practices that could result in the extinction of species and the damage of their habitats. Therefore, further research should be done to determine the proper cultivation practices, including the standardized propagation protocols. Moreover, the expansion of the MAPSs sector is associated with the mechanization of cultivation practices that will allow the reduction of production cost and further increase the added value of the final products. Organizing and educating the individuals involved in the entire production chain (production, harvesting, processing, and marketing) should also be a priority to ensure the safety of the final products and the compliance to market and consumer needs. Finally, to prevent the extinction and genetic erosion of several native plants as well as to protect the consumer, there should be adequate national coverage and reinforcement of crops.

REFERENCES

Abbasi, Bilal Haider, Praveen K. Saxena, Susan J. Murch, and Chun Zhao Liu. 2007. "*Echinacea* Biotechnology: Challenges and Opportunities." *In Vitro Cellular and Developmental Biology - Plant* 43 (6): 481–92. https://doi.org/10.1007/s11627-007-9057-2.

Abbaszadeh, Bohlool, Hossein Aliabadi Farahani, and Elham Morteza. 2009. "Effects of Irrigation Levels on Essential Oil of Balm (*Melissa officinalis* L.)." *American-Eurasian Journal of Sustainable Agriculture* 3 (1): 53–56.

Aguilar, Ramiro, Edson Jacob Cristóbal-Pérez, Francisco Javier Balvino-Olvera, María de Jesús Aguilar-Aguilar, Natalia Aguirre-Acosta, Lorena Ashworth, Jorge A. Lobo, et al. 2019. "Habitat Fragmentation Reduces Plant Progeny Quality: A Global Synthesis." *Ecology Letters* 22 (7): 1163–73. https://doi.org/10.1111/ele.13272.

Ahuja, A., Verma, M, and Grewal, Simranjot. 1982. "Clonal Propagation of *Ocimum* Species by Tissue Culture." *Indian Journal of Experimental Biology* 20: 455–58.

Alan, Ali R., Susan J. Murch, and Praveen K. Saxena. 2015. "Evaluation of Ploidy Variations in *Hypericum perforatum* L. (St. John's Wort) Germplasm from Seeds, *in Vitro* Germplasm Collection, and Regenerants from Floral Cultures." *In Vitro Cellular and Developmental Biology - Plant* 51 (4): 452–62. https://doi.org/10.1007/s11627-015-9708-7.

Albrecht, Sebastian, and Lars-Gernot Otto. 2020. "*Matricaria recutita* L.: True Chamomile." In *Medicinal, Aromatic and Stimulant Plants, Handbook of Plant Breeding 12*, edited by J. Novak and W.-D. Blüthner, 313–31. Cham: Springer Nature Switzerland AG. https://doi.org/10.1007/978-3-030-38792-1_7.

Alcántara-Flores, Ela, Alicia E. Brechú-Franco, Angel Villegas-Monter, Guillermo Laguna-Hernández, and Armando Gómez-Campos. 2017. "Sexual and Vegetative Propagation of the Medicinal Mexican Species *Phyllonoma laticuspis* (Phyllonomaceae)." *Revista de Biologia Tropical* 65 (March): 9–19.

Alexieva, I. N., A. T. Popova, and D. Sp Mihaylova. 2020. "Trends in Herbal Usage – A Survey Study." *Food Research* 4 (2): 500–6. https://doi.org/10.26656/fr.2017.4(2).346.

Alexopoulos, A., A. C. Kimbaris, S. Plessas, I. Mantzourani, I. Theodoridou, E. Stavropoulou, M. G. Polissiou, and E. Bezirtzoglou. 2011. "Antibacterial Activities of Essential Oils from Eight Greek Aromatic Plants against Clinical Isolates of *Staphylococcus aureus*." *Anaerobe* 17 (6): 399–402. https://doi.org/10.1016/j.anaerobe.2011.03.024.

Ali, Muzamil, A. Mujib, Dipti Tonk, and Nadia Zafar. 2017. "Plant Regeneration through Somatic Embryogenesis and Genome Size Analysis of *Coriandrum sativum* L." *Protoplasma* 254 (1): 343–52. https://doi.org/10.1007/s00709-016-0954-2.

Allison, V. J. 2002. "Nutrients, Arbuscular Mycorrhizas and Competition Interact to Influence Seed Production and Germination Success in *Achillea millefolium*." *Functional Ecology* 16 (6): 742–49. https://doi.org/10.1046/j.1365-2435.2002.00675.x.

Amer, Ahmed, and Hanaa Omar. 2019. "In-Vitro Propagation of the Multipurpose Egyptian Medicinal Plant *Pimpinella anisum*." *Egyptian Pharmaceutical Journal* 18 (3): 254. https://doi.org/10.4103/epj.epj_12_19.

Anwar, M., D. D. Patra, S. Chand, Kumar Alpesh, A. A. Naqvi, and S. P. S. Khanuja. 2005. "Effect of Organic Manures and Inorganic Fertilizer on Growth, Herb and Oil Yield, Nutrient Accumulation, and Oil Quality of French Basil." *Communications in Soil Science and Plant Analysis* 36 (13–14): 1737–46. https://doi.org/10.1081/CSS-200062434.

Atia, Abdallah, Zouhaier Barhoumi, Rabhi Mokded, Chedly Abdelly, and Abderrazak Smaoui. 2011. "Environmental Eco-Physiology and Economical Potential of the Halophyte *Crithmum maritimum* L. (Apiaceae)." *Journal of Medicinal Plants Research* 5 (16): 3564–71.

Axiotis, Evangelos, Maria Halabalaki, and Leandros A. Skaltsounis. 2018. "An Ethnobotanical Study of Medicinal Plants in the Greek Islands of North Aegean Region." *Frontiers in Pharmacology* 9 (MAY): 1–6. https://doi.org/10.3389/fphar.2018.00409.

Baatour, Olfa, R. Kaddour, W. Aidi Wannes, M. Lachaâl, and B. Marzouk. 2010. "Salt Effects on the Growth, Mineral Nutrition, Essential Oil Yield and Composition of Marjoram (*Origanum majorana*)." *Acta Physiologiae Plantarum* 32 (1): 45–51. https://doi.org/10.1007/s11738-009-0374-4.

Baranauskiene, Renata, Petras Rimantas Venskutonis, Pranas Viškelis, and Edita Dambrauskiene. 2003. "Influence of Nitrogen Fertilizers on the Yield and Composition of Thyme (*Thymus vulgaris*)." *Journal of Agricultural and Food Chemistry* 51 (26): 7751–58. https://doi.org/10.1021/jf0303316.

Bekmirzaev, Gulom, Jose Beltrao, and Baghdad Ouddane. 2019. "Effect of Irrigation Water Regimes on Yield of *Tetragonia tetragonioides*." *Agriculture (Switzerland)* 9 (22): 1–9. https://doi.org/10.3390/agriculture9010022.

Bakoumé, C. 2016. "Genetic Diversity, Erosion, and Conservation in Oil Palm (*Elaeis guineensis* Jacq.)." In M. Ahuja and S. Jain (eds), *Genetic Diversity and Erosion in Plants. Sustainable Development and Biodiversity*, vol. 8, Springer. https://doi.org/10.1007/978-3-319-25954-3_1

Brutovská, Renáta, Eva Čellárová, and Jaroslav Doležel. 1998. "Cytogenetic Variability of in Vitro Regenerated *Hypericum perforatum* L. Plants and Their Seed Progenies." *Plant Science* 133 (2): 221–29. https://doi.org/10.1016/S0168-9452(98)00041-7.

Canter, Peter H., Howard Thomas, and Edzard Ernst. 2005. "Bringing Medicinal Plants into Cultivation: Opportunities and Challenges for Biotechnology." *Trends in Biotechnology* 23 (4): 180–85. https://doi.org/10.1016/j.tibtech.2005.02.002.

Čellárová, E., K. Kimáková, J. Halušková, and Z. Daxnerová. 1994. "The Variability of the Hypericin Content in the Regenerants of *Hypericum perforatum*." *Acta Biotechnologica* 14 (3): 267–74. https://doi.org/10.1002/abio.370140309.

Chen, Shi Lin, Hua Yu, Hong Mei Luo, Qiong Wu, Chun Fang Li, and André Steinmetz. 2016. "Conservation and Sustainable Use of Medicinal Plants: Problems, Progress, and Prospects." *Chinese Medicine (United Kingdom)* 11 (1): 1–10. https://doi.org/10.1186/s13020-016-0108-7.

Christaki, Efterpi, Eleftherios Bonos, Ilias Giannenas, and Panagiota Florou-Paneri. 2012. "Aromatic Plants as a Source of Bioactive Compounds." *Agriculture (Switzerland)* 2 (3): 228–43. https://doi.org/10.3390/agriculture2030228.

Chrysargyris, Antonios, Charalampia Kloukina, Rea Vassiliou, Ekaterina Michaela Tomou, Helen Skaltsa, and Nikolaos Tzortzakis. 2019. "Cultivation Strategy to Improve Chemical Profile and Anti-Oxidant Activity of *Sideritis perfoliata* L. Subsp. *Perfoliata*." *Industrial Crops and Products* 140 (July): 111694. https://doi.org/10.1016/j.indcrop.2019.111694.

Chrysargyris, Antonios, Spyriidon A. Petropoulos, and Nikolaos Tzortzakis. 2022. "Essential Oil Composition and Bioactive Properties of Lemon Balm Aerial Parts as Affected by Cropping System And." *Agronomy* 12 (649): 1–17.

Clapa, Doina, and Monica Hârţa. 2021. "Establishment of an Efficient Micropropagation System for *Humulus lupulus* L. Cv. Cascade and Confirmation of Genetic Uniformity of the Regenerated Plants through DNA Markers." *Agronomy* 11 (2268): 1–16. https://doi.org/10.3390/agronomy11112268.

Copetta, Andrea, Guido Lingua, and Graziella Berta. 2006. "Effects of Three AM Fungi on Growth, Distribution of Glandular Hairs, and Essential Oil Production in *Ocimum Basilicum* L. Var. Genovese." *Mycorrhiza* 16 (7): 485–94. https://doi.org/10.1007/s00572-006-0065-6.

Craker, Lyle E., Zoë Gardner, and Selma C. Etter. 2003. "Herbs in American Fields: A Horticultural Perspective of Herb and Medicinal Plant Production in the United States, 1903 to 2003." *HortScience* 38 (5): 977–83. https://doi.org/10.21273/hortsci.38.5.977.

Dagar, J. C., R. K. Yadav, S. R. Dar, and Sharif Ahamad. 2015. "Liquorice (*Glycyrrhiza glabra*): A Potential Salt-Tolerant, Highly Remunerative Medicinal Crop for Remediation of Alkali Soils." *Current Science* 108 (9): 1683–88.

Dastanpoor, Nasrollah, Hamid Fahimi, Mansour Shariati, Saeid Davazdahemami, Sayed Mojtaba, and Modarres Hashemi. 2013. "Effects of Hydropriming on Seed Germination and Seedling Growth in Sage (*Salvia officinalis* L.)." *African Journal of Biotechnology* 12 (11): 1223–28. https://doi.org/10.5897/AJB12.1941.

Dehghanpour Farashah, H., R. Tavakkol Afshari, F. Sharifzadeh, and S. Chavoshinasab. 2011. "Germination Improvement and α-Amylase and β-1,3-Glucanase Activity in Dormant and Nondormant Seeds of Oregano (*Origanum vulgare*)." *Australian Journal of Crop Science* 5 (4): 421–27.

DeKalb, Courtney D., Brian A. Kahn, Bruce L. Dunn, Mark E. Payton, and Allen V. Barker. 2014. "Substitution of a Soilless Medium with Yard Waste Compost for Basil Transplant Production." *HortTechnology* 24 (6): 668–75. https://doi.org/10.21273/horttech.24.6.668.

Echeverrigaray, S., R. Basso, and L. B. Andrade. 2005. "Micropropagation of *Lavandula dentata* from Axillary Buds of Field-Grown Adult Plants." *Biologia Plantarum* 49 (3): 439–42. https://doi.org/10.1007/s10535-005-0024-7.

EL-Keltawi, N. E., R. F. Abdou, and D. W. Bishay. 1985. "Comparative Studies on Growth and Volatile Oil Contents of Some Induced Mutants of *Origanum majorana*." In *Essential Oils and Aromatic Plants*, edited by A. Baerheim Svendsen and J. J. C. Scheffer, 191–97. Dordrecht: Junk Publishers. https://doi.org/10.1007/978-94-009-5137-2_21.

El-Khateeb, M.A., A.B. El-Attar, and R.M. Nour. 2017. "Application of Plant Biostimulants to Improve the Biological Responses and Essential Oil Production of Marjoram (*Majorana hortensis*, Moench) Plants." *Middle East Journal of Agriculture Research* 6 (4): 928–41.

Eshaghi Gorgi, Olia, Hormoz Fallah, Yosoof Niknejad, and Davood Barari Tari. 2022. "Effect of Plant Growth Promoting Rhizobacteria (PGPR) and Mycorrhizal Fungi Inoculations on Essential Oil in *Melissa officinalis* L. under Drought Stress." *Biologia* 77: 11–20. https://doi.org/10.1007/s11756-021-00919-2.

Fernandes, R. P. P., M. A. Trindade, F. G. Tonin, C. G. Lima, S. M.P. Pugine, P. E.S. Munekata, J. M. Lorenzo, and M. P. de Melo. 2016. "Evaluation of Antioxidant Capacity of 13 Plant Extracts by Three Different Methods: Cluster Analyses Applied for Selection of the Natural Extracts with Higher Antioxidant Capacity to Replace Synthetic Antioxidant in Lamb Burgers." *Journal of Food Science and Technology* 53 (1): 451–60. https://doi.org/10.1007/s13197-015-1994-x.

Geerts, Sam, and Dirk Raes. 2009. "Deficit Irrigation as an On-Farm Strategy to Maximize Crop Water Productivity in Dry Areas." *Agricultural Water Management* 96 (9): 1275–84. https://doi.org/10.1016/j.agwat.2009.04.009.

Geneva, Maria P., Ira V. Stancheva, Madlen M. Boychinova, Nadezhda H. Mincheva, and Petranka A. Yonova. 2010. "Effects of Foliar Fertilization and Arbuscular Mycorrhizal Colonization on *Salvia officinalis* L. Growth, Antioxidant Capacity, and Essential Oil Composition." *Journal of the Science of Food and Agriculture* 90 (4): 696–702. https://doi.org/10.1002/jsfa.3871.

George, E. F., M. A. Hall, and G. J. D. Klerk. 2008. "Micropropagation: Uses and Methods." In *Plant Propagation by Tissue Culture*, edited by E. F. George, M. A. Hall, and G. J. D. Klerk, 29–64. Dordrecht: Springer.

Giannoulis, Kyriakos D., Vasileios Evangelopoulos, Nikolaos Gougoulias, and Eleni Wogiatzi. 2020a. "Lavender Organic Cultivation Yield and Essential Oil Can Be Improved by Using Bio-Stimulants." *Acta Agriculturae Scandinavica Section B: Soil and Plant Science* 70 (8): 648–56. https://doi.org/10.1080/09064710.2020.1833974.

———. 2020b. "Could Bio-Stimulators Affect Flower, Essential Oil Yield, and Its Composition in Organic Lavender (*Lavandula angustifolia*) Cultivation?" *Industrial Crops and Products* 154 (April). https://doi.org/10.1016/j.indcrop.2020.112611.

Giannoulis, Kyriakos D., Christina Anna Kamvoukou, Nikolaos Gougoulias, and Eleni Wogiatzi. 2020a. "Irrigation and Nitrogen Application Affect Greek Oregano (*Origanum vulgare* ssp. *hirtum*) Dry Biomass, Essential Oil Yield and Composition." *Industrial Crops and Products* 150: 112392. https://doi.org/10.1016/j.indcrop.2020.112392.

———. 2020b. "*Matricaria chamomilla* L. (German Chamomile) Flower Yield and Essential Oil Affected by Irrigation and Nitrogen Fertilization." *Emirates Journal of Food and Agriculture* 32 (5): 328–35. https://doi.org/10.9755/ejfa.2020.v32.i5.2099.

Giannoulis, Kyriakos D., Elpiniki Skoufogianni, Dimitrios Bartzialis, Alexandra D. Solomou, and Nicholaos G. Danalatos. 2021. "Growth and Productivity of *Salvia officinalis* L. under Mediterranean Climatic

Conditions Depends on Biofertilizer, Nitrogen Fertilization, and Sowing Density." *Industrial Crops and Products* 160 (May 2020). https://doi.org/10.1016/j.indcrop.2020.113136.

Goldman, P. 2001. "Herbal Medicines Today and the Roots of Modern Pharmacology." *Annals of Internal* 135 (8): 595–600.

Grigoriadou, K., N. Krigas, V. Sarropoulou, K. Papanastasi, G. Tsoktouridis, and E. Maloupa. 2019. "In Vitro Propagation of Medicinal and Aromatic Plants : The Case of Selected Greek Species with Conservation Priority." *In Vitro and Cellular Development Biology-Plant* 55: 635–46.

Gurib-Fakim, Ameenah. 2006. "Medicinal Plants: Traditions of Yesterday and Drugs of Tomorrow." *Molecular Aspects of Medicine* 27 (1): 1–93. https://doi.org/10.1016/j.mam.2005.07.008.

Hammer, M., and W. Junghanns. 2020. "*Rosmarinus Officinalis* L.: Rosemary." In J. Novak and W.D. Blüthner (eds), *Medicinal, Aromatic and Stimulant Plants. Handbook of Plant Breeding, Vol 12*, 501–22. Cham: Springer Nature Switzerland AG.

Hanafy, M. S., H. A. Ashour, and F. M. Sedek. 2018. "Effect of Some Bio-Stimulants and Micronutrients on Growth, Yield and Essential Oil Production of *Majorana hortensis* Plants." *International Journal of Environment* 7 (1): 37–52.

Hayden, Anita L. 2006. "Aeroponic and Hydroponic Systems for Medicinal Herb, Rhizome, and Root Crops." *HortScience* 41 (3): 536–38. https://doi.org/10.21273/hortsci.41.3.536.

He, Ke. 2015. "Traditional Chinese and Thai Medicine in a Comparative Perspective." *Complementary Therapies in Medicine* 23 (6): 821–26. https://doi.org/10.1016/j.ctim.2015.10.003.

Helgason, Thorunn, and Alastair Fitter. 2005. "The Ecology and Evolution of the Arbuscular Mycorrhizal Fungi." *Mycologist* 19 (3): 96–101. https://doi.org/10.1017/S0269-915X(05)00302-2.

Idrees, Mohd, M. Masroor, A. Khan, Tariq Aftab, M. Naeem, and Nadeem Hashmi. 2010. "Salicylic Acid-Induced Physiological and Biochemical Changes in Lemongrass Varieties under Water Stress." *Journal of Plant Interactions* 5 (4): 293–303. https://doi.org/10.1080/17429145.2010.508566.

Jamshidi-Kia, Fatemeh, Zahra Lorigooini, and Hossein Amini-Khoei. 2018. "Medicinal Plants: Past History and Future Perspective." *Journal of HerbMed Pharmacology* 7 (1): 1–7. https://doi.org/10.15171/jhp.2018.01.

Jeelani, Syed Mudassir, Gulzar A. Rather, Arti Sharma, and Surrinder K. Lattoo. 2018. "In Perspective: Potential Medicinal Plant Resources of Kashmir Himalayas, Their Domestication and Cultivation for Commercial Exploitation." *Journal of Applied Research on Medicinal and Aromatic Plants* 8 (December 2017): 10–25. https://doi.org/10.1016/j.jarmap.2017.11.001.

Kala, Chandra Prakash, Pitamber Prasad Dhyani, and Bikram Singh Sajwan. 2006. "Developing the Medicinal Plants Sector in Northern India: Challenges and Opportunities." *Journal of Ethnobiology and Ethnomedicine* 2 (32): 1–15. https://doi.org/10.1186/1746-4269-2-32.

Karioti, A., H. Skaltsa, C. Demetzos, D. Perdetzoglou, C. D. Economakis, and A. B. Salem. 2003. "Effect of Nitrogen Concentration of the Nutrient Solution on the Volatile Constituents of Leaves of *Salvia fruticosa* Mill. in Solution Culture." *Journal of Agricultural and Food Chemistry* 51 (22): 6505–8. https://doi.org/10.1021/jf030308k.

Karkanis, A., N. Martins, S. A. Petropoulos, and I. C. F. R. Ferreira. 2016. "Phytochemical Composition, Health Effects, and Crop Management of Liquorice (*Glycyrrhiza glabra* L.): A Medicinal Plant." *Food Reviews International* 00 (00): 1–22. https://doi.org/10.1080/87559129.2016.1261300.

Karunamoorthi, Kaliyaperumal, Kaliyaperumal Jegajeevanram, Jegajeevanram Vijayalakshmi, and Embialle Mengistie. 2013. "Traditional Medicinal Plants: A Source of Phytotherapeutic Modality in Resource-Constrained Health Care Settings." *Journal of Evidence-Based Complementary and Alternative Medicine* 18 (1): 67–74. https://doi.org/10.1177/2156587212460241.

Kaur, R., M. Sood, S. Chander, R. Mahajan, V. Kumar, and D. R. Sharma. 1999. "*In Vitro* Propagation of *Valeriana jatamansi*." *Plant Cell, Tissue and Organ Culture* 59 (3): 227–29. https://doi.org/10.1023/A:1006425230046.

Khalil, M. Y., M. A. M. Kandil, and M. F. Swaefy Hend. 2008. "Effect of Three Different Compost Levels on Fennel and Salvia Growth Character and Their Essential Oils." *Research Journal of Agriculture and Biological Sciences* 4 (1): 34–39.

Khan, H., and A. Rauf. 2014. "Medicinal Plants: Economic Perspective and Recent Developments." *World Applied Sciences Journal* 31 (11): 1925–29. https://doi.org/10.5829/idosi.wasj.2014.31.11.14494.

Khaosaad, T., H. Vierheilig, M. Nell, K. Zitterl-Eglseer, and J. Novak. 2006. "Arbuscular Mycorrhiza Alter the Concentration of Essential Oils in Oregano (*Origanum* sp., Lamiaceae)." *Mycorrhiza* 16 (6): 443–46. https://doi.org/10.1007/s00572-006-0062-9.

Konstantinidou, Elissavet, Ioannis Takos, and Theodora Merou. 2008. "Desiccation and Storage Behavior of Bay Laurel (*Laurus nobilis* L.) Seeds." *European Journal of Forest Research* 127 (2): 125–31. https://doi.org/10.1007/s10342-007-0189-z.

Kumar, Birendra. 2012. "Prediction of Germination Potential in Seeds of Indian Basil (*Ocimum basilicum* L.)." *Journal of Crop Improvement* 26 (4): 532–39. https://doi.org/10.1080/15427528.2012.659418.

Kumar, Peeyush, Sapna Mishra, Anushree Malik, and Santosh Satya. 2011. "Insecticidal Properties of *Mentha* Species: A Review." *Industrial Crops and Products* 34 (1): 802–17. https://doi.org/10.1016/j.indcrop.2011.02.019.

Kumari, M., M. Kumar, and S. S. Solankey. 2020. "*Zingiber officinale* Roscoe: Ginger." In *Medicinal, Aromatic and Stimulant Plants, Handbook of Plant Breeding 12*, edited by J. Novak and W.-D. Blüthner, 605–21. Cham: Springer Nature Switzerland AG. https://doi.org/10.1007/978-3-030-38792-1.

Kunwar, Ripu M., Bal K. Nepal, Hari B. Kshhetri, Sanjeev K. Rai, and Rainer W. Bussmann. 2006. "Ethnomedicine in Himalaya: A Case Study from Dolpa, Humla, Jumla and Mustang Districts of Nepal." *Journal of Ethnobiology and Ethnomedicine* 2: 1–6. https://doi.org/10.1186/1746-4269-2-27.

Kuris, A., A. Altman, and E. Putievsky. 1981. "Vegetative Propagation of Spice-Plants: Root Formation in Oregano Stem Cuttings." *Scientia Horticulturae* 14 (2): 151–56. https://doi.org/10.1016/0304-4238(81)90007-8.

Kwiatkowski, Cezary A., and Jolanta Juszczak. 2012. "The Response of Sweet Basil (*Ocimum basilicum* L.) to the Application of Growth Stimulators and Forecrops." *Acta Agrobotanica* 64 (2): 69–76. https://doi.org/10.5586/aa.2011.019.

Laghmouchi, Yousif, Omar Belmehdi, Abdelhakim Bouyahya, Nadia Skali Senhaji, and Jamal Abrini. 2017. "Effect of Temperature, Salt Stress and PH on Seed Germination of Medicinal Plant *Origanum compactum*." *Biocatalysis and Agricultural Biotechnology* 10 (January): 156–60. https://doi.org/10.1016/j.bcab.2017.03.002.

Lakhiar, Imran Ali, Jianmin Gao, Tabinda Naz Syed, Farman Ali Chandio, and Noman Ali Buttar. 2018. "Modern Plant Cultivation Technologies in Agriculture under Controlled Environment: A Review on Aeroponics." *Journal of Plant Interactions* 13 (1): 338–52. https://doi.org/10.1080/17429145.2018.1472308.

Leaman, Danna J. 2008. "The International Standard for Sustainable Wild Collection of Medicinal and Aromatic Plants (ISSC-MAP)." International Expert Workshop on CITES Non-Detriment Findings Perennial Plant Working Group (Ornamentals, Medicinal and Aromatic Plants Cancun, Mexico, November 2008. https://cites.org/sites/default/files/ndf_material/THE%20INTERNATIONAL%20STANDARD%20FOR%20SUSTAINABLE%20WILD%20COLLECTION%20OF%20MEDICINAL%20AND%20AROMATIC%20PLANTS.pdf.

Lim, You Jin, and Seok Hyun Eom. 2013. "Effects of Different Light Types on Root Formation of *Ocimum basilicum* L. Cuttings." *Scientia Horticulturae* 164: 552–55. https://doi.org/10.1016/j.scienta.2013.09.057.

Lubbe, Andrea, and Robert Verpoorte. 2011. "Cultivation of Medicinal and Aromatic Plants for Specialty Industrial Materials." *Industrial Crops and Products* 34 (1): 785–801. https://doi.org/10.1016/j.indcrop.2011.01.019.

Mackay, W. A., and S. L. Kitto. 1988. "Factors Affecting *in vitro* Shoot Proliferation of French Tarragon." *Journal of the American Society for Horticultural Science* 113 (2): 282–87.

Mahdieh, Majid, Mojtaba Yazdani, and Shahla Mahdieh. 2013. "The High Potential of *Pelargonium Roseum* Plant for Phytoremediation of Heavy Metals." *Environmental Monitoring and Assessment* 185 (9): 7877–81. https://doi.org/10.1007/s10661-013-3141-3.

Maluin, Farhatun Najat, Mohd Zobir Hussein, Nik Nor Liyana Nik Ibrahim, Aimrun Wayayok, and Norhayati Hashim. 2021. "Some Emerging Opportunities of Nanotechnology Development for Soilless and Microgreen Farming." *Agronomy* 11 (6): 1–28. https://doi.org/10.3390/agronomy11061213.

Marques, Patricia Angélica A., Lineu Bernardi Filho, and José Antônio Frizzone. 2012. "Economic Analysis for Oregano under Irrigation Considering Economic Risk Factors." *Horticultura Brasileira* 30 (2): 234–39. https://doi.org/10.1590/s0102-05362012000200009.

Marthe, F. 2021. "*Petroselinum crispum* (Mill.) Nyman (Parsley)." In *Medicinal, Aromatic and Stimulant Plants, Handbook of Plant Breeding 12*, edited by J. Novak and W.-D. Blüthner, 435–66. Cham: Springer Nature Switzerland AG. https://doi.org/10.1007/978-3-030-38792-1.

Mathur, Jaideep. 1992. "Plantlet Regeneration from Suspension Cultures of *Valeriana wallichii* DC." *Plant Science* 81 (1): 111–15. https://doi.org/10.1016/0168-9452(92)90030-P.

McGaw, L., A. Jager, C. Fennel, and J. van Staden. 2005. "Medicinal Plants." In *Ethics in Agriculture-An African Perspective*, edited by Alvin van Niekerk, Vol. 10:1–11 Cham: Springer. https://doi.org/10.3390/plants10071355.

Mehalaine, Souad, and Haroun Chenchouni. 2021. "New Insights for the Production of Medicinal Plant Materials: *Ex vitro* and *in vitro* Propagation of Valuable Lamiaceae Species from Northern Africa." *Current Plant Biology* 27: 100216. https://doi.org/10.1016/j.cpb.2021.100216.

Miclea, Ileana, Andreea Suhani, Marius Zahan, and Andrea Bunea. 2020. "Effect of Jasmonic Acid and Salicylic Acid on Growth and Biochemical Composition of *In-Vitro*-Propagated *Lavandula angustifolia* Mill." *Agronomy* 10 (11): 1722. https://doi.org/10.3390/agronomy10111722.

Moghimipour, Zohreh, Mohammad Mahmoodi Sourestani, Naser Alemzadeh Ansari, and Zahra Ramezani. 2017. "The Effect of Foliar Application of Zinc on Essential Oil Content and Composition of Holy Basil [*Ocimum sanctum*] at First and Second Harvests." *Journal of Essential Oil-Bearing Plants* 20 (2): 449–58. https://doi.org/10.1080/0972060X.2017.1284609.

Mohammad, Sharrif Moghaddasi, Hamed Haddad Kashani, and Zohre Azarbad. 2012. "*Capparis spinosa* L. Propagation and Medicinal Uses." *Life Science Journal* 9 (4): 684–86.

Mohammadipour, Ehsan, Ahmad Golchin, Jafar Mohammadi, Naser Negahdar, and Mohammad Zarchini. 2012. "Effect of Humic Acid on Yield and Quality of Marigold (*Calendula officinalis* L.)." *Availa Scholars Research Library Annals of Biological Research* 2012 (11): 5095–98.

Momeni, Maryam, Abdollah Ghasemi Pirbalouti, Amir Mousavi, and Hassanali Naghdi Badi. 2020. "Effect of Foliar Applications of Salicylic Acid and Chitosan on the Essential Oil of *Thymbra spicata* L. under Different Soil Moisture Conditions." *Journal of Essential Oil-Bearing Plants* 23 (5): 1142–53. https://doi.org/10.1080/0972060X.2020.1801519.

Morone-Fortunato, Irene, and Pinarosa Avato. 2008. "Plant Development and Synthesis of Essential Oils in Micropropagated and Mycorrhiza Inoculated Plants of *Origanum vulgare* L. ssp. *hirtum* (Link) Ietswaart." *Plant Cell, Tissue and Organ Culture* 93 (2): 139–49. https://doi.org/10.1007/s11240-008-9353-5.

Mukherjee, Pulok K., M. Venkatesh, and Venkatesh V. Kumar. 2007. "An Overview on the Development in Regulation and Control of Medicinal and Aromatic Plants in the Indian System of Medicine." *Boletin Latinoamericano y Del Caribe de Plantas Medicinales y Aromaticas* 6 (4): 129–36.

Narnoliya, Lokesh Kumar, Jyoti Singh Jadaun, and Sudhir P. Singh. 2019. "The Phytochemical Composition, Biological Effects and Biotechnological Approaches to the Production of High- Value Essential Oil from Geranium." In *Essential Oil Research*, edited by S. Malik, 327–52. Cham: Springer Nature Switzerland AG. https://doi.org/10.1007/978-3-030-16546-8.

Nasiri, Yousef, Saeid Zehtab-Salmasi, Safar Nasrullahzadeh, Nosratollah Najafi, and Kazem Ghassemi-Golezani. 2010. "Effects of Foliar Application of Micronutrients (Fe and Zn) on Flower Yield and Essential Oil of Chamomile (*Matricaria chamomilla* L.)." *Journal of Medicinal Plants Research* 4 (17): 1733–37. https://doi.org/10.5897/JMPR10.083.

Negbi, M., D. Dagan, A. Dror, and D Basker. 1989. "Growth, Flowering, Vegetative Reproduction and Dormancy in the Saffron Crocus (*Crocus sativus* L.)." *Israel Journal of Botany* 38 (2–3): 95–113.

Nejad, Somayeh Rohani, Zohreh Emami Bistgani, and Allen V. Barker. 2022. "Enhancement of Seed Germination of Yarrow with Gibberellic Acid, Potassium Nitrate, Scarification, or Hydropriming." *Journal of Crop Improvement* 36 (3): 335–49. https://doi.org/10.1080/15427528.2021.1968553.

Nikopoulos, Dimitrios, and Alexios A. Alexopoulos. 2008. "In Vitro Propagation of an Endangered Medicinal Plant: *Pancratium Maritimum* L." *Journal of Food, Agriculture and Environment* 6 (2): 393–98.

Nikopoulos, Dimitrios, Despina Nikopoulou, and Alexios A. Alexopoulos. 2008. "Methods for the Preservation of Genetic Material of *Pancratium maritimum* (Amaryllidaceae)." *Journal of Food, Agriculture and Environment* 6 (3–4): 538–46.

Novak, J., and W. D. Blüthner. 2020. *Medicinal, Aromatic and Stimulant Plants*. Cham: Springer Nature Switzerland AG.

Okigbo, R. N., U. E. Eme, and S. Ogbogu. 2008. "Biodiversity and Conservation of Medicinal and Aromatic Plants in Africa." *Biotechnology and Molecular Biology Reviews* 3 (December): 127–34.

Okwany, R. O., R. T. Peters, K. L. Ringer, and D. B. Walsh. 2012. "Sustained Deficit Irrigation Effects on Peppermint Yield and Oil Quality in the Semi-Arid Pacific Northwest, USA." *Applied Engineering in Agriculture* 28 (4): 551–58.

Okwany, Romulus O., Troy R. Peters, Kerry L. Ringer, Douglas B. Walsh, and Maria Rubio. 2012. "Impact of Sustained Deficit Irrigation on Spearmint (*Mentha spicata* L.) Biomass Production, Oil Yield, and Oil Quality." *Irrigation Science* 30 (3): 213–19. https://doi.org/10.1007/s00271-011-0282-4.

Omer, Elsayed A. 1999. "Response of Wild Egyptian Oregano to Nitrogen Fertilization in a Sandy Soil." *Journal of Plant Nutrition* 22 (1): 103–14. https://doi.org/10.1080/01904169909365610.

Ozguven, Mensure, Filliz Ayanoglu, and Abdulhabib Ozel. 2006. "Effects of Nitrogen Rates and Cutting Times on the Essential Oil Yield and Components of *Origanum syriacum* L. var. *Bevanii*." *Journal of Agronomy*. https://doi.org/10.3923/ja.2006.101.105.

Paatre Shashikanthalu, Sharanyakanth, Lokeswari Ramireddy, and Mahendran Radhakrishnan. 2020. "Stimulation of the Germination and Seedling Growth of *Cuminum cyminum* L. Seeds by Cold Plasma."

Journal of Applied Research on Medicinal and Aromatic Plants 18 (May): 100259. https://doi.org/10.1016/j.jarmap.2020.100259.

Papadatou, Marilena, Catherine Argyropoulou, Catherine Grigoriadou, Eleni Maloupa, and Helen Skaltsa. 2015. "Essential Oil Content of Cultivated *Satureja* spp. in Northern Greece." *Natural Volatiles and Essential Oils* 2 (1): 37–48.

Parlak, Salih, and Devrim Semizer-Cuming. 2012. "Anatomical Examination of Root Formation on Bay Laurel (*Laurus nobilis* L.) Cuttings." *Journal of Plant Biology Research* 1 (4): 145–50.

Petrakou, Kassiani, Gregoris Iatrou, and Fotini N. Lamari. 2020. "Ethnopharmacological Survey of Medicinal Plants Traded in Herbal Markets in the Peloponnisos, Greece." *Journal of Herbal Medicine* 19 (July 2018): 100305. https://doi.org/10.1016/j.hermed.2019.100305.

Pimm, Stuart L., and Lucas N. Joppa. 2015. "How Many Plant Species Are There, Where Are They, and at What Rate Are They Going Extinct?" *Annals of the Missouri Botanical Garden* 100 (3): 170–76. https://doi.org/10.3417/2012018.

Pirbalouti, Abdollah Ghasemi, Mehdi Rahimmalek, Ladan Elikaei-Nejhad, and Behzad Hamedi. 2014. "Essential Oil Compositions of Summer Savory under Foliar Application of Jasmonic Acid and Salicylic Acid." *Journal of Essential Oil Research* 26 (5): 342–47. https://doi.org/10.1080/10412905.2014.922508.

Posmyk, Małgorzata M., and Katarzyna Szafrańska. 2016. "Biostimulators: A New Trend towards Solving an Old Problem." *Frontiers in Plant Science* 7 (May 2016): 1–6. https://doi.org/10.3389/fpls.2016.00748.

Pratap, Prawal, Singh Verma, R. C. Padalia, and V. R. Singh. 2017. "Influence of Vermicompost with FYM and Soil on Propagation of Marjoram (*Majorana Hortensis* L.) and Oregano (*Origanum vulgare*) with Green Cuttings." *Journal of Medicinal Plants Studies* 5 (3): 284–87.

Ravindran, P. N., S. Pillai, and M. Divakaran. 2012. "Other Herbs and Spices: Mango Ginger to Wasabi." In K.V. Peter (ed.), *Handbook of Herbs and Spices*, 557–82. Woodhead Publishing Limited.

Rout, G. R., S. Samantaray, and P. Das. 2000. "*In vitro* Manipulation and Propagation of Medicinal Plants." *Biotechnology Advances* 18 (2): 91–120. https://doi.org/10.1016/S0734-9750(99)00026-9.

Ruzzi, Maurizio, and Ricardo Aroca. 2015. "Plant Growth-Promoting Rhizobacteria Act as Biostimulants in Horticulture." *Scientia Horticulturae* 196: 124–34. https://doi.org/10.1016/j.scienta.2015.08.042.

Safari, Fateme, Morteza Akramian, Hossein Salehi-Arjmand, and Ali Khadivi. 2019. "Physiological and Molecular Mechanisms Underlying Salicylic Acid-Mitigated Mercury Toxicity in Lemon Balm (*Melissa officinalis* L.)." *Ecotoxicology and Environmental Safety* 183 (August): 109542. https://doi.org/10.1016/j.ecoenv.2019.109542.

Saglam, C., I. Atakisi, H. Turhan, S. Kaba, F. Arslanoglu, and F. Onemli. 2004. "Effect of Propagation Method, Plant Density, and Age on Lemon Balm (*Melissa officinalis*) Herb and Oil Yield." *New Zealand Journal of Crop and Horticultural Science* 32 (4): 419–23. https://doi.org/10.1080/01140671.2004.9514323.

Said-Al Ahl, Hah, A. G. El Gendy, and Omer Ea. 2016. "Humic Acid and Indole Acetic Acid Affect Yield and Essential Oil of Dill Grown under Two Different Locations in Egypt." *International Journal of Pharmacy and Pharmaceutical Sciences* 8 (8): 146–57.

Salmerón-Manzano, Esther, Jose Antonio Garrido-Cardenas, and Francisco Manzano-Agugliaro. 2020. "Worldwide Research Trends on Medicinal Plants." *International Journal of Environmental Research and Public Health* 17 (10). https://doi.org/10.3390/ijerph17103376.

Schippmann, U., A. B. Cunningham, D. J. Leaman, and S. Walter. 2005. "Impact of Cultivation and Collection on the Conservation of Medicinal Plants: Global Trends and Issues." *Acta Horticulturae* 676: 31–44. https://doi.org/10.17660/actahortic.2005.676.3.

Schippmann, Uwe, Danna Leaman, and A.B. Cunningham. 2006. "A Comparison of Cultivation and Wild Collection of Medicinal and Aromatic Plants Under Sustainability Aspects." In *Medicinal and Aromatic Plants*, edited by R. J. Bogers, L. E. Craker, and D. Lange, 75–95. Heidelberg: Springer. https://doi.org/10.1007/1-4020-5449-1_6.

Seidler-Łożykowska, Katarzyna, Romuald Mordalski, Wojciech Kucharski, Elżbieta Kędzia, Kamila Nowosad, and Jan Bocianowski. 2015. "Effect of Organic Cultivation on Yield and Quality of Lemon Balm Herb (*Melissa officinalis* L.)." *Acta Scientiarum Polonorum, Hortorum Cultus* 14 (5): 55–67.

Sifola, Maria Isabella, and G. Barbieri. 2006. "Growth, Yield and Essential Oil Content of Three Cultivars of Basil Grown under Different Levels of Nitrogen in the Field." *Scientia Horticulturae* 108 (4): 408–13. https://doi.org/10.1016/j.scienta.2006.02.002.

Skendi, Adriana, Dimitrios N. Katsantonis, Paschalina Chatzopoulou, Maria Irakli, and Maria Papageorgiou. 2020. "Antifungal Activity of Aromatic Plants of the Lamiaceae Family in Bread." *Foods* 9 (11): 8–12. https://doi.org/10.3390/foods9111642.

Slikkerveer, L. Jan. 2006. "The Challenge of Non-Experimental Validation of Mac Plants." In *Medicinal and Aromatic Plants*, edited by R. J. Bogers, L. E. Craker, and D. Lange, 1–28. Heidelberg: Springer. https://doi.org/10.1007/1-4020-5449-1_1.

Smith-Hall, Carsten, Helle Overgaard Larsen, and Mariève Pouliot. 2012. "People, Plants and Health: A Conceptual Framework for Assessing Changes in Medicinal Plant Consumption." *Journal of Ethnobiology and Ethnomedicine* 8: 1–11. https://doi.org/10.1186/1746-4269-8-43.

Smitha, G R. 2013. "Vegetative Propagation of Ashoka [*Saraca asoca* (Roxb.) de Wilde) - An Endangered Medicinal Plant." *Research on Crops* 14 (1): 274–83.

Sotiropoulou, D. E., and A. J. Karamanos. 2010. "Field Studies of Nitrogen Application on Growth and Yield of Greek Oregano (*Origanum vulgare* ssp. *Hirtum* (Link) Ietswaart)." *Industrial Crops and Products* 32 (3): 450–57. https://doi.org/10.1016/j.indcrop.2010.06.014.

Steinhoff, B. 2005. "Laws and Regulation on Medicinal and Aromatic Plants in Europe." *Acta Horticulturae* 678 (February): 13–22. https://doi.org/10.17660/ActaHortic.2005.678.1.

Tahaei, Amirreza, Ali Soleymani, and Majid Shams. 2016. "Seed Germination of Medicinal Plant, Fennel (*Foeniculum vulgare* Mill), as Affected by Different Priming Techniques." *Applied Biochemistry and Biotechnology* 180 (1): 26–40. https://doi.org/10.1007/s12010-016-2082-z.

Tajkarimi, M. M., S. A. Ibrahim, and D. O. Cliver. 2010. "Antimicrobial Herb and Spice Compounds in Food." *Food Control* 21 (9): 1199–1218. https://doi.org/10.1016/j.foodcont.2010.02.003.

Tariq, Saika, Saira Wani, Waseem Rasool, Khushboo Shafi, Muzzaffar Ahmad Bhat, Anil Prabhakar, Aabid Hussain Shalla, and Manzoor A. Rather. 2019. "A Comprehensive Review of the Antibacterial, Antifungal and Antiviral Potential of Essential Oils and Their Chemical Constituents against Drug-Resistant Microbial Pathogens." *Microbial Pathogenesis* 134 (June): 103580. https://doi.org/10.1016/j.micpath.2019.103580.

Tarraf, Waed, Claudia Ruta, Francesca De Cillis, Anna Tagarelli, Luigi Tedone, and Giuseppe De Mastro. 2015. "Effects of Mycorrhiza on Growth and Essential Oil Production in Selected Aromatic Plants." *Italian Journal of Agronomy* 10 (3): 160–62. https://doi.org/10.4081/ija.2015.633.

Thanos, Costas A., and Maria A. Doussi. 1995. "Ecophysiology of Seed Germination in Endemic Labiates of Crete." *Israel Journal of Plant Sciences* 43 (3): 227–37. https://doi.org/10.1080/07929978.1995.10676607.

Traykova, Boryanka, Marina Stanilova, Milena Nikolova, and Strahil Berkov. 2019. "Growth and Essential Oils of *Salvia officinalis* Plants Derived from Conventional or Aeroponic Produced Seedlings." *Agriculturae Conspectus Scientificus* 84 (1): 77–81.

Tucker, A. O. 2012. "Genetics and Breeding of the Genus *Mentha*: A Model for Other Polyploid Species with Secondary Constituents." *Journal of Medicinally Active Plants* 1 (1): 19–29.

Urwin, Nigel A. R., Jennie Horsnell, and Therese Moon. 2007. "Generation and Characterisation of Colchicine-Induced Autotetraploid *Lavandula angustifolia*." *Euphytica* 156 (1–2): 257–66. https://doi.org/10.1007/s10681-007-9373-y.

Vining, Kelly J., Kim E. Hummer, Nahla V. Bassil, B. Markus Lange, Colin K. Khoury, and Dan Carver. 2020. "Crop Wild Relatives as Germplasm Resource for Cultivar Improvement in Mint (*Mentha* L.)." *Frontiers in Plant Science* 11 (Article 1217): 1–15. https://doi.org/10.3389/fpls.2020.01217.

Vleeshouwers, L. M., H. J. Bouwmeester, and C. M. Karssen. 1995. "Redefining Seed Dormancy: An Attempt to Integrate Physiology and Ecology." *The Journal of Ecology* 83 (6): 1031. https://doi.org/10.2307/2261184.

Vouillamoz, J. F., and B. Christ. 2020. "*Thymus vulgaris* L.: Thyme." In *Medicinal, Aromatic and Stimulant Plants, Handbook of Plant Breeding 12*, edited by J. Novak and W. D. Blüthner, 547–57. Cham: Springer Nature Switzerland AG.

Williams, V. L., J. E. Victor, and N. R. Crouch. 2013. "Red Listed Medicinal Plants of South Africa: Status, Trends, and Assessment Challenges." *South African Journal of Botany* 86: 23–35. https://doi.org/10.1016/j.sajb.2013.01.006.

Wiśniewski, J., M. Szczepanik, B. Kołodziej, and B. Król. 2016. "Plantation Methods Effects on Common Valerian (*Valeriana officinalis*) Yield and Quality." *Journal of Animal and Plant Sciences* 26 (1): 177–84.

Yin, Heng, Xavier C. Fretté, Lars P. Christensen, and Kai Grevsen. 2012. "Chitosan Oligosaccharides Promote the Content of Polyphenols in Greek Oregano (*Origanum vulgare* ssp. *hirtum*)." *Journal of Agricultural and Food Chemistry* 60 (1): 136–43. https://doi.org/10.1021/jf204376j.

Yu, F., T. Takahashi, J. Moriya, K. Kawaura, J. Yamakawa, K. Kusaka, T. Itoh, S. Morimoto, N. Yamaguchi, and T. Kanda. 2006. "Traditional Chinese Medicine and Kampo: A Review from the Distant Past for the Future." *Journal of International Medical Research* 34 (3): 231–39. https://doi.org/10.1177/147323000603400301.

Zheljazkov, Valcho, Boris Yankov, and Venelin Topalov. 1996. "Comparison of Three Methods of Mint Propagation and Their Effect on the Yield of Fresh Material and Essential Oil." *Journal of Essential Oil Research* 8 (1): 35–45. https://doi.org/10.1080/10412905.1996.9700551.

Zimnitskaya, S. A. 2009. "State of the Reproductive System of Populations of Species of the Genus *Glycyrrhiza* L. (Fabaceae)." *Contemporary Problems of Ecology* 2 (4): 392–95. https://doi.org/10.1134/S1995425509040146.

Zolfaghari, Mayam, Vahideh Nazeri, Fatemeh Sefidkon, and Farhad Rejali. 2013. "Effect of Arbuscular Mycorrhizal Fungi on Plant Growth and Essential Oil Content and Composition of *Ocimum basilicum* L." *Iranian Journal of Plant Physiology* 3 (2): 643–50.

Zulfiqar, Faisal, Andrea Casadesús, Henry Brockman, and Sergi Munné-Bosch. 2020. "An Overview of Plant-Based Natural Biostimulants for Sustainable Horticulture with a Particular Focus on Moringa Leaf Extracts." *Plant Science* 295: 110194.

4 Threatened and Endangered Medicinal and Aromatic Plants

Sohail Ahmad, Maria Faraz, and Arshad Farid
Gomal University

Shakira Ghazanfar
National Institute for Genomics Advanced Biotechnology
(NIGAB) National Agricultural Research Centre

CONTENTS

DOI: 10.1201/b22842-4

4.1 INTRODUCTION

Plants are sources of non-food industrial goods in addition to being a source of food. Such plants are often grown on a large scale to make fine chemicals or specialized goods. One group of plants that falls under this category is medicinal plants, which are a veritable chemical treasure trove with numerous uses in human wellness. Any plant that, as a whole or in one or more of its organs, either contains such substances that can be used for the synthesis of useful drugs or contains such substances with medicinal properties that can be used directly for therapeutic purposes or can affect human health is considered a medicinal plant by the World Health Organization (Farnsworth and Soejarto 1991). The whole plant or any of its components, including the roots, stem, leaves, stem bark, fruits, or seeds, may be used to treat or control a disease. These chemical elements are therapeutically useful, non-nutrient molecules, which are frequently referred to as phytochemicals, bioactive chemicals, or active compounds (Doughari et al. 2009; Liu 2004). Because of their usage in traditional medicine, a remarkable number of modern medications have also been identified from these natural plant species. Numerous phytochemicals act as vital medications that are currently employed worldwide to treat a wide range of serious disorders. Each plant contains a variety of phytochemicals that can be used to treat a variety of illness conditions. Asthma, fever, constipation, infections of the urinary tract, gastrointestinal, and biliary systems, as well as skin infections, have all been treated with traditionally used medicinal herbs and their constituents (Saganuwan and Onyeyili 2010). The plants are used in a variety of methods, including poultices, formulations of various plant mixtures, infusions as teas or tinctures, or as component mixtures in porridges and soups that are administered in different ways either orally or topically.

Utilizing medicinal plants for therapeutic purposes has a long history that reaches back to antiquity. Herbal medicines are mostly made from medicinal herbs. These plants have played a significant role in historical systems of traditional medicine, including Egyptian, Chinese, and Ayurvedic (Sarker, Nahar, and Kumarasamy 2007). Over the years, medicinal plant products have taken on a very central role in the healthcare system and as a substitute source of drugs that are good for your health. The growing ineffectiveness and adverse consequences of many contemporary synthetic medications, such as the rise in bacterial resistance, are the main causes of the increased use of medicinal plants (Smolinski and Pestka 2003). About two-thirds of the global population in developing nations rely on plant-based traditional medicines and herbal treatments for their basic healthcare needs, according to the World Health Organization (WHO). In addition to being utilized as a source of raw materials for the synthesis of a variety of products ranging from traditional to modern medications, medicinal plants have the ability to treat both serious and minor human diseases (Oksman-Caldentey 2007). Numerous plant families, including the ones listed below, have been shown to be significant and useful by regular scientific research. The therapeutic applications of the Asteraceae, Apocynaceae, Liliaceae, Rutaceae, Caesalpiniaceae, Solanaceae, Piperaceae, Ranunculaceae, Apiaceae, Sapotaceae, etc., and their bioactive chemicals play a significant role in the natural wealth. These plants with medicinally significant bioactive components have a wide range of therapeutic effects, including the successful prevention and treatment of diseases, relief of disease symptoms, and advantageous control of the body's physical and mental status. It is generally known and well accepted that using medicinal plants to make herbal medicines is both safe and effective. A medical strategy that prioritizes policy, safety/quality/efficacy, access, and rational use of traditional medicine should be in place to encourage the right use of traditional medicine and medicinal plant products (WHO).

According to Bhalodia and Shukla (2011), medicinal plants exhibit a wide spectrum of pharmacological activities including anti-inflammatory, anti-bacterial, and anti-fungal characteristics, which has highlighted their significance for human health (Bhalodia and Shukla 2011). The existence of certain compounds in medicinal plants, which cause the body to react physiologically in a certain way, is what gives them their value. These phytochemicals are found across the plant kingdom and serve a variety of physiological and ecological purposes. These bioactive secondary products' primary function in plants is to aid them in coping with diverse abiotic and biotic challenges, such as chemical defense against pathogens, predators, diseases, and allopathic agents. Secondary metabolites are the medicinally significant active components of pharmacological plants. They are comparatively small chemical molecules that are widely distributed across the plant kingdom; although it is not always clear how they affect the plant life. These phytochemicals, which include phenolic compounds, flavonoids, alkaloids, tannins, and terpenes, are produced and accumulated by medicinal plants. They are used therapeutically or as building blocks for the creation of effective pharmaceuticals (Altemimi et al. 2017). There is a wide variety of these chemical components that are biologically active, and each one has a specific physiological effect on the human body (Atanasov et al. 2015). Stems and leaves of most medicinal plants are discovered to be abundant in a wide range of secondary metabolites with noticeable physiological activity (Hussein and El-Anssary 2019).

4.2 DIVERSITY OF MEDICINAL AND AROMATIC PLANTS USED WORLDWIDE

Up to 70,000 species are thought to be employed in traditional medicine globally. According to the WHO, about 21,000 plant species are utilized as medicines (Rajaei and Mohamadi 2012). Unfortunately, it is impossible to determine the precise number of medicinal and aromatic plant species utilized globally, because it is unknown that how many species are in use in other industrial uses, such as cosmetics, spirits, or aromas. However, based on the projected total number of 300–350,000 flowering plants, it can be said that at least every fourth plant is in use. There are a staggering variety of aromatic and medicinal plant species that are employed in various regions. About 7,500 species, or half of India's 17,000 native plant species, are utilized in ethnomedicines, making it one of the oldest, richest, and most diverse cultural traditions in the world for the use of medicinal plants. According to Pei-Gen (1991), there are about 6,000 kinds of medicinal plants utilized throughout China, and Shan-An He estimated that there are about 10,000 species. Of these, 1,000 plant species are regularly employed in Chinese medicine, and about half of these are regarded as the primary medicinal plants (He 1998). These two nations are followed by Vietnam, Sri Lanka, Indonesia, and other nations, with medicinal plant usage rates ranging from 6.1% in Pakistan to 20% in India. In comparison to other plant families, some have a higher proportion of vulnerable species as well as a higher density of therapeutic plants (Wakdikar 2004). Only some medicinal plants have been classified as threatened (Deeb et al. 2013; Schipmann et al. 2005) because of genetic erosion and resource damage (Table 4.1).

4.2.1 THE STATUS OF MEDICINAL AND AROMATIC PLANTS IN CHINA

With around 8,000 medicinal plants, China has a lengthy history of employing plants for healing dating back to the Paleolithic era (Van Alfen 2014). At least 6,000 species have been identified in the literature, including 4,300 angiosperm species and about 100 gymnosperm species. Different species in each family of gymnosperms, which make up nearly all gymnosperms in China, have been employed as herbal treatments. Many families and genera of angiosperms have been studied

TABLE 4.1
Numbers of Plants Species Used Worldwide for Medicinal Purposes

Country	Number of Plants Species	Medicinal Plant Species	Percentage (%)
China	26,092	4,941	18.9
India	17,000	7,500	20.0
Indonesia	22,500	1000	4.4
Malaysia	15,500	1,200	7.7
Nepal	6,973	700	10.0
Pakistan	4,950	300	6.1
Philippines	8,931	850	9.5
Sri Lanka	3,314	550	16.6
Thailand	11,625	1,800	15.5
USA	21,641	2,564	11.8
Vietnam	10,500	1,800	17.1
Average	13,366	1,700	12.5
World	422,000	70,000	

for their potential as medicinal plants. Examples include the 39-genus family Ranunculaceae and c. 700 species are used in China for medicinal purposes, including plants from the families Berberidaceae (260 species in 11 genera) and Papaveraceae (300 species in 19 genera), as well as Verbenaceae (180 species in 21 genera), of which 80 species in 14 genera are utilized in herbal medicine. Campanulaceae, with more than 150 species in 17 genera native to China, of which 50 species in 11 genera have been documented in herbal recipes; Lamiaceae, with 800 species in 99 genera native to China, of which 250 species in 50 genera are used medicinally; Apiaceae, with 500 species in 90 genera native to China, of which 150 species in 45 genera have been used medicinally; and Zingiberaceae, with more than 100 species in 19 genera native to China, of which China has 560 species in 60 genera, the bulk of which are plants used for medicine (Huang 2011). Some important medicinal and aromatic plants in China include *Glycyrrhiza uralensis*, *Epimedium brevicorum*, *Gastrodia alata*, *Gentiana macrophylla*, *Fritillaria* spp., *Dendrobium* spp., *Bergeria purpurascens*, *Eucommia ulmoides*, *Taxus chinensis*, and *Erigeron breviscapus*.

4.2.2 The Status of Medicinal and Aromatic Plants in India

India, which has ten bio-geographic areas, is one of the top 12 countries in the world for mega diversity. Two of the eight biodiversity hot spots in the world are located in India alone. The country's diverse ecological habitats, along with the country's climatic and altitudinal variations, have helped to develop incredibly rich vegetation, including a unique diversity of medicinal plants. These plants serve as an important source of medicinal raw materials for both domestic and international pharmaceutical industries, as well as traditional medical systems. Over 21,000 plant species used for therapeutic purposes are listed by the WHO. About 2,500 plant species are used in India's traditional medical practices. There are 427 Indian medicinal plant entries on the Red Data Book's list of endangered species, of which 28 are thought to be extinct, 124 are endangered, 81 are rare, and 34 are not well enough known. The distribution of these species over several bio-geographical zones, diversified habitats, and landscape features has been discovered. Around 18% of these species are only found in the Himalayan and Trans-Himalayan region, which includes North East India; 4% are only found in the Western Ghats; and 0.5% are only found in the Desert region. The remaining species (about 77%) are widely distributed throughout the other bio-geographic zones of the nation (Chandra 2016). Examples of medicinal and aromatic plants of India include *Rauwolfia Serpentina*,

Glycyrrhiza Glabra, Salacia Oblonga, Holostemma Ada-Kodien, Picrorhiza kurroa, Ginkgo biloba, Celastrus paniculatis, Bacopa monnieri, Oroxylum indicum, and *Tinospora cordifolia*.

4.2.3 The Status of Medicinal and Aromatic Plants in Indonesia

One of the nations with the greatest amount of mega biodiversity and abundant biological resources is Indonesia. A total of 30,000 different plant species, out of the 40,000 different plant species in the globe, are found in this nation; 940 of them are medicinal plants. Because medicinal plants can be a different option for treating a variety of conditions, the use of local plants as a source of medicine is an alternative that can be created (Rahayu and Andini 2019). Important MAPS in Indonesia are *Graptophyllum pictum* (L.) Griff, *Amaranthus spinosus* L., *Impatiens balsamina* L., *Symphytum officinale* L., *Laurentia longiflora* (Linn.) Peterm, *Caesalpinia sappan* L., *Panguim edule, Persea americana, Dracaena angustifloia*, and *Lagerstromeia speciosa*.

4.2.4 The Status of Medicinal and Aromatic Plants in Malaysia

There are 14,500 kinds of flowering plants in Malaysia, 1,300 of which are said to have therapeutic properties, and only approximately 100 have had their full potential thoroughly explored. Around 1,200 plant species, or about 80% of all the plant species used for medicinal reasons by local communities in Sabah (excluding Bryophyta, Algae, and Fungi), have been identified so far (Kulip et al. 2010). Malaysia includes following important medicinal and aromatic plants: *Alstonia angustiloba, Aquilaria malaccensis, Bauhinia kockiana, Curculigo latifolia, Dinochloa scandens, Ficus recurve, Fordia splendidissima, Globba pendula, Polyalthia bullata*, and *Fagraea cuspidate*.

4.2.5 The Status of Medicinal and Aromatic Plants in Nepal

There are 246 indigenous flowering plant species among the 7,000 documented from the nation. The nation ranks 27th in the world for flowering plant diversity. From 70 species in a 1968 survey to 1,463 species in 1997, the list of documented medicinal plants used in Nepal has increased. This growth over time shows that the total number of species utilized as medicine in the nation is still unknown. In Nepal, 14% of the land is protected. According to reports, Nepal is home to about 100 different medicinal plants that are used in the production of herbal medicines. Some of the most significant medicinal plants are *Asparagus racemosus* Willd., *Bergenia ciliata* (Haw.) Sternb., *Cinnamomum glaucescens* Nees, *Picrorhiza scrophulariiflora* Pennell, *Swertia chirayita* (Roxb. ex Fleming) H., and *Zanthollum armatum* DC. S. Other plants include *Rheum australe* D., *Piper* spp., and *Acorus calamus* L. *Rubia cordifolia* L., Don and *Valeriana jatamansi* Jones exhibit a similar pattern in demand and distribution. Although their total quantities collected were lower than those of the first group, their utility value was higher (Tiwari, Uprety, and Rana 2019). In Nepal, some important medicinal and aromatic plants include *Adiantum capillus-veneris, Rhus parviflora, Angelica archangelica, Pleumeria rubra L, Ageratum conyzoides, Inula racemosa, Xanthium strumarium, Chenopodium album, Cordyceps sinensis*, and *Dioscorea deltoidea*.

4.2.6 The Status of Medicinal and Aromatic Plants in Pakistan

With an area of 87.98 million hectares, Pakistan is a developing nation in South Asia. Of this region, 88% is dry and semi-arid, while just 12% is humid and sub-humid, both of which are found in the Himalayan ranges. It has the ninth-highest population density in the world. About 2,000 species of medicinal plants are thought to exist in Pakistan, 400 of which are widely used in traditional medicine. Commercially exploited medicinal plants are most common in alpine and high-altitude regions, temperate montane forests, sub-tropical foothill forests, and semi-arid scrub lands (Baig and Al-Subaiee 2009). Important medicinal and aromatic plants of Pakistan are *Trianthema portulacastrum* Linn, *Amaranthus viridis Linn. Aerva persica, Alternanthera sessilis* Linn, *Cataranthus roseus* Linn, *Thevetia aperuviana* (Pers.) Schum, *Leptadenia pyrotechnica* (Forssk) Decne, *Calotropis procera, Ageratum houstonianum*, and *Ageratum conzeoides.*

4.2.7 The Status of Medicinal and Aromatic Plants in Philippines

The Philippines has a long history of using herbal medicines, from rural traditional healers to housewives who create cures in their backyard gardens. The primary source of medicines in conventional medical procedures is medicinal plants. More than 1,500 medicinal plants utilized by traditional healers in the Philippines have been documented and 120 plants have received scientific validation for their efficacy and safety. Out of all known Philippine medicinal plants, the majority have undergone evaluation in order to corroborate traditional claims (Rondilla et al. 2021). Philippines includes following medicinal and aromatic plants: *Hemigraphis colorata, Justicia gendarussa, Eurycles amboinensis, Spondias purpurea, Carica papaya, Brophyllum pinnatum, Jathropa curcas, Arachis hypogaea, Coleus blumei*, and *Corchorus olitorius.*

4.2.8 The Status of Medicinal and Aromatic Plants in Sri Lanka

Sri Lanka has a high amount of biodiversity because to its diverse topography and climate. Sri Lanka, one of the most ecologically varied nations in Asia, has 4,143 plant species spread over 214 groups and a forest cover of 29.7%. 1,025 of these plant species are indigenous to the nation. As a result, Sri Lanka is acknowledged as a biodiversity hot spot of both global and local significance. Ayurveda, Siddha, Unani, and Deshiya Chikitsa are the four traditional medical systems in Sri Lanka that have been using plants to cure various illnesses for thousands of years. By meeting 60–70% of the primary healthcare needs of the rural population in Sri Lanka, traditional medical systems have a significant impact on their lives. 1,430 species belonging to 181 families and 838 genera are regarded as medicinal plants. 174 (12%) of the total number of medicinal plant species are indigenous to the nation. There are 250 species of medicinal plants that are frequently employed in traditional medicine, of which 50 species are heavily used. Thus, it blatantly emphasizes the significance of medicinal plants in Sri Lanka's various medical systems (Dharmadasa et al. 2016). Some important medicinal and aromatic plants of Sri Lanka include *Barleria, lupulina* Lindl, *Acorus calamus* L., *Berberis aristata* DC, *Bombax ceiba* L., *Garcinia mangostana* L., *Asparagus racemosus, Heliotropium indicum, Mesua ferrea*, and *Garcinia mangostana.*

4.2.9 The Status of Medicinal and Aromatic Plants in Thailand

Thailand has a wealth of natural resources, a variety of ecological settings, a wide range of ethnic groups, including ancient civilizations. There are more than 10,000 different plant species, and 1,400 of them are classified as native medicinal and aromatic plants. Global production and use of plant-based raw materials in the pharmaceutical, cosmetics, fragrance, and allied industries are enormous. Several medicinal and aromatic plants are consumed domestically and exported to Europe, the USA, and Japan as raw materials or intermediary chemicals. Included among the candidate species are Senna (*Cassia angustifolia* Vahl.), Chilli pepper (*Capsicum frutescens* L.), Pepper (*Capsicum* spp.), Sweet Basil (*Ocimum basilicum* L.), Citronella (*Cymbopogon nardus* (L.) Rendle), Jasmine (*Jasminum* spp.), Champaca (*Michelia* spp.), Ylang-Ylang (*Cananga odorata* (Lam.), and *Cymbopogon citratus* (DC.) Stapf. (Phumthum et al. 2020). Important medicinal and aromatic plants of Thailand include *Amaranthus spinosus, Amomum krervanh* Pierre ex Gagnep, *Amomum krervanh* Pierre ex Gagnep, *Amomum xanthioides* Wall, *Anacardium occidentale* L., *Anethum graveolens* L., *Belamcanda chinensis* DC., *Abrus precatorius, Alpinia galanga, Acorus calamus*, and *Aegle marmelos.*

4.2.10 The Status of Medicinal and Aromatic Plants in United States

The use of medicinal plants as a medication to address medical conditions is common in the United States, with about one-fourth of individuals reporting doing so in the previous year. It has been discovered that more than 3,000 medicinal plants have been used for many years in the United States. 10% of plants in the United States, according to the survey, are used medicinally. Patients with specific chronic medical illnesses, such as breast cancer (12%), liver disease (21%), human immunodeficiency virus (22%), asthma (24%), and rheumatic diseases (26%), frequently utilize herbal medications (Bent and Ko 2004). United States of America includes some following important medicinal and aromatic plants: *Taraxacum officinale, Plantago major, Arctium minus, Urtica dioica, Achillea millefolium, Prunella vulgaris, Rubus* spp., *Salix* spp., *Echinacea angustifolia*, and *Stellaria media.*

4.2.11 The Status of Medicinal and Aromatic Plants in Vietnam

Vietnam is a tropical nation with a wide range of geographical and climatic characteristics. It boasts a wealth of various natural resources, including aromatic and medicinal plants. In 350 districts, province capitals, and 2,795 villages and quarters, an assessment of the genetic resources of medicinal and aromatic plants utilized in traditional remedies was carried out: 1,863 species and sub-species altogether (now about 2,000 species). Three quarters are said to be growing wild and are dispersed everywhere, belonging to 238 families. About 10,000 herbarium specimens of the 1,296 species that occur in Vietnam have been collected, together with data on the flowering and fruiting times of 1,423 species, and close to 1,000 traditional and ethnomedical treatments. Asteraceae (105 species), Fabaceae (137 species), Euphorbiaceae (91 species), Rubiaceae (65 species), and Lamiaceae (45 species) are the top five families of medicinal plants (Van Sam, Bass, and Keßler 2008). Examples of some important medicinal and aromatic plants of Vietnam are *Alstonia scholaris, Careya arborea, Medinilla septentrionalis, Elephantopus mollis, Calamus palustris, Paris polyphylla* Smith, *Drynaria bonii* Christ, *Ardisia silvestris* Pitard, *Chukrasia tabularis*, and *Curculigo orchioides* Gaertn.

4.3 THREATS TO MEDICINAL PLANTS BIODIVERSITY

Natural resource consumers and conservationists are both concerned about the ongoing threat to the world's supply of therapeutic plants. Due to the rising demand for medicinal plants in the pharmaceutical industry, for herbal treatments, and for other natural goods in recent years, plants are being harvested from the wild and are exploited in an unsustainable way (Prance, Chadwick, and Marsh 1994; Kumar, Kumar, and Khan 2011). Threats to genetic diversity and species survival have risen in the case of medicinal plant resources as a result of overexploitation, habitat degradation, land use changes, and other related causes (Arora and Engels 1993). Aside from that, medicinal plants are also traded for economical practices on an unsustainable scale (Natesh 1999).

In addition to endangering plant resources, habitat modification or destruction poses a serious risk to traditional community life, cultural diversity, and the understanding of the therapeutic value of some endemic species. With the growing human population and plant consumption, over 20% of the wild medicinal plants have been consumed (Ross 2005). The nature, scope, and influence of these elements, as well as the gravity of the threat, varied according to the nation. In order to adopt a workable conservation approach to preserve them, it is crucial to understand the biological properties and geographic distribution of medicinal plants.

4.3.1 Habitat Destruction

Habitat loss is regarded as a serious issue that could lead to the extinction of living species, including plants used as medicines worldwide. In the modern world, human influence may be found in all natural habitats of animals and plants, including high alpine regions, coasts, rainforests, and deserts

(TRAFFIC 1999). The diversity of medicinal plants may be impacted by this danger (habitat degradation) in a variety of ways. The rapid extinction of significant plant species caused by ecological damage is one of the biggest worries. One of the finest examples is the rise in slash-and-burn clearings in the rainforests of South America, Africa, and Southeast Asia (Touchell 1997). Additionally, uncontrollable soil erosions might occur if the soil in such locations, which has lost its native flora, is not allowed to regenerate back to its usual position. As a result, the natural greenery and its healing herbs permanently vanish. In many developing nations, mangrove forests are destroyed for aquaculture, farming, and other similar objectives, which results in the loss of wetlands and coastal plant species (Intergovernmental Panel on Climate Change 2007).

Changes in habitat architecture caused by anthropogenic activity and climate change over the past century finally led to plant species having insufficient tolerance to shifting climatic conditions (Intergovernmental Panel on Climate Change 2007). One of the world's 34 biodiversity hotspots, the Himalayan region, has lost more than 70% of its natural habitat (Hachfeld and Schippmann 2000). Due to a rise in population in these areas, more forests, marshes, and grasslands have been converted to agricultural land and human settlements. The construction of roads and highways across natural environments has eventually led to habitat fragmentation, increased the spread of dangerous illnesses, insects, and invasive species, and resulted in the loss of precious medicinal flora (Davis et al. 1997). As a result, there is an urgent need for a better understanding and threat assessment of various plant species as well as their habitats, with a particular emphasis on their adaptation range.

4.3.2 Over-exploitation

Over-exploitation of medicinal plant resources has become a serious threat to the survival of the world's medicinal plant diversity in this age of humans, where no one is aware of the bankruptcy of Earth's resource capital (Sodhi and Ehrlich 2010). Humans exploit the resources on the surface of the earth and also peer into its deeper layers, which is a completely unsustainable activity. Over-harvesting of medicinal plant species is reducing access to traditional treatments and endangers some of the most precious wild species and their habitats. Approximately 15,000 medical plant species may be threatened by over-harvesting, according to the World Conservation Union (IUCN), which estimates that 50,000–80,000 medicinal plants are used worldwide in traditional healthcare systems (Chen et al. 2016). Numerous species are on the verge of extinction as a result of the increased pressure on wild medicinal plants brought on by over-harvesting. Even with the smallest levels of collection, some species of medicinal plants can go extinct. However, many of these species are not vulnerable to harvesting pressure (Sodhi and Ehrlich 2010). As a result, human activity is depleting the earth's precious resources, endangering future generations.

4.3.3 Genetic Erosion

In addition to other dangers, genetic erosion poses a serious risk to species of medicinal plants. In addition to the genetic erosion that occurs naturally, humans also cause genetic modifications through different mutations. The cultivation and enhancement of many varieties of medicinal plant species cause genetic alterations that frequently increase the concentration of specific chemicals. In contrast to the changes that occur naturally over a few years, these alterations happen relatively quickly. Although these modifications initially produce the desired outcomes, they upset the plants' normal equilibrium. Although these desirable components can be extracted and used to treat a variety of disorders, the long-term efficacy of these plants as medicines is still debatable (Walter and Gillett 1998). Furthermore, nothing is known about genetic degradation in wild plants. There is some uncertainty based on theoretical concerns regarding the possibility that many medicinal plant species not designated as threatened are currently experiencing or will experience genetic erosion in the near future (Holsinger and Gottlieb 1991). 10,000 medicinal plant species are considered to be under danger (Vorhies 2000).

4.3.4 Bioprospecting and Biopiracy

The medicinal plant industry is quite lucrative. According to a 1995 analysis, each new plant-based medication is worth an average of $449 million to society and $94 million to pharmaceutical businesses (Mendelsohn and Balick 1995). According to other projections, non-prescription medicinal plant sales in the United States range from $1.5 to $5.7 billion yearly, with $24.4 billion in global sales. In 1985, the market for prescription and over-the-counter plant-based medications was estimated to be worth $84.3 billion globally and $19.8 billion in the United States (Fearce and Moran 1994; Tuxill 1999). Resources from medicinal plants that have historically been used by societies that have relied on them for millennia may become less readily available as a result of this profitability. The practice of private firms patenting folk cures from the wild and selling them at a significant profit, frequently allowing little to no of that wealth to go back to the nation or the indigenous and local communities of origin, has been termed as "biopiracy". Occasionally, after a medicine is patented, indigenous and local communities that have utilized it for millennia may no longer be able to use it or gain from it. For instance, the neem tree, a relative of mahogany from South Asia, has produced highly powerful and lucrative germicides, fungicides, and other substances. At least 50 patents for goods made from this species had been given to American and forcign businesses by 1995. International humanitarian organizations and environmentalists have opposed these patents (Torrance 2000).

4.3.5 Lack of Effective Regulation

The threat to medicinal plants has increased as a result of the erosion of traditional regulations that have historically controlled the use of natural resources (Pant 2002). These laws are easily undermined by contemporary socio-economic factors. Many European nations have created regional and national laws in order to address various conservation issues related to medicinal plants, in addition to being required to abide by the Convention on International Trade in Endangered Species of Wild Fauna and Flora (CITES) and the Convention on Biological Diversity (CBD). The UK Wildlife and Countryside Act, for instance, makes it illegal for anybody other than landowners or other authorized individuals to uproot any type of wild plant. Additionally, nations like Poland have identified medicinal plant species that cannot be gathered without appropriate authorization, while Italy created a law that regulated the gathering of medicinal plants in 1931 (Lange 1998). However, most developing nations lack this efficient regulation, making it challenging to effectively conserve endangered species. Some of the community norms that historically protected medicinal plants have been superseded by forest laws, which are still not well-implemented and enforced, particularly in the remote regions of the greater Himalayas. Because of the difficulty in enforcing rules, the population of several valued aromatic plants, such as *Nardostachys grandiflora* and *Picrorhiza kurroa*, has decreased owing to illegal trading in Nepal and Indian borders. In addition to these species, a number of emerging nations also lack suitable speciation practices (Mulliken 2000). Furthermore, it is not stated that which threats are predominant in which societal perspectives or why.

4.4 EXAMPLES OF THREATENED SPECIES

The plants listed below are just a few of the rare species that are highly prized for their medicinal properties.

4.4.1 Black Cohosh (*Cimicifuga racemosa*)

Native Americans employed black cohosh root tea to treat women's health problems, including menopause and monthly irregularities. Inflammatory disorders, tinnitus, and potentially arthritic and rheumatism symptoms have all been linked to it. Harvesting pressures, habitat loss, and invasive species competition are reasons for its loss.

4.4.2 Goldenseal (Hydrastis canadensis)

Native Americans who lived along the East Coast of North America highly valued the root of the goldenseal plant, because it is an essential immune-supporting herb. It can be used to treat congestion, infections of the skin and other kinds, and to assist the digestive system. It is included in red list because of its over-harvesting.

4.4.3 Hawaiian Sandalwood (Santalum ellipticum)

Only in Hawaii is this type of sandalwood to be found. It has a long history of traditional usage, including those for immune support, astringent, anti-inflammatory, natural perfume, anti-bacterial, incense in religious and spiritual acts, and general body-toning. Over-harvesting due to high demand, a 40-year period of maturity, and a lack of regulations on harvest are reasons for its loss.

4.4.4 *American Ginseng* (Panax quinquefolius, Panacis quinquefolis)

This woody plant, a perennial herb native to Eastern North America, is valued for its root. Numerous health benefits of American ginseng include adaptogenic, cardiotonic, sedative, immune-supportive, tonic, and stomachic effects. Additionally, American ginseng promotes the endocrine system and aids in the body's ability to adapt to various stressors. Over-harvesting and habitat loss are reasons for its rarity.

4.4.5 Asian Ginseng (Panax ginseng)

Chinese have used Asian ginseng for thousands of years as a medicine, and the emperors once had exclusive access to the root. It is regarded as an adaptogenic herb and functions as a general tonic for the body even in stressful situations. It also helps to maintain energy, boost the immune system, reduce inflammation in the body, and has anti-oxidant characteristics. The reason for its loss is over-harvesting. The majority of Asian Ginseng available on the market today is now cultivated.

4.4.6 Bloodroot (Sanguinaria candadensis)

The sap of the bloodroot plant, a perennial blooming plant that is indigenous to the Eastern North American woodlands, is typically advised for external use only. Bloodroot is mostly applied topically and is well recognized for treating eczema, moles, and other skin disorders like skin tags. Over-harvesting is the main reason for its rarity.

4.4.7 Wild Yam (Dioscorea villosa)

The roots and stems of this plant, which is native to North America, Mexico, and Asia, are used medicinally to reduce inflammation, aid digestion, support healthy blood sugar levels, act as an anti-oxidant, ease muscle cramps, aid in maintaining a healthy metabolism, boost virility, and support healthy cholesterol levels. It is lost because of over-harvesting.

4.4.8 *Rosy Periwinkle* (Catharanthus rosea)

This miraculous herb, which is indigenous to the forest, woodlands, and grasslands of the island of Madagascar, has been shown to strengthen the body's defenses against cancer, diabetes, and

Hodgkin's lymphoma as well as to aid in the body's recovery from malaria. This plant is endangered due to over-harvesting and habitat loss.

4.4.9 Eyebright (Euphrasia officinalis)

The Eyebright plant's stem, leaves, and flowers are edible and have therapeutic qualities. This plant is used to treat skin conditions such as acne and stretch marks, respiratory conditions like bronchitis, colds, and allergies, as well as eye health issues such as eye inflammation, conjunctivitis, cataracts, and styes. The eyebright plant might also aid with memory enhancement. This plant is endangered due to over-harvesting and habitat loss.

4.4.10 Slippery Elm (Ulmus rubra)

The mucilaginous inner bark of this tall flowering tree, which is native to the Eastern and Central United States and into Canada, has a variety of therapeutic benefits, including relieving coughs and sore throats, reducing inflammation, and supporting the digestive, glandular, and urinary tract systems. As a result of Dutch elm disease and unsustainable harvesting practices, this tree is seriously endangered.

4.5 NEED FOR CONSERVATION

As long as forest resources are being destroyed, over-exploitation of medicinal plant diversity and their natural habitats will continue to be a threat (Walter and Gillett 1998). Conservation is therefore crucial in maintaining the natural habitats of vulnerable medicinal plant species and achieving sustainable utilization in less vulnerable places (Cunningham 1997). Numerous significant medicinal plant species are being destroyed at an alarming rate due to the destruction of forests, and some of the raw materials used to make medicines are already in limited supply in various pharmaceutical businesses worldwide. A mass extinction is expected to occur if deforestation keeps up its current pace, and a sizable portion of genetic diversity is thought to be at risk. Deforestation and habitat fragmentation have a negative impact on the diversity of rare species (Phartyal et al. 2002). Several conservation strategies, including commercial cultivation, habitat preservation, creation of natural reserves, and enforcement of regulations, should be implemented in order to maintain the diversity of medicinal plants (Rao, Rabinowitz, and Khaing 2002). Numerous plant species are in danger of going extinct, according to recent studies, while others have already lost their habitats. Currently, a number of initiatives have been proposed by both government and non-governmental organizations to preserve the diversity of medicinal plants that are deteriorating (Phartyal et al. 2002). Despite their acknowledged significance and usefulness, medicinal plants are endangered in their native habitats due to the degradation of their natural ecosystems. With the opening of forest regions for farming and trans-migration, this issue is getting worse (Hikmat et al. 2001 2001). Numerous therapeutic plants have been exhausted, and numerous fresh species identities and advantages would be lost to humankind, since they were not fully studied and documented (Mir et al. 2021). There is a need to collect and record these plants with their traditional applications in order to safeguard the medicinal plant resources.

4.6 CONSERVATION STRATEGIES FOR MEDICINAL PLANT RESOURCES

Over-harvesting of medicinal plant resources, mostly from wild populations, has been caused by recent increases in demand for wild resources (Ross 2005). At least one potential medication is lost every two years, according to estimates, and the current rate of plant species extinction is higher than projected (Pimm et al. 1995). The International Union for Conservation of Nature and the World Wild Life Fund estimate that between 50,000 and 80,000 plants are utilized worldwide for therapeutic purposes, with 15,000 of those species facing various threats (Bent and Ko 2004). Due to these dangers, medicinal plant extinction risk has increased globally, particularly in China, India, Kenya, Nepal, and Uganda (Heywood and Iriondo 2003; Zerabruk and Yirga 2012). Consequently, the proper study of medicinal plant conservation has taken place (Larsen and Olsen 2006). The creation of systems for species inventory, improvement of cultivation efforts, improved management of wild populations, public awareness, law enforcement, and the requirement for conservation practices based on in situ and ex situ strategies are just a few of the recommendations made for the conservation of medicinal plants (Hamilton 2004). While botanical gardens and seed banks are significant examples of ex situ conservation and sustainable development, wild nurseries and natural habitats, for instance, are typical examples of maintaining the medicinal efficacy of plants in their natural settings (Sheikh et al. 2002; Coley et al. 2003).

4.6.1 In Situ Conservation Approach

To guarantee that plant species continue to flourish in the wild and in their natural habitat, in situ conservation is used. This method of conservation comprises the creation of plant resources in areas where they naturally occur and the safeguarding of wild populations (Cunningham 1997). By ensuring that as many plant species continue to flourish in regulated habitats as possible, in situ conservation is offered in some protected areas where they develop naturally. These resources are exploited at a very high rate without any effective management due to strong market demand and poverty, because some communities are entirely dependent on the sale of medicinal plant varieties. A variety of actions are included in a strategy for the maintenance of medicinal plants, such as the identification and protection of regions with a high density of fragile species and a high rate of commercial exploitation (Srivastava 2000).

Traditional medicinal plants are protected by in situ conservation, which also upholds the integrity of natural communities and their intricate network relationships (Gepts 2006). In situ conservation can improve the diversity of species that need to be saved in addition to establishing a relationship between resource conservation and sustainable use (Lange 1998). The goal of in situ conservation is to create protected areas, and this strategy is more ecosystem-focused than species-focused. Rules, restrictions, and prospective agreements between medicinal plant species and growth habitats are necessary for this form of protection (Ma, Rong, and Cheng 2012). This kind of conservation promotes genetic diversity, a species' ability to adapt to changing environmental conditions, and the potential for ecological and evolutionary processes to occur. In order to preserve biodiversity, natural environments are preferred (Heywood and Watson 1995).

4.6.1.1 Protected Areas

By creating protected area systems, biodiversity protection is frequently accomplished. These are geographically isolated places that are kept up and under control to achieve certain conservation goals (Gillison and Boyle 1996). The preservation of medicinal plant resources in their natural habitat is greatly aided by protected areas. It is vital to put enough human and financial resources in place for protected areas to be effective in maintaining plant resources. Locals should receive an equal share of benefits from the conservation and sustainable use of plant resources. Numerous nations (including African nations) have demonstrated that they are unable to achieve their conservation goals mainly because local people are not included in the process (Zegeye, Teketay, and Kelbessa 2006; Jeffries 1997).

They consist of places like wild nurseries and natural reserves. Natural reserves were established as protected places to preserve and replenish declining biodiversity. 12,700 have been built globally, covering 13.2 million km^2 (Huang et al. 2002). To conserve medicinal plant species by preserving important natural habitats, it is necessary to examine the contributions and ecological functions of each unique habitat (Liu et al. 2001). In wild nurseries, medicinal plant species that are in danger of extinction are preserved by being domesticated and grown in their original habitats or only a short distance from them (Hamilton 2004; Schipmann et al. 2005). Wild nurseries' approach to in situ conservation is the greatest way to preserve these endemic and endangered medicinal plant species, because wild medicinal plants are under heavy threat from over-harvesting, habitat degradation, and exotic species (Liu et al. 2001).

4.6.1.2 On-Farm Conservation

Crops, their wild relatives, and the agroecosystems in which they exist are all protected through on-farm conservation. Agroecosystems include backyard gardens, crop fields, grazing grounds, fallow fields, and agroforestry systems. The preservation of the variety of medicinal plants depends on on-farm in situ conservation. Indigenous resource management strategies and farming practices are of utmost importance for the preservation and diversification of domestic medicinal plants (McNeely et al. 1995). Worldwide, farmers and pastoralists cultivate a staggering variety of therapeutic plant species. Additionally, indigenous farming practices, knowledge, and skills are crucial for managing and conserving agrobiodiversity (Deribe et al. 2002). Therefore, it has been discovered that farmer-based on-farm conservation of medicinal plant biodiversity is a more effective strategy.

4.6.2 Ex Situ Conservation Approach

Ex situ conservation is the process of preserving biological diversity away from where it naturally occurs (Antofie 2011). It guarantees the preservation of therapeutic plants with sluggish development, low quantity, and disease susceptibility that are over-exploited and endangered (Hamilton 2004; Havens et al. 2006; Yu et al. 2010). Ex situ conservation aims to nurture and naturalize endangered species in order to ensure their ongoing survival and to supply the necessary quantity of material for the production of prospective medications. This method is frequently used as an emergency measure to preserve medicinal plant resources (Swarts and Dixon 2009).

Establishing plantations and maintaining live collections in farm fields, backyard gardens, and botanical gardens away from their natural habitats are all included in this conservation strategy (Roche 1975). When medicinal plant species are grown far from their natural environments, they may help to keep their high potency as well as the ability to select and store their reproductive resources in gene banks for future replanting. Additionally, ex situ conservation serves as a source of data for research and ecological restoration (Hamilton 2004). In order to compete with the prices received by gatherers of medicinal plants from the wild, it is the kind of quick development of alternative supply of medicinal plant resources by cultivation in large quantities at reduced rates. As a result, the market's demands will be met, which will also lead to more secure jobs (Cunningham 1997).

4.6.2.1 Botanical Gardens

Herbariums, lecture rooms, labs, libraries, museums, and paintings created for experimentation or research are all included in botanical gardens. According to the Institute of Biodiversity Conservation, it could be a collection of a specific family, genus, or species of fragrant or medicinal plants. Botanical gardens play a key part in ex situ conservation, because they preserve the environment to help rare and endangered plant species survive (Christie 1998). Many botanical gardens have a taxonomically and ecologically diverse flora as well as a diversity of plant species that are all grown together in similar environments (Primack and Miller-Rushing 2009). Arable medicinal plants can be produced in controlled and modified environments on a platform provided by botanical gardens. Botanical gardens, which are thought of as living labs, encourage and carry out scientific research on medicinal plants in particular and biological diversity in general (Brütting, Hensen, and Wesche 2013). Therefore, botanical gardens are the organizations that maintain the recorded collections of living plants for scientific research, protection, display, and education.

4.6.2.2 Gene Banks

Another method for preserving the resources of medicinal plants is genome banking. *In vitro* gene banks for plant tissues and cells, seed banks for seeds, field gene banks for live plants, pollen, chromosomal, and deoxyribonucleic acid (DNA) banks for the conservation of plant resources kept in laboratory storage for short- or long term are all examples of genome banking (Clarke 2009). The goal of a gene bank is to maintain genetic variety in order to protect medicinal plant species that are in danger of extinction and to lessen the rate of regeneration. With the advancement of molecular genetics and genomics, the need for DNA for molecular investigations is rising quickly from gene banks (Dulloo, Hunter, and Borelli 2010). Seeds are the best material for conservation in gene banks since they are the most convenient and easy to keep in a viable form for extended periods of time. Dried seeds of endangered and other medicinal plants are kept in seed banks at temperatures between 18°C and 28°C for long-term preservation and between 3% and 7% moisture content at 4°C for short-term conservation (Dulloo, Hunter, and Borelli 2010; Brütting, Hensen, and Wesche 2013).

4.6.3 Cultivation Practice

Domestication is typically a widespread and accepted activity, despite the fact that many people believe wild medicinal plant resources to be more effective than those that have been grown (Gepts 2006; Joshi and Joshi 2014). Cultivation offers the chance to employ novel methods to address issues with medicinal plant production, such as pesticide contamination, poisonous components, low concentrations of active compounds, and incorrect identification of botanical origin (Raina et al. 2011). Cultivation can increase the generation of active secondary metabolite chemicals and assure stability, when growing conditions are controlled. Due to this, cultivation techniques are created to give the ideal amounts of nutrients, water, additives, and environmental elements like temperature, light, and humidity to produce increased yields of pharmaceuticals (Liu et al. 2001; Wong et al. 2014). Additionally, as cultivation increases, the amount of medicinal plant resources harvested declines, which supports the recovery of their wild relatives and brings down costs to a fair range (Larsen and Olsen 2006; Gepts 2006).

4.6.3.1 Good Agricultural Practices (GAP)

Good agricultural practices have been developed for medicinal plants to regulate the production and facilitate the standardization of herbal medicines, ensuring high-quality, safe, and pollution-free herbal medicines (Chang, Hua, and Shi-Lin 2011). Good agricultural practices cover a wide range of topics, such as the ecological environment of production sites, germplasm cultivation, collection and quality aspect of pesticide detection, macroscopic or microscopic authentication, chemical characterization of bioactive compounds, and assessment of metal elements (Makunga, Philander, and Smith 2008). In regions where these medicinal plants are being historically grown, many nations, including China, have aggressively pushed the implementation of GAP for the development of regularly used herbal medicines (Ma, Rong, and Cheng 2012). Organic farming has drawn a lot of attention for its capacity to develop production systems that are both economically and environmentally viable. Regular use of organic fertilizers enhances soil stability, while also supplying soil nutrients, which has an impact on the development of medicinal plants. Therefore, organic farming is gaining more attention as a means of ensuring the growth and sustainability of medicinal plants (Macilwain 2004).

4.7 CONCLUSION

The ability of natural ecosystems to support human health and life depends on biodiversity. As a component of this biodiversity, medicinal plants are the backbone of the country's economy and healthcare system. Herbal remedies represent a significant natural resource that is close to extinction and have a significant role in the healthcare and environmental sectors. Since the majority of people worldwide depend directly or indirectly on medicinal plant resources for their primary healthcare and prominent pharmaceutical industries use these resources to their advantage, our medicinal plant resources are being over-harvested, which poses a serious threat to their extinction. This threat is also brought on by the market's rising demand for these resources. Due to habitat degradation and other damaging anthropogenic activities, a number of medicinal plant species are disappearing at an alarming rate. They are in danger as a result of the uncontrolled trading of medicinal plant resources and the unsustainable use of those resources for grazing, firewood, food, and medicine, among other uses. In order to preserve this disappearing heritage, medicinal plant resources require significant preservation, management, research, and raising public awareness. These therapeutic plant resources and their environments must both be preserved. It is important to encourage the creation of natural protected areas such as botanical gardens, wild nurseries, and wild reserves. Enhancing culture is necessary for medicinal plant preservation and bulk manufacturing. Additionally, national and international non-governmental organizations working with the governments can play a significant role in saving the biotic diversity, which will benefit society in a number of ways. The ability to end the current devastation on these resources and to support alternative agricultural development of these resources will determine the future of medicinal plants around the world. Many scholars emphasized the necessity to preserve medicinal plant resources through sustainable management strategies, and as such, these strategies must explore the current pattern of medicinal plant distribution, collecting, and consumption. Investigation of the variables that can serve as preventative measures to stop these actions or reveal the degree to which they are detrimental to the sustainability of medicinal plants in a particular location is urgently needed. In order to preserve this priceless heritage, it is crucial to fight for the conservation and sustainable utilization of medicinal plant resources.

REFERENCES

Alfen, Neal K Van. 2014. *Encyclopedia of Agriculture and Food Systems*. Amsterdam: Elsevier.

Altemimi, Ammar, Naoufal Lakhssassi, Azam Baharlouei, Dennis G Watson, and David A. Lightfoot 2017. "Phytochemicals: Extraction, Isolation, and Identification of Bioactive Compounds from Plant Extracts" *Plants* 6(4): 42.

Antofie, Maria-Mihaela. 2011. "Current Political Commitments' Challenges for Ex Situ Conservation of Plant Genetic Resources for Food and Agriculture" *Analele Universitatii din Oradea, Fascicula Biologie* 18(2): 157–163.

Arora, R K, and J M M Engels. 1993. "Genetic Resources in Medicinal and Aromatic Plants: Their Conservation and Use." *Acta Horticulturae* 330: 21–38.

Atanasov, Atanas G, Birgit Waltenberger, Eva-Maria Pferschy-Wenzig, Thomas Linder, Christoph Wawrosch, Pavel Uhrin, Veronika Temml, Limei Wang, Stefan Schwaiger, and Elke H. Heiss. 2015. "Discovery and Resupply of Pharmacologically Active Plant-Derived Natural Products: A Review" *Biotechnology Advances* 33(8): 1582–614.

Baig, Mirza B, and Faisal Sultan Al-Subaiee. 2009. "Biodiversity in Pakistan: Key Issues" *Biodiversity* 10(4): 20–29.

Bent, Stephen, and Richard Ko. 2004. "Commonly Used Herbal Medicines in the United States: A Review" *The American Journal of Medicine* 116(7): 478–85.

Bhalodia, Nayan R, and V J Shukla 2011. "Antibacterial and Antifungal Activities from Leaf Extracts of Cassia Fistula l.: An Ethnomedicinal Plant" *Journal of Advanced Pharmaceutical Technology and Research* 2(2): 104.

Brütting, C, I Hensen, and K Wesche. 2013. "Ex Situ Cultivation Affects Genetic Structure and Diversity in Arable Plants" *Plant Biology* 15(3): 505–13.

Chandra, L D 2016. "Bio-Diversity and Conservation of Medicinal and Aromatic Plants" *Advances in Plants and Agricultural Research* 5(4): 186.

Chang, L I U, Y U Hua, and Chen Shi-Lin. 2011. "Framework for Sustainable Use of Medicinal Plants in China" *Plant Diversity* 33(01): 65.

Chen, Shi Lin, Hua Yu, Hong Mei Luo, Qiong Wu, Chun Fang Li, and André Steinmetz. 2016. "Conservation and Sustainable Use of Medicinal Plants: Problems, Progress, and Prospects." *Chinese Medicine (United Kingdom)* 11(1): 1–10. https://doi.org/10.1186/s13020-016-0108-7.

Christie, S. 1998. "Why Keep Tigers in Zoos?" In: Tilson R, Nyhus P (Eds) *Tigers of the World: Theb Science, Politics and Conservation of Panthera Tigris*. Amsterdam: Elsevier Inc, 205–14.

Clarke, A G. 2009. "The Frozen Ark Project: The Role of Zoos and Aquariums in Preserving the Genetic Material of Threatened Animals" *International Zoo Yearbook* 43(1): 222–30.

Coley, Phyllis D, Maria V Heller, Rafael Aizprua, Blanca Araúz, Nayda Flores, Mireya Correa, Mahabir Gupta, et al. 2003. "Using Ecological Criteria to Design Plant Collection Strategies for Drug Discovery" *Frontiers in the Ecology Environment* 1(8): 421–28.

Cunningham, A B. 1997. "An Africa-Wide Overview of Medicinal Plant Harvesting" *Medicinal Plants for Forest Conservation, and Health Care* 92: 116.

Davis, S D, V H Heywood, O Herrera-MacBryde, J Villa-Lobos, and A C Hamilton. 1997. "Centres of Plant Diversity: A Guide and Strategy for Their Conservation. Vol. 3." *The Americas*. The Worldwide Fund for Nature (WWF)/The World Conservation Union (IUCN). ISBN: 2-8317-0199-6.

Deeb, Taha, Khouzama Knio, Zabta K Shinwari, Sawsan Kreydiyyeh, and Elias Baydoun. 2013. "Survey of Medicinal Plants Currently Used by Herbalists in Lebanon" *Pakistan Journal of Botany* 45(2): 543–55.

Deribe, Shewaye, Zemede Asfaw, Awegechew Teshome, and Sebsebe Demissew. 2002. "Management of Agrobiodiversity in the Borkena Watershed, South Wollo, Ethiopia: Farmers Allocate Crops/Landraces to Farm Types" *Ethiopian Journal of Biological Sciences* 1(1): 13–36.

Dharmadasa, R M, G C Akalanka, P R M Muthukumarana, and R G S Wijesekara. 2016. "Ethnopharmacological Survey on Medicinal Plants Used in Snakebite Treatments in Western and Sabaragamuwa Provinces in Sri Lanka" *Journal of Ethnopharmacology* 179: 110–27.

Doughari, James Hamuel, I S Human, A J S Benadé, and Patrick Alois Ndakidemi. 2009. "Phytochemicals as Chemotherapeutic Agents and Antioxidants: Possible Solution to the Control of Antibiotic Resistant Verocytotoxin Producing Bacteria." *Journal of Medicinal Plants Research* 3(11): 839–48.

Dulloo, Mohammad Ehsan, Danny Hunter, and Teresa Borelli. 2010. "Ex Situ and in Situ Conservation of Agricultural Biodiversity: Major Advances and Research Needs" *Notulae Botanicae Horti Agrobotanici Cluj-Napoca* 38(2): 123–35.

Farnsworth, Norman R, and Djaja D Soejarto. 1991. "Global Importance of Medicinal Plants" *The Conservation of Medicinal Plants* 26: 25–51.

Fearce, D, and D Moran. 1994. *The Economic Value of Biodiversity*. London: Earthscan Publication.

Gepts, Paul. 2006. "Plant Genetic Resources Conservation and Utilization: The Accomplishments and Future of a Societal Insurance Policy" *Crop Science* 46(5): 2278–92.

Gillison, A N and T J B Boyle. 1996. "Measures for Conservation of Biodiversity and Sustainable Use of Its Components, section 13.3.3: Managing Biodiversity in Forestry." In: V H Heywood and R T Watson (eds.), *Global Biodiversity Assessment* (pp. 952–960).

Hachfeld, B, and U Schippmann. 2000. "Conservation Data Sheet 2: Exploitation, Trade and Population Status of Harpagophytum Procumbens in Southern Africa" *Medicinal Plant Conservation* 6: 4–9.

Hamilton, Alan C. 2004. "Medicinal Plants, Conservation and Livelihoods" *Biodiversity and Conservation* 13(8): 1477–1517.

Havens, Kayri, Pati Vitt, Mike Maunder, Edward O Guerrant, and Kingsley Dixon. 2006. "Ex Situ Plant Conservation and Beyond" *BioScience* 56(6): 525–31.

He, Shan-An. 1998. "Utilization and Conservation of Medicinal Plants in China with Special Reference to Atractylodes lancea." In T. Tomlinson and O. Akerele (eds.), *Medicinal Plants: Their Role in Health and Biodiversity* (pp. 161–8). Philadelphia: University of Pennsylvania Press. https://doi.org/10.9783/9780812292633-017

Heywood, Vernon Hilton, and José M Iriondo. 2003. "Plant Conservation: Old Problems, New Perspectives" *Biological Conservation* 113(3): 321–35.

Heywood, Vernon Hilton, and Robert Tony Watson. 1995. *Global Biodiversity Assessment*. Vol. 1140. Cambridge: Cambridge University Press.

Hikmat A, E A Zuhud, S E Siswoyo, and R K Sari. 2001. "Revitalisasi konservasi tumbuhan obat keluarga (toga) guna meningkatkan kesehatan dan ekonomi keluarga mandiri di desa Contoh Lingkar Kampus IPB Darmaga Bogor." *Jurnal Ilmu Pertanian Indonesia*, *16*(2), 71–80.

Holsinger, Kent E, and L D Gottlieb, 1991. "Conservation of Rare and Endangered Plants: Principles and Prospects." In D A Falk and K E Holsinger (eds.), *Geneticsand Conservation of Rare Plants* (pp. 195–208). New York: Oxford University Press.

Huang, H, X Han, L Kang, P Raven, P W Jackson, and Y Chen. 2002. "Conserving Native Plants in China" *Science* 297(5583): 935–36.

Huang, Hongwen. 2011. "Plant Diversity and Conservation in China: Planning a Strategic Bioresource for a Sustainable Future" *Botanical Journal of the Linnean Society* 166(3): 282–300.

Hussein, Rehab A, and Amira A El-Anssary. 2019. "Plants Secondary Metabolites: The Key Drivers of the Pharmacological Actions of Medicinal Plants" *Herbal Medicine* 1(3). IntechOpen. doi: 10.5772/intechopen.76139

Intergovernmental Panel on Climate Change. 2007. "Climate Change 2007: The Physical Science Basis." Contribution of Working Group I to the Fourth Assessment Report of the Intergovernmental Panel on Climate Change. In S Solomon, D Qin, M Manning, Z Chen, M Marquis, K B Averyt, M Tignor and H L Miller (eds.). Cambridge: Cambridge University Press, 996 p.

Jeffries, Michael 1997. *Biodiversity and Conservation (Routledge Introductions to Environment)*. London, and New York: Routledge. 208p.

Joshi, Bipin Chandra, and Rakesh K Joshi. 2014. "The Role of Medicinal Plants in Livelihood Improvement in Uttarakhand" *International Journal of Herbal Medicine* 1(6): 55–58.

Kulip, Julius, Lam Nyee Fan, Nurhuda Manshoor, Avelinah Julius, Idris Mohd Said, Johnny Gisil, Julianah A Joseph, Welly Frederick Tukin. 2010. "Medicinal Plants in Maliau Basin, Sabah, Malaysia." *Journal of Tropical Biology and Conservation* 6: 21–33.

Kumar, Suresh, R Kumar, and A Khan. 2011. "Medicinal Plant Resources: Manifestation and Prospects of Life-Sustaining Healthcare System" *Continental Journal of Biological Science* 4(1): 19–29.

Lange, Dagmar. 1998. *Europe's Medicinal and Aromatic Plants: Their Use, Trade and Conservation*. Cambridge: Traffic International.

Larsen, Helle Overgaard, and Carsten Smith Olsen. 2006. "Unsustainable Collection and Unfair Trade? Uncovering and Assessing Assumptions Regarding Central Himalayan Medicinal Plant Conservation" In: D L Hawksworth and A T Bull (eds.), *Plant Conservation and Biodiversity*, vol. 6. Dordrecht: Springer, 105–23. https://doi.org/10.1007/978-1-4020-6444-9_8

Liu, Jianguo, Marc Linderman, Zhiyun Ouyang, Li An, Jian Yang, and Hemin Zhang. 2001. "Ecological Degradation in Protected Areas: The Case of Wolong Nature Reserve for Giant Pandas" *Science* 292 (5514): 98–101.

Liu, Rui Hai. 2004. "Potential Synergy of Phytochemicals in Cancer Prevention: Mechanism of Action" *The Journal of Nutrition* 134(12): 3479S–85S.

Ma, Jianzhang, Ke Rong, and Kun Cheng. 2012. "Research and Practice on Biodiversity in Situ Conservation in China: Progress and Prospect" *Biodiversity Science* 20(5): 551–58.

Macilwain, Colin. 2004. "Organic: Is It the Future of Farming?" *Nature* 428(6985): 792–94.

Makunga, N P, L E Philander, and M Smith. 2008. "Current Perspectives on an Emerging Formal Natural Products Sector in South Africa" *Journal of Ethnopharmacology* 119(3): 365–75.

McNeely, J, M Gadgil, C Leveque, C Padoch, and Redford, K. 1995. Human Influences on Biodiversity." In V H Heywood (ed.), *Global Biodiversity Assessment*. Published for the United Nations Environment Programme, Cambridge University Press.

Mendelsohn, Robert, and Michael J. Balick 1995. "The Value of Undiscovered Pharmaceuticals in Tropical Forests" *Economic Botany* 49(2): 223–28.

Mir, Tawseef Ahmad, Muatasim Jan, Rakesh Kumar Khare, and Musadiq Hussain Bhat. 2021. "Medicinal Plant Resources: Threat to Its Biodiversity and Conservation Strategies" In T Aftab and K R Hakeem (eds.), *Medicinal and Aromatic Plants*. Cham: Springer, 717–39.

Mulliken, T A. 2000. "Implementing CITES for Himalayan Medicinal Plants Nardostachys Grandiflora and Picrorhiza Kurrooa" *Traffic Bulletin-Cambridge-Traffic International* 18(2): 63–72.

Natesh, S. 1999. "Conservation of Medicinal and Aromatic Plants in India-an Overview." In S Natesh, A Osman, and A K Azizol (eds.), *Medicinal, Aromatic Plants: Strategies, and Technologies for Conservation* 1–11. Kuala Lumpur: Forest Research Institute.

Oksman-Caldentey, Kirsi-Marja. 2007. "Tropane and Nicotine Alkaloid Biosynthesis-Novel Approaches towards Biotechnological Production of Plant-Derived Pharmaceuticals" *Current Pharmaceutical Biotechnology* 8(4): 203–10.

Pant, Ruchi. 2002. *Customs and Conservation: Cases of Traditional and Modern Law in India and Nepal*. Maharashtra,India: Kalpavriksh & International Institute of Environment and Development.

Pei-Gen, Xiao. 1991. "The Chinese Approach to Medicinal Plants: Their Utilization and Conservation." In O. Akerele, V. Heywood, and H. Synge (eds.), *The Conservation of Medicinal Plants* (pp. 305–13). Proceedings of an International Consultation. Cambridge: Cambridge University Press.

Phartyal, Shyam S, R C Thapliyal, Nico Koedam, and Sandrine Godefroid. 2002. "Ex Situ Conservation of Rare and Valuable Forest Tree Species through Seed-Gene Bank," *Current Science* 1351–57.

Phumthum, Methee, Henrik Balslev, Rapeeporn Kantasrila, Sukhumaabhorn Kaewsangsai, and Angkhana Inta. 2020. "Ethnomedicinal Plant Knowledge of the Karen in Thailand" *Plants* 9(7): 813.

Pimm, Stuart L, Gareth J Russell, John L Gittleman, and Thomas M Brooks. 1995. "The Future of Biodiversity" *Science* 269(5222): 347–50.

Prance, Ghillean T, Derek Chadwick, and Joan Marsh. 1994. "Ethnobotany and the Search for New Drugs" In: *Ciba Foundation Symposium (USA). No. 185*. Hotel Praia Centro, Fortaleza, Brazil.

Primack, Richard B, and Abraham J Miller-Rushing. 2009. "The Role of Botanical Gardens in Climate Change Research" *New Phytologist* 182(2): 303–13.

Rahayu, Slamet Mardiyanto, Arista Suci Andini. 2019. "Ethnobotanical Study on Medicinal Plants in Sesaot Forest, Narmada, West Lombok, Indonesia" *Biosaintifika: Journal of Biology, and Biology Education* 11(2): 234–42.

Raina, R, Romesh Chand, Yash Pal. 2011. "Conservation Strategies of Some Important Medicinal Plants" *International Journal of Medicinal Sharma, and Aromatic Plants* 1(3): 342–47.

Rajaei, Peyman, and Neda Mohamadi. 2012. "Ethnobotanical Study of Medicinal Plants of Hezar Mountain Allocated in South East of Iran" *Iranian Journal of Pharmaceutical Research: IJPR* 11(4): 1153.

Rao, Madhu, Alan Rabinowitz, and Saw Tun Khaing. 2002. "Status Review of the Protected-area System in Myanmar, with Recommendations for Conservation Planning" *Conservation Biology* 16(2): 360–68.

Roche, L. F A O Rome. 1975. "Guidelines for the Methodology of Conservation of Forest Genetic Resources," *The Methodology of Conservation of Forest Genetic Resources-Report on a Pilot Project* 201–3. FAO/UNEP.

Rondilla, Nadine Angela, Ian Christopher N Rocha, Shannon Jean Roque, Ricardo Martin Lu, Nica Lois B Apolinar, Alyssa A Solaiman-Balt, Theorell Joshua Abion, Pauline Bianca Banatin, and Carina Viktoria Javier. 2021. "Folk Medicine in the Philippines: A Phenomenological Study of Health-Seeking Individuals" *International Journal of Medical Students* 9(1): 25–32.

Ross, Ivan A. 2005. *Medicinal Plants of the World, Volume 3: Chemical Constituents, Traditional and Modern Medicinal Uses*. Heidelberg, Germany: Springer.

Saganuwan, Saganuwan Alhaji, and Patrick Azubuike Onyeyili. 2010. "Biochemical Effects of Aequous Leaf Extract of Abrus Precatorius (Jecquirity Bean) in Swiss Albino Mice" *Herba Polonica* 56(3): 63–80.

Sam, Hoang Van, Pieter Baas, Paul J A Keßler. 2008. "Traditional Medicinal Plants in Ben En National Park, Vietnam" *Blumea-Biodiversity Evolution, and Biogeography of Plants* 53(3): 569–601.

Sarker, Satyajit D, Lutfun Nahar, and Yashodharan Kumarasamy. 2007. "Microtitre Plate-Based Antibacterial Assay Incorporating Resazurin as an Indicator of Cell Growth, and Its Application in the in Vitro Antibacterial Screening of Phytochemicals" *Methods* 42(4): 321–24.

Schipmann, U, D J Leaman, A B Cunningham, and S Walter. 2005. "Impact of Cultivation and Collection on the Conservation of Medicinal Plants: Global Trends and Issues." *Acta Horticulturae* 676: 31–44.

Sheikh, Kashif, Tahira Ahmad, Mir Ajab Khan. 2002. "Use, Exploitation and Prospects for Conservation: People and Plant Biodiversity of Naltar Valley, Northwestern Karakorums, Pakistan" *Biodiversity and Conservation* 11(4): 715–42.

Smolinski, Alexa T, and James Pestka. 2003. "Modulation of Lipopolysaccharide-Induced Proinflammatory Cytokine Production in Vitro and in Vivo by the Herbal Constituents Apigenin (Chamomile), Ginsenoside Rb1 (Ginseng) and Parthenolide (Feverfew)" *Food and Chemical Toxicology* 41(10): 1381–90.

Sodhi, Navjot S, and Paul R Ehrlich. 2010. *Conservation Biology for All*. Oxford: Oxford University Press.

Srivastava, R. 2000. "Studying the Information Needs of Medicinal Plant Stakeholders in Europe" *Traffic Dispatches* 15(5): 13.

Swarts, Nigel D, and Kingsley W Dixon. 2009. "Terrestrial Orchid Conservation in the Age of Extinction" *Annals of Botany* 104(3): 543–56.

Tiwari, Achyut, Yadav Uprety, and Santosh Kumar Rana. 2019. "Plant Endemism in the Nepal Himalayas and Phytogeographical Implications" *Plant Diversity* 41(3): 174–82.

Torrance, Andrew W. 2000. "Bioprospecting and the Convention on Biological Diversity." Harvard Library Student Paper, Harvard University Press.

Touchell, D H. 1997. *Conservation Into the 21st Century: Proceedings of the 4th International Botanic Gardens Conservation Congress, Perth, Western Australia; [from 25 to 29 September 1996]*. Kings Park and Botanic Garden.

TRAFFIC. 1999. "Conservation of Medicinal Plants Trade in Europe" *Traffic Bullet*, 1999.

Tuxill, John. 1999. *Nature's Cornucopia: Our Stake in Plant Diversity*. Washington, DC: Worldwatch Inst.

Vorhies F. 2000. The global dimension of threatened medicinal plants from a conservation point of view. In S Honnef and R Melisch (eds.), *Medicinal Utilization of Wild Species: Challenge for Man and Nature in the New Millennium* (pp. 26–29). WWF Germany/TRAFFIC Europe-Germany, EXPO 2000, Hannover, Germany.

Wakdikar, Sandhya. 2004. "Global Health Care Challenge: Indian Experiences and New Prescriptions" *Electronic Journal of Biotechnology* 7(3): 2–3.

Walter, Kerry S, and Harriet J Gillett. 1998. *1997 IUCN Red List of Threatened Plants*. Rue Mauverney, Gland Switzerland: IUCN.

Wong, Kam Lok, Ricky Ngok Shun Wong, Liang Zhang, Wing Keung Liu, Tzi Bun Ng, Pang Chui Shaw, Philip Chi Lip Kwok, Yau Ming Lai, Zhang Jin Zhang, and Yanbo Zhang. 2014. "Bioactive Proteins and Peptides Isolated from Chinese Medicines with Pharmaceutical Potential" *Chinese Medicine* 9(1): 1–14.

Yu, Hua, Caixiang Xie, Jingyuan Song, Yingqun Zhou, and Shilin Chen. 2010. "TCMGIS-II Based Prediction of Medicinal Plant Distribution for Conservation Planning: A Case Study of Rheum Tanguticum" *Chinese Medicine* 5(1): 1–9.

Zegeye, Haileab, Demel Teketay, and Ensermu Kelbessa. 2006. "Diversity, Regeneration Status and Socio-Economic Importance of the Vegetation in the Islands of Lake Ziway, South-Central Ethiopia" *Flora-Morphology Distribution, Functional Ecology of Plants* 201(6): 483–98.

Zerabruk, Samuel, and Gidey Yirga. 2012. "Traditional Knowledge of Medicinal Plants in Gindeberet District, Western Ethiopia" *South African Journal of Botany* 78: 165–69.

5 Ethnobotany, Ethnopharmacology, and Traditional Uses of Medicinal and Aromatic Plants

Noureddine Chaachouay and Abdelhamid Azeroual
Hassan First University

Lahcen Zidane
Ibn Tofail University

CONTENTS

5.1 INTRODUCTION

Humans have relied on plants for their necessities and survival since the beginning of time (Allard 1999). Exploring human life on the planet requires comprehending the function of medicinal and aromatic plants in historical and current-day practices. Over time, indigenous communities and societies have attempted to use beneficial plants. Even today, we rely on plants and their essential pollinators for our survival.

Ethnobotany is the study of the historical and cross-cultural interactions between plants and people, focusing on the significance of plants in human culture and customs, as well as how humans have used and transformed plants and how they represent them in their knowledge systems (Leonti 2011; Nolan and Turner 2011). Ethnobotany has also made crucial contributions to human well-being and health, particularly in less developed parts of the world, where plant-based popular remedies are the mainstay of sickness remedies (Moret 2008; Barboza et al. 2009).

Ethnopharmacology is a scientific approach for examining the pharmacological effects of any human-administered substance that is known to have either beneficial or detrimental qualities or other fundamental biological features (Holmstedt and Bruhn 1983; Heinrich 2010). As a result, the

DOI: 10.1201/b22842-5

focus is on a broad range of anthropological, belief, pharmacological, and toxicological research of these formulations rather than on explaining (usually local or traditional) practices (Mosihuzzaman and Choudhary 2008). However, as a unique area of research, ethnopharmacology has a limited history, stretching back around 50 years (Yeung, Heinrich, and Atanasov 2018).

Today, ethnobotanical and ethnopharmacological investigations encompass all aspects of culture, including advanced communities that use plants in folk medicine or as common remedies to treat diseases and wounds. Traditional plant-based healing modalities continue to play a critical role in health care, with around 80% of the world's population depending on herbal medicine for their primary health care (Zizka et al. 2015).

The conservation status of medicinal and aromatic plants is determined by taking into account a variety of threats, including socio-economic and biological problems such as loss of local livelihoods, species extinction, destructive harvesting, and habitat change, all of which have contributed to critical trends in species extinction (Okigbo, Eme, and Ogbogu 2008). Furthermore, a lack of cooperation in the herb sector leads to the over-exploitation of these plants, which is putting them in jeopardy. The present chapter discusses the ethnobotanical and ethnopharmacological approaches to traditional benefits of medicinal and aromatic plants, prioritizing the preservation of widespread medical knowledge and medical plant resources.

5.2 ETHNOBOTANY

Ethnobotany is a term derived from the combination of two concepts (Ethno and Botany): Ethno (as in "ethnic") is a term that refers to people, culture, and the aggregate body of a culture's beliefs, language, aesthetics, knowledge, and practice. Botany studies plants, beginning with the tiny fern or blade of grass and ending with the tallest or oldest tree. It encompasses both wild and cultivated plants. In an article published by the Philadelphia Evening Telegram, the term ethnobotany was born under the pen of the American professor John William Harsherberg, botanist and agro-botanist on December 5, 1895 (Brousse 2011). He defined it as "the practice of plant species by aboriginal and primitive peoples." In 1941, Volney Jones described it as "ethnobotany is a science that studies the relationships between human groups, their environment, and plants, namely the use and development of plants in different cultural and temporal spaces" (Brousse 2011). This discipline defines the role of plants in human societies (especially in tribal communities). Numerous sub-fields of ethnobotany exist, including ethnopharmacology, ethnotoxicology, ethnopharmacognosy, ethnotaxonomy, ethnophytotaxonomy, ethnomycology, ethnomedicobotany, ethnoagriculture, ethnoveterinary medicine, ethnoecology, and ethnogynecology. Oral folklore, ceramics, literary sources, images, other mythologies, the study of reference species in herbaria and museums, and plants unearthed in jars or midden heaps (trash dumps) at archaeological sites have all been used to document ethnobotanical knowledge (Posnansky 2013). However, ethnobotanists have been critical in understanding and recording these plant–human relationships and unlocking the data via various surveys and interviews (Saha, Sundriyal, and Sundriyal 2014). Ethnobotanical and ethnopharmacological studies have long been a priority for investigators. Ethnobotanists have mainly used numerous ethnobotany studies globally to disseminate knowledge on plant usage (Moret 2008; Heinrich 2010; Mesfin et al. 2009; Abdurhman, n.d.; Chaachouay et al. 2019, 2020; Ahmad et al. 2014; Rahmatullah et al. 2010; Agra et al. 2008; Tene et al. 2007; Cakilcioglu et al. 2011; Cakilcioglu and Turkoglu 2010; Morton 1968; Manandhar 1995; Hirschmann and de Arias 1990; Simbo 2010). Ethnobotanical inventories and documentations were scientifically systematized with local and scientific names, cultural interpretations, medical usages, and data on the phenology, ecology, distribution, botany, managing, harvesting, and preservation of medicinal plants (Sheng-Ji 2001). Ethnobotany is the study of plants by humans; it encompasses any traditional usage of plants (Kufer 2005). Ethnobotanical studies may be classified according to their intended use (rosaries, agricultural implements, ceremonial services, beads, containers, beverages, cultural benefits, condiments, detergent dye, edible equipment, fatty oils, essential oils, feed and fodder, incense, fragrances, gum, insect repellent, manure, fibre, intoxicant, medicine, insecticide, religious uses, musical instruments, resin, spices, tools, tan, weapons, toxicants).

5.3 ETHNOPHARMACOLOGY

Ethnopharmacology is a distinct subject of study that has only been around for a short time. The term was initially used in the title of a hallucinogen book in 1967 (Ghorbani, Naghibi, and Mosadegh 2006). Ethnopharmacology is described as "the study of the usage, mechanisms of action, and biological consequences of herbal remedies, stimulants, and psychotropic substances by various ethnic or cultural groups" (Holmstedt and Bruhn 1983; Houghton 1995; Schultes and Swain 1976). This definition emphasizes the scientific research of indigenous medications, although it does not explicitly address the question of drug discovery. Ethnopharmacology is inextricably tied to ethnobotany, the study of plants, because plants are the primary route of drug administration. It is also often associated with ethnopharmacy. However, ethnopharmacology varies from ethnopharmacy in that it is concerned with the biological evaluation of traditional remedies, whereas ethnopharmacy is concerned with far broader issues about medication benefits (Kumar et al. 2021). These factors are connected to how drugs are perceived, used, and managed in a given human community. Pharmacoepidemiology, the study of how medications affect large groups of people, is also related to ethnopharmacology. When researching a natural substance utilized as a medication by culture, the collection, extraction, and preparation processes must be similar to those used by the ethnic group. This is to ensure that the experimentation is consistent and legitimate. Ethnopharmacology has resurfaced as a novel method of medication development. Most ethnopharmacology research is conducted on botanicals, which provide natural libraries for chemical structures and mixtures of varying complexity (Chaguturu and Patwardhan 2017). Hundreds of chemicals and bioactive substances can be found in a single plant.

5.4 MEDICINAL AND AROMATIC PLANTS

Medicinal and aromatic plants are botanical raw materials, also referred to as herbal pharmaceuticals, used in health foods, cosmetics, therapeutic goods, and other natural health products for their aromatic, medicinal, and gastronomic properties. The term "medicinal plant" does not refer to a taxonomic classification but rather a sub-category of plants used for therapeutic purposes (Raj Paroda et al. 2014). A medicinal plant, also called a medicinal herb, is any plant that, in one or more of its parts (roots, leaves, fruits, seeds, flowers, bulbs, bark), contains chemical compounds that can be used for therapeutic, purgative, tonic, or other health-promoting benefits or that are precursors for the synthesis of valuable remedies (Meuss 2000). Any plant that is used medicinally or therapeutically, whether in advanced or in organized systems of ancestral medicine such as Homeopathy, Acupuncture, Allopathy, Ayurveda, Acupressure, the Unani system, Hydrotherapy, Yoga, Kampo, Naturopathy, or the widely practiced autochthonous and folk systems, can be classified as a medicinal plant. On the other hand, aromatic plants, usually referred to as herbs and spices, contain aromatic composites that are essentially volatile essential oils at room temperature (Kunwar et al. 2011). These essential oils are odorous, hydrophobic, volatile, and highly concentrated. Both classes are now referred to as "medicinal" and "aromatic."

5.5 HERBAL MEDICINE

Herbal medicine is also known as phytotherapy, phytomedicine, botanical medicine, or vegetable medicine (Figure 5.1). It is a term used by the World Health Organization (WHO) to refer to herbs, herbal preparations, herbal materials, and finished herbal products that contain whole plants, plant parts, or other plant products (World Health Organization 2001). According to herbal remedy professionals, a complete plant has more influence than its constituent components. Additionally,

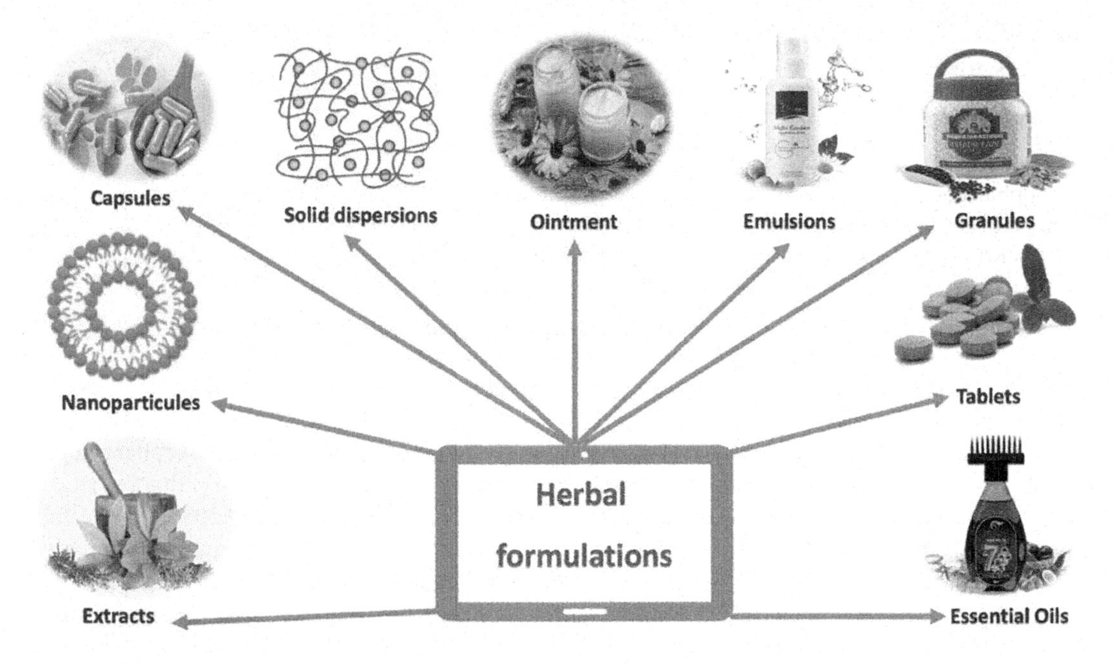

FIGURE 5.1 Herbal remedies are available in a variety of formulations (Byeon et al. 2019).

practitioners think that an active ingredient might lose its efficacy, when administered in isolation from the rest of the plant or become less safe. Numerous other variables, including gender, age, ethnic origin, education, and socio-economic standing, are also associated with herbal remedy usage (Bishop and Lewith 2010). Herbal medicine use is more prevalent in indigenous communities when members suffer from chronic ailments such as asthma, type 2 diabetes, cancer, stroke, depression, and chronic kidney disease (Burstein et al. 1999; Egede et al. 2002). According to the WHO, around 80% of the indigenous societies of some African and Asian peoples now utilize herbal medicine for some component of primary health care. Herbal medicine seems to have universal applicability across several civilizations and cultures. However, the medicinal plants used to cure the same diseases and treatment modalities may vary per civilization. Because herbal medications are natural, they may be misinterpreted as entirely safe (Meuss 2000). This is not always the case, though. Herbal remedies may have various adverse effects, ranging from moderate to severe, such as asthma, allergic reactions, rashes, nausea, headaches, diarrhoea, and vomiting (Edwards and Aronson 2000). A qualified and trained practitioner should only prescribe herbal medicine as with other prescription treatments.

5.6 TRADITIONAL HEALERS

However, the phrase "traditional healer" is deceptive since it often refers to all non-Western-trained healers. It is characterized as a person living in a primitive culture who uses long-established practices handed down from healer to healer to cure various disorders, many of which have psychological bases (Cheetham and Griffiths 1982; Meissner 2004; Nelms and Gorski 2006). Traditional healers are present worldwide and will continue to be an essential source of health care for large sectors of the population for decades, if not millennia (Jenkins 2003). Traditional healers practice culturally acceptable forms of health-care coverage in virtually all areas of the globe, dealing with differing degrees of satisfaction with many of the health problems of the local population. Their approaches vary, and although some are unquestionably detrimental, it is also possible that some of the herbal remedies employed have beneficial psychotropic qualities and that some treatments provide vital psychosocial support to individuals, families, and societies (Jenkins 2003; Gureje et al. 2015). The preaching of super-natural powers and malevolent spirits, the conduct of religious rites, preventive measures, cauterization, surgery, psychotherapy, and the use of mineral and animal products, among other things, form the fundamental foundation of the majority of widespread remedies (Chaachouay and Zidane 2022).

5.7 MEDICINAL AND AROMATIC PLANTS IN MODERN MEDICINE

Indigenous communities worldwide gather and employ medicinal and aromatic plants for preventing and treating diseases. The WHO has recognized the significance of traditional therapies, especially in numerous developing nations, where about 70–95% of the local people depend on these drugs for primary care (Zhang et al. 2005). An emphasis on integrated medicine currently gives a wide range of plant and plant component options for ailment treatment. Modern therapy and the use of medicinal and aromatic plants, crude drugs, aromatherapy, and several other medicines have been adopted in the hospital and in the home (Inoue, Hayashi, and Craker 2019). The basis of these health-care systems is medicinal and aromatic herbs. The majority of active mechanisms in current pharmaceutical medications are obtained directly or indirectly from natural compounds found in plants and other living things (Shakya 2016). Despite the introduction of synthetic and combinatorial chemistry, this remains the case. The application of plants is seen to be relatively safe, with no or minor adverse effects. Plants and their specimens are employed in cosmetics, food, and high-end items. Medicinal plants are valuable for developing novel medications and treating mental and physical ailments. The promise of medicinal and aromatic plants has long provided humans with hope. Typically, an alkaloid from a medicinal and aromatic plant has a noticeable effect on the central nervous system of humans, even if they are taking a small dose of the alkaloid (Hesse 2002). Medicinal and aromatic plants with an alkaloid have limitations of use or are on the list of poisonous plants.

On the other hand, those alkaloids are commonly employed as components in drug development, either pharmacopoeial, non-pharmacopoeial, or synthetic drugs. There have been numerous excellent investigations to find beneficial alkaloids (Hadi and Bremner 2001; Roeder 1995, 2000; Shamsa et al. 2008). The magical ingredients of medicinal plants support protecting humans until the present, such as food, medicine, recreation, and healing (Inoue, Hayashi, and Craker 2019). Many harsh ailments, such as infectious diseases, diabetes, tuberculosis, cancer, Ebola, and AIDS, could be treated using medicinal and aromatic plant species. These herbal remedies are now seen as a symbol of safety, in contrast to synthetic pharmaceuticals, which are seen to be harmful to both humans and the environment (Table 5.1).

5.8 TRADE OF MEDICINAL AND AROMATIC PLANTS

Teas, herbal treatments, spirits, sweets, nutritional supplements, varnishes, pesticides, and cosmetics are just a few of the items made from medicinal and aromatic plants. The new fair practices, strategy and product measures, and risk control tools allow us to enhance the trade, processing, and manufacturing of plants, natural components, and traditional therapy (Dürbeck and Hüttenhofer 2015). Plants supply the basis for a considerable number of current pharmacological drugs. As a result, the medical plant-based industry is a bright spot with plenty of possibilities for growth (Dhar et al. 2002; Kala, Dhyani, and Sajwan 2006). Many rural households benefit from plant-growing and harvesting activities. Numerous rural families benefit from the cultivation and harvesting of medicinal and aromatic plants. These medicinal and aromatic plants may provide a significant source of income for disadvantaged people living in remote areas and contribute significantly to the economies of the source countries (Klingenstein et al. 2006). The global income from traditional Chinese medicine, for example, was USD 83 billion in 2012 (CITES 2020). In Morocco, the sector of medicinal and aromatic products generated average annual earnings of USD 59.2 million in 2015 (Dupree 1977). In 2009, the Republic of Korea spent USD 7.4 billion on traditional medicine,

TABLE 5.1
Plant-derived Anticancer Pharmaceuticals Have Been Approved for Commercial Manufacturing by the FDA

Drug Name	Plant Resources	Feature
Etoposide	*Podophyllum peltatum* L.	A semi-synthetic derivative of a plant chemical epipodophyllotoxin.
Irinotecan	*Camptotheca acuminata* Decne.	The Food and Drug Administration has been authorized to treat metastatic colorectal cancer.
Taxol/paclitaxel	*Taxus brevifolia* Nutt.	It is now the first-line treatment for various tumourous malignancies, including breast cancer.
Vinblastine	*Catharanthus roseus* (L.) G.Don	It is the first-line treatment for many types of leukaemia and has boosted the survival rate of children's leukaemia by 80% since the 1950s.

TABLE 5.2
The Twelve Countries with the Highest Export and Import Volumes of Medicinal Plants Classified as Pharmaceutical Plants

Country of Import	Quantity (tonnes)	Value (US$)	Country of Export	Quantity (tonnes)	Value (US$)
China	15,550	41,602,800	Egypt	11,800	13,476,000
France	21,800	51,975,000	USA	13,050	104,572,000
Germany	44,750	104,457,200	Mexico	37,600	14,257,500
Hong Kong	59,950	263,484,200	China	150,600	266,038,500
Italy	11,950	43,006,600	Bulgaria	10,300	14,355,500
Japan	46,450	131,031,500	India	40,400	61,665,500
Malaysia	7,050	38,685,400	Singapore	7,950	52,620,700
Pakistan	10,650	9,813,800	Chile	9,850	26,352,000
Republic Korea	33,500	49,889,200	Germany	15,100	68,243,200
Spain	9,850	27,648,300	Morocco	8,500	13,685,400
UK	7,950	29,551,000	Albania	8,050	11,693,300
USA	51,200	139,379,500	Hong Kong	55,000	201,021,200
Total	320,550	930,524,400	Total	368,100	847,980,800

UN Comtrade (2014).

whereas the United States spent USD 14.8 billion on natural products privately. The herbal supplement and pharmaceutical industries in Europe are expected to reach 7.4 billion dollars annually. Between 2001 and 2014, yearly average growth rates of 2.4% in volume and 9.2% in export value of medicinal plant material were seen, resulting in a three-fold increase in international commerce in medicinal and aromatic plants since 1999 (CITES 2020). The demand for wild-collected medicinal plants is steadily growing. On the other hand, the wild collection has become a risk to nature and humans due to a lack of respect for the approach, product standards, and resource and risk management tools, among other things (Table 5.2).

5.9 CONSERVATION AND SUSTAINABLE USES OF MEDICINAL AND AROMATIC PLANTS

Worldwide, about 60,000 species are used for medicinal, nutritional, and aromatic purposes, with over 500,000 tons of goods derived from these species sold each year (World Health Organization

2013). However, 80% of individuals in developing countries depend exclusively on herbal remedies as their primary source of therapy, and over 25% of medications prescribed in developed regions are derived from wild plant species (Hamilton 2004). Medicinal plant usage is rapidly growing globally, owing to the growing demand for herbal, pharmaceuticals, natural health products, and secondary metabolites of plant species (Vanisree et al. 2004; Taylor et al. 2001). The World Wildlife Fund (WWF) and the International Union for the Conservation of Nature (IUCN) estimate that between 50,000 and 80,000 medicinal and aromatic plant species are used therapeutically. Approximately 15,000 species face extinction due to over-harvesting, indiscriminate collecting, uncontrolled deforestation, habitat degradation, unregulated or illicit international commerce, and climate change. All these affect species rarity but do not illustrate an individual species' susceptibility or resistance to accumulating pressure (Chen et al. 2016; Huxley 1992; Srivastava 2018). Plant species are progressively being harvested in large numbers, mostly from wild populations. In recent decades, demand for wild resources has grown by 8–15% per year in Asia, Europe, and North America (Heywood and Iriondo 2003). There is a point where a plant species' reproductive ability irrevocably diminishes (Semwal et al. 2007). Thus, in situ and ex situ preservation and complementary conservation strategies are being used to conserve plant genetic resources in general and medicinal and aromatic species in Europe and other parts of the world mainland (Huxley 1992; CITES 2020; IUCN 2001).

5.10 CONCLUSION

This chapter emphasized the significance, potential, and the role of ethnobotany, ethnopharmacology, and the traditional benefits of medicinal and aromatic plants. We consider that ethnobotany-investigation institutions, organizations, and societies around the world should form cross-disciplinary collaborations and alliances with other fields to attain sustainable development objectives in the most suitable interests of people and the globe, and with a particular priority on the conservation of indigenous ethnomedical knowledge as well as medicinal and aromatic plant resources. We also think that preserving indigenous ethnomedical knowledge, and medicinal and aromatic plant resources should be a top priority.

REFERENCES

Abdurhman, Nurya. n.d. *Ethnobotanical Study of Medicinal Plants Used by Local People in Ofla Wereda, Southern Zone of Tigray Region Ethiopia*. Addis Ababa University.

Agra, Maria de Fátima, Kiriaki Nurit Silva, Ionaldo José Lima Diniz Basílio, Patrícia França de Freitas, and José Maria Barbosa-Filho. 2008. "Survey of Medicinal Plants Used in the Region Northeast of Brazil" *Revista brasileira de farmacognosia* 18: 472–508.

Ahmad, Mushtaq, Shazia Sultana, Syed Fazl-i-Hadi, Taibi Ben Hadda, Sofia Rashid, Muhammad Zafar, Mir Ajab Khan, Muhammad Pukhtoon Zada Khan, Ghulam Yaseen. 2014. "An Ethnobotanical Study of Medicinal Plants in High Mountainous Region of Chail Valley (District Swat-Pakistan)" *Journal of Ethnobiology and Ethnomedicine* 10(1): 1–18.

Allard, Robert W. 1999. *Principles of Plant Breeding*. John Wiley & Sons.

Barboza, Gloria E, Juan J Cantero, César Núñez, Adriana Pacciaroni, and Luis Ariza Espinar. 2009. "Medicinal Plants: A General Review and a Phytochemical and Ethnopharmacological Screening of the Native Argentine Flora" *Kurtziana* 34 (1–2): 7–365.

Bishop, F L, and G T Lewith. 2010. "Who uses CAM? A narrative review of demographic characteristics and health factors associated with CAM use" *Evidence-Based Complementary Alternative Medicine* 7: 11–28.

Brousse, C. 2011. "Une Analyse Historique et Ethnobotanique Des Relations Entre Les Activités Humaines et La Végétation Prairiale" *Fourrages* 208: 245–51.

Burstein, Harold J, Shari Gelber, Edward Guadagnoli, and Jane C. 1999. "Use of Alternative Medicine by Women with Early-Stage Breast Cancer" *New England Journal of Medicine Weeks* 340(22): 1733–39.

Byeon, J C, J B Ahn, W S Jang, et al. 2019. "Recent Formulation Approaches to Oral Delivery of Herbal Medicines. *Journal of Pharmaceutical Investigation* 49: 17–26. https://doi.org/10.1007/s40005-018-0394-4

Cakilcioglu, Ugur, Selima Khatun, Ismail Turkoglu, and Sukru Hayta. 2011. "Ethnopharmacological Survey of Medicinal Plants in Maden (Elazig-Turkey)" *Journal of Ethnopharmacology* 137(1): 469–86.

Cakilcioglu, Ugur, and Ismail Turkoglu. 2010. "An Ethnobotanical Survey of Medicinal Plants in Sivrice (Elazığ-Turkey)" *Journal of Ethnopharmacology* 132(1): 165–75.

Chaachouay, Noureddine, Ouafae Benkhnigue, Mohamed Fadli, Hamid El Ibaoui, and Lahcen Zidane. 2019. "Ethnobotanical and Ethnopharmacological Studies of Medicinal and Aromatic Plants Used in the Treatment of Metabolic Diseases in the Moroccan Rif" *Heliyon* 5(10): e02191.

Chaachouay, Noureddine, Allal Douira, Rachida Hassikou, Najiba Brhadda, Jamila Dahmani, Nadia Belahbib, Rabea Ziri, and Lahcen Zidane. 2020. "Mr Chaachouay Noureddine Sous Le Thème" Etude Floristique et Ethnomédicinale Des Plantes Aromatiques et Médicinales Dans Le Rif (Nord Du Maroc)"." Département de Biologie-Université Ibn Tofail-Kénitra.

Chaachouay, Noureddine, Lahcen Zidane. 2022. "The Symbolic Efficacy of Plants in Rituals and Socio-Religious Ceremonies in Morocco, Northwest of Africa" *Journal of Religious and Theological Information* 21(1–2): 34–53.

Chaguturu Rathnam, Bhushan Patwardhan. 2017. "Chapter 1: Drug Discovery Impasse: Pharmacognosy Holds the Key." In Bhushan Patwardhan and Rathnam Chaguturu (eds.), *Innovative Approaches in Drug Discovery* (pp. 1–22). Academic Press, ISBN 9780128018149, https://doi.org/10.1016/B978-0-12-801814-9.00001-5.

Cheetham, R W S, and J A Griffiths. 1982. "The Traditional Healer/Diviner as Psychotherapist" *South African Medical Journal* 62(25): 957–58.

Chen, Shi Lin, Hua Yu, Hong Mei Luo, Qiong Wu, Chun Fang Li, and André Steinmetz. 2016. "Conservation and Sustainable Use of Medicinal Plants: Problems, Progress, and Prospects." *Chinese Medicine (United Kingdom)* 11(1): 1–10. https://doi.org/10.1186/s13020-016-0108-7.

CITES. 2020. "Convention on International Trade in Endangered Species of Wild Fauna and Flora." 11 Chemin des Anémones CH-1219 Châtelaine, Geneva Switzerland. 2020. https://cites.org/eng/prog/medplants.

Dhar, Uppeandra, Sumit Manjkhola, Mitali Joshi, Arvind Bhatt, A K Bisht, and Meena Joshi. 2002. "Current Status and Future Strategy for Development of Medicinal Plants Sector in Uttaranchal, India" *Current Science* 83(8): 956–64.

Dupree, Louis. 1977. *USAID [United States Agency for International Development] and Social Scientists Discuss Afghanistan's Development Prospects*. American University Field Staff Report. Southwest Asia Series 21 (2).

Dürbeck, Klaus, and Teresa Hüttenhofer. 2015. "International Trade of Medicinal and Aromatic Plants" In Á Máthé (ed.), *Medicinal and Aromatic Plants of the World*, vol. 1, 375–82. Dordrecht: Springer.

Edwards, I Ralph, and Jeffrey K Aronson. 2000. "Adverse Drug Reactions: Definitions, Diagnosis, and Management" *The Lancet* 356(9237): 1255–59.

Egede, Leonard E, Xiaobou Ye, Deyi Zheng, and Marc D Silverstein. 2002. "The Prevalence and Pattern of Complementary and Alternative Medicine Use in Individuals with Diabetes" *Diabetes Care* 25(2): 324–29.

Ghorbani, Abdolbaset, Farzaneh Naghibi, and M Mosadegh. 2006. "Ethnobotany, Ethnopharmacology and Drug Discovery." *Iranian Journal of Pharmaceutical Sciences*, 2(2): 109–118.

Gureje, Oye, Gareth Nortje, Victor Makanjuola, Bibilola D Oladeji, Soraya Seedat, and Rachel Jenkins. 2015. "The Role of Global Traditional and Complementary Systems of Medicine in the Treatment of Mental Health Disorders" *The Lancet Psychiatry* 2(2): 168–77.

Hadi, Surya, and John B Bremner. 2001. "Initial Studies on Alkaloids from Lombok Medicinal Plants" *Molecules* 6(2): 117–29.

Hamilton, Alan C. 2004. "Medicinal Plants, Conservation and Livelihoods" *Biodiversity, and Conservation* 13(8): 1477–517.

Heinrich, Michael. 2010. "Ethnopharmacology in the 21st Century-Grand Challenges" *Frontiers in Pharmacology* 1: 8.

Hesse, Manfred. 2002. *Alkaloids: Nature's Curse or Blessing?* John Wiley & Sons.

Heywood, Vernon H, and José M Iriondo. 2003. "Plant Conservation: Old Problems, New Perspectives" *Biological Conservation* 113(3): 321–35.

Hirschmann, Guillermo Schmeda, and Antonieta Rojas de Arias. 1990. "A Survey of Medicinal Plants of Minas Gerais, Brazil" *Journal of Ethnopharmacology* 29(2): 159–72.

Holmstedt, B O, and Jan G Bruhn. 1983. "Ethnopharmacology—a Challenge" *Journal of Ethnopharmacology* 8(3): 251–56.

Houghton, Peter J. 1995. "The Role of Plants in Traditional Medicine and Current Therapy" *The Journal of Alternative and Complementary Medicine* 1(2): 131–43.

Huxley, Anthony. 1992. *Green Inheritance: The World Wildlife Fund Book of Plants*. Four Walls Eight Windows.

Inoue, Maiko, Shinichiro Hayashi, and Lyle E Craker. 2019. "Role of Medicinal and Aromatic Plants: Past, Present, and Future," *Pharmacognosy-Medicinal Plants* 1–13. IntechOpen. doi: 10.5772/intechopen.82497

IUCN, Species Survival Commision Prepared by the IUCN Species Survival Commission. 2001. "IUCN Red List Categories and Criteria: Version 3.1."

Jenkins, Rachel. 2003. "Supporting Governments to Adopt Mental Health Policies" *World Psychiatry* 2(1): 14.

Kala, Chandra Prakash, Pitamber Prasad Dhyani, and Bikram Singh Sajwan. 2006. "Developing the Medicinal Plants Sector in Northern India: Challenges and Opportunities." *Journal of Ethnobiology and Ethnomedicine* 2(32): 1–15. https://doi.org/10.1186/1746-4269-2-32.

Klingenstein, Frank, Hartmut Vogtmann, Susanne Honnef, and Danna Leaman. 2006. "Sustainable Wild Collection of Medicinal and Aromatic Plants: Practice Standards and Performance Criteria" *IUCN* 141: 97–107.

Kufer, Johanna Kathrin. 2005. *Plants Used as Medicine and Food by the Ch'orti'Maya: Ethnobotanical Studies in Eastern Guatemala*. University of London, University College London (United Kingdom).

Kumar, Ajay, Sushil Kumar, Nirala Ramchiary, and Pardeep Singh. 2021. "Role of Traditional Ethnobotanical Knowledge and Indigenous Communities in Achieving Sustainable Development Goals" *Sustainability* 13(6): 3062.

Kunwar, R M, K P Thapa, R Shrestha, P R Shrestha, N K Bhattarai, N N Tiwari, and K K Shrestha. 2011. "Medicinal and Aromatic Plants Network (MAPs-Net) Nepal: An Open Access Digital Database" *Banko Janakari* 21(1): 48–50.

Leonti, Marco. 2011. "The Future Is Written: Impact of Scripts on the Cognition, Selection, Knowledge and Transmission of Medicinal Plant Use and Its Implications for Ethnobotany and Ethnopharmacology" *Journal of Ethnopharmacology* 134(3): 542–55.

Manandhar, Narayan P. 1995. "A Survey of Medicinal Plants of Jajarkot District, Nepal" *Journal of Ethnopharmacology* 48(1): 1–6.

Meissner, Ortrun. 2004. "The Traditional Healer as Part of the Primary Health Care Team?" *South African Medical Journal* 94(11): 901–2.

Mesfin, Fisseha, Sebsebe Demissew, Tilahun. 2009. "An Ethnobotanical Study of Medicinal Plants in Wonago Woreda, SNNPR, Ethiopia" *Journal of Ethnobiology Teklehaymanot, and Ethnomedicine* 5(1): 1–18.

Meuss, A 2000. "Herbal Medicine" *Current Science* 78: 35–39.

Moret, Erica. 2008. "Afro-Cuban Religion, Ethnobotany and Healthcare in the Context of Global Political and Economic Change" *Bulletin of Latin American Research* 27(3): 333–50.

Morton, Julia F. 1968. "A Survey of Medicinal Plants of Curacao" *Economic Botany* 22(1): 87–102.

Mosihuzzaman, M, and M Iqbal Choudhary. 2008. "Protocols on Safety, Efficacy, Standardization, and Documentation of Herbal Medicine (IUPAC Technical Report)" *Pure and Applied Chemistry* 80(10): 2195–2230.

Nelms, Linda W, and June Gorski. 2006. "The Role of the African Traditional Healer in Women's Health" *Journal of Transcultural Nursing* 17(2): 184–89.

Nolan, Justin M, and Nancy J Turner. 2011. "Ethnobotany: The Study of People-Plant Relationships" *Ethnobiology* 9: 135–41.

Okigbo, R N, U E Eme, and S Ogbogu. 2008. "Biodiversity and Conservation of Medicinal and Aromatic Plants in Africa." *Biotechnology and Molecular Biology Reviews* 3(December): 127–34.

Posnansky, Merrick. 2013. "Digging through Twentieth-Century Rubbish at Hani, Ghana" *Historical Archaeology* 47(2): 64–75.

Rahmatullah, Mohammed, Dilara Ferdausi, A Mollik, Rownak Jahan, Majeedul H Chowdhury, Wahid Mozammel Haque. 2010. "A Survey of Medicinal Plants Used by Kavirajes of Chalna Area, Khulna District, Bangladesh" *African Journal of Traditional, Complementary and Alternative Medicines* 7(2): 91–7.

Raj Paroda, S. Dasgupta, Bhag Mal, S.P. Ghosh and S.K. Pareek. 2014. Expert Consultation on Promotion of Medicinal and Aromatic Plants in the Asia-Pacific Region: Proceedings, Bangkok, Thailand; 2-3 December, 2013. 259 p.

Roeder, E. 1995. "Medicinal Plants in Europe Containing Pyrrolizidine Alkaloids." *Die Pharmazie* 50(2): 83–98.

———. 2000. "Medicinal Plants in China Containing Pyrrolizidine Alkaloids." *Die Pharmazie* 55(10): 711–26.

Saha, Debabrata, Manju Sundriyal, and R C Sundriyal. 2014. "Diversity of Food Composition and Nutritive Analysis of Edible Wild Plants in a Multi-Ethnic Tribal Land, Northeast India: An Important Facet for Food Supply." *Indian Journal of Traditional Knowledge* 13: 698–705.

Schultes, RICHARD E, and T Swain. 1976. "The Plant Kingdom: A Virgin Field for New Biodynamic Constituents." In *2. Phillip Morris Science Symposium, Richmond, VA.(USA), 1975*. Phillip Morris, Inc.

Semwal, D P, P Pardha Saradhi, B P Nautiyal, and A B Bhatt. 2007. "Current Status, Distribution and Conservation of Rare and Endangered Medicinal Plants of Kedarnath Wildlife Sanctuary, Central Himalayas, India," *Current Science* 92: 1733–38.

Shakya, Arvind Kumar. 2016. "Medicinal Plants: Future Source of New Drugs" *International Journal of Herbal Medicine* 4(4): 59–64.

Shamsa, Fazel, Hamidreza Monsef, Rouhollah Ghamooshi, and Mohammadreza Verdian-rizi. 2008. "Spectrophotometric Determination of Total Alkaloids in Some Iranian Medicinal Plants" *Thai Journal of Pharmaceutical Sciences* 32: 17–20.

Sheng-Ji, Pei. 2001. "Ethnobotanical Approaches of Traditional Medicine Studies: Some Experiences from Asia" *Pharmaceutical Biology* 39 (supl): 74–79.

Simbo, David J. 2010. "An Ethnobotanical Survey of Medicinal Plants in Babungo, Northwest Region, Cameroon" *Journal of Ethnobiology, and Ethnomedicine* 6(1): 1–7.

Srivastava, Akhileshwar Kumar. 2018. "Significance of Medicinal Plants in Human Life." In Ashish Tewari and Supriya Tiwar (eds.), *Synthesis of Medicinal Agents from Plants*, 1–24. Elsevier.

Taylor, J L S, T Rabe, L J McGaw, A K Jäger, and J Van Staden. 2001. "Towards the Scientific Validation of Traditional Medicinal Plants" *Plant Growth Regulation* 34(1): 23–37.

Tene, Vicente, Omar Malagon, Paola Vita Finzi, Giovanni Vidari, Chabaco Armijos, and Tomas Zaragoza. 2007. "An Ethnobotanical Survey of Medicinal Plants Used in Loja and Zamora-Chinchipe, Ecuador" *Journal of Ethnopharmacology* 111(1): 63–81.

UN Comtrade. 2014. United Nations Commodity Trade Statistics Database. SITC Rev. 3. http://comtrade.un.org/

Vanisree, Mulabagal, Chen-Yue Lee, Shu-Fung Lo, Satish Manohar Nalawade, Chien Yih Lin, and Hsin-Sheng Tsay. 2004. "Studies on the Production of Some Important Secondary Metabolites from Medicinal Plants by Plant Tissue Cultures" *Botanical Bulletin of Academica Sinica* 45(1): 1–22.

World Health Organization. 2001. "Legal Status of Traditional Medicine and Complementary/Alternative Medicine: A Worldwide Review." Geneva: World Health Organization.

———. 2013. *WHO Traditional Medicine Strategy: 2014–2023*. World Health Organization.

Yeung, Andy Wai Kan, Michael Heinrich, and Atanas G Atanasov. 2018. "Ethnopharmacology—a Bibliometric Analysis of a Field of Research Meandering between Medicine and Food Science?" *Frontiers in Pharmacology* 9: 215.

Zhang, Jin-lan, Ming Cui, Yun He, Hai-lan Yu, De-an Guo. 2005. "Chemical Fingerprint and Metabolic Fingerprint Analysis of Danshen Injection by HPLC–UV and HPLC–MS Methods" *Journal of Pharmaceutical, and Biomedical Analysis* 36(5): 1029–35.

Zizka, Alexander, Adjima Thiombiano, Stefan Dressler, Blandine M I Nacoulma, Amadé Ouédraogo, Issaka Ouédraogo, Oumarou Ouédraogo, et al. 2015. "Traditional Plant Use in Burkina Faso (West Africa): A National-Scale Analysis with Focus on Traditional Medicine" *Journal of Ethnobiology and Ethnomedicine* 11(1): 1–10.

6 Wild Edible Medicinal and Aromatic Plants in Ancient Traditions

Mahima Verma, Shireen Fatima, Prakriti Mishra, and Irfan Ahmad Ansari
Integral University

CONTENTS

6.1 INTRODUCTION

Medicinal and aromatic plants (MAPs) contain aromatic compounds used before history was documented for traditional medicinal uses worldwide. Humans and plants have had a symbiotic relationship since the dawn of time. Since its inception, humans have relied on wild plants for food, housing, energy, and health. Plants and plant products have gradually gained additional intangible niches of value, such as cultural and religious significance as humans have evolved and civilizations have developed. Aromatic plants include odorous volatile chemicals such as essential oil, exudate gum, balsam, and oleo gum resin in many parts of the plant, including bark, wood, root, leaves, stem, flower, and fruit. There are many complicated chemical compounds present in wild edible and medicinal plants responsible for the characteristic aroma. In the case of citrus fruits, essential oils can be obtained using various physical and chemical techniques like hydrodistillation, steam distillation, and expression. To separate volatile components, enfleurage, a procedure that utilizes odorless fats that are stable at 37 degrees, is used to catch fragranced molecules released by plants. Flavors and fragrances are the most common applications for volatile compounds and essential oils as they reproduce the active components of the plants. Essential oils account for approximately 17% of the global flavoring and fragrance market. The global output of essential oils ranges from 40,000 to 60,000 tons per year. The demand for spice oils is 2,000 tons annually (Sadanandan, Peter, and Hamza 2002).

The number of industries has increased the use of medicinal plants for human consumption, including cosmetics, pharma industry, healthcare system, organic food industry, and pesticides. Natural compounds have been used extensively to formulate approximately 40% of newly approved drugs in the last two decades. More and more patents are getting filed for medicinal plants from

DOI: 10.1201/b22842-6

pharma companies to use medicinal plants and their derivatives. Plants have been used for therapeutic purposes in Asian countries since ancient times. In recent years, MAP production has emerged as a new export business in many Asian countries. India has one of the richest sources of many types of MAPs; however, because farmers are unaware of their potential and rewards, they have had limited success in harnessing the potential of these plants (Riaz et al. 2021).

India has a well-documented and practical understanding of ancient herbal therapy (Kamboj 2000). Medicinal plants have been used in some form or another by traditional medical systems such as Ayurveda, Siddha, and Unani. India is a significant supplier of raw medicinal herbs and processed plant-based pharmaceuticals. The biodiversity of the Indian Himalayan Region has traditionally offered diverse commodities and services to the indigenous mountain inhabitants, influencing their traditional food and medical systems. India has a diverse landscape and climate, impacting its flora and floristic diversity. This subcontinent is among the world's 12 leading biodiversity centers, with 16 distinct agro-climatic regions, 25 biotic provinces, 10 vegetation zones, and 426 species habitats. The Indian subcontinent is home to over 45,000 plants (almost 20% of the world's species). Approximately 3,500 plant species of both upper and lower groupings have therapeutic properties.

"Pen T'Sao," a Chinese treatise on roots and grasses published around 2500 BC by Emperor Shen Nung, describes 365 medications derived from dried sections of plant extracts (Wiart 2006). For almost 2,200 years, medicinal herbs and concoctions have been utilized in China. The truth that they are so popular in Western nations may be ascribed to the customers' fear of unintended side effects from synthetic pharmaceuticals and the "Green Movement" that has been resurrected in North America, Europe, and Australia during the last 20 years. Plant components make up more than 80% of herbal medications used in traditional Chinese medicine (TCM). Chinese medicine is among the most extensive and well-documented traditional and folk medicine systems in human history. Since ancient times, Chinese and several other Asian people have utilized it to treat various ailments and preserve health due to the limits of current Western medicine and the need for alternative and complete medications.

The Arabian Peninsula is rich in medicinal, aromatic plants with life-saving, disease-fighting potential against diseases such as acquired immunodeficiency syndrome, diabetes, hypertension, and asthma. The therapeutic approaches are being industrialized against these ailments from the medicinal herbs of African countries like Libya, Somalia, Gambia, Egypt, and Nigeria (Gunjan et al. 2015). According to Grover, Yadav, and Vats (2002), there are 150 significant plants, which have been identified from the literature survey, that have the most medicinal properties and have been used since ancient times for treatment purposes. Some of these plants and their medicinal properties are listed in Table 6.1. A literature survey revealed that plants have been contributing to genetic engineering, drug development, and the food and beverage industry by producing secondary metabolites, sugars, protein, clothes, gums, etc. In a recent *in silico* study conducted by Lim et al. (2021), shreds of evidence of the antiviral effect of medicinal plants on severe acute respiratory syndrome coronavirus-2 (SARS-CoV-2) were found. The study was conducted on four medicinal plants (viz., *Vernonia amygdalina*, *Nigella sativa*, *Eurycoma Longifolia*, and *Azadirachta indica*) of which SARS-CoV-2-specific antiviral effects were shown by *Azadirachta indica*. These findings suggest the need for further investigation of MAPs to assess their astonishing therapeutic effects (Lim, Teh, and Tan 2021). Because billions of people worldwide are dependent on medicinal plants for their health, sustainability, and survival, it must be our principal focus. We must conserve the richness of plant environments, particularly in tropical rainforests. This chapter is focused on sharing knowledge and addressing the influence of plants on human life, especially on Indian, Chinese, and Arabian traditions.

6.2 THE WORLDWIDE HISTORY OF THE USE OF AROMATIC PLANTS

The World Health Organization (WHO) reported that nearly 60–80% of the world's population has depended on plants for remedial strategies in developing countries since ancient times due to limited or lack of access to modern therapy (Awoyemi et al. 2012; Zollman and Vickers 1999; Giacometti et al. 2018). The value of medicinal plants is rapidly being acknowledged from social, ecological, and economic standpoints (Arnold and Pérez 2001). Mostly, all ancient cultures showed some understanding of the medical uses of plants. Aromatic plants and herbs were employed by ancient cultures to treat physiological and psychic ailments. People thought that disease had a spiritual source or arose from evil during the period. As a result, therapists were held in great regard and played a significant role in their communities. A person's bones were enclosed by six to seven medicinal plants containing Ephedra at a Neanderthal floral burial site in Northern Iraq around 60,000 years ago, indicating that people used plants for the first time (Solecki 1975). The Sumerian culture came after, displaying written formulae on clay rocks unearthed 5,000 years ago in the Nagpur district, which includes plant preparations of over 200 herbs comprising secondary metabolites such as alkaloids (licorice, henbane, poppy, and mandrake) (Kelly 2009; Duke 2002). Traditional practices and beliefs were combined and altered by past civilizations, including Babylonians, Assyrians, and other Mesopotamian residents, as is widely known. This is reinforced by discovering a cuneiform inscription on clay tablets, alluding to 1,000 plants used to cure various illnesses in Mesopotamia around 2500 BCE (Elgood 2010).

Aromatic medicinal herbs were the first pharmacological chemicals utilized in ancient times to cure illnesses or other aberrant states. They are still employed in folk or ethnic-type therapeutic applications. According to the Bible and the sacred Jewish book "Talmud," cinnamon oil, myrtle, marjoram, dill, anise, fenugreek, frankincense, mint, juniper, garlic, mustard, saffron, myrrh, and other fragrant plants were utilized as holy ointments used in rites (Smith et al. 2005). Furthermore, Iranian medicine was industrialized and classified in the 8th and 9th centuries by the physicians Razi and Avicenna, as demonstrated by their medical texts, "Canon of Medicine" (Jamshidi-Kia, Lorigooini, and Amini-Khoei 2018).

Plants and spices that have been used as medicine in India date back to around 1500 BCE, as evidenced by transcripts of Atharva and Rig Veda's sacred scriptures and the Sushruta Samhita (Sumner 2000). *Rauvolfia serpentina* (snakeroot), used to sedate and cure snake bites and mental illnesses for generations, was one of the most popular herbs. However, reserpine, one of its main active components, was only identified in the last decade by Western pharmacies. Turmeric, an essential polyphenolic component of curcumin and basil (*Ocimum* spp.), also known as tulsi in India and Nepal, has long been used in Ayurvedic medicine to treat a variety of respiratory, gastrointestinal, skin, renal, blood, and other problems (Henrotin, Priem, and Mobasheri 2013; Singletary 2018). Moreover, oral disorders have long been treated with the most common spices in the Indian kitchen, such as haldi (*Curcuma longa*) and clove (*Syzygium aromaticum*) (Hongal et al. 2014). After Ayurveda, Unani is another established therapeutic approach originating from "Greek" and conveyed to Persia by most leading philosophers Aristotle. Moreover, it was recognized as a preventive medication in India and the Asia-Islamic World in the 13th century. It was further improved and practiced by Indian medical procedures such as Charaka and Sushruta and in the United States of America (USA), Germany, Canada, United Kingdom (UK), Bangladesh, and Pakistan till now (Rahman 1994).

In Europe, Saffron (*Crocus sativus* L.), used to cure cardiovascular illnesses and as a flavoring agent since ancient times, was one of the spices of interest in the Mediterranean basin (Khorasanchi et al. 2018). The evident discovery of the anticancer activity of garlic and hellebore supports the usefulness and utilization of these plants from ancient times (Blowman et al. 2018). During the second year BCE, Nikander was the first to report on poisons and their plant-based medicinal antidotes, followed by King Eupator of Pontus. He made significant contributions to that learning. One more fantastic finding was the use of alkaloid (hyoscine) as a sedative before surgery was discovered in the mandrake of potatoes. Despite its analgesic qualities, its usage was limited due to its toxicity, as an overdose might be lethal (Jones 1996).

On the other hand, the discovery and isolation of chemicals extracted from medicinal plants resulted in pharmaceuticals such as cocaine, opium, codeine, quinine, morphine, and digitoxin in

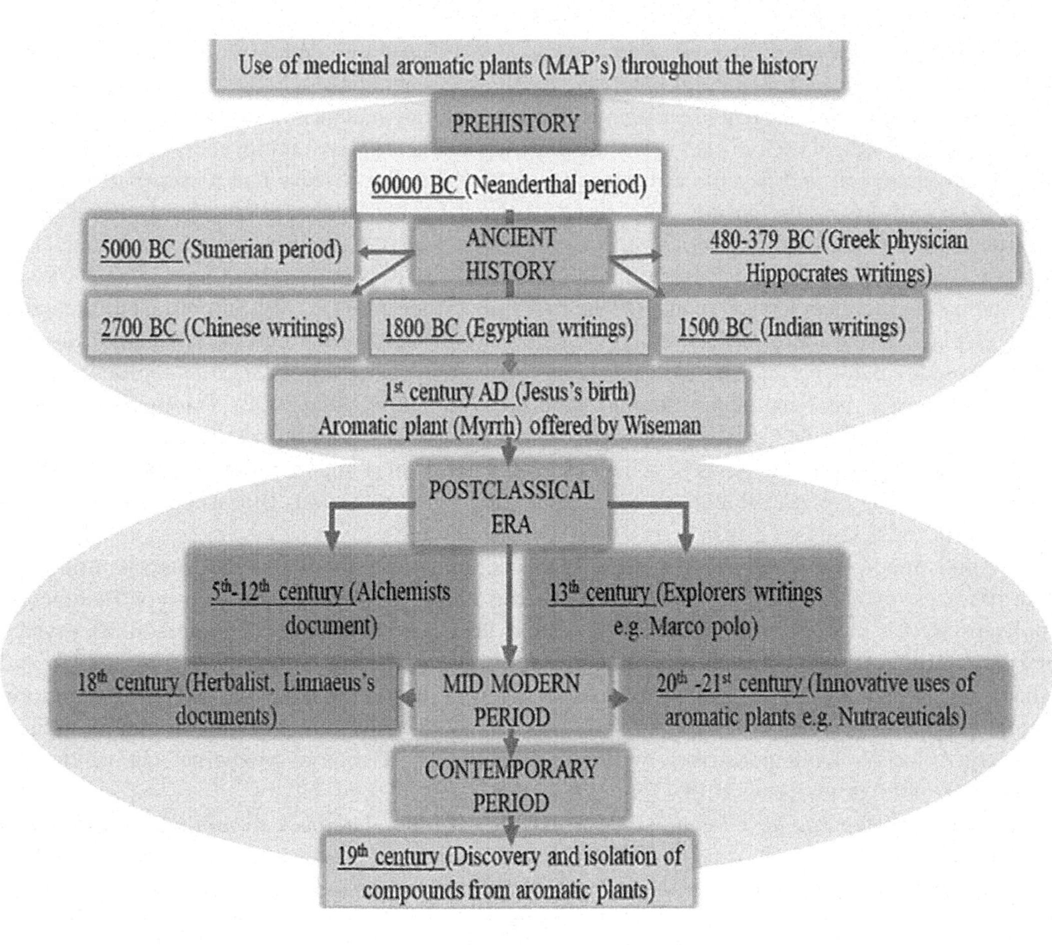

FIGURE 6.1 Use of MAPs throughout the world history. MAP, medicinal aromatic plant.

the 19th century. Additionally, the significance of medicinal plants was assessed by another discovery of plant isolate artemisinin, which was reported to considerably decrease the malaria patients' death rate (Balunas and Kinghorn 2005; Tu 2011). The good establishment of herbal medicine was evidenced, when claimed by Russia that the use of mold for the treatment of infected wounds was in the application even before the discovery of penicillin (Solovieva 2005). The Chinese Materia Medica (1100) included approximately 300 plants. By contrast, Chinese medicine, a well-known proponent of natural therapy, offers its first documented writings from Emperor Shen Nong Ben Jing's existence (c.2700 BCE).

6.3 WILD EDIBLE PLANTS IN THE INDIAN TRADITION

Plants have been used as medicine for thousands of years; traditionally, these remedies began as rudimentary pharmaceuticals, such as teas, tinctures, powders, poultices, pills, and other herbal formulations (Lim, Teh, and Tan 2021). India has well-documented and well-practiced expertise in ancient herbal remedies under indigenous medical systems such as Ayurveda, Siddha, and Unani. Ayurveda, which means "science of life," is one of the oldest Indian therapeutic systems. This comprehensive therapy was founded on Rig Veda teachings and is being used today, not just in Hindu communities in India but also in the West. Basil (*Ocimum* spp.), also known as Tulsi in India and Nepal, has long been used in Ayurvedic medicine to treat a variety of respiratory, gastrointestinal, renal, blood, skin, and other problems (Jaiswal and Williams 2017; Singletary 2018).

Withania somnifera (Solanaceae family), which is also known as Ashwagandha in common language in India, have been found to have a strong effect against inflammatory, diuretic, tumor, hypotension, and immunomodulatory diseases (Anbalagan and Sadique 1981; Devi 1996; Mishra, Singh, and Dagenais 2000). On the other hand, *Tinospora cordifolia*, which belongs to family Menispermaceae also called as Guduchi, have shown anticancer, anti-allergic, antioxidant, anti-leprotic, and anti-malarial effects (Sultana et al. 1995; Singh et al. 2003). Makoy (*Solanum nigrum*) is also reported to be a rich source of therapeutic compounds as it shows efficacy against numerous diseases, such as fever, piles, dysentery, and stomach ulcer, along with hepatoprotective properties (Chauhan et al. 2012; Jainu and Devi 2004). Additionally, Ashoka (*Saraca indica*) have long been documented to treat gynecological disorders (Krishnaraju et al. 2005).

It is well-known in Ayurveda that many plants are reported to have historic bioactivity, and amla (*Embelica officinalis*) is one of them that have anticancer, anti-inflammatory, laxative, diuretic, anti-fungal, and hepatoprotective properties (Variya et al., 2016). Turmeric, which includes the poly-phenolic component curcumin, is another essential plant that is still utilized to create numerous pharmacological regimens across the world (Henrotin, Priem, and Mobasheri 2013). On the other end, with around 6,000 plants comprising approximately 75% of the medical needs of third-world nations, India is a significant global exporter of wild MAPs and processed plant-based pharmaceuticals.

The Indian government has taken several steps to strengthen herbal medication standards and develop the Indian medical system by prioritizing quality standards (Mukherjee 2002). With this historical basis, India needs to enhance its global market share. The US market is estimated to reach $5 trillion by 2050 (Joshi et al. 2004). Indian states Gujarat, Rajasthan, Haryana, Andhra Pradesh, Tamil Nadu, and Uttarakhand are the leading producers of medicinal herbal plants in India. After India, China is the most significant producer of medicinal plants, accounting for more than 40% of the world's variety (plants have been used as medicine for thousands of years).

In the Indian legal system, the Department of AYUSH issued additional regulations to update the Drugs and Cosmetics Provisions, 1945. According to the proposal, a certificate of good manu-facturing practices (GMP) to producers of Ayurveda, Siddha, or Unani pharmaceuticals would be provided to licensees who meet the required standards of GMP of Ayurveda, Siddha, and Unani medicines as laid forth in Schedule (Table 6.2) (Mukherjee 2002).

6.4 WILD EDIBLE PLANTS IN CHINESE TRADITION

Herbal medicine has been practiced in China for over 5,000 years. Despite the fact that each plant has its own set of indications, ailments are routinely treated in traditional Chinese herbal medicine by mixing herbs into formulations. While TCM is extensively practiced in Asian communities such as China, Hong Kong, Taiwan, and Singapore, many non-Asian nations have realized the enor-mous therapeutic potential of this ancient treatment in recent decades and have been aggressively using its advantages to give patients with an extra option in their health care. Over 60,000 herbal prescriptions/formulas were documented during the early Ming dynasty (CE 1368–1644) (Florou-Paneri, Christaki, and Giannenas 2019). In North-western China, there are around 8,300 recognized medicinal plant species, of which 400 are employed in TCM or by drug production facilities in China. The majority of these medicinal plants may be found in North-western China's arid and semi-arid fringe mountainous zones, as well as desert areas (Dongling, Yinquan, and Ling 2017).

TCM systems of pharmacology database and analytic platform was established in China as a comprehensive repository of herb pharmacochemistry, pharmacokinetic characteristics, and other information to aid drug development from herbal sources. Artesunate produced from *Artemisia annua*, which is used to treat severe malaria, and arsenic trioxide, which is used to treat acute pro-myelocytic leukemia, are two well-known examples of TCM active components that have been therapeutically successful (Fung and Linn 2015). A renowned Chinese herbalist, Li Shizhen (1518–1593), produced the pharmacopoeia "Pen t'sao kang mu," which translates as "The Great Herbal." He summarized everything was understood about herbal medicine up to the late 16th century, and he

TABLE 6.1
Major MAP used in Indian tradition for treatment of different ailments

N	Botanical Name of the Plant	Common Name of the Plant	Major Portion Used	Family	Bioactivity	References
1	*Ficus religiosa*	Peepal	Seeds, bark, leaves, fruit, latex	Moraceae	Skin-related ailments, gynecological diseases, constipation, neuralgia	(Samuelsson 2004)
2	*Trigonella foenum*	Methi	Seeds	Fabaceae	Antidiabetic, constipation	(Chakraborty, Dam, and Abraham 2016)
3	*Phoenix dactylifera*	Khajoor	Fruit	Arecaceae	Diarrhea, urinary tract ailments	(Yasin, El-Fawal, and Mousa 2015)
4	*Azadirachta indica*	Neem	Bark, root, flower	Meliaceae	Cough, antidiabetic, bronchitis, arthritis	(Hashmat, Azad, and Ahmed 2012)
5	*Nigella sativa*	Kalonji	Seeds	Ranunculaceae	Dysentery, diarrhea	(Ahmad et al. 2013)
6	*Datura stramonium*	Datura	Leaves, fruit	Solanaceae	Asthma, cardiac pain	(Neeraj, Ayesha, and Balu 2013)
7	*Ocimum sanctum*	Tulsi	Leaves	Lamiaceae	Antidiabetic, antiallergic	(Siva et al. 2016)
8	*Punica granatum*	Anar	Flower, seeds	Punicaceae	Bronchitis, syphilis, stomachic	(Ranjha et al. 2023)
9	*Thymus vulgaris*	Ajwain	Seeds	Apiaceae	antispasmodic, antiseptic	(Hosseinzadeh et al. 2015)
10	*Cinnamomum verum*	Dalcheeni	Bark	Lauraceae		(Singh et al. 2021)
11	*Amomum subulatum*	Ilayachi	Fruit	Zingiberaceae	Stomachic, skin-related ailments, aphrodisiac, nausea	(Bisht et al. 2011)
12	*Curcuma longa*	Haldi	Root	Zingiberaceae	Anti-inflammatory, antiparasitic, anticancer	(Verma et al. 2018)
13	*Cymbopogon citratus*	Lemon grass	Leaf	Gramineae	Antibacterial, antifungal, antimycobacterial, antiamebic, antimalarial	(Manvitha and Bidya 2014)
14	*Jasminum officinale*	Jasmine	Flower	Oleaceae	Antibacterial, antifungal, antioxidant	(Prakkash, Ragunathan, and Jesteena 2019)
15	*Plantago ovata*	Isabgol	Leaves, seeds	Plantaginaceae	Anti-inflammatory, diarrhea, laxative, demulcent, emollient	(Dhar et al. 2005)
16	*Saussurea costus*	Kuth	Root	Asteraceae	Anti-arthritic, aphrodisiac, anti-inflammatory, antioxidant, antiseptic, cytotoxic	(Pandey, Rastogi, and Rawat 2007)
17	*Myristica fragrans*	Jaiphal	Fruit, seed, essential oil	Myristicaceae	Anti-inflammatory, anti-carcinogenic, hepatoprotective, antibacterial, antifungal	(Al-Qahtani et al. 2022)

(Continued)

TABLE 6.1 (*Continued*)
Major MAP used in Indian tradition for treatment of different ailments

N	Botanical Name of the Plant	Common Name of the Plant	Major Portion Used	Family	Bioactivity	References
18	*Plumbago zeylanica*	Chitrak	Root	Plumbaginaceae	Antibacterial, anti-hyperglycemic, anti-inflammatory, anticancer, antiplasmodial, antifungal, anti-atherosclerotic	(Roy and Bharadvaja 2017)
19	*Syzygium aromaticum*	Laung	Flower bud	Myrtaceae	Antifungal, antibacterial, anticancer, antiviral, antidepressant, antiprotozoal, antiulcer	(El-Saber Batiha et al. 2020)
20	*Nardostachys jatamansi*	Jatamansi	Root	Caprifoliaceae	Antispasmodic, laxative, hepatoprotective, neuroprotective	(U. M. Singh et al. 2013)
21	*Asparagus recemosus*	Satavatri	Root	Asparagaceae	Aphrodisiac	(Zhang et al. 2019)

MAP, medicinal and aromatic plant.

described almost 1,800 plants, their therapeutic properties, and uses. *Rhei rhizoma*, camphor, *Theeae folium*, Podophyllum, the big yellow gentian, cinnamon bark, ginseng, jimson weed, and ephedra are among the 365 dried medicinal herbs utilized in ancient times (Florou-Paneri, Christaki, and Giannenas 2019). Synergy was considered essential for the success of herbal therapy in TCM. Many traditional herbal formulations are being utilized nowadays. For example, one formula called "Shi Quan Da Bu Tang" has 11 herbs and is used to treat arthritic joint pain; another formula called "Huo Luo Xiao Ling Dan" contains 10 herbs and is used to cure exhaustion and energy, especially in the elderly (Yamada and Saiki 2005). "Maxingshigan-Yinqiaosan," a TCM concoction, was used to treat uncomplicated H1N1 influenza (Jaiswal and Williams 2017). To treat respiratory tract infection, "Maxingshigan- Yinqiaosan" is a 12-herb combination that combines two ancient TCM formulations (TCMFs) for their "diaphoretic and heat-clearing" actions. MLC601 (Trade name: Neuroaid) is a TCMF called "Danqipiantan jiaonang" that contains nine herbs and five animal components. It may help improve functional independence and motor recovery as a supplement to standard treatment and is safe for patients with non-acute stable stroke. After a 12-month therapy, a Tianqi capsule comprising ten TCM herbs was found to dramatically reduce the prevalence of type 2 diabetes mellitus development in participants with impaired glucose tolerance (Fung and Linn 2015).

Traditional medicine, according to the WHO, is one of the key sources of health care. Prof. Youyou Tu was also awarded the Nobel Prize in Physiology or Medicine in 2015 for her exceptional success in employing artemisinin therapy. As a result, TCM is gaining popularity among medical professionals all over the world. The TCMF is a measure of Chinese medicine therapy of disorders that contains two or more types of herbs, is developed for relatively specific symptoms, and comprises two or more kinds of herbs as components (Chen et al. 2019). A few TCMFs are now undergoing FDA clinical studies, taking on the task of meeting the demanding standards of good clinical practice. Danshen dripping pill (*Salvia miltiorrhiza*, *Radix notoginseng*, and Borneol) was used in a phase III study for stable angina, Kanglaite injection (coix seed oil and excipients) for cancer, Xuezhikang capsule (red yeast) for hyperlipidemia, and FuzhengHuayu capsule (*Salvia miltiorrhiza*, peach seed, pine pollen, Gynostemm (Xu 2011).

TABLE 6.2
Regulatory Considerations of Natural Goods based on the Drugs and Cosmetics Provision of India

Item	Section	Criteria
Misbranded drugs	33 E	ASU drugs are deemed to be misbranded: If colored or coated to conceal the damage or made appear better than therapeutic value If it is not labelled in prescribed manner If label or container accompanying drug bears any false, claim or misleading
Adulterated drugs	33EE	ASU drugs are deemed to be adulterated: • If it consists filthy or decomposed material • If prepared, packed, or stored under insanitary conditions • If its container contains any poisonous or deleterious substance color other than the one that is prescribed harmful or toxic substance
Spurious drugs	33EEA	ASU drugs are deemed to be spurious: • If it is sold or offered under another name • If it is an imitation or substitute for another drug • If the label or container bears the name of an individual or company which is fictions. • If it has been substituted by another drug
Regulation of manufacture for sale of ASU drugs	33EEB	No person shall manufacture for sale or distribution, any ASU drugs except in accordance with prescribed standards.
Prohibition of manufacture for sale of ASU drugs	33EC	No person shall manufacture ASU drugs as follows: • Misbranded, adulterated, or spurious • Patent or proprietary medicine • Contravention to any of the provisions of the act
Power of central government to prohibit manufacture of ASU drugs in public interest	33ED	Central government can prohibit manufacture of ASU if: • Drugs involve any risk to human beings or animals • Drugs does not have the therapeutic value claimed
Government analysts	33F	Central or state government can appoint any person with the prescribed qualification and do not have any financial interest in ASU drug
Inspectors	33G	Central or state government can appoint any person with the prescribed qualification and do not have any financial interest in ASU drug
Penalty for manufacture, sale, etc., of Ayurvedic, Siddha, or Unani drug in contravention of this chapter	33I	Manufacturers for sale or for distribution of any ASU drugs deemed to be adulterated or without a valid license as required
Penalty for subsequent offences	33J	Shall be punishable with imprisonment for a term not less than two years, but which may extend to six years and with fine not less than 5,000 INR
Confiscation	33K	Any person convicted under the act, the respective stock of ASU drug can be confiscated
Cognizance of offences	33M	No prosecution under this chapter shall be instituted except by an inspector. No court inferior to that of a [Metropolitan Magistrate] or of a [Judicial Magistrate] of the first class shall try an offence punishable under this chapter.
Power of central government to make rules	33N	The Central Government may, after consultation with, or on the recommendation of, the Board, and after previous publication by notification in the official gazette, make rules for the purpose of giving effect to the provisions of this chapter

(*Continued*)

TABLE 6.2 (*Continued*)
Regulatory Considerations of Natural Goods based on the Drugs and Cosmetics Provision of India

Item	Section	Criteria
Application for license to manufacture Ayurvedic (including Siddha) or Unani drugs	153	An application for the grant or renewal of a license to manufacture for sale any ASU drugs shall be made in Form 24-D to the licensing authority along with a fee of rupees 60
Form of license to manufacture Ayurvedic (including Siddha) or Unani drugs	154	Subject to the conditions of rule 157 being fulfilled, a license to manufacture for sale any ASU drugs shall be issued in Form 25-D
Loan license	153A	An application for the grant of renewal of a loan license to manufacture for sale of any ASU drugs shall be made in Form 25-E to the licensing authority along with a fee of rupees 60
Certificate of award of GMP Ayurveda, Siddha, and Unani drugs	155B	Shall be issued to licensees who comply with the requirements of GMP of ASU drugs as laid down in Schedule T
Standards to be complied with in manufactures for sale or for distribution of Ayurvedic, Siddha, and Unani drugs	168	Single drugs: The standards for identity, purity and strength as given in editions of Ayurvedic Pharmacopoeia of India. Asavas and Arishtas: The upper limit of alcohol as self-generated alcohol should not exceed 12% v/v

GMP, good manufacturing practices.

Currently in the COVID-19 pandemic, 102 cases with moderate symptoms were medicated with TCM, the clinical symptom disappeared with in reduced time, the body temperature recovery time was reduced by 1.7 days, the average length of stay in hospital was shortened by 2.2 days, the CT image improvement rate was improved by 22%, the clinical cure rate was increased by 33%, the rate of common to severe cases was reduced by 27.4 percent, and the lymphocyte count rose by 70%. TCM has proposed prescribing prescriptions that are likely to be effective, such as qingfei paidu decoction, sheganmahuang decoction, gancaoganjiang decoction, and qingfei touxie fuzheng recipe, based on current treatment of COVID-19 in China (Ren, Zhang, and Wang 2020). All these data are summarized in Table 6.3.

6.5 WILD EDIBLE PLANTS IN ARABIAN TRADITION

The Peninsula is a massive territory bordered on three sides by the Arabian Sea on the east, the Red Sea on the West, and the Mediterranean Sea on the north. Syria, Jordan, Iraq, and Kuwait border it on the north, while the Indian Ocean surrounds it on the south (Ash 2005; Ahmad et al. 1983). In the Gulf, herbal medicine is widely used, and medicinal plants can be attained from the local markets. More than 4,500 species have been found in Syria, Palestine, Jordan, and Lebanon, of which Lebanon was the primary source of these plants. Based on the data provided from the history and case study of inappropriate treatment of various diseases from hospitals and folk medical practitioners, the number of naturopaths cannot be assumed. Still, the utilization of medicinal, aromatic

TABLE 6.3
Some Commonly Used MAP of China

Common Name	Pharmaceutical	Chinese in Traditional Script	Listed in 2020 TCM Pharmacopeia	IUCN Red List Assessment	References
Eucommia bark	Cortex Eucommiae	dù zhòng	Yes	Vulnerable	(Lange 2002)
Golden thread	Rhizoma Coptidis	huáng lián	Yes	Not evaluated	
Storax	Styrax	sū hé xiāng	Yes	Endangered	
Officinal Dendrobium	Herba Dendrobii Officinalis	Tiě pí shí hú	Yes	Critically endangered	
Tall Gastrodia Tuber	Rhizoma Gastrodiae	tiān má	Yes	Vulnerable	
Black Cohosh Rhizome	Black cohosh	Shengma	Yes	Endangered	
Ginkgo seed & Ginkgo leaf	Semen Ginkgo	yín xìng yè piàn	Yes	Endangered	
Borneol	Borneolum	méi huā bīng piàn	No	Vulnerable	
Golden larch bark	Cortex Pseudolaricis	Tǔ jīng pí	Yes	Vulnerable	

MAP, medicinal and aromatic plant; TCM, traditional Chinese medicine.

plants is massive. Moreover, many Arabic medicinal plants are facing extinction due to global climate change and manmade complications (Pan et al. 2014; Farooqi 1998).

In the Arabian Peninsula, MAPs have been in practice to treat multiple ailments since the beginning of humankind for good health and care. For example, *Citrus sinensis*, *Pistachio lentiscus*, and *Cicer arietinum* are reportedly used to treat cancer. On the contrary, *Cocos nucifera*, *Coriandrum sativum*, *Cinnamomum zeylanicum*, and *Borago officinalis* are used to treat psychosis. Additionally, Pistachio, *Olea europaea*, *Phaseolus vulgaris*, *Lycopersicon esculentum*, *Glycyrrhiza glabra*, *Allium cepa*, and many more are used to treat illnesses like diabetes, hypertension, hepatic disorders, tuberculosis, and asthma (Ashur 1986; Robertson 2008). Moreover, burns and wounds have long been treated with aloe (*Aloe* species), whereas aspirin (*Salix* species) is being used as a muscle relaxant, as well as in blood dilution (Tuxill 1999; Marinelli 2004). Some of the significant plants of the Arabian Peninsula are listed in Table 6.4.

6.6 USE OF MAPs: CURRENT PERSPECTIVE IN HUMAN AND VETERINARY MEDICATION, PLANT LONGEVITY, AND SAFETY CONCERNS

According to history textbooks and other recorded evidence from throughout the world, plants have been utilized as medicines for millennia. They are currently used as model substances in the pharmaceutical business (Schmidt et al. 2008). The use of plants for healing precedes history and is at the root of contemporary medicine. Today, the creation of novel plant-origin medications and plant-extracted nutritional supplements must integrate herbal medical understanding with cutting-edge technology because several traditional pharmaceuticals are obtained from plants. A century ago, the majority of the few successful drugs were plant-based, for example, digoxin (derived from foxglove), aspirin (derived from willow bark), quinine (derived from cinchona bark), and morphine (from the opium poppy) (Shen 2015).

TABLE 6.4
Major MAPs of Arabian Peninsula with Their Bioactivity

N°	Botanical Name	Common Name	English Name	Major Portion Used	Family Name	Bioactivity	References
1	*Thymus serpyllum*	Saatar	Wild thyme	Whole plant	Labiatae	Jaundice, fatigue, digestive, helminthiasis, carminative, stimulant, sudorific, mucolytic, antidote for poisoning from snake venom and other poisons	(Nikolić et al. 2014)
2	*Luffa acutangula*	Silq	Silk guard	Whole plant, juice	Cucurbitaceae	Diuretic, emollient, aesthesia, edema, gout, alopecia, warts, leprosy, herpetic eruptions, splenitis, cholagogue, eryspelax, migraine, headache, facial paralysis	(Partap et al. 2012)
3	*Sesamum indicum*	Simsim	Sesame	-	Pedaliaceae	Ecbolic arteriosclerosis, snake bite, aesthetic emmenagoge, voice clearing, cholagogue, infertility, nephritis, diuretic	(Wu et al. 2019)
4	*Lycopersicum esculentum*	Tamatim	Tomato	Fruits	Solanaceae	Arthritis, diuretic, foot cracking, renal and urinary calculi, aesthetic, rheumatism, scurvy	(Khan et al. 2014)
5	*Helianthus annuus*	Ibadusshams	Sunflower	Roots, seeds	Compositae	Night blindness, gingivitis, consolation, diuretic, malaria, mucolytic, arteriosclerosis, hypercholestrolemia	(Guo, Ge, and Na Jom 2017)
6	*Juniperus oxycedrus*	Arar	Juniper	Flowers	Cupressaceae	Skin diseases, rheumatism, urinary purifier, diuretic, anti-sweating, expels rodents, antipoison, diabetes, hemorrhoids, colic, stomachic, digestive, chest pain, chronic cough	(Chaouche et al. 2013)
7	*Citrus sinensis*	Inab	Grape	Stem, bark, leaves, fruits, roots	Rutaceae	Bladder and renal pains, cancer, laxative, diuretic, sedative, gastritis, hepatic and adrenal problems, polycythemia, thirst, inflammations, cough	(Sahib, Al-Shareefi, and Hameed 2019)
8	*Zizyphus vulgaris*	Anbar	Jujube	Leaves, seeds, oil	Rhamnaceae	Mucolytic, cough, sudorific, resin, volatile oils, cinnamic, sterol, gum, stimulant, perfume, digestive, tonic	(Asgarpanah and Haghighat 2012)
9	*Dactylis glomerata*	Najiyl	Dactylis	Roots	Gramineae, Poaceae	Cystitis, gout, rheumatism, skin rashes, fever, diaphoretic, diabetes, blood tonic, diuretic	(Hauck et al. 2014)

(*Continued*)

TABLE 6.4 (*Continued*)
Major MAPs of Arabian Peninsula with Their Bioactivity

N°	Botanical Name	Common Name	English Name	Major Portion Used	Family Name	Bioactivity	References
10	*Ricinus communis*	Kharu	Castor oil tree	Leaves, oil	Euphorbiaceae	HIV/AIDS, constipation, bloat, callus, hemorrhoids, dystonia, catarrhs, aesthetic, umbilitis, contraceptive, warts, carminative, cancer, ecbolic, emmenagogue	(Abdul et al. 2018)
11	*Punica granatum*	Ruman	Pomegranate	Roots, barks	Punicaceae	Laxative, anthelmintic, jaundice, pain, chest, vomiting, suppresses bile diuretic, emollient, cough	(Ranjha et al. 2023)
12	*Persea Americana*	Zabadiyya	Avocado	Fruits	Lauraceae	Growth promoter, stimulant, antipathogenic, digestive	(Dabas et al. 2013)
13	*Citrus aurantifolia*	Naranj	Bitter lemon	Leaves, flowers, peels	Rutaceae	Antispasmodic, carminative, colic, ecbolic, anthelmintic, aesthesia, eczema, cold, malaria, fevers	(Enejoh et al. 2015)
14	*Cucurbita maxima*	Yaqtin	Pumpkin	Seeds	Cucurbitaceae	Urethritis, skin eruption, cystitis, diabetes, insomnia, soothner, diuretic, digestive, enteritis, constipation, impotence	(Men et al. 2021)
15	*Sorghum bicolor*	Shair	Guinea corn	Leaves, seeds, flower	Gramineae/Poaceae	Cough, diuretic, diarrhea, antipyretic, euphoric, emollient, general debility, cystitis, laxative, purgative hypertension, tuberculosis	(Espitia-Hernández et al. 2022)
16	*Avena sativa*	Shawfan	Oat	Seed, bark	Urticaceae	Renal calculi, hemorrhoids, gall bladder, inflammations, renal colic, whooping cough, emollient, diuretic, pimples, gout, chronic cough	(R. Singh, De, and Belkheir 2013)
17	*Citrullus colocynthis*	Hanzwal	Colocynth	Fruits, oil	Cucurbitaceae	Pains of ligaments, joint and sciatic nerve, gout, skin diseases, rheumatism, cold, gastrointestinal problem, nose bleeding, scorpion sting	(Hussain et al. 2014)
18	*Brassica nigra*	Khardalsauda'a	Black mustard	Leaves, seeds	Brassicaceae	Acanthosis, stomatitis, migraine, emmenagogue, asthma, dizziness, headache, numbness, dyspnea, jaundice, hypertension	(Rajamurugan et al. 2012)

(*Continued*)

TABLE 6.4 (*Continued*)
Major MAPs of Arabian Peninsula with Their Bioactivity

N°	Botanical Name	Common Name	English Name	Major Portion Used	Family Name	Bioactivity	References
19	*Lawsonia inermis*	Hinaa	Henna plant	Leaves	Lythraceae	Headache, histological stain, dye, aesthetic, antibiotic, antifungal, seborrhea, baldness, sudorific, aromatic acanthosis	(Chaudhary, Goyal, and Poonia 2010)
20	*Santalum album*	Sandal	Sandal wood	Leaves, oil	Santalaceae	Perfumery, cold, gout, gastric and cardiac problems, heart palpitation, astringent, carminative, colic, deodorant, thirst, fevers, stimulant, diaphoretic	(Choudhary and Chaudhary 2021)
21	*Plantago major*	Lisanulhamal	Plantain	Leaves, fruits	Musaceae	Thrombosis, heart problem, elephantiasis, dropsy, epilepsy, strangury, hemophilia, pneumonia, earache, gingivitis, splenitis, stomatitis, hemorrhage, cancer, tuberculosis	(Najafian et al. 2018)
22	*Brassica oleracea var capitata*	Karnab	Cabbage	Roots, leaves	Brassicaceae	Myopia, wound and kidney diseases, cough, obesity, impotence, dejection, rheumatism, fevers, vermifuge emmenagogue, diuretic, jaundice, splenitis, gout, coarseness	(Šamec, Pavlović, and Salopek-Sondi 2017)
23	*Piper cubeba*	Kababatussinniyya	Chinese cubeb	Seeds	Piperaceae	Dental caries, clear throat, headache, diuretic, emollient, gonorrhea, diarrhea, colic, carminative, digestive, appetizer	(Takooree et al. 2019)
24	*Apium graveolens*	Karfasalhumadl-ahmar	Celery	Flowers	Umbeliferae	Jaundice, arthritis, chest ailments, laxative, body tonic, antacid, digestive	(Fazal and Singla 2012)
25	*Coriandrum sativum*	Kazbara	Corriander	Whole plant	Umbeliferae	Diarrhea, scabies, itching, hypertension, edema, antipsychotic, antispasmodic, analgesic, arteriosclerosis, antialchoholic, flavoring agent in candies, beverages and tobacco products	(Pathak Nimish et al. 2011)
26	*Pyrus communis*	Kuthra	Pear	Fruits	Cucurbitaceae	Choleretic, diuretic, hypertension, ashma, eruptions, astringents, arteriosclerosis, heart kidney and liver diseases	(Kolniak-Ostek et al. 2020)

(*Continued*)

TABLE 6.4 (*Continued*)
Major MAPs of Arabian Peninsula with Their Bioactivity

N°	Botanical Name	Common Name	English Name	Major Portion Used	Family Name	Bioactivity	References
27	*Brassica rapa*	Lift	Turnip	Oil seeds	Brassicaceae	Foot cracking, cold, catarrh, fatigue, antiobesity, diuretic, eczema, calculi, aphrodisiac, cough, cholagogue, galactogogue	(Xue et al. 2020)
28	*Origanum majorana*	Mardaqush	Marjoram	Leaves, flowers	Labiatae	Isomnia, annuria, appetizer, liver renal and gastric problems, catarrh, diuretic, antideacy, antipoison, emmenagogue	(Joshi, Lekhak, and Sharma 2009)
29	*Citrus limon*	LaimunBanzahir	Lemon	Leaves, Fruits	Rutaceae	Hepatoprotective, antipoison, fatigue, antivomiting, vermifuge, appetizer, fever, eczema, malaria, herpetic, rheumatism	(Klimek-Szczykutowicz, Szopa, and Ekiert 2020)
30	*Prunus americana*	Mishmish	Apricot	Seeds	Rosaceae	Cancer, piles, anemia, fever, growth promoter, cholagogue, insomnia, thirst, diarrhea, digestive, basic acids, blood ailments, body tonic, appetizer	(Poonam et al. 2011)
31	*Musa sapientum*	Mauz, Talh	Banana, Philosphers food, fruit of wise people	Fruits, Leaves	Musaceae	banana yields 1.5% of calcium, cholagogue, good for pregnant/breast feeding women, heartburn, self-balance, diuresis, good vision, increases sperm, anemia, dental caries	(Siddique et al. 2018)
32	*Myrtus communis*	Habbulas, Aas, Alhabbal-as	Myrtle	Leaves, Juice, Flowers	Myrtaceae	Vaginal secretion, antidiarrhoeic, cystitis, antisudorific, myrtilblenorrhalgia, cough	(Asgarpanah and Haghighat 2012)
33	*Agaricus campestris*	Aqrasulmali, Aishul-gurab	Mushroom	Whole plant (Some species are very toxic)	Halvallaceae	Dizziness, headache	(Muszynska et al. 2017)
	Ananas comosus	Ananas	Pineapple	Leaves, fruits	Bromeliaceae	Arthritis, diuretic, obesity, digestive, growth promoter, antipoison, bromeline, anemia	(Debnath, Singh, and Manna 2021)
34	*Citrus aurantium*	Burtuqal	Orange	Leaves, peels,Seeds, Juice	Rutaceae	Antipoison, digestive ecbolic, antiemtic, scorpion sting, fever, migraine, headache, purgative, nervous conditions, meningitis, diaphoretic, numbness	(Suryawanshi 2011)

(*Continued*)

TABLE 6.4 (*Continued*)
Major MAPs of Arabian Peninsula with Their Bioactivity

N°	Botanical Name	Common Name	English Name	Major Portion Used	Family Name	Bioactivity	References
35	*Allium cepa*	Basl	Onion	Bulbs	Liliaceae/Aliaceae	Scorpion sting, diuretic, hemorrhoid, appetizer, pneumonia, antimicrobial, cough, jaundice, crack, splenitis, cancer, emmenogogue, splenitis	(Alam, Hoq, and Uddin 2016)
36	*Balsamodendron myrrha*	Balsam	Balsam	Barks, gum, roots, resin, leaves	Burseraceae	Infalmmations, eruption, vermifuge, rheumatism chronic gastric ulcer, dysentery	(Cao et al. 2019)
37	*Securidaca longepedunculata*	Banafsaj	Violet tree	Whole plant, flowers, leaves, roots	Polygalaceae	Whooping cough, diuretic, ophthalmia, analgesic, chest pain, gastric and urinary tract diseases, renal, hepatic	(Alafe et al. 2014)
38	*Malus sylvestris*	Tufah	Apple	Fruits	Cucurbitaceae	Hepatic disease, anemia, pneumonia, antipoison, diarrhea, emetic, neuralgia, anthelmintic, cholagogue, wound, sialogogue, boils, fever, acidic urine	(Stojiljković, Arsić, and Tadić 2016)
39	*Ficus carica*	Tin	Fig	Leaves	Moraceae	Dysmenorrhea, hemorrhoid, intestinal statis, jaundice, nose bleeding, herpetic eruptions, chronic cough, boils, ear and esophageal pains	(Salma et al. 2020)
40	*Syzygium cumini*	Jawa	Java	Roots, Leaves, whole plant	Myrtaceae	Lymphangitis, sudorific, antipoison, diuretic, rheumatism, tranquilizer, asthma, digestive, stimulant	(Ahmed et al. 2019)
41	*Daucus carota*	Jizr	Carrot	Leaves, Roots	Apiaceae	Skin diseases, mucolytic, chest pain, nervousness, cough, hypertension, hepatic and gastric problems, fatigue, diuretic, digestive, brain stimulant, coarseness of voice	(Bahrami et al. 2018)
42	*Solanum nigrum*	Jawaffa	Common night shade, Black night shade	Roots, Fruits, Leaves	Solanaceae	Antidote for the plant poisoning. It is poisonous, caution should be exercised	(Medha and Patki 2009)
43	*Myristica fragrance*	Jawztib	Nutmeg	Fruits, leaves	Myristicaceae	Digestive, strangury, food sweetening, splenic, aphrodisiac, hepatic and gastric diseases, rheumatism, anesthetic	(Ha et al. 2020)

(*Continued*)

TABLE 6.4 (*Continued*)
Major MAPs of Arabian Peninsula with Their Bioactivity

N°	Botanical Name	Common Name	English Name	Major Portion Used	Family Name	Bioactivity	References
44	*Nigella sativa*	Habbatul Baraka, HabbatusSauda, Shunaiz	Black cumin, Black caraway	Leaves, seeds, oil	Ranunculaceae	Expel rodents, general ailments, tinea pedis, carminative, tinea capitis, mucolytic, hemorrhoid, ulcers, headache, mange, migraine, diabetes, antiobesity, antidote, hepatoprotective, hypertension, asthma	(Ahmad et al. 2013)
45	*Humulus lupulus*	Hashishatu dinar	Hops, Humulus	Flowers, Whole plant	Moraceae	Insomnia, tranquilizer, antiaphrodisiac, appetizer, nervous disorders, wounds, oil, gum, ulcers	(Zanoli and Zavatti 2008)
46	*Rosmarinus officinalis*	Haslaban	Rosemary	Flowers, whole plant	Rosaceae	Cough suppressant, tinnitus, anemia, ankylosis, emmenagogue, diuretic, stimulant, tranquilizer, cholagogue	(Hamidpour, Hamidpour, and Elias 2017)
47	*Cinnamomum zeylanicum*	Sinamik	Cinnamon	Leaves, whole plant	Lauraceae	Laxative, asthma, cinnamon cough, jaundice, arthritis, aesthetic, cinnamic lumbago, piles, migraine, chronic headache	(Singh et al. 2021)
48	*Hibiscus sabdariffa*	Himad/hamid	Sorrel	Leaves, flowers	Malvaceae	Diarrhea gout, diabetes, hypertension, gall bladder problems, constipation	(Riaz and Chopra 2018)
49	*Ferula asafoetida*	Hiltit/ abukabir	Devil's drug, asafoetida	Roots, leaves	Umbelliferae	Mucolytic, vermifuge, piles, abortifacient, antidote, jaundice, hepatoprotective, carminative, chronic gum deafness, edema, ophthalmia	(Iranshahy and Iranshahi 2011)
50	*Trygonella foenumgraecum fenugreek*	Hulbah	Sickle fruit	Seeds, roots, leaves,oil	Legumnosae	Cough suppressant, antiasthmatic, uterine, edema, colic, diuretic, asthma, pneumonia, anti-inflammation, galactogogue, tuberculosis, emmenagogue anemia	(Goyal, Gupta, and Chatterjee 2016)
51	*Salvadora persica*	Siwak/ Arak	Tooth brush tree	Fresh root	Salvadoraceae	Voice clearance, antipyretic, laxative, pimples, strangury, pain, psychosis, asthma, liver and heart disease, eczema, catarrh, diuretic	(Aumeeruddy, Zengin, and Mahomoodally 2018)

(*Continued*)

TABLE 6.4 (*Continued*)
Major MAPs of Arabian Peninsula with Their Bioactivity

N°	Botanical Name	Common Name	English Name	Major Portion Used	Family Name	Bioactivity	References
52	*Cinnamomum zeylanicum*	Sinamik	Cinnamic	Bark	Lauraceae	Blurred vision, antipyretic, cough, asthma, liver and heart disease, laxative, uterine disease	(Dorri, Hashemitabar, and Hosseinzadeh 2018)
53	*Atropa belladona*	Sittulhiss	Deadly night shade	Roots, leaves	Solanaceae	Ulcer, gastritis, local anesthetic	(Almubayedh et al. 2018)
54	*Raphanus sativus*	Fajal	Wild radish	Seeds, juice	Cruciferae	Aphrodisiac, aesthetic, deafness, joint pain, cough vomiting, diuretic, stomach tonic	(Mohammed and Hameed 2018)
55	*Vigna unguiculata*	Ful	Common bean	Flower	Papilionaceae	Vomiting, kidney inflammation, urinary and gall bladder disease, digestive stimulant diuresis	(Chandrasekaran, Rajkishore, and Ramalingam 2015)
56	*Phaseolus vulgaris*	Fasuliyya	Common European bean	Seeds, roots, leaves	Papilionaceae	Maldigestion, hypertension, sedative, kidney and liver diseases, inflammations, burns, heart failure	(Poudel et al. 2020)
57	*Eugenia caryophyllata*	Qaranfal/ Mismar	---	Seeds, oil	Myrtaceae	Antipyretic, disinfectant, diarrhea, antipoison, toothache, vomiting, liver tonic, digestive assistance, powerful CNS stimulant, impaired vision, renal incontinence, antipyretic, infertility, antipoison, toothache, disinfectant, diarrhea, vomiting	(Nejad, Özgüneş, and Başaran 2017)
58	*Trapa bispinosa*	Qustul	chestnut	Barks, leaves	Onagraceae	Inflammation, aesthesia, ulcers, hemorrhoids, colic, diabetes, whooping cough	(Adkar et al. 2014)
59	*Heliotropium indicum*	Bone setter	Qasiyn	Flower roots, leaves	Vitaceae	Foot cracking, anuria, TB, rheumatism, joint problems, stimulants, vomiting, diarrhea	(Roy 2015)
60	*Triticum sativum*	Wheat	Qamh	Seeds	Gramineae	Fevers, cough, constipation, diarrhea, maldigestion, rheumatism, colic, prophylaxis, whitlow	(Zilani et al. 2017)

(*Continued*)

TABLE 6.4 (*Continued*)
Major MAPs of Arabian Peninsula with Their Bioactivity

N°	Botanical Name	Common Name	English Name	Major Portion Used	Family Name	Bioactivity	References
	Colocasia antiquorum	Elephants ear	Qalqas	Root, stem	Araceae	Anticancer, CNS stimulant, voice clearing, cough	(Prajapati et al. 2011)
61	*Coffea arabica*	Coffee	Qahawa	Fruit	Coffea	Fevers, diarrhea, stimulant, digestive (overconsumption causes insomnia), cardiotoxic effect	(ALAsmari, Zeid, and Al-Attar 2020)
62	*Zingiber officinale*	Ginger	Zanjebil	Rhizomes	Zingiberaceae	Sudorific, antipyretic, antiscorbutic, food condiment, hepatoprotective, digestive, aphrodisiac, gout, rheumatism, voice clearance	(Kumar, Karthik, and Rao 2011)
63	*Linum usitatissimum*	Flax	Kitan	Bark	Linaceae	Ulcer, diarrhea, aesthesia, renal disease, inflammation, pain	(Akhtar et al. 2017)
64	*Olea europaea*	Olive tree	Zytun	Leaves, bark, oil	Palmae	Esophageal swelling, ulcers, edemas, emollient, cholagogue, calculi, diabetes, wound, demulcent	(Hashmi et al. 2015)
65	*Ruta graveolens*	Common rue	Sazaab	Oil, leaves	Rutaceae	Jaundice, hiccup, urinary tract disease, strangury, anemia, inflammation, vermifuge, facial paralysis, headache, arthritis, lumbago	(Parray et al. 2012)
66	*Daucus carota*	Carrot	Jizr	Leaves, roots	Apiaceae	Gastric and hepatic disease, cough, chest pain, diuretic, digestive problems, fatigue, nervousness, hypertension, mucolytic, brain stimulant, voice clearance, skin disease	(Akhtar et al. 2017)
67	*Phoenix dactylifera*	Date palm	Annakhlatu	Leaves, fruits, stems, spadix, roots	Palmae	Diabetes, renal disease, blennorrhagia, diarrhea, hypertension, hepatitis, vermifuge, antipoison	(Qadir et al. 2020)

MAP, medicinal and aromatic plant.

Aromatic plants and herbs are regarded as natural and safer than standard synthetic medications based on hereditary knowledge and long-term use to treat numerous disorders over generations. There is evidence that the toxicity of many plants was understood in ancient times. Significant adverse effects have been documented following the intake of herbal remedies, and current scientific research has revealed that many plants thought to be therapeutic are potentially poisonous, mutagenic, and malignant (Fennell et al. 2004).

Moreover, various limits to the scientific data support the notion that plant extracts are entirely safe (Raskin et al. 2002). According to Frohne and Pfander (2005), over 750 toxic components are naturally available from more than 1,000 species of plants. However, only a small number of plants may cause severe poisoning once a modest amount is consumed. Most hazardous plants, in most cases, can only induce poisoning under specified conditions like inappropriate administration, misidentification, and wrong preparation of dosage (Kankara et al. 2015). Therefore, researchers proposed that food and medication interactions be avoided by altering pharmacokinetics, such as drug absorption, distribution, metabolism, and excretion, or by compromising the nutritional benefits of essential nutrients.

Since they are not synthetic, aromatic herb and herbal blends may become popular in veterinary medication. Before medicinal herbs are widely used in animal diets, extra attention should be paid to extraction, manufacturing, toxicity assessment, side effects, and safety standardization (Chang 2000). Animals do not intentionally ingest dangerous plants since they taste bitter (Sumner 2000). However, recorded animal intoxications are caused mainly by accidental absorption of pharmaceuticals, metals, or other hazardous chemicals. The restriction on the use of antibiotics frequently added to diets in Europe and its possible prohibition globally heightened interest in discovering antibiotic alternatives for agricultural animals, particularly broiler chickens and pigs. Herbs and spices were employed in the chicken diet to boost growth performance in pioneer research around 40 years ago. On the other hand, an increase in the usage of aromatic plants as animal feed additives has been perceived as medicinal herbs that have been demonstrated to heal ailments and improve commercial poultry's productive efficiency (Windisch et al. 2008).

Traditionally, medicinal plants and herbs are derived from natural sources, where they grow organically. However, due to various detrimental human activity and natural causes, including overexploitation, logging, soil erosion, and climate change, medicinal plants have faced a significant threat of extinction, particularly in the last decade. According to reports, over 15,000 medicinal species of plants are on the verge of extinction owing to habitat loss, overharvesting, and big business around the world (Osman 2011). Pharmaceuticals and industries have put pressure on the systematic production of wild aromatic species of plants due to rising demand and use. The reliability of botanical identification, regularity in quantity supply, control of favored genotypes, and management of quality and safety concerns that may arise from wild plants are only a few of the advantages of producing wild plant species (Posadzki, Watson, and Ernst 2013).

6.7 CONCLUSION

Traditional medicines, usually plant remedies, are used by more than 80% of the population in underdeveloped nations for primary health care. Herbal treatments may be found in most pharmacies and supermarket stores. MAPs will be sustained in all forms, as powder, plant extract, or essential oil by the food industry, pharmaceuticals, and many more. Ayurveda, Siddha, and Unani are examples of Indian medical systems that may compete in worldwide markets. Ayurvedic products are affordable and well received by patients with almost negligible side effects. The Indian government has taken several steps to strengthen herbal medication standards and promote the Indian medical system. In addition, the need for new goods and uses in the feed sector will extend their usage in the coming years. The administration of herbal ingredients and their preparations in domesticated animals as an essential antibacterial agent will be based on scientific knowledge obtained through peer-reviewed literature. To get the most out of natural resources, international coordination should lead to harmonizing different nations' rules. Faced with a multibillion-dollar worldwide herbal medicine trade, several Asian countries are gearing to increase their portion of the market while implementing necessary regulatory steps to protect human safety.

REFERENCES

Abdul, Waseem Mohammed, Nahid H Hajrah, Jamal S M Sabir, Saleh M Al-Garni, Meshaal J Sabir, Saleh A Kabli, Kulvinder Singh Saini, and Roop Singh Bora. 2018. "Therapeutic Role of Ricinus Communis L. and Its Bioactive Compounds in Disease Prevention and Treatment." *Asian Pacific Journal of Tropical Medicine* 11 (3): 177.

Adkar, Prafulla, Amita Dongare, Shirishkumar Ambavade, and V H Bhaskar. 2014. "Trapa Bispinosa Roxb.: A Review on Nutritional and Pharmacological Aspects." *Advances in Pharmacological Sciences* 2014, 959830.

Ahmad, A, A Aliyu, A Abdulazeez, A Ahmadu, A Ahmad. 1983. "Explanation on Historical Literature of Arabs for Secondary Schools." In: Part 1:212. Alexander (p. 212), Egypt.

Ahmad, Aftab, Asif Husain, Mohd Mujeeb, Shah Alam Khan, Abul Kalam Najmi, Nasir Ali Siddique, Zoheir A Damanhouri, and Firoz Anwar. 2013. "A Review on Therapeutic Potential of Nigella Sativa: A Miracle Herb." *Asian Pacific Journal of Tropical Biomedicine* 3 (5): 337–52.

Ahmed, Rashid, Muhammad Tariq, Maria Hussain, Anisa Andleeb, Muhammad Shareef Masoud, Imran Ali, Fatima Mraiche, and Anwarul Hasan. 2019. "Phenolic Contents-Based Assessment of Therapeutic Potential of Syzygium Cumini Leaves Extract." *Plos One* 14 (8): e0221318.

Akhtar, Saeed, Abdur Rauf, Muhammad Imran, Muhammad Qamar, Muhammad Riaz, and Mohammad S Mubarak. 2017. "Black Carrot (Daucus Carota L.), Dietary and Health Promoting Perspectives of Its Polyphenols: A Review." *Trends in Food Science & Technology* 66: 36–47.

Al-Qahtani, Wahidah H, Yuvaraj Dinakarkumar, Selvaraj Arokiyaraj, Vigneshwar Saravanakumar, Jothi Ramalingam Rajabathar, Kowsalya Arjun, P K Gayathri, and Jimmy Nelson Appaturi. 2022. "Phyto-Chemical and Biological Activity of Myristica Fragrans, an Ayurvedic Medicinal Plant in Southern India and Its Ingredient Analysis." *Saudi Journal of Biological Sciences* 29 (5): 3815–21.

Alafe, A O, T O Elufioye, O S Faborode, and J O Moody. 2014. "Anti-Inflamatory and Analgesic Activities of Securidaca Longepedunculata Fers (Polygalaceae) Leaf and Stem Bark Methanolic Extract." *African Journal of Biomedical Research* 17 (3): 187–91.

Alam, K, O Hoq, and S Uddin. 2016. "Medicinal Plant Allium Sativum. A Review." *Journal of Medicinal Plant Studies* 4 (6): 72–79.

ALAsmari, Khalid Mushabbab, Isam M Abu Zeid, and Atef M Al-Attar. 2020. "Medicinal Properties of Arabica Coffee (Coffea Arabica) Oil: An Overview." *Advancements in Life Sciences* 8 (1): 20–29.

Almubayedh, Hanine, Reem Albannay, Kawthar Alelq, Rizwan Ahmad, Niyaz Ahmad, and Atta Abbas Naqvi. 2018. "Clinical Uses and Toxicity of Atropa Belladonna; an Evidence Based Comprehensive Retrospective Review." *Bioscience Biotechnology Research Communications* 11: 41–48.

Anbalagan, K, and J Sadique. 1981. "Influence of an Indian Medicine (Ashwagandha) on Acute-Phase Reactants in Inflammation." *Indian Journal of Experimental Biology* 19 (3): 245–49.

Arnold, J E Michael, and M Ruiz Pérez. 2001. "Can Non-Timber Forest Products Match Tropical Forest Conservation and Development Objectives?" *Ecological Economics* 39 (3): 437–47.

Asgarpanah, Jinous, and Elaheh Haghighat. 2012. "A Review of Phytochemistry and Medicinal Properties of Jujube (Ziziphus Vulgaris L.)." *Journal of Pharmaceutical and Health Sciences* 1(4): 89–97.

Ash, R. 2005. "Every Subject on Earth." In Russell Ash (Ed.) *Whitaker's World of Facts*. London: A & C Black Publishers Ltd., 320.

Ashur, A. 1986. "Herbs Are Your Natural Doctor; Treatment with Herbs and Plants." *Written in Arabic (Translated from Arabic into English by SA Sagamuwan). Ibn Sina Bookshop, Printing–Publishing–Distributing-Exporting, Heliopolis, Cairo, Egypt*, 192.

Aumeeruddy, Muhammad Zakariyyah, Gokhan Zengin, and Mohamad Fawzi Mahomoodally. 2018. "A Review of the Traditional and Modern Uses of Salvadora Persica L.(Miswak): Toothbrush Tree of Prophet Muhammad." *Journal of Ethnopharmacology* 213: 409–44.

Awoyemi, O K, I A Abdulkarim, E E Ewa, and A R Aduloju. 2012. "Ethnobotanical Assessment of Herbal Plants in South-Western Nigeria." *Academic Research International* 2 (3): 50.

Bahrami, Rosita, Ali Ghobadi, Nasim Behnoud, and Elham Akhtari. 2018. "Medicinal Properties of Daucus Carota in Traditional Persian Medicine and Modern Phytotherapy." *Journal of Biochemical Technology* 9 (2): 107–14.

Balunas, Marcy J, and A Douglas Kinghorn. 2005. "Drug Discovery from Medicinal Plants." *Life Sciences* 78 (5): 431–41.

Bisht, V K, J S Negi, A K Bhandari, and R C Sundriyal. 2011. "Amomum Subulatum Roxb: Traditional, Phytochemical and Biological Activities-An Overview." *African Journal of Agricultural Research* 6 (24): 5386–90.

Blowman, K, M Magalhães, M F L Lemos, C Cabral, and I M Pires. 2018. "Anticancer Properties of Essential Oils and Other Natural Products." *Evidence-Based Complementary and Alternative Medicine* 2018, 3149362. https://doi.org/10.1155/2018/3149362

Cao, Bo, Xi-Chuan Wei, Xiao-Rong Xu, Hai-Zhu Zhang, Chuan-Hong Luo, Bi Feng, Run-Chun Xu, Sheng-Yu Zhao, Xiao-Juan Du, and Li Han. 2019. "Seeing the Unseen of the Combination of Two Natural Resins, Frankincense and Myrrh: Changes in Chemical Constituents and Pharmacological Activities." *Molecules* 24 (17): 3076.

Chakraborty, Pritha, Deblina Dam, and Jayanthi Abraham. 2016. "Bioactivity of Lanthanum Nanoparticle Synthesized Using Trigonella Foenum-Graecum Seed Extract." *Journal of Pharmaceutical Sciences and Research* 8 (11): 1253.

Chandrasekaran, Suruthi, Vijaya Bharathi Rajkishore, and Radha Ramalingam. 2015. "Vigna Unguiculata-an Overall Review." *Research Journal of Pharmacognosy and Phytochemistry* 7 (4): 219.

Chang, Joseph. 2000. "Medicinal Herbs: Drugs or Dietary Supplements?" *Biochemical Pharmacology* 59 (3): 211–19.

Chaouche, T M, F Haddouchi, R Ksouri, F Medini, and F Atik-Bekara. 2013. "In Vitro Evaluation of Antioxidant Activity of the Hydro-Methanolic Extracts of Juniperus Oxycedrus Subsp. Oxycedrus." *Phytothérapie* 11 (4): 244–49.

Chaudhary, Gagandeep, Sandeep Goyal, and Priyanka Poonia. 2010. "Lawsonia Inermis Linnaeus: A Phytopharmacological Review." *International Journal of Pharmaceutical Sciences and Drug Research* 2 (2): 91–98.

Chauhan, Rajani, K M Ruby, Aastha Shori, and Jaya Dwivedi. 2012. "Solanum Nigrum with Dynamic Therapeutic Role: A Review." *International Journal of Pharmaceutical Sciences Review and Research* 15 (1): 65–71.

Chen, Yi-Bing, Xiao-Fang Tong, Junge Ren, Chun-Quan Yu, and Yuan-Lu Cui. 2019. "Current Research Trends in Traditional Chinese Medicine Formula: A Bibliometric Review from 2000 to 2016." *Evidence-Based Complementary and Alternative Medicine* 2019, 3961395. https://doi.org/10.1155/2019/3961395.

Choudhary, Shailja, and Gitika Chaudhary. 2021. "Sandalwood (Santalum Album): Ancient Tree with Significant Medicinal Benefits." *International Journal of Ayurveda and Pharma Research* 9(4): 90–99.

Dabas, Deepti, Rachel M Shegog, Gregory R Ziegler, and Joshua D Lambert. 2013. "Avocado (Persea Americana) Seed as a Source of Bioactive Phytochemicals." *Current Pharmaceutical Design* 19 (34): 6133–40.

Debnath, Bikash, Waikhom Somraj Singh, and Kuntal Manna. 2021. "A Phytopharmacological Review on Ananas Comosus." *Advances in Traditional Medicine*, 1–8. https://doi.org/10.1007/s13596-021-00563-w

Devi, P Uma. 1996. "Withania Somnifera Dunal (Ashwagandha): Potential Plant Source of a Promising Drug for Cancer Chemotherapy and Radiosensitization." *Indian Journal of Experimental Biology* 34 (10): 927–32.

Dhar, M K, S Kaul, S Sareen, and A K Koul. 2005. "Plantago Ovata: Genetic Diversity, Cultivation, Utilization and Chemistry." *Plant Genetic Resources* 3 (2): 252–63.

Dongling, Liu, Wang Yinquan, and Tian Ling. 2017. "Medicinal Plants in the Northwestern China and Their Medicinal Uses." *Aromatic Medicinal Plants Back to Nature*, 215–18. InTech. DOI: 10.5772/66739.

Dorri, Mahyar, Shirin Hashemitabar, and Hossein Hosseinzadeh. 2018. "Cinnamon (Cinnamomum Zeylanicum) as an Antidote or a Protective Agent against Natural or Chemical Toxicities: A Review." *Drug and Chemical Toxicology* 41 (3): 338–51.

Duke, James A. 2002. *Handbook of Medicinal Herbs*. CRC Press.

El-Saber Batiha, Gaber, Amany Magdy Beshbishy, Amany El-Mleeh, Mohamed M. Abdel-Daim, and Hari Prasad Devkota. 2020. "Traditional Uses, Bioactive Chemical Constituents, and Pharmacological and Toxicological Activities of Glycyrrhiza Glabra L.(Fabaceae)." *Biomolecules* 10 (3): 352.

Elgood, Cyril. 2010. *A Medical History of Persia and the Eastern Caliphate: From the Earliest Times until the Year AD 1932*. Cambridge University Press.

Enejoh, Onyilofe Sunday, Ibukun Oladejo Ogunyemi, Madu Smart Bala, Isaiah Sotonye Oruene, Mohammed Musa Suleiman, and Suleiman Folorunsho Ambali. 2015. "Ethnomedical Importance of Citrus Aurantifolia (Christm) Swingle." *The Pharma Innovation* 4 (8, Part A): 1.

Espitia-Hernández, Pilar, Monica L Chavez Gonzalez, Juan A Ascacio-Valdés, Desiree Dávila-Medina, Antonio Flores-Naveda, Teresinha Silva, Xochitl Ruelas Chacon, and Leonardo Sepúlveda. 2022. "Sorghum

(Sorghum Bicolor L.) as a Potential Source of Bioactive Substances and Their Biological Properties." *Critical Reviews in Food Science and Nutrition* 62 (8): 2269–80.

Farooqi, Mohammad I H. 1998. *Medicinal Plants in the Traditions of Prophet Muhammad: Medicinal, Aromatic and Food Plants Mentioned in the Traditions of Prophet Muhammad (SAAS.* Sidrah Publishers.

Fazal, Syed Sufiyan, and Rajeev K Singla. 2012. "Review on the Pharmacognostical & Pharmacological Characterization of Apium Graveolens Linn." *Indo Global Journal of Pharmaceutical Sciences* 2 (1): 36–42.

Fennell, C W, K L Lindsey, L J McGaw, S G Sparg, G I Stafford, E E Elgorashi, O M Grace, and J Van Staden. 2004. "Assessing African Medicinal Plants for Efficacy and Safety: Pharmacological Screening and Toxicology." *Journal of Ethnopharmacology* 94 (2–3): 205–17.

Florou-Paneri, Panagiota, Efterpi Christaki, and Ilias Giannenas. 2019. *Feed Additives: Aromatic Plants and Herbs in Animal Nutrition and Health.* Academic Press.

Fung, Foon Yin, and Yeh Ching Linn. 2015. "Developing Traditional Chinese Medicine in the Era of Evidence-Based Medicine: Current Evidences and Challenges." *Evidence-Based Complementary and Alternative Medicine* 2015, 425037. https://doi.org/10.1155/2015/425037.

Giacometti, Jasminka, Danijela Bursać Kovačević, Predrag Putnik, Domagoj Gabrić, Tea Bilušić, Greta Krešić, Višnja Stulić, Francisco J Barba, Farid Chemat, and Gustavo Barbosa-Cánovas. 2018. "Extraction of Bioactive Compounds and Essential Oils from Mediterranean Herbs by Conventional and Green Innovative Techniques: A Review." *Food Research International* 113: 245–62.

Goyal, Shivangi, Nidhi Gupta, and Sreemoyee Chatterjee. 2016. "Investigating Therapeutic Potential of Trigonella Foenum-Graecum L. as Our Defense Mechanism against Several Human Diseases." *Journal of Toxicology* 2016, 1250387. https://doi.org/10.1155/2016/1250387.

Grover, J K, S Yadav, and V Vats. 2002. "Medicinal Plants of India with Anti-Diabetic Potential." *Journal of Ethnopharmacology* 81 (1): 81–100.

Gunjan, Manish, Thein Win Naing, Rahul Singh Saini, A Ahmad, Jegathambigai Rameshwar Naidu, and Ishab Kumar. 2015. "Marketing Trends & Future Prospects of Herbal Medicine in the Treatment of Various Disease." *World Journal of Pharmaceutical Research* 4 (9): 132–55.

Guo, Shuangshuang, Yan Ge, and Kriskamol Na Jom. 2017. "A Review of Phytochemistry, Metabolite Changes, and Medicinal Uses of the Common Sunflower Seed and Sprouts (Helianthus Annuus L.)." *Chemistry Central Journal* 11 (1): 1–10.

Ha, Manh Tuan, Ngoc Khanh Vu, Thu Huong Tran, Jeong Ah Kim, Mi Hee Woo, and Byung Sun Min. 2020. "Phytochemical and Pharmacological Properties of Myristica Fragrans Houtt.: An Updated Review." *Archives of Pharmacal Research* 43 (11): 1067–92.

Hamidpour, Rafie, Soheila Hamidpour, and Grant Elias. 2017. "Rosmarinus Officinalis (Rosemary): A Novel Therapeutic Agent for Antioxidant, Antimicrobial, Anticancer, Antidiabetic, Antidepressant, Neuroprotective, Anti-Inflammatory, and Anti-Obesity Treatment." *Biomedical Journal of Scientific and Technical Research* 1 (4): 1–6.

Hashmat, Imam, Hussain Azad, and Ajij Ahmed. 2012. "Neem (Azadirachta Indica A. Juss)-A Nature's Drugstore: An Overview." *International Research Journal of Biological Sciences* 1 (6): 76–79.

Hashmi, Muhammad Ali, Afsar Khan, Muhammad Hanif, Umar Farooq, and Shagufta Perveen. 2015. "Traditional Uses, Phytochemistry, and Pharmacology of Olea Europaea (Olive)." *Evidence-Based Complementary and Alternative Medicine* 2015, 541591. https://doi.org/10.1155/2015/541591.

Hauck, Barbara, Joe A Gallagher, S Michael Morris, David Leemans, and Ana L Winters. 2014. "Soluble Phenolic Compounds in Fresh and Ensiled Orchard Grass (Dactylis Glomerata L.), a Common Species in Permanent Pastures with Potential as a Biomass Feedstock." *Journal of Agricultural and Food Chemistry* 62 (2): 468–75.

Henrotin, Yves, Fabian Priem, and Ali Mobasheri. 2013. "Curcumin: A New Paradigm and Therapeutic Opportunity for the Treatment of Osteoarthritis: Curcumin for Osteoarthritis Management." *Springerplus* 2 (1): 1–9.

Hongal, Sudhir, Nilesh Arjun Torwane, Goel Pankaj, B R Chandrashekhar, and Abhishek Gouraha. 2014. "Role of Unani System of Medicine in Management of Orofacial Diseases: A Review." *Journal of Clinical and Diagnostic Research: JCDR* 8 (10): ZE12.

Hosseinzadeh, Saleh, Azizollah Jafarikukhdan, Ahmadreza Hosseini, and Raham Armand. 2015. "The Application of Medicinal Plants in Traditional and Modern Medicine: A Review of Thymus Vulgaris." *International Journal of Clinical Medicine* 6 (09): 635.

Hussain, Abdullah I, Hassaan A Rathore, Munavvar Z A Sattar, Shahzad A S Chatha, Satyajit D Sarker, and Anwar H Gilani. 2014. "Citrullus Colocynthis (L.) Schrad (Bitter Apple Fruit): A Review of Its Phytochemistry, Pharmacology, Traditional Uses and Nutritional Potential." *Journal of Ethnopharmacology* 155 (1): 54–66.

Iranshahy, Milad, and Mehrdad Iranshahi. 2011. "Traditional Uses, Phytochemistry and Pharmacology of Asafoetida (Ferula Assa-Foetida Oleo-Gum-Resin)—A Review." *Journal of Ethnopharmacology* 134 (1): 1–10.

Jainu, Mallika, and C S Devi. 2004. "Antioxidant Effect of Methanolic Extract of Solanum Nigrum Berries on Aspirin Induced Gastric Mucosal Injury." *Indian Journal of Clinical Biochemistry* 19 (1): 57–61.

Jaiswal, Yogini S, and Leonard L Williams. 2017. "A Glimpse of Ayurveda–The Forgotten History and Principles of Indian Traditional Medicine." *Journal of Traditional and Complementary Medicine* 7 (1): 50–53.

Jamshidi-Kia, Fatemeh, Zahra Lorigooini, and Hossein Amini-Khoei. 2018. "Medicinal Plants: Past History and Future Perspective." *Journal of Herbmed Pharmacology* 7 (1): 1–7.

Jones, Francis Avery. 1996. "Herbs–Useful Plants. Their Role in History and Today." *European Journal of Gastroenterology & Hepatology* 8 (12): 1227–31.

Joshi, Bishnu, Sunil Lekhak, and Anuja Sharma. 2009. "Antibacterial Property of Different Medicinal Plants: Ocimum Sanctum, Cinnamomum Zeylanicum, Xanthoxylum Armatum and Origanum Majorana." *Kathmandu University Journal of Science, Engineering and Technology* 5 (1): 143–50.

Joshi, Kalpana, Preeti Chavan, Dnyaneshwar Warude, and Bhushan Patwardhan. 2004. "Molecular Markers in Herbal Drug Technology." *Current Science* 87: 159–65.

Kamboj, Ved P. 2000. "Herbal Medicine." *Current Science* 78 (1): 35–39.

Kankara, Sulaiman Sani, Mohd H Ibrahim, Muskhazli Mustafa, and Rusea Go. 2015. "Ethnobotanical Survey of Medicinal Plants Used for Traditional Maternal Healthcare in Katsina State, Nigeria." *South African Journal of Botany* 97: 165–75.

Kelly, Kate. 2009. *Early Civilizations: Prehistoric Times to 500 CE*. Infobase Publishing.

Khan, M R, K Ranjini, T K Godan, and S N Suresh. 2014. "Pharmacognostic Study and Phytochemical Investigation of Lycopersicon Esculentum (Tomato) Flower Extracts." *Research Journal of Pharmaceutical, Biological and Chemical Sciences* 5: 1691–98.

Khorasanchi, Zahra, Mojtaba Shafiee, Farnoush Kermanshahi, Majid Khazaei, Mikhail Ryzhikov, Mohammad Reza Parizadeh, Behnoush Kermanshahi, Gordon A Ferns, Amir Avan, and Seyed Mahdi Hassanian. 2018. "Crocus Sativus a Natural Food Coloring and Flavoring Has Potent Anti-Tumor Properties." *Phytomedicine* 43: 21–27.

Klimek-Szczykutowicz, Marta, Agnieszka Szopa, and Halina Ekiert. 2020. "Citrus Limon (Lemon) Phenomenon—a Review of the Chemistry, Pharmacological Properties, Applications in the Modern Pharmaceutical, Food, and Cosmetics Industries, and Biotechnological Studies." *Plants* 9 (1): 119.

Kolniak-Ostek, Joanna, Dagmara Kłopotowska, Krzysztof P Rutkowski, Anna Skorupińska, and Dorota E Kruczyńska. 2020. "Bioactive Compounds and Health-Promoting Properties of Pear (Pyrus Communis L.) Fruits." *Molecules* 25 (19): 4444.

Krishnaraju, Alluri V, Tayi V N Rao, Dodda Sundararaju, Mulabagal Vanisree, Hsin-Sheng Tsay, and Gottumukkala V Subbaraju. 2005. "Assessment of Bioactivity of Indian Medicinal Plants Using Brine Shrimp (Artemia Salina) Lethality Assay." *International Journal of Applied Science and Engineering* 3 (2): 125–34.

Kumar, Gaurav, L Karthik, and K V Bhaskara Rao. 2011. "A Review on Pharmacological and Phytochemical Properties of Zingiber Officinale Roscoe (Zingiberaceae)." *Journal of Pharmacy Research* 4 (9): 2963–66.

Lange, Dagmar. 2002. "Medicinal and Aromatic Plants: Trade, Production, and Management of Botanical Resources." *Acta Horticulturae* 629: 177–197.

Lim, Xin Yi, Bee Ping Teh, and Terence Yew Chin Tan. 2021. "Medicinal Plants in COVID-19: Potential and Limitations." *Frontiers in Pharmacology* 12: 611408.

Manvitha, Karkala, and Bhushan Bidya. 2014. "Review on Pharmacological Activity of Cymbopogon Citratus." *International Journal of Herbal Medicine* 6: 7.

Marinelli, Janet. 2004. *Plant: The Ultimate Visual Reference to Plants and Flowers of the World*. Dorling Kindersley.

Medha, Kshirsagar, and P S Patki. 2009. "Solanum Nigrum-a Review." *Biomed* 4 (2): 99–108.

Men, Xiao, Sun-Il Choi, Xionggao Han, Hee-Yeon Kwon, Gill-Woong Jang, Ye-Eun Choi, Sung-Min Park, and Ok-Hwan Lee. 2021. "Physicochemical, Nutritional and Functional Properties of Cucurbita Moschata." *Food Science and Biotechnology* 30 (2): 171–83.

Mishra, Lakshmi-Chandra, Betsy B Singh, and Simon Dagenais. 2000. "Scientific Basis for the Therapeutic Use of Withania Somnifera (Ashwagandha): A Review." *Alternative Medicine Review* 5 (4): 334–46.

Mohammed, Ghaidaa Jihadi, and Imad Hadi Hameed. 2018. "Pharmacological Activities: Hepatoprotective, Cardio Protective, Anti-Cancer and Anti-Microbial Activity of (Raphanus Raphanistrum Subsp. Sativus): A Review." *Indian Journal of Public Health Research and Development* 9 (3): 212–17.

Mukherjee, Pulok K. 2002. "Problems and Prospects for Good Manufacturing Practice for Herbal Drugs in Indian Systems of Medicine." *Drug Information Journal: DIJ/Drug Information Association* 36 (3): 635–44.

Muszynska, Bozena, Katarzyna Kala, Jacek Rojowski, Agata Grzywacz, and Włodzimierz Opoka. 2017. "Composition and Biological Properties of Agaricus Bisporus Fruiting Bodies-a Review." *Polish Journal of Food and Nutrition Sciences* 67 (3): 173–81.

Najafian, Younes, Shokouh Sadat Hamedi, Masoumeh Kaboli Farshchi, and Zohre Feyzabadi. 2018. "Plantago Major in Traditional Persian Medicine and Modern Phytotherapy: A Narrative Review." *Electronic Physician* 10 (2): 6390.

Neeraj, O Maheshwari, Khan Ayesha, and A Chopade Balu. 2013. "Rediscovering the Medicinal Properties of Datura Sp.: A Review." *Journal of Medicinal Plants Research* 7 (39): 2885–97.

Nejad, Solmaz Mohammadi, Hilal Özgüneş, and Nursen Başaran. 2017. "Pharmacological and Toxicological Properties of Eugenol." *Turkish Journal of Pharmaceutical Sciences* 14 (2): 201.

Nikolić, Miloš, Jasmina Glamočlija, Isabel C F R Ferreira, Ricardo C Calhelha, Ângela Fernandes, Tatjana Marković, Dejan Marković, Abdulhamed Giweli, and Marina Soković. 2014. "Chemical Composition, Antimicrobial, Antioxidant and Antitumor Activity of Thymus Serpyllum L., Thymus Algeriensis Boiss. and Reut and Thymus Vulgaris L. Essential Oils." *Industrial Crops and Products* 52: 183–90.

Osman Y. 2011. "Organic vs Chemical Fertilization of Medicinal Plants: A Concise Review of Researches." *Advances in Environmental Biology* 5 (2): 394–400.

Pan, Si-Yuan, Gerhard Litscher, Si-Hua Gao, Shu-Feng Zhou, Zhi-Ling Yu, Hou-Qi Chen, Shuo-Feng Zhang, Min-Ke Tang, Jian-Ning Sun, and Kam-Ming Ko. 2014. "Historical Perspective of Traditional Indigenous Medical Practices: The Current Renaissance and Conservation of Herbal Resources." *Evidence-Based Complementary and Alternative Medicine* 2014, 525340. https://doi.org/10.1155/2014/525340.

Pandey, Madan Mohan, Subha Rastogi, and Ajay Kumar Singh Rawat. 2007. "Saussurea Costus: Botanical, Chemical and Pharmacological Review of an Ayurvedic Medicinal Plant." *Journal of Ethnopharmacology* 110 (3): 379–90.

Parray, Shabir Ahmad, J U Bhat, Ghufran Ahmad, Najeeb Jahan, G Sofi, and M IFS. 2012. "Ruta Graveolens: From Traditional System of Medicine to Modern Pharmacology: An Overview." *American Journal of Pharm Tech Research* 2 (2): 239–52.

Partap, Sangh, Amit Kumar, Neeraj Kant Sharma, and K K Jha. 2012. "Luffa Cylindrica: An Important Medicinal Plant." *Journal of Natural Product and Plant Resources* 2 (1): 127–34.

Pathak Nimish, L, B Kasture Sanjay, M Bhatt Nayna, and D Rathod Jaimik. 2011. "Phytopharmacological Properties of Coriander Sativum as a Potential Medicinal Tree: An Overview." *Journal of Applied Pharmaceutical Science* 1 (4): 20–25.

Poonam, V, G Kumar, C S Reddy L, R Jain, S K Sharma, A K Prasad, and V S Parmar. 2011. "Chemical Constituents of the Genus Prunus and Their Medicinal Properties." *Current Medicinal Chemistry* 18 (25): 3758–824.

Posadzki, Paul, Leala Watson, and Edzard Ernst. 2013. "Contamination and Adulteration of Herbal Medicinal Products (HMPs): An Overview of Systematic Reviews." *European Journal of Clinical Pharmacology* 69 (3): 295–307.

Poudel, Saroj, Y Jyothi, R Narendra, and V Gowthami. 2020. "An Updated Review on Various Pharmacological Activities of Phaseolus Vulgaris Linn." *Drug Invention Today* 14 (3): 387–91.

Prajapati, Rakesh, Manisha Kalariya, Rahul Umbarkar, Sachin Parmar, and Navin Sheth. 2011. "Colocasia Esculenta: A Potent Indigenous Plant." *International Journal of Nutrition, Pharmacology, Neurological Diseases* 1 (2): 90.

Prakkash, M A Jaya, R Ragunathan, and Johney Jesteena. 2019. "Evaluation of Bioactive Compounds from Jasminum Polyanthum and Its Medicinal Properties." *Journal of Drug Delivery and Therapeutics* 9 (2): 303–10.

Qadir, Abdul, Faiyaz Shakeel, Athar Ali, and Md Faiyazuddin. 2020. "Phytotherapeutic Potential and Pharmaceutical Impact of Phoenix Dactylifera (Date Palm): Current Research and Future Prospects." *Journal of Food Science and Technology* 57 (4): 1191–1204.

Rahman, Syed Zillur. 1994. "Unani Medicine in India during 1901–1947." *New Delhi* 13 (1): 97–112.

Rajamurugan, R, N Selvaganabathy, S Kumaravel, C H Ramamurthy, V Sujatha, and C Thirunavukkarasu. 2012. "Polyphenol Contents and Antioxidant Activity of Brassica Nigra (L.) Koch. Leaf Extract." *Natural Product Research* 26 (23): 2208–10.

Ranjha, Muhammad Modassar Ali Nawaz, Bakhtawar Shafique, Lufeng Wang, Shafeeqa Irfan, Muhammad Naeem Safdar, Mian Anjum Murtaza, Muhammad Nadeem, Shahid Mahmood, Ghulam Mueen-ud-Din, and Hafiz Rehan Nadeem. 2023. "A Comprehensive Review on Phytochemistry, Bioactivity and

Medicinal Value of Bioactive Compounds of Pomegranate (Punica Granatum)." *Advances in Traditional Medicine* 23: 37–57.

Raskin, Ilya, David M Ribnicky, Slavko Komarnytsky, Nebojsa Ilic, Alexander Poulev, Nikolai Borisjuk, Anita Brinker, Diego A Moreno, Christophe Ripoll, and Nir Yakoby. 2002. "Plants and Human Health in the Twenty-First Century." *TRENDS in Biotechnology* 20 (12): 522–31.

Ren, Jun-ling, Ai-Hua Zhang, and Xi-Jun Wang. 2020. "Traditional Chinese Medicine for COVID-19 Treatment." *Pharmacological Research* 155: 104743.

Riaz, Ghazala, and Rajni Chopra. 2018. "A Review on Phytochemistry and Therapeutic Uses of Hibiscus Sabdariffa L." *Biomedicine & Pharmacotherapy* 102: 575–86.

Riaz, U, S Iqbal, M I Sohail, T Samreen, M Ashraf, F Akmal, A Siddiqui, I Ahmad, M Naveed, and N I Khan. 2021. "A Comprehensive Review on Emerging Importance and Economical Potential of Medicinal and Aromatic Plants (MAPs) in Current Scenario." *Pakistan Journal of Agricultural Research* 34: 381–92.

Robertson, E. 2008. *Medicinal Plants at Risk. Nature's Pharmacy, Our Treasure Chest: Why We Must Conserve Our Natural Heritage*. Center for Biological Diversity.

Roy, Anupam. 2015. "Pharmacological Activities of Indian Heliotrope (Heliotropium Indicum L.): A Review." *Journal of Pharmacognosy and Phytochemistry* 4 (3): 101–4.

Roy, Arpita, and Navneeta Bharadvaja. 2017. "A Review on Pharmaceutically Important Medical Plant: Plumbago Zeylanica." *Journal of Ayurvedic and Herbal Medicine* 3 (4): 225–28.

Sadanandan, A K, K V Peter, and S Hamza. 2002. "Role of Potassium Nutrition in Improving Yield and Quality of Spice Crops in India." International Potash Institute, Switzerland, 445–54.

Sahib, Ahmed Hadi Abdal, Ekhlas Al-Shareefi, and Imad Hadi Hameed. 2019. "Detection of Bioactive Compounds of Vitex Agnus-Castus and Citrus Sinensis Using Fourier-Transform Infrared Spectroscopic Profile and Evaluation of Its Anti-Microbial Activity." *Indian Journal of Public Health Research & Development* 10 (1): 954–59.

Salma, Salma, Yasmeen Shamsi, Saba Ansari, and Sadia Nikhat. 2020. "Ficus Carica L.: A Panacea of Nutritional and Medicinal Benefits." *Cellmed* 10 (1): 1.1–1.6.

Šamec, Dunja, Iva Pavlović, and Branka Salopek-Sondi. 2017. "White Cabbage (Brassica Oleracea Var. Capitata f. Alba): Botanical, Phytochemical and Pharmacological Overview." *Phytochemistry Reviews* 16 (1): 117–35.

Samuelsson, G. 2004. *Drugs of Natural Origin: A Textbook of Pharmacognosy*. 5th Swedish Pharmaceutical Press.

Schmidt, Barbara, David M Ribnicky, Alexander Poulev, Sithes Logendra, William T Cefalu, and Ilya Raskin. 2008. "A Natural History of Botanical Therapeutics." *Metabolism* 57: S3–9.

Shen, Ben. 2015. "A New Golden Age of Natural Products Drug Discovery." *Cell* 163 (6): 1297–300.

Siddique, Sarmad, Shamsa Nawaz, Faqir Muhammad, Bushra Akhtar, and Bilal Aslam. 2018. "Phytochemical Screening and In-Vitro Evaluation of Pharmacological Activities of Peels of Musa Sapientum and Carica Papaya Fruit." *Natural Product Research* 32 (11): 1333–36.

Singh, Neetu, Amrender Singh Rao, Abhishek Nandal, Sanjiv Kumar, Surender Singh Yadav, Showkat Ahmad Ganaie, and Balasubramanian Narasimhan. 2021. "Phytochemical and Pharmacological Review of Cinnamomum Verum J. Presl-a Versatile Spice Used in Food and Nutrition." *Food Chemistry* 338: 127773.

Singh, Rajinder, Subrata De, and Asma Belkheir. 2013. "Avena Sativa (Oat), a Potential Neutraceutical and Therapeutic Agent: An Overview." *Critical Reviews in Food Science and Nutrition* 53 (2): 126–44.

Singh, S S, S C Pandey, S Srivastava, V S Gupta, and B Patro. 2003. "Chemistry and Medicinal Properties of Tinospora Cordifolia (Guduchi)." *Indian Journal of Pharmacology* 35 (2): 83.

Singh, Uma M, Vijayta Gupta, V P Rao, Rakesh S Sengar, and M K Yadav. 2013. "A Review on Biological Activities and Conservation of Endangered Medicinal Herb Nardostachys Jatamansi." *International Journal of Medicinal and Aromatic Plants* 3 (1): 113–24.

Singletary, Keith W. 2018. "Basil: A Brief Summary of Potential Health Benefits." *Nutrition Today* 53 (2): 92–97.

Siva, M, K R Shanmugam, B Shanmugam, Subbaiah G Venkata, S Ravi, R K Sathyavelu, and K Mallikarjuna. 2016. "Ocimum Sanctum: A Review on the Pharmacological Properties." *International Journal of Basic Clinical Pharmacology* 5: 558–65.

Smith, R L, Samuel Monroe Cohen, J Doull, V J Feron, J I Goodman, L J Marnett, P S Portoghese, W J Waddell, B M Wagner, and R L Hall. 2005. "A Procedure for the Safety Evaluation of Natural Flavor Complexes Used as Ingredients in Food: Essential Oils." *Food and Chemical Toxicology* 43 (3): 345–63.

Solecki, Ralph S. 1975. "Shanidar IV, a Neanderthal Flower Burial in Northern Iraq." *Science* 190 (4217): 880–81.

Solovieva, V A. 2005. "Traditional Methods of Health Promotion [Narodnye Metody Ukreplenija Zdorovja]." *Izdatelsky Dom" Neva", St-Petersburg*, 352.

Stojiljković, Dragana, Ivana Arsić, and Vanja Tadić. 2016. "Extracts of Wild Apple Fruit (Malus Sylvestris (L.) Mill., Rosaceae) as a Source of Antioxidant Substances for Use in Production of Nutraceuticals and Cosmeceuticals." *Industrial Crops and Products* 80: 165–76.

Sultana, Sarwat, Shahid Perwaiz, Mohammad Iqbal, and Mohammad Athar. 1995. "Crude Extracts of Hepatoprotective Plants, Solanum Nigrum and Cichorium Intybus Inhibit Free Radical-Mediated DNA Damage." *Journal of Ethnopharmacology* 45 (3): 189–92.

Sumner, Judith. 2000. *The Natural History of Medicinal Plants*. Timber press.

Suryawanshi, Jyotsna A Saonere. 2011. "An Overview of Citrus Aurantium Used in Treatment of Various Diseases." *African Journal of Plant Science* 5 (7): 390–95.

Takooree, Heerasing, Muhammad Z Aumeeruddy, Kannan R R Rengasamy, Katharigatta N Venugopala, Rajesh Jeewon, Gokhan Zengin, and Mohamad F Mahomoodally. 2019. "A Systematic Review on Black Pepper (Piper Nigrum L.): From Folk Uses to Pharmacological Applications." *Critical Reviews in Food Science and Nutrition* 59 (sup1): S210–43.

Tu, Youyou. 2011. "The Discovery of Artemisinin (Qinghaosu) and Gifts from Chinese Medicine." *Nature Medicine* 17 (10): 1217–20.

Tuxill, John. 1999. *Nature's Cornucopia: Our Stake in Plant Diversity*. Washington, DC: Worldwatch Inst.

Variya, B C, A K Bakrania, & S S Patel. 2016. "Emblica officinalis (Amla): A Review for Its Phytochemistry, Ethnomedicinal Uses and Medicinal Potentials with Respect to Molecular Mechanisms." *Pharmacological Research*, 111: 180–200.

Verma, Rahul Kumar, Preeti Kumari, Rohit Kumar Maurya, Vijay Kumar, R B Verma, and Rahul Kumar Singh. 2018. "Medicinal Properties of Turmeric (Curcuma Longa L.): A Review." *International Journal of Chemical Studies* 6 (4): 1354–57.

Wiart, Christophe. 2006. *Medicinal Plants of Asia and the Pacific*. CRC Press.

Windisch, W, K Schedle, Ch Plitzner, and A Kroismayr. 2008. "Use of Phytogenic Products as Feed Additives for Swine and Poultry." *Journal of Animal Science* 86 (suppl_14): E140–48.

Wu, Ming-Shun, Levent Bless B Aquino, Marjette Ylreb U Barbaza, Chieh-Lun Hsieh, Kathlia A De Castro-Cruz, Ling-Ling Yang, and Po-Wei Tsai. 2019. "Anti-Inflammatory and Anticancer Properties of Bioactive Compounds from Sesamum Indicum L.—A Review." *Molecules* 24 (24): 4426.

Xu, Zhiguo. 2011. "Modernization: One Step at a Time." *Nature* 480 (7378): S90–92.

Xue, You-Lin, Jia-Nan Chen, Hao-Ting Han, Chun-Ju Liu, Qi Gao, Jia-Heng Li, Da-Jing Li, Masaru Tanokura, and Chun-Quan Liu. 2020. "Multivariate Analyses of the Physicochemical Properties of Turnip (Brassica Rapa L.) Chips Dried Using Different Methods." *Drying Technology* 38 (4): 411–19.

Yamada, Haruki, and Ikuo Saiki. 2005. *Juzen-Taiho-to (Shi-Quan-Da-Bu-Tang): Scientific Evaluation and Clinical Applications*. CRC Press.

Yasin, Bibi R, Hassan A N El-Fawal, and Shaker A Mousa. 2015. "Date (Phoenix Dactylifera) Polyphenolics and Other Bioactive Compounds: A Traditional Islamic Remedy's Potential in Prevention of Cell Damage, Cancer Therapeutics and Beyond." *International Journal of Molecular Sciences* 16 (12): 30075–90.

Zanoli, Paola, and Manuela Zavatti. 2008. "Pharmacognostic and Pharmacological Profile of Humulus Lupulus L." *Journal of Ethnopharmacology* 116 (3): 383–96.

Zhang, Hongxia, John Birch, Jinjin Pei, Zheng Feei Ma, and Alaa El-Din Bekhit. 2019. "Phytochemical Compounds and Biological Activity in Asparagus Roots: A Review." *International Journal of Food Science & Technology* 54 (4): 966–77.

Zilani, Md Nazmul Hasan, Tamanna Sultana, S M Asabur Rahman, Md Anisuzzman, Md Amirul Islam, Jamil A Shilpi, and Md Golam Hossain. 2017. "Chemical Composition and Pharmacological Activities of Pisum Sativum." *BMC Complementary and Alternative Medicine* 17 (1): 1–9.

Zollman, Catherine, and Andrew Vickers. 1999. "Complementary Medicine in Conventional Practice." *BMJ* 319 (7214): 901–4.

7 Ethnopharmacology and Ethnopharmacognosy

Current Perspectives and Future Prospects

Riadh Badraoui, Arif Jamal Siddiqui, and Fevzi Bardakci
University of Ha'il

Hmed Ben-Nasr
University of Gafsa

CONTENTS

7.1 INTRODUCTION

Archeological records predict the use of medicinal plants since 3500 BCE (Nair et al., 2012). Since then, nature always leads to drug discovery by preparing from natural sources to discover new compounds. Furthermore, plants showed much better effects than chemically synthesized compounds, which justify the interest in both ethnopharmacology and ethnopharmacognosy throughout the history of human civilization. Usually, scientists use two major approaches to find an appropriate plant medication, which can be categorized either as a random search or a targeted search (Sen et al., 2015). The major goal of this search is to come up with a suitable plant or one of its parts for a particular disease/pain/discomfort. Random search is based on the collection and the screening of a set of plants from a specific area regardless of their taxonomic status, e.g. paclitaxel (Taxol®), which can be considered an important natural drug used to treat breast and ovarian cancer. Taxol®, as a life-saving compound (Von Hoff et al. 2013), was discovered accidentally by the National Cancer Institute while searching for something else in a randomized study. Taxol is produced from the bark of the Pacific yew tree (*Taxus brevifolia*) and expands the treatment options for several cancerous diseases. This compound is also used in pancreatic cancer, non-small cell lung cancer, and Kaposi's sarcoma. In fact, it results in cell death by blocking the cancer cell division (Weaver 2014).

DOI: 10.1201/b22842-7

Targeted search is considered a systematic approach based on the targeted selection of both the disease(s) and the studied plants. The selection of potential plants can be (i) based on the production of specific compounds; (ii) based on particular biological properties (antioxidant, antibacterial, and antiviral) of some plants that live in a particular area/ecosystem; or (iii) based on use by indigenous people in traditional medications. Thus, it could be deduced that the selection can be phylogenic, ecological, or ethnobotanical, respectively. The latter often resulted in positive and significant findings (Garima et al., 2020). In fact, evidence of use for medicinal purposes already exists. Furthermore, the design of new drugs and future medicines usually sounds promising.

7.2 SAFETY IN ETHNOPHARMACOLOGY AND ETHNOPHARMACOGNOSY AND IMPORTANCE OF DOUBLE-BLIND CLINICAL TRIALS

The world contains not only medicinal plants but also toxic plants, consumable or quite for maintaining/conservation of the earth. Several phytochemicals/extracts have been reported to show moderate-to-severe toxicity (Shi et al. 2021). Biochemical, histopathological, radiographic, and many other investigation methods exhibited urinary, dermal, and digestive disruptions following the consumption of some traditional and/or ethnobotanical products (Badraoui et al. 2007; He et al. 2021). As several pharmacological reports are only performed *in vitro-* or *in silico*-based studies, additional *in vivo* assays in experimental animal models are usually required to assess the side effects for the safety approval of phyto-therapeutic applications (Zhang et al., 2015). Thus, double-blind clinical trials sounds crucial for safe applications of phyto-therapy and to avoid any drug failure at advanced stages.

7.3 USEFUL PLANT SUBSTANCES IN ETHNOPHARMACOLOGY AND ETHNOPHARMACOGNOSY: PRIMARY AND SECONDARY METABOLITES

Metabolites are intermediate products of cellular metabolism catalyzed by various enzymes that naturally occur within the cells. Medicinal plants synthesize substances, which are useful in health promotion (Li et al., 2020). Usually, these beneficial compounds are aromatic substances, specifically phenols or ox-substituted derivatives (tannins). If the active compound that is responsible for the medicinal properties is the primary molecule or a parental molecule, it is categorized as a primary metabolite. Usually, primary metabolites include sugars (carbohydrates and glycoproteins) and fats, which are required for the plant's basic metabolic activities, such as growth and production. However, if the active compound is not directly involved in the medicinal properties rather than its metabolite, the compound is categorized as a secondary metabolite. Functions of the secondary metabolites within the plants include toxins and pheromones. They possess deleterious properties for insects or their attraction for pollination, respectively (Erb et al., 2020; Lohaus, 2022).

7.4 FROM ETHNOPHARMACOLOGY AND ETHNOPHARMACOGNOSY TO DRUG DESIGN

Nature's diversity is one of the biggest resources of therapeutic lead compounds. Traditionally used herbal remedies harbor a variety of bioactive compounds providing researchers with starting points for drug development. Ethnopharmacological investigations of plant preparations showed several health-promoting compounds. Traditional medicine strongly relies on the beneficial effects of herbal remedies.

The bioactive constituents of such plant preparations display a rich source for the discovery of novel compounds with significant potential for pharmacological applications and drug development. Besides small organic molecules, secondary plant metabolites have been identified and express a variety of beneficial effects, hence constituting a rich source for the discovery of bioactive compounds (Badraoui et al. 2022; Siddiqui et al. 2022). A range of intrinsic bioactivities has been reported for plant phytochemicals. Yet, some underlying molecular mechanisms, in

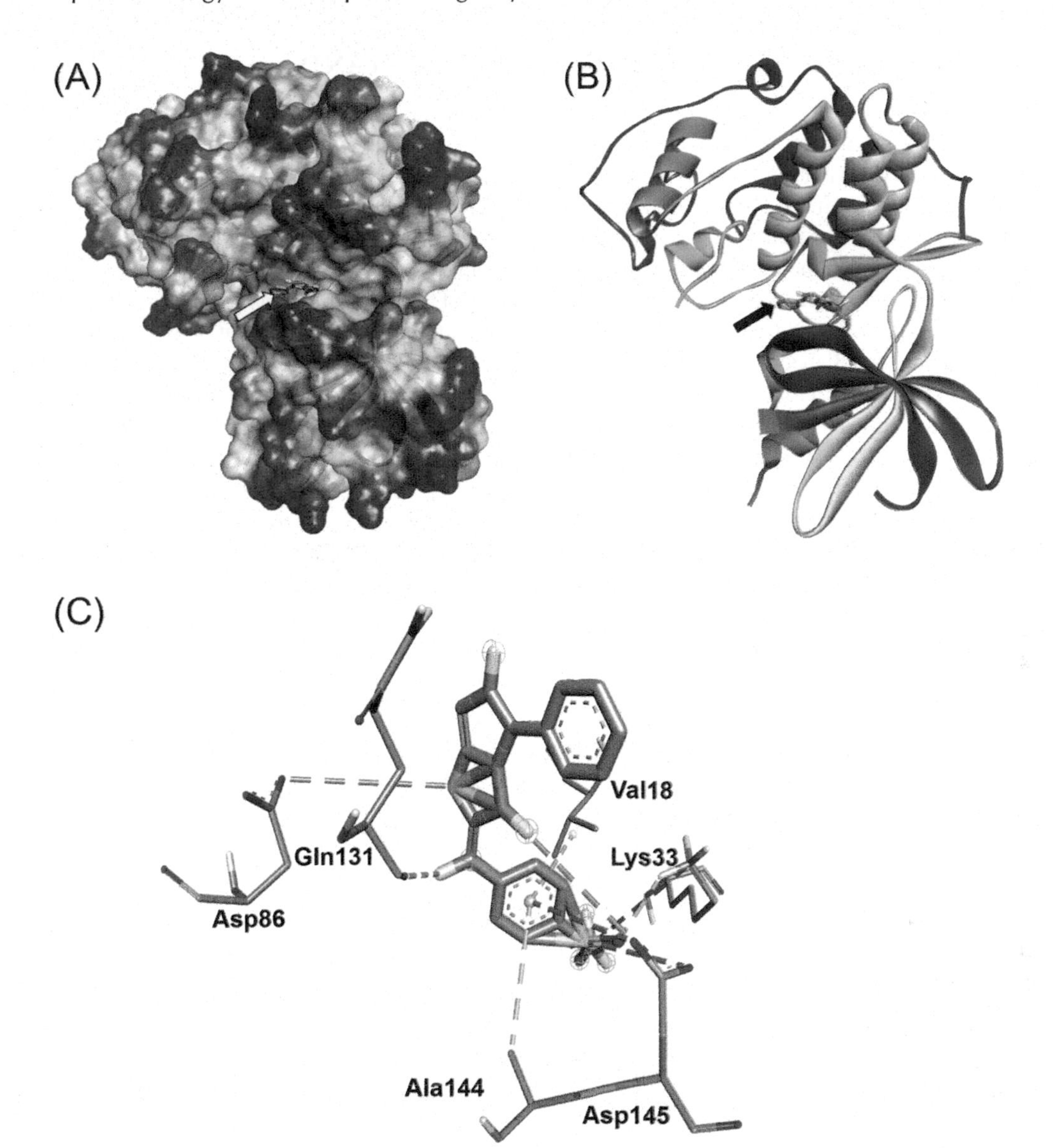

FIGURE 7.1 Illustration of a ligand/phytochemical-bound crystal structure of a targeted receptor as a result of a drug design approach. Tridimensional illustrations of the ligand receptor complex that exhibit hydrophobicity (A) and the ribbon structure of the protein (B) and the resulting molecular interactions (C). The arrow indicates the ligand docked to the pocket region of the targeted receptor.

particular anticancer, antidiabetic, antioxidant, and anti-inflammatory, have been reported. This includes several appropriate signaling pathways via targeting specific receptors (Koehbach et al., 2013; Cock, 2015). In fact, their actions closely relay to a ligand-bound receptor, which usually fits into the pocket region with a significant affinity, molecular interactions, and deep embedding. Analyses such as microscale thermophoresis, surface plasmon resonance, and isothermal titration calorimetry may provide additional values, support the drug design approaches, permit a better understanding of the thermodynamic parameters of the binding interactions (affinity, enthalpy, and stoichiometry), and detect the studied phytochemical movement and its molecular interactions in real time. Figure 7.1 illustrates a ligand-bound crystal structure of the receptor and the resultant molecular interactions.

7.5 PHARMACEUTICAL EFFECTS OF MEDICINAL PLANTS

Phytochemicals, called bioactive compounds, are produced as metabolites in plants and possess beneficial and health-promoting effects once consumed as nutrients. Phytochemicals contribute effectively to the formation of the color, smell, and taste of each plant. The latter is related to the phytochemical combinations. The interest in using plant phytochemicals and their synthetic materials as an alternative in the treatment of many diseases has increased (Nagulapalli Venkata et al., 2017). This trend has led to the development of new ethnobotanical and para-pharmaceutical markets. In this chapter's section, we will discuss the major confirmed heath-promoting and biological activities of plant preparations and phytochemicals. This includes anticancer, antioxidant, anti-inflammatory, antimicrobial, antidiabetic, and antihypertensive properties. These properties depend first of all on the drug-likeness, bioavailability, and pharmacokinetic properties of the phytochemicals themselves, which in turn depend on the physicochemical attributes of the phytocompounds and their interactions with the major cytochrome P450 (CYP) isoforms, specifically CYP1A2, CYP2C9, CYP2C19, CYP2D6, and CYP3A4 (Badraoui et al. 2022; Zammel, Oudadesse, et al. 2021). This can, at least, explain the common exploration of these parameters in several ethnopharmacological and ethnopharmacognostic investigations (Rahmouni et al. 2022; Jedli et al. 2022). Table 7.1 below illustrates the major pharmacokinetic assessment of drug-likeness of phytochemicals.

TABLE 7.1
Overview of Major Explored Lipophilicity, Bioavailability, and Pharmacokinetic Parameters based on the ADMET Properties of the Phytochemicals Used for Ethnopharmacological and Ethnopharmacognostic Purposes

Entry
Medicinal Chemistry
Consensus Log *Po/w*
Lipinski's Rule
Synthetic Accessibility
Bioavailability
Bioavailability Score
Lipophilicity
Molecular Size
Polarity
Insolubility
Insaturation
Flexibility
Pharmacokinetics
Gastrointestinal Absorption
Blood–Brain Barrier Permeant
Cytochrome P450 (CYP) Inhibition
CYP1A2
CYP2C19
CYP2C9
CYP2D6
CYP3A4
Log Kp (cm/s)
P-Glycoprotein Substrate

Numerous herbs have been reported to offer a variety of biological functions, including the ability to treat cancer, rheumatism, exhalation channel infections, menstrual abnormalities, wounds, skin illnesses, and monthly irregularities (Talib et al. 2020). About 250,000 plant species are known to exist in the plant world, but only approximately 10% have been studied and confirmed to be used as medicines for the treatment of different diseases (Ijaz et al. 2018; Iqbal et al. 2017). Table 7.2 illustrates some well-known medicinal plants, their major active compounds, and their confirmed beneficial effects.

7.5.1 Anticancer Effects

As one of the major causes of mortality, cancer was largely treated and tested for decades using ethnobotanical products. Many plant derivatives have been recognized for their anticancer properties (Siddiqui et al. 2022). Plant extracts and/or phytochemicals are also recognized as effective against different types of cancers. Recent reports confirmed their chemotherapeutic potentials. Despite oncology's achievements, cancer is still one of the deadliest diseases in the world. Recent studies estimated about 10 million deaths from cancer and around 19.5 million new cases (Sung et al. 2021). In the beginning, cancer is categorized as a localized disease, but via the metastatic process which includes invasion and migration, it can colonize different organs and/or the whole parts of the body (Fares et al. 2020; Badraoui et al. 2009; Badraoui, Ben-Nasr, et al. 2014). Through lymphatic or blood vessels, metastasis distributes cancer cells from the initial spot to several regions of the body (Badraoui, Boubakri, et al. 2014). The complex process of metastasis involves a number of phases that start with detachment followed by amassing and motility of malignant cells, which leads to their attachment to the endothelial cells, their extravasation, and their proliferation in some localized foci (Badraoui, Boubakri, et al. 2014; Badraoui et al. 2009). Due to apoptosis and tolerance to several cytotoxic treatments, metastasis remains one of the important causes of cancer-resulted mortality. Because current chemotherapy medicines showed ineffectiveness in completely eliminating cancer cells without harming healthy ones, patients with metastatic cancer have a high death and morbidity rate (Hchicha et al. 2021; Mhadhbi et al. 2022).

Humans get medicinal plants as a gift from nature to aid in their need for improved health. Mankind has always recognized and utilized natural resources as the primary source of therapeutic medications, which continue to be a source of powerful and effective bioactive compounds that may be employed as drugs (Akacha et al. 2022; Rahmouni et al. 2022; Mzid et al. 2017).

Many plant parts, including fruits, leaves, seeds, and flowers, contain various phytochemicals and/or their bioactive compounds with a range of pharmacological properties, including antidiabetic, antioxidant, anticancer, anti-osteoporotic, antimicrobial, hepatoprotective, antimalarial, anti-aging, immunomodulator, anti-inflammatory, antihypertensive, and others (Table 7.2).

Regarding cancer, numerous plant-based substances have been proven to possess pharmacological properties, either by activation of DNA repair mechanisms or inhibition of cancer-promoting proteins, cancer-related enzymes, and signaling pathways, which include topoisomerase enzyme, cyclooxygenase (COX), mitogen-activated protein kinase (MAPK)/extracellular signal-regulated kinase pathway (also known as Ras-Raf-MEK-ERK), protein kinase B (PKB, also known as Akt), cytokines, B-cell lymphoma 2 (Bcl-2), phosphoinositide 3-kinase (PI3K), cyclin-dependent kinases (CDK) –4, –2, CDC2, and mechanistic target of rapamycin (mTOR) (Singh et al. 2018).

Some plants are reported in this chapter together with their mode of action and a probable mechanism that might be involved in their ethnopharmacological effects. These plants have been explored and their use in the treatment and/or prevention of some diseases has been previously confirmed in several studies (Akacha et al., 2022; Jebahi et al., 2022; Mzid et al., 2017). Numerous studies have shown that phytochemicals, such as Taxol, can inhibit microtubule depolymerization and promote microtubule polymerization, which resulted in apoptosis and cell cycle arrest (Liu et al. 2015). As an antiproliferative drug, Taxol affects cancer cells in a distinctive way through its interactions with tubulin (El-Sayed et al. 2020). Additionally, some phytochemicals' unusual structures and

TABLE 7.2
Some Well-Known Medicinal Plants, Their Major Active Phytochemicals, and the Confirmed Beneficial Effects

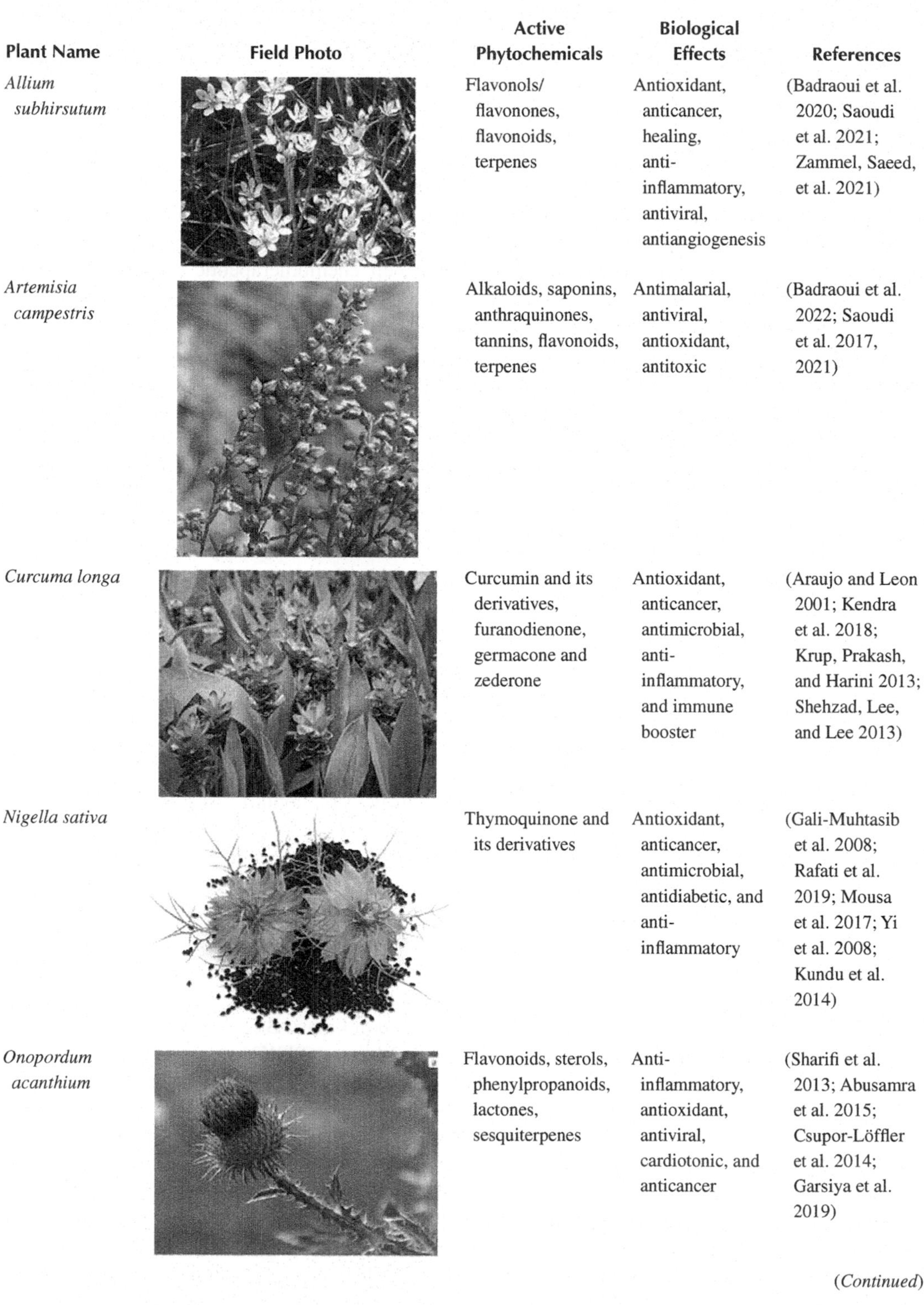

Plant Name	Field Photo	Active Phytochemicals	Biological Effects	References
Allium subhirsutum		Flavonols/ flavonones, flavonoids, terpenes	Antioxidant, anticancer, healing, anti-inflammatory, antiviral, antiangiogenesis	(Badraoui et al. 2020; Saoudi et al. 2021; Zammel, Saeed, et al. 2021)
Artemisia campestris		Alkaloids, saponins, anthraquinones, tannins, flavonoids, terpenes	Antimalarial, antiviral, antioxidant, antitoxic	(Badraoui et al. 2022; Saoudi et al. 2017, 2021)
Curcuma longa		Curcumin and its derivatives, furanodienone, germacone and zederone	Antioxidant, anticancer, antimicrobial, anti-inflammatory, and immune booster	(Araujo and Leon 2001; Kendra et al. 2018; Krup, Prakash, and Harini 2013; Shehzad, Lee, and Lee 2013)
Nigella sativa		Thymoquinone and its derivatives	Antioxidant, anticancer, antimicrobial, antidiabetic, and anti-inflammatory	(Gali-Muhtasib et al. 2008; Rafati et al. 2019; Mousa et al. 2017; Yi et al. 2008; Kundu et al. 2014)
Onopordum acanthium		Flavonoids, sterols, phenylpropanoids, lactones, sesquiterpenes	Anti-inflammatory, antioxidant, antiviral, cardiotonic, and anticancer	(Sharifi et al. 2013; Abusamra et al. 2015; Csupor-Löffler et al. 2014; Garsiya et al. 2019)

(*Continued*)

TABLE 7.2 (*Continued*)
Some Well-Known Medicinal Plants, Their Major Active Phytochemicals, and the Confirmed Beneficial Effects

Plant Name	Field Photo	Active Phytochemicals	Biological Effects	References
Opuntia ficus-indica		Flavonoids: kaempferol, and quercetin	Antitoxic, antiulcerogenic, chemoprotective, antioxidant, anticancer, and neuroprotective	(Akacha et al. 2022)
Psoralea corylifolia		Coumarins, psoralidin, flavonoids, and terpenes	Antioxidant, anticancer, antimicrobial, anti-inflammatory, and antidepressant	(Chopra, Dhingra, and Dhar 2013; Alam, Khan, and Asad 2018; Sharifi-Rad et al. 2020)
Taxus baccata		Paclitaxel/Taxol, taxine, taxusin, baccatin, phenols, and flavonoids	Anti-inflammatory, antirheumatic, anticancer, antimalarial, and antinociceptive	(Ahmadi et al. 2020; Asif et al. 2016)
Taxus chinensis (Chinese yew)		Paclitaxel/Taxol	Antioxidant and anticancer	(Shu et al. 2014; Expósito et al. 2009; Liu et al. 2015; Li et al. 2017; Diwaker and Gunjan 2012)
Teucrium polium		Flavonoids, isoprenoids, coumarin derivatives, phenols	Antioxidant, anticancer, anti-inflammatory, healing promotion, hepatoprotective, anti-genotoxic	(Alreshidi et al. 2020; Noumi et al. 2020; Rahmouni et al. 2018, 2022)

(*Continued*)

TABLE 7.2 (*Continued*)
Some Well-Known Medicinal Plants, Their Major Active Phytochemicals, and the Confirmed Beneficial Effects

Plant Name	Field Photo	Active Phytochemicals	Biological Effects	References
Urtica urens		Phenols, flavonoid	Anti-osteoporotic, antioxidant, antitoxic, antiviral, chemoprotective, anti-inflammatory	(Mzid et al. 2017; Gaafar et al. 2020)
Zingiber officinale Roscoe		Gingerols, phenols, and terpenes	Antioxidant, anticancer, antimicrobial, anti-inflammatory	(Banerjee et al. 2011; Ghafoor et al. 2020; Ansari et al. 2016; Kumar Gupta and Sharma 2014; Zammel, Saeed, et al. 2021; Jedli et al. 2022)

antitumoral potentials have attracted great interest from researchers from all over the world. In fact, some plant-based products enhance the autophagy of cancer cells, stimulate apoptosis via the mitochondrial signaling pathway or by blocking tumor growth, and interfere with NF- κB or MAPK/ERK kinase phosphorylation (Osafo, Mensah, and Yeboah 2017; Ansha and Mensah 2013; Alam, Khan, and Asad 2018; Koul et al. 2019) or G2/M phase of the cell cycle (Garsiya et al. 2019; Cho et al. 2017).

7.5.2 Antioxidant Effects

Free radicals are examples of reactive oxygen and reactive nitrogen species (ROS and RNS, respectively) that can lead to oxidative damage to DNA, lipids, and proteins of the cells (Badraoui et al. 2009; Nasr et al. 2009; Mzid et al. 2017; Akacha et al. 2022). In healthy conditions, the body's antioxidant system is able to scavenge these radicals, thus keeping a balance between oxidation and anti-oxidation. Oxidative stress is categorized as one of the major pathogenetic principles for the development of almost all diseases, particularly in old age. For reliable anti-oxidative purposes, dietary supplementation of a natural mixture of plant-based products, specifically flavonoids, tannins, and vitamins A, C, and E, has been recommended, as they are free from drawbacks once compared with the synthesized ones (Hässig et al. 1999).

7.5.3 Anti-Inflammatory Effects

The effectiveness of some natural products and their beneficial effects against inflammation associated with several diseases have recently attracted a lot of attention. In fact, plant-based natural products might be a suitable alternative treatment to prevent and treat inflammatory diseases.

Particularly, it was reported that phytochemicals possessed anti-inflammatory potentials and targeted both acute and chronic inflammations and their associated disorders (Saoudi et al. 2021; Zammel, Oudadesse, et al. 2021; Jedli et al. 2022). They offer a valuable source for developing and designing inflammatory pharmaceuticals (Badraoui et al. 2020). This is mainly due to the abundance of promising phytochemicals, such as polyphenols, flavonoids, and tannins, in plant-based products. The suppression of toll-like receptor (TLR)-6, IL-1, nuclear factor kappa B (NF-κB), and tumor necrosis factor (TNF) signaling is frequently linked to an anti-inflammatory potential (Barros et al. 2020). In fact, medicinal herbs suppressed TNF and MAPK and decreased the production of pro-inflammatory cytokines (Nguyen and Kim 2020; Zammel, Oudadesse, et al. 2021; Jedli et al. 2022).

TLRs are frequently targeted during anti-inflammatory therapy and are just one of the many receptors and pathways that make up the inflammatory cascade. Inflammation is a relatively prevalent illness. Human diseases such as enteritis, gastritis, bronchitis, and others ending in "-itis" are serious conditions, which associate with inflammation. Conventional anti-inflammatory medications have a disheartening profile of side effects, despite their critical function in the management of pain and inflammatory conditions (James and Hawkey 2003; Zammel, Oudadesse, et al. 2021). This indicates the requirement to secure anti-inflammatory medications based on bioactive components like flavonoids that are present in the majority of plants (Rengasamy et al. 2019; Rahmouni et al. 2022). Plants are considered promising alternatives to counteract oxidative injury and alleviate inflammation and several diseases due to the biological activities of their rich phytochemical contents such as flavonoids, polyphenols, and tannins (Rengasamy et al. 2019; Akacha et al. 2022; Jedli et al. 2022). These chemicals have a considerable effect on the inflammatory process, according to earlier research reports.

7.6 FUTURE DIRECTIONS IN ETHNOPHARMACOLOGY AND ETHNOPHARMACOGNOSY

Phytochemicals, which possess antioxidant, antiproliferative, antiangiogenic, proapoptotic, and antitumor activities, are currently providing huge breakthroughs and providing benefits for human health, preventing and treating several diseases. These plant products may not only be possible therapeutic agents/drugs with antioxidant, antiproliferative, anti-inflammatory, antimicrobial, and anticancer activities but also may act as antioxidants in healthy cells while acting as a pro-oxidant in pathological/abnormal cells. In fact, they may cause pro-oxidant-induced oxidative DNA damage and trigger apoptotic tumor cell killing via a downstream pathway. The beneficial properties appear to be mainly linked to the antioxidant potentials, which can be attributed to the phytochemical compounds/structures and free phenolic hydroxyls.

It has been reported that plant components have potent pharmacological effects in a variety of diseases, suggesting the possibility to be employed as therapeutic agents or drugs either alone or as a complementary medication. In cancer diseases, some phytochemicals stimulate apoptosis via the intrinsic apoptotic route, hence playing a significant role in cancer progression development and metastasis. They have been shown also to be effective against a variety of malignancies by decreasing the overproduction of ROS.

In pharmaceutical industries, combination therapy is an appealing alternative to drug development for resolving drug resistance, reducing unpleasant drug responses, and improving medication efficiency. In several diseases, due to associated complicated biology and course, the therapeutic scenarios necessitate a combined therapy. However, for a better translational result in clinical trials, medication procedure and development still need further research activities to provide appropriate efficacies and satisfactions. More in-depth clinical trials on ethnopharmacology and ethnopharmacognosy knowledge and application will be critical in the development of new therapeutic drugs. The clinical and preclinical findings satisfactorily showed a wide variety of antioxidant, anti-inflammatory, and antitumor effects against a variety of diseases through a wide number of biological processes. Phytochemicals and/or plant extracts, however, can either be used alone or in conjunction with other therapeutic agents/drugs for better control and management of targeted diseases.

REFERENCES

Abusamra, Yousef Abdel-Kareem, Michele Scuruchi, Sofiane Habibatni, Zenib Maammeri, Samir Benayache, Angela D'Ascola, Angela Avenoso, Giuseppe Maurizio Campo, and Edoardo Spina. 2015. "Evaluation of Putative Cytotoxic Activity of Crude Extracts from Onopordum Acanthium Leaves and Spartium Junceum Flowers against the U-373 Glioblastoma Cell Line" *Pakistan Journal of Pharmaceutical Sciences* 28(4): 1225–32.

Ahmadi, Kourosh, Seyed Jalil Alavi, Ghavamudin Zahedi Amiri, Seyed Mohsen Hosseini, Josep M Serra-Diaz, and Jens-Christian Svenning. 2020. "Patterns of Density and Structure of Natural Populations of Taxus Baccata in the Hyrcanian Forests of Iran" *Nordic Journal of Botany* 38(3): 1–10.

Akacha, Amira, Riadh Badraoui, Tarek Rebai, Lazhar Zourgui. 2022. "Effect of Opuntia Ficus Indica Extract on Methotrexate-Induced Testicular Injury: A Biochemical, Docking and Histological Study" *Journal of Biomolecular Structure and Dynamics* 40(10): 4341–51.

Alam, Fiaz, Gul Nawaz Khan, and Muhammad Hassham Hassan Bin Asad. 2018. "Psoralea Corylifolia L: Ethnobotanical, Biological, and Chemical Aspects: A Review" *Phytotherapy Research* 32(4): 597–615.

Alreshidi, Mousa, Emira Noumi, Lamjed Bouslama, Ozgur Ceylan, Vajid N Veettil, Mohd Adnan, Corina Danciu, Salem Elkahoui, Riadh Badraoui, and Khalid A Al-Motair. 2020. "Phytochemical Screening, Antibacterial, Antifungal, Antiviral, Cytotoxic, and Anti-Quorum-Sensing Properties of Teucrium Polium L. Aerial Parts Methanolic Extract" *Plants* 9(11): 1418.

Ansari, Jamal Akhtar, Mohammad Kaleem Ahmad, Abdul Rahman Khan, Nishat Fatima, Homa Jilani Khan, Namrata Rastogi, Durga Prasad Mishra, and Abbas Ali Mahdi. 2016. "Anticancer and Antioxidant Activity of Zingiber Officinale Roscoe Rhizome." *Indian Journal of Experimental Biology*, 54(11): 767–73.

Ansha, C, and K B Mensah. 2013. "A Review of the Anticancer Potential of the Antimalarial Herbal Cryptolepis Sanguinolenta and Its Major Alkaloid Cryptolepine" *Ghana Medical Journal* 47(3): 137–47.

Araujo, C A C, and L L Cruz Leon. 2001. "Biological Activities of Curcuma Longa L" *Memórias do Instituto Oswaldo* 96: 723–28.

Asif, Muhammad, Ghazala H Rizwani, Hina Zahid, Zahid Khan, and Rao Qasim. 2016. "Pharmacognostic Studies on Taxus Baccata L.: A Brilliant Source of Anti-Cancer Agents" *Pakistan Journal of Pharmaceutical Sciences* 29(1): 105–9.

Badraoui, Riadh, Hmed Ben-Nasr, Selma Amamou, Michèle Véronique El-May, Tarek Rebai. 2014. "Walker 256/B Malignant Breast Cancer Cells Disrupt Osteoclast Cytomorphometry and Activity in Rats: Modulation by α-Tocopherol Acetate" *Pathology-Research and Practice* 210(3): 135–41.

Badraoui, Riadh, Stéphane Blouin, Marie Françoise Moreau, Yves Gallois, Tarek Rebai, Zouhaier Sahnoun, Michel Baslé, and Daniel Chappard. 2009. "Effect of Alpha Tocopherol Acetate in Walker 256/B Cells-Induced Oxidative Damage in a Rat Model of Breast Cancer Skeletal Metastases" *Chemico-Biological Interactions* 182(2–3): 98–105.

Badraoui, Riadh, Mariem Boubakri, Maissa Bedbabiss, Hmed Ben-Nasr, and Tarek Rebai. 2014. "Walker 256/B Malignant Breast Cancer Cells Improve Femur Angioarchitecture and Disrupt Hematological Parameters in a Rat Model of Tumor Osteolysis" *Tumor Biology* 35(4): 3663–70.

Badraoui, Riadh, Tarek Rebai, Salem Elkahoui, Mousa Alreshidi, Vajid N. Veettil, Emira Noumi, Khaled A. Al-Motair, Kaïss Aouadi, Adel Kadri, and Vincenzo De Feo. 2020. "Allium Subhirsutum L. as a Potential Source of Antioxidant and Anticancer Bioactive Molecules: HR-LCMS Phytochemical Profiling, in Vitro and in Vivo Pharmacological Study" *Antioxidants* 9(10): 1003.

Badraoui, Riadh, Zouhaier Sahnoun, Nouha Bouayed Abdelmoula, Ahmed Hakim, Moncef Fki, Tarek Rebaï. 2007. "May Antioxidants Status Depletion by Tetradifon Induce Secondary Genotoxicity in Female Wistar Rats via Oxidative Stress?" *Pesticide Biochemistry and Physiology* 88(2): 149–55.

Badraoui, Riadh, Mongi Saoudi, Walid S Hamadou, Salem Elkahoui, Arif J Siddiqui, Jahoor M Alam, Arshad Jamal, Mohd Adnan, Abdel M E Suliemen, and Mousa M Alreshidi. 2022. "Antiviral Effects of Artemisinin and Its Derivatives against SARS-CoV-2 Main Protease: Computational Evidences and Interactions with ACE2 Allelic Variants" *Pharmaceuticals* 15(2): 129.

Banerjee, S, H I Mullick, J Banerjee, and A Ghosh. 2011. "Zingiber Officinale: 'A Natural Gold'" *International Journal of Pharmaceutical and Bio-Science* 2: 283–94.

Barros, Silvana, Ana Paula D Ribeiro, Steven Offenbacher, and Zvi G Loewy. 2020. "Anti-Inflammatory Effects of Vitamin E in Response to Candida Albicans" *Microorganisms* 8(6): 804.

Cho, Jung Hae, Young Hoon Joo, Eun Young Shin, Eun Ji Park, and Min Sik Kim. 2017. "Anticancer Effects of Colchicine on Hypopharyngeal Cancer" *Anticancer Research* 37(11): 6269–80.

Chopra, Bhawna, Ashwani Kumar Dhingra, and Kanaya Lal Dhar. 2013. "Psoralea Corylifolia L.(Buguchi)—Folklore to Modern Evidence" *Fitoterapia* 90: 44–56.

Cock IE. 2015. "The medicinal properties and phytochemistry of plants of the genus Terminalia (Combretaceae)" *Inflammopharmacology* 23(5): 203–29. doi: 10.1007/s10787-015-0246-z.

Csupor-Löffler, Boglárka, István Zupkó, Judit Molnár, Peter Forgo, and Judit Hohmann. 2014. "Bioactivity-Guided Isolation of Antiproliferative Compounds from the Roots of Onopordum Acanthium" *Natural Product Communications* 9(3): 1934578X1400900313.

Diwaker, A K, and Jadon Gunjan. 2012. "Plant-Based Anticancer Molecules: A Chemical and Biological Profile of Some Important Leads" *International Journal of Advanced Research in Pharmaceutical and Bio Sciences* 1(2): 16–25.

El-Sayed, Ashraf S A, Manal T El-Sayed, Amgad M Rady, Nabila Zein, Gamal Enan, Ahmed Shindia, Sara El-Hefnawy, Mahmoud Sitohy, and Basel Sitohy. 2020. "Exploiting the Biosynthetic Potency of Taxol from Fungal Endophytes of Conifers Plants; Genome Mining and Metabolic Manipulation" *Molecules* 25(13): 3000.

Erb, Matthias, and Daniel J Kliebenstein. 2020. "Plant Secondary Metabolites as Defenses, Regulators, and Primary Metabolites: The Blurred Functional Trichotomy" *Plant Physiology* 184(1):39–52. doi: 10.1104/pp.20.00433.

Expósito, O, M Bonfill, E Moyano, M Onrubia, M H Mirjalili, R M Cusido, and J Palazon. 2009. "Biotechnological Production of Taxol and Related Taxoids: Current State and Prospects" *Anti-Cancer Agents in Medicinal Chemistry* 9(1): 109–21.

Fares, Jawad, Mohamad Y Fares, Hussein H Khachfe, Hamza A Salhab, Youssef Fares. 2020. "Molecular Principles of Metastasis: A Hallmark of Cancer Revisited" *Signal Transduction and Targeted Therapy* 5(1): 1–17.

Gaafar, Alaa A, Sami I Ali, Omnia Kutkat, Ahmed M Kandeil, and Salwa M El-Hallouty. 2020. "Bioactive Ingredients and Anti-Influenza (H5N1), Anticancer, and Antioxidant Properties of Urtica Urens L" *Jordan Journal of Biological Sciences* 13: 647–57.

Garima, Singh, Passari Ajit Kumar, D Momin Marcy, Ravi Sakthivel, Singh Bhim Pratap and Kumar Nachimuthu Senthil. 2020. "Ethnobotanical survey of medicinal plants used in the management of cancer and diabetes" *Journal of traditional Chinese medicine = Chung i tsa chih ying wen pan* 40(6):1007–1017. doi: 10.19852/j.cnki.jtcm.2020.06.012.

Gali-Muhtasib, Hala, Matthias Ocker, Doerthe Kuester, Sabine Krueger, Zeina El-Hajj, Antje Diestel, Matthias Evert, et al. 2008. "Thymoquinone Reduces Mouse Colon Tumor Cell Invasion and Inhibits Tumor Growth in Murine Colon Cancer Models" *Journal of Cellular and Molecular Medicine* 12(1): 330–42.

Garsiya, Ekaterina Robertovna, Dmitryi Alexeevich Konovalov, Arnold Alexeevich Shamilov, Margarita Petrovna Glushko, and Kulpan Kenzhebaevna Orynbasarova. 2019. "Traditional Medicine Plant, Onopordum Acanthium L.(Asteraceae): Chemical Composition and Pharmacological Research" *Plants* 8(2): 40.

Ghafoor, Kashif, Fahad Al Juhaimi, Mehmet Musa Özcan, Nurhan Uslu, Elfadıl E Babiker, and Isam A Mohamed Ahmed. 2020. "Total Phenolics, Total Carotenoids, Individual Phenolics and Antioxidant Activity of Ginger (Zingiber Officinale) Rhizome as Affected by Drying Methods" *LWT* 126: 109354.

Hässig, A, W X Linag, H Schwabl, and K Stampfli. 1999. "Flavonoids and Tannins: Plant-Based Antioxidants with Vitamin Character" *Medical hypotheses* 52(5): 479–81.

Hchicha, Khouloud, Marcus Korb, Riadh Badraoui, and Houcine Naïli. 2021. "A Novel Sulfate-Bridged Binuclear Copper (II) Complex: Structure, Optical, ADMET and in Vivo Approach in a Murine Model of Bone Metastasis" *New Journal of Chemistry* 45(31): 13775–84.

He, Yisheng, Lin Zhu, Jiang Ma, and Ge Lin. 2021. "Metabolism-Mediated Cytotoxicity and Genotoxicity of Pyrrolizidine Alkaloids" *Archives of Toxicology* 95(6): 1917–42.

Hoff, Daniel D Von, Thomas Ervin, Francis P Arena, E Gabriela Chiorean, Jeffrey Infante, Malcolm Moore, Thomas Seay, Sergei A Tjulandin, Wen Wee Ma, and Mansoor N Saleh. 2013. "Increased Survival in Pancreatic Cancer with Nab-Paclitaxel Plus Gemcitabine" *New England Journal of Medicine* 369(18): 1691–1703.

Ijaz, Shakeel, Naveed Akhtar, Muhammad Shoaib Khan, Abdul Hameed, Muhammad Irfan, Muhammad Adeel Arshad, Sajid Ali, Muhammad Asrar. 2018. "Plant Derived Anticancer Agents: A Green Approach towards Skin Cancers" *Biomedicine and Pharmacotherapy* 103: 1643–51.

Iqbal, Javed, Banzeer Ahsan Abbasi, Tariq Mahmood, Sobia Kanwal, Barkat Ali, Sayed Afzal Shah, and Ali Talha Khalil. 2017. "Plant-Derived Anticancer Agents: A Green Anticancer Approach" *Asian Pacific Journal of Tropical Biomedicine* 7(12): 1129–50.

James, Martin W, and Christopher J Hawkey. 2003. "Assessment of Non-steroidal Anti-inflammatory Drug (NSAID) Damage in the Human Gastrointestinal Tract" *British Journal of Clinical Pharmacology* 56(-2): 146–55.

Jebahi, Samira, Ghada Ben Salah, Soufien Jarray, Mounir Naffati, Mohammad Ayaz Ahmad, Faten Brahmi, Mohd Saeed, Arif J Siddiqui, Khabir Abdelmajid and Riadh Badraoui. 2022. "Chitosan-Based Gastric Dressing Materials Loaded with Pomegranate Peel as Bioactive Agents: Pharmacokinetics and Effects on Experimentally Induced Gastric Ulcers in Rabbits" *Metabolites* 12(12), 1158.

Jedli, Olfa, Hmed Ben-Nasr, Nourhène Zammel, Tarek Rebai, Mongi Saoudi, Salem Elkahoui, Arshad Jamal, Arif J Siddiqui, Abdelmoneim E Sulieman, and Mousa M Alreshidi. 2022. "Attenuation of Ovalbumin-Induced Inflammation and Lung Oxidative Injury in Asthmatic Rats by Zingiber Officinale Extract: Combined in Silico and in Vivo Study on Antioxidant Potential, STAT6 and TNF-α Pathways" *3 Biotech* 12(9): 1–13.

Kendra, K Vigyan, I Preeti Kumari, R Kumar Maurya Scholar, V Kumar, R Kumar Verma, P Kumari, R Kumar Maurya, R Verma, and R Kumar Singh. 2018. "Medicinal Properties of Turmeric (Curcuma Longa L.): A Review" *International Journal of Chemical Studies* 6(4): 1354–57.

Koehbach, Johannes and Christian W Gruber. 2013. "From ethnopharmacology to drug design" *Communicative & integrative biology* 6(6):e27583. doi: 10.4161/cib.27583.

Koul, Bhupendra, Pooja Taak, Arvind Kumar, Anil Kumar, and Indraneel Sanyal. 2019. "Genus Psoralea: A Review of the Traditional and Modern Uses, Phytochemistry and Pharmacology" *Journal of Ethnopharmacology* 232: 201–26.

Krup, Vasavda, L Hedge Prakash, and A Harini. 2013. "Pharmacological Activities of Turmeric (Curcuma Longa Linn): A Review" *Journal of Homeopathy and Ayurvedic Medicine* 2(133): 2167-1206.100 0133.

Kumar Gupta, Subash, and Anand Sharma. 2014. "Medicinal Properties of Zingiber Officinale Roscoe-A Review" *Journal of Pharmaceutical and Biological Sciences* 9: 124–29.

Kundu, Juthika, Kyung-Soo Chun, Okezie I Aruoma, Joydeb Kumar Kundu. 2014. "Mechanistic Perspectives on Cancer Chemoprevention/Chemotherapeutic Effects of Thymoquinone" *Mutation Research/Fundamental and Molecular Mechanisms of Mutagenesis* 768: 22–34.

Li, Fengzhi, Tao Jiang, Qingyong Li, and Xiang Ling. 2017. "Camptothecin (CPT) and Its Derivatives Are Known to Target Topoisomerase I (Topl) as Their Mechanism of Action: Did We Miss Something in CPT Analogue Molecular Targets for Treating Human Disease Such as Cancer?" *American Journal of Cancer Research* 7(12): 2350.

Li, Yanqun, Dexin Kong, Ying Fu, Michael R Sussman, and Hong Wu. 2020. "The effect of developmental and environmental factors on secondary metabolites in medicinal plants" *Plant Physiology and Biochemistry* 148:80–89. doi: 10.1016/j.plaphy.2020.01.006.

Liu, Zhihui, Xiao Zheng, Jiajia Lv, Xiaowen Zhou, Qiong Wang, Xiaozhou Wen, Huan Liu, Jingyi Jiang, and Liling Wang. 2015. "Pharmacokinetic Synergy from the Taxane Extract of Taxus Chinensis Improves the Bioavailability of Paclitaxel" *Phytomedicine* 22(5): 573–78.

Lohaus G. 2022. "Review primary and secondary metabolites in phloem sap collected with aphid stylectomy" *Journal of Plant Physiology* 271:153645. doi: 10.1016/j.jplph.2022.153645.

Mhadhbi, Noureddine, Noureddine Issaoui, Walid S Hamadou, Jahoor M Alam, Abdelmonein S Elhadi, Mohd Adnan, Houcine Naïli, and Riadh Badraoui. 2022. "Physico-Chemical Properties, Pharmacokinetics, Molecular Docking and In-Vitro Pharmacological Study of a Cobalt (II) Complex Based on 2-Aminopyridine" *ChemistrySelect* 7(3): e202103592.

Mousa, HebatAlla Fathi Mohamed, Nesrin Kamal Abd-El-Fatah, Olfat Abdel-Hamid Darwish, Shehata Farag Shehata, and Shady Hassan Fadel. 2017. "Effect of Nigella Sativa Seed Administration on Prevention of Febrile Neutropenia during Chemotherapy among Children with Brain Tumors" *Child's Nervous System* 33(5): 793–800.

Mzid, Massara, Riadh Badraoui, Sameh Ben Khedir, Zouheir Sahnoun, Tarek Rebai. 2017. "Protective Effect of Ethanolic Extract of Urtica Urens L. against the Toxicity of Imidacloprid on Bone Remodeling in Rats and Antioxidant Activities" *Biomedicine and Pharmacotherapy* 91: 1022–41.

Nagulapalli Venkata, Kalyan C, Anand Swaroop, Debasis Bagchi, and Anupam Bishayee. 2017. A small plant with big benefits: Fenugreek (Trigonella foenum-graecum Linn.) for disease prevention and health promotion. *Molecular nutrition & food research* 61(6). doi: 10.1002/mnfr.201600950.

Nair, Rajesh, Senthy Sellaturay, and Seshadri Sriprasad. 2012. "The history of ginseng in the management of erectile dysfunction in ancient China (3500–2600 BCE)" *Indian journal of urology : IJU : journal of the Urological Society of India* 28(1):15–20. doi: 10.4103/0970-1591.94946.

Nasr, Hmed Ben, Hammami Serria, Selma Chaker, Badraoui Riadh, Sahnoun Zouheir, Jamoussi Kamel, Rebai Tarek, Zeghal Khaled. 2009. "Some Biological Effects of Scorpion Envenomation in Late Pregnant Rats" *Experimental and Toxicologic Pathology* 61(6): 573–80.

Nguyen, Anh Thu, and Ki-young Kim. 2020. "Inhibition of Proinflammatory Cytokines in Cutibacterium Acnes-Induced Inflammation in HaCaT Cells by Using Buddleja Davidii Aqueous Extract" *International Journal of Inflammation, 2020*, 8063289. https://doi.org/10.1155/2020/8063289.

Noumi, Emira, Mejdi Snoussi, El Hassane Anouar, Mousa Alreshidi, Vajid N Veettil, Salem Elkahoui, Mohd Adnan, Mitesh Patel, Adel Kadri, and Kaïss Aouadi. 2020. "HR-LCMS-Based Metabolite Profiling, Antioxidant, and Anticancer Properties of Teucrium Polium L. Methanolic Extract: Computational and in Vitro Study" *Antioxidants* 9(11): 1089.

Osafo, Newman, Kwesi Boadu Mensah, and Oduro Kofi Yeboah. 2017. "Phytochemical and Pharmacological Review of Cryptolepis Sanguinolenta (Lindl.) Schlechter" *Advances in Pharmacological Sciences, 2017*, 3026370. https://doi.org/10.1155/2017/3026370.

Rafati, Mohammadreza, Arash Ghasemi, Majid Saeedi, Emran Habibi, Ebrahim Salehifar, Mahmood Mosazadeh, and Monireh Maham. 2019. "Nigella Sativa L. for Prevention of Acute Radiation Dermatitis in Breast Cancer: A Randomized, Double-Blind, Placebo-Controlled, Clinical Trial" *Complementary Therapies in Medicine* 47: 102205.

Rahmouni, Fatma, Riadh Badraoui, Hmed Ben-Nasr, Fevzi Bardakci, Salem Elkahoui, Arif J Siddiqui, Mohd Saeed, Mejdi Snoussi, Mongi Saoudi, and Tarek Rebai. 2022. "Pharmacokinetics and Therapeutic Potential of Teucrium Polium against Liver Damage Associated Hepatotoxicity and Oxidative Injury in Rats: Computational, Biochemical and Histological Studies" *Life* 12(7): 1092.

Rahmouni, Fatma, Mongi Saoudi, Nahed Amri, Abdelfattah El-Feki, Tarek Rebai, Riadh Badraoui. 2018. "Protective Effect of Teucrium Polium on Carbon Tetrachloride Induced Genotoxicity and Oxidative Stress in Rats" *Archives of Physiology and Biochemistry* 124(1): 1–9.

Rengasamy, Kannan R R, Haroon Khan, Shanmugaraj Gowrishankar, Ricardo J L Lagoa, Fawzi M Mahomoodally, Ziyad Khan, Shanoo Suroowan, et al. 2019. "The Role of Flavonoids in Autoimmune Diseases: Therapeutic Updates" *Pharmacology & Therapeutics* 194: 107–31.

Saoudi, Mongi, Riadh Badraoui, Houda Bouhajja, Marwa Ncir, Fatma Rahmouni, Malek Grati, Kamel Jamoussi, Abdelfattah El Feki. 2017. "Deltamethrin Induced Oxidative Stress in Kidney and Brain of Rats: Protective Effect of Artemisia Campestris Essential Oil" *Biomedicine and Pharmacotherapy* 94: 955–63.

Saoudi, Mongi, Riadh Badraoui, Ahlem Chira, Mohd Saeed, Nouha Bouali, Salem Elkahoui, Jahoor M Alam, Choumous Kallel, and Abdelfattah El Feki. 2021. "The Role of Allium Subhirsutum L. in the Attenuation of Dermal Wounds by Modulating Oxidative Stress and Inflammation in Wistar Albino Rats" *Molecules* 26(16): 4875.

Sen, Tuhinadri and Samir Kumar Samanta. 2015. "Medicinal plants, human health and biodiversity: a broad review" *Advances in biochemical engineering/biotechnology* 147:59–110. doi: 10.1007/10_2014_273.

Sharifi, Niusha, Effat Souri, Seyed Ali Ziai, Gholamreza Amin, and Massoud Amanlou. 2013. "Discovery of New Angiotensin Converting Enzyme (ACE) Inhibitors from Medicinal Plants to Treat Hypertension Using an in Vitro Assay" *DARU Journal of Pharmaceutical Sciences* 21(1): 1–8.

Sharifi-Rad, Javad, Senem Kamiloglu, Balakyz Yeskaliyeva, Ahmet Beyatli, Mary Angelia Alfred, Bahare Salehi, Daniela Calina, Anca Oana Docea, Muhammad Imran, and Nanjangud Venaktesh Anil Kumar. 2020. "Pharmacological Activities of Psoralidin: A Comprehensive Review of the Molecular Mechanisms of Action" *Frontiers in Pharmacology* 11: 571459.

Shehzad, Adeeb, Jaetae Lee, and Young Sup Lee. 2013. "Curcumin in Various Cancers" *Biofactors* 39(1): 56–68.

Shi, Yinxian, Min Zhou, Yu Zhang, Yao Fu, Jianwen Li, and Xuefei Yang. 2021. "Poisonous Delicacy: Market-Oriented Surveys of the Consumption of Rhododendron Flowers in Yunnan, China" *Journal of Ethnopharmacology* 265: 113320.

Shu, Qijin, Minhe Shen, Binbin Wang, Qingli Cui, Xiaoying Zhou, and Luming Zhu. 2014. "Aqueous Extract of Taxus Chinensis (Pilger) Rehd Inhibits Lung Carcinoma A549 Cells through the Epidermal Growth Factor Receptor/Mitogen-Activated Protein Kinase Pathway in Vitro and in Vivo" *Journal of Traditional Chinese Medicine* 34(3): 293–301.

Siddiqui, Arif Jamal, Sadaf Jahan, Ritu Singh, Juhi Saxena, Syed Amir Ashraf, Andleeb Khan, Ranjay Kumar Choudhary, Santhanaraj Balakrishnan, Riadh Badraoui, and Fevzi Bardakci. 2022. "Plants in Anticancer Drug Discovery: From Molecular Mechanism to Chemoprevention" *BioMed Research International* 2022, 5425485. https://doi.org/10.1155/2022/5425485.

Singh, Swati, Manika Awasthi, Veda P Pandey, Upendra N Dwivedi. 2018. "Natural Products as Anticancerous Therapeutic Molecules with Special Reference to Enzymatic Targets Topoisomerase, COX, LOX and Aromatase" *Current Protein and Peptide Science* 19(3): 238–74.

Sung, Hyuna, Jacques Ferlay, Rebecca L Siegel, Mathieu Laversanne, Isabelle Soerjomataram, Ahmedin Jemal, and Freddie Bray. 2021. "Global Cancer Statistics 2020: GLOBOCAN Estimates of Incidence and Mortality Worldwide for 36 Cancers in 185 Countries." *CA: A Cancer Journal for Clinicians* 71(3): 209–49.

Talib, Wamidh H, Izzeddin Alsalahat, Safa Daoud, Reem Fawaz Abutayeh, and Asma Ismail Mahmod. 2020. "Plant-Derived Natural Products in Cancer Research: Extraction, Mechanism of Action, and Drug Formulation" *Molecules* 25(22): 5319.

Weaver, Beth A. 2014. "How Taxol/Paclitaxel Kills Cancer Cells" *Molecular Biology of the Cell* 25(18): 2677–81.

Yi, Tingfang, Sung-Gook Cho, Zhengfang Yi, Xiufeng Pang, Melissa Rodriguez, Ying Wang, Gautam Sethi, Bharat B Aggarwal, and Mingyao Liu. 2008. "Thymoquinone Inhibits Tumor Angiogenesis and Tumor Growth through Suppressing AKT and Extracellular Signal-Regulated Kinase Signaling Pathways" *Molecular Cancer Therapeutics* 7(7): 1789–96.

Zammel, Nourhene, Hassane Oudadesse, Ikram Allagui, Bertrand Lefeuvre, Tarek Rebai, and Riadh Badraoui. 2021. "Evaluation of Lumbar Vertebrae Mineral Composition in Rat Model of Severe Osteopenia: A Fourier Transform Infrared Spectroscopy (FTIR) Analysis" *Vibrational Spectroscopy* 115: 103279.

Zammel, Nourhene, Mohd Saeed, Nouha Bouali, Salem Elkahoui, Jahoor M Alam, Tarek Rebai, Mohd A Kausar, Mohd Adnan, Arif J Siddiqui, and Riadh Badraoui. 2021. "Antioxidant and Anti-Inflammatory Effects of Zingiber Officinale Roscoe and Allium Subhirsutum: In Silico, Biochemical and Histological Study" *Foods* 10(6): 1383.

Zhang, Jian-Ping, Guo-Wei Wang, Xin-Hui Tian , Yong-Xun Yang, Qing-Xin Liu, Li-Ping Chen, Hui-Liang Li, and Wei-Dong Zhang. 2015. "The genus Carpesium: a review of its ethnopharmacology, phytochemistry and pharmacology" *Journal of ethnopharmacology* 163:173–91. doi: 10.1016/j.jep.2015.01.027.

8 Phytochemistry and Biosynthesis of Phytochemicals

Stephano Hanolo Mlozi
University of Dar es Salaam

CONTENTS

8.1 INTRODUCTION

Plants have the miraculous natural ability to synthesize a voluminous and diverse range of chemical substances called phytochemicals. The scientific study of phytochemicals is called phytochemistry. Phytochemicals are commonly also known as natural products or secondary metabolites or bioactive compounds. Typically, plants biosynthesize phytochemicals ceaselessly for their own defence and ecological adaption. Phytochemicals are usually biosynthesized as an adaptation mechanism for self-defence against ultraviolet exposure, herbivores, pathogens, pests, and other geographical environmental threats. Accordingly, geographical location and ecological factors have a great influence on the type and structure of phytochemicals. This entails that phytochemical biosynthesized in different regions even of the same species may vary in terms of type and quantity. The biosynthesis of phytochemicals involves three major biological processes called biosynthetic pathways; these are the acetate pathway, shikimate pathway, and mevalonate pathway (Dewick 2009). Biosynthetic pathways result in a wide variety of phytochemicals, which are grouped based on their origin of precursors. Plant secondary metabolites are many but can be grouped into major three classes namely; terpenoids, alkaloids, and phenolics (Francesca et al. 2019; Hussein and El-anssary 2018; Mendoza and Silva 2018; Dewick 2009). For instance, terpenoids are derived from the mevalonate pathway, alkaloids are derived from the shikimate pathway, and phenolics are derived from acetate or a combination of pathways. These phytochemicals have a wide range of bioactivities, and many of them are important pharmaceutically.

Terpenoids (terpenes) are the most diverse group of phytochemicals found in plants, with over 50,000 distinct compounds (Ramawat and Mérillon 2013). They are non-saponifiable lipids from a chemical standpoint since fatty acids do not play a role in their synthesis. They're also called isoprenoids as isoprene is the basic chemical structural unit that makes them up (Ramawat and Mérillon 2013). Terpenoids are categorized depending on how many isoprene units they contain.

DOI: 10.1201/b22842-8

Hemiterpenes, which have a single isoprene unit and five carbons in their structure, are the simplest of all the classes. Hemiterpene is a common volatile product that emerges from photosynthetically active tissues. The nomenclature of terpenes depends on the number of isoprene units used to make up a compound. For instance, one made up of two units of isoprene is called monoterpenes, one with three units is sesquiterpenes, one with four units is diterpenes, one with six units is triterpenes, and one with eight units is tetraterpenes.

Phenolic compounds are one more diverse group of phytochemicals. Their chemically diverse group of phytochemicals is due to the several classes they comprise. These are chemical substances with a hydroxyl group directly linked to aromatic hydrocarbons. In most cases, phenol is the basic unit member of this group of phytocompounds. They include simple phenols, acidic phenols, acetophenones, benzophenones, xanthones, phenylacetic acids, hydroxycinnamic acids, flavonoids, coumarins, stilbenes, quinones, lignans, neolignans, and tannins (Cartea et al. 2011; Bhuyan and Basu 2017). The phenolic chemicals are generated in plant cells via the shikimate or acetate pathway, or both pathways. For example, orsellinic acid (originated as polyketides) is biosynthesized by the acetate pathway while flavonoids are biosynthesized through both shikimate and acetate pathways.

On the other hand, alkaloids are another large group of secondary metabolites that consist of molecules obtained from vascular aromatic plants. They are alkali by nature, and that's the origin of this name. Alkaloids are characterized as having a nitrogen atom in their structure. Aromatic plants biosynthesize a complex mixture of alkaloids, with one or more major constituents predominating. Even though their structures differ significantly, the metabolic origin of alkaloids present in a given plant is common. Nevertheless, some amino acids are essentially involved as precursors in the biosynthesis of alkaloids, whereby the foremost ones known as tyrosine, ornithine, tryptophan, nicotinic acid, lysine, and anthranilic acid (Dewick 2009). The concentration of alkaloids varies greatly from one area of the plant to the next, and some parts may be devoid of them. Relatively, mammals and microbes such as fungi, all contain alkaloids. Of course, in most cases, alkaloids have a bitter taste, stimulant activity, and potential anti-fatigue, pain relieving, and poisonous properties (Dewick 2009). Examples of alkaloids with such properties are morphine, tropane, and cocaine.

Phytochemistry involves several stages (procedures), and it is a fundamental arena of bioprospecting. It involves stages such as the collection of plant samples, extraction of plant materials using selected solvents, isolation and purification of compounds, phytochemical structure elucidation, study of the biosynthesis of secondary metabolites and biological assay of secondary metabolites, as well as drug development and clinical trials. The ethnomedical properties of plants are primary informative (criteria) in the selection and collection of plant materials for phytochemistry. Bio-guided assays are the best approach in searching bioactive compounds for the discovery and development of drugs. The extraction method of the intent plant species depends on the interested compounds. For instance, polar compounds are commonly extracted using polar solvents such as acetone, methanol, and ethanol. On the other hand, dichloromethane and ethyl acetate are used to extract less polar compounds while petroleum ether or n-hexane is used to extract the least polar compounds. The extracts are commonly subjected to a biological assay (bioassay) so as to identify the ones with active phytocompounds. Extracts with active phytochemicals are then subjected to isolation. Chromatographic techniques are employed during isolation and purification. The isolated secondary metabolites are then evaluated for their usefulness through a biological assay in order to determine their potency in drug development. On the other hand, a combination of spectroscopic techniques such as ultraviolet visible spectroscopy, nuclear magnetic resonance, infrared, and mass spectroscopy are used to identify the structure of isolated and purified compounds.

Accordingly, drug discovery and development depend on the bioactivity of the phytochemical constituents. Bioactivity is assessed *in vitro* when the bioassay is conducted on a small scale and *in vivo* when the bioassay is conducted on a large scale. For the reason of advanced technology, nowadays, it is easy for drug discovery and development, as it is possible to predict interactions of drug(s) with cells through nanotechnology and computational chemistry. Discovered active extracts or pure compounds are extended to drug formulation or development. Certainly, the developed drug is approved by the relevant authority for global therapeutic usage after the clinical trials have shown that it has less or tolerable toxicity levels.

8.2 BIOSYNTHESIS OF PHYTOCHEMICALS

Plants have natural biochemical and biosynthetic laboratories that use air, water, minerals, and sunlight to produce both primary and secondary metabolites. Biomolecules such as sugars, amino acids, and fatty acids are fundamental (primary) metabolites that are required for plant growth and physiological development; also, they are used as food by humans and other living things. Nevertheless, terpenoids, glycosides, flavonoids, alkaloids, volatile oils, steroids, and other secondary metabolites are biosynthetically produced from primary metabolites. As a result, biosynthesis of phytochemicals refers to metabolic processes in plants that result in the formation of secondary metabolites with different structures for different roles. The metabolism takes place in the presence of enzyme catalysis where the glucose breaks down into the intermediate products such as acetic acid, shikimic acid, and mevalonic acid. These intermediates are used as precursors (starting materials) for secondary metabolism called acetate pathway, shikimate pathway, and mevalonic pathway (Figure 8.1). Since biosynthesis of phytochemicals takes place in the cell, the enzymes play a great role in their biochemical conversion.

8.2.1 ACETATE PATHWAY

Fatty acids and polyketides are biosynthesized from acetic acid through acetate pathways under the mediation of coenzymes. The building unit of fatty acid and polyketide phytocompounds is two carbons (C_2), which is acetate commonly known as acetyl-CoA. Aldol and Claisen reactions are responsible for joining together the C_2 biochemical units (acetyl-CoA). By first turning acetyl-CoA into malonyl-CoA via a carboxylation reaction with CO_2 using ATP and the coenzyme biotin, reactions involving acetyl-CoA become even more favourable (Figure 8.2). In a biotin–enzyme

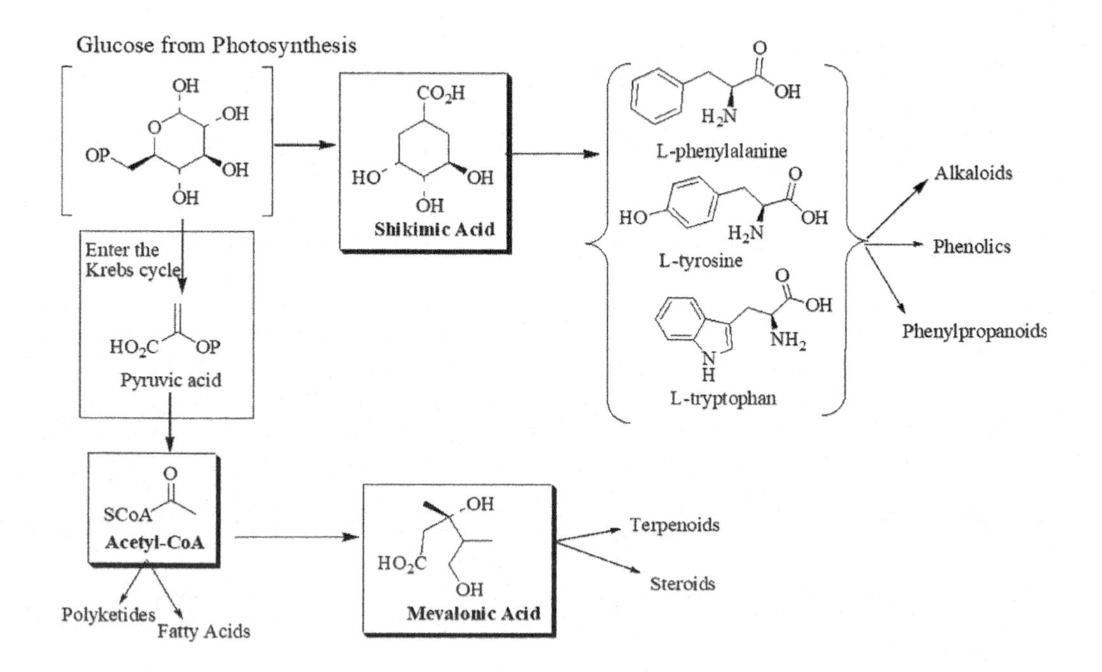

FIGURE 8.1 Schematic showing summary of precursors and products in the biosynthetic pathways.

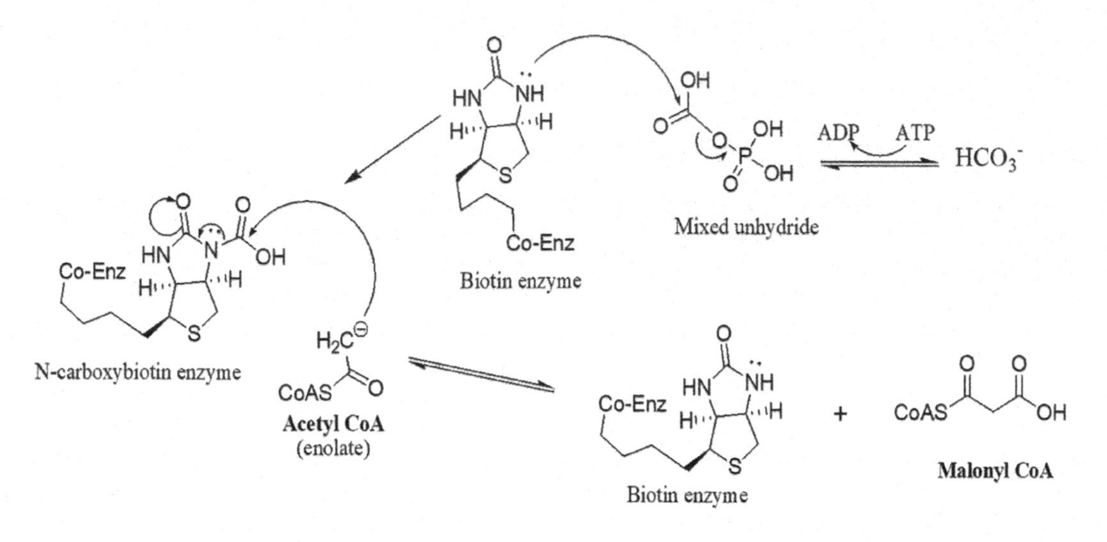

FIGURE 8.2 Schematic reactions for the formation of malonyl-CoA from acetyl-CoA.

combination, ATP and CO_2 (as bicarbonate, HCO_3^-) produce a mixed anhydride, which carboxylates the coenzyme. Enhanced acidity is manifested with α-hydrogens that are flanked by two carbonyl groups during the conversion of acetyl-CoA to malonyl-CoA. The acidity environment creates a further favourable nucleophile for the Claisen reactions.

Fatty acid biosynthesis is commonly catalyzed by the enzyme fatty acid synthase. Usually, malonyl-CoA and acetyl-CoA do not involve in the condensation phase themselves; they are instead converted into the enzyme-bound thioesters, ordinarily known as malonyl ester via an *acyl carrier protein* (ACP). Fatty acid synthase contains an ACP-binding site and the active site cysteine residue in the β-ketoacyl synthase domain. The acetyl and malonyl groups are continuously transferred from coenzyme A esters and attached to the thiol groups of Cys and ACP (Figure 8.2). Claisen condensation reactions occur, and then processes of reduction, dehydration, and reduction occur, whereas the growing chain is attached to ACP. For example, a C_{16} fatty acid (palmitic acid) is produced by combining one acetate starter unit with seven malonates while a C_{18} fatty acid is produced by combining one acetate starter unit with eight malonates (stearic acid). As a result, the two carbons at the head of the chain (methyl end) come from acetate rather than malonate while the rest comes from malonate, which is made by the carboxylation of acetate. In actual fact, this explains why typical fatty acids have an even number of carbon atoms and are straight-chained. Natural fatty acids can have 4 to 30, or possibly more carbon atoms, and those with 16 or 18 carbon atoms being the most common (Figure 8.3).

Fatty acids are usually found in ester combination with glycerol forming triglycerides. Animal fats are solids because they include a significant proportion of saturated fatty acid glycerides, whereas plant and fish fats are liquids because they contain mostly unsaturated fatty acid esters. Major groups of unsaturated fatty acids are designated as ω-3 (omega-3), ω-6 (omega-6), and ω-9 (omega-9). The stereochemistry of the double bond is almost always Z (Cis), resulting in a 'bend' in the alkyl chain. This serves to preserve the fluidity of oils and cellular membranes by interfering with the intimate interaction and aggregation of molecules that is possible in saturated structures. Fats and oils, which are commercially generated for use as meals, toiletries, medicines, or pharmaceutical formulation aids, are long-term energy reserves for most organisms and can be subjected to oxidative metabolism as needed. The naming of fatty acids depends on the number of carbons, number of double bonds, and position of double bonds in a molecule as well as the stereochemistry of the double bond.

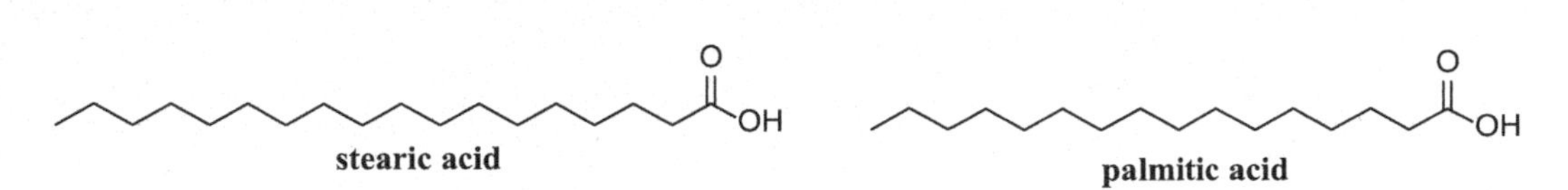

FIGURE 8.3 Chemical structure of stearic acid and palmitic acid.

Plant oils differ significantly depending on the environmental circumstances in which they are grown. In colder regions, the plant produces a higher proportion of polyunsaturated fatty acids to keep the fluidity of its stored lipids and membranes. The melting points of fats are determined by the relative quantities of various fatty acids, which are principally determined by the chain length and unsaturation. At room temperature, fatty acids that are saturated and have longer chains become more solid. Cocoa butter, for example, is solid because it includes a large concentration of saturated fatty acids with long chains. Palm kernel and coconut oils are semisolids with a high saturated C_{12} concentration (lauric acid). Ricinoleic acid is another example of a major fatty acid that is found in seeds of the castor oil plant (*Ricinus communis*) and is the 12-hydroxy derivative of oleic acid. Ricinoleic acid is formed by direct hydroxylation of oleic acid through the action of an O_2- and NADPH-dependent mixed function oxidase. Castor oil has a longstanding experience of use as a purgative agent, and it is now used as a cream basis in the same way. Furthermore, olive oil contains a significant amount of oleic acid (18:1), but rapeseed oil contains a high amount of long-chain C_{20} and C_{22} fatty acids, such as erucic acid (22:1).

On the other hand, oleic acid can be converted into docosahexaenoate and prostaglandins which constitute lipids in sperm, the retina, and the brain (Dewick 2009). This is most important for brain development, and its deficiency is associated with abnormalities in brain function. The importance of essential fatty acids supplied from plant sources, such as linoleic and γ-linolenic acids, in the diet is well appreciated. The synthesis of prostaglandins would be impaired without a source of arachidonic acid or chemicals that can be converted into arachidonic acid and this would have a significant impact on many typical metabolic processes. Prostaglandins (steroids) display a wide range of pharmacological activities, including contraction and relaxation of smooth muscles of the uterus, the cardiovascular system, the intestinal tract, and of the bronchial tissues. They may also inhibit gastric acid secretion, control blood pressure, and suppress blood platelet aggregation (Alamgir 2017; Dewick 2009).

In such case of fatty acid biosynthesis, where the reduction after each condensation step results in developing a hydrocarbon chain, the expanding poly-β-keto chain must be stabilized on the enzyme surface until the chain length is complete, at which point cyclization or other processes can take place. Polyketide synthases catalyse these reactions. Poly-β-keto ester is extremely reactive, and there are a variety of ways for it to undergo intramolecular Claisen or aldol reactions, depending on the enzyme and how the substrate is folded. Normally, two carbonyls flank methylenes, permitting the production of carbanions/enolates and subsequent interaction with ketone or ester carbonyl groups, with a natural inclination to build strain-free six-membered rings (Figure 8.4). The formed rings are polyketide systems which are mainly aromatics and macrolides (Dewick 2009). A distinguishing feature of an aromatic ring system formed via the acetate pathway is that:

i. Numerous carbonyl oxygen of the poly-β-keto systems are reserved in the end product.
ii. The meta-oxygenation pattern is apparent, indicating that the molecule is biosynthetic.
iii. Around the ring system(s), the carbonyl group ends up on alternative carbons.
iv. One or more might be used in forming carbon–carbon bond as viewed in orsellinic acid.

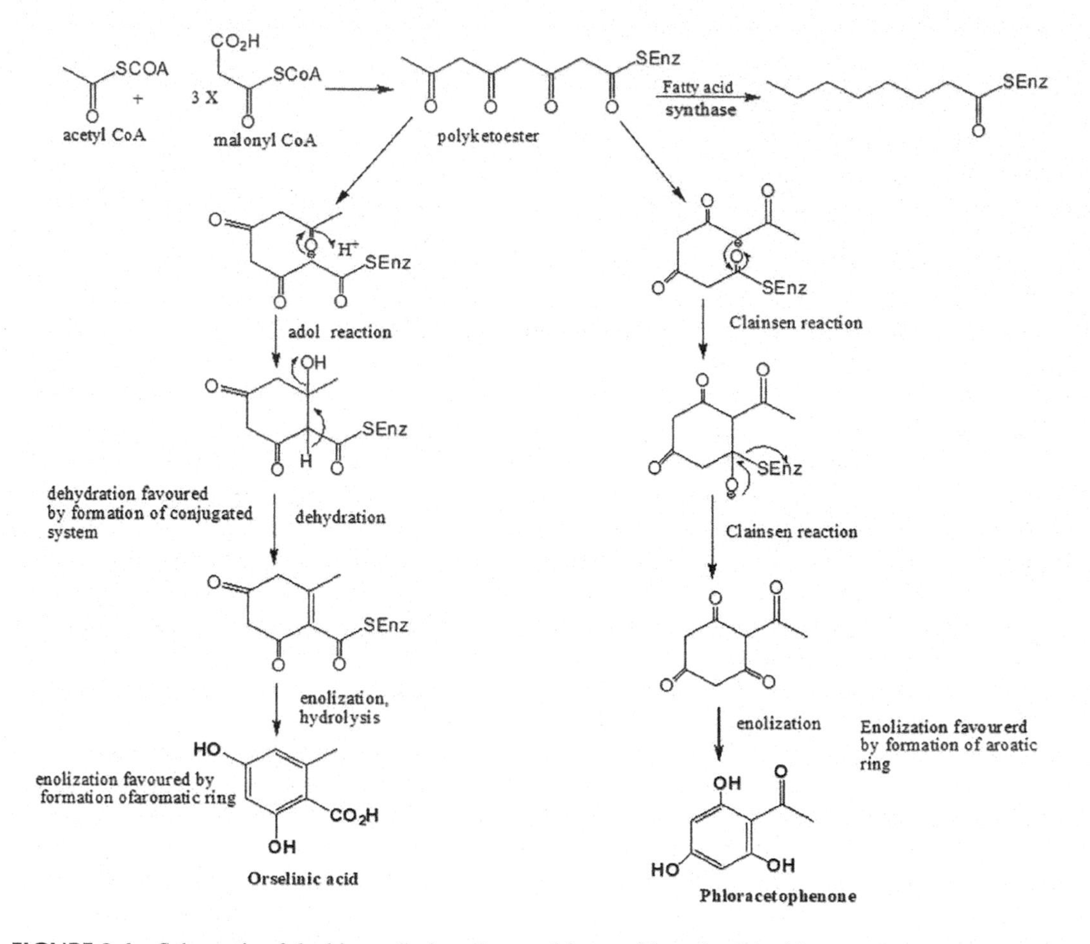

FIGURE 8.4 Schematic of the biosynthetic pathway of fatty acids and polyketides.

8.2.2 Mevalonate Pathway

In addition to acetyl-CoA, mevalonic acid is a product of three molecules of acetate. Compounds such as terpenoids and steroids are primarily biosynthesized from mevalonic acid through mevalonate pathways (Figure 8.5). The pathway starts with the reaction of three molecules of acetates to form mevalonic acid, a compound of six carbons (C_6). Decarboxylation of mevalonic acid gives a building block of five carbons (C_5), commonly known as isoprene or isoprenoids. The bioenergetic molecule, adenosine triphosphate (ATP), supplies energy during the biosynthesis of these phytochemicals. During the biosynthetic process (Figure 8.4), the mevalonic acid diphosphate is formed by ATP-mediated phosphorylation of mevalonic acid, which is then decarboxylated to yield the first isoprene unit called isopentyl pyrophosphate (IPP). The isomerase enzyme converts the IPP to dimethyl allyl pyrophosphate (DMAPP), which is the second isoprene unit. The C_{10}unit, geranyl pyrophosphate (GPP), is produced by the electrophilic addition of IPP with DMAPP via the enzyme prenyl transferase, which is the precursor for monoterpene synthesis. The addition of the IPP unit to GPP results in C_{15}, farnesyl pyrophosphate (FPP), a C_{15}unit that works as a precursor for the production of sesquiterpene. The addition of the IPP unit to FPP results in C_{20}, geranyl geraniol pyrophosphate (GGPP), which can be used to make a variety of diterpenes. Sesterterpenes

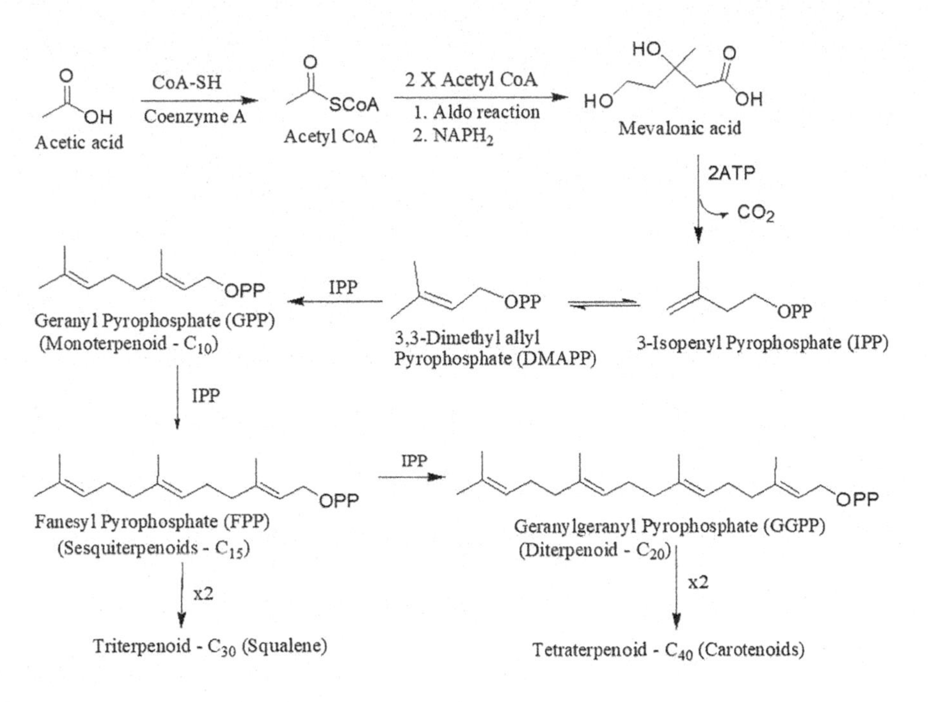

FIGURE 8.5 Schematic showing the mevalonate pathway.

are formed by adding an IPP unit to C_{25} geranylfarnesyl pyrophosphate. The C_{30}unit triterpene is obtained by adding two FPP units tail to tail. In the same way, two units of GGPP yield C_{40}, a tetraterpene. The acetate/mevalonate pathway thus produces two distinct skeleton-containing molecules, steroids and triterpenoids, via IPP and DMAPP via squalene. In conjunction with other routes, it also creates a wide range of monoterpenoids, sesquiterpenoids, diterpenoids, carotenoids, polyprenols, glycosides, and sometimes alkaloids when there is a combination with shikimate pathways (Dewick 2009).

8.2.3 Shikimate Pathway

The shikimate pathway provides an alternative for the synthesis of aromatic compounds apart from aromatic polyketides biosynthesized through acetate pathways. Besides aromatic polyketides that may be grouped as phenolic compounds (polyphenols and phenylpropanoids), and alkaloids are biosynthesized through this pathway because the intermediate precursors carry nitrogen by nature. A combination of other pathways results in the versatility of alkaloids and aromatic compounds. Various enzymes are involved in different biosyntheses of phytochemicals in the shikimate pathway as well. For instance, some enzymes in the formation of chalcones and stilbenes are chalcone synthase and stilbene synthase (resveratrol synthase), respectively.

Shikimic acid was first isolated from the highly toxic Japanese shikimi (*Illicium anisatum*) flower. Its mean of derivation (biosynthetic pathway) became shikimate, whereby the shikimic acid being the central precursor produces L-phenylalanine, L-tyrosine, and L-tryptophan as intermediates. The shikimate pathway employs the aromatic chemicals (intermediates) in a different method to react with acetates to form a long chain, which under enzyme actions results in various aromatic compounds such as chalcones, stilbenes, and flavonoids (Figure 8.6). This pathway is used

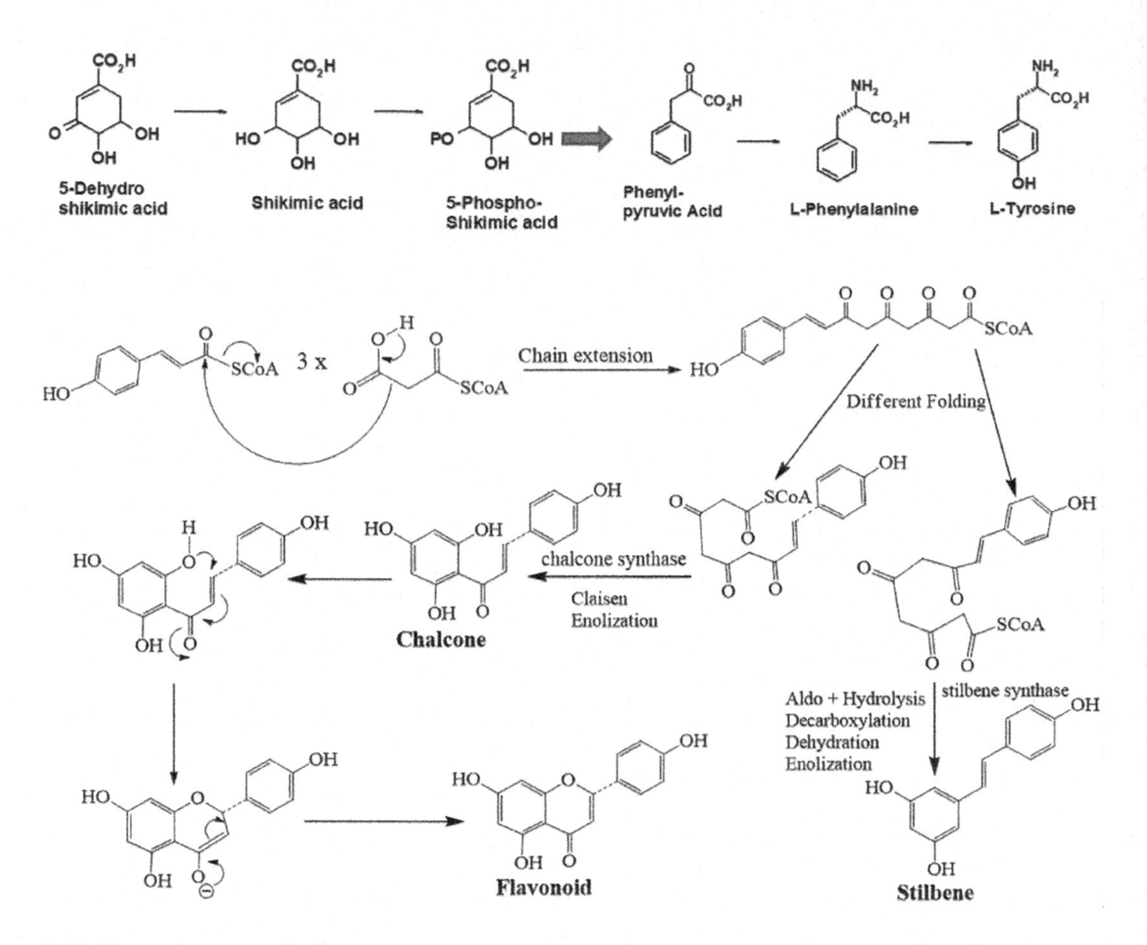

FIGURE 8.6 Example intermediates and products of biosynthesis in the shikimate pathway.

by microbes and plants but not by animals; hence, aromatic amino acids are among the necessary amino acids that must be supplied through diet of humans and other animals.

8.3 DRUGS FROM PHYTOCHEMICALS

Since the immemorial of human history, medicinal plants have been a vital source of drugs for the treatment of various communicable and noncommunicable diseases. Records for the role played by medicinal plants in health settings cannot be overemphasized as several therapeutical drugs in medical services have been developed from phytochemicals (Alamgir 2017). According to estimates, natural compounds or their semisynthetic derivatives account for around 40% of available medicines globally (Alamgir 2017). Equally, a large number of people especially in developing countries depend on traditional medicine systems, predominantly medicinal plants. On the other hand, approximately 30% of modern pharmaceuticals in the United States are derived from plants while in several European nations, there are thousands of herbal and related items on the market (Newman, Cragg, and Kingston 2003; Alamgir 2017). Examples of those drugs are quinine, artemisinin, morphine, and Taxol.

8.3.1 Antimalarial from Phytochemicals

Cinchona tree's bark is the natural source where quinine is derived (Scholar 2007). The tree can be found in Central and South America, as well as various Caribbean islands and western Africa. For

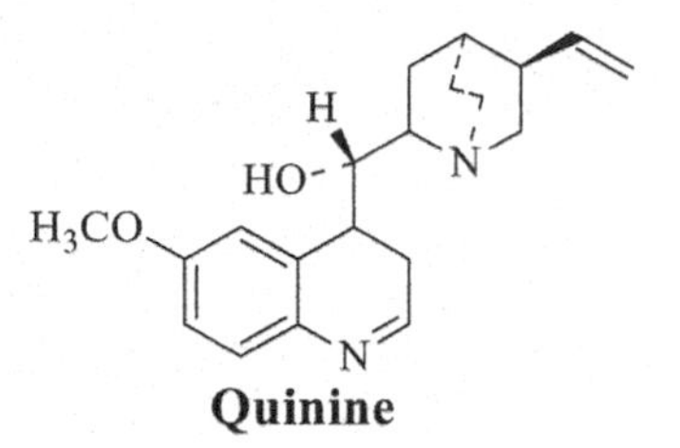

FIGURE 8.7 Chemical structure of quinine.

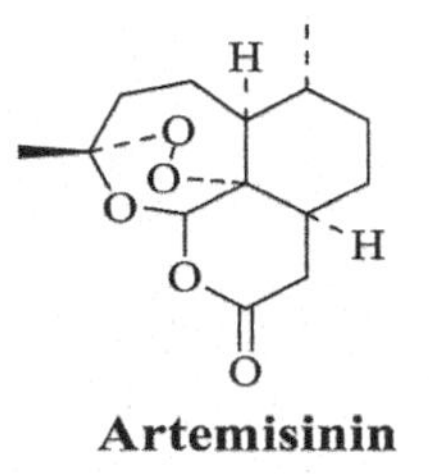

FIGURE 8.8 Chemical structure of artemisinin.

ages, people have used quinine in tonic water for the treatment of malaria. Quinine (Figure 8.7) has been used in traditional, as well as conventional medicine systems (Alamgir 2017). It is an alkaloid that is available in capsules for oral administration, and it was mostly used for the treatment of severe malaria. Apart from being pharmaceutically effective against *Plasmodium falciparum,* quinine is reported as a potential in dermatology for skin treatments (Gelfman 2021). Such medicinal properties necessitate more scientific researches to isolate bioactive phytochemicals.

Due to the increase in drug resistance of *Plasmodium falciparum* for the treatment of malaria, artemisinin is the current medicine of choice (Mannan et al. 2010). Artemisinin (Figure 8.8) is a terpenoid extracted and isolated from the annual herb *Artemisia annua*. It is less toxic compared to quinine, though it is obtained in small amounts relative to bulk sources. Apart from the antimalarial activity of artemisinin, it is reported also as a potential drug for a variety of other ailments such as cancers and also as a natural herbicide (Mannan et al. 2010). Widespread fear of the present artemisinin production that will not be sufficient to fulfil the growing demand from the pharmaceutical industry necessitated a search for alternative sources. The enquiry was carried out in order to find alternatives such as a new artemisinin-producing plant, laboratory synthesis, and genetic engineering using microorganisms (Ikram and Simonsen 2017; Singh and Sarin 2010; Mannan et al. 2010). Accordingly, some of the findings reported that *Artemisia scoparia* contains the antimalarial medication artemisinin (Ding et al. 2021).

8.3.2 Anticancer from Phytochemicals

Taxol is well known as paclitaxel; it is the foremost natural-source cancer drug and is still one of the most widely used as an anticancer. Taxol (Figure 8.9) is a terpenoid that was discovered in the United States, and it was derived from the bark of the Pacific yew tree plant scientifically known as *Taxus brevifolia* (Wani et al. 1970). This drug is used in the treatment of various cancers, especially breast, lung, and ovarian cancers, as well as Kaposi's sarcoma. Taxol has been approved as the most favourable anticancer drug (Sabzehzari, Zeinali, and Naghavi 2020). Taxol has the ability to kill cancer cells by stabilizing microtubules and preventing them from degrading (Schiff, Fant, and Horwitz 1979). Its activity is attributed to the presence of several functional groups in it. Due to its limited water solubility, it is made up of a blend of dehydrated ethanol and Cremophor EL. Its global market is predicted to expand by 8.2% from 2020 to 2025, reaching $152 million in 2025,

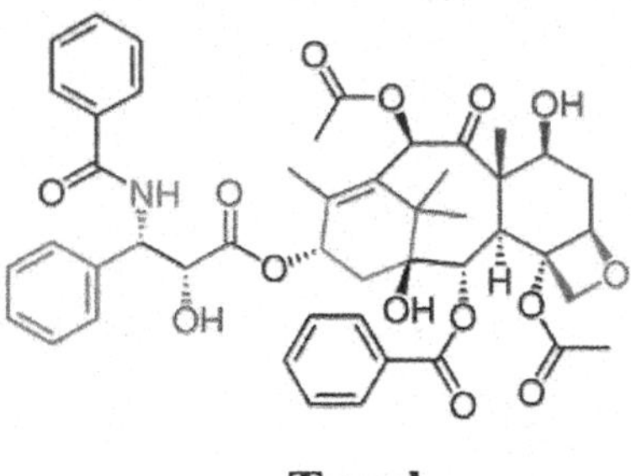

FIGURE 8.9 Chemical structure of Taxol.

up from $110 million in 2019 (Ning et al., 2019). The anticancer properties of paclitaxel have been discovered to be distinct and novel when compared to other anti-tumour drugs. Despite its inability to polymerize tubulins into microtubules, this anticancer drug can bind into the microtubules and eventually impede their depolymerization, hence preventing mitosis (Zhu and Chen 2019). According to the findings, paclitaxel exerts its anticancer effect by limiting microtubule dynamics, preventing the formation of mitotic spindles and chromosomal separation during cell division at concentrations of 1–10 nM (Zhu and Chen 2019). The halting of the cell cycle is linked to cytotoxicity, which could lead to apoptosis induction and death.

Despite the potentiality of Taxol, it is found at a low concentration of 0.01–0.05%, and takes 3,000 trees or 10,000 kg of yew barks to produce 1 kg of the medication (Liu, Gong, and Zhu 2016; Sabzehzari, Zeinali, and Naghavi 2020). As a result of difficulties in harvesting Taxol and the challenges in manufacturing the molecule, the progress towards the clinic has been slow. Since the medicine was found to be effective against breast tumours and ovarian cancer, it is prompting researchers to look for an alternative way to isolate significant amounts for clinical usage. Polysciences, Inc. was the first company to produce Taxol on a significant scale. A method for extracting a precursor of Taxol, 10-deacetyl-baccatin III from the common yew *Taxus baccata*, was developed, and then chemical synthesis was used to transform the precursor to Taxol. As a result, a variety of other procedures, such as tissue culture (genetic engineering using microorganisms), have been proposed as solutions to these challenges. After a few years, *Taxus* cultures produced more Taxol, which is a useful medicine (Liu, Gong, and Zhu 2016). At the moment, a cell culture method created by Bristol-Myers Squibb produces Taxol.

8.3.3 Painkiller from Phytochemicals

Morphine is one of the most significant alkaloids generated from *Papaver somniferum*, commonly known as a poppy plant (Dewey 2007). It is a class of opioid analgesic, which means that it relieves pain. Morphine is one of the earliest bioactive phytocompounds then synthesised from medicinal plants in the laboratory (Figure 8.10). It is a highly effective medicine for the alleviation

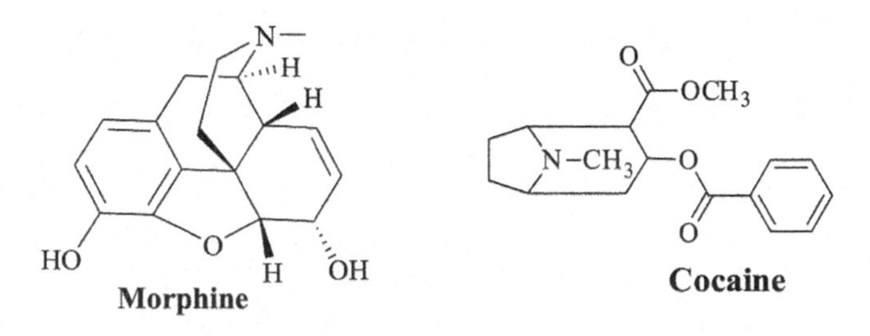

FIGURE 8.10 Chemical structure of morphine and cocaine.

of moderate-to-severe pain and serves as the gold standard against which all other pain relievers are judged (Dewey 2007; Alamgir 2017). It alters the way your body perceives and responds to pain by acting on the brain. Morphine is also used in hospitals before surgery to alleviate anxiety, induce sleep, and lower the anaesthetic dose. Sometimes, it can be used to treat myocardial infarction because of its vasodilatory and bradycardic properties. It can also be used to cure pulmonary oedema. At therapeutic levels, morphine causes respiratory depression and gastrointestinal consequences (Dewey 2007). Consequently, at some point, this chemical is abused and regulated by national and international regulatory bodies. In addition to morphine, codeine derived from the opium poppy and cocaine (Figure 8.10) derived from the coca plant are essential therapeutic alkaloids. These alkaloids are pain relievers that act on the neurological system. Anaesthesia and ophthalmology both use atropine, a neurotoxin produced from the deadly nightshade plant.

8.3.4 Potential Phytochemicals for Drug Development

Apart from the aforementioned drugs developed from phytochemicals, there are some bioactives that are potential phytochemicals for drug development as well. For instance, rotenoids (isoflavonoids such as rotenone and deguelin) (Figure 8.11) from *Tephrosia* species are the potential for the development or formulation of pesticides (Caboni et al. 2004).

Podophyllotoxin (Figure 8.12) is another potential bioactive which is isolated from May apple, *Podophyllum peltatum,* but was at first shelved regarding its highly intolerable toxicity (Newman, Cragg, and Kingston 2003). A few years later, the drug was developed from the same compound and subjected to pharmacological bioassay as an anticancer agent. Equally, two useful semisynthetic compounds, etoposide and teniposide, playing a role as DNA topoisomerase II inhibitors were derived from podophyllotoxin (Newman, Cragg, and Kingston 2003). Hence, they indicate that the phytocompounds can be modified and subjected to other medicinal purposes.

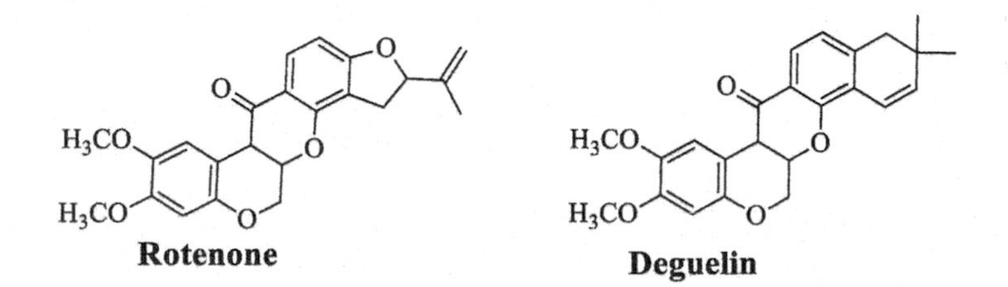

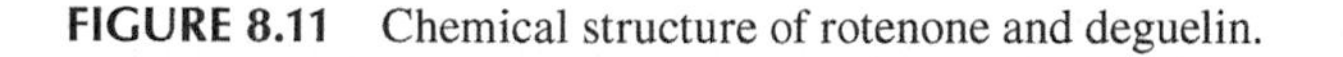
FIGURE 8.11 Chemical structure of rotenone and deguelin.

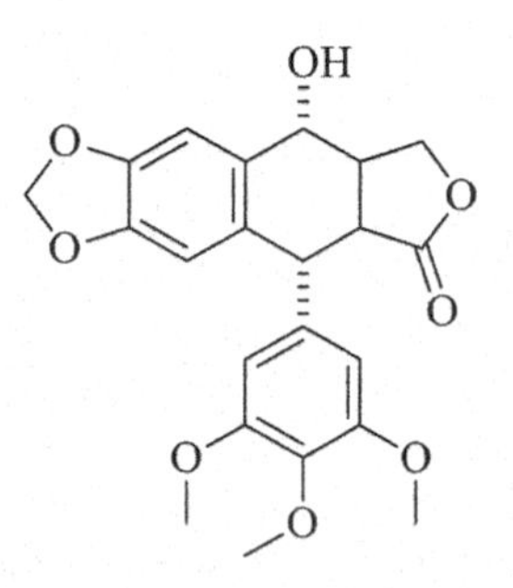

FIGURE 8.12 Chemical structure of podophyllotoxin.

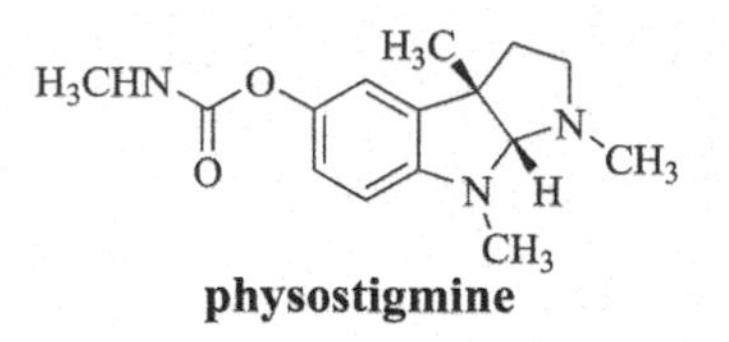

FIGURE 8.13 Chemical structure of physostigmine.

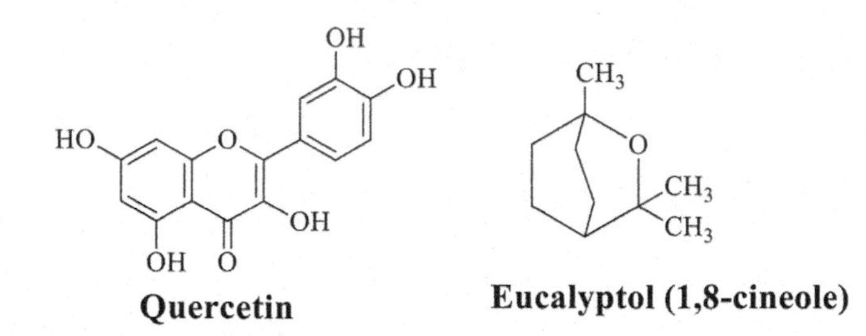

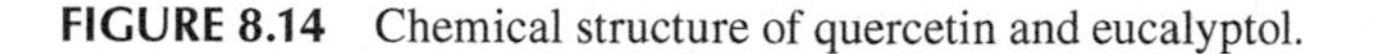

FIGURE 8.14 Chemical structure of quercetin and eucalyptol.

Physostigmine (Figure 8.13) is derived from *Physostigma venenosum* seeds, commonly known as Calabar bean. It's a type of reversible anticholinesterase that boosts acetylcholine levels at cholinergic transmission sites (Alamgir 2017). Acetylcholine is rapidly hydrolyzed by the enzyme anticholinesterase; hence, its activity is usually brief. Thus, anticholinesterase is inhibited by physostigmine, which prolongs and increases the effects of acetylcholine.

Flavonoids such as quercetin and essential oils such as eucalyptol (Figure 8.14) have been reported to fight against various viruses (Galan et al. 2020; Manuel et al. 2020). Traditionally, the plants biosynthesizing these phytocompounds have been reported to combat SARS-CoV-2 too. Therefore, they are potential bioactives for drug development against COVID-19. These are a few examples of bioactive phytochemicals among many that are in clinical routine from medicinal plants.

In the view of that, more effort in research on phytochemicals is needed because the natural product compounds are still among the key source of novel pharmaceuticals against microbial infectious diseases as well as noninfectious diseases. These may include anticancer, anti-hypertensive, anti-infective, immunosuppression, and neurological disorder agents.

8.4 CONCLUSION

As afore-discussed though not mentioned exhaustively about phytochemicals for pharmaceutical drugs for the treatment of human diseases, there are many others for example perfume formulation as well as pesticide development. The global demand for plant-derived goods has increased daily. This is assessed for the fact that more than 85% of people in the Middle East, Latin America, Africa, and Asia rely on traditional medicine, particularly herbal medications, for their health care. Despite considerable experience with medicinal plants in traditional medicine, scientific research and identification of bioactives and their effects may also eventually lead to the discovery of novel therapeutic benefits and the manufacturing of natural goods. Numerous secondary metabolites found in plants are multifunctional chemicals that defend them from microorganisms such as bacterial, viral, and protozoal infections, as well as herbivores like insects, worms, and animals. This means that once phytochemicals show activity in one situation, more research should be done, so they are subjected to many other areas in order to hunt for their possible potential. As a result, substantial research is required to ensure the quality of raw medications and formulations in order to justify their use in the current health settings.

Nevertheless, a limited amount of bioactives in relation to the volume of plant materials collected as a source is the most challenging and difficult for the use of phytochemicals in drug development. It is common to isolate minute active secondary metabolites from plants, which compels us to find other means (alternative) of getting a large amount of the same phytochemicals. Such an alternative includes searching for the same phytochemicals from plants in the same genus, the family far away from the family. Another alternative is looking at the possibility of totally synthesizing the compound or using it as the lead compound for semisynthesis in the laboratory. Moreover, tissue culture technology may necessarily be applied in industries where it is possible to obtain a bulk of active phytocompounds.

REFERENCES

Alamgir, A. N. M. 2017. *Progress in Drug Research: Therapeutic Use of Medicinal Plants and Their Extracts.* Springer. Vol. 73. https://doi.org/10.1007/978-3-319-63862-1_4.

Bhuyan, Deep Jyoti, and Amrita Basu. 2017. "Phenolic Compounds: Potential Health Benefits and Toxicity." In *Utilisation of Bioactive Compounds from Agricultural and Food Production Waste*, 27–59. CRC Press, Taylor & Francis Group.

Caboni, Pierluigi, Todd B. Sherer, Nanjing Zhang, Georgia Taylor, Hye Me Na, J. Timothy Greenamyre, and John E. Casida. 2004. "Rotenone, Deguelin, Their Metabolites, and the Rat Model of Parkinson's Disease." *Chem. Res. Toxicol* 17 (11): 1540–48. https://doi.org/10.1021/tx049867r.

Cartea, María Elena, Marta Francisco, Pilar Soengas, and Pablo Velasco. 2011. "Phenolic Compounds in Brassica Vegetables." *Molecules* 16: 251–80. https://doi.org/10.3390/molecules16010251.

Dewey, William. 2007. "Morphine." *Virginia Commonwealth University, Richmond, USA*. Elsevier Inc.

Dewick, Paul M. 2009. *Medicinal Natural Product: A Biosynthetic Approach*, 3rd Edition. John Wiley and Sons, Ltd. https://doi.org/10.1002/9780470742761.

Ding, Jiwei, Linlin Wang, Chunnian He, Jun Zhao, Lijun Si, and Hua Huang. 2021. "Artemisia Scoparia: Traditional Uses, Active Constituents and Pharmacological Effects." *Journal of Ethnopharmacology* 273: 1–15. https://doi.org/10.1016/j.jcp.2021.113960.

Francesca, Irina, González Mera, Daniela Estefanía, González Falconí, and Vivian Morera Córdova. 2019. "Secondary Metabolites in Plants: Main Classes, Phytochemical Analysis and Pharmacological Activities." *Revista Bionatura* 4: 1000–1009. https://doi.org/10.21931/RB/2019.04.04.11.

Galan, Derick M., Ngozi E. Ezeudu, Jasmine Garcia, Chalice A. Geronimo, Nicholas M. Berry, and Benjamin J. Malcolm. 2020. "Respiratory Disorders?" *Journal of Essential Oil Research* 32 (2): 103–10. https://doi.org/10.1080/10412905.2020.1716867.

Gelfman, Daniel M. 2021. "Reflections on Quinine and Its Importance in Dermatology Today." *Clinics in Dermatology* 21: 8–11. https://doi.org/10.1016/j.clindermatol.2021.08.017.

Hussein, Rehab A., and Amira A. El-anssary. 2018. "Plants Secondary Metabolites: The Key Drivers of the Pharmacological Actions of of Medicinal Plants." In *Intechopen*, 11–30. https://doi.org/10.5772/intechopen.76139.

Ikram, Nur K. B. K., and Henrik T. Simonsen. 2017. "A Review of Biotechnological Artemisinin Production in Plants." *Frontiers in Life Science* 8 (November): 1–10. https://doi.org/10.3389/fpls.2017.01966.

Liu, W. C., T. Gong, and P. Zhu. 2016. "RSC Advances Advances in Exploring Alternative Taxol Sources." *RSC Advances* 6: 48800–809. https://doi.org/10.1039/C6RA06640B.

Mannan, Abdul, Ibrar Ahmed, Waheed Arshad, Muhammad F. Asim, Rizwana A. Qureshi, and Izhar Hussain. 2010. "Survey of Artemisinin Production by Diverse Artemisia Species in Northern Pakistan." *Malaria Journal* 9: 1–9. https://doi.org/doi:10.1186/1475-2875-9-310.

Manuel, Ruben, Luciano Colunga, Max Berrill, John D. Catravas, and Paul E. Marik. 2020. "Quercetin and Vitamin C: An Experimental, Synergistic Therapy for the Prevention and Treatment of SARS-CoV-2 Related Disease (COVID-19)." *Frontiers in Immunology* 11 (June): 1–11. https://doi.org/10.3389/fimmu.2020.01451.

Mendoza, Nadia, and Eleazar M. Escamilla Silva. 2018. "Introduction to Phytochemicals: Secondary Metabolites Metabolites from Active Plants Principles with Active Principles for from Plants with for Pharmacological Pharmacological Importance Importance." In *Intechopen*, 27–47. https://doi.org/10.5772/intechopen.78226.

Newman, David J, Gordon M Cragg, and David G I Kingston. 2003. *Natural Products as Pharmaceuticals and Sources for Lead Structures. The Practice of Medicinal Chemistry*, 2nd Edition. Elsevier Inc. https://doi.org/10.1016/B978-0-12-744481-9.50010-6.

Ning, Like, Chaoqun You, Yu Zhang, Xun Li, and Fei Wang. 2019. "Synthesis and Biological Evaluation of Surface-Modified Nanocellulose Hydrogel Loaded with Paclitaxel." *Life Sciences*, 117137. https://doi.org/10.1016/j.lfs.2019.117137.

Ramawat, Kishan Gopal, and Jean Michel Mérillon. 2013. "Terpenes: Chemistry, Biological Role, and Therapeutic Applications." In *Natural Products: Phytochemistry, Botany and Metabolism of Alkaloids, Phenolics and Terpenes*, 2667–91. Springer-Verlag Berlin Heidelberg. https://doi.org/10.1007/978-3-642-22144-6_120.

Sabzehzari, Mohammad, Masoumeh Zeinali, and Mohammad Reza Naghavi. 2020. "Alternative Sources and Metabolic Engineering of Taxol: Advances and Future Perspectives." *Biotechnology Advances*, 1–35. https://doi.org/10.1016/j.biotechadv.2020.107569.

Schiff, P. B., J. Fant, and S. B. Horwitz. 1979. "Promotion of Microtubule Assembly in Vitro by Taxol." *Nature* 277: 665–67.

Scholar, Eric. 2007. "Quinine." In S J Enna and D B Bylund (eds.), *The Comprehensive Pharmacology Reference* (pp. 1–6). *University of Nebraska Medical Center*, 1–6. Elsevier

Singh, Aditi, and Renu Sarin. 2010. "Artemisia Scoparia – A New Source of Artemisinin." *Journal of the Bangladesh Pharmacological Society* 5: 17–20. https://doi.org/10.3329/bjp.v5i1.4901.

Wani, M. C., J. A. Kepler, J. B. Thompson, E. Wall, and Chem. S. G. Levine. 1970. "Plant Anti-Cancer Agents. VI. The Isolation and Structure of Taxol, a Novel Antileukemic and Anti-Cancer Agent from Taxus Brevifolia." *Journal of the American Chemical Society* 243 (9): 2325–27. https://doi.org/10.1021/ja00738a045.

Zhu, Linyan, and Liqun Chen. 2019. "Progress in Research on Paclitaxel and Tumor Immunotherapy." *Cellular & Molecular Biology Letters* 24: 1–11. https://doi.org/10.1186/s11658-019-0164-y.

9 Phytochemicals as Immunomodulators, Nutraceuticals, and Pharma Foods

Syed Amir Ashraf, Abd Elmoneim O. Elkhalifa, Arif Jamal Siddiqui, and Ashanul Haque
University of Ha'il

Danish Mahmood
Qassim University

CONTENTS

9.1 INTRODUCTION

The term "phytochemicals" denotes a wide variety of non-nutritive components derived from the plant sources such as vegetables, herbs, beans, fruits, and grains, which provide a defensive mechanism or have the preventative action against diseases. Since phytochemicals are non-nutritive nutrients, they are not essential by the human beings for sustaining the life, but for supporting life (Abbasi, Shah, and Khan 2015). However, phytochemicals display numerous health beneficial effects, such as protecting against chronic degenerative disorders; coronary heart disease; osteoporosis; cancers; diabetes; psychotic diseases; inflammatory disorders; microbial, viral, and parasitic infections; and other related ailments (Singh et al. 2014, Sullivan 2014, Prof, Gupta, and Sharma 2012). Phytochemicals' health effects are linked to the presence of various bioactive components in food, since plant-based foods are complex mixtures of bioactive compounds and they are categorized into various classes.

DOI: 10.1201/b22842-9

9.2 CLASSIFICATION OF PHYTOCHEMICALS

Phytochemicals have been classified based on their structural heterogeneity, biochemical activities, and a wide range of dissemination among plant species. Moreover, phytochemicals are in general classified into six major categories based on their chemical characteristics and structures. These classes include alkaloids, glycosides, flavonoids, phenolic compounds, vitamins, and carotenoids (Campos-Vega and Oomah 2013).

9.2.1 Alkaloids

Alkaloids are a huge collection of naturally occurring chemical components, which contain one or more nitrogen atoms (amino or amido group in some cases). These nitrogen atoms make these compounds alkaline. Normally, these nitrogen atoms are arranged in a ring (cyclic) structure. Alkaloids (whose name is derived from "alkali-like"), like inorganic alkalis, can react with acids to generate salts (Joanna 2019). Many plant species, mostly blooming plants as well as certain animals synthesize, produce, and store by-products and are classified as secondary metabolites. These metabolites are stored in various amounts in different parts of the plant, including the leaves, stem, root, and fruits (Joanna 2019). Alkaloids are considered as one of the key active components in Chinese herbal medicine due to their significant biological activities. Alkaloids have been instrumental for the development of various therapeutically effective anticancer drugs. Among the catharanthus alkaloids, vinblastine is known for its antimitotic activity that prevents tubulin polymerization (Chagas and Alisaraie 2019). Additionally, curcumin has been reported for its cytotoxic effects on medulloblastoma cells, suppressing cell growth and causing cell cycle arrest at the G(2)/M phase (Elamin et al. 2010). Sotoing Taïwe et al. (2012) investigated the antipsychotic and sedative activities of an alkaloid fraction produced from the *Crassocephalum bauchiense* (Hutch.) Milne-Redh leaves in mice. They found that the fraction of alkaloid from *C. bauchiense* might limit rearing activities in a dose-dependent manner and lower body temperature significantly (Sotoing Taïwe et al. 2012, Jachak 2001). Huperzine A supplementation can help the patients with cognitive impairments, oxidative stress, and the apoptotic cascade that is triggered by acute hypobaric hypoxia, since it improves oxygen availability to the brain and reduces the chances of neurological impairment (Vessal, Hemmati, and Vasei 2003). The chief alkaloid of *Piper nigrum* (piperine) was investigated for blood levels of glucose in alloxan-induced diabetic mice. It was found that piperine has a statistically significant antihyperglycemic effect, but at higher concentration, it increases blood glucose levels. Nuciferine derived from *Nelumbo nucifera* was found to increase both stages of insulin production in isolated islets in the study of Nguyen et al. (2012). The benzo[c] phenanthridine alkaloids efficiently reduced the growth of *Mycrocystis aeruginosa*, according to the study by Meng et al. (2009). *Chelidonium majus* L. has a wide range of biological activity and has been used to treat a variety of infectious disorders. Furthermore, alkaloids enhance the immune system by controlling the growth of thymic and splenic lymphocytes and the release of cytokines (Zhou et al. 2020). The primary active ingredient of *Euodia rutaecarpa* (Juss) Benth, evodiamine, has been found to control mouse immunity (Jiang et al. 2021).

9.2.2 Flavonoids and Glycosides

Flavonoids are a family of natural compounds present in fruits, bark, roots, vegetables, flowers, tea, grains, stems, and wine. Flavonoids are the pigments that give most flowers, fruits, and seeds their color. Moreover, naturally occurring flavonoids are very well known for their health benefits, and researchers are continuously working to extract these components (Panche, Diwan, and Chandra 2016). In recent years, the role of flavonoids is recognized as an essential component for the development of pharmaceutical, nutraceutical, and cosmetic products. Numerous flavonoids are now widely used in the realms of nutrition, food safety, and health because of their features. Flavonoids are also known for their antifungal, antiviral, and antibacterial characteristics. Several flavonoids such as quercetin, naringin, hesperetin, catechin, apigenin, and many more possess various biological activities. The flavonoids quercetin and naringin in particular possess an antioxidant activity, which may help to prevent oxidative stress in rats, which is induced by sodium fluoride (Nabavi et al. 2012). Kaempferol, one of the important flavonoids, has been reported to possess a chemopreventive activity, due to its ability to promote an apoptotic activity in HepG2 cells (Guo et al. 2016). Antiviral activity varies among flavonoids such as quercetin, naringin, hesperetin, and catechin. Quercetin and apigenin are two flavonoids that have been extensively investigated for their antibacterial properties (Wu et al. 2008). According to Li and Xu (2008), quercetin derived from lotus leaves could be a promising antibacterial agent. Flavonoids have shown to alter arachidonic acid metabolism by inhibiting cyclo-oxygenase (COX) and lipoxygenase activities. It has also been suggested that anti-inflammatory and anti-allergic characteristics of flavonoids are due to their inhibitory effects on arachidonic acid metabolism (Panche, Diwan, and Chandra 2016). Studies report that flavonoids can inhibit the growth of various cancer cell lines (Devavrat 2021). Furthermore, among the flavonoids, quercetin is one of the most studied flavonoid members of the polyphenolic group. Quercetin has anticancer, antiviral, anti-inflammatory, antiobesity, cardiovascular, hepatoprotective, neuroprotective, and antidepressant activities (Güleç and Demirel 2016, Zhang, Ning, and Wang 2020). Rutin is also a prominent flavonoid with anti-inflammatory, anti-allergic, antispasmodic, cytoprotective, antitumor, antibacterial, and antiviral properties, including the potential inhibition of SARS-CoV-2 vital proteins (Ekaette and Saldaña 2021).

Flavonoids are regarded as promising novel anti-inflammatory medicines and flavonoids may play a vita role in the cascade of events that could cause a progressive neuronal damage in neurodegenerative illnesses. As a result, therapeutic regimens that target neuroinflammatory processes may help to delay the progression of these debilitating brain diseases. Biological effects of flavonoids reflect their different ways of action in inflammation. The inhibitory effect on inflammatory cells, particularly mast cells, appears to be superior to that of any other clinically available drug (Rathee et al. 2009). The ability of flavonoids to affect the activities of numerous inflammatory mediators suggests that they have the potential to influence inflammation. This could lead to the creation of new pharmacological anti-inflammatory drugs and new insights into the regulation of the inflammatory process (Rathee et al. 2009).

Glycosides are also one of the secondary metabolites of plant, which has a sugar moiety attached to a nonsugar moiety. Meanwhile, many plants retain glycosides in the inactive form that can be stimulated by enzyme hydrolysis. Glycosides play numerous important roles in living organisms. Glycosides are classified as alcoholic glycosides (such as salicin found in the genus Salix), cardiac glycosides (found in Ouabain, Squill, and Thevetia species), anthraquinone glycosides (found in Senna, Aloe, Cascara, and Rhubarb species), and flavonoid glycosides (rutin and quercetin) (Onaolapo and Onaolapo 2019).

Therefore, flavonoids have many immunomodulating properties and can influence immune system response. For instance, dendritic cells (DCs) taken from mouse bone marrow were treated by quercetin in an experimental study. Quercetin has been shown to have impacts on the immune system. This flavonoid has the ability to significantly reduce the expression of MHC class II and co-stimulatory molecules, as well as the synthesis of pro-inflammatory cytokines and chemokines. These circumstances stop DCs from becoming activated by LPS. Treatment with quercetin also reduces DC migration mediated by LPS and DC endocytosis (Sun et al. 2015).

9.2.3 Phenolic Acids

The phrase "phenolic acids" refers to phenolic compounds that include a single carboxylic acid group. Phenolic acids are usually present in a wide variety of plant-based foods, with the largest amounts in fruit skins, seeds, and vegetable leaves (Pereira et al. 2009). Phenolic acids are aromatic phytochemicals with characteristic carbon frameworks, such as hydroxycinnamic and hydroxybenzoic acid (Stalikas 2007, Khoddami, Wilkes, and Roberts 2013). In humans, phenolic acids have been reported to be beneficial due to their possible antioxidant activity, and they also help to deter the damaged cells resulting from free radical oxidation reactions (Piazzon et al. 2012, Saxena, Saxena, and Pradhan 2012). Fraisse et al. (2011) discovered that chicoric acid and dicaffeoylquinic acid contribute to 68.96% and 48.92% of antioxidant activity in *Taraxacum officinale* (Fraisse et al. 2011). Caffeic acid analogs such as dicaffeoylquinic acid and chicoric acid have been reported for their anti-HIV efficacy (Stoszko et al. 2016). Chicoric acid was also found to posses anti-inflammatory and antidiabetic characteristics, as it improved insulin resistance, increased glucose absorption, and reduced glucosamine-induced inflammation (Zhu et al. 2015).

Phenolic acids possess a wide range of biological applications and are used in various pharmaceuticals, cosmetics, and food industries (Shahidi and Ambigaipalan 2015). The capacity of natural bacteria to metabolize phenolic acids is a crucially important, since they provide an important alternative to man-made chemicals that are similarly damaging to the environment. The role of phenolic acids in diabetes cannot be overstated. The action of glucose and insulin receptors is influenced by phenolic acids. Both chlorogenic and ferulic acids have been shown to stimulate transporters and serve as antidiabetic drugs (Jung et al. 2006, Prasad et al. 2010). The chemical structure of phenolic acids has been used to evaluate their antibacterial properties, particularly saturated chain length, its position, and substitutions at core aromatic ring. The antibacterial activity of phenolic acids is lower than that of their methyl and butyl esters (Cueva et al. 2010). The activity of phenolic acid oligomers is significantly higher than that of the monomers (Elegir et al. 2008, Merkl et al. 2010).

According to epidemiological studies, a diet rich in antioxidant-rich fruits and vegetables lowers the threat of various cancers, implying that particular dietetic antioxidants could be useful agents for the prevention of cancer incidences. Phenolic acids and their derivatives, such as hydroxybenzoic and hydroxycinnamic acids, are important in the prevention and treatment of cancer (Kumar et al. 2019, Kumar et al. 2017, Badhani, Sharma, and Kakkar 2015). Plant phenolics may play an important role, as natural products or their derivatives accounted for more than half of all anticancer prescription medications approved internationally between the 1940s and 2006, and numerous clinical investigations are ongoing (Efferth et al. 2007). They limit the creation of genotoxic compounds and restrict the action of mutagen-transforming enzymes (Słoczyńska et al. 2014), regulate heme-containing phase I enzymes (Basheer and Kerem 2015) and carcinogen-detoxifying phase II enzymes, and stop the synthesis of DNA adducts. The majority of phenolics work at different times to treat or suppress various cancers (Choi et al. 2014).

Exogenous medicinal plants and foods high in phenolic compounds can help to prevent numerous chronic diseases and improve health. The enormous range of health advantages and industrial applications of phenolic compound(s) has prompted scientists to improve extraction and purification strategies for these naturally given compounds. Nonflavonoid polyphenols, such as phenolic acids, are a common type of polyphenol found in the human diet. Consumption of vegetables and fruits, such as apples, plums, cherries, kiwis, citrus fruits, onions, whole grains, tea, coffee, and flour produced from whole corn, wheat, rice, or oats, could provide roughly 200 mg of phenolic acids per day or more (Scalbert and Williamson 2000).

Published studies established the immunomodulatory effects of polyphenols both *in vivo* and *in vitro*. These findings demonstrate the prospective of polyphenols in the treatment and prevention of illnesses with underlying inflammatory disorders, such as cancer, neurological diseases, obesity, type II diabetes, and cardiovascular illnesses. Moreover, it has been suggested that nuclear factor kappa B (NF-κB) signaling pathways are intimately connected to the spread of cancer. By preventing NF-κB activation, polyphenols can interfere with the propensity for cancer to metastasize (Xia, Shen, and Verma 2014). By reducing NF-κB expression and downregulating the production of VEGF (vascular endothelial growth factor), COX-2, and MMP-9 (matrix metallopeptidase-9) in tissues of the breast, brain, lung, liver, and spleen, curcumin is a solid example of how to reduce cancer metastasis in mice (Kim et al. 2012).

9.2.4 Carotenoids

Carotenoids are a cluster of more than 700 colors naturally synthesized by algae, plants, and photosynthetic microorganisms. Many plants' colors like yellow, orange, and red come from these brightly colored carotenoid molecules. Among the various carotenoids, 40–50 different types of carotenoids present in fruits and veggies are a part of human diet. Moreover, frequently present dietary carotenoids include carotene, lutein, cryptoxanthin, and lycopene (Maoka 2019). Pro-vitamin A carotenoids such as carotene and cryptoxanthin can be converted into retinol. In humans, carotenoids has a vital role in scavenging peroxyl radicals and singlet oxygen. Dietary carotenoids help to prevent a variety of ailments, including some malignancies, cardiovascular disease, and eye disease. Carotenoids are commonly utilized in nutritional supplements, skincare cosmetics, and medicinal goods, as well as acting as a natural color for meals and feeds, due to their bioactivities (Stahl and Sies 2005). Carotenoids may act as antioxidants and promote oxidative stress tolerance in all species. Carotenoids act as an antioxidant defense system in the human body (Saini et al. 2022). Tocopherols and carotenoids both can quench single oxygen, and they are known to be effective antioxidants that may scavenge reactive oxygen species produced during photooxidative stress (Stahl et al. 2000). Carotenoids (astaxanthin, α-carotene, zeaxanthin, lycopene, β-cryptoxanthin, lutein, fucoxanthin) and β-carotene, as well as α-carotene, may help prevent cancer (Nishino et al. 2002).

Yamaguchi et al. (2006) found that drinking fortified juice, which contains more β-cryptoxanthin than conventional juice, helps to prevent bone loss as people get older (Yamaguchi et al. 2006). Carotenoids and several of their metabolites are thought to protect against a variety of ROS-mediated diseases, including cardiovascular disease, cancer, neurological disorders, and photosensitive or eye-related ailments (Fiedor and Burda 2014). Lowering blood pressure, reducing pro-inflammatory cytokines and indicators of inflammation (such as C-reactive protein), and improving insulin sensitivity in the muscle, liver, and adipose tissues are all features of carotenoids that may help to reduce cardiovascular risk (Gammone, Riccioni, and D'Orazio 2015). Due to their chemical structure and interaction with biological membranes, carotenoids have been shown to have antioxidant effects.

Carotenoids have received particular interest as protective agents in skin photo-related illnesses due to their outstanding $1O_2$ quenching and other ROS scavenging characteristics. Carotenoids (particularly carotene and canthaxanthin) have been proposed as efficient scavengers of excited triplet forms of endogenous photosensitizers like protoporphyrin, which is deposited in the blood and skin of people with hereditary erythropoietic protoporphyria (Meléndez-Martínez, Stinco, and Mapelli-Brahm 2019).

According to reports, β-carotene and other carotenoids have immunomodulatory effects on both human beings and animals. Carotenoids have been reported to offer protection against a variety of cancers (Elkhalifa et al. 2021). Bendich and Shapiro (1986) were the first to identify a specific function for carotenoids in the immune system. They demonstrated that canthaxanthin, a carotenoid with negligible pro-vitamin A activity, increased mitogen-induced lymphocyte proliferation in rats and that dietary carotene had a comparable effect. Later research has confirmed the immunostimulatory effects of carotenoids lacking provitamin A activity, including lutein, lycopene, astaxanthin, and canthaxanthin (Chew and Park 2004). There is evidence linking regular tomato product consumption to beneficial immunomodulatory outcomes. Tomato extracts have also been demonstrated to exhibit antioxidant, anticarcinogenic, and antithrombotic activities *in vitro*, and lycopene, a form of carotenoids that gives tomatoes their red color, plays a part in the immune system modulation (Bessler et al. 2008).

9.3 ROLE OF PHYTOCHEMICALS IN IMMUNOMODULATION

The immunity status of any host cell is to maintain an equilibrium state and to provide enough defense material to fight against the infection, disease, or other unwanted foreign matter invasion. The immune system plays a key role in the maintenance of immunological or physiological functions of the host cell along with its internal environment. Furthermore, any issue with the functioning of immune activities may lead to several disorders in human beings, such as inflammatory disorders, arthritis, and cancer. Therefore, the immune system plays an important role in the survival of human being. Moreover, immune responses are the result of an effective interaction between innate (natural and nonspecific) and acquired (adaptive and specific) components of the immune system (Chaplin 2010). Immune system comprises mainly white blood cells (neutrophils, monocytes, lymphocytes, and macrophages) and some of the very specific immune substances such as proteins, antibodies, and cytokines. These components improve the defense and resistance system of the host by protecting against various infections or diseases arising from foreign substances like viruses, bacteria, and parasites. The interaction between the immune cells and the cell mediators prompts the generation of ideal immune responses (Behl et al. 2021). Human immune system has been categorized into two parts: natural (or innate) and adaptive (or acquired or specific), as presented in Figure 9.1. Both of them comprises various blood-borne factors (like cytokines, antibodies, complement) and cells (Calder and Kew 2002), which give the immunity to the host cells.

Furthermore, the development of first immune response occurs when a foreign substance enters into the host body and causes the production of innate immune system. These responses are quick, but they are not very much specialized, since they seem to be carry low impact paralleled to the acquired immune response. Consequently, these immune cells directly destroy the foreign substances or pathogens by liberating toxic materials (like hydrogen peroxide or superoxide radical) from phagocytes or by emancipating toxic proteins produced by natural killer cells.

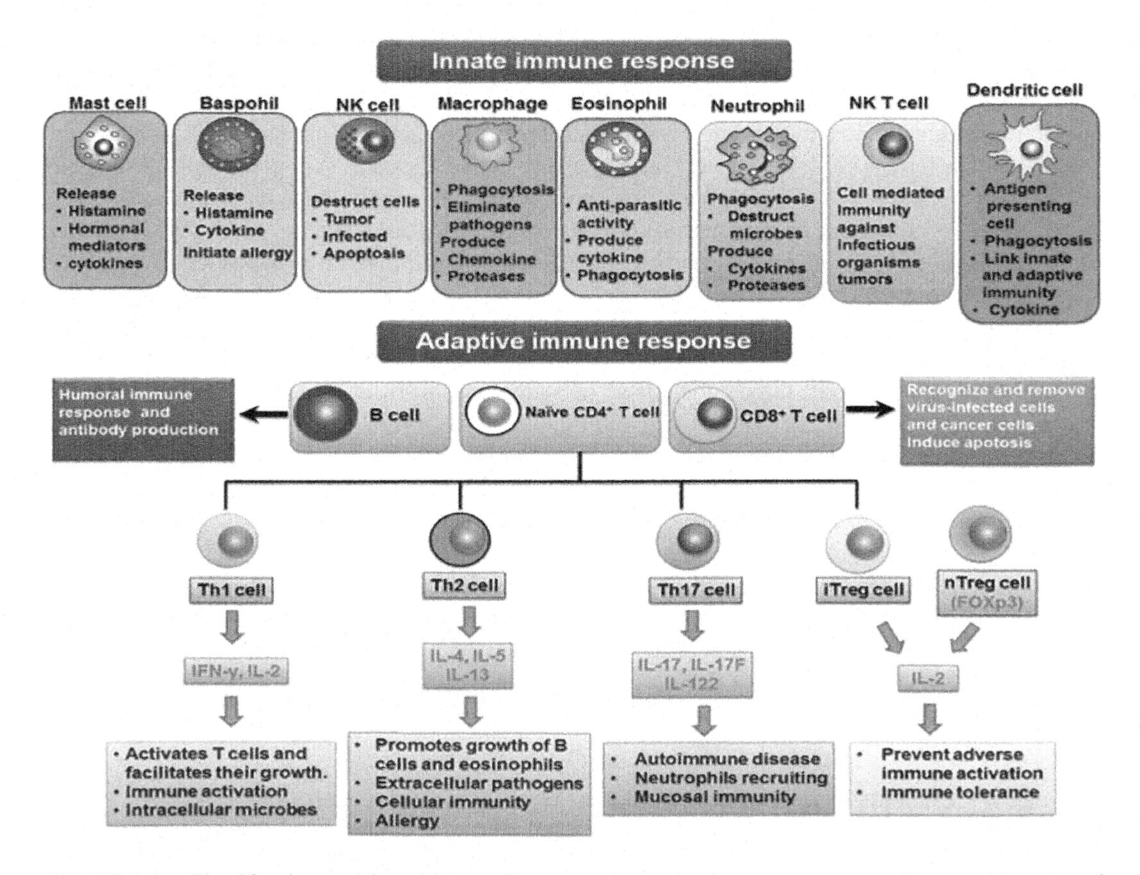

FIGURE 9.1 Classification and functioning of innate and adaptive immune responses (Jantan, Ahmad, and Bukhari 2015).

Phagocytosis is one of the processes of the immune system, which works on by engulfing pathogens responsible for invading the host cell (Venkatalakshmi et al. 2016). The acquired immune system controls and exclusively identifies any foreign material, and be aware of it in case our body comes across it in the future. In such scenario, immune responses start the production of T-lymphocytes, which have a significant role in the recognition of antigen. Additionally, T-lymphocytes have been well known for their cytotoxic activity, and the helper T-lymphocytes help in monitoring the functioning of other cells involved in the immune response. The production of interleukin (IL)-2 and interferon gamma (IFN-γ) arises from Type 1 T helper (Th1), which plays an important role in the antiviral and cellular immune response. Furthermore, the other interleukins such as IL-4, IL-5, and IL-13 are produced from Type 2 T helper (Th2), which possess allergic, antiparasitic, and humoral immune responses (Maheshwari et al. 2022). The other lymphocytes of our adaptive immune system include B cell, which produces antibody or immunoglobulin (Ig), and these Igs usually respond to the antigen that enters into the host cells. Ig is very much distinctive for the pathogens as it assists our immune system to recognize and kill them. Therefore, B-lymphocytes are responsible for humoral immunity (Schroeder and Cavacini 2010). Moreover, host immune system works by three ways; among the three ways, the first way includes surveillance, where immune cells get involved in response to the surrounding foreign invasion by moving in and out of the tissues, blood, and lymph. Furthermore, once these surveillance immune cells are not in a position to combat the foreign invasion, the second mode of the operation starts, i.e., activation or response (Percival 2011). These immune responses or cell activation causes the synthesis of cytokines, leukotrienes, and prostaglandins, leading to communication to the other cells associated with immune response. Subsequently, it leads to the proliferation of cell and ultimately causes the destruction of foreign substances by cytotoxicity activity. Furthermore, when such reaction occurs in the host cell, symptoms like inflammation or fever appears and causes illness.

Finally, the third mode of immune response gets activated, leading to the production of $CD8^+$ T cells, macrophages, γδ T cells, and natural killer cells, which try to kill activated immune cells to end the response. Once all the stages of immune response completed, again the immune system comes back to the regular surveillance action to receive any further foreign substance (Percival 2011). Immunodeficiencies usually arise, once a host immune system becomes inactive due to the absence of one or more of the components of the immune system. Additionally, several factors such as age, sex, lifestyle, malnutrition, genetic variability, alcohol/drug abuse, stress, and environmental pollution play an important role in altering the immune response or competence. Furthermore, several exogenous as well as endogenous factors causing either suppression or stimulation of immune system affect the competence and function of the immune system.

These bioactive components improve host immunity by scavenging free radical species, reducing oxidative stress, supporting the proliferation of lymphocytes and aggregation of platelets, improving anti-inflammatory as well as immunomodulatory mechanism. In healthy individual body, the host cell maintains homeostasis with the help of immune system. Thus, the supplementation of food intake with an appropriate quantity of phytochemicals could improve body's natural defense mechanism by strengthening our immune response. Therefore, any compounds with a potential to normalize or modulate pathophysiological processes are called immunomodulators. The biomolecules of synthetic or biological origin capable of modulating, suppressing, and stimulating any components of adaptive or innate immunity are known as immunomodulators, immune-restoratives, immune-augmenters, or biological response modifiers (Jantan, Ahmad, and Bukhari 2015). Immunomodulation is a very comprehensive term, which refers to any alterations in the immune response, and the development of these responses could be attributable to various physiological processes such as induction, expression, amplification, or inhibition of any part or phase in the immune response. Moreover, immunomodulators can be classified as immune-stimulants, immune-adjuvants, or immune-suppressants as presented in Figure 9.2.

Over the past few decades, the understanding and knowledge about these phytochemicals have significantly increased not only about their gastronomic values, but also about their medicinal

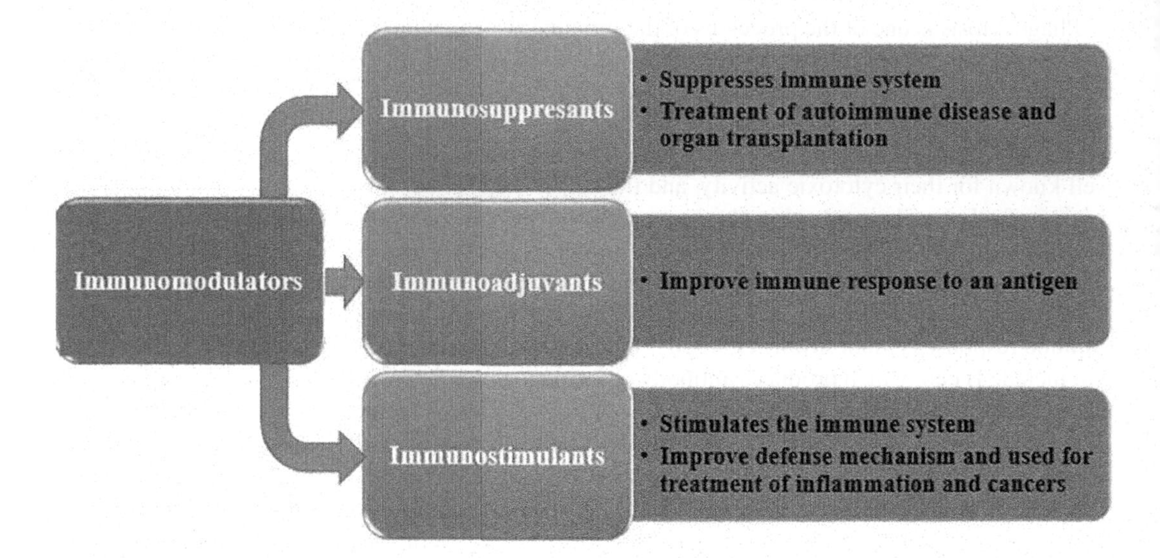

FIGURE 9.2 Classification of immunomodulators and their possible mechanisms of action.

properties. Furthermore, multifold surge in researches involving phytochemicals at both preclinical and clinical levels has led their potential medicinal values in various disorders including metabolic disorders (Francini-Pesenti, Spinella, and Calò 2019). It is a well-known fact that to keep healthy immune system, plant foods containing a rich supply of phytochemicals have a very significant role. Therefore, in recent years, scientific communities have explored these phytochemicals at various levels to put forward the scientific evidence to confirm the efficacy of such phytochemicals for improving the immune system of the host cells. Furthermore, considering the scientific fact regarding such phytochemicals (alkaloids, phenolics, flavonoids, anthocyanins, terpenoids, volatile oils, saponin, tannins, anthocyanin, sterols, resins, polysaccharides, lectins, and many more) as presented in Table 9.1, the consumption of phytochemicals has become so ubiquitous and it could be considered as the need of the hours to combat various bacterial, viral, and metabolic diseases (Maheshwari et al. 2022). Phytochemicals derived from plant sources such as glycosides, alkaloids, phenolics, flavonoids, volatile oils, saponin, tannins, anthocyanin, sterols, resins, polysaccharides, and lectins are well known for their immune-modulating properties. Various phytochemicals and their sources along with possible mechanism of actions are presented in Table 9.1.

9.4 PHYTOCHEMICALS AS NUTRACEUTICALS AND PHARMA FOOD

In recent years, there is a global shift in the treatment modalities, as well as preventive approaches, as awareness regarding the consumption of natural bioactive compounds such as phytochemicals has increased. Phytochemicals have been regarded as one of the important bioactive components for the prevention of metabolic disorders, as well as infectious diseases. Therefore, the role of these phytochemicals in recent years has largely taken the form of nutraceuticals, pharma foods, and functional foods. Among all these forms, nowadays nutraceuticals are gaining more attention due to their natural occurrence in food and their preventive, maintenance, and health-promoting effect. These nutraceuticals have been reported for their therapeutic values in several diseases such as diabetes, common cold, cancer, hypertension, inflammatory bowel disease, arthritis, dyslipidemia, and increased life span by postponing aging, integrity of the body, and support of smooth normal functioning. Nutraceuticals were first presented by Dr. Stephen DeFelice, founder and chairperson of the Foundation for Innovation in Medicine, Cranford, New Jersey, in 1989 (Kalra 2003). DeFelice introduces the "nutraceuticals" term by merging these two words "nutrition" and "pharmaceuticals,"

TABLE 9.1
List of Various Phytochemicals, Sources, and their Possible Mechanism of Action

Phytochemicals	Source	Possible Mechanism of Action	References
Curcumin	*Curcuma longa* L.	Enhance bone marrow cellularity, alpha esterase-positive cells, and phagocytic activity.	(Bhattacharya and Paul 2021)
Genistein	Soybeans		
Allicin	*Allium sativum*		
Ginseng	*Panax ginseng*		
Piperine	*Piper longum*	Increases total WBC count, bone marrow cellularity, and total antibody production	(Venkatalakshmi et al. 2016)
Tetrandrine	*Stephania tetrandra*	Suppresses cytokine synthesis and prevents NF-kB-mediated release of inflammatory factors	(Venkatalakshmi et al. 2016)
Aucubin	*Plantago major*	Enhances lymphocyte propagation and secretion of IFN-γ	(Chiang et al. 2003)
Mangiferin	*Mangefera indica*	Enhances IgG1 and IgG2b production	(García et al. 2003)
Ellagic acid	*Punica granatum*	Antioxidant and antiproliferative effects	(Venkatalakshmi et al. 2016)
Orientin	*Jatropha curcas*	Stimulation of humoral and cell-mediated immune response	
Vitexin		Stimulation of humoral and cell-mediated immune response	(Venkatalakshmi et al. 2016)
Quercetin-3-O-rutinoside	*Urtica dioica*	Immunomodulation	(Akbay et al. 2003)
Kaempherol-3-O-rutinoside		Immunomodulation	
Monoterpenes	*Ocimum sanctum*	Activation of innate and adaptive immunity/immunostimulant	(Antonioli and Fornai 2020)
Polysaccharides	*Astragalus membranaceous* (Fisch.) Bge.	Activation of cellular immunity/immunostimulant	(Antonioli and Fornai 2020)
Berberine	*Coptis chinensis* Franch	T-helper cells' cytokines [Th1 (TNF-α, IL-2), and Th2 (IL-4)] production down-regulated	(Jantan, Ahmad, and Bukhari 2015)
Sophocarpine	*Sophora alopecuroides* L.	Inhibits the production of NO and pro-inflammatory cytokines TNF-α and IL-6. Impedes the expression of iNOS and COX-2	(Jantan, Ahmad, and Bukhari 2015)
Luteolin	*Lonicera japonica*	Reduces inflammatory mediators (IFN-γ, IL-6) secretions and decreased COX-2 and ICAM-1 expression. Blocks heat shock protein 90 activity	(Jantan, Ahmad, and Bukhari 2015)
Wogonin	*Scutellaria baicalensis* Georgi	Inhibits adhesion and relocation of leukocytes by preventing cell adhesion molecules' expression. Decreases allergic airway inflammation by inducing eosinophil apoptosis through the activation of caspase-3.	(Jantan, Ahmad, and Bukhari 2015)
Thymoquinone	*Nigella sativa* L.	Inhibits LPS-induced fibroblast proliferation and H_2O_2-induced 4-hydroxynonenal generation. Improvement of number of circulating and thymus-homing CD4+ and CD8+. Inhibits IL-1β, TNF-α, MMP-13, COX-2, and PGE2 while blocking phosphorylation of MAPK p38, ERK1/2, and NF-kBp65	(Vaillancourt et al. 2011, Ahmad et al. 2021)
Vitamin C	Fruits	Antioxidant, immunomodulatory properties, free radical scavenging property	(Maheshwari et al. 2022)

and he described the term nutraceuticals as "food or a part of food which not only imparts health benefits but also contributes in preventing or treating various diseases" (Ashraf and Elkhalifa 2020). Moreover, nutraceuticals have no regulatory definitions; however, it can be summarized as any food bioactive compounds, which play an important role in maintaining the normal physiological functions in human beings. More importantly, these naturally derived nutraceutical product has a minimal or no side effect compared to other synthetic biomolecule-based drugs. Therefore, nutraceuticals could play a significant role in upgrading human health and well-being (Ashraf et al. 2020, Ahmad et al. 2013). Nutraceuticals have been categorized in several ways, based upon their mechanisms of action and applications. Naturally, available food sources and their phytoactive ingredients are considered for their nutraceutical values, which can be characterized as probiotic, prebiotic, antioxidants, dietary fiber, polyunsaturated fatty acids, vitamins, polyphenols, and spices (Das et al. 2012).

9.4.1 Role of Phytochemicals as Nutraceuticals and Theirs Health-Promoting Action

Phytochemicals have been known to the mankind since antiquity days in the form of nutrient components of plants. The bioactive constituents from plant or plant-based food have been recognized as phytochemicals. These phytochemicals are present in several plant foods, such as fruits, whole grains, vegetables, nuts, spices, even dark chocolate, and red wine. These phytochemicals are used in foods for their gastronomic and sensorimotor values, such as taste, color, texture, and aroma. Additionally, these phytochemicals have the ability to regulate key pathophysiological processes of the host cell. Meanwhile, several phytochemicals have been found to possess excellent health-promoting properties, which prevent the damage to macromolecules such as DNA, RNA, and protein including lipids, and other cellular components from oxidative stress. In addition, phytochemicals were found to control protein synthesis, act like hormones, and affect biochemical properties of blood (Murakami 2013, Palacios et al. 2020). Various phytochemicals derived from plant foods are currently explored for their nutraceutical properties since they possess a beneficial effect in preventing and/or treating metabolic disorders due to their unique therapeutic properties and safety (Bacanli et al. 2019), since these nutraceuticals are widely explored for the management of various metabolic diseases, diabetes (Moreno-Valdespino et al. 2020), obesity (Li et al. 2019), insulin resistance (Mahdavi et al. 2021), and cardiovascular diseases (Pop et al. 2018). According to a report of 2008, World Health Organization, 80% of diabetic patients were found to use the herbal medicine or nutraceutical product in some or the other forms (Bacanli et al. 2019, Su et al. 2021). Phytochemicals such as alkylresorcinols (phenolic lipids), allicin (organosulfur), carotenoids, capsaicin, curcumin, flavonoids, genistein, indoles (organosulfur compound), isothiocyanates (organosulfur compounds, e.g., sulforaphane), monoterpenes (including limonene), phenolic acids, resveratrol, saponins (glucosides), and tannins are widely known plant-derived compounds, which have been investigated for innumerable medicinal properties.

Metabolic disorders such as obesity, diabetes, non-alcoholic fatty liver disease, and cardiovascular diseases like hypertension, renal disease and cognitive deficits and hypertriglyceridemia, low high-density lipoprotein cholesterol, and cancer are considered as one of the biggest health problems of the 21st century. Moreover, it has been also noted that dietary and lifestyle modifications could prove a way to encounter these issues and decrease the occurrence of metabolic disorders. Significant evidence has confirmed that dietary bioactive compounds, which are also referred as phytochemicals, are beneficial for preventing and managing metabolic disorders (Xiao and Bai 2019). Dietary intake of whole-grain foods, which contain alkylresorcinols, has been linked to lowering the risk of chronic diseases such as diabetes, obesity, and heart diseases (Bondia-Pons et al. 2009). Studies have found that alkylresorcinols, which are phenolic lipids, found in whole grain, wheat, and rye grains, lower the risks of diabetes, coronary heart diseases, hypertension, and also several types of cancer (Rejman and Kozubek 2004, Bondia-Pons et al. 2009, Lillioja et al. 2013). Allicin, a known organosulfur compound, which is commonly found in leeks, chives, garlic, and onion, possesses excellent antimicrobial properties, thus reducing ulcers and lowering blood cholesterol (Shi et al. 2019, Miękus et al. 2020, Panyod and Wu 2022).

Carotenoids such as β-carotene, lutein, and lycopene are present in mostly pigmented fruits and vegetables. These carotenoids have been reported to function as potent free radical scavengers. Studies have reported that the consumption of carotenoids reduces the risks of cancer and other diseases (Raju et al. 2021, Jianjun Wu 2022). Additionally, carotenoids have been reported for reducing obesity and its associated diseases as low serum carotenoids are linked to the development of obesity and associated diseases (Marcelino, Machate, and Freitas 2020, Iqbal et al. 2021).

Capsaicin is a major active phytoconstituent found in chilli and peppers, and has been reported to lower the risk of fatal clots in the heart and arteries by preventing blood coagulation. Capsaicin has been reported to activate vanilloid receptors, since these receptors help to detect a wide range of temperatures from hot and cold noxious temperatures to innocuous thermal stimuli (Messeguer, Planells-Cases, and Ferrer Montiel 2006). Preclinical as well as *in vitro* studies have established the effectiveness of low-dose dietary capsaicin in attenuating metabolic disorders (Panchal, Bliss, and Brown 2018).

Curcumin is a polyphenol compound predominantly found in the rhizome of *Curcuma longa*. It is also the active ingredient of the dietary spice turmeric. It has been extensively studied for its antioxidant, antihepatotoxic, and anti-inflammatory properties, including its ability to reverse insulin resistance, hyperglycemia, and hyperlipidemia, and in treating various other metabolic disorders (Aggarwal 2010, Jabczyk, Nowak, and Hudzik 2021). In high-fat diet-induced obese and leptin-deficient ob/ob mice, a nutraceutical formulation containing curcumin has been found to lower glucose intolerance and insulin resistance (Weisberg, Leibel, and Tortoriello 2008, Lee et al. 2019). Additionally, it has been reported that curcumin could downregulate obesity and reduce the impact of associated problems (Alappat and Awad 2010). The metabolic benefits of curcumin was associated with the reduction in macrophage infiltration of white adipose tissue, an increase in the production of adiponectin in the adipose tissue, and a decrease in NF-κB activity in the liver (Akbari et al. 2019, Kasprzak-Drozd and Oniszczuk 2022). In a genetically modified diabetic db/db mice model, curcumin was found to increase the expression of AMP-activated protein kinase (AMPK) and peroxisome proliferator-activated receptor gamma (PPAR-γ), and diminished NF-κB protein in db/db mice (Jiménez-Flores et al. 2014). Also, one of the studies reported that beneficial effects of curcumin were due to the activation of PPAR-γ (Mazidi et al. 2016). PPAR-α and PPAR-γ have been found to regulate lipid metabolism, insulin sensitivity, and glucose homeostasis, and their agonists are found to be beneficial in the treatment of hyperlipidemia and diabetes mellitus (Monsalve et al. 2013).

Epigallocatechin gallate is a polyphenol flavonoid primarily found in green tea; however, some other variety of foods also contain this important bioactive nutraceutical, such as avocados, kiwis, pears, strawberries, cherries, peaches, pistachios, blackberries, and hazelnuts (Harnly et al. 2006). Furthermore, epigallocatechin gallate has been explored by scientific communities for various health benefits, such as lowering the levels of circulating cholesterol, cardiovascular protection, weight loss, attenuation of inflammation as well as an antidiabetic agent (Legeay et al. 2015). Epigallocatechin gallate has been reported to improve wound healing in STZ-induced diabetic mice, as it has been noted that it reduces the pro-inflammatory markers (such as IL-6, TNF-α, and IL-1β) along with causing the inhibition of notch signaling and macrophage accumulation at a wound site (Huang et al. 2019). Moreover, the supplementation of epigallocatechin gallate at 0.1% in diet has been observed to suppress hyperglycemia-evoked inflammation in the adipose tissue of Goto-Kakizaki rats, a polygenic nonobese strain that develops adult-onset type 2 diabetes early in life (Uchiyama et al. 2013). Epigallocatechin gallate is also reported to act as an antioxidant in human hepatocellular liver carcinoma cells (HepG2) to attenuate oxidative stress by preventing the synthesis of ROS and endorsing antioxidant enzyme activities such as superoxide dismutase and glutathione peroxidase. It has also been reported that epigallocatechin gallate has potential to suppress vascular inflammation by preventing NF-κB activation (Yang et al. 2013). Therefore, based upon several potential benefits of epigallocatechin gallate, this can be considered as one of the phytochemicals with its efficacy for the developments of nutraceuticals.

Naringenin is one of the important flavanones largely present in grapefruit and other citrus fruits (Den Hartogh and Tsiani 2019). Studies reported that naringenin has shown a positive response as antidiabetic agent since it has the capability to ameliorate hyperlipidemia, hyperglycemia, and insulin resistance in the rat model of T2DM (Ren et al. 2016). It has been also reported that naringenin was effective in reducing the glucose intolerance in mice during pregnancy via improving the membrane translocation of insulin-responding glucose transporter GLUT4 and glucose uptake through the AMPK pathway in skeletal muscle cells (Li et al. 2019). Naringenin has shown a strong prospective to reduce vascular complications of diabetes and improve abnormally altered lipid profile in a mouse model of diabetic nephropathy. Furthermore, naringenin administration in diabetic rat reduces the hyperglycemia-mediated inflammation in hepatic and pancreatic tissue models (Annadurai, Thomas, and Geraldine 2013). Henceforth, naringenin is notably capable of suppressing circulating pro-inflammatory cytokine levels, including IL-6, TGF-β, TNF-α, and ICAM-1, in pregnant diabetic rats (Zhao and Yin 2020). *In vitro* studies suggested that incubation with naringenin significantly inhibited the expression of chemokines and pro-inflammatory cytokines in human placenta and visceral adipose tissue instigated by TNF-α, an inflammation inducer (Nguyen-Ngo, Willcox, and Lappas 2019). More precisely, naringenin is reported to reduce the inflammatory response of macrophages and neutrophil cells that play an important role in modulating the innate immune response of adipose and other metabolic tissues (Tsuhako et al. 2020). Furthermore, based upon the scientific evidence found by several authors, naringenin possesses very unique properties to become an important nutraceutical.

Saponin possesses structurally a diverse range of phytochemicals with several functional characteristics. Several food items such as soybean, green vegetables, tomatoes, alfalfa sprouts, other lentils, sprouts, and potatoes are found to be a rich source of saponin. Saponin and its secondary metabolites have shown a preventive effect on metabolic syndrome, and their major site of action is in the human intestine. Studies report that saponin displays an antioxidant activity, induction of apoptosis, cell cycle arrest, cellular invasion and inhibition, and autophagy (Elekofehinti and Iwaloye 2021, Luo et al. 2020).

Hesperetin is the aglycone of hesperidin flavanone, which is mainly present in orange, lemon, and other citrus fruits (Choi and Ahn 2008). Hesperetin has been reported to have antidiabetic, anti-inflammatory, antihypertensive, antioxidative, antihyperlipidemic, and cardioprotective properties (Iranshahi et al. 2015). Hesperetin reduces hyperglycemia by decreasing oxidative stress and quashing the production of pro-inflammatory cytokines IL-6 and TNF-α in the rat model of STZ-induced diabetes (Jayaraman et al. 2018). Moreover, oral administration of hesperetin in streptozotocin-induced diabetic rats reduces the morphological and functional abnormalities in the kidney via modulatory effects on the AGEs/RAGE axis and TGF-β1-ILK-Akt signaling pathway (Chen et al. 2019). Hence, hesperetin may serve as a dietary supplement or nutraceutical for the treatment of diabetes and other metabolic disorders.

Chrysin is one of the most prominent flavonoids present in honey (Wollenweber et al. 1985). Studies reported that chrysin has emerged as one noticeable bioactive nutraceutical for the management and treatment of diabetes along with anti-inflammatory activity. In addition, chrysin has been reported for various metabolic functions due to its PPAR-γ agonist activity. Chrysin has been reported to increase systolic blood pressure *in vivo* and alleviate vasoconstriction *in vitro* in the rat model of fructose-induced insulin resistance and vascular complications, and the whole mechanism is largely dependent on PPAR-γ action (El-Bassossy, Abo-Warda, and Fahmy 2014). Chrysin has shown a neuroprotective effect accompanied by the inhibition of inflammatory signals, via a blunted activation of NF-κB and inhibition of pro-inflammatory cytokines (IL-1β, IL-6, and TNF-α) in the cerebral cortex and hippocampus of STZ-induced diabetic rats (Li et al. 2014).

Kaempferol is one of the typical flavonoids present in our diet, including tea, fruits, and vegetables like raspberry, onion, chives, peach, cucumber, leek, beans, broccoli, tomato, grapes, strawberries, cowberries, and apples. It is more prominent in various Chinese medicinal herbs such as *Crocus sativus*, *Centella asiatica*, *Lycium barbarum*, and Aloe vera (Calderón-Montaño et al. 2011). Among the various health-promoting effects, kaempferol has potential to combat diabetes. Studies reported that kaempferol glycosides fractions lead to a reduction in body weight gain, triglycerides level, blood glucose levels along with improved insulin resistance in mice fed a high-fat diet (Zang et al. 2015). Additionally, the administration of kaempferol in diabetic-induced rats ameliorated insulin resistance via the inhibition of NF-κB-mediated hepatic inflammation (Luo et al. 2015). Along with antidiabetic activity, kaempferol has also been reported for its cardioprotective effect in diabetic rats by attenuating oxidative stress and inflammation in heart tissue (Suchal et al. 2017).

Apigenin (4′,5,7-trihydroxyflavone) belongs to the flavonoid family and is abundantly present in fruits (oranges, onions, grapefruit), vegetables, herbs, tea, and chamomile (Shukla and Gupta 2010). Apigenin possesses antidiabetic, anti-inflammatory, antiapoptotic, antioxidant, and antifibrotic properties. Apigenin has been reported to act as a pharmacological inhibitor of CD38, the master regulator of extracellular NAD+ pools. Apigenin administration to obese mice augments NAD+ levels and reduces the global protein acetylation, and by virtue of that, it restores homeostasis of lipid and glucose metabolism, which concludes apigenin potential as antidiabetic agent and its therapeutic effects in other metabolic diseases via NAD+-dependent pathways (Escande et al. 2013). Moreover, apigenin treatment in diabetic nephropathy of the rat model alleviates renal dysfunction, glomerulosclerosis, inflammation, and oxidative stress in association with substantial blockade of MAPK in the kidney (Malik et al. 2017).

Ellagic acid is a hydrolyzed tannin, is majorly present in pomegranate, berries, and dried fruit. Ellagic acid has been reported for various metabolic disorders such as cardiovascular disease (Jordão et al. 2017) and neurodegenerative diseases (Firdaus et al. 2018), along with other biological effects such as anticancer, antioxidant, and anti-inflammatory activities (Baradaran Rahimi et al. 2020). Interestingly, a recent report suggested that ellagic acid is a remarkable antidiabetic compound. Moreover, in type 2 diabetes and its related kidney injury, ellagic acid treatment significantly combated insulin resistance, hyperglycemia, and diabetic nephropathy in rat model by inhibiting the activation of NF-κB. Therefore, the protective effect of ellagic acid was closely associated with the suppression of inflammatory response, evidenced by notably lowered serum levels of pro-inflammatory cytokines (IL-1β, IL-6, and TNF-α) (Gratas-Delamarche et al. 2014). Specifically, ellagic acid treatment significantly alleviated hyperglycemia-evoked renal injury in DKD (diabetic kidney disease) mice by the inhibition of the high-mobility group box 1-toll-like receptor 4-NF-κB pathway (Zhou et al. 2019).

Glucosinolates in recent years have been ranked very important phytochemical due to their biological activities, especially in cancer management. Glucosinolates are present in food sources such as cauliflower, broccoli, horseradish, mustard greens, cabbage, kale, and Brussels sprouts. Since they contain organosulfur compound (indoles) as phytochemicals, which may induce enzymatic reaction, they could inhibit carcinogen-mediated damage to macromolecules like proteins, RNA, DNA, and other enzymes and also exhibit an estrogen inhibitory activity (Ruhee et al. 2020). The above-mentioned vegetables also contain another phytochemical isothiocyanates (organosulfur compounds that include sulforaphane), which have excellent antioxidant potentials, inhibit enzymes that activate carcinogens, and trigger enzymes that detoxify carcinogens, reducing the risk of breast cancer and prostate cancer. Also, glucosinolates are another phytochemicals, exclusively present in cruciferous vegetables, that exhibit anti-inflammatory, chemoprotective, and antioxidant effects (de Figueiredo et al. 2015, Connolly et al. 2021).

9.5 CONCLUSION

Phytochemicals have been proved as one of the most potent natural biomolecules for providing alternative therapeutics for the management of various chronic diseases such as cancer, rheumatoid arthritis, cold and flu, immunomodulatory, diabetes, asthma, and cardiovascular diseases. This chapter highlights the importance of phytochemicals as immunomodulators, nutraceuticals, and pharma foods. Various micronutrients such as vitamins A, C, and E in addition to non-nutritive phytochemicals like carotenoids, flavonoids, phenolic acids, and tannins are essential for antibody production, stimulating phagocytosis, inhibition of activity of pro-inflammatory cytokines, lymphocyte proliferation, and to improve the activity of natural killer cells. This keep our immune system in surveillance mode and play a key role in enhancing cells involved in both innate and acquired immune responses. Phytochemicals of nutraceuticals importance are bioactive constituents that sustain or promote health and occur at the intersection of food and pharmaceutical industries. They have a tremendous impact on the health care system and may provide medical health benefits including the prevention and/or treatment of diseases and physiological disorders. They play specific pharmacological effects in human health as anti-inflammatory, antioxidants, antispasmodic, hepatoprotective, hypolipidemic, neuroprotective, hypotensive, antiaging, diabetes, osteoporosis, DNA damage, cancer, heart diseases, and many more. These identified bioactive phytochemicals have been currently employed in making pharma food as well, which could pave the way to prevent various noncommunicable diseases. Moreover, *in vivo* and *in vitro* studies are needed to understand the activity of these phytochemicals and to further explore the immunomodulatory mechanism, nutraceuticals, and pharma foods. This comprehensive chapter throws light on various phytochemicals present in fruits and vegetables, and this evidence could be further used for the development of novel immune booster functional foods or pharma foods with natural ingredients.

REFERENCES

Abbasi, Arshad Mehmood, Munir Hussain Shah, and Mir Ajab Khan. 2015. "Phytochemicals and nutraceuticals." In *Wild Edible Vegetables of Lesser Himalayas: Ethnobotanical and Nutraceutical Aspects, Volume 1*, edited by Arshad Mehmood Abbasi, Munir Hussain Shah and Mir Ajab Khan, 31–65. Cham: Springer International Publishing.

Aggarwal, B. B. 2010. "Targeting inflammation-induced obesity and metabolic diseases by curcumin and other nutraceuticals." *Annu Rev Nutr* no. 30:173–99 doi: 10.1146/annurev.nutr.012809.104755.

Ahmad, Faruque, Fakhruddin Ali Ahmad, Zr Azaz, Ahmad Azad, Sarfaraz Alam, and Syed Amir Ashraf. 2013. "Nutraceutical Is the Need of Hour." *World Journal of Pharmacy and Pharmaceutical Sciences* 2(5): 2516–25.

Ahmad, Md Faruque, Fakhruddin Ali Ahmad, Syed Amir Ashraf, Hisham H Saad, Shadma Wahab, Mohammed Idreesh Khan, M. Ali, Syam Mohan, Khalid Rehman Hakeem, and Md Tanwir Athar. 2021. "An updated knowledge of Black seed (Nigella sativa Linn.): Review of phytochemical constituents and pharmacological properties." *J Herb Med* no. 25:100404. doi: 10.1016/j.hermed.2020.100404.

Akbari, M., K. B. Lankarani, R. Tabrizi, M. Ghayour-Mobarhan, P. Peymani, G. Ferns, A. Ghaderi, and Z. Asemi. 2019. "The effects of curcumin on weight loss among patients with metabolic syndrome and related disorders: A systematic review and meta-analysis of randomized controlled trials." *Front Pharmacol* no. 10:649. doi: 10.3389/fphar.2019.00649.

Akbay, P., A. A. Basaran, U. Undeger, and N. Basaran. 2003. "In vitro immunomodulatory activity of flavonoid glycosides from Urtica dioica L." *Phytother Res* no. 17 (1):34–37. doi: 10.1002/ptr.1068.

Alappat, L., and A. B. Awad. 2010. "Curcumin and obesity: Evidence and mechanisms." *Nutr Rev* no. 68 (12):729–38. doi: 10.1111/j.1753-4887.2010.00341.x.

Annadurai, T., P. A. Thomas, and P. Geraldine. 2013. "Ameliorative effect of naringenin on hyperglycemia-mediated inflammation in hepatic and pancreatic tissues of Wistar rats with streptozotocin- nicotinamide-induced experimental diabetes mellitus." *Free Radic Res* no. 47 (10):793–803. doi: 10.3109/10715762.2013.823643.

Antonioli, L., and M. Fornai. 2020. "NKG2A and COVID-19: Another brick in the wall." no. 17 (6):672–74. doi: 10.1038/s41423-020-0450-7.

Ashraf, Syed A., Mohd Adnan, Mitesh Patel, Arif J. Siddiqui, Manojkumar Sachidanandan, Mejdi Snoussi, and Sibte Hadi. 2020. Fish-based bioactives as potent nutraceuticals: Exploring the therapeutic perspective of sustainable food from the sea. *Marine Drugs* 18 (5): 265.

Ashraf, S. A., and A. E. O. Elkhalifa. 2020. "Cordycepin for health and wellbeing: A potent bioactive metabolite of an entomopathogenic cordyceps medicinal fungus and its nutraceutical and therapeutic potential." no. 25 (12). doi: 10.3390/molecules25122735.

Bacanli, M., S. A. Dilsiz, N. Başaran, and A. A. Başaran. 2019. "Effects of phytochemicals against diabetes." *Adv Food Nutr Res* no. 89:209–38 doi: 10.1016/bs.afnr.2019.02.006.

Badhani, Bharti, Neha Sharma, and Rita Kakkar. 2015. "Gallic acid: a versatile antioxidant with promising therapeutic and industrial applications." *RSC Adv* no. 5 (35):27540–57. doi: 10.1039/C5RA01911G.

Baradaran Rahimi, V., M. Ghadiri, M. Ramezani, and V. R. Askari. 2020. "Antiinflammatory and anti-cancer activities of pomegranate and its constituent, ellagic acid: Evidence from cellular, animal, and clinical studies." no. 34 (4):685–720. doi: 10.1002/ptr.6565.

Basheer, L., and Z. Kerem. 2015. "Interactions between CYP3A4 and dietary polyphenols." *Oxid Med Cell Longev* no. 2015:854015. doi: 10.1155/2015/854015.

Behl, Tapan, Keshav Kumar, Ciprian Brisc, Marius Rus, Delia Carmen Nistor-Cseppento, Cristiana Bustea, Raluca Anca Corb Aron, Carmen Pantis, Gokhan Zengin, Aayush Sehgal, Rajwinder Kaur, Arun Kumar, Sandeep Arora, Dhruv Setia, Deepak Chandel, and Simona Bungau. 2021. "Exploring the multifocal role of phytochemicals as immunomodulators." *Biomed Pharmacother* no. 133:110959. doi: https://doi.org/10.1016/j.biopha.2020.110959.

Bendich A, S S Shapiro. 1986. Effect of beta-carotene and canthaxanthin on the immune responses of the rat. *The Journal of Nutrition* 116(11): 2254–62. doi:10.1093/jn/116.11.2254

Bessler, H., H. Salman, M. Bergman, Y. Alcalay, and M. Djaldetti. 2008. "In vitro effect of lycopene on cytokine production by human peripheral blood mononuclear cells." *Immunol Invest* no. 37 (3):183–90. doi: 10.1080/08820130801967809.

Bhattacharya, Sonali, and Sudipta Majumdar Nee Paul. 2021. "Efficacy of phytochemicals as immunomodulators in managing COVID-19: A comprehensive view." *VirusDisease* no. 32 (3):435–45. doi: 10.1007/s13337-021-00706–2.

Bondia-Pons, Isabel, Anna-Marja Aura, Satu Vuorela, Marjukka Kolehmainen, Hannu Mykkänen, and Kaisa Poutanen. 2009. "Rye phenolics in nutrition and health." *J Cereal Sci* no. 49 (3):323–36. doi: https://doi.org/10.1016/j.jcs.2009.01.007.

Calder, P. C., and S. Kew. 2002. "The immune system: A target for functional foods?" *Br J Nutr* no. 88 (Suppl 2):S165–77. doi: 10.1079/bjn2002682.

Calderón-Montaño, J. M., E. Burgos-Morón, C. Pérez-Guerrero, and M. López-Lázaro. 2011. "A review on the dietary flavonoid kaempferol." *Mini Rev Med Chem* no. 11 (4):298–344. doi: 10.2174/138955711795305335.

Campos-Vega, R., and B. D. Oomah. 2013. "Chemistry and classification of phytochemicals." *Handbook of Plant Food Phytochemicals: Sources, Stability and Extraction* (pp. 5–48), edited by N. P. Brunton B. K. Tiwari, and C. S. Brennan. Lewiston, New York: John Wiley and Sons Ltd.

Chagas, Caroline Manto, and Laleh Alisaraie. 2019. "Metabolites of vinca alkaloid vinblastine: Tubulin binding and activation of nausea-associated receptors." *ACS Omega* no. 4 (6):9784–99. doi: 10.1021/acsomega.9b00652.

Chaplin, D. D. 2010. "Overview of the immune response." *J Allergy Clin Immunol* no. 125 (2 Suppl 2):S3–23. doi: 10.1016/j.jaci.2009.12.980.

Chen, Y. J., L. Kong, Z. Z. Tang, Y. M. Zhang, Y. Liu, T. Y. Wang, and Y. W. Liu. 2019. "Hesperetin ameliorates diabetic nephropathy in rats by activating Nrf2/ARE/glyoxalase 1 pathway." *Biomed Pharmacother* no. 111:1166–75 doi: 10.1016/j.biopha.2019.01.030.

Chew, B. P., and J. S. Park. 2004. "Carotenoid action on the immune response." *J Nutr* no. 134 (1):257s–61s. doi: 10.1093/jn/134.1.257S.

Chiang, L. C., L. T. Ng, W. Chiang, M. Y. Chang, and C. C. Lin. 2003. "Immunomodulatory activities of flavonoids, monoterpenoids, triterpenoids, iridoid glycosides and phenolic compounds of Plantago species." *Planta Med* no. 69 (7):600–4. doi: 10.1055/s-2003-41113.

Choi, E. J., and W. S. Ahn. 2008. "Neuroprotective effects of chronic hesperetin administration in mice." *Arch Pharm Res* no. 31 (11):1457–62. doi: 10.1007/s12272-001-2130-1.

Choi, J., X. Jiang, J. B. Jeong, and S. H. Lee. 2014. "Anticancer activity of protocatechualdehyde in human breast cancer cells." *J Med Food* no. 17 (8):842–48. doi: 10.1089/jmf.2013.0159.

Connolly, E. L., M. Sim, N. Travica, W. Marx, G. Beasy, G. S. Lynch, C. P. Bondonno, J. R. Lewis, J. M. Hodgson, and L. C. Blekkenhorst. 2021. "Glucosinolates from cruciferous vegetables and their potential role in chronic disease: Investigating the preclinical and clinical evidence." *Front Pharmacol* no. 12:767975. doi: 10.3389/fphar.2021.767975.

Cueva, C., M. V. Moreno-Arribas, P. J. Martín-Alvarez, G. Bills, M. F. Vicente, A. Basilio, C. L. Rivas, T. Requena, J. M. Rodríguez, and B. Bartolomé. 2010. "Antimicrobial activity of phenolic acids against commensal, probiotic and pathogenic bacteria." *Res Microbiol* no. 161 (5):372–82. doi: 10.1016/j.resmic.2010.04.006.

Das, L., E. Bhaumik, U. Raychaudhuri, and R. Chakraborty. 2012. "Role of nutraceuticals in human health." *J Food Sci Technol* no. 49 (2):173–83. doi: 10.1007/s13197-011-0269-4.

de Figueiredo, S. M., N. S. Binda, J. A. Nogueira-Machado, S. A. Vieira-Filho, and R. B. Caligiorne. 2015. "The antioxidant properties of organosulfur compounds (sulforaphane)." *Recent Pat Endocr Metab Immune Drug Discov* no. 9 (1):24–39. doi: 10.2174/1872214809666150505164138.

Den Hartogh, D. J., and E. Tsiani. 2019. "Antidiabetic properties of naringenin: A citrus fruit polyphenol." *Biomolecules* no. 9 (3). doi: 10.3390/biom9030099.

Devavrat, Tripathi. 2021. "Quercetin induces proteolysis of mesenchymal marker vimentin through activation of caspase-3, and decreases cancer stem cell population in human papillary thyroid cancer cell line." *Phytomedicine Plus* no. 1 (4):100108. doi: 10.1016/j.phyplu.2021.100108.

Efferth, T., P. C. Li, V. S. Konkimalla, and B. Kaina. 2007. "From traditional Chinese medicine to rational cancer therapy." *Trends Mol Med* no. 13 (8):353–61. doi: 10.1016/j.molmed.2007.07.001.

Ekaette, I., and M. D. A. Saldaña. 2021. "Ultrasound processing of rutin in food-grade solvents: Derivative compounds, antioxidant activities and optical rotation." *Food Chem* no. 344:128629. doi: 10.1016/j.foodchem.2020.128629.

El-Bassossy, H. M., S. M. Abo-Warda, and A. Fahmy. 2014. "Chrysin and luteolin alleviate vascular complications associated with insulin resistance mainly through PPAR-γ activation." *Am J Chin Med* no. 42 (5):1153–67. doi: 10.1142/s0192415x14500724.

Elamin, M. H., Z. Shinwari, S. F. Hendrayani, H. Al-Hindi, E. Al-Shail, Y. Khafaga, A. Al-Kofide, and A. Aboussekhra. 2010. "Curcumin inhibits the Sonic Hedgehog signaling pathway and triggers apoptosis in medulloblastoma cells." *Mol Carcinog* no. 49 (3):302–14. doi: 10.1002/mc.20604.

Elegir, G., A. Kindl, P. Sadocco, and M. Orlandi. 2008. "Development of antimicrobial cellulose packaging through laccase-mediated grafting of phenolic compounds." *Enzyme Microbial Technology* no. 43 (2):84–92. doi: https://doi.org/10.1016/j.enzmictec.2007.10.003.

Elekofehinti, O. O., and O. Iwaloye. 2021. "Saponins in cancer treatment: Current progress and future prospects." no. 28 (2):250–72. doi: 10.3390/pathophysiology28020017.

Elkhalifa, A. E. O., Al-Shammari, E., Alam, M. J., Alcantara, J. C., Khan, M. A., Eltoum, N. E. and Ashraf, S. A. 2021. "Okra-derived dietary Carotenoid lutein against breast cancer, with an approach towards developing a nutraceutical product: A meta-analysis study." *J Pharm Res Int* no. 33 (40A):135–42. doi: 10.9734/jpri/2021/v33i40A32230.

Escande, C., V. Nin, N. L. Price, V. Capellini, A. P. Gomes, M. T. Barbosa, L. O'Neil, T. A. White, D. A. Sinclair, and E. N. Chini. 2013. "Flavonoid apigenin is an inhibitor of the NAD+ ase CD38: Implications for cellular NAD+ metabolism, protein acetylation, and treatment of metabolic syndrome." *Diabetes* no. 62 (4):1084–93. doi: 10.2337/db12-1139.

Fiedor, J., and K. Burda. 2014. "Potential role of carotenoids as antioxidants in human health and disease." *Nutrients* no. 6 (2):466–88. doi: 10.3390/nu6020466.

Firdaus, F., M. F. Zafeer, E. Anis, M. Ahmad, and M. Afzal. 2018. "Ellagic acid attenuates arsenic induced neuro-inflammation and mitochondrial dysfunction associated apoptosis." *Toxicol Rep* no. 5:411–17. doi: 10.1016/j.toxrep.2018.02.017.

Fraisse, Didier, Catherine Felgines, Odile Texier, and Jean-Louis Lamaison. 2011. "Caffeoyl derivatives: Major antioxidant compounds of some wild herbs of the Asteraceae family." *Food Nutr Sci* no. 2: 181–192. doi: 10.4236/fns.2011.230025.

Francini-Pesenti, F., P. Spinella, and L. A. Calò. 2019. "Potential role of phytochemicals in metabolic syndrome prevention and therapy." *Diabetes Metab Syndr Obes* no. 12:1987–2002. doi: 10.2147/dmso.s214550.

Gammone, M. A., G. Riccioni, and N. D'Orazio. 2015. "Carotenoids: Potential allies of cardiovascular health?" *Food Nutr Res* no. 59:26762. doi: 10.3402/fnr.v59.26762.

García, D., J. Leiro, R. Delgado, M. L. Sanmartín, and F. M. Ubeira. 2003. "Mangifera indica L. extract (Vimang) and mangiferin modulate mouse humoral immune responses." *Phytother Res* no. 17 (10):1182–87. doi: 10.1002/ptr.1338.

Gratas-Delamarche, A., F. Derbré, S. Vincent, and J. Cillard. 2014. "Physical inactivity, insulin resistance, and the oxidative-inflammatory loop." *Free Radic Res* no. 48 (1):93–108. doi: 10.3109/10715762.2013.847528.

Güleç, K., and M. Demirel. 2016. "Characterization and antioxidant activity of quercetin/methyl-β-cyclodextrin complexes." *Curr Drug Deliv* no. 13 (3):444–51. doi: 10.2174/1567201813666151207112514.

Guo, Haiqing, Feng Ren, Li Zhang, Xiangying Zhang, Rongrong Yang, Bangxiang Xie, Zhuo Li, Zhongjie Hu, Zhongping Duan, and Jing Zhang. 2016. "Kaempferol induces apoptosis in HepG2 cells via activation of the endoplasmic reticulum stress pathway." *Mol Med Rep* no. 13 (3):2791–800. doi: 10.3892/mmr.2016.4845.

Harnly, J. M., R. F. Doherty, G. R. Beecher, J. M. Holden, D. B. Haytowitz, S. Bhagwat, and S. Gebhardt. 2006. "Flavonoid content of U.S. fruits, vegetables, and nuts." *J Agric Food Chem* no. 54 (26):9966–77. doi: 10.1021/jf061478a.

Huang, Y. W., Q. Q. Zhu, X. Y. Yang, H. H. Xu, B. Sun, X. J. Wang, and J. Sheng. 2019. "Wound healing can be improved by (-)-epigallocatechin gallate through targeting Notch in streptozotocin-induced diabetic mice." *Faseb J* no. 33 (1):953–64. doi: 10.1096/fj.201800337R.

Iqbal, W. A., I. Mendes, K. Finney, A. Oxley, and G. Lietz. 2021. "Reduced plasma carotenoids in individuals suffering from metabolic diseases with disturbances in lipid metabolism: a systematic review and meta-analysis of observational studies." *Int J Food Sci Nutr* no. 72 (7):879–91. doi: 10.1080/09637486.2021.1882962.

Iranshahi, M., R. Rezaee, H. Parhiz, A. Roohbakhsh, and F. Soltani. 2015. "Protective effects of flavonoids against microbes and toxins: The cases of hesperidin and hesperetin." *Life Sci* no. 137:125–32. doi: 10.1016/j.lfs.2015.07.014.

Jabczyk, M., J. Nowak, and B. Hudzik. 2021. "Curcumin in metabolic health and disease." *Nutrients* no. 13 (12):4440. doi: 10.3390/nu13124440.

Jachak, S.M. 2001. "Natural products: Potential source of COX inhibitors." *CRIPS* no. 2:12–15.

Jantan, Ibrahim, Waqas Ahmad, and Syed Nasir Abbas Bukhari. 2015. "Plant-derived immunomodulators: An insight on their preclinical evaluation and clinical trials." *Front Plant Sci* no. 6:655. doi: 10.3389/fpls.2015.00655.

Jayaraman, R., S. Subramani, S. H. Sheik Abdullah, and M. Udaiyar. 2018. "Antihyperglycemic effect of hesperetin, a citrus flavonoid, extenuates hyperglycemia and exploring the potential role in antioxidant and antihyperlipidemic in streptozotocin-induced diabetic rats." *Biomed Pharmacother* no. 97:98–106. doi: 10.1016/j.biopha.2017.10.102.

Jiang, L., G. Zhang, Y. Li, G. Shi, and M. Li. 2021. "Potential application of plant-based functional foods in the development of immune boosters." *Front Pharmacol* no. 12:637782. doi: 10.3389/fphar.2021.637782.

Jianjun Wu, Yinan Zhou and Hanqing Hu 2022. "Effects of β-carotene on glucose metabolism dysfunction in humans and type 2 diabetic rats." *Acta Mater Medica* no. 1 (1):138–53. doi: 10.15212/AMM-2021-0009.

Jiménez-Flores, L. M., S. López-Briones, M. H. Macías-Cervantes, J. Ramírez-Emiliano, and V. Pérez-Vázquez. 2014. "A PPARγ, NF-κB and AMPK-dependent mechanism may be involved in the beneficial effects of curcumin in the diabetic db/db mice liver." *Molecules* no. 19 (6):8289–302. doi: 10.3390/molecules19068289.

Joanna, Kurek. 2019. "Introductory chapter: Alkaloids - their importance in nature and for human life." In *Alkaloids* (1–7), edited by Kurek Joanna, Ch. 1. Rijeka: IntechOpen.

Jordão, J. B. R., H. K. P. Porto, F. M. Lopes, A. C. Batista, and M. L. Rocha. 2017. "Protective effects of ellagic acid on cardiovascular injuries caused by hypertension in rats." *Planta Med* no. 83 (10):830–36. doi: 10.1055/s-0043-103281.

Jung, U. J., M. K. Lee, Y. B. Park, S. M. Jeon, and M. S. Choi. 2006. "Antihyperglycemic and antioxidant properties of caffeic acid in db/db mice." *J Pharmacol Exp Ther* no. 318 (2):476–83. doi: 10.1124/jpet.106.105163.

Kalra, E. K. 2003. "Nutraceutical--definition and introduction." *AAPS PharmSci* no. 5 (3):E25. doi: 10.1208/ps050325.

Kasprzak-Drozd, K., and T. Oniszczuk. 2022. "Curcumin and weight loss: Does it work?" no. 23 (2). doi: 10.3390/ijms23020639.

Khoddami, A., M. A. Wilkes, and T. H. Roberts. 2013. "Techniques for analysis of plant phenolic compounds." *Molecules* no. 18 (2):2328–75. doi: 10.3390/molecules18022328.

Kim, J. M., E. M. Noh, K. B. Kwon, J. S. Kim, Y. O. You, J. K. Hwang, B. M. Hwang, B. S. Kim, S. H. Lee, S. J. Lee, S. H. Jung, H. J. Youn, and Y. R. Lee. 2012. "Curcumin suppresses the TPA-induced invasion through inhibition of PKCα-dependent MMP-expression in MCF-7 human breast cancer cells." *Phytomedicine* no. 19 (12):1085–92. doi: 10.1016/j.phymed.2012.07.002.

Kumar, Naresh, Nidhi Goel, Tara Chand Yadav, and Vikas Pruthi. 2017. "Quantum chemical, ADMET and molecular docking studies of ferulic acid amide derivatives with a novel anticancer drug target." *Med Chem Res* no. 26 (8):1822–34. doi: 10.1007/s00044-017-1893-y.

Kumar, N., S. Gupta, T. Chand Yadav, V. Pruthi, P. Kumar Varadwaj, and N. Goel. 2019. "Extrapolation of phenolic compounds as multi-target agents against cancer and inflammation." *J Biomol Struct Dyn* no. 37 (9):2355–69. doi: 10.1080/07391102.2018.1481457.

Lee, E., E. L. Miedzybrodzka, X. Zhang, and R. Hatano. 2019. "Diet-induced obese mice and leptin-deficient Lep(ob/ob) mice exhibit increased circulating gip levels produced by different mechanisms." *Int J Mol Sci* no. 20 (18):4448. doi: 10.3390/ijms20184448.

Legeay, S., M. Rodier, L. Fillon, S. Faure, and N. Clere. 2015. "Epigallocatechin gallate: A review of its beneficial properties to prevent metabolic syndrome." *Nutrients* no. 7 (7):5443–68. doi: 10.3390/nu7075230.

Li, Mingyu, and Zhuting Xu. 2008. "Quercetin in a lotus leaves extract may be responsible for antibacterial activity." *Arch Pharm Res* no. 31 (5):640–44. doi: 10.1007/s12272-001-1206-5.

Li, R., A. Zang, L. Zhang, H. Zhang, L. Zhao, Z. Qi, and H. Wang. 2014. "Chrysin ameliorates diabetes-associated cognitive deficits in Wistar rats." *Neurol Sci* no. 35 (10):1527–32. doi: 10.1007/s10072-014-1784-7.

Li, S., Y. Zhang, Y. Sun, G. Zhang, J. Bai, J. Guo, X. Su, H. Du, X. Cao, J. Yang, and T. Wang. 2019. "Naringenin improves insulin sensitivity in gestational diabetes mellitus mice through AMPK." *Nutr Diabetes* no. 9 (1):28. doi: 10.1038/s41387-019-0095-8.

Lillioja, S., A. L. Neal, L. Tapsell, and D. R. Jacobs, Jr. 2013. "Whole grains, type 2 diabetes, coronary heart disease, and hypertension: Links to the aleurone preferred over indigestible fiber." *Biofactors* no. 39 (3):242–58. doi: 10.1002/biof.1077.

Luo, C., H. Yang, C. Tang, G. Yao, L. Kong, H. He, and Y. Zhou. 2015. "Kaempferol alleviates insulin resistance via hepatic IKK/NF-κB signal in type 2 diabetic rats." *Int Immunopharmacol* no. 28 (1):744–50. doi: 10.1016/j.intimp.2015.07.018.

Luo, Z., W. Xu, Y. Zhang, L. Di, and J. Shan. 2020. "A review of saponin intervention in metabolic syndrome suggests further study on intestinal microbiota." *Pharmacol Res* no. 160:105088. doi: 10.1016/j.phrs.2020.105088.

Mahdavi, A., M. Bagherniya, M. S. Mirenayat, S. L. Atkin, and A. Sahebkar. 2021. "Medicinal plants and phytochemicals regulating insulin resistance and glucose homeostasis in type 2 diabetic patients: A clinical review." *Adv Exp Med Biol* no. 1308:161–83. doi: 10.1007/978-3-030-64872-5_13.

Maheshwari, Shruti, Vivek Kumar, Geeta Bhadauria, and Abhinandan Mishra. 2022. "Immunomodulatory potential of phytochemicals and other bioactive compounds of fruits: A review." *Food Frontiers* no. 3 (2):221–38. doi: https://doi.org/10.1002/fft2.129.

Malik, S., K. Suchal, S. I. Khan, J. Bhatia, K. Kishore, A. K. Dinda, and D. S. Arya. 2017. "Apigenin ameliorates streptozotocin-induced diabetic nephropathy in rats via MAPK-NF-κB-TNF-α and TGF-β1-MAPK-fibronectin pathways." *Am J Physiol Renal Physiol* no. 313 (2):F414–22. doi: 10.1152/ajprenal.00393.2016.

Maoka, Takashi. 2019. "Carotenoids as natural functional pigments." *J Nat Med* no. 74:1–16. doi: 10.1007/s11418-019-01364-x.

Marcelino, G., D. J. Machate, and K. C. Freitas. 2020. "β-carotene: Preventive role for type 2 diabetes mellitus and obesity: A review." no. 25 (24): 5803. doi: 10.3390/molecules25245803.

Mazidi, Mohsen, Ehsan Karimi, Mohsen Meydani, Majid Ghayour-Mobarhan, and Gordon Ferns. 2016. "Potential effects of curcumin on peroxisome proliferator-activated receptor-γ in vitro and in vivo." *World J Methodol* no. 6:112. doi: 10.5662/wjm.v6.i1.112.

Meléndez-Martínez, A. J., C. M. Stinco, and P. Mapelli-Brahm. 2019. "Skin carotenoids in public health and nutricosmetics: The emerging roles and applications of the UV radiation-absorbing colourless carotenoids phytoene and phytofluene." *Nutrients* no. 11 (5): 1093. doi: 10.3390/nu11051093.

Meng, F., G. Zuo, X. Hao, G. Wang, H. Xiao, J. Zhang, and G. Xu. 2009. "Antifungal activity of the benzo[c] phenanthridine alkaloids from Chelidonium majus Linn against resistant clinical yeast isolates." *J Ethnopharmacol* no. 125 (3):494–96. doi: 10.1016/j.jep.2009.07.029.

Merkl R., Hrádková I., Filip V., Šmidrkal J.. 2010. "Antimicrobial and antioxidant properties of phenolic acids alkyl esters." *Czech J Food Sci.* no. 28:275–79. doi: https://doi.org/10.17221/132/2010-CJFS.

Messeguer, Angel, Rosa Planells-Cases, and Antonio Ferrer Montiel. 2006. "Physiology and pharmacology of the vanilloid receptor." *Curr Neuropharmacol* no. 4:1–15. doi: 10.2174/157015906775202995.

Miękus, N., K. Marszałek, M. Podlacha, A. Iqbal, C. Puchalski, and A.H. Świergiel. 2020. "Health benefits of plant-derived sulfur compounds, glucosinolates, and organosulfur compounds." *Molecules* no. 25 (17):3804. doi: 10.3390/molecules25173804.

Monsalve, F. A., R. D. Pyarasani, F. Delgado-Lopez, and R. Moore-Carrasco. 2013. "Peroxisome proliferator-activated receptor targets for the treatment of metabolic diseases." *Mediat Inflamm* no. 2013:549627. doi: 10.1155/2013/549627.

Moreno-Valdespino, C. A., D. Luna-Vital, R. M. Camacho-Ruiz, and L. Mojica. 2020. "Bioactive proteins and phytochemicals from legumes: Mechanisms of action preventing obesity and type-2 diabetes." *Food Res Int* no. 130:108905. doi: 10.1016/j.foodres.2019.108905.

Murakami, A. 2013. "Modulation of protein quality control systems by food phytochemicals." *J Clin Biochem Nutr* no. 52 (3):215–27. doi: 10.3164/jcbn.12-126.

Nabavi, S. F., S. M. Nabavi, M. Mirzaei, and A. H. Moghaddam. 2012. "Protective effect of quercetin against sodium fluoride induced oxidative stress in rat's heart." *Food Funct* no. 3 (4):437–41. doi: 10.1039/c2fo10264a.

Nguyen, K. H., T. N. Ta, T. H. Pham, Q. T. Nguyen, H. D. Pham, S. Mishra, and B. L. Nyomba. 2012. "Nuciferine stimulates insulin secretion from beta cells-an in vitro comparison with glibenclamide." *J Ethnopharmacol* no. 142 (2):488–95. doi: 10.1016/j.jep.2012.05.024.

Nguyen-Ngo, C., J. C. Willcox, and M. Lappas. 2019. "Anti-diabetic, anti-inflammatory, and anti-oxidant effects of naringenin in an in vitro human model and an in vivo murine model of gestational diabetes mellitus." *Mol Nutr Food Res* no. 63 (19):e1900224. doi: 10.1002/mnfr.201900224.

Nishino, H., M. Murakosh, T. Ii, M. Takemura, M. Kuchide, M. Kanazawa, X. Y. Mou, S. Wada, M. Masuda, Y. Ohsaka, S. Yogosawa, Y. Satomi, and K. Jinno. 2002. "Carotenoids in cancer chemoprevention." *Cancer Metastasis Rev* no. 21 (3–4):257–64. doi: 10.1023/a:1021206826750.

Onaolapo, A. Y., and O. J. Onaolapo. 2019. "9-herbal beverages and brain function in health and disease." In *Functional and Medicinal Beverages*, edited by Alexandru Mihai Grumezescu and Alina Maria Holban, 313–49. Cambridge, MA: Academic Press.

Palacios, Orsolya M., Heather Nelson Cortes, Belinda H. Jenks, and Kevin C. Maki. 2020. "Naturally occurring hormones in foods and potential health effects." *Toxicol Res Appl* no. 4:2397847320936281. doi: 10.1177/2397847320936281.

Panchal, S. K., E. Bliss, and L. Brown. 2018. "Capsaicin in metabolic syndrome." *Nutrients* no. 10 (5):630. doi: 10.3390/nu10050630.

Panche, A. N., A. D. Diwan, and S. R. Chandra. 2016. "Flavonoids: An overview." *J Nutr Sci* no. 5:e47. doi: 10.1017/jns.2016.41.

Panyod, S., and W. K. Wu. 2022. "Atherosclerosis amelioration by allicin in raw garlic through gut microbiota and trimethylamine-N-oxide modulation." no. 8 (1):4. doi: 10.1038/s41522-022-00266–3.

Percival, S. S.. 2011. "Nutrition and immunity: Balancing diet and immune function." *Nutrition Today* no. 46 (1):12–17. doi: https://doi.org/10.1097/NT.

Pereira, David, Patrícia Valentão, José Pereira, and Paula Andrade. 2009. "Phenolics: From chemistry to biology." *Molecules* no. 14(6): 2202–11. doi: 10.3390/molecules14062202.

Piazzon, A., U. Vrhovsek, D. Masuero, F. Mattivi, F. Mandoj, and M. Nardini. 2012. "Antioxidant activity of phenolic acids and their metabolites: Synthesis and antioxidant properties of the sulfate derivatives of ferulic and caffeic acids and of the acyl glucuronide of ferulic acid." *J Agric Food Chem* no. 60 (50):12312–23. doi: 10.1021/jf304076z.

Pop, Raluca, Ada Popolo, Adrian Trifa, and Luminita Stanciu. 2018. "Phytochemicals in cardiovascular and respiratory diseases: Evidence in oxidative stress and inflammation." *Oxid Med Cell Longev* no. 2018:1–3. doi: 10.1155/2018/1603872.

Prasad, C. N., T. Anjana, A. Banerji, and A. Gopalakrishnapillai. 2010. "Gallic acid induces GLUT4 translocation and glucose uptake activity in 3T3-L1 cells." *FEBS Lett* no. 584 (3):531–36. doi: 10.1016/j.febslet.2009.11.092.

Prof, Dhan, Charu Gupta, and Prof Girish Sharma. 2012. "Importance of phytochemicals in nutraceuticals." *J Chin Med Res Dev* no. 1:70–78.

Raju, Marisiddaiah, Poorigali Raghavendra-Rao Sowmya, Rudrappa Ambedkar, Bangalore Prabhashankar Arathi, and Rangaswamy Lakshminarayana. 2021. "Chapter 32- Carotenoid metabolic pathways and their functional role in health and diseases." In *Global Perspectives on Astaxanthin*, edited by Gokare A. Ravishankar and Ambati Ranga Rao, 671–91. Cambridge, MA: Academic Press.

Rathee, P., H. Chaudhary, S. Rathee, D. Rathee, V. Kumar, and K. Kohli. 2009. "Mechanism of action of flavonoids as anti-inflammatory agents: A review." *Inflamm Allergy Drug Targets* no. 8 (3):229–35. doi: 10.2174/187152809788681029.

Rejman, Joanna, and Arkadiusz Kozubek. 2004. "Inhibitory effect of natural phenolic lipids upon NAD-dependent dehydrogenases and on triglyceride accumulation in 3T3-L1 cells in culture." *J Agric Food Chem* no. 52 (2):246–50. doi: 10.1021/jf034745a.

Ren, B., W. Qin, F. Wu, S. Wang, C. Pan, L. Wang, B. Zeng, S. Ma, and J. Liang. 2016. "Apigenin and naringenin regulate glucose and lipid metabolism, and ameliorate vascular dysfunction in type 2 diabetic rats." *Eur J Pharmacol* no. 773:13–23. doi: 10.1016/j.ejphar.2016.01.002.

Ruhee, R. T., L. A. Roberts, S. Ma, and K. Suzuki. 2020. "Organosulfur compounds: A review of their anti-inflammatory effects in human health." *Front Nutr* no. 7:64. doi: 10.3389/fnut.2020.00064.

Saini, Ramesh K., Parchuri Prasad, Veeresh Lokesh, Xiaomin Shang, Juhyun Shin, Young-Soo Keum, and Ji-Ho Lee. 2022. Carotenoids: Dietary sources, extraction, encapsulation, bioavailability, and health benefits—A review of recent advancements. *Antioxidants* 11 (4):795.

Saxena, Mehul, J. Saxena, and A. Pradhan. 2012. "Flavonoids and phenolic acids as antioxidants in plants and human health." *Int J Pharm Sci Rev Res* no. 16:130–34.

Scalbert, A., and G. Williamson. 2000. "Dietary intake and bioavailability of polyphenols." *J Nutr* no. 130 (8S Suppl):2073s–85s. doi: 10.1093/jn/130.8.2073S.

Schroeder, H. W., Jr., and L. Cavacini. 2010. "Structure and function of immunoglobulins." *J Allergy Clin Immunol* no. 125 (2 Suppl 2):S41–52. doi: 10.1016/j.jaci.2009.09.046.

Shahidi, Fereidoon, and Priyatharini Ambigaipalan. 2015. "Phenolics and polyphenolics in foods, beverages and spices: Antioxidant activity and health effects – A review." *J Funct Foods* no. 18:820–97. doi: 10.1016/j.jff.2015.06.018.

Shi, X., X. Zhou, X. Chu, J. Wang, B. Xie, J. Ge, Y. Guo, and X. Li. 2019. "Allicin improves metabolism in high-fat diet-induced obese mice by modulating the gut microbiota." *Nutrients* no. 11 (12):2909. doi: 10.3390/nu11122909.

Shukla, S., and S. Gupta. 2010. "Apigenin: A promising molecule for cancer prevention." *Pharm Res* no. 27 (6):962–78. doi: 10.1007/s11095-010-0089-7.

Singh, Dr Rambir, Tasleem Arif, Imran Khan, and Dr Poonam Sharma. 2014. "Phytochemicals in antidiabetic drug discovery." *J Biomed Therapeut Sci* no. 1:1–33.

Słoczyńska, K., B. Powroźnik, E. Pękala, and A. M. Waszkielewicz. 2014. "Antimutagenic compounds and their possible mechanisms of action." *J Appl Genet* no. 55 (2):273–85. doi: 10.1007/s13353-014-0198-9.

Sotoing Taïwe, G., E. Ngo Bum, E. Talla, A. Dawe, F. C. Okomolo Moto, G. Temkou Ngoupaye, N. Sidiki, B. Dabole, P. D. Djomeni Dzeufiet, T. Dimo, and M. De Waard. 2012. "Antipsychotic and sedative effects of the leaf extract of Crassocephalum bauchiense (Hutch.) Milne-Redh (Asteraceae) in rodents." *J Ethnopharmacol* no. 143 (1):213–20. doi: 10.1016/j.jep.2012.06.026.

Stahl, W., U. Heinrich, H. Jungmann, H. Sies, and H. Tronnier. 2000. "Carotenoids and carotenoids plus vitamin E protect against ultraviolet light-induced erythema in humans." *Am J Clin Nutr* no. 71 (3):795–98. doi: 10.1093/ajcn/71.3.795.

Stahl, Wilhelm, and Helmut Sies. 2005. "Bioactivity and protective effects of natural carotenoids." *Biochimica et Biophysica Acta (BBA) – Mol Basis Di* no. 1740 (2):101–7. doi: https://doi.org/10.1016/j.bbadis.2004.12.006.

Stalikas, C. D. 2007. "Extraction, separation, and detection methods for phenolic acids and flavonoids." *J Sep Sci* no. 30 (18):3268–95. doi: 10.1002/jssc.200700261.

Stoszko, M., E. De Crignis, C. Rokx, M. M. Khalid, C. Lungu, R. J. Palstra, T. W. Kan, C. Boucher, A. Verbon, E. C. Dykhuizen, and T. Mahmoudi. 2016. "Small molecule inhibitors of BAF; a promising family of compounds in HIV-1 latency reversal." *EBioMedicine* no. 3:108–21. doi: 10.1016/j.ebiom.2015.11.047.

Su, Z., Y. Guo, X. Huang, B. Feng, L. Tang, G. Zheng, and Y. Zhu. 2021. "Phytochemicals: targeting mitophagy to treat metabolic disorders." *Front Cell Dev Biol* no. 9:686820. doi: 10.3389/fcell.2021.686820.

Suchal, K., S. Malik, S. I. Khan, R. K. Malhotra, S. N. Goyal, J. Bhatia, S. Ojha, and D. S. Arya. 2017. "Molecular pathways involved in the amelioration of myocardial injury in diabetic rats by kaempferol." *Int J Mol Sci* no. 18 (5). doi: 10.3390/ijms18051001.

Sullivan, M. L. 2014. "Perennial peanut (Arachis glabrata Benth.) leaves contain hydroxycinnamoyl-CoA:tartaric acid hydroxycinnamoyl transferase activity and accumulate hydroxycinnamoyl-tartaric acid esters." *Planta* no. 239 (5):1091–100. doi: 10.1007/s00425-014-2038-x.

Sun, Xufeng, Masayuki Yamasaki, Takuya Katsube, and Kuninori Shiwaku. 2015. "Effects of quercetin derivatives from mulberry leaves: Improved gene expression related hepatic lipid and glucose metabolism in short-term high-fat fed mice." *Nutr Res Pract* no. 9:137–43. doi: 10.4162/nrp.2015.9.2.137.

Tsuhako, R., H. Yoshida, C. Sugita, and M. Kurokawa. 2020. "Naringenin suppresses neutrophil infiltration into adipose tissue in high-fat diet-induced obese mice." *J Nat Med* no. 74 (1):229–37. doi: 10.1007/s11418-019-01332-5.

Uchiyama, Y., T. Suzuki, K. Mochizuki, and T. Goda. 2013. "Dietary supplementation with (-)-epigallocatechin-3-gallate reduces inflammatory response in adipose tissue of non-obese type 2 diabetic Goto-Kakizaki (GK) rats." *J Agric Food Chem* no. 61 (47):11410–17. doi: 10.1021/jf401635w.

Vaillancourt, F., P. Silva, Q. Shi, H. Fahmi, J. C. Fernandes, and M. Benderdour. 2011. "Elucidation of molecular mechanisms underlying the protective effects of thymoquinone against rheumatoid arthritis." *J Cell Biochem* no. 112 (1):107–17. doi: 10.1002/jcb.22884.

Venkatalakshmi, P., V. Vadivel, P. Brindha. 2016. "Role of phytochemicals as immunomodulatory agents: A review." *Int J Green Pharm* no. 10 (1):1–18.

Vessal, M., M. Hemmati, and M. Vasei. 2003. "Antidiabetic effects of quercetin in streptozocin-induced diabetic rats." *Comp Biochem Physiol C Toxicol Pharmacol* no. 135c (3):357–64. doi: 10.1016/s1532-0456(03)00140-6.

Weisberg, S. P., R. Leibel, and D. V. Tortoriello. 2008. "Dietary curcumin significantly improves obesity-associated inflammation and diabetes in mouse models of diabesity." *Endocrinology* no. 149 (7):3549–58. doi: 10.1210/en.2008-0262.

Wollenweber, E., G. Kohorst, K. Mann, and J. M. Bell. 1985. "Leaf gland flavonoids in comptonia peregrina and myrica pensylvanica (myricaceae)." *J Plant Physiol* no. 117 (5):423–30. doi: 10.1016/s0176-1617(85)80049-3.

Wu, D., Y. Kong, C. Han, J. Chen, L. Hu, H. Jiang, and X. Shen. 2008. "D-Alanine:D-alanine ligase as a new target for the flavonoids quercetin and apigenin." *Int J Antimicrob Agents* no. 32 (5):421–26. doi: 10.1016/j.ijantimicag.2008.06.010.

Xia, Y., S. Shen, and I. M. Verma. 2014. "NF-κB, an active player in human cancers." *Cancer Immunol Res* no. 2 (9):823–30. doi: 10.1158/2326-6066.cir-14-0112.

Xiao, Jianbo, and Weibin Bai. 2019. "Bioactive phytochemicals." *Crit Rev Food Sci Nutr* no. 59:827–29. doi: 10.1080/10408398.2019.1601848.

Yamaguchi, Masayoshi, Aki Igarashi, Satoshi Uchiyama, Kuniaki Sugawara, Takashi Sumida, Seiichi Morita, Hiroshi Ogawa, Masahito Nishitani, and Yoshitaka Kajimoto. 2006. "Effect of β-crytoxanthin on circulating bone metabolic markers: Intake of juice (Citrus Unshiu) supplemented with β-cryptoxanthin has an effect in menopausal women." *J Health Sci* no. 52:758–68. doi: 10.1248/jhs.52.758.

Yang, J., Y. Han, C. Chen, H. Sun, D. He, J. Guo, B. Jiang, L. Zhou, and C. Zeng. 2013. "EGCG attenuates high glucose-induced endothelial cell inflammation by suppression of PKC and NF-κB signaling in human umbilical vein endothelial cells." *Life Sci* no. 92 (10):589–97. doi: 10.1016/j.lfs.2013.01.025.

Zang, Y., L. Zhang, K. Igarashi, and C. Yu. 2015. "The anti-obesity and anti-diabetic effects of kaempferol glycosides from unripe soybean leaves in high-fat-diet mice." *Food Funct* no. 6 (3):834–41. doi: 10.1039/c4fo00844h.

Zhang, J., L. Ning, and J. Wang. 2020. "Dietary quercetin attenuates depressive-like behaviors by inhibiting astrocyte reactivation in response to stress." *Biochem Biophys Res Commun* no. 533 (4):1338–46. doi: 10.1016/j.bbrc.2020.10.016.

Zhao, C., and X. Yin. 2020. "The renoprotective effects of naringenin (NGN) in gestational pregnancy." *Diabetes Metab Syndr Obes* no. 13:53–63. doi: 10.2147/dmso.s231851.

Zhou, B., Q. Li, J. Wang, P. Chen, and S. Jiang. 2019. "Ellagic acid attenuates streptozocin induced diabetic nephropathy via the regulation of oxidative stress and inflammatory signaling." *Food Chem Toxicol* no. 123:16–27. doi: 10.1016/j.fct.2018.10.036.

Zhou, L. H., Yao, T., Guo, A. L., Lin, J. J., Pan, S. Q., and Chang, Y. M. 2020. "Progress in the study of two-way immune regulation of traditional Chinese medicine in recent ten years." *J Basic Chin Med* no. 26:1016–33.

Zhu, D., Y. Wang, Q. Du, Z. Liu, and X. Liu. 2015. "Cichoric acid reverses insulin resistance and suppresses inflammatory responses in the glucosamine-induced HepG2 cells." *J Agric Food Chem* no. 63 (51):10903–13. doi: 10.1021/acs.jafc.5b04533.

10 Phytochemotaxonomy
Role of Phytochemicals in Plant Classification

Fevzi Bardakci, Riadh Badraoui, Mohd Saeed, Arif Jamal Siddiqui, Walid Sabri Hamadou, and Mousa Alreshidi
University of Ha'il

Bektas Tepe
Kilis 7 Aralik University

CONTENTS

10.1 INTRODUCTION

Man has used categorization in everyday life since the dawn of the living world. Man's propensity for categorizing creatures into groups can be traced back to prehistoric periods (Sivarajan 1991) when distinguishing between different types of plants and animals was crucial for survival. The primary purposes of plant taxonomy are to classify them using a system that makes it useful for everyone interested in the observation, collection, identification, description, and characteristics of plants based on scientifically accepted rules. At the time, group recognition was mainly based on their

DOI: 10.1201/b22842-10

gross physical similarities and differences. In both primitive and developed communities, "folk systematics" provides categories that arose out of necessity and without the influence of science. On the other hand, taxonomists believe that morphology and geography alone are insufficient to arrive at meaningful classifications. Plant taxonomy now incorporates information from anatomy, genetics, cytology, chemistry, reproductive biology, ecology, physiology, and molecular biology.

The term "chemotaxonomy," first introduced by A. P. de Candolle (1816), is a classification of organisms based on their chemical composition differences. Chemical studies of plants for chemotaxonomic purposes began at the beginning of the 20th century. The branch of chemotaxonomy that deals with the taxonomy of plants is called phytochemotaxonomy. Perhaps the most ancient of plant classification by humanity, dating back to "folk taxonomies," is based on specific typical characteristics that could be used to make an instinctive differentiation among plants, such as taste, color, odor, edibility, and toxicity, which were based subjectively on such chemical properties. The biogenetic classification was developed by Mentzer (1966) based on natural interactions between various elements.

Chemosystematics/chemotaxonomy appeared to be a viable approach for tackling issues in developing a natural classification of higher plants for a limited period. However, this hope vanished as it became clear that chemosystematic characteristics in higher plants are affected by environmental stressors, just like all other phenetic variables, and hence are prone to parallel evolution (Wink and Waterman 1999).

Because of the advent of much more potent (macro-) molecular techniques and new data analysis methods formed in parallel with these techniques, phytochemotaxonomic studies based solely on the profiles of small molecules have become largely redundant as tools for studying phylogenetic relationships of higher plants. Plant chemophenetics is a new term coined by Zidorn (2019) to describe the field of research focused on exploiting plant taxa's distinctive arrays of specialized natural products. Chemophenetic studies are investigations into a taxon's variety of specialized secondary metabolites. Chemophenetic research, like anatomical, morphological, and karyological techniques, contributes to taxa's phenetic description (Zidorn 2019).

Secondary metabolites, formed from primary metabolites produced by all organisms, are essential for identification and classification, because their chemical structures are usually unique to taxa. The main driver of attention to secondary plant metabolites is their medicinal value, an essential source of medicine for centuries. Analysis of these chemical compositions in different plant taxa showed that some secondary metabolites were specific to the plant taxa, demonstrating a close phylogenetic relationship. Unlike artificial taxonomy systems, which aim to improve identification, phylogenetic classification is considered "natural," because it primarily reflects the phylogeny of plants. Classifications based on shared ancestral characters are considered "natural" because they primarily reflect the evolutionary history of the characters and taxa. From this perspective, if phytochemicals evolved from a common ancestor over time, a group of plants would have unique phytochemicals useful in their classification. An exciting new interdisciplinary topic called pharmacophylogeny has been developed to comprehensively explore the correlation between phylogeny and medicinal plants' chemical composition and therapeutic effects (Chen et al. 2005).

On the other hand, various studies have clarified that genetically close medicinal plants have similar chemical or biologically active compounds with similar therapeutic effects (Gong et al. 2020). Today, it is generally accepted that certain plant compounds and related substances are specific to particular taxonomic groups (Mannheimer 1999). Chemotaxonomy is applicable at all taxonomic levels, where the ability to form chemicals is maintained through metabolic processes supported by a group of its ancestors. This chapter will examine and compare the use of phytochemicals in classifying plants and their utility to classical molecular systematic methods.

10.2 PHYTOCHEMICALS IN PHYTOCHEMOTAXONOMY

Plants contain thousands of chemical compounds involved in their metabolism called metabolites, intermediates, and products of metabolism. These are categorized based on their chemical and physical characteristics (Arts and Hollman 2005; Heneman and Zidenberg-Cherr 2008). Taxonomically useful natural phytochemicals are divided into two main groups based on their molecular weight: (1) micromolecules are chemical compounds with molecular weight less than or equal to 1,000, for example, fatty acids, flavonoids, amino acids, alkaloids, terpenoids, anthocyanins, betalains, etc., and (2) macromolecules include macromolecular compounds with a molecular weight greater than 1000, for example, proteins, DNA, RNA, complex polysaccharides, etc. (Jones and Luchsinger 1987). Various plant chemical compounds are produced through various metabolic pathways. Some of these plant compounds are directly visible particles and are considered morphological characteristics. Some of these macromolecular compounds are not involved in the transfer of genetic information called non-semantides, e.g., cellulose and starch. Others are known as semantides and carry genetic information such as protein, RNA, and DNA. Micromolecules include both primary and secondary metabolites involved in essential metabolic pathways that act as catalysts, inhibitors, signaling molecules, or simulators in any metabolic activity in the cell. Micromolecules in plants were also categorized into three groups based on their biosynthetic pathway, chemical structure, compounds' composition, and solubility in various solvents. They are categorized into three groups based on their biosynthetic origin: (1) phenolics (without nitrogen), the flavonoids, allied phenolics, and polyphenolics class of compounds that a hydroxyl functional group attached to aromatic rings; (2) terpenes: the non-nitrogen-containing saponins, cardiac glycosides, terpenoids, and terpenoid-derived compounds, which include several steroids containing five-carbon isoprenoid units, carbon, and hydrogen molecule; (3) alkaloids: the nitrogen-containing alkaloids or the sulfur-containing compounds (Mazid, Khan, and Mohammad 2011) (Figure 10.1).

One of the oldest classifications proposed by Mentzer (1966), called biogenetic classification for phytochemical research, is based on natural relationships between the various plant constituents and has been widely used from a taxonomic point of view until now. The three broad categories of constituents used in phytotaxonomy are (1) basic or primary metabolites, e.g., proteins, nucleic acids, chlorophyll, and polysaccharides, etc.; (2) secondary metabolites not involved directly in plant metabolism, i.e., simple phenolic compounds such as flavonoids, terpenes, coumarins, alkaloids, etc.; and (3) miscellaneous substances of which their use in taxonomy is not well understood yet (Mazid, Khan, and Mohammad 2011).

10.2.1 Primary Metabolites

Naturally occurring compounds, primary metabolites, are directly involved in the essential metabolic pathways and often occur in seeds and vegetative storage organs, like intrinsic functions,

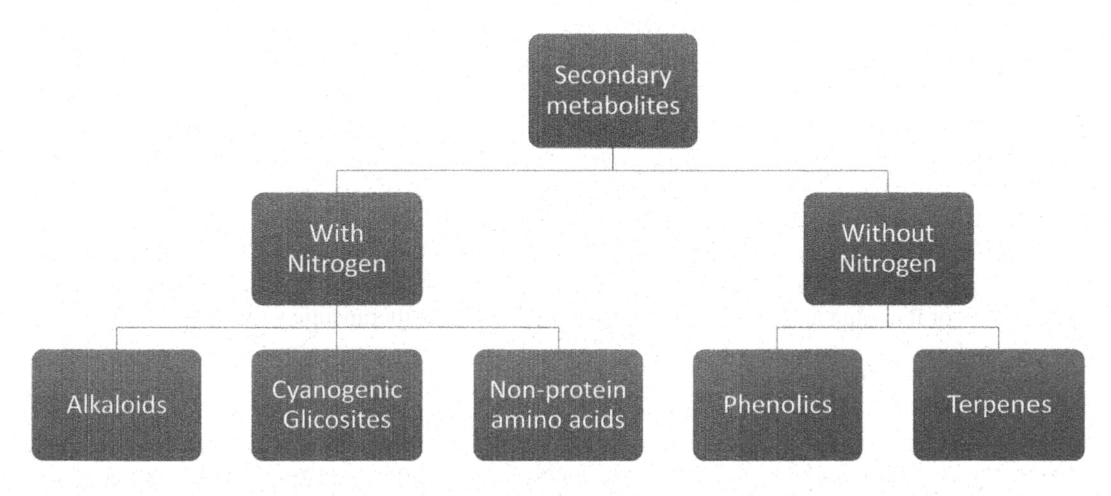

FIGURE 10.1 Classification of secondary metabolites.

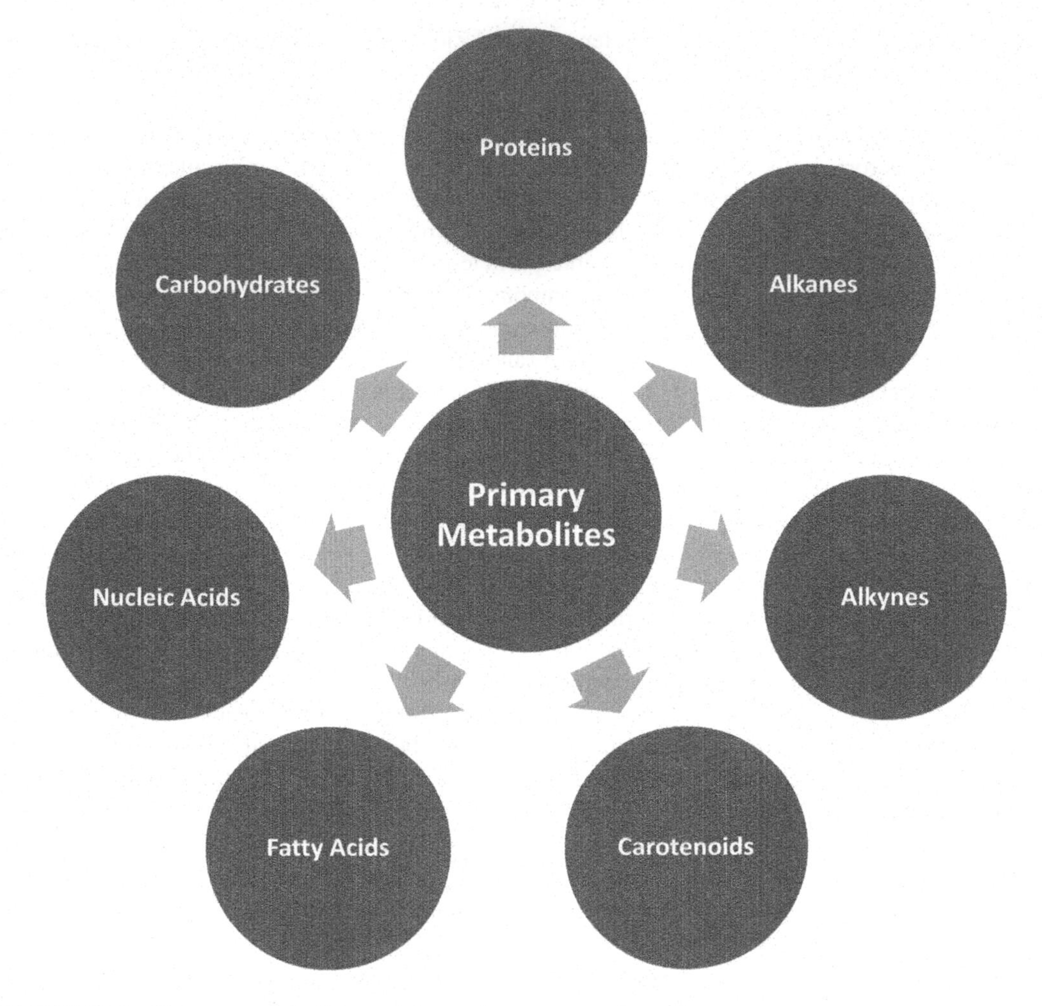

FIGURE 10.2 Primary metabolites useful as a chemotaxonomic marker.

reproduction, growth, and development in basic cell metabolism (Rasool et al. 2010). Primary metabolites exist in all living things and are fundamental in various metabolic pathways involved in synthesizing, degrading, and modifying energy-rich compounds such as fats, carbohydrates, and others (proteins, lignin, and cellulose), are the components of cellular structures. Therefore, primary metabolites are less useful in the phytochemotaxonomic classification of plants (Singh 2016). However, these molecules are useful in plant classification, since their quantity varies among the plants. For example, the accumulation of carbohydrate sedoheptulose in a large quantity is a characteristic of the genus *Sedum*, thus serving as an identification tool for species of this genus. Primary metabolites of higher plants include amino acids, organic acids, and nucleosides, commonly used as industrial raw materials for fermentation and chemical synthesis. They are also widely used as food and nutritional supplements or flavoring agents and do not have a high value of therapeutic value (Figure 10.2).

10.2.1.1 Fatty Acids

Fatty acids are the most abundant class of plant lipids containing a long hydrocarbon chain and a terminal carboxylic group. The commonly occurring fatty acids are either unsaturated or saturated. The unsaturated fatty acids contain up to five double bonds. Fatty acids are aliphatic monocarboxylic acids liberated by hydrolysis and vary in chain length and number, position, and configuration

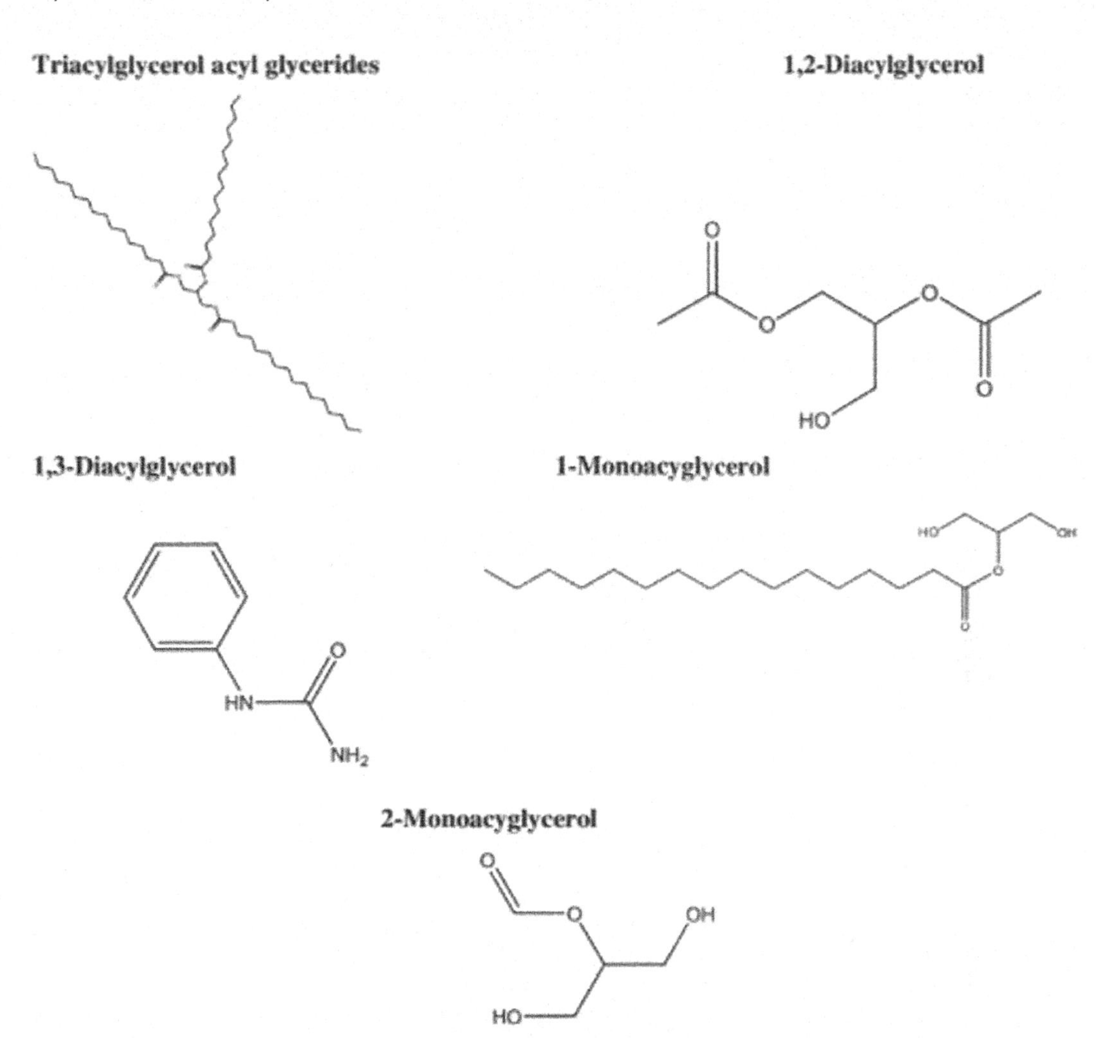

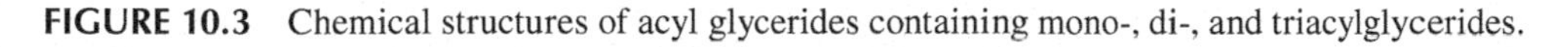

FIGURE 10.3 Chemical structures of acyl glycerides containing mono-, di-, and triacylglycerides.

of double bonds. Essential fatty acids and their derivatives, known as acylglycerols, such as monoacylglycerols, diacylglycerols, and triacylglycerols, are shown in Figure 10.3 (Murphy 2020). They have two major roles in the building blocks of phospholipids and glycolipids in biological membranes and fuel molecules. Overall, over 450 different fatty acid structures have been discovered merely in vascular plants. However, it is believed that many plant families are still being explored for their fatty acid composition, which may lead to the discovery of many other types of structures (Ohlrogge et al. 2018). The occurrence of some fatty acids in plants reviewed by Avato and Tava (2021) is summarized in Table 10.1.

Essential oils characterized by gas chromatography-flame ionization detection (GC-FID) and gas chromatography-mass spectrometry (GC/MS) analyses in *Geranium* (Geraniaceae) and closely related taxa *Erodium* from Macedonia and Serbia proved that they mainly consist of fatty acids and its derivative compounds (45.4–81.3%), including hexadecanoic acid and (E)-phytol as the major components. Since these species are phylogenetically close relatives, not much difference has been found in their intergeneric oil composition variability (Radulović and Dekić 2013). Analysis of the fatty acid composition of 12 *Brassica* species (Brassicaceae) by GC-FID, GC-MS arranged into three groups based on the C18:1 (*n*-7)/(*n*-9) ratios for chemotaxonomy (Barthet 2008). The first clade

TABLE 10.1
Occurrence of Rare Fatty Acids and Lipids Detected in Oilseeds of Different Plants

Fatty Acid/Lipid	Plant Species	References
Tartaric	Santalaceae, Oleaceae	(Spencer et al. 1970; Stuhlfauth et al. 1985)
Crepenynic acid (*cis*-9-octadecen-12-ynoic acid)	*Afzelia cuanzensis* Welw.	(Gunstone and Harwood 2007)
	Atractylodes lancea Thunb. and *A. macrocephala* Koidz.	(Sun, Guo, and Smith 2017)
Enynoic fatty acids	*Onguekoa gore* Engler	(Badami and Gunstone 1963; Miller et al. 1977)
Allenic fatty acids	Sterculiaceae, Malvaceae, Tiliaceae, and Bombacaceae	(Bohannon and Kleiman 1978; Bao-Lin, De-Yuan, and Pei-Gen 2008; Ahmad 2017)
Epoxy fatty acids	*Vernonia* (Asteraceae), Euphorbiaceae, Dipsacaceae, Onagraceae, and Valerianaceae	(Smith Jr 1971; Spitzer 1996; Tsevegsuren, Aitzetmüller, and Vosmann 2004; Gasparetto et al. 2012)
Hydroxy fatty acids	*Ricinus communis* L. (Euphorbiaceae), *Strophanthus* (Apocynaceae), *Lesquerella* (Cruciferae)	(Weselake 2016)
Hydroxy olefinic fatty acids	*Lesquerella* (Cruciferae)	(Smith 1976; Kleiman et al. 1972)
Dienoic fatty acids	Euphorbiaceae	(Chisholm and Hopkins 1964; Smith Jr et al. 1960)
2-Hydroxy fatty acids	Lamiaceae	(Smith Jr and Wolff 1969; Smith Jr 1971; Bohannon and Kleiman 1975; Galliard 1974)
Fluorinated fatty acids	*Dichapetalum* (Dichapetaleceae)	(Msami 1999; Smith and Wolff 1969; Smith 1971; Bohannon and Kleiman 1975; Galliard 1978)
Brominated fatty acids	Eremostachys molucelloides Bunge (Lamiaceae)	(Dembitsky and Srebnik 2002)
γ-Linolenic acid (gamolenic acid)	Onagraceae (e.g., Oenothera biennis L., evening primrose), Grossulariaceae (Ribes nigrum L. black currant), Boraginaceae (Borago officinalis L., borage), Cannabaceae (Cannabis sativa L., hemp), Malvaceae (Durio graveolens Becc.)	(Ohlrogge et al. 2018; Kapoor and Nair 2023; Guil-Guerrero, García Maroto, and Giménez Giménez 2001; Alonso-Esteban et al. 2020)
Cyanolipids	Sapindaceae, Hippocastanaceae, Boraginaceae	(Møller and Seigler 1999)
Acylglycerols	Sapindaceae	(Mikolajczak 1977; Avato et al. 2003, 2005, 2006)
Cyanogenic lipids contain eicosanoic and eicosenoic fatty acids	Sapindaceae	(Tava and Avato 2014)
Petroselinic acid	Geraniaceae, Lamiaceae, Picramiaceae,	(Tsevegsuren, Aitzetmüller, and Vosmann 2004; Ohlrogge et al. 2018; Plant FadFAdb 2016)
	Araliaceae, Apiaceae	(Ohlrogge 2018; Sayed-Ahmad et al. 2017; Ohlrogge et al. 2018; Ušjak et al. 2019)
Erucic acid	Brassicaceae	(Ohlrogge 2018; Avato and Argentieri 2015; Lu et al. 2020)
α-Eleostearic acid	*Vernicia fordii* (Hemsl.),	(Zhang et al. 2019)
	Momordica charantia L. (Cucurbitaceae)	(Chisholm and Hopkins 1964; Yoshime et al. 2016; Jia et al. 2017)

(*Continued*)

TABLE 10.1 (*Continued*)
Occurrence of Rare Fatty Acids and Lipids Detected in Oilseeds of Different Plants

Fatty Acid/Lipid	Plant Species	References
	Schinziophyton rautanenii (Schinz) Radcl.-Sm. (Euphorbiaceae), *Ricinocarpus tuberculatus* Muell. Arg. (Euphorbiaceae), *Parinari montana* Aubl. (Chrysobalanaceae), and *Prunus mahaleb* L.	(Keskin Çavdar 2019)
Trienoic punicic acid (trichosanic acid)	*Punica granatum* L., Trichosantes (T. kirilowii Maxim (TK), and T. anguina L.)	(Hopkins and Chrisholm 1962; Joh et al. 1995; Pereira de Melo et al. 2014; Aruna et al. 2015; Hennessy et al. 2016)
Jacaric and calendic acids	*Jacaranda* (Bignoniaceae)	(Gunstone et al. 2007)
Ximenynic acid	*Ximenia* (Olacaceae), *Santalum* (Santalaceae), Opiliaceae	(Aitzemüller 2012)
Cyclopropenoid fatty acids	Malvales	(Pasha and Ahmad 1992; Ohlrogge 2018)
Cyclopentenyl fatty acids (hydnocarpic acid and chaulmoogric acid)	Flacurtiaceae, Hydnocarpus	(Santos et al. 2008)
Ricinoleic acid	*Ricinus communis* L. (Euphorbiaceae)	(Teomim et al. 1999)

includes *B. rapa*, *Brassica napus*, and *B. tournefortii*, but only *B. napus* branches with a related *Eruca sativa*. The second group includes *Sinapis alba*, *B. tournefortii*, and *Raphanus sativus*. The last group includes three species *B. juncea*, *B. nigra*, and *B. carinata*, which have no similarities or relationships with other species and between them in the C18:1 (*n*-7)/(*n*-9) ratios.

Seed oil content and fatty acid composition of the genus *Crambe* (Brassicaceae) from Turkey were determined for their usefulness as a taxonomic marker (Subasi 2020). Cluster analysis based on a total of 17 fatty acids formed two major classes of *Crambe* species. While the first cluster includes *C. orientalis*, *C. grandiflora*, and *C. tataria*, the second clade consists of only *C. maritima*. Although fatty acid value-based dendrogram displayed a clear discrimination between the groups, there were no close relationships between *C. tataria* and *C. maritima* species, contrary to molecular classification.

Fatty acid components of seed oil of 23 *Stachys* (Labiatae) taxa analyzed using GC-MS were stearic (trace to 5.2%), linoleic (27.1–64.3%), palmitic (4.3–9.1%), and oleic (20.2–48.1%) acids. Among the fatty acids, 6-octadecynoic acid (2.2–34.1%) was the most useful one as a chemotaxonomic marker for species of *Stachys* (2.2–34.1%) (Gören et al. 2012). Species of *Stachys* were classified into four subgroups analysis based on fatty acid composition using the numerical cluster method. This study revealed that 6-octadecynoic acid (2.2–34.1%) is the most important chemotaxonomic marker.

Fatty acids composition of 17 *Coffea* species (Rubiaceae) showed a significant relationship with their ecology and geographic distribution. When the clusters of the species were obtained based on their fatty acid composition compared with the clades inferred from previous molecular phylogenetic studies, they were remarkably congruent (Dussert et al. 2008). Mongrand et al. (2005) have reported the usefulness of leaf fatty acid content in the chemotaxonomy of Rubiaceae. Moreover, principal component analysis (PCA) based on the fatty acid composition discriminated against *Coffeae*, *Psychotrieae*, and *Rubieae*.

Analysis of fatty acids composition in *Pinus abies* from different geographic regions has shown considerably lower intraspecific dissimilarities than interspecific dissimilarities among other *Picea* species, supporting the use as chemotaxonomic markers as they are not affected by edaphic or climatic conditions. Moreover, although a comparison of *Picea* and *Larix* spp. displayed minor differences in seed fatty acid composition, PCA has indicated that such minor differences in fatty acids compositions are sufficient to discriminate between these two genera (Wolff et al. 2001). Moreover, seed fatty acid composition from distinct locations in France (Wolff et al. 2001) and Finland (Tillman-Sutela et al. 1995) showed no significant variations in their fatty acid content.

FGC-EIMS analysis of the fatty acid composition of 16 species of native Vochysiaceae has been proposed that the patterns of fatty acids are seemingly characteristic for most of the species with lower intraspecific variability and relatively higher intraspecific variability, and thus, they have been proposed to be useful for species characterization with one exception *S. convallariodora* (Mayworm and Salatino 2002).

Zhang et al. (2015) have investigated 747 species from 207 localities in China and the relative effects of phylogeny and climate on the fatty acid composition of seed oil. The results showed that fatty acid content of seed oil varied considerably across plant species with large-scale spatial patterns indicating the evolutionary association of fatty acid composition with the climate at the family level proposing adaptive significance of fatty acid unsaturation in seed oil in colder climates through evolution (Zhang et al. 2015).

10.2.1.2 Protein and Amino Acids

Proteins are translated products of genetic information and are therefore considered tertiary semantides. The albumin fractions of Leguminosae seeds were useful for taxonomic purposes (Buchanan 1971). Moreover, globulin patterns of the genus *Crotalaria* were also reasonably correlated with their taxonomy (Buchanan and Arnon 1970). A study on the seed storage proteins from 20 taxa of the genus *Lasthenia* using protein gel electrophoresis suggested that nearly all species of this genus have specific arrays of seed proteins, but are not congruent with their taxonomy based on the morphological, cytological, and other biochemical markers (Altosaar 1974).

Amino acid sequences of cytochrome c were generally useful in plant taxonomy. A comparison of phylogenetic relationships of 25 species of vascular plants belonging to 12 families based on their amino acid sequences of cytochrome c was generally congruent with their morphologically based phylogeny (Fairbrothers et al. 1975).

The complete amino acid sequence of [2Fe–2S] ferredoxin from *Panax ginseng* (Araliaceae) (Mino 2006) was investigated using phylogenetic analysis based on the amino acid sequence of ferredoxin. Results suggested a close taxonomic relationship between that *P. ginseng* and umbelliferous plants. Another study on *Datura* ferredoxins showed that three were three amino acid differences between *D. quercifolia* and *D. fastuosa* amino acid sequences belonging to different sections, but exhibited identical amino acid sequences to *D. stramonium* (section *Stramonium*) and *D. metal* (section *Dutra*) ferredoxins, respectively (MINO 1995). Moreover, phylogenetic analysis of the amino acid sequence of ferredoxins from four species of the genus *Solanum* (*S. nigrum*, *S. lyratum*, *S. indicum*, and *S. abutiloides*) showed that these species are distantly related and clustered with *Datura, Lycium*, *Scopolia*, *Physalis*, *Nicotiana*, and *Capsicum,* indicating a discrepancy between protein chemotaxonomy and classical taxonomy.

10.2.1.3 Carbohydrates

Cellulose and pectin are the most extensively studied polysaccharides by infrared spectroscopy. This method was used to identify individual polysaccharide types and their mixture in higher plants. The constituent of monosaccharides of the studied hemicellulosic and pectic polysaccharides revealed some characteristic bands (Kacurakova et al. 2000), but was not discussed for their taxonomic significance. Nevertheless, it is promising that sugar components and their structures provide the basis for the chemotaxonomy of flowering plants. Although the significance of carbohydrates as a chemotaxonomic marker has not been well documented in any higher plant species, several studies put forward the usefulness of sugar components in the cell wall in algal (e.g., *Chlorella*, Chlorophyceae) (Takeda 1991) and fungal (Ruiz-Herrera and Ortiz-Castellanos 2010) taxonomy.

10.2.1.4 Alkanes and Alkynes

Leaf wax alkanes were determined in several species of the grass family (Gramineae) and used as chemotaxonomic characteristics. GC-MS analysis of leaf wax alkanes of 93 species belonging to five subfamilies (Bambusoideae, Arundinoideae, Panicoideae, Chloridoideae, and Pooideae) showed a significant morphological data at subfamilial and tribal levels, conforming the usefulness of leaf wax alkanes as chemotaxonomic markers (Maffei 1996).

The occurrence of odd carbon number isoalkanes is relevant as a chemotaxonomic characteristic in Lamiaceae, while the presence of even carbon number *n*-alkanes and isoalkanes has been confirmed to be common in taxa belonging to different families (Mecklenburg 1966; Herbin and Robins 1969; Osborne, Salatino, and Salatino 1989; Osborne et al. 1993). Leaf wax *n*-alkanes and isoalkanes were determined in several species of Lamiaceae and compared with Boraginaceae, Verbenaceae, Solanaceae, and Scrophulariaceae. Analysis of some *n*-alkanes and isoalkanes provided a good partition inside Lamiaceae and among the five different families examined, and a much deeper insight into pattern partition confirmed the chemotaxonomic usefulness of leaf wax alkanes (Maffei 1994).

Investigation of alkanes' foliar epicuticular waxes of eight species of *Huberia* DC. (Melastomataceae) has shown a substantial variation at the species level and below, proving the usefulness of alkaline distribution as a chemotaxonomic characteristic. The results of this study have displayed two pattern distributions, one characterized by samples with C31 or C33 and the other with C29 as the main homolog supported by UPGMA showing intraspecific differences in alkanes pattern of *Huberia glazioviana* and *H. semiserrata*, while *H. ovalifolia* and *H. nettoana* had conservative distributions, irrespective of variations in habitats (Mimura et al. 1998).

Comparison of *n*-alkane composition and the nonacosan-10-ol content in the needle cuticular wax of Serbian spruce (*Picea omorika*), Macedonian pine (*Pinus peuce*), and Bosnian pine (*Pinus heldreichii*) showed that the amounts of nonacosan 10-ol content in *P. heldreichii* and *P. peuce* were lower than that in *P. omorika*. However, the most common n-alkanes found in the needle waxes of *P. amorika* (C_{18}-C_{35}) were C_{29}, C_{23}, C_{27}, and C_{25} in those of *P. heldreichi* (C_{18}-C33) and C_{29}, C_{25}, C_{27}, and C_{23} in those of *P. peuce* (C_{18}-C_{33}). PCA of nine *n*-alkanes based on their content discriminates *P. amorika* from those of the other two species studied with a partial overlap. The distinction among these species was also approved by cluster analysis, where the *P. heldreichii* and *P. peuce* populations were in the same cluster. Overall results of the *n*-alkalene composition of *Pinus* species were found useful in their chemotaxonomy (Nikolić et al. 2013).

Essential oils in dry fruits of *Scandix balansae* (Apiaceae) using GC-MS analysis showed that the major volatile compounds were a long-chain homolog nonacosane (7.6%) and medium-chain length *n*-alkalenes heptadecane (19.3%), pentadecane (13.4%), and tridecane (6.7%) (Radulović and Dekić 2013). Several minor oils such as octadecyl 2-methyl propanoate, pentanoate, 3-methyl butanoate, and tetradecyl 3-methyl butanoate have limited occurrence only in umbellifers, but also in the plant kingdom. These rare plant constituents' identity presents excellent chemotaxonomic marker candidates for *Scandix*. *n*-alkane chain length distribution patterns in the leaf cuticular *n*-alkane have also been useful as a chemotaxonomic marker for eggplant (Solanaceae) and its related species (Haliński, Szafranek, and Stepnowski 2011).

GC and GC-MS analysis have shown that leaf and roots oils composition of *Artemisia* (Asteraceae) is reached in phenyl alkynes: *A. capillaris* (Joshi, Padalia, and Mathela 2010; Joshi 2013), *A. scoparia,* and *A. capetris* var. *glutinosa* (Juteau et al. 2002). Although no chemotaxonomic study has explicitly been conducted on the usefulness of alkynes in plant taxonomy, it is evident from the studies carried out on some plant species given above that they have potential use in plant chemotaxonomy and identification.

10.2.1.5 Carotenoids

Carotenoids, a group of isoprenoid metabolites, are essential for all photosynthetic organisms, including cyanobacteria, algae, and plants, with various functions in photosynthesis photoprotection, pigmentation, phytohormone synthesis, and signaling (DellaPenna and Pogson 2006). The attractive coloration of carotenoids in many flowers plays a significant role in attracting pollinators and seed dispersal (Lightbourn et al. 2008). *Capsicum* genus contains substrates such as β-carotene, zeaxanthin, and β-cryptoxanthin for capsanthin-capsorubin synthase (Kevrešan et al. 2009), essential for the synthesis of major carotenoids, e.g., capsorubin and capsanthin in red-fruited peppers (Wahyuni et al. 2013).

HPLC-PAD, GC-MS, and UHPLC-PAD-ESI-MS analyses were used to determine carotenoid content in leaves and berries of six varieties of Carpathians' Sea buckthorn (*Hippophae rhamnoides* L., ssp. *carpatica*) (Pop et al. 2014). GC-MS analysis showed a specific fatty acid profile for each berry variety, and the main fraction among them were carotenoid diesters, of which zeaxanthin dipalmitate was the major compound. However, only free carotenoids like β-carotene, neoxanthin, and violaxanthin existed in the leaves. Suitable carotenoids were specific for the Carpathians' Sea buckthorn from Romania, identified by PCA, and suggested to be a useful taxonomic marker for their identification and classification.

10.2.2 Secondary Metabolites

Secondary metabolites, derivatives of primary metabolites, are usually not directly involved in metabolism and play a significant role in environmental conditions such as adaptation to environmental stresses, defense against pathogens and predators, and mechanical damage (Osawa et al. 1994). Secondary metabolites, unlike primary metabolites, are produced in small quantities at the particular developmental stages of the plants by specialized cells, but restricted to a specific taxon of plant, e.g., alkaloids, terpenoids, essential oils, toxins, peptides, growth factors, lectins, polymeric substances, and pigments. The phytochemical content of closely related plant species is typically similar (Raghuveer et al. 2015) and is most widely studied concerning plant systematics. Secondary metabolites of plants have been a new source for drug discovery and utilized in the food industry, e.g., food additives, flavor, and others.

Despite the differences given above, there is not always a clear distinction between these two groups of metabolites, such as some rare types of sugars and fatty acids restricted to only some plant species, unlike primary metabolites. Plant growth regulators are considered intermediates between primary and secondary metabolites due to their role in plant development and growth. However, from a functionality point of view, while primary metabolites are involved in metabolic processes in the cell, secondary metabolites play a role in interaction with the environment and other organisms (Raghuveer et al. 2015). Secondary metabolites widely used in plant classification are described as follows:

10.2.2.1 Phenolics

Phenolic compounds are among the most common class of secondary metabolites in terrestrial vascular plants (Ertel and Hedges 1984) and exhibit a significant qualitative and quantitative variation at different genetic levels between and within species and clones (Nichols-Orians, Fritz, and Clausen 1993; Hakulinen, Julkunen-Tiitto, and Tahvanainen 1995; Witzell, Gref, and Näsholm 2003). They are localized mainly in a vacuole of plant cells in free and conjugated forms, including simple phenols and quinones, phenol carboxylic acids and their derivatives, flavones, flavonols, catechins, and leucoanthocyanins (Zaprometov 1996). However, phenolic compounds are chiefly in chloroplasts and nuclei in young, intensively growing plant tissues, whose cells have weekly developed vacuoles (Khlestkina 2013).

Many naturally occurring phenolics are generated from phenylpropanoid and phenylpropanoid-acetate pathways (Croteau and Johnson 1985). Concerning their chemical structure, phenolic

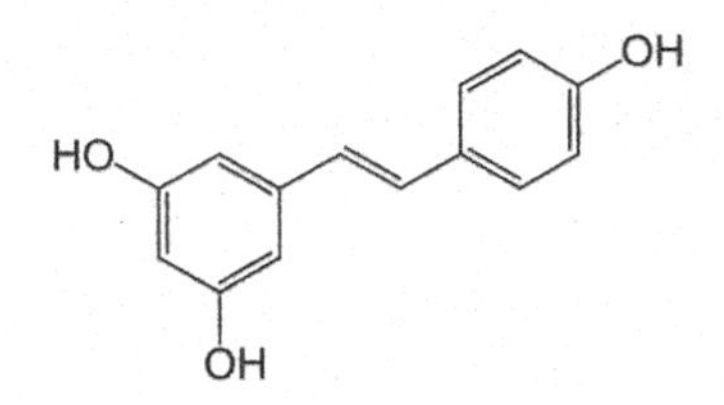

FIGURE 10.4 Resveratrol is a biologically important stilbenoid.

compounds are composed of an aromatic hydrocarbon with one or more "acidic" hydroxyl (–OH) functional groups. The classification of phenols is based primarily on the number of phenolic rings and structural elements connecting those rings. Phenolic compounds are categorized simply as flavonoids and non-flavonoids and further subdivided into many subcategories based on the number of phenolic units, substituent groups, and/or the nature of the bond between the phenolic moieties. Flavonoids are classified into six subgroups, namely, flavones, flavonols, flavonols, flavanones, isoflavones, and anthocyanins, while non-flavonoids include phenolic acids, stilbenes (Figure 10.4), and lignans (Raghuveer et al. 2015). These phenolic compounds are, in general, involved in defense against pathogens. Moreover, some phenolic compounds give plants colors, flavors, and astringency (Chitindingu et al. 2007). They also have various biological activities, most notably being an antioxidant involved in the oxidation of potent inhibitors of polyunsaturated fatty acids and low-density lipoproteins; thus, they are considered effective in reducing the risk of cardiovascular diseases (Morton et al. 2000).

Many studies have been reported on phenolic compounds' usefulness in the identification, taxonomy, and phylogeny of various plant species. Over 100 compounds have been reported from 121 species belonging to 21 genera of Gentianaceae. Of these compounds, four were taxonomically important, namely, xanthones, iridoids, mangiferin, and C-glucoflavones, and xanthones are not present in all taxa Gentianaceae. Categorization genera of this family according to the presence of the type of xanthones generated four groups: Group 1, the taxa containing only a few, and biosynthetically primitive xanthones, including *Anthocleista*, *Blackstonia*, *Macrocarpaea*, *Gentianopsis*, and *Orphium*; Group 2, contains xanthones with an intermediate degree of biosynthetic advancement *Comastoma*, *Lomatogonium*, *Swertia*, *Gentiana*, *Gentianella*, and tentatively, *Tripterospermum*. Group has the most advanced compounds, with the xanthones found in Group 2 being the biosynthetic precursors 3 including *Frasera*, *Veratrill*, and *Halenia*; Group 4 contains another set of biosynthetically advanced compounds including *Canscora*, *Centaurium*, *Hoppea*, *Eustoma*, *Chironia*, *Ixanthus*, and, with some reservation, *Schultesia*. The molecular phylogeny of Gentianaceae based on the sequences of *trn*L intron and *mat*K and the most recent classification proposed by Struwe et al. (2002) was well in accord with the grouping given above.

Phenylethanoid glycosides (forsythoside B, verbascoside, samioside, alyssonoside, isoverbascoside, leucosceptosides A and B, and martynoside) have been found exclusively in certain species of the genus *Phlomis* (Lamiaceae) from Turkey, indicating a good photochemical taxonomic marker (Kirmizibekmez et al. 2005). Mitreski et al. (2014) examined the polyphenolic contents and profiles of four *Teucrium* species (*T. polium* L., *T. chamaedrys* L., *T. scordium* L., and *T. montanum* L.) from Macedonia by LC/DAD/ESI-MS^n chromatographic method. A total of 31 phenolic compounds identified were divided into four categories: 2 hydroxycinnamic acid derivatives, 12 phenylethanoid glycosides, flavonoid 11 glycosides, and 6 flavonoid aglycones. Systematic analysis of *Teucrium* species based on their phenolic content and profiles displayed their chemodiversity potential to contribute to their chemotaxonomy, approximately 90% for *T. montanum* and *T. chamaedrys* and 60% for *T. polium* and *T. scordium*. Similarly, the composition and quantification of phenolic acids and flavonoid compounds in *Achillea millefolium* aggregate (Asteraceae) from Middle Europe by SPE-HPLC/UV analysis revealed differences, especially between distantly related species (Benedek et al. 2007). In another study done on the content and composition of phenolic acids and flavonoids, the complex genus *Plantago* (Plantaginaceae) from Croatia displayed a significant difference between its species, thus indicating to be a chemotaxonomic marker for this genus (Jurišić Grubešić et al. 2013).

Chemotaxonomic significance of paeonol and analogs has also been shown for 14 species and two subspecies of *Paeonia* (Ranunculaceae) (Bao-Lin, De-Yuan, and Pei-Gen 2008). The occurrence and content of these phenolic compounds were evaluated in three sections: (1) sect. *Moutan*, high content of paeonol and its analogs in all species; (2) sect. *Paeonia*, low content of paeonol and its analogs in plants of four taxa, *P. anomala* ssp. *veitchii*, *P. lactiflora*, *P. mairei* and *P. intermedia*; and (3) sect. *Onaepia*, none of these compounds found.

Twenty *Salvia* species (Lamiaceae) grown and cultivated in Poland have been examined for their chemical content using conjugated (i.e., binary) chromatographic fingerprints. The method proposed was successfully useful in defining chemical similarities and differences between *Salvia* species that was found to be useful for their chemotaxonomy and identification of new species (Ciesla et al. 2010).

Polyphenols of 89 wild, cultivated, and hybrid tea plants from Japan and China were examined for their potential use as a chemotaxonomic marker (Li et al. 2010). PCA of 13 polyphenol patterns ((–)-Epigallocatechin 3-O-gallate (EGCG) (1); (–)-epigallocatechin (EGC) (2); (–)-epicatechin 3-O-gallate (ECG) (3); (–)-epicatechin (EC) (4); and (+)-catechin, strictinin, and gallic acid) were used as a chemotaxonomic marker (Li et al. 2010). Ward's minimum-variance cluster analysis is based on those taxonomical information, the structure-types of the flavonoid B-rings, such as the pyrogallol-(EGCG (**1**) and EGC (**2**)) and catechol-(ECG (**3**) and EC (**4**)) types displayed three subclusters, namely, **A**, **B**, and **C**. Cluster **A** contained old tea trees cv "Taidi" cha and "Gushu" cha (*C. sinensis* var. *assamica*) with higher number of compounds 1 and 3 and lower amounts of compounds 1 and 2, indicating relatively primitive trees. Subclusters **B** and **C** include Chinese hybrids and Japanese and Taiwanese tea trees, respectively, with relatively lower amounts of **3** and **4** than subcluster **A**. Cluster and PCAs showed that the content of polyphenols indicates the most recent origin of tea plant supporting the current hypothesis proposing the Xishuangbanna district and Puer City are among the center of origin for tea plants.

Míka et al. (2005) investigated non-structural phenolic compounds in the polyploid complex of *Dactylis* L. (Poaceae) as a phytochemical marker. They found a clear-cut difference between the diploid ($2n=14$) subspecies of *Dactylis* and the tetraploid ($2n=28$) ones. However, lower similarities were found between octoploid brome species and those belonging to various sections. Overall results suggest that phenolic profiles of grasses are useful chemical markers for discrimination at the low taxonomic level when their environmental conditions for plant growth, growth stages of the plant, etc., were standardized (Míka et al. 2005).

Apart from the studies mentioned above, there have been more reports on the taxonomic value of phenolic compounds in various plant species, etc., Moraceae (Royer et al. 2010), Crassulaceae (Z. Liu et al. 2013), Gesneriaceae (Bai et al. 2013) (*Veronica*; Plantaginaceae) (Taskova, Gotfredsen, and Jensen 2006), and *Cannabis* (Cannabaceae) (Hillig and Mahlberg 2004).

10.2.2.2 Non-flavonoids and Flavonoids

Flavonoids are widespread secondary metabolites in plants, seeds, and fruits, which are vital for the defense and the physiological development of plants. Non-flavonoid secondary metabolites include phenolic acids, polyphenolic stilbenes, and lignans. Benzoic acid and cinnamic acid derivatives are the main class of these compounds, mainly in conjugation with other polyphenols, quinic acid, glucose, or other plant structural components (Chang, Reiner, and Xie 2005) (Figure 10.5).

Another small category of non-flavonoids, the stilbenes (1,2-diarylethenes), are characterized by two phenyl groups linked by a two-carbon methylene group. One of the best known is resveratrol (3,5,4'-trihydroxystilbene) produced by various plants such as grapes, peanuts, and berries (Han, Shen, and Lou 2007). Stilbenes are polyphenols produced by plants as a defensive response to microbial attacks. Salicylic acid, produced by some plant species in response to fungal attacks, has

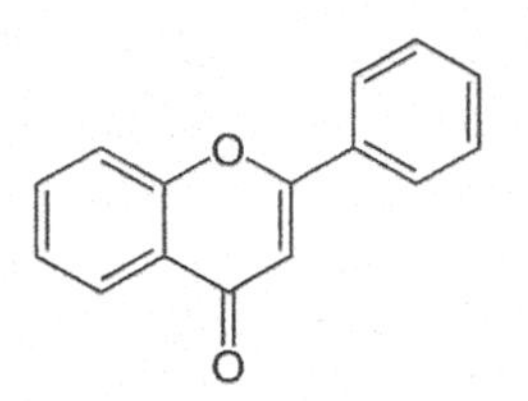

FIGURE 10.5 Molecular structure of the flavone backbone (2-phenyl-1,4-benzopyrone).

been synthesized and modified to serve as an anti-inflammatory, analgesic, and antipyretic agent (Raghuveer et al. 2015).

Lignans are phenolic dimers having a 2,3-dibenzylbutane structure. Such compounds are recognized to exist as minor components of many plants, in which they shape the construction blocks for the formation of lignin in the plant cell wall. The compounds occur mainly in the glycosidic form. Lignans are found in legumes, seeds, and vegetable oils; they are mainly present in their free forms, even as the glycosylated structure is not always abundant (Axelson et al. 1982; MacRae and Towers 1984).

Flavonoids are part of a large family of phenolic compounds or polyphenols and include more than 6,000 different structures (Šamec et al. 2021). They are very abundant secondary metabolites in plants, fruits, and seeds, responsible for the color, fragrance, and aromatic characteristics. In plants, flavonoids regulate cell growth, attract pollinating insects, and protect against biotic and abiotic stresses (Rodríguez De Luna, Ramírez-Garza, and Serna Saldívar 2020). Flavonoids subclasses include chalcones, anthocyanidins, isoflavanones, flavones, flavanones, isoflavonoids, and flavonols. These compounds have been associated with the antioxidant properties of many medicinal plants (Pandey and Rizvi 2009).

Flavonoids have been beneficial as a chemotaxonomic marker for numerous plant species. Using high-performance liquid chromatography (HPLC), the flavonoid content analysis in four different buckthorn berries origins using HPLC displayed 12 flavonoid compounds. Relative peaks of characteristic flavonoid compounds and correlation coefficient of similarity in chromatograms revealed a clear distinction between different species of buckthorn berries, except for closely related two subspecies (Chen et al. 2007).

HPLC chromatogram profiles of icariin and related compounds in 35 species and one variety of the Chinese *Epimedium* (Berberidaceae), most of which belong to the *Diphyllon* sect and subgen. *Epimedium* was divided into four main types and nine subtypes labeled as the "ABCI" peak group; II-3 has been proposed as the most primitive type; II-1, IV, and I-3 were primitive and closely related to II-3; I-1 was of the basic type; I-2, I-4, III, and II-2 were types derived from correlation analysis with flower morphology. This division was consistent with W. T. Stearn's classification of sect. *Diphyllon* with four series in 2002 (Guo, Li-Kuan, and Pei-Gen 2008).

A detailed chemotaxonomic study was conducted to determine the qualitative profile and actual amounts of the pharmacologically active isoflavone aglycones genistein, daidzein, formononetin, and biochanin A in aerial parts of 13 *Trifolium* L. (clover) species originating from Poland. Chemotaxonomic differences combined with flower color variability showed that the levels of certain isoflavones ranged from ~3 to ~3,300 μg/g dry weight in ten *Trifolium* species with pink, white, or purple-red corolla. In contrast, the isoflavone compounds were not detected at all in three yellow-flowered ones (*T. aureum*, *T. dubium*, and *T. campestre*) (Zgórka 2009). The genus *Iris* (Iridaceae) is recognized as rich in seconder metabolite content, of which flavonoids are predominant. As reviewed by Wang et al. (2010), over 90 flavonoid compounds were identified in *Iris* species, of which 38 were new compounds from 15 species. Since features and restricted occurrence of some flavonoid compounds, in particular *Iris* species put forward their significance as a chemotaxonomic marker (Wang, Cui, and Zhao 2010).

Two genera of the family Annonaceae, namely, *Dasymaschalon* and *Desmos*, containing formyl-substituted with A-ring and unsubstituted B-ring flavonoids supported their taxonomy based on their molecular phylogeny, gross morphology, and leaf anatomy so that they could be considered as chemotaxonomic markers for these two independent genera (Xiao-Lei et al. 2012).

HPLC fingerprint of flavonoids from the six commonly used clinical Chinese materia medicas revealed two types named naringin and hesperidin according to their phytochemotaxonomy. It was concluded that the six popular traditional Chinese medicines' flavonoid fingerprints pattern could be used as a chemotaxonomic marker (Chen and Lin 2011).

The LC/MS profiling using the CID/MS^n experiments showed that structures of 175 flavonoid glycoconjugates from roots and leaves of three Mediterranean species (*Lupinus luteus*, *L. albus*, and *L. angustifolius*), eight North America lupine species (*L. elegans*, *L. hintonii*, *L. exaltatus*, *L. montanus*, *L. stipulates*, *L. mexicanus*, *L. rotundiflorus*, and *Lupinus* sp.), and one species from South America domesticated in Europe (*L. mutabilis*) discriminate between North American and the Mediterranean lupines (Wojakowska et al. 2013).

A correlation between leaf flavonoids and taxonomy and geographic distribution was demonstrated in the genus *Disporum*. Quercetin and kaempferol were the main flavonoids constituting five North American species. At the same time, the flavone luteolin, together with some apigenin and chrysoeriol, was the major component in the five Asian taxa supporting the geographical division of *Disporum* into two sections. This geographic distinction between North American and Asian plants was further supported by 16 chemical differences in the flavonoid glycosides in four taxa. Most species within Asian taxa have some characteristic flavonoid profiles, which were suggested to be good evidence for chemical polymorphism in the *D. cantoniense*-*D. sessile* group. Although three distinct flavonoid profiles were found among different accession of *D. cantoniense* from the Himalayas and China, surprisingly, *D. sessile* from Japan was chemically indistinguishable from some morphologically similar specimens of *D. cantoniense* (Williams et al. 1993).

10.2.2.3 Essential Oils and Volatile Terpenes

Terpenes constitute a highly diverse class of secondary metabolites with the formula $(C_5H_8)_n$, predominantly produced by plants, especially conifers (Breitmaier 2006) (Figure 10.6). While they are simple hydrocarbons, terpenoids (also known as isoprenoids) are oxygenated hydrocarbons. They have been further categorized based on the number of isoprene units in the structure like (C_{10}), sesquiterpenes (C_{15}), diterpenes (C_{20}), etc. (Figure 10.7) (Martin et al. 2003) (Ashour, Wink, and Gershenzon 2010). A major component of turpentine alpha-pinene is the well-known monoterpene. Some terpenoids, such as sterols, are considered primary metabolites, and others are produced as secondary (plant) products involved in fundamental biological processes such as electron transport in plastids and mitochondria and plant hormones such as gibberellins, strigolactones, and brassinosteroids.

Moreover, various terpenoids in defense against abiotic and biotic stress factors are signal molecules for attracting pollinator insects (Boncan et al. 2020). Terpenoids are exclusively stored in various plant tissues with special structures such as latex canals, serin canals, secretory cavities, and glandular trichomes (Holopainen et al. 2013). Chemical properties of terpenoids, such as hydrophobicity and vapor pressure, affect their storage and volatility. At the same time, their emission rates and patterns are influenced by seasonality, temperature, humidity, irradiance, and their interaction with other organisms (Yazaki, Arimura, and Ohnishi 2017).

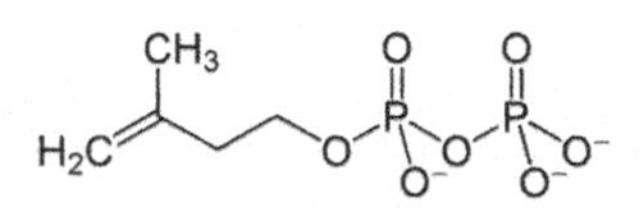

FIGURE 10.6 The structural formula for isopentenyl pyrophosphate (IPP), the carbon atoms of which constitute building blocks in terpenes.

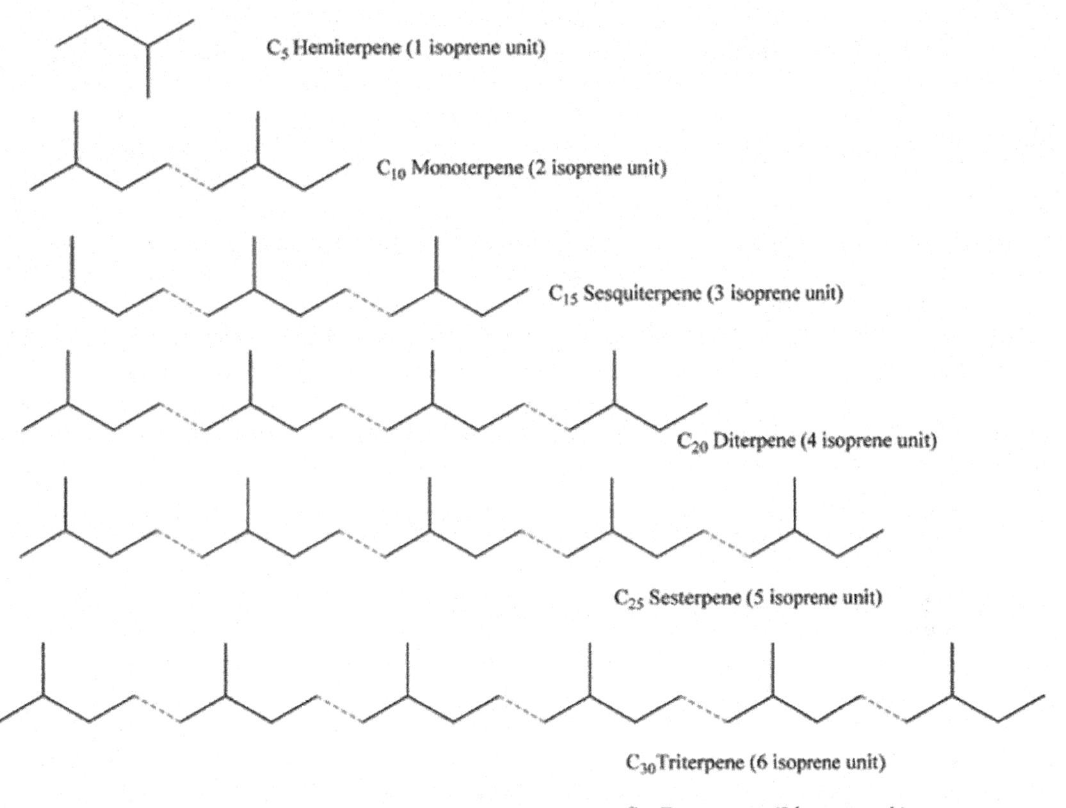

FIGURE 10.7 Classification of terpenes.

β-Carotene is present in all green cells. Terpenoid glycosides (cardiac glycosides) are present in Apocynaceae, Liliac, Moraceae, etc. Monoterpenes are also known as essential oils. It depends on odors and essence, e.g., Lamiaceae, Apiaceae, Rutaceae, etc. Sesquiterpene lactones are a group of bitter-tasting compounds. Out of 1,400 in Asteraceae, 1,340 are present. Cronquist has shown the evolution of Asteraceae on this basis, i.e., Rubiales → Dipsacales → Asteraceae, rather than through Campanulales, where lactones are altogether absent.

A chemotaxonomic study was conducted on essential oils of the flowers and leaves of *Alpinia allughas* Rosc., *Alpinia speciosa* K. Schum., *Alpinia galanga* (L.) Willd., and *Alpinia calcarata* Rosc. by capillary GC and GC/MS which revealed that monoterpenoids such as 1,8-cineole, terpinene-4-ol, α-terpineol, (*E*)-methyl cinnamate, α- and β-pinenes, and camphor were the main constituents. However, *endo*-fenchyl acetate, *endo*-fenchol, and *exo*-fenchyl acetate essential oils were only determined in rhizomes of *A. galanga*, *A. speciosa*, and *A. calcarata*, while β-pinene was the dominant rhizome oil in *A. allughas*. Moreover, essential oil components of Alpine species native to subtropical and subtemperate regions of Northern India were significantly different in qualitative and quantitative contents. Cluster analysis based on the oil composition of these species put forwarded those similarities and differences, which were due to their molecular skeletons, of which monoterpenoids, *viz*., 1,8-cineole, terpinene-4-ol, camphor, pinenes, (*E*)-methyl cinnamate, and fenchyl derivatives, were considered as chemotaxonomic markers (Padalia et al. 2010).

Chemotaxonomic significance of essential oils of *Ferula* (Umbelliferae) species from Iran using cluster analysis categorized *Ferula* species into four groups containing either (i) oxygenated monoterpenes, (ii) organosulfur compounds, (iii) monoterpene hydrocarbons, or (iv) aliphatic hydrocarbons, monoterpene, and sesquiterpene as the principal classes of compounds (Kanani et al. 2011).

Another study on the volatile compounds of flowers and leaves of *Alpinia purpurata* and *A. speciosa* from North Brazil using GC-MS identified that terpinene-4-ol (22.7%), limonene (25.1%), and γ-terpinene (17.4%) are the main constituents in the leaf of *A. speciosa*, whereas 1,8-cineole (23.1%), terpinene-4-ol (20.4%), and sabinene (14.5%) were the major ones in flowers. On the other hand, the major oil constituents found in flowers of *A. purpurata* were β-pinene (27.8%) and α-pinene (16.9%) (Zoghbi, Andrade, and Maia 1999). Variation in essential oil constitutes within and between six subspecies of *Phebalium squamulosum* (Rutaceae: Boronieae), discriminating distinct chemotypes supporting their classification based on morphological characteristics (Sadgrove et al. 2014). More studies were providing evidence on the chemotaxonomic significance of the specific compositions and contents of essential oils in various plants from different geographic regions, such as two *Hyptis suaveolens* (L.) Poit leaves from Nigeria (Eshilokun, Kasali, and Giwa-Ajeniya 2005), *Stachys inflata* Benth. from Iran (Norouzi-Arasi et al. 2006), *Hypericum triquetrifolium* Turra (Guttiferae, Hypericoideae) from Calabria (Italy), etc. (Bertoli et al. 2003), *Hypericum* (Guttiferae) (Yuce and Bagci 2012), six *Gingidia* (Umbelliferae) species from Australia and New Zealand (Sansom et al. 2013), and two different *Ocimum basilicum* L. varieties from Iran (Pirmoradi, Moghaddam, and Farhadi 2013).

A chemotaxonomic study on the terpene compounds in wild populations of *Pinus nigra* (ssp. *nigra*, var. *gocensis*, ssp. *pallasiana*, and var. *banatica*) was found to be similar to the populations from Greece (ssp. *nigra*) and Central Italy, and the populations are divided into three groups by cluster analysis: group 1, populations I, II, III, IV, and V (ssp. *nigra* group), group 2, Population VI (ssp. *pallasiana* group); and group 3, *Population* VII, with the most distinct oil composition (ssp. *banatica* group) (Šarac et al. 2013).

Phylogenetic analysis of peel oil of volatile profiles of fruit and leaf and 29 other genotypes of *Citrus*, *Fortunella*, and *Poncirus* based on hierarchical cluster analysis (HCA) differentiated between these taxa, as well as between subgroups of *Citrus*, namely, *Citrophorum*, *Sinocitrus*, and *Cephalocitrus*. A wild germplasm *Citrus nobilis* Lauriro (Mangshanyegan) volatile compounds were similar to pomelo (Liu et al. 2013).

10.2.2.4 Sesquiterpene

Sesquiterpenes, often with the molecular formula $C_{15}H_{24}$, are a class of terpenes built from three isoprene units found particularly in higher plants and other organisms. They have a significant structural diversity mainly due to changes in the assembly of their carbon skeleton and the rearrangement of functional groups. Hydrocarbons and oxygenated forms of naturally occurring sesquiterpenes are acids, lactones, alcohols, ketones and aldehydes, aromatic components, and essential oils with various pharmacological activities. A survey of the profiles of 34 sesquiterpene lactones in 25 species of the genera *Centaurea*, *Psephellus*, *Chartolepis*, *Leuzea*, *Zoegea*, and *Stizolophus* (subtribe centaureinae, Family Compositae) discriminated the species belonging to the genera *Psephellus, Chartolepis,* and *Centaurea* subg. *Hyalinella* (Nowak 1992).

Sesquiterpene lactones isolated from the genus *Anthemis* (family Asteraceae, tribe Anthemideae) were investigated for their potential as a chemotaxonomic marker. Two groups of species were defined according to the skeletal type of lactones: the first group includes species containing guaianolides (*A. alpestris, A. aetnensis,* and *A. hydruntina*), and the second one comprises species generating germacranolides (*A. punctata* subsp. *cupaniana* and *A. stribrnyi* subsp. *tracica*). Overall results obtained from this study support the current classification of European flora with some exceptions (Staneva, Todorova, and Evstatieva 2008).

The occurrence of these molecules varies among different plants taxa and has been found useful as chemotaxonomic markers in various plant species to date. Sesquiterpene dialdehyde variants investigated for their potential as a chemotaxonomic marker for four *Pseudowintera* species ((Winteraceae) endemic to New Zealand showed *P. axillaris* with high levels (2.2–6.9%) of paxidal, *P. insperata* individuals with high levels (3.0–6.9% of dry leaf wt.) of the coumarate, and *P. colorata* with varying levels of polygodial (1.4–2.9%) and 9-deoxymuzigadial (0–2.9%) (Wayman et al. 2010).

Sesquiterpene lactones were chemically characterized in one of the most prominent angiosperm families, Asteraceae. Sesquiterpene lactones from three tribes, except Inuleae, of the family Asteraceae were validated using the self-organizing maps to associate with their botanical sources. This method was successfully used in the discrimination of the Asteraceae tribes, and similarities among the Heliantheae, Eupatorieae, and Helenieae tribes, also between the Inuleae and Anthemideae tribes giving support to their previous classification mainly based on morphology and molecular data (Scotti et al. 2012).

From a chemotaxonomic point, two *Salvia* species (*S. stenophylla* and *S. runcinata*) native to South Africa were studied as a new source of a commercially important aroma, α-bisabolol, from a chemotaxonomic point of view. These two species were differentiated by analyzing their oil content using OPLS-DA on GC-MS and MIRS data (Sandasi, Kamatou, and Viljoen 2012).

GC/EM analysis of dichloromethane extracts from leaves of five *Polygonium* species belonging to the *Persicaria* section from the northeast and central lowlands of Argentina was used to investigate the presence of four drimane-type sesquiterpenes, namely, (**1**), isopolygodial (**2**), drimenol (**3**), and confertifolin (**4**), previously isolated from *P. acuminatum*, and the presence of three flavonoids: pinostrobin (**5**), flavokawin B (**6**), and cardamonin (**7**), previously isolated from *P. persicaria*. Results showed that among the five species of *Persicaria* section studied, two species contained sesquiterpenes **1–4** but not flavonoids **5–7**, the other two species included flavonoids **5–7** but not sesquiterpenes **1–4**, and only one species contained compounds **1–7**. These results indicated the potential use of sesquiterpenes as a chemotaxonomic marker for *Polygonum* species (Derita and Zacchino 2011).

10.2.2.5 Diterpenes

Diterpene is a compound consisting of four isoprene units with the molecular formula $C_{20}H_{32}$. The most common biologically important diterpenes are phytol (Figure 10.8), retinol, and retinal, which are well known for their anti-inflammatory and antimicrobial activities.

Plant or fungal diterpenes are determined from gummy exudates, resins, and high boiling fractions of resinous material after distillation of essential oils. The most commonly encountered diterpenes are non-volatile acids from conifers and legumes (Croteau and Johnson 1985; Langenheim 1990).

Phylogenetic and chemotaxonomic studies were conducted on some *Portulaca* species (Portulacaceae) with different chromosome numbers. The clerodane ([6.6]-fused ring), prinziane ([7.6]-fused ring), and portulane ([7.5]-fused ring) were found as major diterpenoid compounds in *P.* cv. Jewel, *P. pilosa*, and *P. grandiflora*, respectively. These species have diterpenoids with 1,4-dihydroxy-2-buten-2-ylethyl side chains that could serve as a phylogenetic and chemotaxonomic marker (Ohsaki et al. 1999).

Plant-derived diterpenoids have long been used as a chemotaxonomic marker for conifers. Recently, Diefendorf et al. (2019) have determined the content of diterpenoids and carbon isotopes of 43 conifer species and *Ginko biloba* from the University of California Botanical Garden at Berkeley, where all extant conifer families and two-thirds of genera are present. Although diterpenoid concentrations were highly variable among families and species, there was only a family-level phylogenetic structuring with a minute concentration in Pinaceae. However, a significant phylogenetic signal for the abietane and tetracyclic structures was obtained for the abietane and tetracyclic structures, when terpenoids were analyzed by the proportion of diterpenoid compound structural

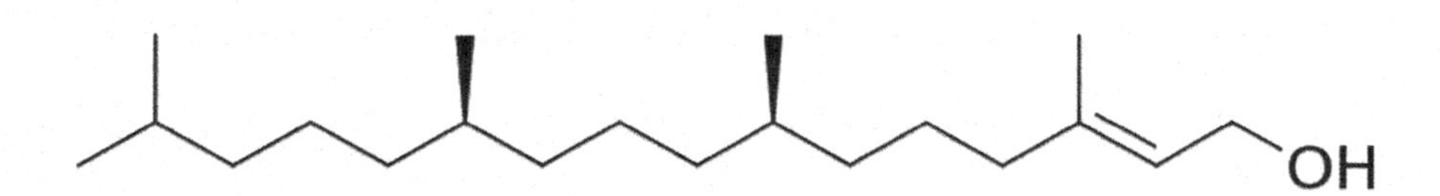

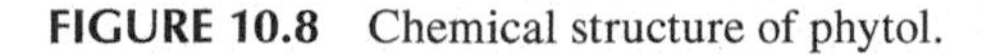
FIGURE 10.8 Chemical structure of phytol.

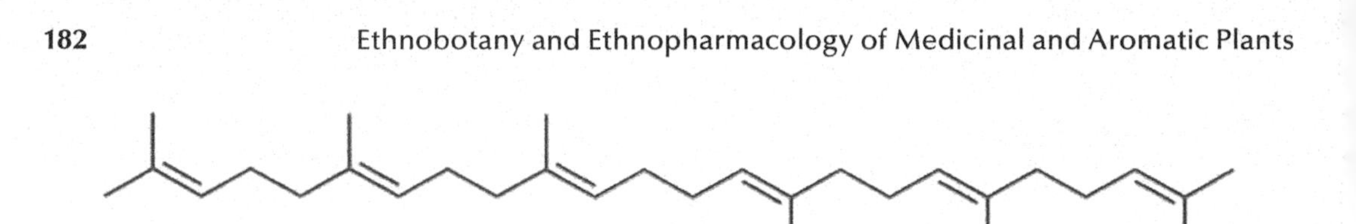

FIGURE 10.9 Chemical structure of squalene.

classes. An investigation of the distribution of lipophilic compounds in the needles defoliated twigs, and the outer bark of the *Pinus thunbergia* shoot system, was proposed that they might be used for chemotaxonomic purposes (Diefendorf, Leslie, and Wing 2019; Shpatov et al. 2013).

10.2.2.6 Triterpene Saponins

The triterpenes are a class of natural plant products composed of six isoprene units derived from the squalene biosynthetic pathway and have the molecular formula $C_{30}H_{48}$. Many methyl groups and oxidation of alcohol, carboxylic acids, and aldehydes in triterpenes make it various molecules with different biological activities. Triterpenes are converted to another big class of compounds called saponins (triterpene glycoside), because of their various active sites for glycosylation (Perveen 2021). Moreover, triterpenes, including squalene (Figure 10.9), are the precursors of all steroids, whereas some simple compounds act as signaling molecules, and more complex glycosylated triterpenes (saponins) are involved in defense against pathogens and pests. Apart from triterpenes' metabolic activities in the cell, they also have many industrial applications in the food and health sectors.

Chemotaxonomic utility of alkaloids and triterpenes was investigated in *Symphytum officinale* agg. (*Boraginaceae*) using a microextraction method (Jaarsma et al. 1989). The triterpene isobauerenol was found in *S. officinale*, *S. officinale* var. *lanceolatum* Weinm, and *S. bohemicum* Schmidt, *S. tanaicense*. An identical pyrrolizidine alkaloid and triterpene pattern were found in *S. bohemicum* ($2n=24$) and *S. officinale* ($2n=24$). These two species are also similar cytologically, morphologically, and phytochemically, and *S. bohemicum* was successfully crossed with the diploid, white-flowered W. European *S. officinale* suggests that they are conspecific. The similarity in pyrrolizidine alkaloid and triterpene profile, cytology, and morphology between *S. tanaicense* and *S. officinale* ($2n=40$) suggested that they are conspecific intraspecific variants.

Some rare triterpenes and sterols such as cardiac glycosides and withanolides in Scrophulariaceae, Apocynaceae, Asclepiadaceae, and quassinoids and limonoids in Rutales have a limited distribution. However, some families like Leguminosae produce triterpenes in the form of saponin (Wink and Waterman 1999).

The ursolic acid triterpene and the diterpene phytol were isolated from *Wendlandia formosana* Cowan (Lakshmana Raju et al. 2004); however, the universal presence of this compound in the family shows no taxonomic significance. Unlike two acyl lupeol from the stem bark and the pentacyclic triterpenes ursolic acid from the leaves of *Augusta longifolia* (Spreng.) Rehder, Rubiaceae-Ixoroideae and flavonoids, and coumarins were significant taxonomic markers in determining its systematic position (Choze, Delprete, and Lião 2010).

The order Ericales comprises around 25 families and 347 genera distributed in 4 clades: Balsaminoids, Ericoids, Polemonioids, and Primuloids. Phylogenetic analyses based on their phytochemical data showed that triterpenes, together with flavonoids, were useful as a taxonomic marker, mainly due to their molecular diversity and general occurrence. Saponins and lignans were considered phytochemotaxonomic markers for Primulaceae and Styracaceae, respectively (do Nascimento Rocha et al. 2015).

10.2.2.7 Alkaloids

Alkaloids are naturally occurring compounds that contain one or more nitrogen atoms making them alkalic. Over 3,000 alkaloid compounds have been described in the roots, leaves, stems, and fruits

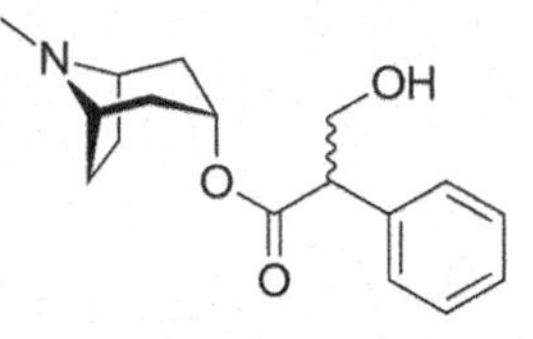

FIGURE 10.10 Chemical structure of an alkaloid, atropine.

of more than 3,000 plant species. According to their structure, alkaloids are classified as quinolines, pyridines, isoquinolines, indoles, pyrrolizidines, pyrrolidines, tropanes, terpenoids, and steroids. Solanaceae, Ranunculaceae, Papaveraceae, Apocynaceae, Berberidaceae, Rubiaceae, and Fabaceae are plants rich in alkaloids. The bitter taste of alkaloid compounds defers herbivorous animals; thus, some are used as a natural pesticide. These biologically active compounds are analgesics, anti-inflammatory, anticancer, local anesthetic, pain relief, antifungal, antimicrobial, and neuropharmacological (Kurek 2019). One of the most known alkaloids (Figure 10.10) is morphine used for various medical purposes.

Geographical distribution and chemotaxonomic utility of tropane alkaloids were investigated within the plant families Solanaceae, Erythroxylaceae, Rhizophoraceae, Proteaceae, Cruciferae, Euphorbiaceae, and Convolvulaceae. Of these families, tropane alkaloids are characteristic of the genera *Datura*, *Duboisia*, and *Brugmansia* of Solanaceae, but distribution in other plant families is more widespread as a novel tropane derivative (Griffin and Lin 2000). Detailed chemical structure of alkaloids in several species of *Datura* and *Brugmansia* revealed that they all have the same type of alkaloid spectrum (Evans 1979). While scopolamine and/or hyoscyamine are the principal alkaloids in the aerial parts of these plants, many forms of esters derivatives of dihydroxytropane and teloidine exist in their roots. The alkaloid content of five *Solandra* species (*S. grandifolia*, *S. hirsute*, *S. guttata*, *S. hartwegii*, and *S. macrantha*) was examined from a taxonomic point of view. Atropine and/or hyoscyamine and their non-derivatives were principal alkaloids in a uniform taxonomic group (Evans, Ghani, and Woolley 1972).

A phytochemical investigation on six *Sceletium* species (Aizoaceae) conducted on the mesembrine-type alkaloids showed that they are usually associated with Sceletium. Their venation patterns were either "emarcidum" or "tortuosum" type, indicating that the emarcidum type consisted of *S. emarcidum*, *S. rigidum*, and *S. exalatum*. In contrast, the tortuosum type consisted of *S. tortuosum*, *S. strictum*, and *S. expansum* (Patnala and Kanfer 2013). This is important for identifying alkaloid content from these plants' materials for the manufacture and quality control of their products.

A study performed on the alkaloid diversity, including 49 compounds of homolycorine, haemantamine, lycorine, tazettine, and narciclasine types, in 14 populations of *Lapiedra martinezii* Lag. (Amaryllidaceae) covering almost its entire distribution area, and their relationship with their phenological changes and biogeographical pattern, proposed a new tool for analyzing their geographic distribution. Alkaloid patterns differed between populations distributed in the north and south margins and those of the central area. The geographic distribution of alkaloid content in the *martinezii* populations was interpreted in a phylogenetic sense (Ríos et al. 2013). A recent review of the alkaloid content and profiles in the family Amaryllidaceae revealed more than 600 diverse alkaloids from 350 species, making 44% of all species in the subfamily, of which the majority are unique to this family. A survey of 636 compounds isolated or identified tentatively as alkaloids categorized into 42 skeleton types put forward that every member of the Amaryllidaceae family contains bioactive alkaloids indicating its chemotaxonomical importance (Berkov et al. 2020).

Chemosystematic identification of 15 new cocaine-bearing *Erythroxylum* cultigens throughout Colombia has been propagated. The alkaloid content of randomly selected five plants/cultigens with the three known varieties showed that *Erythroxylum coca* var. *ipadu* alkaloid profiles of ten

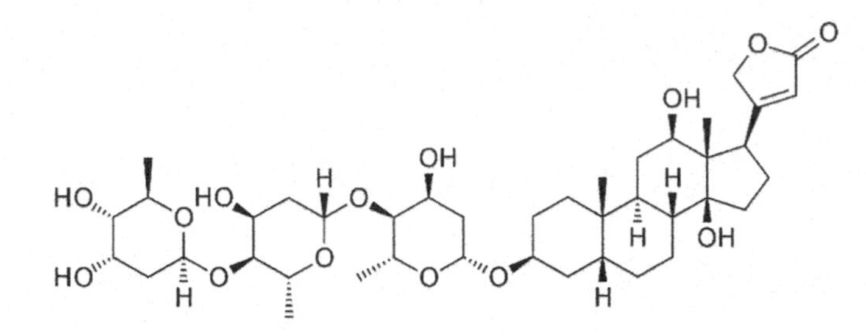

FIGURE 10.11 Chemical structure of digoxin.

cultigens were classic, *Erythroxylum novogranatense* and *Erythroxylum coca* var. *ipadu* were congruent with their hybrids, and one cultigen alkaloid profile could not be characterized due to the heterogeneous alkaloid profiles (Casale, Mallette, and Jones 2014).

10.2.2.8 Glycosides

Glycoside is a molecule in which a sugar molecule is linked to another non-sugar functional group through a glycosidic bond. Glycosides exist in flowers and fruits of many plants, and sometimes in the form of inactive glycosides that are activated by the removal of sugar part by enzyme hydrolysis.

They are highly diverse and categorized as O-glycoside, S-glycoside, C-glycoside, and N-glycoside. O-glycosides like rhein have a wider distribution and taxonomic value of these glycoside compounds. Moreover, flavanol glycoside was regarded as a significant taxonomic marker for the identification of Rosa species belonging to sections Gallicanae, Synstylae, Cinnamomeae, and Caninae, and the *R. rugosa* flavonol glycosides were also reported as a significant chemotaxonomic marker for the classification of Cinnamomeae (Sarangowa et al. 2014). Cardiac glycosides (Figure 10.11) used in the treatment of cardiac diseases are present in *Digitalis* (Scrophulariaceae), *Strophanthus* (Apocynaceae), and Squill bulb (Liliaceae) as digoxin, digitoxin, and digitonin.

Cyanogenic glycosides were found in over 3,000 plant species, especially widely distributed among the members of Asteraceae, Linaceae, Fabaceae, and Rosaceae. These compounds are bioactive toxins derived from amino acids involved in defense against some fungal pathogens and herbivores. Therefore, they have been recognized as natural herbicides. Glycosides derived from leucine amino acids are widespread in the subfamily Amygdaloideae (almond) and Maloideae (apple) of the family Rosaceae. In contrast, those derived from tyrosine are common in the families of the order Mangnoliales and Laurales (Singh 2012). Moreover, *n*-alkyl glycoside derivatives isolated from the glandular trichome exudates of *Geranium carolinianum* were proposed for chemotaxonomic purposes (Asai et al. 2011).

10.2.2.9 Non-protein (Non-proteinogenic) Amino Acids

Non-protein amino acids are those not incorporated in proteins during protein synthesis. These so-called non-protein amino acids are especially abundant in Liliaceae and Leguminosae, but also in very low concentrations in some other plants; e.g., L-azetidine-2-carboxylic acid (Figure 10.12) isolated from *Convallaria majalis* (Liliaceae) is a non-protein amino acid-like proline.

Nevertheless, the distribution of non-protein amino acids may be a valuable tool to establish systematic relationships between plant taxa. Canavanine (2-amino-4-guanidoxybutanoic acid, a chemical analog of arginine) was found restricted to Leguminosae and only within one subfamily of this family Papilionoideae. Also, it does not exist in all genera and species of this subfamily. As supported by cytological evidence, the other two subfamilies of Leguminosae, namely, Caesalpinioideae and Mimosoideae, diverged relatively early during evolution from the stem line now represented by Papilionoideae. Therefore, it was proposed that this non-protein amino acid has a chemotaxonomic significance (Bell et al. 2008). A survey revealed that *m*-carboxyphenylalanine

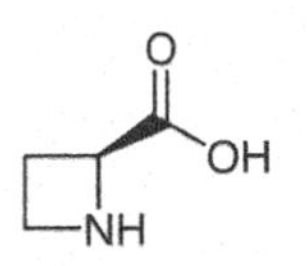

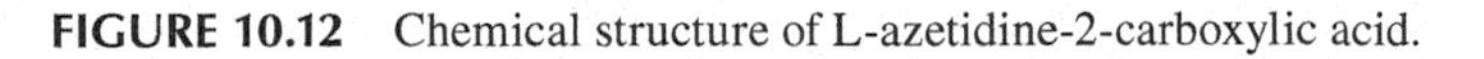

FIGURE 10.12 Chemical structure of L-azetidine-2-carboxylic acid.

was a constituent of some members of Cucurbitaceae (Dunnill and Fowden 1965). On the contrary, the comparison of seed extracts of 120 *Astragalus* species showed no definite chemotaxonomic value (Dunnill and Fowden 1967). To the best of our knowledge, there have not been any studies on the taxonomic value of non-amino acid proteins of the higher plants.

10.2.3 Macromolecules

Macromolecules are high molecular weight compounds, e.g., DNAs, RNAs, proteins, and complex polysaccharides. Semantides are the macromolecules that carry genetic information and classified into three categories in the order of sequential flow of genetic information: the genes themselves are known as "primary semantides," the mRNA transcripts "secondary semantides," and polypeptides "tertiary semantides." These molecules have been widely used for plant taxonomy, since they are offering a better phylogenetic relationship between plant taxa than secondary metabolites and morphological, cytological, and embryological characteristics in plants.

10.2.4 Conclusion

Humankind has long before used plant materials to treat diseases and source of nutrition and flavor. Therefore, they have developed their natural ability to discriminate between valuable plants. Classification is a natural way of learning about matters around us. Analytical thinking led to the scientific investigation of the classification, known as taxonomy. Secondary metabolites of plants have also become the center of interest due to their benefits; therefore, more efforts have been given to their isolation, identification, and biological activities for a wide range of applications in our daily life. Scientific development in taxonomy and systematics has reached a climax based on the natural classification of living things with a phylogenetic approach based on their inherited molecule DNA and its translated products proteins. Such classification is based on their similarities inherited from their common ancestors. Apart from genetic material, organisms have been classified based on taxonomically important characteristics such as morphology, physiology, biochemistry, behavior, cytogenetics, etc. Classification of plants based on their metabolites is a kind of necessity to ease the benefit, due to their significance for a wide range of medical and economic values. As we have reviewed in this chapter, the chemical structure and content of plant metabolites vary among plant taxa. Some characteristics are unique to some defined plant taxa that could be useful as chemotaxonomic characteristics for those plants. The success behind the usefulness of plant metabolites as a chemotaxonomic marker is mainly due to the diversity in the enzymes involved in their metabolic processes. Since these enzymatic proteins are translations of their genetic code, they may differentiate from different ancestral lineages among plants. They could successfully be used for tracing their origin and their phylogenetic relationships and geographic distributions. As seen from the literature review of plant metabolic compounds, they have been successfully used to determine the identification, phylogenetic relationships, and taxonomic positions. Though, the plant metabolic content could show plasticity by epigenetic factors, mainly the biotic and abiotic ones. Therefore, these characteristics could be considered phenetic characteristics because they are the secondary products of their inherited genetic material, such as morphological characteristics, as not all of them are taxonomically significant. Perhaps for these reasons, new term called plant chemophenetics was proposed for plant chemosystematics/plant chemotaxonomy.

As a result, it can be concluded that plants' metabolic constituents were successfully used in plant taxonomy and phylogeny in many cases described above; it seems they would not be an ultimate tool for plant taxonomy, but could be useful for the classification of the plants with specific valuable chemical content that would make easy for their industrial use. Therefore, recently, a new terminology, pharmacophylogeny, suggested that the classification of medicinal plants based on their molecular phylogeny and pharmacological activities might be more useful mainly for practical applications.

REFERENCES

Ahmad, Moghis U. 2017. *Fatty Acids: Chemistry, Synthesis, and Applications*. London, UK: Elsevier.

Aitzetmüller, Kurt. 2012. "Santalbic Acid in the Plant Kingdom." *Plant Systematics and Evolution* 298 (9): 1609–1617. doi: 10.1007/s00606-012-0678-5.

Alonso-Esteban, José Ignacio, María José González-Fernández, Dmitri Fabrikov, Esperanza Torija-Isasa, María de Cortes Sánchez-Mata, and José Luis Guil-Guerrero. 2020. "Hemp (*Cannabis sativa* L.) Varieties: Fatty Acid Profiles and Upgrading of Γ-Linolenic Acid–Containing Hemp Seed Oils." *European Journal of Lipid Science Technology* 122 (7): 1900445.

Altosaar, Illimar. 1974. *Plant Protein Chemotaxonomy: I. Disc Electrophoresis of Lasthenia Seed Albumins and Globulins: II. Partial Characterization and Sequence Studies of Sambucus Ferredoxin*. Vancouver: University of British Columbia.

Arts, Ilja C W, and Peter C H Hollman. 2005. "Polyphenols and Disease Risk in Epidemiologic Studies." *The American Journal of Clinical Nutrition* 81 (1): 317S–25S.

Aruna, P, D Venkataramanamma, Alok Kumar Singh, and R P Singh, 2015. "Health Benefits of Punicic Acid: A Review." *Comprehensive Reviews in Food Science and Food Safety* 15 (1): 16–27. doi: https://doi.org/10.1111/1541-4337.12171.

Asai, Teigo, Takaomi Sakai, Kiyoshi Ohyama, and Yoshinori Fujimoto. 2011. "N-Octyl α-L-Rhamnopyranosyl-(1→ 2)-β-D-Glucopyranoside Derivatives from the Glandular Trichome Exudate of Geranium Carolinianum." *Chemical Bulletin* 59 (6): 747–52.

Ashour, Mohamed, Michael Wink, and Jonathan Gershenzon. 2010. "Biochemistry of Terpenoids: Monoterpenes, Sesquiterpenes and Diterpenes." *Annual Plant Reviews Volume 40: Biochemistry of Plant Secondary Metabolism* 2: 258–303.

Avato, P, M A Pesante, F P Fanizzi, and C Aimbiré de Moraes Santos. 2003. "Seed Oil Composition of Paullinia Cupana Var. Sorbilis (Mart.) Ducke." *Lipids* 38 (7): 773–80.

Avato, P, I Rosito, P Papadia, and F P Fanizzi. 2005. "Cyanolipid-rich Seed Oils from *Allophylus natalensis* and *A. dregeanus*." *Lipids* 40 (10): 1051–56.

Avato, Pinarosa, Isabella Rosito, Paride Papadia, and Francesco P Fanizzi. 2006. "Characterization of Seed Oil Components from *Nephelium lappaceum* L." *Natural Product Communications* 1 (9): 1934578X0600100910.

Avato, P., and M. P. Argentieri. 2015. "Brassicaceae: A Rich Source of Health Improving Phytochemicals." *Phytochemistry Reviews* 14 (6): 1019–1033. doi: 10.1007/s11101-015-9414-4.

Avato, P, and A Tava. 2021. "Rare Fatty Acids and Lipids in Plant Oilseeds: Occurrence and Bioactivity." *Phytochemistry Reviews* 21: 1–28.

Axelson, M, J Sjövall, B E Gustafsson, and K D R Setchell. 1982. "Origin of Lignans in Mammals and Identification of a Precursor from Plants." *Nature* 298 (5875): 659–60.

Badami, R C, and F D Gunstone. 1963. "Vegetable Oils. XIII.—the Component Acids of Isano (Boleko) Oil." *Journal of the Science of Food Agriculture* 14 (12): 863–66.

Bai, Zhen-Fang, Xiao-Qin Wang, Pei-Gen Xiao, and Yong Liu. 2013. "Phenylethanoid Glycosides Distribution in Medicinal Plants of Gesneriaceae." *China Journal of Chinese Materia Medica* 38 (24): 4267–70.

Bao-Lin, G U O, Hong De-Yuan, and Xiao Pei-Gen. 2008. "Further Research on Chemotaxonomy of Paeonol and Analogs in *Paeonia* (Ranunculaceae)." *Journal of Systematics Evolution* 46 (5): 724.

Barthet, Véronique J. 2008. "(N-7) and (n-9) Cis-Monounsaturated Fatty Acid Contents of 12 Brassica Species." *Phytochemistry* 69 (2): 411–17.

Bell, E Arthur, Alison A Watson, and Robert J Nash. 2008. "Non-Protein Amino Acids: A Review of the Biosynthesis and Taxonomic Significance." *Natural Product Communications* 3 (1): 1934578X0800300117.

Benedek, Birgit, Noela Gjoncaj, Johannes Saukel, and Brigitte Kopp. 2007. "Distribution of Phenolic Compounds in Middleeuropean Taxa of the *Achillea millefolium* L. Aggregate." *Chemistry Biodiversity* 4 (5): 849–57.

Berkov, Strahil, Edison Osorio, Francesc Viladomat, and Jaume Bastida. 2020. "Chemodiversity, Chemotaxonomy and Chemoecology of Amaryllidaceae Alkaloids." *The Alkaloids: Chemistry Biology* 83: 113–85.

Bertoli, Alessandra, Francesco Menichini, Michele Mazzetti, Guido Spinelli, and Ivano Morelli. 2003. "Volatile Constituents of the Leaves and Flowers of *Hypericum triquetrifolium* Turra." *Flavour and Fragrance Journal* 18 (2): 91–94.

Bohannon, M B, and R Kleiman. 1975. "Unsaturated C18 α-Hydroxy Acids InSalvia Nilotica." *Lipids* 10 (11): 703–6.

Bohannon, M B, and R Kleiman. 1978. "Cyclopropene Fatty Acids of Selected Seed Oils from Bombacaceae, Malvaceae, and Sterculiaceae." *Lipids* 13 (4): 270–73.

Boncan, Delbert Almerick T, Stacey S K Tsang, Chade Li, Ivy H T Lee, Hon-Ming Lam, Ting-Fung Chan, and Jerome H L Hui.. "Terpenes and Terpenoids in Plants: Interactions with Environment and Insects." *International Journal of Molecular Sciences* 21 (19): 7382.

Breitmaier, Eberhard. 2006. *Terpenes: Flavors, Fragrances, Pharmaca, Pheromones*. New Jersey: John Wiley & Sons.

Buchanan, Bob B. 1971. "Ferredoxins from Photo-Synthetic Bacteria, Algae, and Higher Plants." *Methods in Enzymology* 23: 413–40.

Buchanan, Bob B, and Daniel I Arnon. 1970. "Ferredoxins: Chemistry and Function in Photosynthesis, Nitrogen Fixation, and Fermentative Metabolism." *Advances in Enzymology and Related Areas of Molecular Biology* 33: 119–76.

Casale, John F, Jennifer R Mallette, and Laura M Jones. 2014. "Chemosystematic Identification of Fifteen New Cocaine-Bearing *Erythroxylum cultigens* Grown in Colombia for Illicit Cocaine Production." *Forensic Science International* 237: 30–39.

Chang, Junbiao, John Reiner, and Jingxi Xie. 2005. "Progress on the Chemistry of Dibenzocyclooctadiene Lignans." *Chemical Reviews* 105 (12): 4581–4609.

Chen, Chu, Hao Zhang, Wei Xiao, Zheng-Ping Yong, and Nan Bai. 2007. "High-Performance Liquid Chromatographic Fingerprint Analysis for Different Origins of Sea Buckthorn Berries." *Journal of Chromatography A* 1154 (1–2): 250–59.

Chen, S B, Y Peng, S L Chen, and P G Xiao. 2005. "Introduction of Pharmaphylogeny." *World Science and Technology-Modernization of Traditional Chinese Medicine* 7 (6): 97.

Chen, Yonggang, and Li Lin. 2011. "Study and Comparison on HPLC Fingerprints of Flavonoids of Frequently Used Chinese Materia Medica in Citrus." *J Zhongguo Zhong Yao Za Zhi= Zhongguo Zhongyao Zazhi= China Journal of Chinese Materia Medica* 36 (19): 2660–65.

Chisholm, Mary J, and C Y Hopkins. 1964. "Fatty Acid Composition of Some Cucurbitaceae Seed Oils." *Canadian Journal of Chemistry* 42 (3): 560–64.

Chitindingu, K, A R Ndhlala, C Chapano, M A Benhura, and M Muchuweti. 2007. "Phenolic Compound Content, Profiles and Antioxidant Activities of *Amaranthus hybridus* (Pigweed), Brachiaria Brizantha (Upright Brachiaria) and *Panicum maximum* (Guinea Grass)." *Journal of Food Biochemistry* 31 (2): 206–16.

Choze, Rafael, Piero G Delprete, and Luciano M Lião. 2010. "Chemotaxonomic Significance of Flavonoids, Coumarins and Triterpenes of *Augusta Longifolia* (Spreng.) Rehder, Rubiaceae-Ixoroideae, with New Insights about Its Systematic Position within the Family." *Revista Brasileira de Farmacognosia* 20: 295–99.

Ciesla, Łukasz, Michał Hajnos, Dorota Staszek, Łukasz Wojtal, Teresa Kowalska, and Monika Waksmundzka-Hajnos. 2010. "Validated Binary High-Performance Thin-Layer Chromatographic Fingerprints of Polyphenolics for Distinguishing Different *Salvia* Species." *Journal of Chromatographic Science* 48 (6): 421–27.

Croteau, Rodney, and Mark A Johnson. 1985. "Biosynthesis of Terpenoid Wood Extractives." *Biosynthesis Components, Biodegradation of Wood*, 379–439.

DellaPenna, Dean, and Barry J Pogson. 2006. "Vitamin Synthesis in Plants: Tocopherols and Carotenoids." *Annual Review of Plant Biology* 57 (1): 711–38.

Dembitsky, Valery M, and Morris Srebnik. 2002. "Natural Halogenated Fatty Acids: Their Analogues and Derivatives." *Progress in Lipid Research* 41 (4): 315–67.

Derita, M, and S Zacchino. 2011. "Chemotaxonomic Importance of Sesquiterpenes and Flavonoids in Five Argentinian Species of *Polygonum* Genus." *Journal of Essential Oil Research* 23 (5): 11–14.

Diefendorf, Aaron F, Andrew B Leslie, and Scott L Wing. 2019. "A Phylogenetic Analysis of Conifer Diterpenoids and Their Carbon Isotopes for Chemotaxonomic Applications." *Organic Geochemistry* 127: 50–58.

Dunnill, P M, and L Fowden. 1967. "The Amino Acids of the Genus *Astragalus*." *Phytochemistry* 6 (12): 1659–63.

Dunnill, Patricia M, and L Fowden. 1965. "The Amino Acids of Seeds of the Cucurbitaceae." *Phytochemistry* 4 (6): 933–44.

Dussert, Stéphane, Andréina Laffargue, Alexandre de Kochko, and Thierry Joët. 2008. "Effectiveness of the Fatty Acid and Sterol Composition of Seeds for the Chemotaxonomy of *Coffea* Subgenus Coffea." *Phytochemistry* 69 (17): 2950–60.

Ertel, John R, and John I Hedges. 1984. "The Lignin Component of Humic Substances: Distribution among Soil and Sedimentary Humic, Fulvic, and Base-Insoluble Fractions." *Geochimica et Cosmochimica Acta* 48 (10): 2065–74.

Eshilokun, Adeolu O, Adeleke A Kasali, and Abdullatif O Giwa-Ajeniya. 2005. "Chemical Composition of Essential Oils of Two *Hyptis Suaveolens* (L.) Poit Leaves from Nigeria." *Flavour and Fragrance Journal* 20 (5): 528–30.

Evans, W C, A Ghani, and Valerie A Woolley. 1972. "Alkaloids of Solandra Species." *Phytochemistry* 11 (1): 470–72.

Evans, W. C. 1979. "Tropane Alkaloids of the Solanaceae." In *The Biology and Taxonomy of the Solanaceae*, edited by J. G. Hawkes, R. N. Lester, and A. D. Skelding, 241. London: Linnean Soc. Symp. Series No. 7.

Fairbrothers, D E, T J Mabry, R L Scogin, and B L Turner. 1975. "The Bases of Angiosperm Phylogeny: Chemotaxonomy." *Annals of the Missouri Botanical Garden*, 765–800.

Galliard, T. 1974. "Unusual Fatty Acids in Plants." In John T. Romeo (ed.), *Recent Advances in Phytochemistry*, 8:209–41. London, UK: Elsevier.

Galliard, T. 1978. "Unusual Fatty Acids in Plants." In *Metabolism and Regulation of Secondary Plant Products*, edited by V.C. Runeckles and F.E. Conn, 209–241. New York: Academic Press.

Gasparetto, João Cleverson, Cleverson Antônio Ferreira Martins, Sirlei Sayomi Hayashi, Michel Fleith Otuky, and Roberto Pontarolo. 2012. "Ethnobotanical and Scientific Aspects of *Malva Sylvestris* L.: A Millennial Herbal Medicine." *Journal of Pharmacy and Pharmacology* 64 (2): 172–89.

Gong, Xue, Min Yang, Chun-nian He, Ya-qiong Bi, Chun-hong Zhang, Min-hui Li, and Pei-gen Xiao. 2020. "Plant Pharmacophylogeny: Review and Future Directions." *Chinese Journal of Integrative Medicine* 28: 567–74.

Gören, Ahmet C, Ekrem Akçiçek, Tuncay Dirmenci, Turgut Kılıç, Erkan Mozioğlu, and Hasibe Yilmaz. 2012. "Fatty Acid Composition and Chemotaxonomic Evaluation of Species Of *Stachys*." *Natural Product Research* 26 (1): 84–90. doi: 10.1080/14786419.2010.544025.

Griffin, William J, and G David Lin. 2000. "Chemotaxonomy and Geographical Distribution of Tropane Alkaloids." *Phytochemistry* 53 (6): 623–37.

Guil-Guerrero, J L, F F García Maroto, and A Giménez Giménez. 2001. "Fatty Acid Profiles from Forty-Nine Plant Species That Are Potential New Sources of γ-Linolenic Acid." *Journal of the American Oil Chemists' Society* 78 (7): 677–84.

Gunstone, Frank D, and John L Harwood. 2007. *The Lipid Handbook with CD-ROM*. Florida: CRC Press.

Guo, Bao-Lin, P E I Li-Kuan, and Xiao Pei-Gen. 2008. "Further Research on Taxonomic Significance of Flavonoids in *Epimedium* (Berberidaceae)." *Journal of Systematics Evolution* 46 (6): 874.

Hakulinen, Johanna, Riitta Julkunen-Tiitto, and Jorma Tahvanainen. 1995. "Does Nitrogen Fertilization Have an Impact on the Trade-off between Willow Growth and Defensive Secondary Metabolism?" *Trees* 9 (4): 235–40.

Haliński, Łukasz P, Janusz Szafranek, and Piotr Stepnowski. 2011. "Leaf Cuticular N-alkanes as Markers in the Chemotaxonomy of the Eggplant (*Solanum Melongena* L.) and Related Species." *Plant Biology* 13 (6): 932–39.

Han, Xiuzhen, Tao Shen, and Hongxiang Lou. 2007. "Dietary Polyphenols and Their Biological Significance." *International Journal of Molecular Sciences* 8 (9): 950–88.

Heneman, Karrie, and Sheri Zidenberg-Cherr. 2008. "Nutrition and Health Info Sheet: Phytochemicals." https://nutrition.ucdavis.edu/outreach/nutr-health-info-sheets/pro-phytochemical

Hennessy, Alan A., Paul R. Ross, Gerald F. Fitzgerald, and Catherine Stanton. 2016. "Sources and Bioactive Properties of Conjugated Dietary Fatty Acids." *Lipids* 51 (4): 377–397. doi: 10.1007/s11745-016-4135-z.

Herbin, G A, and P A Robins. 1969. "Patterns of Variation and Development in Leaf Wax Alkanes." *Phytochemistry* 8 (10): 1985–98.

Hillig, Karl W, and Paul G Mahlberg. 2004. "A Chemotaxonomic Analysis of Cannabinoid Variation in *Cannabis* (Cannabaceae)." *American Journal of Botany* 91 (6): 966–75.

Holopainen, Jarmo K, Sari J Himanen, J S Yuan, F Chen, and C Neal Stewart. 2013. "Ecological Functions of Terpenoids in Changing Climates." In K Ramawat and J M Mérillon (eds), *Natural Products*. Berlin/Heidelberg, Germany: Springer.

Hopkins, C Y, and Mary J. Chisholm .1962. "The Conjugated Triene Acid of Catalpa Ovata Seed Oil (Resumed)" *Journal of the Chemical Society (Resumed)* 573–573, doi: https://doi.org/10.1039/jr9620000573.
Jaarsma, Tea A, Elisabeth Lohmanns, Theo W J Gadella, and Theo M Malingré. 1989. "Chemotaxonomy of *The Symphytum officinale* Agg.(Boraginaceae)." *Plant Systematics Evolution* 167 (3): 113–27.
Jia, Shuo, Mingyue Shen, Fan Zhang, and Jianhua Xie. 2017. "Recent Advances in Momordica Charantia: Functional Components and Biological Activities." *International Journal of Molecular Sciences* 18 (12): 2555. doi: 10.3390/ijms18122555
Joh, Yong-Goe, Seung-Jin Kim, William W. Christie. 1995. "The Structure of the Triacylglycerols, Containing Punicic Acid, in the Seed Oil of *Trichosanthes Kirilowii" Journal of the American Oil Chemists' Society* 72: 1037–1042. doi: https://doi.org/10.1007/bf02660718.
Jones, S B, and A E Luchsinger. 1987. *Plant Systematics* (2nd. Ed.). New York: McGraw Hill Inc.
Joshi, Rakesh K. 2013. "*Artemisia capillaris*: Medicinal Uses and Future Source for Commercial Uses from Western Himalaya of Uttrakhand." *Asian Journal of Research in Pharmaceutical Sciences* 3 (3): 137–40.
Joshi, Rakesh K, Rajendra C Padalia, and Chandra S Mathela. 2010. "Phenyl Alkynes Rich Essential Oil of Artemisia Capillaris." *Natural Product Communications* 5 (5): 1934578X1000500528.
Jurišić Grubešić, Renata, Goran Srečnik, Dario Kremer, Jadranka Vuković Rodríguez, Toni Nikolić, and Sanda Vladimir-Knežević. 2013. "Simultaneous RP-HPLC-DAD Separation, and Determination of Flavonoids and Phenolic Acids in Plantago L. Species." *Chemistry Biodiversity* 10 (7): 1305–16.
Juteau, Fabien, Véronique Masotti, Jean-Marie Bessière, and Josette Viano. 2002. "Compositional Characteristics of the Essential Oil of *Artemisia Campestris* Var. Glutinosa." *Biochemical Systematics Ecology* 30 (11): 1065–70.
Kacurakova, M, P Capek, V Sasinkova, N Wellner, and A Ebringerova. 2000. "FT-IR Study of Plant Cell Wall Model Compounds: Pectic Polysaccharides and Hemicelluloses." *Carbohydrate Polymers* 43 (2): 195–203.
Kanani, Mohammad Reza, Mohammad Reza Rahiminejad, Ali Sonboli, Valiollah Mozaffarian, Shahrokh Kazempour Osaloo, and Samad Nejad Ebrahimi. 2011. "Chemotaxonomic Significance of the Essential Oils of 18 *Ferula* Species (Apiaceae) from Iran." *Chemistry Biodiversity* 8 (3): 503–17.
Kapoor, Rakesh, and Harikumar Nair. 2023. "Gamma Linolenic Acid: Sources and Functions." In F. Shahidi (ed.), *Bailey's Industrial Oil and Fat Products*, 1–45. https://doi.org/10.1002/047167849X.bio026.pub2
Keskin Çavdar, Hasene. 2019. "Active Compounds, Health Effects, and Extraction of Unconventional Plant Seed Oils." *Plant and Human Health,* Volume 2: 245–285. doi: 10.1007/978-3-030-03344-6_10
Kevrešan, Žarko S, Anamarija P Mandić, Ksenija N Kuhajda, and Marijana B Sakač Quality. 2009. "Carotenoid Content in Fresh and Dry Pepper (*Capsicum Annuum* L.): Fruits for Paprika Production." *Food Processing Safety* 36 (1–2): 21–27.
Khlestkina, E. 2013. "The Adaptive Role of Flavonoids: Emphasis on Cereals." *Cereal Research Communications* 41 (2): 185–98.
Kirmizibekmez, H, P Montoro, S Piacente, C Pizza, A Dönmez, and I Caliş. 2005. "Identification by HPLC-PAD-MS and Quantification by HPLC-PAD of Phenylethanoid Glycosides of Five Phlomis Species." *Phytochem Anal* 16 (1): 1–6. https://doi.org/10.1002/pca.802.
Kleiman, R, G F Spencer, F R Earle, H J Nieschlag, and A S Barclay. 1972. "Tetra-Acid Triglycerides Containing a New Hydroxy Eicosadienoyl Moiety In *Lesquerella Auriculata* Seed Oil." *Lipids* 7 (10): 660–65.
Kurek, Joanna. 2019. Alkaloids: Their Importance in Nature and Human Life. IntechOpen. doi: 10.5772/intechopen.85400.
Lakshmana Raju, B, Shwu-Jiuan Lin, Wen-Chi Hou, Zhi-Yang Lai, Pan-Chun Liu, and Feng-Lin Hsu. 2004. "Antioxidant Iridoid Glucosides from Wendlandia Formosana." *Natural Product Research* 18 (4): 357–64.
Langenheim, Jean H. 1990. "Plant Resins." *American Scientist* 78 (1): 16–24.
Li, Jia-Hua, Atsushi Nesumi, Keiichi Shimizu, Yusuke Sakata, Ming-Zhi Liang, Qing-Yuan He, Hong-Jie Zhou, and Fumio Hashimoto. 2010. "Chemosystematics of Tea Trees Based on Tea Leaf Polyphenols as Phenetic Markers." *Phytochemistry* 71 (11–12): 1342–49.
Lightbourn, Gordon J, Robert J Griesbach, Janet A Novotny, Beverly A Clevidence, David D Rao, and John R Stommel. 2008. "Effects of Anthocyanin and Carotenoid Combinations on Foliage and Immature Fruit Color of *Capsicum Annuum* L." *Journal of Heredity* 99 (2): 105–11.
Liu, Cuihua, Dong Jiang, Yunjiang Cheng, Xiuxin Deng, Feng Chen, Liu Fang, Zhaocheng Ma, and Juan Xu. 2013. "Chemotaxonomic Study of *Citrus*, *Poncirus* and *Fortunella* Genotypes Based on Peel Oil Volatile Compounds-Deciphering the Genetic Origin of Mangshanyegan (Citrus Nobilis Lauriro)." *PLoS One* 8 (3): e58411.
Liu, Zhenli, Yuanyan Liu, Chunsheng Liu, Zhiqian Song, Qing Li, Qinglin Zha, Cheng Lu, Chun Wang, Zhangchi Ning, and Yuxin Zhang. 2013. "The Chemotaxonomic Classification of *Rhodiola* Plants and Its Correlation with Morphological Characteristics and Genetic Taxonomy." *Chemistry Central Journal* 7 (1): 1–8.

Lu, Shao-Ping, Mina Aziz, Drew Sturtevant, Kent D Chapman, and Liang Guo. 2020. "Heterogeneous Distribution of Erucic Acid in Brassica Napus Seeds." *Frontiers in Plant Science* 10 (January). doi: 10.3389/fpls.2019.01744.

MacRae, W Donald, and G H Neil Towers. 1984. "Biological Activities of Lignans." *Phytochemistry* 23 (6): 1207–20.

Maffei, M. 1994. "Discriminant Analysis of Leaf Wax Alkanes in the Lamiaceae and Four Other Plant Families." *Biochemical Systematics Ecology* 22 (7): 711–28.

Maffei, Massimo. 1996. "Chemotaxonomic Significance of Leaf Wax Alkanes in the Gramineae." *Biochemical Systematics Ecology* 24 (1): 53–64.

Mannheimer, C A. 1999. *An Overview of Chemotaxonomy, and Its Role in Creating a Phylogenic Classification System.* Windhoek: National Botanical Research Institute, Ministry of Agriculture, Water and Rural Development, 87–90.

Martin, Vincent J J, Douglas J Pitera, Sydnor T Withers, Jack D Newman, and Jay D Keasling. 2003. "Engineering a Mevalonate Pathway in *Escherichia Coli* for Production of Terpenoids." *Nature Biotechnology* 21 (7): 796–802.

Mayworm, Marco A S, and Antonio Salatino. 2002. "Distribution of Seed Fatty Acids and the Taxonomy of Vochysiaceae." *Biochemical Systematics Ecology* 30 (10): 961–72.

Mazid, M, T A Khan, and F Mohammad. 2011. "Role of Secondary Metabolites in Defense Mechanisms of Plants." *Biology Medicine* 3 (2): 232–49.

Mecklenburg, Helen Cameron. 1966. "Inflorescence Hydrocarbons of Some Species of Solanum L., and Their Possible Taxonomic Significance." *Phytochemistry* 5 (6): 1201–9.

Mentzer, C. 1966. *Biogenetic Classification of Plant Constituents*. New York: Academic Press, 21–31.

Mĭka, V, V Kuban, B Klejdus, V Odstrcilova, and P Nerusil. 2005. "Phenolic Compounds as Chemical Markers of Low Taxonomic Levels in the Family Poaceae." *Plant Soil Environment* 51 (11): 506.

Mikolajczak, K L. 1977. "Cyanolipids." *Progress in the Chemistry of Fats Lipids, Other* 15 (2): 97–130.

Miller, Roger W, David Weisleder, Robert Kleiman, Ronald D Plattner, and Cecil R Smith Jr. 1977. "Oxygenated Fatty Acids of Isano Oil." *Phytochemistry* 16: 947–51.

Mimura, Maria R M, Maria L F Salatino, Antonio Salatino, and José F A Baumgratz. 1998. "Alkanes from Foliar Epicuticular Waxes of *Huberia* Species: Taxonomic Implications." *Biochemical Systematics Ecology* 26 (5): 581–88.

Mino, Yoshiki. 1995. "Protein Chemotaxonomy of Genus *Datura*. IV. Amino Acid Sequence of Datura Ferredoxins Depends Not on the Species but the Section of Datura Plants from Which It Comes." *Chemical Bulletin, Pharmaceutical* 43 (7): 1186–89.

Mino, Yoshiki. 2006. "Protein Chemotaxonomy. XIII. Amino Acid Sequence of Ferredoxin from *Panax Ginseng*." *Biological Bulletin, Pharmaceutical* 29 (8): 1771–74.

Mitreski, Ilija, Jasmina Petreska Stanoeva, Marina Stefova, Gjoshe Stefkov, and Svetlana Kulevanova. 2014. "Polyphenols in Representative *Teucrium* Species in the Flora of R. Macedonia: LC/DAD/ESI-MS n Profile and Content." *Natural Product Communications* 9 (2): 1934578X1400900211.

Mohammed Khysar Pasha, and F Ahmad. 1992. "Analysis of Triacylglycerols Containing Cyclopropane Fatty Acids in Sterculia Fostida (Linn.) Seed Lipids." *Journal of Agricultural and Food Chemistry* 40 (4): 626–629. doi: 10.1021/jf00016a020.

Møller, B L, and D S Seigler. 1999. "Biosynthesis of Cyanogenic Glycosides, Cyanolipids and Related Compounds." *Plant Amino Acids Biochemistry Biotechnology* 69: 563–609.

Mongrand, Sébastien, Alain Badoc, Brigitte Patouille, Chantal Lacomblez, Marie Chavent, and Jean-Jacques Bessoule. 2005. "Chemotaxonomy of the Rubiaceae Family Based on Leaf Fatty Acid Composition." *Phytochemistry* 66 (5): 549–59.

Morton, Lincoln W, Rima Abu-Amsha Caccetta, Ian B Puddey, and Kevin D Croft. 2000. "Chemistry and Biological Effects of Dietary Phenolic Compounds: Relevance to Cardiovascular Disease." *Clinical Pharmacology, Experimental Physiology* 27 (3): 152–59.

Msami, H.M. 1999. "An Outbreak of Suspected Poisoning of Cattle by *Dichapetalum* sp." *Tropical Animal Health and Production* 31 (1): 1–7. doi:10.1023/a:1005162015366

Murphy, Denis J. 2020. "The Study and Utilisation of Plant Lipids: From Margarine to Lipid Rafts." In Denis J. Murphy (ed.), *Plant Lipids*, 1–26. Florida: CRC Press.

Nascimento Rocha, Marco Eduardo do, Maria Raquel Figueiredo, Maria Auxiliadora Coelho Kaplan, Tony Durst, and John Thor Arnason. 2015. "Chemotaxonomy of the Ericales." *Biochemical Systematics Ecology* 61: 441–49.

Nichols-Orians, Colin M, Robert S Fritz, and Thomas P Clausen. 1993. "The Genetic Basis for Variation in the Concentration of Phenolic Glycosides in *Salix Sericea*: Clonal Variation and Sex-Based Differences." *Biochemical Systematics Ecology Biochemical Systematics Ecology* 21 (5): 535–42.

Nikolić, Biljana, Vele Tešević, Srdjan Bojović, and Petar D Marin. 2013. "Chemotaxonomic Implications of the N-alkane Composition and the Nonacosan-10-ol Content in Picea Omorika, *Pinus Heldreichii*, and *Pinus Peuce*." *Chemistry Biodiversity* 10 (4): 677–86.

Norouzi-Arasi, Hassan, Issa Yavari, Vahid Kia-Rostami, Raoof Jabbari, and Mohammad Ghasvari-Jahromi. 2006. "Volatile Constituents of *Stachys Inflata* Benth. from Iran." *Flavour and Fragrance Journal* 21 (2): 262–64.

Nowak, Gerard. 1992. "A Chemotaxonomic Study of Sesquiterpene Lactones from Subtribe Centaureinae of the Compositae." *Phytochemistry* 31 (7): 2363–68.

Ohlrogge, John, Nick Thrower, Vandana Mhaske, Sten Stymne, Melissa Baxter, Weili Yang, Jinjie Liu, Kathleen Shaw, Basil Shorrosh, and Meng Zhang. 2018. "Plant FA Db: A Resource for Exploring Hundreds of Plant Fatty Acid Structures Synthesized by Thousands of Plants and Their Phylogenetic Relationships." *The Plant Journal* 96 (6): 1299–308.

Ohsaki, Ayumi, Kozo Shibata, Takashi Kubota, and Takashi Tokoroyama. 1999. "Phylogenetic and Chemotaxonomic Significance of Diterpenes in Some *Portulaca* Species (Portulacaceae)." *Biochemical Systematics Ecology* 27 (3): 289–96.

Osawa, Kenji, Hideyuki Yasuda, Takashi Maruyama, Hiroshi Morita, Koichi Takeya, and Hideji Itokawa. 1994. "Antibacterial Trichorabdal Diterpenes from *Rabdosia Trichocarpa*." *Phytochemistry* 36 (5): 1287–91.

Osborne, R, A Salatino, M L F Salatino, C M Sekiya, and M Vazquez Torres. 1993. "Alkanes of Foliar Epicuticular Waxes from Five Cycad Genera in the Zamiaceae." *Phytochemistry* 33 (3): 607–9.

Osborne, R, M L F Salatino, and A Salatino. 1989. "Alkanes of Foliar Epicuticular Waxes of the Genus *Encephalartos*." *Phytochemistry* 28 (11): 3027–30.

Padalia, Rajendra C, Ram S Verma, Velusamy Sundaresan, and Chandan S Chanotiya. 2010. "Chemical Diversity in the Genus *Alpinia* (Zingiberaceae): Comparative Composition of Four *Alpinia* Species Grown in Northern India." *Chemistry Biodiversity* 7 (8): 2076–87.

Pandey, Kanti Bhooshan, and Syed Ibrahim Rizvi. 2009. "Plant Polyphenols as Dietary Antioxidants in Human Health and Disease." *Oxidative Medicine and Cellular Longevity* 2 (5): 270–78.

Patnala, Srinivas, and Isadore Kanfer. 2013. "Chemotaxonomic Studies of Mesembrine-Type Alkaloids in *Sceletium* Plant Species." *South African Journal of Science* 109 (3): 1–5.

Pereira de Melo, Illana Louise, Eliane Carvalho, and Jorge Mancini Filho. 2014. " Pomegranate seed oil (Punica garanatum L.): a source of punicic acid (conjugated α-linolenic acid)."*Journal of Human Nutrition & Food Science* 2: 1024–1035.

Perveen, Shagufta. 2021. "Introductory Chapter: Terpenes and Terpenoids." In S. Perveen and A. Mohammad Al-Taweel (eds.), *Terpenes and Terpenoids-Recent Advances*. Biochemistry. IntechOpen. doi: 10.5772/intechopen.87558.

Pirmoradi, Mohammad Reza, Mohammad Moghaddam, and Nasrin Farhadi. 2013. "Chemotaxonomic Analysis of the Aroma Compounds in Essential Oils of Two Different *Ocimum Basilicum* L. Varieties from Iran." *Chemistry Biodiversity* 10 (7): 1361–71.

PlantFAdb—Exploring Phylogenetic Relathionships between Hundreds of Plant Fatty Acids Synthsized by Thousands of Plants ." 2016. *Plantfadb.org*. Michigan State University. https://plantfadb.org/pages/about.

Pop, Raluca Maria, Yannick Weesepoel, Carmen Socaciu, Adela Pintea, Jean-Paul Vincken, and Harry Gruppen. 2014. "Carotenoid Composition of Berries and Leaves from Six Romanian Sea Buckthorn (*Hippophae Rhamnoides* L.) Varieties." *Food Chemistry* 147: 1–9.

Radulović, Niko S, and Milan S Dekić. 2013. "Volatiles of Geranium Purpureum Vill. and Geranium Phaeum L.: Chemotaxonomy of Balkan *Geranium* and *Erodium* Species (Geraniaceae)." *Chemistry Biodiversity* 10 (11): 2042–52.

Raghuveer, Irchhaiya, Kumar Anurag, Yadav Anumalik, Gupta Nitika, Kumar Swadesh, Gupta Nikhil, Kumar Santosh, Yadav Vinay, Prakash Anuj, and Gurjar Himanshu. 2015. "Metabolites in Plants and Its Classification." *World Journal of Pharmacy and Pharmaceutical Sciences* 4 (1): 287–305.

Rasool, Rafia, Bashir A Ganai, Seema Akbar, Azra N Kamili, and Akbar Masood. 2010. "Phytochemical Screening of Prunella Vulgaris L.-an Important Medicinal Plant of Kashmir." *Pakistan Journal of Pharmaceutical Sciences* 23 (4): 399–402.

Ríos, Segundo, Strahil Berkov, Vanessa Martínez-Francés, and Jaume Bastida. 2013. "Biogeographical Patterns and Phenological Changes in *Lapiedra Martinezii* Lag. Related to Its Alkaloid Diversity." *Chemistry Biodiversity* 10 (7): 1220–38.

Rodríguez De Luna, Sara Luisa, R E Ramírez-Garza, and Sergio O Serna Saldívar. 2020. "Environmentally Friendly Methods for Flavonoid Extraction from Plant Material: Impact of Their Operating Conditions on Yield and Antioxidant Properties." *The Scientific World Journal* 2020: 6792069.

Royer, Mariana, Gaëtan Herbette, Véronique Eparvier, Jacques Beauchêne, Bernard Thibaut, and Didier Stien. 2010. "Secondary Metabolites of Bagassa Guianensis Aubl. Wood: A Study of the Chemotaxonomy of the Moraceae Family." *Phytochemistry* 71 (14–15): 1708–13.

Ruiz-Herrera, José, and Lucila Ortiz-Castellanos. 2010. "Analysis of the Phylogenetic Relationships and Evolution of the Cell Walls from Yeasts and Fungi." *FEMS Yeast Research* 10 (3): 225–43.
Sadgrove, Nicholas J, Ian R H Telford, Ben W Greatrex, and Graham L Jones. 2014. "Composition and Antimicrobial Activity of the Essential Oils from the *Phebalium Squamulosum* Species Complex (Rutaceae) in New South Wales, Australia." *Phytochemistry* 97: 38–45.
Šamec, Dunja, Erna Karalija, Ivana Šola, Valerija Vujčić Bok, and Branka Salopek-Sondi. 2021. "The Role of Polyphenols in Abiotic Stress Response: The Influence of Molecular Structure." *Plants* 10 (1): 118.
Sandasi, Maxleene, Guy P P Kamatou, and Alvaro M Viljoen. 2012. "An Untargeted Metabolomic Approach in the Chemotaxonomic Assessment of Two *Salvia* Species as a Potential Source of α-Bisabolol." *Phytochemistry* 84: 94–101.
Sansom, Catherine E, Peter B Heenan, Nigel B Perry, Bruce M Smallfield, and John W van Klink. 2013. "Chemosystematic Analyses of *Gingidia Volatiles*." *Chemistry & Biodiversity* 10 (12): 2226–34.
Santos, Fernando Sergio Dumas dos, Letícia Pumar Alves de Souza, and Antonio Carlos Siani. 2008. "O Óleo de Chaulmoogra Como Conhecimento Científico: A Construção de Uma Terapêutica Antileprótica." *História, Ciências, Saúde-Manguinhos* 15 (15): 29–46. doi: 10.1590/S0104-59702008000100003.
Šarac, Zorica, Srdjan Bojović, Biljana Nikolić, Vele Tešević, Iris Đorđević, and Petar D Marin. 2013. "Chemotaxonomic Significance of the Terpene Composition in Natural Populations of *Pinus Nigra* JF Arnold from Serbia." *Chemistry & Biodiversity* 10 (8): 1507–20.
Sarangowa, Ochir, Tsutomu Kanazawa, Makoto Nishizawa, Takao Myoda, Changxi Bai, and Takashi Yamagishi. 2014. "Flavonol Glycosides in the Petal of *Rosa* Species as Chemotaxonomic Markers." *Phytochemistry* 107: 61–68.
Sayed-Ahmad, Bouchra, Thierry Talou, Zeinab Saad, Akram Hijazi, and Othmane Merah. 2017. "The Apiaceae: Ethnomedicinal Family as Source for Industrial Uses." *Industrial Crops and Products* 109 (December): 661–671. doi: 10.1016/j.indcrop.2017.09.027.
Scotti, Marcus T, Vicente Emerenciano, Marcelo J P Ferreira, Luciana Scotti, Ricardo Stefani, Marcelo S Da Silva, and Francisco Jaime B Mendonça Junior. 2012. "Self-Organizing Maps of Molecular Descriptors for Sesquiterpene Lactones and Their Application to the Chemotaxonomy of the Asteraceae Family." *Molecules* 17 (4): 4684–702.
Shpatov, Alexander V, Sergey A Popov, Olga I Salnikova, Ekaterina A Khokhrina, Emma N Shmidt, and Byung Hun Um. 2013. "Low-Volatile Lipophilic Compounds in Needles, Defoliated Twigs, and Outer Bark of *Pinus Thunbergii*." *Natural Product Communications* 8 (12): 1934578X1300801227.
Singh, P. 2012. "Plant Taxonomy: Past, Present and Future, Edited by R Gupta." *TERI* 233: 1–376.
Singh, Ram. 2016. "Chemotaxonomy: A Tool for Plant Classification." *Journal of Medicinal Plants Studies* 4 (2): 90–93.
Sivarajan, V V. 1991. *Introduction to the Principles of Plant Taxonomy*. Cambridge: Cambridge University Press.
Smith Jr, C R. 1971. "Occurrence of Unusual Fatty Acids in Plants." *Progress in the Chemistry of Fats and Other Lipids* 11: 137–77.
Smith Jr, C R, T L Wilson, E H Melvin, and I A Wolff. 1960. "Dimorphecolic Acid—A Unique Hydroxydienoid Fatty Acid2." *Journal of the American Chemical Society* 82 (6): 1417–21.
Smith Jr, C R, and I A Wolff. 1969. "Characterization of Naturally Occurring A-Hydroxylinolenic Acid." *Lipids* 4 (1): 9–14.
Smith, Philip M. 1976. "The Chemotaxonomy of Plants." Philip M. Smith Contemporary Biology Series. London: Edward Arnold, 313 pp.
Spencer, G F, R Kleiman, F R Earle, and I A Wolff. 1970. "TheTrans-6 Fatty Acids OfPicramnia Sellowii Seed Oil." *Lipids* 5 (3): 285–87.
Spitzer, Volker. 1996. "Fatty Acid Composition of Some Seed Oils of the Sapindaceae." *Phytochemistry* 42 (5): 1357–60.
Staneva, Jordanka D, Milka N Todorova, and Ljuba N Evstatieva. 2008. "Sesquiterpene Lactones as Chemotaxonomic Markers in Genus *Anthemis*." *Phytochemistry* 69 (3): 607–18.
Struwe, L, J W Kadereit, J Klackenberg, S Nilsson, M Thiv, K B Von Hagen, and V A Albert. 2002. "Systematics, Character Evolution, and Biogeography of Gentianaceae, Including a New Tribal and Subtribal Classification." *Gentianaceae: Systematics and Natural History*, 21–309.
Stuhlfauth, T, H Fock, H Huber, and K Klug. 1985. "The Distribution of Fatty Acids Including Petroselinic and Tariric Acids in the Fruit and Seed Oils of the Pittosporaceae, Araliaceae, Umbelliferae, Simarubaceae and Rutaceae." *Biochemical Systematics Ecology* 13 (4): 447–53.
Subaşi, İlhan. 2020. "Seed Fatty Acid Compositions and Chemotaxonomy of Wild Crambe (Brassicaceae) Taxa in Turkey." *Turkish Journal of Agriculture and Forestry* 44 (6): 662–670. doi: 10.3906/tar-1912-76.

Sun, Jin-Yue, Xu Guo, and Mark A Smith. 2017. "Identification of Crepenynic Acid in the Seed Oil of Atractylodes Lancea and A. Macrocephala." *Journal of the American Oil Chemists' Society* 94 (5): 655–60.

Takeda, Hiroshi. 1991. "Sugar Composition of the Cell Wall and the Taxonomy of *Chlorella* (Chlorophyceae) 1." *Journal of Phycology* 27 (2): 224–32.

Taskova, Rilka Mladenova, Charlotte Held Gotfredsen, and Søren Rosendal Jensen. 2006. "Chemotaxonomy of Veroniceae and Its Allies in the Plantaginaceae." *Phytochemistry* 67 (3): 286–301.

Tava, Aldo, and Pinarosa Avato. 2014. "Analysis of Cyanolipids from Sapindaceae Seed Oils by Gas Chromatography–EI-Mass Spectrometry." *Lipids* 49 (4): 335–45.

Teomim, Doron, Abraham Nyska, and Abraham J. Domb. 1999. "Ricinoleic Acid-Based Biopolymers." Journal of Biomedical Materials Research 45 (3): 258–267. doi: 10.1002/(sici)1097-4636(19990605)45:3%3C258::aid-jbm14%3E3.0.co;2-w.

Tillman-Sutela, Eila, Anu Johansson, Päivi Laakso, Tomi Mattila, and Heikki Kallio. 1995. "Triacylglycerols in the Seeds of Northern Scots Pine, *Pinus Sylvestris* L., and Norway Spruce, *Picea Abies* (L.) Karst." *Trees* 10 (1): 40–45.

Tsevegsuren, N, K Aitzetmuller, and K Vosmann. 2004. "*Geranium Sanguineum* (Geraniaceae) Seed Oil: A New Source of Petroselinic and Vernolic Acid." *Lipids* 39 (6): 571–76.

Ušjak, Ljuboš, Ivana Sofrenić, Vele Tešević, Milica Drobac, Marjan Niketić, and Silvana Petrović. 2019. "Fatty Acids, Sterols, and Triterpenes of the Fruits of 8 *Heracleum* Taxa." *Natural Product Communications* 14 (6): 1934578X1985678. doi: 10.1177/1934578x19856788.

Wahyuni, Yuni, Ana-Rosa Ballester, Enny Sudarmonowati, Raoul J Bino, and Arnaud G Bovy. 2013. "Secondary Metabolites of *Capsicum* Species and Their Importance in the Human Diet." *Journal of Natural Products* 76 (4): 783–93.

Wang, Hui, Yanmei Cui, and Changqi Zhao. 2010. "Flavonoids of the Genus *Iris* (Iridaceae)." *Mini Reviews in Medicinal Chemistry* 10 (7): 643–61.

Wayman, Kjirsten A, Peter J de Lange, Lesley Larsen, Catherine E Sansom, and Nigel B Perry. 2010. "Chemotaxonomy of Pseudowintera: Sesquiterpene Dialdehyde Variants Are Species Markers." *Phytochemistry* 71 (7): 766–72.

Weselake, R J. 2016. "Engineering Oil Accumulation in Vegetative Tissue." In T A McKeon, D G Hayes, D F Hildebrand, and R J Weselake (eds.), *Industrial Oil Crops* (pp. 413–34). New York: AOCS Press. [Google Scholar].

Williams, Christine A, James Richardson, Jenny Greenham, and John Eagles. 1993. "Correlations between Leaf Flavonoids, Taxonomy and Plant Geography in the Genus *Disporum*." *Phytochemistry* 34 (1): 197–203.

Wink, Michael, and Peter G Waterman. 1999. "Chemotaxonomy in Relation to Molecular Phylogeny of Plants." *Annual Plant Reviews* 2: 300–41.

Witzell, Johanna, Rolf Gref, and Torgny Näsholm. 2003. "Plant-Part Specific and Temporal Variation in Phenolic Compounds of Boreal Bilberry (*Vaccinium Myrtillus*) Plants." *Biochemical Systematics Ecology* 31 (2): 115–27.

Wojakowska, Anna, Anna Piasecka, Pedro M García-López, Francisco Zamora-Natera, Paweł Krajewski, Łukasz Marczak, Piotr Kachlicki, and Maciej Stobiecki. 2013. "Structural Analysis and Profiling of Phenolic Secondary Metabolites of Mexican *Lupine* Species Using LC–MS Techniques." *Phytochemistry* 92: 71–86.

Wolff, Robert L, Olivier Lavialle, Frédérique Pédrono, Elodie Pasquier, Laurent G Deluc, Anne M Marpeau, and Kurt Aitzetmüller. 2001. "Fatty Acid Composition of Pinaceae as Taxonomic Markers." *Lipids* 36 (5): 439–51.

Xiao-Lei, Zhou, Shirley Tan Siang Ning, Baixin Jiao, Liu Bing-Tao, and Wu Jiu-Hong. 2012. "Phylogenetic Relationship between Dasymaschalon (Hook. f. et Thoms.) Dalle Torre et Harms and Desmos Lour. as Well as Study on Their Constituents." *Chinese Traditional and Herbal Drugs* 43: 1852–57.

Yazaki, Kazufumi, Gen-ichiro Arimura, and Toshiyuki Ohnishi. 2017. "'Hidden' Terpenoids in Plants: Their Biosynthesis, Localization and Ecological Roles." *Plant and Cell Physiology* 58 (10): 1615–21.

Yoshime, Luciana Tedesco, Illana Louise Pereira de Melo, José Augusto Gasparotto Sattler, Eliane Bonifácio Teixeira de Carvalho, and Jorge Mancini-Filho. 2016. "Bitter Gourd (*Momordica Charantia* L.) Seed Oil as a Naturally Rich Source of Bioactive Compounds for Nutraceutical Purposes." *Nutrire* 41 (1). doi: 10.1186/s41110-016-0013-y.

Yuce, Ebru, and Eyup Bagci. 2012. "The Essential Oils of the Aerial Parts of Two Hypericum Taxa (*Hypericum Triquetrifolium* and *Hypericum Avicularifolium* Subsp. *depilatum* Var. Depilatum (Clusiaceae)) from Turkey." *Natural Product Research* 26 (21): 1985–90.

Zaprometov, M N. 1996. "Phenolic Compounds and Their Role in Plant Life of a plant: 56th Timiryazev reading. Moscow Nauka, 45 p.

Zgórka, Grazyna. 2009. "Ultrasound-assisted Solid-phase Extraction Coupled with Photodiode-array and Fluorescence Detection for Chemotaxonomy of Isoflavone Phytoestrogens in *Trifolium* L.(Clover) Species." *Journal of Separation Science* 32 (7): 965–72.

Zhang, Jiao-Lin, Shi-Bao Zhang, Yi-Ping Zhang, and Kaoru Kitajima. 2015. "Effects of Phylogeny and Climate on Seed Oil Fatty Acid Composition across 747 Plant Species in China." *Industrial Crops Products* 63: 1–8.

Zhang, Lin, Meilan Liu, Hongxu Long, Wei Dong, Asher Pasha, Eddi Esteban, Wen-Ying Li, et al. 2019. "Tung Tree (*Vernicia fordii*) Genome Provides a Resource for Understanding Genome Evolution and Improved Oil Production." *Genomics, Proteomics & Bioinformatics* 17 (6): 558–575. doi: 10.1016/j.gpb.2019.03.006.

Zidorn, Christian. 2019. "Plant Chemophenetics– A New Term for Plant Chemosystematics/Plant Chemotaxonomy in the Macro-Molecular Era." *Phytochemistry* 163: 147–48.

Zoghbi, Maria Das Graças B, Eloisa Helena A Andrade, and José Guilherme S Maia. 1999. "Volatile Constituents from Leaves and Flowers of *Alpinia Speciosa* K. Schum. and *A. Purpurata* (Viell.) Schum." *Flavour and Fragrance Journal* 14 (6): 411–14.

11 Biological Roles and Mechanism of Phytochemicals in Disease Prevention and Treatment

Manzar Alam, Anas Shamsi, and Md Imtaiyaz Hassan
Jamia Millia Islamia

CONTENTS

11.1 INTRODUCTION

Chronic diseases are the leading reasons of death globally (World Health Organization 2014). Recent evidence sturdily indicates that diets affluent in plant foods are linked with a diminished risk of multiple diseases, including cardiovascular disease (CVD) (Dauchet, Amouyel, and Dallongeville 2009), neurodegenerative diseases (NDDs) (D'Onofrio et al. 2017; Jarząb and Kukula-Koch 2018), diabetes (A Stravodimos et al. 2017), and cancer (Key 2011). Oxidative stress and inflammation (OSI) are constantly high in public suffering from chronic diseases (Calder et al. 2009). However, *in vitro*, *in vivo*, and clinical trial data show that the plant-based diet may decrease the risk of cancer, NDDs, and CVD, because of the incidence of biologically dynamic plant phytochemicals (Upadhyay, Dixit, and longevity 2015).

Phytochemicals are considered as bioactive compounds in fruits, grains, vegetables, and other plant foods, which have decreased the risk of numerous diseases (Alam, Ali, et al. 2022; Alam, Ali, Ahmed, et al. 2021). Phytochemicals act as a defense system for combating diseases. Antioxidant phytochemicals occur extensively in fruits, cereal grains, vegetables, edible macrofungi, and medicinal plants (Deng, Shen et al. 2012; Guo et al. 2012; Alam, Ahmed, et al. 2022). However, common fruits, such as berries, grape, pomegranate, guava, sweetsop, persimmon, and plum, are rich in antioxidants and phytochemicals (Alam, Ahmed, et al. 2022; Fu et al. 2011). Besides, fruit wastes also comprise high contents of phytochemicals, such as catechin, cyanidin 3-glucoside, gallic acid, epicatechin, kaempferol, and chlorogenic acid (Alam, Ahmed, et al. 2022; Lin et al. 2021). Vegetables

DOI: 10.1201/b22842-11

have high antioxidant capacity and total phenolic substances (Deng et al. 2013; Alam, Ashraf, et al. 2022).

Phytochemicals are influential compounds belong to secondary metabolites of plants. They include polyphenols, steroidal saponins, flavonoids, organosulfur compounds, and vitamins (Vasanthi, ShriShriMal, and Das 2012; Alam, Alam, et al. 2022). Polyphenols and carotenoids contribute to the antioxidant functions (Alam, Alam, et al. 2022). Natural polyphenols contain a plenty of antioxidants in human food diets, and radical scavenging actions are associated with substituting hydroxyl groups of phenolics (Rokayya et al. 2013; Alam, Hasan, et al. 2022). Flavonoids comprise a huge group of polyphenolic compounds, which have a benzo-γ-pyrone in their structure, and they universally exist in plants. They are synthesized through the phenylpropanoid pathway (Kumar and Pandey 2013).

Phytochemicals play a beneficial role in various diseases by regulating many cellular and molecular pathways, including regulation of inflammation, metabolic disorder, redox potentials, apoptosis, and so forth (Maraldi et al. 2014; Ali, Alam, and Hassan 2022). Polyphenols exhibit extensive range of defensive roles, including hypolipidemic, antiproliferative, antioxidative, and anti-inflammatory, in reducing disease development (Alissa and Ferns 2012; George et al. 2009; Liu 2003). Phytochemicals have antioxidant/antimicrobial activities, modulation of detoxification enzymes, an incentive of the immune system, reduction of platelet aggregation, neuroprotective effects, and anticancer properties. Multiple phytochemicals may defend humans against diseases (Rao 2003; Alam, Ali, Mohammad, et al. 2021; Ali et al. 2022). Phytochemicals exhibit various health-promoting functions, such as preventing and treating CVDs, cancer, neural disorders, and Alzheimer's disease (AD) (Winter et al. 2017; Xue et al. 2022). In this chapter, we summarize the current progress on the human health benefits of phytochemicals and their promising mechanisms in treating and preventing multiple diseases.

11.2 BIOLOGICAL ACTIVITIES OF PHYTOCHEMICALS

The phytochemicals in plants are responsible for preventing diseases and promoting health benefits. They have been studied broadly to demonstrate their efficiency and recognize the basic mechanism of their action. Phytochemicals, as plant constituents with distinct bioactivities towards biochemistry and metabolism, are being extensively studied for their capability to provide health benefits (Dillard and German 2000). Such studies include the identification and isolation of the chemical constituents, the establishment of their biological effectiveness by *in vivo* and *in vitro* studies in experimental animals and by epidemiological as well as clinical-case control (Saxena et al. 2013). Several plant metabolites were studied on the animal and human cells that exhibit extremely exciting biological activities. Hence, they were indicated to be beneficial in pharmaceutical applications, nutrition, and dietary supplements (Hidalgo et al. 2018). They are source of food diet and having great medicinal significance (Phillipson 2001). About 200 species are considered medicinal plants, and approximately 25% of the medicines have plant sources (Gurnani et al. 2014).

Phytochemicals have been used to prevent and treat CVDs, cancer, neural disorders, and AD (Winter et al. 2017). Phytochemicals and their therapeutic role in various diseases are summarized in Table 11.1. Phytochemicals might detoxify substances that cause tumor. They seem to neutralize free radicals, prevent enzymes that stimulate carcinogens, and trigger enzymes, which detoxify carcinogens. Genistein inhibits the formation of new capillaries needed for cancer growth and metastasis (Morrison and Hark 1996). However, the physiologic functions of comparatively limited phytochemicals are finely understood, and considerable research has been made on the probable function of phytochemicals in treating or preventing tumor and heart disease (Mathai 2000). Phytochemicals have been endorsed for treating and preventing high blood pressure, diabetes, and macular degeneration (Saxena et al. 2013).

TABLE 11.1
Phytochemicals and Their Therapeutic Role in Several Diseases

Phytochemicals	Disease	References
Quercetin	Cancer, CVD, AD	(Lara-Guzman et al. 2012; Sak 2014; Ansari et al. 2009)
Catechin	Cancer, AD	(Manikandan et al. 2012; Davinelli et al. 2012)
EGCG	Cancer	(Cordero-Herrera et al. 2013)
Lycopene	CVD, diabetes, cancer	(Weberling, Boehm, and Froehlich 2011; Dembinska-Kiec et al. 2008; Luo and Wu 2011; Gloria et al. 2014)
Genistein	Cancer, diabetes	(Xie et al. 2014; Fu et al. 2012)
Curcumin	Cancer, AD	(Manikandan et al. 2012; Rinwa and Kumar 2012)
Resveratrol	CVD, cancer, AD	(Davinelli et al. 2012; Lin et al. 2021; Bishayee, Politis, and Darvesh 2010; Whitlock and Baek 2012)
Caffeic acid	Cancer, Neurological disorders	(Alam, Ahmed, et al. 2022; Alam, Ashraf, et al. 2022)
Kaempferol	Cancer, diabetes	(Kumar, Kumar, and Kaur 2012; Zhang and Liu 2011)
Naringenin	Diabetes, AD	(Rahigude et al. 2012; Cho et al. 2012)
Ellagitannins	Cancer	(Barrajón-Catalán et al. 2010)
Garlic	Cancer	(Tattelman 2005; Shukla and Kalra 2007)
Turmeric	Cancer	(Jurenka 2009; Hatcher et al. 2008)
Indian gooseberry	Cancer	(Khan 2009)

11.3 BIOAVAILABILITY OF PHYTOCHEMICALS

The bioavailability of distinct compounds of curiosity at the target position is one of the significant challenges or parameters for determining the therapeutic efficacy of the target drug (Aqil et al. 2013). Bioavailability of a compound might not be exactly predicted; though, an investigation by Lipinski's 'rule of five' provides nearly insight: overall, a compound will have excellent bioavailability, when it comprises not more than five H-bond donors, not greater than ten H-bond acceptors, has a molecular mass not more than 500 Daltons, a partition coefficient log *P*-value of not more than five, and comprises less than ten rotatable bonds (Lipinski et al. 1997). It has been documented that quick conjugation of phytochemicals, particularly through glucuronidation in the intestine and liver, in connection with the response of C-P450 enzymes that are identified as significant clearance mechanisms, is principally answerable for their poor bioavailability (Manach et al. 2004). Several researchers have revealed that codelivery of chief target phytochemicals with a drug that may modify the action of glucuronidation or inhibit C-P450 arbitrated clearance mechanisms probable for increasing the bioavailability of dynamic compounds of attention at the target site (Manach et al. 2004; Wu et al. 2011). Numerous phytochemicals were recognized as potential therapeutic agents in the initial *in vitro* examinations. Though, when those phytochemicals were studied *in vivo*, several failed to translate the preclinical or clinical outcomes, because these compounds either were unbalanced in the gut or showed poor bioavailability (Aqil et al. 2013; Rahman, Biswas, and Kirkham 2006). Hence, a detailed preclinical and clinical examination of the bioavailability of active compounds or phytochemicals is instantly needed for understanding their therapeutic limitation and finding out the well-established delivery system for achieving the best efficacy level of the agent on the target organ (Upadhyay, Dixit, and longevity 2015).

Approval of dietary phytochemicals in the human system and bioavailability for targeting cells enable their bioefficiency for protecting health (D'Archivio et al. 2010). Phytochemicals have comparatively poor bioavailability as they are handled via the body as xenobiotics; their existence in the body is transient (Holst and Williamson 2008). After the ingestion of compounds (phytochemicals), some but not all constituents are absorbed into the circulatory system through the small intestine

(D'Archivio et al. 2010). These phytochemicals might be exposed to metabolism in the liver, and the metabolites are liberated back to the circulatory system (D'Archivio et al. 2010; Holst and Williamson 2008; Crozier, Jaganath, and Clifford 2009). The oral bioavailability of phytochemicals might be apprised via the application of *in silico* modeling broadly utilized in pharmaceutical sciences (Van De Waterbeemd and Gifford 2003) as well as drug discovery (Camp et al. 2013). These models associate *in vivo* and/or *in vitro* passive absorption of agents with chemical structures defined via physicochemical functions for predicting the absorption of similar compounds (Van de Waterbeemd 2005).

11.4 ANTI-INFLAMMATORY ACTIVITY OF PHYTOCHEMICALS

Inflammation is a usual biological phenomena in response to tissue damage, microbial pathogen infection, and chemical irritation. It is initiated through the migration of immune cells from blood vessels and the liberation of mediators at the injury site. This procedure is followed via enrolment of inflammatory cells and discharge of RNS, ROS, and proinflammatory cytokines to eliminate foreign pathogens and restore injured tissues (Alam, Ansari, et al. 2022). Overall, inflammation is quick and self-limiting, but abnormal resolution and persistent inflammation cause several chronic diseases (Pan et al. 2010; Alam, Hasan, and Hassan 2021). Phytochemicals with anti-inflammatory actions have been scientifically reviewed (Howes 2018). A broad range of flavonoids with numerous chemical structures have been linked with diverse anti-inflammatory mechanistic results (Gomes et al. 2008). Glycosides of apigenin and luteolin are the greatest diffuse flavones (King 1962). Apigenin represses nitric oxide (NO) and prostaglandin by preventing iNOS and COX-2, correspondingly (Raso et al. 2001). Luteolin inhibited chronic inflammation through *in vitro* co-culture of macrophages, adipocytes, and JNK phosphorylation (Hirai et al. 2010). Hence, quercetin and its glycosides have been confirmed to be potential anti-inflammatory drugs for sarcoidosis patients and *in vivo* models of arthritis and allergic airway inflammation (Boots et al. 2011; Oliveira and Fierro 2018). Genistein might prevent inflammation by obstructing NF-κB activation and less regulating IL-6 and TNF-α expression (Wang et al. 2008). Hence, anthocyanins exhibit their anti-inflammatory results, principally by the MAPK pathway (Vendrame and Klimis-Zacas 2015).

Phytochemicals are found to exert anti-inflammatory results via downregulation of LPS-mediated expression of COX-2, and iNOS in mouse macrophages chiefly through preventing NF-κB pathways (Heiss et al. 2001; Wagner et al. 2013a). *In vivo* examination of C57BL/6 exhibited that the pretreatment of mice with SFN effected a significant decrease of DSS-induced colitis compared with PBS-treated mice (Wagner et al. 2013b). Epigallocatechin-3-Gallate (EGCG), and theaflavins, decreased LPS-induced TNF-α and iNOS expression via inhibiting the NF-κB activation (Lin et al. 1999; Stangl et al. 2007). Additionally, this phenolic component may be considered a potential anti-inflammatory drug as this is capable of preventing the expression of iNOS and COX-2 in LPS-stimulated macrophages (Leger et al. 2005; Maiuri et al. 2005). Hence, polyphenols are capable of inhibiting COX-2 as well as LOX in a dose-dependent way after application for LPS-activated murine RAW264 cells (Ichikawa et al. 2004). Curcumin has been revealed to inhibit LPS-induced making of TNF-α and IL-1β in macrophage cell for preventing LPS-induced NF-κB activation and decrease the biological action of TNF-α (Salminen et al. 2008).

11.5 ANTIOXIDANT ACTIVITY OF PHYTOCHEMICALS

Flavonoids possess several biochemical functions, including antioxidant properties. The antioxidant action of flavonoids depends on the arrangement of functional groups around the nuclear structure. The substitution, configuration, and the number of hydroxyl groups considerably stimulate numerous mechanisms of antioxidant action including radical scavenging and metal ion chelation capability (Heim, Tagliaferro, and Bobilya 2002; Pandey et al. 2012). The ring hydroxyl arrangement is the most significant important determinant of scavenging of RNS and ROS because it gives hydrogen and an electron for hydroxyl, peroxyl, and peroxynitrite radicals, steadying them and providing an increase to a comparatively constant flavonoids radical (Cao et al. 1997). Flavonoids prevent the enzymes involved in ROS making, which are microsomal monooxygenase, GST, mitochondrial succinoxidase, and NADH oxidase (Brown et al. 1998). LPO is a common sign of oxidative stress. Hence, flavonoids defend lipids against oxidative injury via several mechanisms (Kumar, Mishra, and Pandey 2013; Kumar and Pandey 2012).

The antioxidant and pro-oxidant functions of phytochemicals in tumor progression are arguable. Recognizing the molecular mechanism of a specific compound provides evidence of its therapeutic value (Gaikwad and Srivastava 2022). Phytochemicals performance (either enhancing or reducing oxidative stress) and classifying the dose and molecular mechanism of the used bioactive compound are significant parameters in the tumor. Phytochemicals were studied expansively using *in vivo* and *in vitro* methods for their antitumor action (Gaikwad and Srivastava 2022). Phytochemical antioxidants might have paradoxical results that can be cell type-specific and concentration dependent. However, they may either scavenge ROS or illogically make high oxidative stress to inhibit tumor cell proliferation (Loo 2003). Large amounts of H_2O_2 are formed in several tumor cells (Szatrowski and Nathan 1991).

11.6 ANTI-CANCER ACTIVITY OF PHYTOCHEMICALS

Presently, there is evidence concerning cellular mechanisms by which polyphenols might disturb carcinogenesis (Kang et al. 2011), cancer cell proliferation and death, inflammation (Sarkar et al. 2016), angiogenesis (Albini et al. 2012), and drug resistance (Garg et al. 2005). Phytochemicals can modify the initiation of carcinogenesis progression through defense against DNA injury. The suitable lifestyle alterations could inhibit more than two-thirds of human tumors, and the diet involves around 35% of human tumor mortality (Sak 2014). Polyphenols perform a significant role in the antitumor action of phytochemicals. Polyphenols ellagitannins, as well as epicatechin gallate, exhibited anticarcinogenic functions (Cordero-Herrera et al. 2013; Xie et al. 2014).

Polyphenols have a regulatory effect on MAPK signaling. Resveratrol dually controls this signaling pathway by preventing the activation of the MAPK pathway linked to influencing cell proliferation and persuading MAPK-associated apoptosis (Yan et al. 2020; Pramanik et al. 2018). Polyphenols might display a chemopreventive potential against colon carcinogenesis by affecting the mTOR/AKT signaling and miR-143 (Banerjee et al. 2016; Alam, Ali, and Hassan 2022; Alam and Mishra 2020). In this sense, targeting the miRNAs via polyphenols was considered as a novel and potential approach to antitumor chemotherapy (Pandima Devi et al. 2017). Polyphenols might obstruct the IκB kinase action, thus stopping translocation to the nucleus, which NF-κB persuades the expression of genes connected to apoptosis, invasion, and metastasis. Hence, by preventing NF-κB activation, polyphenols repress the expression of diverse cell survival and proliferative genes (Aggarwal et al. 2006; Pramanik et al. 2016).

Phytochemicals significantly stimulate cell death in several cancers via preventing the activity of JAK/STAT pathway and activating apoptosis (Kim et al. 2016). Resveratrol stimulates cell death via preventing a constituent activation of STAT3 and decreasing regulating survivin and Mcl-1 (Quoc Trung et al. 2013). Phytochemicals inhibit the translocation and accretion of β-catenin in the nucleus through the activation of GSK3 (Pramanik et al. 2016, 2018; Taipale and Beachy 2001; Tsai et al. 2013). Phytochemicals exhibit anticancer roles in tissues via regulating inflammation, metastasis, angiogenesis, and invasion (Wung et al. 2005). Phytochemicals suppress cancer development through several pathways (Figure 11.1).

Antioxidant phytochemicals might prevent cell proliferation and enhance tumor apoptosis. Curcumin targeted tumor stem cells for expressing their antitumor capability (Li and Zhang 2014), and it can exert a synergistic antitumor action with catechin against several cancer cell lines (Manikandan et al. 2012). A study by Yang and Liu (2009) reported that genistein, quercetin, and resveratrol revealed higher induction of quinone reductase than others among the 18 phytochemicals

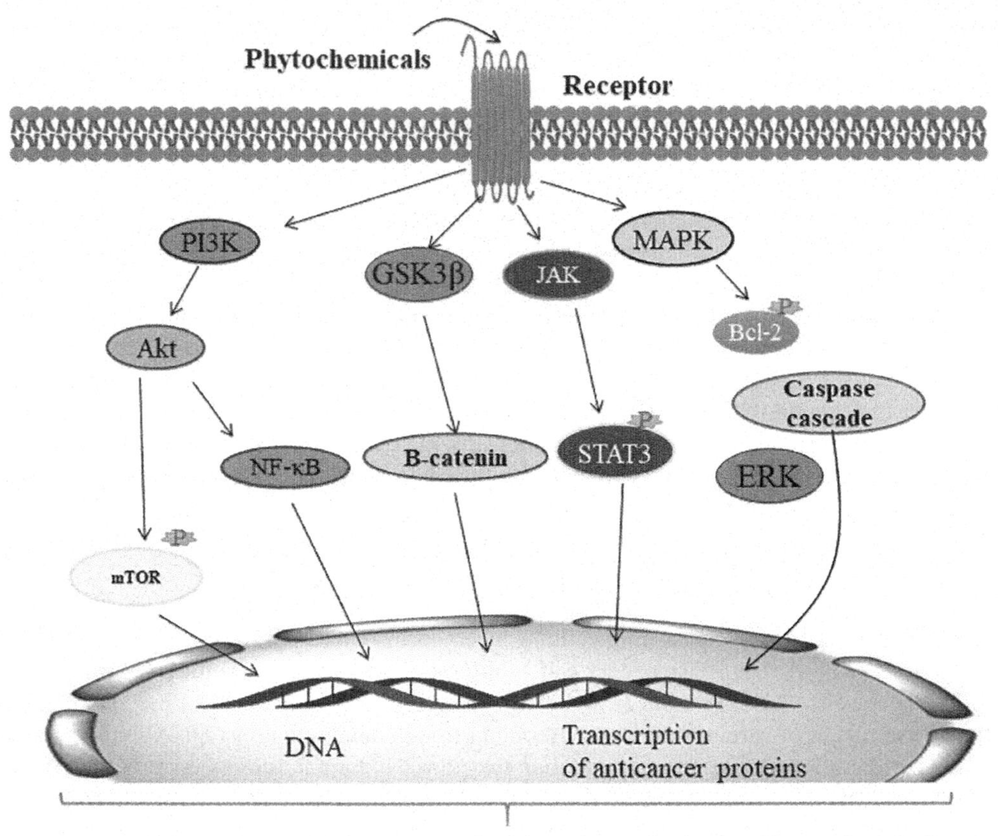

FIGURE 11.1 Interaction of signaling molecules with phytochemicals in cancer chemoprevention. This figure denotes the action of phytochemicals on ERK, MAPK, Akt/PI3K, NF-κB, GSK3β, and JAK/STAT pathways stimulating tumor cell death via the activation of multiple internal signaling molecules. This figure is adapted from George, Chandran, and Abrahamse 2021.

studied. The antitumor results of quercetin in multiple cancers have been validated (Sak 2014). Resveratrol exerted an antitumor activity via preventing cancer initiation, promotion, and development (Bishayee, Politis, and Darvesh 2010; Whitlock and Baek 2012). Furthermore, butein displayed an antitumor activity through persuading cell cycle arrest and cell death in lung tumor cells (Zhang et al. 2015). Additionally, lycopene consumption and serum lycopene levels have been inversely connected to certain types of tumors (Luo and Wu 2011; El-Rouby 2011; Tang et al. 2011). Lycopene and carotene might prevent cell cycle arrest and cell proliferation and enhance apoptosis of breast tumor cells (Gloria et al. 2014). Certain polyphenols enhance cell death by binding to Bcl-2 or Bcl-xL or changing the cellular microtubule cytoskeleton (Mena et al. 2012; Alam et al. 2017; Alam and Mishra 2021) (Figure 11.1). Cocoa polyphenols stimulate cell death by enhancing the expression of Bax and caspase 3and reducing the expression of Bcl-xL (Saadatdoust et al. 2015; Alam et al. 2019).

11.7 NEUROPROTECTIVE EFFECT OF PHYTOCHEMICALS

NDDs are generally illustrious as disorders with loss or damage of neurons. NDDs are a heterogeneous cluster of chronic and deadly states, characterized via a progressive functional impairment of the nervous system, stimulated via the deterioration of neurons, neurotransmission, myelin sheath, and movement control (George, Chandran, and Abrahamse 2021). The damaging consequences of oxidative stress, as well as chronic neuroinflammation in neuronal apoptosis, were associated with the pathogenesis of NDDs, including Parkinson's disease (PD) and AD (Wang et al. 2016). Multiple studies have validated the advantage of phytochemicals for treating NDDs. However, phytochemicals have extensive anti-inflammatory and antioxidant effects, exhibiting an inhibitory function in the oxidative and inflammatory mechanisms linked with NDDs (Joseph et al. 2009; Hashimoto and Hossain 2011).

Resveratrol validates a significant neuroprotective action both *in vivo* and *in vitro*. However, resveratrol is cytoprotective in cells visible to Aβ and/or to Aβ-metal complex through Sirt3-induced mechanisms (Granzotto and Zatta 2011; Yan et al. 2018). Additionally, an animal model of cerebral amyloid deposition after administration of resveratrol dropped microglial activation linked with cortical amyloid plaque creation (Capiralla et al. 2012). Curcumin stimulates a neuroprotective activity via the control of pathogenetic oxidative and inflammatory actions in *in vitro* and *in vivo* models of AD and PD. Further, in Neuro2a mouse neuroblastoma cells, curcumin increases cell viability via reducing ROS and preventing proapoptotic signals (Dutta, Ghosh, and Basu 2009). Additionally, curcumin defends against α-synuclein-persuaded cytotoxicity in SH-SY5Y cells, reducing the cytotoxicity of accumulated α-synuclein, dropping ROS, and preventing caspase-3 activation (Wang et al. 2010). *In vivo*, curcumin meaningfully improved spatial memory deficits in AD's APP/PS1 model, endorsing the cholinergic neuronal role (Quoc Trung et al. 2013). Curcumin decreased the activation of astrocytes and microglia and cytokine making and presented NF-κB pathway, which suggests that the beneficial results of curcumin in AD mice are attributable to the repression of neuroinflammation (Bhullar, Rupasinghe, and longevity 2013). However, in the PD model, stimulated through the neurotoxin MPTP, curcumin is neuroprotective and inhibits glutathione depletion, and LPO is produced via this toxin. Curcumin returned motor deficits and increased the actions of antioxidant enzymes in rotenone-stimulated PD in mice (Walker et al. 2015).

Several research reports have been studied in AD models where phytochemical (EGCG) has been administered. Hence, D-gal has been administered to AD animal models, considerably decreasing amyloid plaques (Chan et al. 2016; Walker et al. 2015). A research group has recommended that via administering EGCG, β- and γ-secretases have been reduced by preventing NF-κβ and ERK, hence inhibiting neuronal apoptosis (Liu et al. 2014). By administering EGCG, amyloid plaques have decreased because of APP cleavage via α-secretase (Smith et al. 2010). Further, in the case of PD, one report has proposed that three or more cups of tea faced a reduced risk of progressing PD. The free radical scavenging scheme is strengthened via enhancing glutathione, which stimulates CREB and Bcl-2 and leads to positive manifestations (Choi et al. 2002). A study has recommended that concurrent intake of EGCG prohibited the loss of these cells in substantia nigra in PD (Koh et al. 2003). Therefore, neuroprotective results of quercetin have been detected at 10 μM and 30 μM concentrations. Quercetin might cross the blood-brain barrier (BBB) and inhibits cytotoxicity stimulated via H_2O_2 (Heo and Lee 2004). Quercetin effects depend on the cell type, and exposure time higher than 100 μM concentrations showed antiproliferative, cytotoxic, and genotoxic actions. NF-κB is initiated to be controlled through quercetin, which might enhance inflammatory procedures linked to NDDs (Dajas et al. 2015). Quercetin defends from neurodegeneration via mitochondria-targeted consequences (Dajas et al. 2015).

11.8 PROTECTIVE ACTION ON CARDIOVASCULAR DISEASES

CVD is the leading cause of death (Reuland et al. 2013). The etiology of CVD is incredibly complex and challenging, and more making of oxidants is the chief pathogenic reason. Modifying vascular role is a critical pathogenic procedure common to several significant and extremely wordy human pathologies (Van Gaal, Mertens, and De Block 2006). The oxidative injury might cause endothelial cell damage and deleterious vasodilator results. Antioxidant polyphenols might change molecular incidents towards an enhancement in the endothelial role and play a significant role in the inhibition of CVD. Flavonoids exhibited atheroprotective results (Costa et al. 2013). Hence, the morphofunctional integrity of the vascular endothelium is a highly homeostatic procedure connecting maintenance of vasorelaxation capability and anti-inflammatory, and barrier roles with significant results on atherogenesis and enhanced risk of CVD (Grassi, Desideri, and Ferri 2011; Fonseca et al. 2019).

Polyphenols might defend the cardiovascular system against oxidative stress and other injuries. They have different physiological results, including a decrease in blood pressure and inflammation-reducing activity (Prahalathan, Saravanakumar, and Raja 2012). Dehydroglyasperin C diminished proliferation, and migration of human arterial smooth muscle cells that have been stimulated via platelet-derived growth factor (Kim et al. 2013). Accretion of vascular smooth muscle cells performs a vital function in the creation and progression of lesions in atherosclerosis; hence, dehydroglyasperin C might be beneficial for CVD. Furthermore, research exhibited that seven phenolic acids exerted promising atheroprotective effects (Xie et al. 2011). However, platelet accumulation, as well as adhesion under pathophysiological situations, may be the reason for thrombosis as well as blockage of coronary arteries that are connected with the CVD pathogenesis. Polyphenols can change molecular incidents toward preventing platelet aggregation (Costa et al. 2013). Stilbenoids possess anti-inflammatory and antioxidant activities and exhibit inhibitory effects in human platelet adhesion and aggregation (Kloypan et al. 2012). Additionally, phlorizin possesses an antioxidant action and reduces diabetic macrovascular problems in diabetic mice (Shen et al. 2012). These are the important reasons for CVD; phlorizin might be beneficial. Moreover, anthocyanins were confirmed to have a defensive activity against numerous cardiovascular risk factors (Kruger et al. 2014).

The cardioprotective results of resveratrol in human examinations were documented (Hung et al. 2000). Nevertheless, the *in vivo* data and the clinical trials are less conclusive, frequently due to the poor intestinal absorption of flavonoids and the widespread degradation of several phenolic acids that might retain various antioxidant actions (Vitaglione et al. 2005). One extra role of resveratrol has been remarkably connected to an increased SIRT-1 expression and action, and has been broadly associated with longevity (Kao et al. 2010). In contrast, a pro-oxidant action of resveratrol under certain states has been described (De La Lastra and Villegas 2007). Phytochemicals with antioxidant results were documented to play a protective activity in the progression of CVD. Consequently, they were planned as vital factors in the diet, alike, for example, β-carotene, curcumin, and others (Gammone et al. 2018; Zhang et al. 2018). Some studies emphasize that the anti-atherogenic result of lycopene is linked with the prevention of proinflammatory cytokines secretion (Gammone et al. 2017).

11.9 CONCLUSIONS AND FUTURE DIRECTIONS

Nature is an exclusive source of structures of more phytochemical varieties possessing potent biological actions and medicinal properties. Association between phytochemicals and chronic disease inhibition has been the focus of health studies for nearly half a century. Hence, clinical and epidemiological studies show that phytochemicals decrease the risk of chronic diseases. They have been revealed to have a promising effect on cancer, neurodegeneration, and CVD pathogenesis. Accordingly, phytochemicals are the most promising agents for treating these diseases. Further, they possess numerous biological activities and health benefits, including antioxidant, anti-inflammatory, antitumor, and protective activities for CVD and NDDs.

Despite the potential and beneficial therapeutic effect of phytochemicals on several diseases, including cancer, neurodegenerative, and CVD, the field of phytochemicals and their defensive effect would grow effectively that preclinical or clinical study is capable of integrating reliable science with systematic user understanding and suitability, along with consciousness about the defensive function of phytochemicals on the progression of chronic diseases. Consequently, it must be highlighted that the biological activity is regulated by the product of many cellular results exerted via a phytochemical. A better understanding of these cellular results is pivotal for accurately exploiting the phytochemicals as potential agents that encourage health and inhibit several diseases. In addition, the comprehensive study of phytochemicals based on clinical studies would benefit the approach of novel therapeutic drugs.

REFERENCES

Aggarwal, Sita, Haruyo Ichikawa, Yasunari Takada, Santosh K Sandur, Shishir Shishodia, and Bharat B Aggarwal. 2006. "Curcumin (Diferuloylmethane) Down-Regulates Expression of Cell Proliferation and Antiapoptotic and Metastatic Gene Products through Suppression of IκBα Kinase and Akt Activation" *Molecular Pharmacology* 69(1): 195–206.

Alam, Manzar, Sarfraz Ahmed, Abdelbaset Mohamed Elasbali, Mohd Adnan, Shoaib Alam, Md Imtaiyaz Hassan, and Visweswara Rao Pasupuleti. 2022. "Therapeutic Implications of Caffeic Acid in Cancer and Neurological Diseases" *Frontiers in Oncology* 12. https://doi.org/10.3389/fonc.2022.860508

Alam, Manzar, Shoaib Alam, Anas Shamsi, Mohd Adnan, Abdelbaset Mohamed Elasbali, Waleed Abu Al-Soud, Mousa Alreshidi, Yousef Hawsawi, Anitha Tippana, and Visweswara Rao Pasupuleti. 2022. "Bax/Bcl-2 Cascade Is Regulated by EGFR Pathway: Therapeutic Targeting of Non-Small Cell Lung Cancer" *Frontiers in Oncology* 12: 869672.

Alam, Manzar, Sabeeha Ali, Sarfraz Ahmed, Abdelbaset Mohamed Elasbali, Mohd Adnan, Asimul Islam, Md Imtaiyaz Hassan, and Dharmendra Kumar Yadav. 2021. "Therapeutic Potential of Ursolic Acid in Cancer and Diabetic Neuropathy Diseases" *International Journal of Molecular Sciences* 22(22): 12162.

Alam, Manzar, Sabeeha Ali, Gulam M Ashraf, Anwar L Bilgrami, Dharmendra Kumar Yadav, and Md Imtaiyaz Hassan. 2022. "Epigallocatechin 3-Gallate: From Green Tea to Cancer Therapeutics" *Food Chemistry* 132135. doi:10.1016/j.foodchem.2022.132135

Alam, Manzar, Sabeeha Ali, and Md Imtaiyaz Hassan. 2022. "Akt Inhibitors in Cancer Therapy." In Md. Imtaiyaz Hassan and Saba Noor (eds.), *Protein Kinase Inhibitors*, 239–60. Academic Press.

Alam, Manzar, Sabeeha Ali, Taj Mohammad, Gulam Mustafa Hasan, Dharmendra Kumar Yadav, and Md Imtaiyaz Hassan. 2021. "B Cell Lymphoma 2: A Potential Therapeutic Target for Cancer Therapy" *International Journal of Molecular Sciences* 22(19): 10442.

Alam, Manzar, Md Meraj Ansari, Saba Noor, Taj Mohammad, Gulam Mustafa Hasan, Syed Naqui Kazim, and Md Imtaiyaz Hassan. 2022. "Therapeutic Targeting of TANK-Binding Kinase Signaling towards Anticancer Drug Development: Challenges and Opportunities." *International Journal of Biological Macromolecules* 207: 1022–37.

Alam, Manzar, Ghulam Md Ashraf, Kayenat Sheikh, Anish Khan, Sabeeha Ali, Md Meraj Ansari, Mohd Adnan, Visweswara Rao Pasupuleti, and Md Imtaiyaz Hassan. 2022. "Potential Therapeutic Implications of Caffeic Acid in Cancer Signaling: Past, Present, and Future" *Frontiers in Pharmacology* 13: 845871.

Alam, Manzar, Gulam Mustafa Hasan, Md Meraj Ansari, Rishi Sharma, Dharmendra Kumar Yadav, and Md Imtaiyaz Hassan. 2022. "Therapeutic Implications and Clinical Manifestations of Thymoquinone" *Phytochemistry* 13: 113213.

Alam, Manzar, Gulam Mustafa Hasan, and Md Imtaiyaz Hassan. 2021. "A Review on the Role of TANK-Binding Kinase 1 Signaling in Cancer" *International Journal of Biological Macromolecules* 183: 2364–75.

Alam, Manzar, Tanushree Kashyap, Prajna Mishra, Aditya K Panda, Siddavaram Nagini, Rajakishore Mishra. 2019. "Role and Regulation of Proapoptotic Bax in Oral Squamous Cell Carcinoma and Drug Resistance" *Head and Neck* 41(1): 185–97.

Alam, Manzar, Tanushree Kashyap, Kamdeo K Pramanik, Abhay K Singh, Siddavaram Nagini, and Rajakishore Mishra. 2017. "The Elevated Activation of NFκB and AP-1 Is Correlated with Differential Regulation of Bcl-2 and Associated with Oral Squamous Cell Carcinoma Progression and Resistance" *Clinical Oral Investigations* 21(9): 2721–31.

Alam, Manzar, Rajakishore Mishra. 2020. "Role of PI3K and EGFR in Oral Cancer Progression and Drug Resistance" *International Journal for Research in Applied Sciences and Biotechnology* 7(6): 85–89.

Alam, Manzar, and Rajakishore Mishra. 2021. "Bcl-XL Expression and Regulation in the Progression, Recurrence, and Cisplatin Resistance of Oral Cancer" *Life Sciences* 280: 119705.

Albini, Adriana, Francesca Tosetti, Vincent W Li, Douglas M Noonan, and William W Li. 2012. "Cancer Prevention by Targeting Angiogenesis" *Nature Reviews Clinical Oncology* 9(9): 498–509.

Ali, Sabeeha, Manzar Alam, and Md Imtaiyaz Hassan. 2022. "Kinase Inhibitors: An Overview" *Protein Kinase Inhibitors* 1–22. Doi: 10.1016/B978-0-323-91287-7.00026-0

Ali, Sabeeha, Manzar Alam, Fatima Khatoon, Urooj Fatima, Abdelbaset Mohamed Elasbali, Mohd Adnan, Asimul Islam, et al. 2022. "Natural Products Can Be Used in Therapeutic Management of COVID-19: Probable Mechanistic Insights" 147: 112658.

Alissa, Eman M, and Gordon A Ferns. 2012. "Functional Foods and Nutraceuticals in the Primary Prevention of Cardiovascular Diseases" *Journal of Nutrition Ferns, and Metabolism* 2012: 569486.

Ansari, Mubeen Ahmad, Hafiz Mohammad Abdul, Gururaj Joshi, Wycliffe O Opii, and D Allan Butterfield. 2009. "Protective Effect of Quercetin in Primary Neurons against Aβ (1–42): Relevance to Alzheimer's Disease" *The Journal of Nutritional Biochemistry* 20(4): 269–75.

Aqil, Farrukh, Radha Munagala, Jeyaprakash Jeyabalan, and Manicka V Vadhanam. 2013. "Bioavailability of Phytochemicals and Its Enhancement by Drug Delivery Systems" *Cancer Letters* 334(1): 133–41.

Banerjee, Nivedita, Hyemee Kim, Stephen T Talcott, Nancy D Turner, David H Byrne, and Susanne U Mertens-Talcott. 2016. "Plum Polyphenols Inhibit Colorectal Aberrant Crypt Foci Formation in Rats: Potential Role of the MiR-143/Protein Kinase B/Mammalian Target of Rapamycin Axis" *Nutrition Research* 36 (10): 1105–13.

Barrajón-Catalán, Enrique, Salvador Fernández-Arroyo, Domingo Saura, Emilio Guillén, Alberto Fernández-Gutiérrez, Antonio Segura-Carretero, Vicente Micol. 2010. "Cistaceae Aqueous Extracts Containing Ellagitannins Show Antioxidant and Antimicrobial Capacity, and Cytotoxic Activity against Human Cancer Cells" *Food and Chemical Toxicology* 48(8–9): 2273–82.

Bhullar, Khushwant S, H P Rupasinghe. 2013. "Polyphenols: Multipotent Therapeutic Agents in Neurodegenerative Diseases" *Oxidative Medicine and Cellular Longevity* 2013: 891748.

Bishayee, Anupam, Themos Politis, and Altaf S Darvesh. 2010. "Resveratrol in the Chemoprevention and Treatment of Hepatocellular Carcinoma" *Cancer Treatment Reviews* 36(1): 43–53.

Boots, Agnes W, Marjolein Drent, Vincent C J de Boer, Aalt Bast, and Guido R M M Haenen. 2011. "Quercetin Reduces Markers of Oxidative Stress and Inflammation in Sarcoidosis" *Clinical Nutrition* 30(4): 506–12.

Brown, E Jonathan, Hicham Khodr, C Robert HIDER, and Catherine A Rice-Evans. 1998. "Structural Dependence of Flavonoid Interactions with Cu2+ Ions: Implications for Their Antioxidant Properties" *Biochemical Journal* 330(3): 1173–78.

Calder, Philip C, R Albers, J-M Antoine, S Blum, R Bourdet-Sicard, G A Ferns, G Folkerts, P S Friedmann, G S Frost, and F Guarner. 2009. "Inflammatory Disease Processes and Interactions with Nutrition" *British Journal of Nutrition* 101(S1): 1–45.

Camp, David, Marc Campitelli, Anthony R Carroll, Rohan A Davis, Ronald J Quinn. 2013. "Front-Loading Natural-Product-Screening Libraries for Log P: Background, Development, and Implementation" *Chemistry and Biodiversity* 10(4): 524–37.

Cao, Guohua, Emin Sofic, Ronald L Prior. 1997. "Antioxidant and Prooxidant Behavior of Flavonoids: Structure-Activity Relationships" *Free Radical Biology and Medicine* 22(5): 749–60.

Capiralla, Hemachander, Valérie Vingtdeux, Haitian Zhao, Roman Sankowski, Yousef Al-Abed, Peter Davies, and Philippe Marambaud. 2012. "Resveratrol Mitigates Lipopolysaccharide-and Aβ-mediated Microglial Inflammation by Inhibiting the TLR4/NF-κB/STAT Signaling Cascade" *Journal of Neurochemistry* 120 (3): 461–72.

Chan, Stephen, Srinivas Kantham, Venkatesan M Rao, Manoj Kumar Palanivelu, Hoang L Pham, P Nicholas Shaw, Ross P McGeary, and Benjamin P Ross. 2016. "Metal Chelation, Radical Scavenging and Inhibition of Aβ42 Fibrillation by Food Constituents in Relation to Alzheimer's Disease" *Food Chemistry* 199: 185–94.

Cho, Jung Keun, Young Bae Ryu, Marcus J Curtis-Long, Hyung Won Ryu, Heung Joo Yuk, Dae Wook Kim, Hye Jin Kim, Woo Song Lee, Ki Hun Park. 2012. "Cholinestrase Inhibitory Effects of Geranylated Flavonoids from Paulownia Tomentosa Fruits" *Bioorganic and Medicinal Chemistry* 20(8): 2595–602.

Choi, Ji-Young, Chang-Shin Park, Dae-Joong Kim, Myung-Haeng Cho, Byung-Kwan Jin, Jae-Eun Pie, and Woon-Gye Chung. 2002. "Prevention of Nitric Oxide-Mediated 1-Methyl-4-Phenyl-1, 2, 3, 6-Tetrahydropyridine-Induced Parkinson's Disease in Mice by Tea Phenolic Epigallocatechin 3-Gallate" *Neurotoxicology* 23(3): 367–74.

Cordero-Herrera, Isabel, María Angeles Martín, Laura Bravo, Luis Goya, Sonia Ramos. 2013. "Epicatechin Gallate Induces Cell Death via P53 Activation and Stimulation of P38 and JNK in Human Colon Cancer SW480 Cells" *Nutrition and Cancer* 65(5): 718–28.

Costa, Andre Gustavo Vasconcelos, Diego F Garcia-Diaz, Paula Jimenez, and Pollyanna Ibrahim Silva. 2013. "Bioactive Compounds and Health Benefits of Exotic Tropical Red–Black Berries" *Journal of Functional Foods* 5(2): 539–49.

Crozier, Alan, Indu B Jaganath, and Michael N Clifford. 2009. "Dietary Phenolics: Chemistry, Bioavailability and Effects on Health" *Natural Product Reports* 26(8): 1001–43.

D'Archivio, Massimo, Carmelina Filesi, Rosaria Varì, Beatrice Scazzocchio, and Roberta Masella. 2010. "Bioavailability of the Polyphenols: Status and Controversies" *International Journal of Molecular Sciences* 11(4): 1321–42.

D'Onofrio, Grazia, Daniele Sancarlo, Qingwei Ruan, Zhuowei Yu, Francesco Panza, Antonio Daniele, Antonio Greco, and Davide Seripa. 2017. "Phytochemicals in the Treatment of Alzheimer's Disease: A Systematic Review" *Current Drug Targets* 18(13): 1487–98.

Dajas, Federico, Juan Andrés Abin-Carriquiry, Florencia Arredondo, Fernanda Blasina, Carolina Echeverry, Marcela Martínez, Felicia Rivera, and Lucía Vaamonde. 2015. "Quercetin in Brain Diseases: Potential and Limits" *Neurochemistry International* 89: 140–48.

Dauchet, Luc, Philippe Amouyel, and Jean Dallongeville. 2009. "Fruits, Vegetables and Coronary Heart Disease" *Nature Reviews Cardiology* 6(9): 599–608.

Davinelli, Sergio, Nadia Sapere, Davide Zella, Renata Bracale, Mariano Intrieri, Giovanni Scapagnini. 2012. "Pleiotropic Protective Effects of Phytochemicals in Alzheimer's Disease" *Oxidative Medicine and Cellular Longevity* 2012: 386527.

Dembinska-Kiec, Aldona, Otto Mykkänen, Beata Kiec-Wilk, and Hannu Mykkänen. 2008. "Antioxidant Phytochemicals against Type 2 Diabetes" *British Journal of Nutrition* 99(E-S1): ES109–17.

Deng, Gui-Fang, Xi Lin, Xiang-Rong Xu, Li-Li Gao, Jie-Feng Xie, and Hua-Bin Li. 2013. "Antioxidant Capacities and Total Phenolic Contents of 56 Vegetables" *Journal of Functional Foods* 5(1): 260–66.

Deng, Gui-Fang, Chen Shen, Xiang-Rong Xu, Ru-Dan Kuang, Ya-Jun Guo, Li-Shan Zeng, Li-Li Gao, Xi Lin, Jie-Feng Xie, and En-Qin Xia. 2012. "Potential of Fruit Wastes as Natural Resources of Bioactive Compounds" *International Journal of Molecular Sciences* 13(7): 8308–23.

Dillard, Cora J, and J Bruce German. 2000. "Phytochemicals: Nutraceuticals and Human Health" *Journal of the Science of Food and Agriculture* 80(12): 1744–56.

Dutta, Kallol, Debapriya Ghosh, and Anirban Basu. 2009. "Curcumin Protects Neuronal Cells from Japanese Encephalitis Virus-Mediated Cell Death and Also Inhibits Infective Viral Particle Formation by Dysregulation of Ubiquitin–Proteasome System" *Journal of Neuroimmune Pharmacology* 4(3): 328–37.

El-Rouby, Dalia Hussein. 2011. "Histological and Immunohistochemical Evaluation of the Chemopreventive Role of Lycopene in Tongue Carcinogenesis Induced by 4-Nitroquinoline-1-Oxide" *Archives of Oral Biology* 56(7): 664–71.

Fonseca, Lucas José Sá da, Valéria Nunes-Souza, Marília Oliveira Fonseca Goulart, Luiza Antas Rabelo. 2019. "Oxidative Stress in Rheumatoid Arthritis: What the Future Might Hold Regarding Novel Biomarkers and Add-on Therapies" *Oxidative Medicine and Cellular Longevity* 2019: 7536805.

Fu, Li, Bo-Tao Xu, Xiang-Rong Xu, Ren-You Gan, Yuan Zhang, En-Qin Xia, and Hua-Bin Li. 2011. "Antioxidant Capacities and Total Phenolic Contents of 62 Fruits" *Food Chemistry* 129(2): 345–50.

Fu, Zhuo, Elizabeth R Gilbert, Liliane Pfeiffer, Yanling Zhang, Yu Fu, Dongmin Liu. 2012. "Genistein Ameliorates Hyperglycemia in a Mouse Model of Nongenetic Type 2 Diabetes" *Applied Physiology Nutrition, and Metabolism* 37(3): 480–88.

Gaal, Luc F Van, Ilse L Mertens, and Christophe E De Block. 2006. "Mechanisms Linking Obesity with Cardiovascular Disease" *Nature* 444(7121): 875–80.

Gaikwad, Shreyas R, and Sanjay K Srivastava. 2022. "Antioxidant Activity of Phytochemicals in Cancer." In *Handbook of Oxidative Stress in Cancer: Therapeutic Aspects*, 1–17. Singapore: Springer.

Gammone, Maria Alessandra, Konstantinos Efthymakis, Francesca Romana Pluchinotta, Sonia Bergante, Guido Tettamanti, Graziano Riccioni, and Nicolantonio D'Orazio. 2018. "Impact of Chocolate on the Cardiovascular Health" *Frontiers in Bioscience-Landmark* 23(5): 852–64.

Gammone, Maria Alessandra, Francesca Romana Pluchinotta, Sonia Bergante, Guido Tettamanti, and Nicolantonio D'Orazio. 2017. "Prevention of Cardiovascular Diseases with Carotenoids" *Frontiers in Bioscience-Scholar* 9(1): 165–71.

Garg, Amit K, Thomas A Buchholz, Bharat B Aggarwal. 2005. "Chemosensitization and Radiosensitization of Tumors by Plant Polyphenols" *Antioxidants and Redox Signaling* 7(11–12): 1630–47.

George, Blassan P, Rahul Chandran, and Heidi Abrahamse. 2021. "Role of Phytochemicals in Cancer Chemoprevention: Insights" *Antioxidants* 10(9): 1455.

George, Trevor W, Chutamat Niwat, Saran Waroonphan, Michael H Gordon, and Julie A Lovegrove. 2009. "Effects of Chronic and Acute Consumption of Fruit-and Vegetable-Puree-Based Drinks on Vasodilation, Risk Factors for CVD and the Response as a Result of the ENOS G298T Polymorphism: Conference on 'Multidisciplinary Approaches to Nutritional Problems'" *Proceedings of the Nutrition Society* 68(2): 148–61.

Gloria, Nathalie Fonseca, Nathalia Soares, Camila Brand, Felipe Leite Oliveira, Radovan Borojevic, and Anderson Junger Teodoro. 2014. "Lycopene and Beta-Carotene Induce Cell-Cycle Arrest and Apoptosis in Human Breast Cancer Cell Lines" *Anticancer Research* 34(3): 1377–86.

Gomes, Ana, Eduarda Fernandes, Jose L F C Lima, Lurdes Mira, and M Luísa Corvo. 2008. "Molecular Mechanisms of Anti-Inflammatory Activity Mediated by Flavonoids" *Current Medicinal Chemistry* 15 (16): 1586–605.

Granzotto, Alberto, and Paolo Zatta. 2011. "Resveratrol Acts Not through Anti-Aggregative Pathways but Mainly via Its Scavenging Properties against Aβ and Aβ-Metal Complexes Toxicity" *PloS One* 6(6): e21565.

Grassi, Davide, Giovambattista Desideri, and Claudio Ferri. 2011. "Cardiovascular Risk and Endothelial Dysfunction: The Preferential Route for Atherosclerosis" *Current Pharmaceutical Biotechnology* 12(9): 1343–53.

Guo, Ya-Jun, Gui-Fang Deng, Xiang-Rong Xu, Shan Wu, Sha Li, En-Qin Xia, Fang Li, et al. 2012. "Antioxidant Capacities, Phenolic Compounds and Polysaccharide Contents of 49 Edible Macro-Fungi" 3(11): 1195–205.

Gurnani, N, D Mehta, M Gupta, and B K Mehta. 2014. "Natural Products: Source of Potential Drugs" *African Journal of Basic and Applied Science* 6(6): 171–86.

Hashimoto, Michio, and Shahdat Hossain. 2011. "Neuroprotective and Ameliorative Actions of Polyunsaturated Fatty Acids against Neuronal Diseases: Beneficial Effect of Docosahexaenoic Acid on Cognitive Decline in Alzheimer's Disease" *Journal of Pharmacological Sciences* 116(2): 150–62.

Hatcher, H, R Planalp, J Cho, F M Torti, S V Torti. 2008. "Curcumin: From Ancient Medicine to Current Clinical Trials" *Cellular and Molecular Life Sciences* 65(11): 1631–52.

Heim, Kelly E, Anthony R Tagliaferro, and Dennis J Bobilya. 2002. "Flavonoid Antioxidants: Chemistry, Metabolism and Structure-Activity Relationships" *The Journal of Nutritional Biochemistry* 13(10): 572–84.

Heiss, Elke, Christian Herhaus, Karin Klimo, Helmut Bartsch, and Clarissa Gerhäuser. 2001. "Nuclear Factor KB Is a Molecular Target for Sulforaphane-Mediated Anti-Inflammatory Mechanisms" *Journal of Biological Chemistry* 276(34): 32008–15.

Heo, Ho Jin, Chang Yong Lee. 2004. "Protective Effects of Quercetin and Vitamin C against Oxidative Stress-Induced Neurodegeneration" *Journal of Agricultural and Food Chemistry* 52(25): 7514–17.

Hidalgo, Diego, Raul Sanchez, Liliana Lalaleo, Mercedes Bonfill, Purificacion Corchete, and Javier Palazon. 2018. "Biotechnological Production of Pharmaceuticals and Biopharmaceuticals in Plant Cell and Organ Cultures" *Current Medicinal Chemistry* 25(30): 3577–96.

Hirai, Shizuka, Nobuyuki Takahashi, Tsuyoshi Goto, Shan Lin, Taku Uemura, Rina Yu, and Teruo Kawada. 2010. "Functional Food Targeting the Regulation of Obesity-Induced Inflammatory Responses and Pathologies" *Mediators of Inflammation* 2010: 367838.

Holst, Birgit, and Gary Williamson. 2008. "Nutrients and Phytochemicals: From Bioavailability to Bioefficacy beyond Antioxidants" *Current Opinion in Biotechnology* 19(2): 73–82.

Howes, Melanie-Jayne R. 2018. "Phytochemicals as Anti-Inflammatory Nutraceuticals and Phytopharmaceuticals." In Shampa Chatterjee, Wolfgang Jungraithmay, and Debasis Bagchi (eds.), *Immunity and Inflammation in Health and Disease*, 363–88. Elsevier.

Hung, Li-Man, Jan-Kan Chen, Shiang-Suo Huang, Ren-Shen Lee, and Ming-Jai Su. 2000. "Cardioprotective Effect of Resveratrol, a Natural Antioxidant Derived from Grapes" *Cardiovascular Research* 47(3): 549–55.

Ichikawa, Daiju, Ayako Matsui, Miwa Imai, Yoshiko Sonoda, Tadashi Kasahara. 2004. "Effect of Various Catechins on the IL-12p40 Production by Murine Peritoneal Macrophages and a Macrophage Cell Line, J774. 1" *Biological and Pharmaceutical Bulletin* 27(9): 1353–58.

Jarząb, Agata, and Wirginia Kukula-Koch. 2018. "Recent Advances in Obesity: The Role of Turmeric Tuber and Its Metabolites in the Prophylaxis and Therapeutical Strategies" *Current Medicinal Chemistry* 25 (37): 4837–53.

Joseph, James, Greg Cole, Elizabeth Head, and Donald Ingram. 2009. "Nutrition, Brain Aging, and Neurodegeneration" *Journal of Neuroscience* 29(41): 12795–801.

Jurenka, Julie S. 2009. "Anti-Inflammatory Properties of Curcumin, a Major Constituent of Curcuma Longa: A Review of Preclinical and Clinical Research" *Alternative Medicine Review* 14(2): 141–53.

Kang, Nam Joo, Seung Ho Shin, Hyong Joo Lee, Ki Won Lee. 2011. "Polyphenols as Small Molecular Inhibitors of Signaling Cascades in Carcinogenesis" *Pharmacology and Therapeutics* 130(3): 310–24.

Kao, Chung-Lan, Liang-Kung Chen, Yuh-Lih Chang, Ming-Chih Yung, Chuan-Chih Hsu, Yu-Chih Chen, Wen-Liang Lo, et al. 2010. "Resveratrol Protects Human Endothelium from H2O2-Induced Oxidative Stress and Senescence via SirT1 Activation" 17(9): 970–79.

Key, T J. 2011. "Fruit and Vegetables and Cancer Risk" *British Journal of Cancer* 104(1): 6–11.

Khan, Kishwar Hayat. 2009. "Roles of Emblica Officinalis in Medicine-A Review" *Botany Research International* 2(4): 218–28.

Kim, Do-Hee, Ki-Woong Park, In Gyeong Chae, Juthika Kundu, Eun-Hee Kim, Joydeb Kumar Kundu, and Kyung-Soo Chun. 2016. "Carnosic Acid Inhibits STAT3 Signaling and Induces Apoptosis through Generation of ROS in Human Colon Cancer HCT116 Cells" *Molecular Carcinogenesis* 55(6): 1096–110.

Kim, Hyo Jung, Byung-Yoon Cha, In Sil Park, Ji Sun Lim, Je-Tae Woo, and Jong-Sang Kim. 2013. "Dehydroglyasperin C, a Component of Liquorice, Attenuates Proliferation and Migration Induced by Platelet-Derived Growth Factor in Human Arterial Smooth Muscle Cells" *British Journal of Nutrition* 110(3): 391–400.

King, H G C. 1962. "Phenolic Compounds of Commercial Wheat Germ" *Journal of Food Science* 27(5): 446–54.

Kloypan, Chiraphat, Rattima Jeenapongsa, Piyawit Sri-in, Surin Chanta, Dech Dokpuang, Santi Tip-pyang, and Nattanan Surapinit. 2012. "Stilbenoids from Gnetum Macrostachyum Attenuate Human Platelet Aggregation and Adhesion" *Phytotherapy Research* 26(10): 1564–68.

Koh, Seong-Ho, Seung H Kim, Hyugsung Kwon, Younjoo Park, Ki Sok Kim, Chi Won Song, Juhan Kim, Myung-Ho Kim, Hyun-Jeung Yu, and Jenny S Henkel. 2003. "Epigallocatechin Gallate Protects Nerve Growth Factor Differentiated PC12 Cells from Oxidative-Radical-Stress-Induced Apoptosis through Its Effect on Phosphoinositide 3-Kinase/Akt and Glycogen Synthase Kinase-3" *Molecular Brain Research* 118(1–2): 72–81.

Kruger, Maria J, Neil Davies, Kathryn H Myburgh, and Sandrine Lecour. 2014. "Proanthocyanidins, Anthocyanins and Cardiovascular Diseases" *Food Research International* 59: 41–52.

Kumar, Manish, Subodh Kumar, and Satwinderjeet Kaur. 2012. "Role of ROS and COX-2/INOS Inhibition in Cancer Chemoprevention: A Review" *Phytochemistry Reviews* 11(2): 309–37.

Kumar, Shashank, Amita Mishra, Abhay K Pandey. 2013. "Antioxidant Mediated Protective Effect of Parthenium Hysterophorus against Oxidative Damage Using in Vitro Models" *BMC Complementary and Alternative Medicine* 13(1): 1–9.

Kumar, S, and A K Pandey. 2012. "Antioxidant, Lipo-Protective and Antibacterial Activities of Phytoconstituents Present in Solanum Xanthocarpum Root" *International Review of Biophysical Chemistry* 3(3): 42–47.

Kumar, Shashank, and Abhay K Pandey. 2013. "Chemistry and Biological Activities of Flavonoids: An Overview" *The Scientific World Journal* 2013: 162750.

La Lastra, C Alarcón De, and Isabel Villegas. 2007. "Resveratrol as an Antioxidant and Pro-Oxidant Agent: Mechanisms and Clinical Implications" *Biochemical Society Transactions* 35(5): 1156–60.

Lara-Guzman, Oscar J, Jorge H Tabares-Guevara, Yudy M Leon-Varela, Rafael M Alvarez, Miguel Roldan, Jelver A Sierra, Julian A Londono-Londono, Jose R Ramirez-Pineda. 2012. "Proatherogenic Macrophage Activities Are Targeted by the Flavonoid Quercetin" *Journal of Pharmacology, and Experimental Therapeutics* 343(2): 296–306.

Leger, C L, M A Carbonneau, F Michel, Emilie Mas, L Monnier, J P Cristol, and B Descomps. 2005. "A Thromboxane Effect of a Hydroxytyrosol-Rich Olive Oil Wastewater Extract in Patients with Uncomplicated Type I Diabetes" *European Journal of Clinical Nutrition* 59(5): 727–30.

Li, Yanyan, and Tao Zhang. 2014. "Targeting Cancer Stem Cells by Curcumin and Clinical Applications" *Cancer Letters* 346(2): 197–205.

Lin, Yu-Li, Shu-Huei Tsai, Shoei-Yn Lin-Shiau, Chi-Tang Ho, and Jen-Kun Lin. 1999. "Theaflavin-3, 3′-Digallate from Black Tea Blocks the Nitric Oxide Synthase by down-Regulating the Activation of NF-KB in Macrophages" *European Journal of Pharmacology* 367(2–3): 379–88.

Lin, Ming-Hsien, Chi-Feng Hung, Hsin-Ching Sung, Shih-Chun Yang, Huang-Ping Yu, Jia-You Fang. 2021. "The bioactivities of resveratrol and its naturally occurring derivatives on skin" *Journal of food and drug analysis* 29(1):15-38. doi: 10.38212/2224-6614.1151. PMID: 35696226; PMCID: PMC9261849.

Lipinski, Christopher A, Franco Lombardo, Beryl W Dominy, and Paul J Feeney. 1997. "Experimental and Computational Approaches to Estimate Solubility and Permeability in Drug Discovery and Development Settings" *Advanced Drug Delivery Reviews* 23(1–3): 3–25.

Liu, Mingyan, Fujun Chen, Lei Sha, Shuang Wang, Lin Tao, Lutian Yao, Miao He, Zhimin Yao, Hang Liu, and Zheng Zhu. 2014. "(−)-Epigallocatechin-3-Gallate Ameliorates Learning and Memory Deficits by Adjusting the Balance of TrkA/P75NTR Signaling in APP/PS1 Transgenic Mice" *Molecular Neurobiology* 49(3): 1350–63.

Liu, Rui Hai. 2003. "Health Benefits of Fruit and Vegetables Are from Additive and Synergistic Combinations of Phytochemicals" *The American Journal of Clinical Nutrition* 78(3): 517S–520S.

Loo, George. 2003. "Redox-Sensitive Mechanisms of Phytochemical-Mediated Inhibition of Cancer Cell Proliferation" *The Journal of Nutritional Biochemistry* 14(2): 64–73.

Luo, Cong, and Xian-Guo Wu. 2011. "Lycopene Enhances Antioxidant Enzyme Activities and Immunity Function in N-Methyl-N′-Nitro-N-Nitrosoguanidine–Induced Gastric Cancer Rats" *International Journal of Molecular Sciences* 12(5): 3340–51.

Maiuri, Maria Chiara, Daniela De Stefano, Paola Di Meglio, Carlo Irace, Maria Savarese, Raffaele Sacchi, Maria Pia Cinelli, and Rosa Carnuccio. 2005. "Hydroxytyrosol, a Phenolic Compound from Virgin Olive Oil, Prevents Macrophage Activation" *Naunyn-Schmiedeberg's Archives of Pharmacology* 371(6): 457–65.

Manach, Claudine, Augustin Scalbert, Christine Morand, Christian Rémésy, and Liliana Jiménez. 2004. "Polyphenols: Food Sources and Bioavailability" *The American Journal of Clinical Nutrition* 79(5): 727–47.

Manikandan, R, M Beulaja, C Arulvasu, S Sellamuthu, D Dinesh, D Prabhu, G Babu, B Vaseeharan, N M Prabhu. 2012. "Synergistic Anticancer Activity of Curcumin and Catechin: An in Vitro Study Using Human Cancer Cell Lines" *Microscopy Research and Technique* 75(2): 112–16.

Maraldi, Tullia, David Vauzour, Cristina Angeloni. 2014. "Dietary Polyphenols and Their Effects on Cell Biochemistry and Pathophysiology 2013." *Oxidative Medicine and Cellular Longevity* 2014: 576363.

Mathai, Kimberly. 2000. "Nutrition in the Adult Years" In LK Mahan, and S Escott-Stump (Eds.). *Krause's Food, Nutrition, and Diet Therapy*, 10th edn. 271: 274–75. Pennsylvania: W.B. Saunders Co.

Mena, Salvador, Maria L Rodriguez, Xavier Ponsoda, Jose M Estrela, Marja Jäättela, and Angel L Ortega. 2012. "Pterostilbene-Induced Tumor Cytotoxicity: A Lysosomal Membrane Permeabilization-Dependent Mechanism." *PLoS ONE* 7(9): 1–12.

Morrison, Gail, and Lisa Hark. 1996. *Medical Nutrition and Disease*. New Jersey, United States: Blackwell Science.

Oliveira, Rodrigo Ayres de, and Iolanda M Fierro. 2018. "New Strategies for Patenting Biological Medicines Used in Rheumatoid Arthritis Treatment" *Expert Opinion on Therapeutic Patents* 28(8): 635–46.

Pan, Min-Hsiung, Ching-Shu Lai, Chi-Tang Ho. 2010. "Anti-Inflammatory Activity of Natural Dietary Flavonoids" *Food and Function* 1(1): 15–31.

Pandima Devi K, T Rajavel, M Daglia, S F Nabavi, A Bishayee, S M Nabavi. 2017. "Targeting miRNAs by polyphenols: Novel therapeutic strategy for cancer." *Seminars in Cancer Biology* 46:146–157.

Pandey, A K, A K Mishra, A Mishra. 2012. "Antifungal and Antioxidative Potential of Oil and Extracts Derived from Leaves of Indian Spice Plant Cinnamomum Tamala" *Cellular and Molecular Biology* 58(1): 142–47.

Phillipson, J David. 2001. "Phytochemistry and Medicinal Plants" *Phytochemistry* 56(3): 237–43.

Prahalathan, Pichavaram, Murugesan Saravanakumar, and Boobalan Raja. 2012. "The Flavonoid Morin Restores Blood Pressure and Lipid Metabolism in DOCA-Salt Hypertensive Rats" *Redox Report* 17(4): 167–75.

Pramanik, Kamdeo K, Siddavaram Nagini, Abhay K Singh, Prajna Mishra, Tanushree Kashyap, Nidhi Nath, Manzar Alam, Ajay Rana, and Rajakishore Mishra. 2018. "Glycogen Synthase Kinase-3β Mediated Regulation of Matrix Metalloproteinase-9 and Its Involvement in Oral Squamous Cell Carcinoma Progression and Invasion" *Cellular Oncology* 41(1): 47–60.

Pramanik, Kamdeo K, Abhay K Singh, Manzar Alam, Tanushree Kashyap, Prajna Mishra, Aditya K Panda, Ratan K Dey, Ajay Rana, Siddavaram Nagini, and Rajakishore Mishra. 2016. "Reversion-Inducing Cysteine-Rich Protein with Kazal Motifs and Its Regulation by Glycogen Synthase Kinase 3 Signaling in Oral Cancer" *Tumor Biology* 37(11): 15253–64.

Quoc Trung, Ly, J Luis Espinoza, Akiyoshi Takami, and Shinji Nakao. 2013. "Resveratrol Induces Cell Cycle Arrest and Apoptosis in Malignant NK Cells via JAK2/STAT3 Pathway Inhibition" *PloS One* 8(1): e55183.

Rahigude, A, P Bhutada, S Kaulaskar, M Aswar, and K Otari. 2012. "Participation of Antioxidant and Cholinergic System in Protective Effect of Naringenin against Type-2 Diabetes-Induced Memory Dysfunction in Rats" *Neuroscience* 226: 62–72.

Rahman, Irfan, Saibal K Biswas, and Paul A Kirkham. 2006. "Regulation of Inflammation and Redox Signaling by Dietary Polyphenols" *Biochemical Pharmacology* 72(11): 1439–52.

Rao, Bs Narasinga. 2003. "Bioactive Phytochemicals in Indian Foods and Their Potential in Health Promotion and Disease Prevention" *Asia Pacific Journal of Clinical Nutrition* 12(1): 9–22.

Raso, Giuseppina Mattace, Rosaria Meli, Giulia Di Carlo, Maria Pacilio, and Raffaele Di Carlo. 2001. "Inhibition of Inducible Nitric Oxide Synthase and Cyclooxygenase-2 Expression by Flavonoids in Macrophage J774A. 1" *Life Sciences* 68(8): 921–31.

Reuland, Danielle J, Joe M McCord, Karyn L Hamilton. 2013. "The Role of Nrf2 in the Attenuation of Cardiovascular Disease" *Exercise and Sport Sciences Reviews* 41(3): 162–68.

Rinwa, Puneet, and Anil Kumar. 2012. "Piperine Potentiates the Protective Effects of Curcumin against Chronic Unpredictable Stress-Induced Cognitive Impairment and Oxidative Damage in Mice" *Brain Research* 1488: 38–50.

Rokayya, Sami, Chun-Juan Li, Yan Zhao, Ying Li, and Chang-Hao Sun. 2013. "Cabbage (Brassica Oleracea L. Var. Capitata) Phytochemicals with Antioxidant and Anti-Inflammatory Potential" *Asian Pacific Journal of Cancer Prevention* 14(11): 6657–62.

Saadatdoust, Zeinab, Ashok Kumar Pandurangan, Suresh Kumar Ananda Sadagopan, Norhaizan Mohd Esa, Amin Ismail, and Mohd Rais Mustafa. 2015. "Dietary Cocoa Inhibits Colitis Associated Cancer: A Crucial Involvement of the IL-6/STAT3 Pathway" *The Journal of Nutritional Biochemistry* 26(12): 1547–58.

Sak, Katrin. 2014. "Site-Specific Anticancer Effects of Dietary Flavonoid Quercetin" *Nutrition and Cancer* 66(2): 177–93.

Salminen, A, M Lehtonen, T Suuronen, K Kaarniranta, and J Huuskonen. 2008. "Terpenoids: Natural Inhibitors of NF-KB Signaling with Anti-Inflammatory and Anticancer Potential" *Cellular and Molecular Life Sciences* 65(19): 2979–99.

Sarkar, Souvik, Somnath Mazumder, Shubhra J Saha, and Uday Bandyopadhyay. 2016. "Management of Inflammation by Natural Polyphenols: A Comprehensive Mechanistic Update" *Current Medicinal Chemistry* 23(16): 1657–95.

Saxena, Mamta, Jyoti Saxena, Rajeev Nema, Dharmendra Singh, and Abhishek Gupta. 2013. "Phytochemistry of Medicinal Plants" *Journal of Pharmacognosy and Phytochemistry* 1(6): 168–82.

Shen, Lin, Bei-an You, Hai-qing Gao, Bao-Ying Li, Fei Yu, and Fei Pei. 2012. "Effects of Phlorizin on Vascular Complications in Diabetes Db/Db Mice" *Chinese Medical Journal* 125(20): 3692–96.

Shukla, Yogeshwer, and Neetu Kalra. 2007. "Cancer Chemoprevention with Garlic and Its Constituents" *Cancer Letters* 247(2): 167–81.

Smith, Adam, Brian Giunta, Paula C Bickford, Michael Fountain, Jun Tan, and R Douglas Shytle. 2010. "Nanolipidic Particles Improve the Bioavailability and α-Secretase Inducing Ability of Epigallocatechin-3-Gallate (EGCG) for the Treatment of Alzheimer's Disease" *International Journal of Pharmaceutics* 389(1–2): 207–12.

Stangl, Verena, Henryk Dreger, Karl Stangl, and Mario Lorenz. 2007. "Molecular Targets of Tea Polyphenols in the Cardiovascular System" *Cardiovascular Research* 73(2): 348–58.

Stravodimos, George A, Ben A Chetter, Efthimios Kyriakis, Anastassia L Kantsadi, Demetra SM Chatzileontiadou, Vassiliki T Skamnaki, Atsushi Kato, Joseph M Hayes, and Demetres D Leonidas. 2017. "Phytogenic Polyphenols as Glycogen Phosphorylase Inhibitors: The Potential of Triterpenes and Flavonoids for Glycaemic Control in Type 2 Diabetes" *Current Medicinal Chemistry* 24(4): 384–403.

Szatrowski, Ted P, and Carl F Nathan. 1991. "Production of Large Amounts of Hydrogen Peroxide by Human Tumor Cells" *Cancer Research* 51(3): 794–98.

Taipale, Jussi, and Philip A Beachy. 2001. "The Hedgehog and Wnt Signalling Pathways in Cancer" *Nature* 411(6835): 349–54.

Tang, Feng-Yao, Man-Hui Pai, Xiang-Dong Wang. 2011. "Consumption of Lycopene Inhibits the Growth and Progression of Colon Cancer in a Mouse Xenograft Model" *Journal of Agricultural and Food Chemistry* 59(16): 9011–21.

Tattelman, Ellen. 2005. "Health Effects of Garlic" *American Family Physician* 72(1): 103–6.

Tsai, Jie-Heng, Li-Sung Hsu, Chih-Li Lin, Hui-Mei Hong, Min-Hsiung Pan, Tzong-Der Way, Wei-Jen Chen. 2013. "3, 5, 4′-Trimethoxystilbene, a Natural Methoxylated Analog of Resveratrol, Inhibits Breast Cancer Cell Invasiveness by Downregulation of PI3K/Akt and Wnt/β-Catenin Signaling Cascades and Reversal of Epithelial–Mesenchymal Transition" *Toxicology and Applied Pharmacology* 272(3): 746–56.

Upadhyay, Swapna, Madhulika Dixit. 2015. "Role of Polyphenols and Other Phytochemicals on Molecular Signaling" *Oxidative Medicine and Cellular Longevity* 2015: 504253.

Vasanthi, HR, N ShriShriMal, and DK Das. 2012. "Phytochemicals from Plants to Combat Cardiovascular Disease" *Current Medicinal Chemistry* 19(14): 2242–51.

Vendrame, Stefano, and Dorothy Klimis-Zacas. 2015. "Anti-Inflammatory Effect of Anthocyanins via Modulation of Nuclear Factor-KB and Mitogen-Activated Protein Kinase Signaling Cascades" *Nutrition Reviews* 73(6): 348–58.

Vitaglione, Paola, Stefano Sforza, Gianni Galaverna, Cristiana Ghidini, Nicola Caporaso, Pier Paolo Vescovi, Vincenzo Fogliano, Rosangela Marchelli. 2005. "Bioavailability of Trans-resveratrol from Red Wine in Humans" *Molecular Nutrition and Food Research* 49(5): 495–504.

Wagner, Anika Eva, Anna Maria Terschluesen, Gerald Rimbach. 2013a. "Health Promoting Effects of Brassica-Derived Phytochemicals: From Chemopreventive and Anti-Inflammatory Activities to Epigenetic Regulation" *Oxidative Medicine and Cellular Longevity* 2013: 964539. https://doi.org/10.1155/2013/964539

Wagner, Anika E, Olga Will, Christine Sturm, Simone Lipinski, Philip Rosenstiel, and Gerald Rimbach. 2013b. "DSS-Induced Acute Colitis in C57BL/6 Mice Is Mitigated by Sulforaphane Pre-Treatment" *The Journal of Nutritional Biochemistry* 24(12): 2085–91.

Walker, Jennifer M, Diana Klakotskaia, Deepa Ajit, Gary A Weisman, W Gibson Wood, Grace Y Sun, Peter Serfozo, Agnes Simonyi, and Todd R Schachtman. 2015. "Beneficial Effects of Dietary EGCG and

Voluntary Exercise on Behavior in an Alzheimer's Disease Mouse Model" *Journal of Alzheimer's Disease* 44(2): 561–72.

Wang, Jintang, Yuetao Song, Maolong Gao, Xujing Bai, and Zheng Chen. 2016. "Neuroprotective Effect of Several Phytochemicals and Its Potential Application in the Prevention of Neurodegenerative Diseases" *Geriatrics* 1(4): 29.

Wang, Juejin, Rongjian Zhang, Youhua Xu, Hong Zhou, Bin Wang, Shengnan Li. 2008. "Genistein Inhibits the Development of Atherosclerosis via Inhibiting NF-κ B and VCAM-1 Expression in LDLR Knockout Mice" *Canadian Journal of Physiology and Pharmacology* 86(11): 777–84.

Wang, Min S, Shanta Boddapati, Sharareh Emadi, and Michael R Sierks. 2010. "Curcumin Reduces α-Synuclein Induced Cytotoxicity in Parkinson's Disease Cell Model" *BMC Neuroscience* 11(1): 1–10.

Waterbeemd, Han Van de. 2005. "From in Vivo to in Vitro/in Silico ADME: Progress and Challenges" *Expert Opinion on Drug Metabolism and Toxicology* 1(1): 1–4.

Waterbeemd, Han Van De, and Eric Gifford. 2003. "ADMET in Silico Modelling: Towards Prediction Paradise?" *Nature Reviews Drug Discovery* 2(3): 192–204.

Weberling, Anke, Volker Boehm, and Kati Froehlich. 2011. "Nutraceuticals-The Relation between Lycopene, Tomato Products and Cardiovascular Diseases" *Agro Food Industry Hi Tech* 22(4): 21.

Whitlock, Nichelle C, Seung Joon Baek. 2012. "The Anticancer Effects of Resveratrol: Modulation of Transcription Factors" *Nutrition and Cancer* 64(4): 493–502.

Winter, Aimee N, Matthew C Brenner, Noelle Punessen, Michael Snodgrass, Caleb Byars, Yingyot Arora, Daniel A Linseman. 2017. "Comparison of the Neuroprotective and Anti-Inflammatory Effects of the Anthocyanin Metabolites, Protocatechuic Acid and 4-Hydroxybenzoic Acid" *Oxidative Medicine and Cellular Longevity* 2017: 6297080.

World Health Organization. 2014. *Global Status Report on Noncommunicable Diseases 2014*. World Health Organization.

Wu, Baojian, Kaustubh Kulkarni, Sumit Basu, Shuxing Zhang, and Ming Hu. 2011. "First-Pass Metabolism via UDP-Glucuronosyltransferase: A Barrier to Oral Bioavailability of Phenolics" *Journal of Pharmaceutical Sciences* 100(9): 3655–81.

Wung, B S, M C Hsu, C C Wu, and C W Hsieh. 2005. "Resveratrol Suppresses IL-6-Induced ICAM-1 Gene Expression in Endothelial Cells: Effects on the Inhibition of STAT3 Phosphorylation" *Life Sciences* 78(4): 389–97.

Xie, Chenghui, Jie Kang, Jin-Ran Chen, Shanmugam Nagarajan, Thomas M Badger, Xianli Wu. 2011. "Phenolic Acids Are in Vivo Atheroprotective Compounds Appearing in the Serum of Rats after Blueberry Consumption" *Journal of Agricultural and Food Chemistry* 59(18): 10381–87.

Xie, Qi, Qian Bai, Ling-Yun Zou, Qian-Yong Zhang, Yong Zhou, Hui Chang, Long Yi, Jun-Dong Zhu, and Man-Tian Mi. 2014. "Genistein Inhibits DNA Methylation and Increases Expression of Tumor Suppressor Genes in Human Breast Cancer Cells" *Genes Chromosomes and Cancer* 53(5): 422–31.

Xue, Bin, Debarati DasGupta, Manzar Alam, Mohd Shahnawaz Khan, Shuo Wang, Anas Shamsi, Asimul Islam, and Md Imtaiyaz Hassan. 2022. "Investigating Binding Mechanism of Thymoquinone to Human Transferrin, Targeting Alzheimer's Disease Therapy" *Journal of Cellular Biochemistry* 123(8): 1381–93.

Yan, Wen-Jun, Ruo-Bin Liu, Ling-Kai Wang, Ya-Bing Ma, Shao-Li Ding, Fei Deng, Zhong-Yuan Hu, and Da-Bin Wang. 2018. "Sirt3-Mediated Autophagy Contributes to Resveratrol-Induced Protection against ER Stress in HT22 Cells" *Frontiers in Neuroscience* 12: 116.

Yan, Zhaoming, Yinzhao Zhong, Yehui Duan, Qinghua Chen, and Fengna Li. 2020. "Antioxidant Mechanism of Tea Polyphenols and Its Impact on Health Benefits" *Animal Nutrition* 6(2): 115–23.

Yang, Jun, and Rui Hai Liu. 2009. "Induction of Phase II Enzyme, Quinone Reductase, in Murine Hepatoma Cells in Vitro by Grape Extracts and Selected Phytochemicals" *Food Chemistry* 114(3): 898–904.

Zhang, Han, Huanhuan Liu, Yulong Chen, and Yan Zhang. 2018. "The Curcumin-Induced Vasorelaxation in Rat Superior Mesenteric Arteries" *Annals of Vascular Surgery* 48: 233–40.

Zhang, Yanling, and Dongmin Liu. 2011. "Flavonol Kaempferol Improves Chronic Hyperglycemia-Impaired Pancreatic Beta-Cell Viability and Insulin Secretory Function" *European Journal of Pharmacology* 670(1): 325–32.

Zhang, Yu-Jie, Ren-You Gan, Sha Li, Yue Zhou, An-Na Li, Dong-Ping Xu, and Hua-Bin Li. 2015. "Antioxidant Phytochemicals for the Prevention and Treatment of Chronic Diseases" *Molecules* 20(12): 21138–56.

12 Metabolomics of Medicinal and Aromatic Plants

Mohd Nehal, Iqra Khan, Jahanarah Khatoon, Salman Akhtar, and Mohammad Kalim Ahmad Khan
Integral University

CONTENTS

DOI: 10.1201/b22842-12

12.1 INTRODUCTION

Over the centuries, the plants were the sole source of all the medicaments and health care for humans and domestic animals (Porwal et al. 2020). Medicinal and aromatic plants (MAPs) were the source of various bioactive components, which have been used for treating countless diseases over a long period. Since ancient times, spices and MAPs have played crucial roles in human nutrition, conferring aroma, flavor, and color to foods (Sousa et al. 2019). Recently, an estimated increase has been found in synthetic substances in various sectors of the economy, including agriculture, food, and pharmaceutics (Adel Mahmoodabad et al. 2014). In the pharmaceutical industry, secondary metabolites of medicinal plants are equally beneficial as lead compounds for the design and production of effective drugs. During this process, vast amounts of harmful solvents were often used to prepare desired drugs. The phytotherapeutic nature of medicinal plants is due to specific secondary metabolites and biologically active components formed during secondary metabolism. They provide many compounds exhibiting various therapeutic effects (Salmerón-Manzano, Garrido-Cardenas, and Manzano-Agugliaro 2020).

While aromatic plants produce aromatic substances exhibiting therapeutic properties like antimicrobial, anti-inflammatory, and antioxidant activities, generally used for culinary purposes, food, and liqueur industries (Jamshidi-Kia, Lorigooini, and Amini-Khoei 2018), the MAPs are the plant's raw materials, rich in secondary metabolites, and play various functions in the growth and nourishment of the body in pigmentation (Cadar et al. 2021). In a recent study, it has been proved that the antimicrobial substance found in herbal extracts can preserve food safety; based on these properties, plants can be used as natural antimicrobials. By using MAPs as raw material, a series of intermediate and final products can be obtained, including allopathic medicines, phytotherapeutic products, botanical supplements, cosmetics and personal care, paints, and others (Enioutina et al. 2020). This chapter envisions promoting and creating awareness about MAPs as a good diversification for small-scale farmers.

The thermal decomposition of MAPs derivative under an oxygen-deficient environment called pyrolysis was used earlier for charcoal preparation and tar production for sealing boats (Rollag et al. 2022). The growth in pyrolysis could be considered an effective way of transforming biomass into bio-oils. Different methodologies, including thermo-chemical biological processes, have converted biomass into valuable products. Among them, pyrolysis is the most common, since it has many advantages, including storing and flexibility in adjuring for transportation through turbines, boilers, or engines (Demirbas 2005). The eventual objective of the pyrolysis is to provide high-yield energy products by gradually ousting non-renewable fossil fuels (Zaman et al. 2017). Typically, a pyrolysis system unit contains the equipment for pre-processing target residues, the reactor, and a unit for downstream processing.

12.2 VALUE OF MAPs AS RAW MATERIALS

The MAPs play an essential role in disease prevention and daily lifestyle-based health management. They are an integral part of the traditional medicine system and vital component of various local trade supply chains found in numerous local communities worldwide. The MAPs are comprised of a wide range of species that have different sources, uses, and characterization, where their activities can be valuable for livelihood as MAPs are one of the most convenient options for harvests trade, as it requires access to natural assets, volunteer labor, and primary species knowledge (Urquiza-Haas and Cloatre 2019). More than 70% of MAPs trade is being used in herbal products, and other plant-based products are harvested from the forest, and their demand increases daily. The Medicinal Plant Specialist Group of the World Conservation Union (IUCN-International Union for Conservation of Nature) predicts that nearly about 15,000 plant species used in the herbal product could be endangered due to unsustainable collection and unfair trade.

The rapid growth of interest in MAPs for both conservation and development organizations was found over the past 30 years; as a result, the indispensable livelihoods contribution by these valuables makes a large number of rural communities for their trade and subsistence. Approximately 1,000 plant species contribute significantly to modern medicines; the anticancer compound like taxol extracted from the *Taxus brevifolia* and phytochemicals such as *Pterocarpus osum* are used to treat sickle cell disease (Amanuel et al. 2019). An estimate of 80% of the population in Africa and Asia generally depends on these plant-based drugs for their health care. Similar statistics were reported earlier by World Health Organization (WHO), revealing that the world population is closely dependent on plant-based medicines to cure their diseases. A few common MAPs and their therapeutic merits are tabulated in Table 12.1, followed by their detailed discussions.

12.2.1 Aegle Marmelos

Aegle marmelos (A. marmelos) belong to the Rutaceae family, a.k.a. Beal (English-Bengal Quince), commonly found throughout India and native to Northern India, Burma, Bangladesh, Sri Lanka, and China. It plays an essential role in the indigenous system of Indian medicine due to its various medicinal properties. *A. marmelos* is a medium- to large-sized tree with 2.5 cm leaves, short flowers and globular fruits. In India, the essential oil of *A. marmelos* is isolated from the leaves, which have an antifungal activity against *Aspergillus niger* and *Aspergillus flavus.* The gastrointestinal disorders are also treated by using *A. marmelos.* The therapeutic properties of *A. marmelos* are illustrated in Figure 12.1.

12.2.2 Prunus Africana

Prunus africana (P. africana) also known as red cherry, pygeum, or winter cherry. It belongs to the Rosaceae family and is used to treat the prostate problem in Africa. *P. africana* is a medium to large-sized evergreen tree, distributed throughout Kenya and Africa. The bark of *P. africana* is used in prostate cancer, and the leaves are used as an inhalant for fever. In Kenya, this plant is used as traditional medicine to treat chest pain, malaria, and fever. The therapeutic properties of *P. africana* are illustrated in Figure 12.2.

12.2.3 Ficus Religiosa

Ficus religiosa (F. religiosa), also known as Bodhi tree, peepal, or sacred fig, belongs to the Moraceae family. *F. religiosa* is distributed throughout Southeast Asia, Southwest China, India,

TABLE 12.1
Examples of Common MAPs and Their Possible Medicinal Properties

Name of Plants	Role in Diseases	References
Agele marmelos	Fever, cholesterol reduction, blood purification	(Ayoub et al. 2018)
Runus Africana	Cardiovascular complaints, circulatory disease	(Komakech et al. 2017)
Ficus religiosa	Cough, fever, headache, genitourinary complaints	(Nallasamy et al. 2020)
Mangifera indica	Gastrointestinal disorders, cuts, wounds, and other uses	(Sekar et al. 2019)
Ananas comosus	Heart disease, arthritis, cancer	(Karam et al. 2016)
Ocimum sanctum	Fever, headache, coughing	(Sayyad, Patil, and Dhumal 2022)
Allium sativum	Diabetes, fever, liver disorder	(Tariq et al. 2021)
Curcuma longa	Heart disease, cancer, indigestion	(Witkin and Li 2013)

FIGURE 12.1 *Aegle marmelos* and its therapeutic properties.

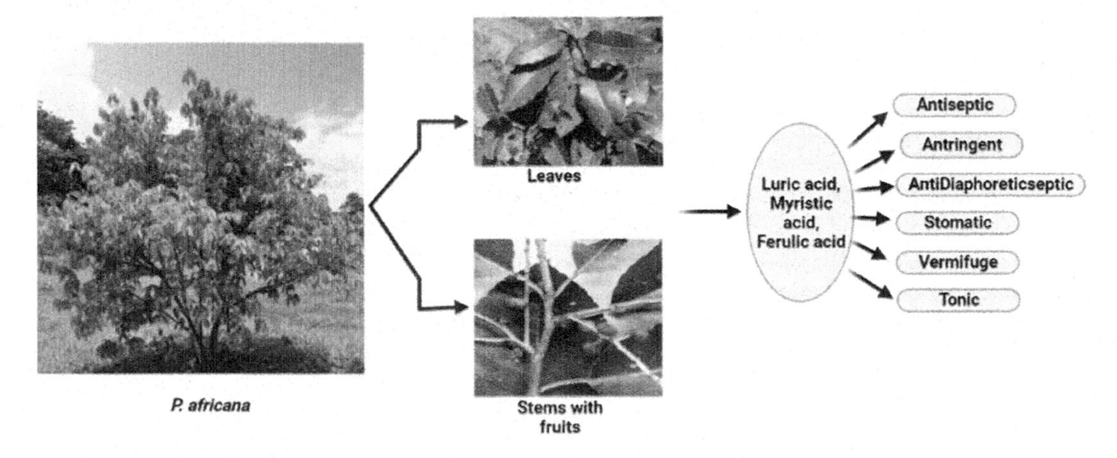

FIGURE 12.2 *Prunus africana* and its therapeutic properties.

and Bangladesh. This plant bark in India is used as antibacterial, antiviral, and antiprotozoal agents for gonorrhea treatment, and the leaves are used for skin diseases, whereas in Bangladesh, this plant is used in treating various diseases such as cancer and inflammation. The therapeutic properties of *F. religiosa* are shown in Figure 12.3.

12.2.4 Mangifera Indica

Mangifera indica (M. indica), commonly known as mango, is the king of fruits. This plant belongs to the Anacardiaceae family, and it is the national fruit of India and the Philippines, while it is considered the national tree of Bangladesh. The mango is native to India and Southeast Asia. It is evergreen, fast-growing and about 30 m to 40 m in height. In traditional medicine, different parts of the *M. indica* are used for enhancing health, such as mango kernel used to treat diarrhea and the unripe fruit, leaves, and bark exhibiting an antibiotic activity. The *M. indica* fruit is a rich source of fibers, polyphenols, vitamins, and potassium. The therapeutic properties of mango are shown in Figure 12.4.

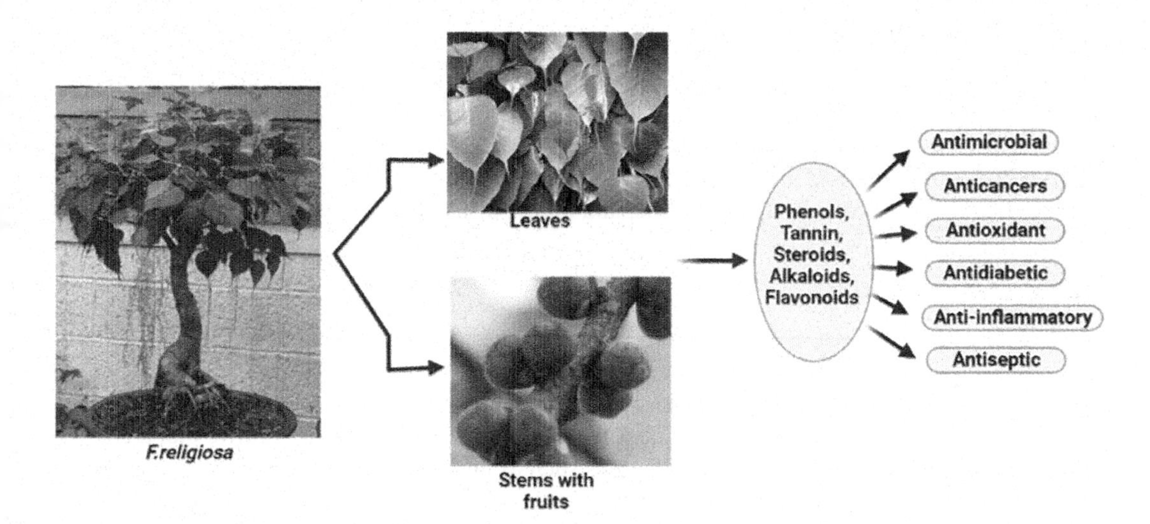

FIGURE 12.3 *Ficus religiosa* and its therapeutic properties.

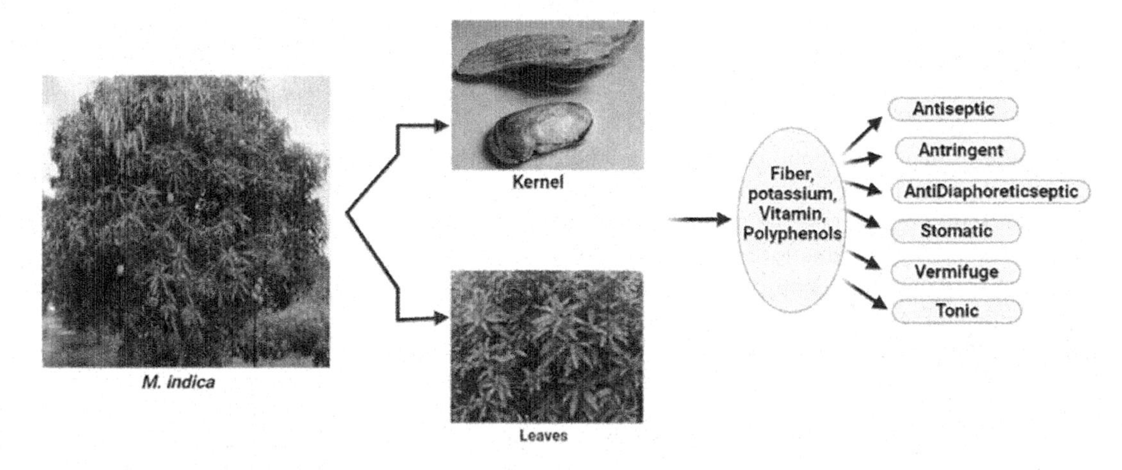

FIGURE 12.4 *Mangifera indica* and its therapeutic properties.

12.2.5 Ananas Comosus

Ananas comosus (A. comosus) is commonly known as Ananas or pineapple. *A. comosus* is native to tropical and subtropical America and belongs to the Bromeliaceae family. The pineapple plant grows worldwide in South Africa, Indonesia, the Philippines, Brazil, Kenya, and India. The fruit of this plant is highly nutritive as it contains essential nutrients such as vitamins, minerals, antioxidants, fiber, and carbohydrates, which improve the quality of the diet (Corrêa et al. 2017).

12.2.6 Ocimum Sanctum

Ocimum sanctum (O. sanctum) is generally utilized in Ayurveda, a traditional form of medicine with a powerful aroma and astringent taste. *O. sanctum*, commonly known as Tulsi or holy basil, belongs to the Lamiaceae family. Tulsi or basil has been used to treat coughing, headache, diarrhea, warts, and worms. It is also used to treat various ailments, including common cold, malaria, inflammation, and heart disease. The Tulsi leaves contain many biologically active components exhibiting nematocidal, fungistatic, insecticidal, and antibacterial activities (Sayyad, Patil, and Dhumal 2022).

12.2.7 Allium Sativum

Allium sativum (*A. sativum*), a.k.a. garlic or lahsun, belongs to *the Alliaceae* family. Garlic has been used to treat pulmonary issues and uterine growths, and throat and fungal infections. Recently, garlic has been used to treat cardiovascular disease, diabetes, and exhibits anticancer and anti-aging effects. Garlic may also help in reducing lipid quantities in the blood. Garlic contains various active compounds like alliin, allicin, diallyl sulfide, and diallyl disulfide (Tariq et al. 2021).

12.2.8 Curcuma Longa

Curcuma longa (C. longa), commonly known as turmeric, belongs to the *Zingiberaceae* family. Turmeric is commonly cultivated in tropical Asian regions and is used as a coloring and flavoring agent for foods. Turmeric contains various active constituents such as curcumin, ketones, alcohol, and p-hydroxycinnamoyl (feruloyl) methane. *C. longa* is commonly used to improve blood circulation. *C. longa* exhibits many biological activities such as anti-inflammatory, antifungal, antibacterial, and antioxidant activities (Witkin and Li 2013).

12.3 BIOCHEMICAL AND PHYSIOLOGICAL FEATURES

The metabolites produced by living organisms are divided into two major groups, viz., primary and secondary. Primary metabolites are considered crucial for functional and operating components of all living organisms as they perform basic metabolic processes such as photosynthesis and respiration. In contrast, the secondary metabolites are mainly derived from the precursors produced during the primary metabolic processes. In other words, the production of secondary metabolites is restricted to only one plant species or related group of species, while the primary metabolites are often found throughout the plant kingdom. Secondary metabolites are propounded to have a momentous ecological function, such as protecting plants against being eaten by primary consumers or infected by a microbial pathogen. They function as plant-microbial symbioses, and the secondary metabolites may serve as an attractant. Also, secondary metabolites (such as amino acids, nucleotides, and sugars) have a limited distribution in the plant kingdom (Guerriero et al. 2018).

12.4 ACTIVE SUBSTANCES OF MAPs

Biological active substances contain different chemical compositions and activities, implying their effect on the living organism. These substances play an important physiological and ecological role for plants, so they act as the basis for their utilization as MAPs in the scientific literature. Commonly known biologically active substances are alkaloids, glycosides, essential oils, and miscellaneous substances.

12.4.1 Alkaloids

Alkaloids are a group of nitrogen-containing chemical substances with complex molecular structures and are bitter in taste. Alkaloids mainly form salts and have strong physiological effects on living organisms; they may function as plant simulator or growth, metabolism, and reproduction regulators. Alkaloids are colorless, crystalline, non-volatile liquids and insoluble in water. Nowadays, many alkaloid compounds are known, viz., betanin (also known as phytolaccamin, which is a water-soluble red pigment commonly used to color food and pharmaceutical products) (Samanta, Awwad, and Algarni 2020), matrine (an anticancer alkaloid used for cancer inhibitions by cancer cell proliferation, inducing apoptosis, arresting the cell cycle, and inhibiting cancer cell metastasis), and brucine (an anti-inflammatory and anti-tumor analgesic drug) (Li et al. 2018), which is widely used in many South Asian countries. However, it is bitter and has high toxicity.

12.4.2 Glycosides

Glycosides are chemical compounds that have physiological effects and divergent metabolic origins. Glycoside's commonly feature in any of sugar derivatives that contain a non-sugar group (aglycone) bound to an oxygen or a nitrogen atom and, on hydrolysis, yields sugar (glucose). The aglycone, in turn, can also be more varied. Various glycosides are derived from plants, such as cardiac glycosides, flavones glycosides, and xanthone glycosides. Cardiac glycoside is medicine for treating heart failure and certain irregular heartbeats. The most common plants producing cardiac glycosides are *Nerium oleander, Digitalis lanata*, and *Digitalis purpurea* (Karaś et al. 2020).

12.4.3 Essential Oils

The locution "essential oils" refer to a mixture of various compounds, chiefly terpenes and their derivatives, which can evaporate at room temperature without residues. Generally, they are isolated by steam distillation and are water insoluble or poorly dissolved in water (Celano et al. 2017). The essential oils mainly produced and traded are orange oil, lemon oil, clove oil, and rose oil, and it is noted that about 65% of the production of essential oils comes from developing countries because of the availability of raw materials. Mostly, they have strong characteristic odor and taste (aroma). In India, essential oil is mainly isolated from mint, peppermint, lemongrass, and citronella plants. These oils have antifungal, antibacterial, and anti-inflammatory properties (Samarth, Samarth, and Matsumoto 2017).

12.4.4 Miscellaneous Substances

Miscellaneous substances are of diverse chemical composition and physiological effectiveness, like aromatic acids, bitter substances, carbohydrates, and mucilaginous plant pigments.

12.5 BIOACTIVE PROPERTIES OF MAPs

12.5.1 Antioxidant Activity

The antioxidant activity of the extract of a mixture was assessed by determining its ability to inhibit lipid peroxidation and oxidative hemolysis. The lower the IC_{50} value, the greater the antioxidant activity of the extract. It is mainly associated with phenolic compounds, organic acid, and tocopherols. Antioxidant might function to terminate free radicals, initiate chain retarding and repair damaged biomolecules. For example, phenolic compounds with more than one hydroxyl group (-OH) can terminate free radical by donating H-atoms, and synthetic phenolic compounds such as BHA (butylated hydroxyanisole) and BHT (butylated hydroxytoluene) are chain-breaking antioxidant (Lockowandt et al. 2019). Common examples of MAPs having antioxidant properties are shown in Table 12.2.

12.5.2 Antimicrobial Activity

The naturally occurring microorganism present in the environment could cause food spoilage, widely affecting all food types, which is the primary reason for food waste in developing countries. Traditionally, the crude extract from different parts of medicinal plants, including root, stem, fruit and twinges, were widely used to treat some human diseases (Khan, Akhtar, and Arif 2018). In medical plants, such as alkaloids, flavonoids, and terpenoids, phytochemicals possess antimicrobial and antioxidant properties (Samarth, Samarth, and Matsumoto 2017). It has been shown that the extract from the medicinal plant can effectively help in reducing bacterial growth during the storage of meat, juices and milk (Das et al. 2021). For example, crude extract of ginger, garlic, curry,

TABLE 12.2
Common Examples of MAPs Having Antioxidant Properties

Botanical Name	COMMON NAME	Type of Plant	Chemical Constituent	Chemical Structure
Solanum lycopersicum	Tomato	Herb	Lycopene	
Allium sativum	Garlic	Herb	Allicin, ajoene	
Glycine max	Soybean	Herb	Genistein	
Daucus carota	Carrot	Shrub	Beta- carotenes	
Camellia sinensis	Tea	Shrub	Catechins	

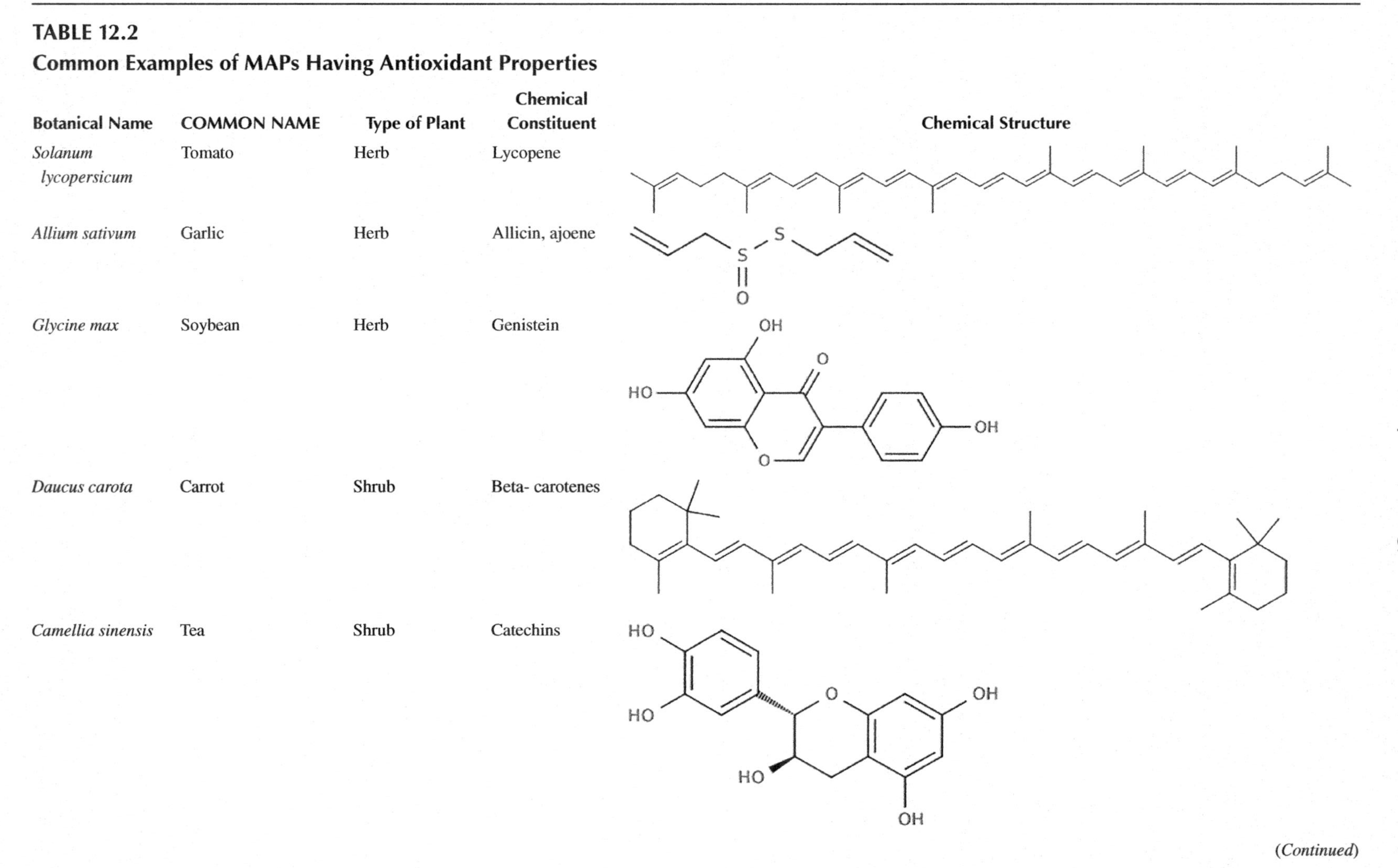

(Continued)

TABLE 12.2 (*Continued*)
Common Examples of MAPs Having Antioxidant Properties

Botanical Name	COMMON NAME	Type of Plant	Chemical Constituent	Chemical Structure
Capsicum	Red chilli	Shrub like herb	Capsaicin	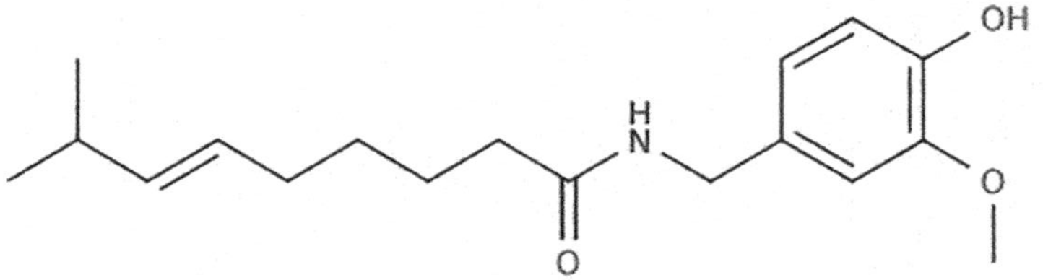
Curcuma longa	Turmeric	Herb	Curcumin	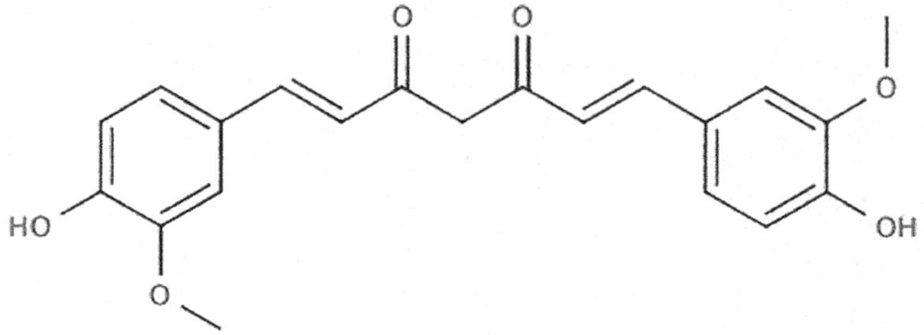
Syzygium aromaticum	Cloves	Tree	Eugenol, isoeugenol	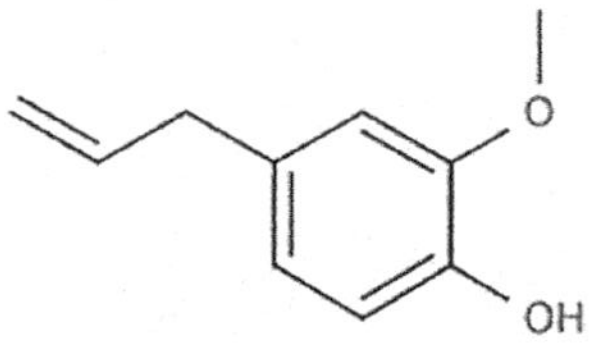
Aloe	Aloe vera	Herb	Emodin	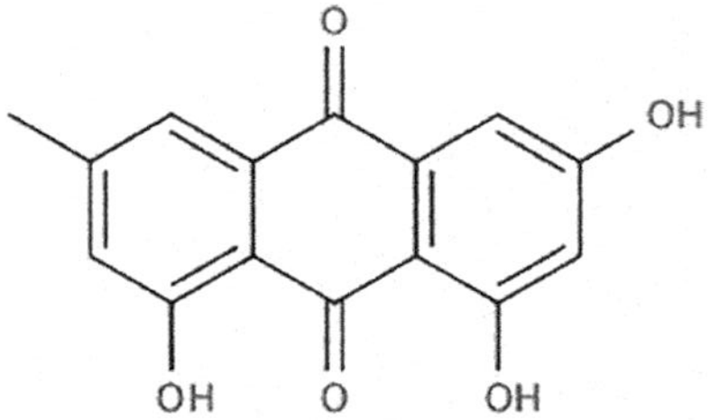

mustard and other herbs exhibits antimicrobial properties against a broad spectrum of bacteria, such as gram-negative and gram-positive bacteria.

12.5.3 Antitumor Activity

The lifestyle-based diseases have been increasing day by day, which are associated with malignancies. Nowadays, working with plants extracts by determining their biological activity and then using them as natural products in cancer therapy has great importance, such as *Colchicum sanguicolle* extract, which showed the most effective tumor activity (Abreu et al. 2011). Common examples of MAPs having anticancer properties are shown in Table 12.3. Moreover, example of MAPs used in traditional medicines is shown in Figure 12.5.

12.6 THERMO-CHEMICAL ROUTES FOR BIOMASS CONVERSION TO FUEL

Different types of thermo-chemical processes have converted biomass into biofuel (bio-oil, biochar, and biogas), viz., combustion, gasification, and pyrolysis. Among them, pyrolysis is highlighted as it has several advantages such as storing, flexibility, and transportation. Sometimes solid biomass and wastes research is challenging through pyrolysis. During combustion, release of heat through the exothermal heat of reaction happens, which is for the oxidation of the fuel and is an interaction of fuel, energy, and environment. Combustion is an ingredient of biomass and oxygen together in a high-temperature environment to form vapors, carbon dioxide, and heat (Lehto et al. 2014). The water content in biomass is an essential factor, and the best burning fuel is dry. To find out combustion is efficient and clean, the ingredients (biomass and oxygen) are well mixed at a fixed temperature for the desired period. Gasification is defined as converting a solid or liquid into a gaseous fuel without leaving any carbonaceous residues, pyrolysis applies to the thermal decomposition of solid waste in an inert atmosphere (Liu et al. 2022). An overview of thermo-chemical routes for biomass conversion to fuel is illustrated in Figure 12.6.

12.7 PYROLYSIS

The process of thermal decomposition of solid waste in an inert environment is called pyrolysis (McNamara, Zitomer, and Liu 2019; Liu et al. 2016). The term "pyrolysis" is taken from the Greek words "pyro", which means fire, and "lysis", which means disintegration into integral parts. The pyrolysis process mainly contains methane, hydrogen, carbon monoxide, and carbon dioxide (Liu et al. 2022). The organic material present in biomass is decomposed around 350–550°C, and it can proceed until the 700–800°C in an inert atmosphere. Biomass contains a mixture of cellulose, hemicelluloses, lignin, pectin, and others. Moisture plays a significant role in the pyrolysis process. During pyrolysis, the larger molecules disrupt smaller molecules, released from the process stream as gases, oils, tars, and solid chars. The yield of gaseous products is irregular; the yield of dry ash-free basis is about ~25% by weight of the refuse, and the yield of char is about ~15–25% by weight of the refuse (Zaman et al. 2017). The final pyrolysis products depend on the temperature, time, heating rate, pressure, precursors, and reactor design. Several risks are associated with pyrolysis; e.g., if the moisture content is high, the primary product becomes liquid, whereas if the water level is low, the surplus dust production is more than oil production. Based on these, there are many types of products produced. If the heating rate is high and the temperature exceeds 800°C, then a more significant fraction of ash and gaseous products are produced. At slow heating rate and temperature less than 450°C mainly biochar is produced, whereas if the temperature is between 250°C and 300°C, the volatile materials are produced. Pyrolysis of a biomass fuel mainly has three objectives: (i) the production of clean-burning fuel without generating char, (ii) high calorific fuel production compared to the initial feed, and (iii) production of a more reactive fuel.

TABLE 12.3
Common Examples of MAPs Having Anticancer Properties

Botanical Name	Common Name	Types of Plant	Chemical Constituent	Effects	Chemical Structure
Tinospora cardifolia	Giloya	Shrub	Cordioside, columbin	Cytotoxic against HeLa cells	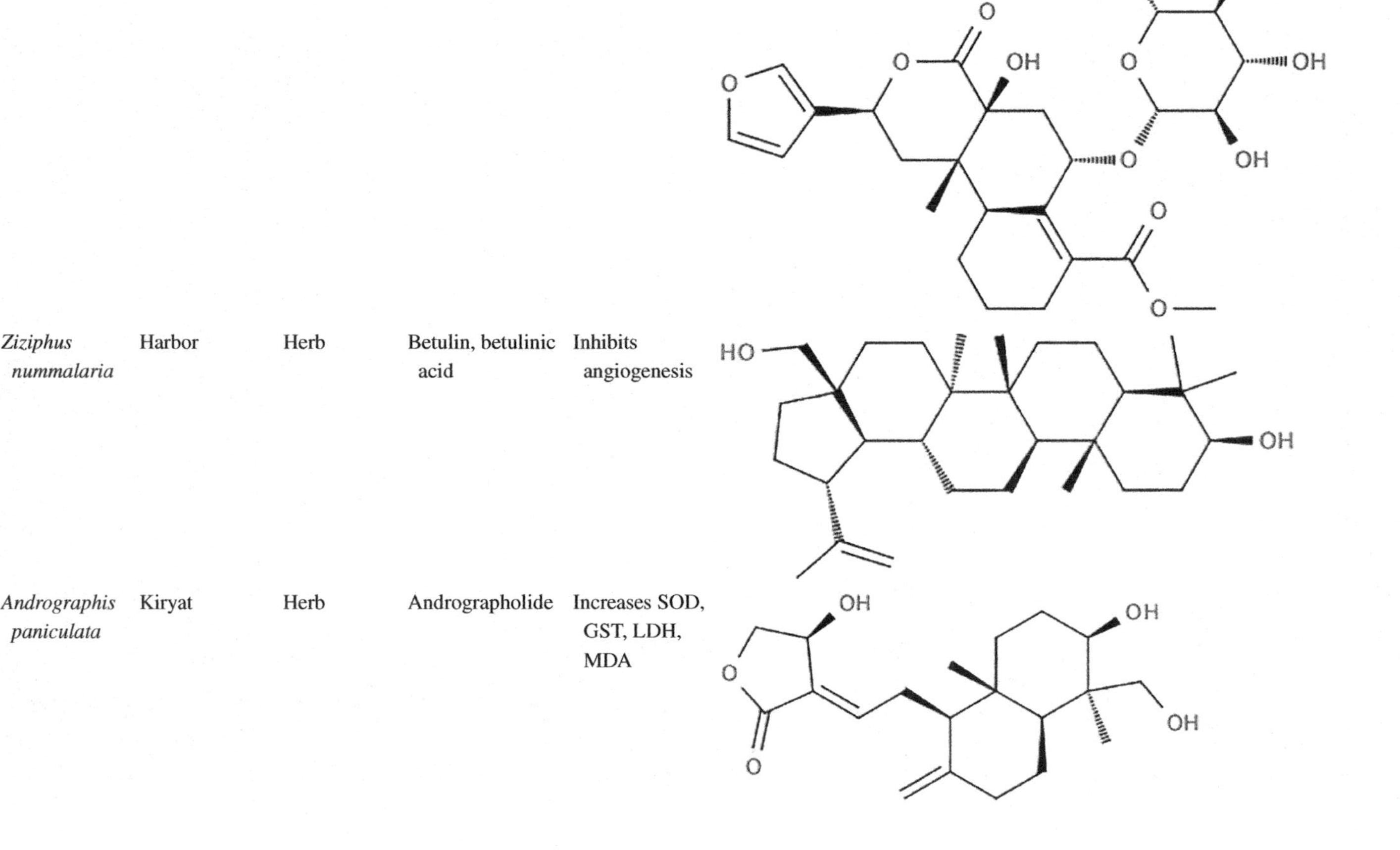
Ziziphus nummalaria	Harbor	Herb	Betulin, betulinic acid	Inhibits angiogenesis	
Andrographis paniculata	Kiryat	Herb	Andrographolide	Increases SOD, GST, LDH, MDA	

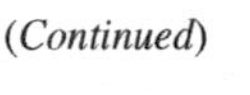

(Continued)

TABLE 12.3 (*Continued*)
Common Examples of MAPs Having Anticancer Properties

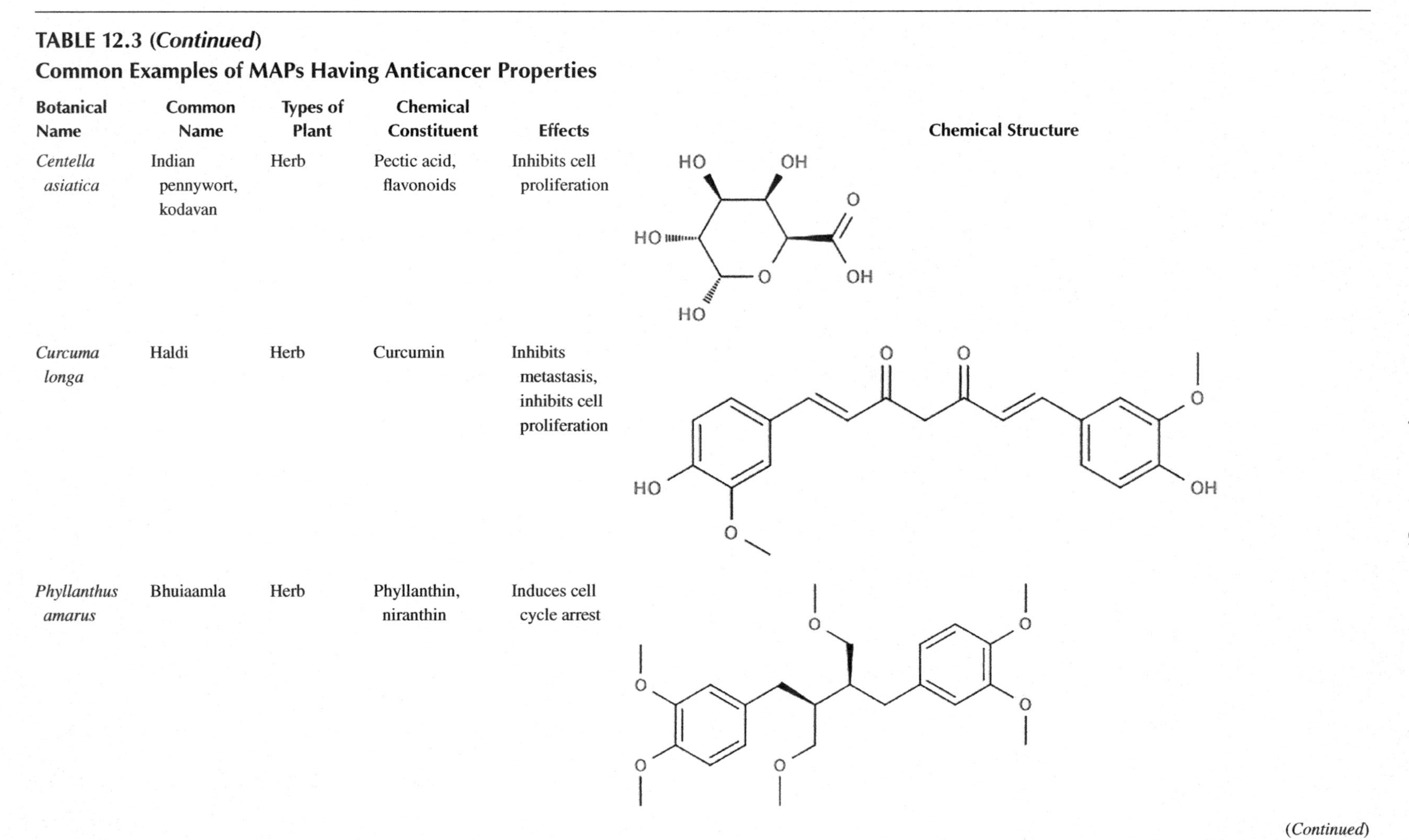

Botanical Name	Common Name	Types of Plant	Chemical Constituent	Effects	Chemical Structure
Centella asiatica	Indian pennywort, kodavan	Herb	Pectic acid, flavonoids	Inhibits cell proliferation	
Curcuma longa	Haldi	Herb	Curcumin	Inhibits metastasis, inhibits cell proliferation	
Phyllanthus amarus	Bhuiaamla	Herb	Phyllanthin, niranthin	Induces cell cycle arrest	

(*Continued*)

TABLE 12.3 (*Continued*)
Common Examples of MAPs Having Anticancer Properties

Botanical Name	Common Name	Types of Plant	Chemical Constituent	Effects	Chemical Structure
Annona atemoya	Sarifa	Tropical tree or shrub	Bullatacin	Induces cell death	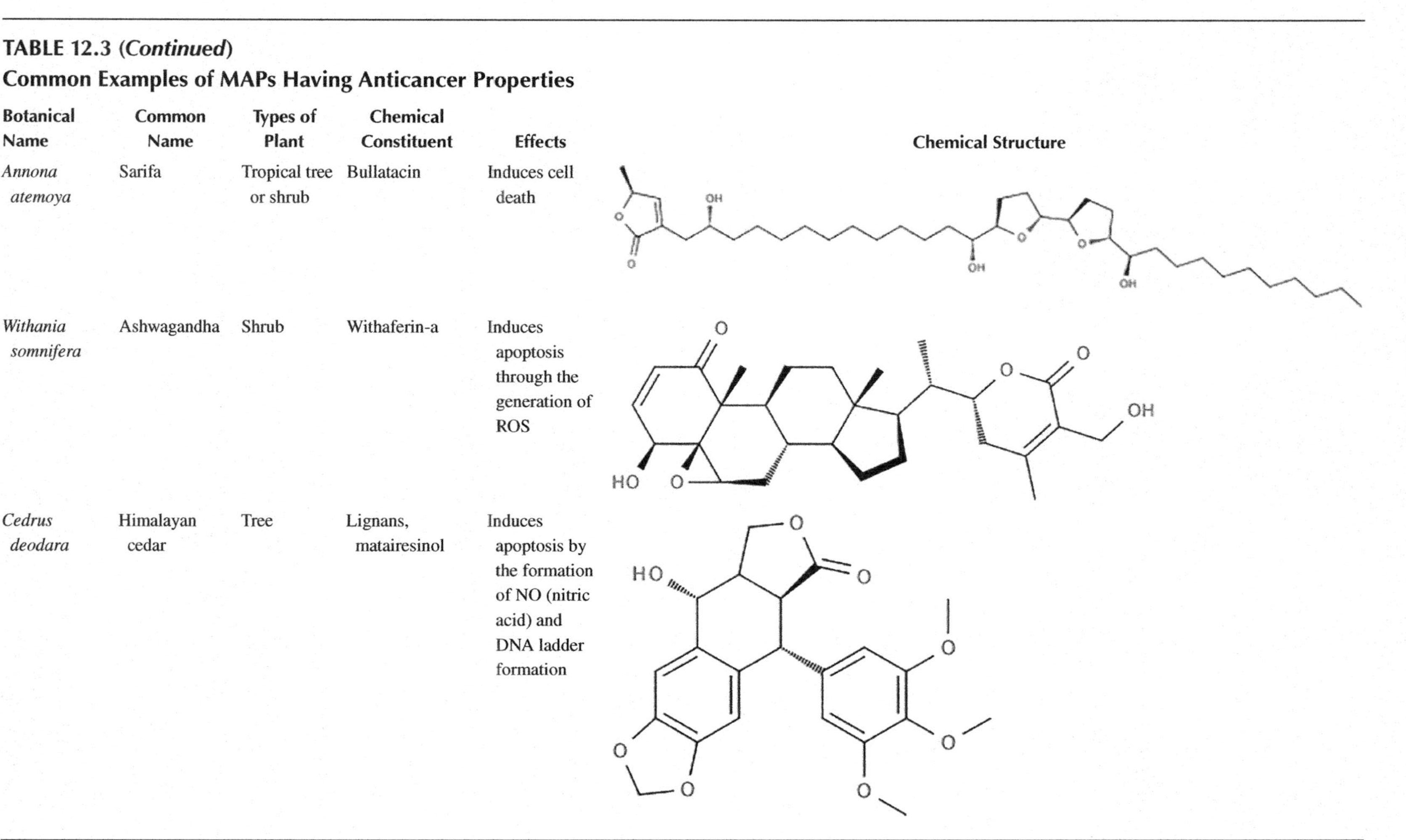
Withania somnifera	Ashwagandha	Shrub	Withaferin-a	Induces apoptosis through the generation of ROS	
Cedrus deodara	Himalayan cedar	Tree	Lignans, matairesinol	Induces apoptosis by the formation of NO (nitric acid) and DNA ladder formation	

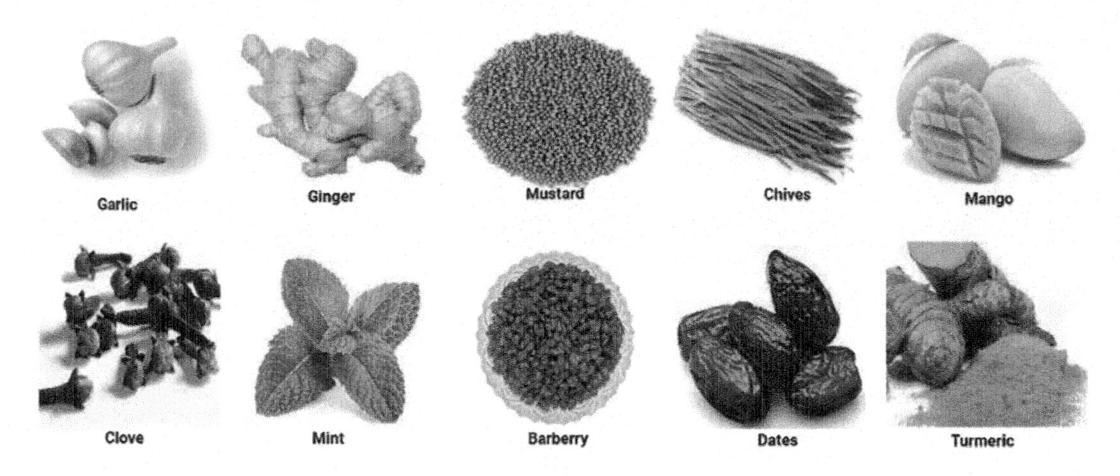

FIGURE 12.5 Common MAPs used in traditional medicine.

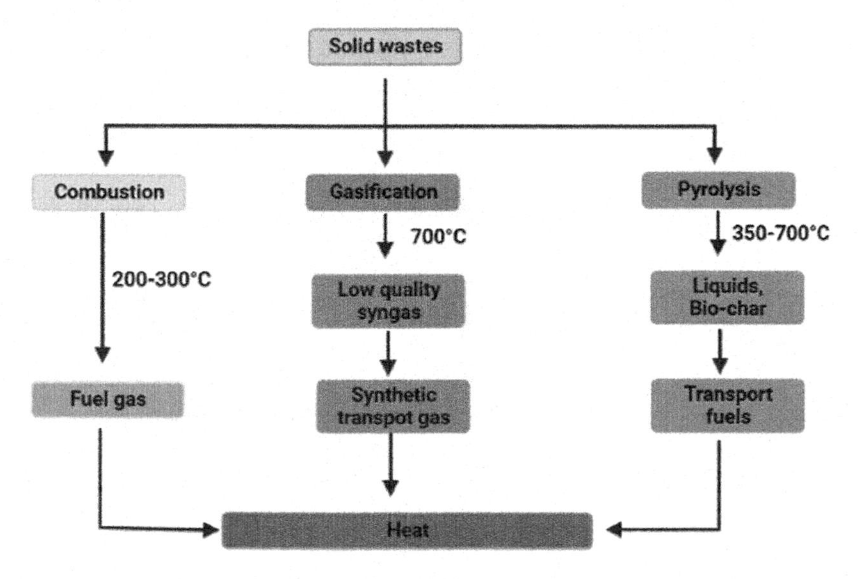

FIGURE 12.6 An overview of thermo-chemical routes for biomass conversion to fuel.

12.8 TYPES OF PYROLYSIS

Depending upon the heat rate, pyrolysis can be categorized into fast pyrolysis and slow pyrolysis. However, on the basis of medium, they are classified as hydrous pyrolysis and hydro-pyrolysis. The hydrous pyrolysis is water-dependent process, whereas the hydro-pyrolysis is hydrogen-dependent process. Usually, slow and fast pyrolysis is carried out in an inert atmosphere.

12.8.1 Slow Pyrolysis

In slow pyrolysis, the time taken by biomass substrate to reach pyrolysis temperature is longer than the final retention time of the substrate. In slow pyrolysis, fine-quality charcoal is produced using low temperature and heating rate. The quality of bio-oil produced during this process is meager, whereas the retention time can be around 5–30 minutes. Slow pyrolysis enhances the expenditure by high input of energy as slow pyrolysis suffers from a low heat transfer value with a longer retention time (Xia et al. 2019).

12.8.2 Fast Pyrolysis

In fast pyrolysis, the initial time taken for biomass substrate heating is smaller than the final retention time of the substrate and, biomass residues are heated in an inert atmosphere at a very high temperature with higher heating rate. On the basis of weight of the biomass, there are several products formed during fast pyrolysis, such as 60–75% liquid biofuels, 10–20% gaseous phase, and about 15–25% biochar residues produced (Perkins, Bhaskar, and Konarova 2018). At a temperature of ~500°C, a maximum yield of ~50% liquid and ~30% gas is expected. Based on the reactor height and gas flow rates, the holding time of carbonization is calculated but usually made of several seconds. It can provide liquid biofuels for turbines, boilers, and engines, and can also supply power for industrial applications.

12.8.3 Flash Pyrolysis

The flash pyrolysis process includes devolatilization using higher heating rates and high pyrolysis temperature (450°C and 1,000°C) under an inert atmosphere. Flash pyrolysis can increase bio-oil production up to 75%, and the gaseous residue time is less than 1s (Xin et al. 2019). This process of biomass can give solid, gaseous, and liquid products.

12.9 PRODUCTS OF PYROLYSIS

12.9.1 Bio-char

A solid carbonaceous material generally obtained from degrading medicinal, aromatic plants, lignin, and hemicelluloses polymer during pyrolysis with high temperature and slow heating rate is known as biochar. However, the higher the temperature, (~650°C) lower the biochar yield, and the lower the temperature, (400°C–500°C) higher will be the yield of biochar (Ahmed et al. 2020), while the yield of biochar decreases at 550–650°C (Dawei et al. 2006). Biochar contains moisture along with fixed carbon, hydrogen, and other constituents, and an aromatic portion of biochar contains H, N, P, O, and S. The percentage of carbon, hydrogen, and other constituents depends upon the types of biomass and pyrolysis processes, respectively (Saxena, Rawat, and Kumar 2017). Biochar can be used as solid fuel and an activator such as carbon, carbon nanotubes, and gaseous fractions.

12.9.2 Syngas

Usually, secondary reactions such as dehydrogenation, deoxygenation, decarboxylation, decarbonylation, and cracking produce syngas. At higher temperatures, around 800°C, tar decomposition is initiated, and the gaseous products are formed, decreasing the oil yield (Kan, Strezov, and Evans 2016). The gaseous product obtained from dry biomass is higher at an early stage of pyrolysis, while at a later stage, the maximum quantity production occurs with wet biomass (Kan, Strezov, and Evans 2016). Generally, the gas obtained after the pyrolysis process at high temperature leads to an endothermic reaction. The gaseous product comprises H_2O, H_2, and CO, due to the presence of oxygen in the biomass CO_2. In Addition, the cracking of hydrocarbon produces hydrogen at high temperature (Zaman et al. 2017). The production of gaseous products at different temperature is given in Table 12.4.

TABLE 12.4
Production of Gaseous Products at Different Temperature

	Temperature (°C)				
% of Products	**200°C**	**300°C**	**400°C**	**500°C**	**600°C**
C	32	28	27	27	25.2
CO	20	0	1.2	1.2	4.5
CO_2	23.9	28	35	35	36
H_2O	36.5	32.5	27	27	22.5

12.9.3 Bio-oil

After pyrolysis, the oil obtained contains a mixture of about 350–450 compounds. These compounds could be viscous due to physical and chemical changes with substituent loss of volatile matters in them. Previously, it was found that the produced oil contains high metal/ash content and water, which will make phase separation difficult as well as lower the heating rate. Thus, the commercial application monitored carefully, so the quality, quantity, and constancy of pyrolysis oil can be improved by method variables such as, heating rate, retention time, and temperature. If the oil contains low molecular weight compounds, it is likely to be used commercially.

12.10 USE OF MAPs

MAPs have a high content of active substances with particular chemical and biochemical properties, used for therapeutic, aromatic, and gastronomic uses. Their use in food, cosmetics, medicines, domestics, and industries products is practically unlimited. There are several products derived from MAPs for this purpose: essential oils, primarily present in leaves, such as rosemary and patchouli aromatic plants. Rosemary is an evergreen herb that belongs to *the Lamiaceae* family, which is used as body perfume, antioxidant, anti-inflammatory, digestive improvement, memory enhancement, and concentration. Spices like cinnamon and nutmeg have antiseptic and flavoring properties. Moreover, cinnamon is an essential age-old spice, and its consumptive health benefits include antimicrobial actions, antioxidant activity, inhibition of cancerous cell escalation (Saha and Basak 2020), and glucose control in diabetes. Bitter substances are used in spirits industries, for example, quina and cuassai bitter plants and chemical substances such as carotenes and anthocyanins are used in food and cosmetics. Besides these MAPs, plants are also used in industrial sectors, for example, powdered plants, essential oils, and extract for pharmacological and industrial raw material purposes.

REFERENCES

Abreu, Rui M V, Isabel C F R Ferreira, Ricardo C Calhelha, Raquel T Lima, M Helena Vasconcelos, Filomena Adega, Raquel Chaves, and Maria-João R P Queiroz. 2011. "Anti-Hepatocellular Carcinoma Activity Using Human HepG2 Cells and Hepatotoxicity of 6-Substituted Methyl 3-Aminothieno [3, 2-b] Pyridine-2-Carboxylate Derivatives: In Vitro Evaluation, Cell Cycle Analysis and QSAR Studies." *European Journal of Medicinal Chemistry* 46(12): 5800–5806.

Adel Mahmoodabad, Hamid, Saeed Hokmalipoor, Morad Shaban, and Reza Ashrafi Parchin. 2014. "Effect of Foliar Spray of Urea and Soil Application of Vermicompost on Essential Oil and Chlorophyll Content of Green Mint (Mentha Spicata L.)." *International Journal of Advanced Biological and Biomedical Research* 2(6): 2104–8.

Ahmed, Ashfaq, Muhammad S Abu Bakar, Rahayu S Sukri, Murid Hussain, Abid Farooq, Surendar Moogi, and Young-Kwon Park. 2020. "Sawdust Pyrolysis from the Furniture Industry in an Auger Pyrolysis Reactor System for Biochar and Bio-Oil Production." *Energy Conversion and Management* 226: 113502.

Amanuel, Wondimagegn, Musse Tesfaye, Adefires Worku, Gezahegne Seyoum, and Zenebe Mekonnen. 2019. "The Role of Dry Land Forests for Climate Change Adaptation: The Case of Liben Woreda, Southern Oromia, Ethiopia." *Journal of Ecology and Environment* 43(1): 1–13.

Ayoub, Zeenat, Archana Mehta, Siddhartha Kumar Mishra, and Laxmi Ahirwal. 2018. "Medicinal Plants as Natural Antioxidants: A Review." *Journal of the Botanical Society, University of Saugor* 1–15: 100–7.

Cadar, Roxana-Larisa, Antonio Amuza, Diana Elena Dumitras, Mihaela Mihai, and Cristina Bianca Pocol. 2021. "Analysing Clusters of Consumers Who Use Medicinal and Aromatic Plant Products." *Sustainability* 13 (15): 8648.

Celano, Rita, Anna Lisa Piccinelli, Imma Pagano, Graziana Roscigno, Luca Campone, Enrica De Falco, Mariateresa Russo, and Luca Rastrelli. 2017. "Oil Distillation Wastewaters from Aromatic Herbs as New Natural Source of Antioxidant Compounds." *Food Research International* 99: 298–307.

Corrêa, J L G, M C Rasia, A Mulet, and J A Cárcel. 2017. "Influence of Ultrasound Application on Both the Osmotic Pretreatment and Subsequent Convective Drying of Pineapple (Ananas Comosus)." *Innovative Food Science & Emerging Technologies* 41: 284–91.

Das, Nibedita, Sanowar Hossain, Jaytirmoy Barmon, Shahnaj Parvin, Mahadi Hasan, Masuma Akter, and Ekramul Islam. 2021. "Evaluation of Leea Rubra Leaf Extract for Oxidative Damage Protection and Antitumor and Antimicrobial Potential." *Journal of Tropical Medicine* 2021: 7239291.

Dawei, An, Wang Zhimin, Zhang Shuting, and Yang Hongxing. 2006. "Low-temperature Pyrolysis of Municipal Solid Waste: Influence of Pyrolysis Temperature on the Characteristics of Solid Fuel." *International Journal of Energy Research* 30(5): 349–57.

Demirbas, Ayhan. 2005. "Pyrolysis of Ground Beech Wood in Irregular Heating Rate Conditions." *Journal of Analytical and Applied Pyrolysis* 73(1): 39–43.

Enioutina, Elena Y, Kathleen M Job, Lubov V Krepkova, Michael D Reed, and Catherine M Sherwin. 2020. "How Can We Improve the Safe Use of Herbal Medicine and Other Natural Products? A Clinical Pharmacologist Mission." *Expert Review of Clinical Pharmacology* 13(9): 935–44.

Guerriero, Gea, Roberto Berni, J Armando Muñoz-Sanchez, Fabio Apone, Eslam M Abdel-Salam, Ahmad A Qahtan, Abdulrahman A Alatar, Claudio Cantini, Giampiero Cai, and Jean-Francois Hausman. 2018. "Production of Plant Secondary Metabolites: Examples, Tips and Suggestions for Biotechnologists." *Genes* 9(6): 309.

Jamshidi-Kia, Fatemeh, Zahra Lorigooini, and Hossein Amini-Khoei. 2018. "Medicinal Plants: Past History and Future Perspective." *Journal of HerbMed Pharmacology* 7(1): 1–7. https://doi.org/10.15171/jhp.2018.01.

Kan, Tao, Vladimir Strezov, and Tim J Evans. 2016. "Lignocellulosic Biomass Pyrolysis: A Review of Product Properties and Effects of Pyrolysis Parameters." *Renewable and Sustainable Energy Reviews* 57: 1126–40.

Karam, Marie Céleste, Jeremy Petit, David Zimmer, Elie Baudelaire Djantou, and Joël Scher. 2016. "Effects of Drying and Grinding in Production of Fruit and Vegetable Powders: A Review." *Journal of Food Engineering* 188: 32–49.

Karaś, Kaja, Anna Sałkowska, Jarosław Dastych, Rafał A Bachorz, and Marcin Ratajewski. 2020. "Cardiac Glycosides with Target at Direct and Indirect Interactions with Nuclear Receptors." *Biomedicine & Pharmacotherapy* 127: 110106.

Khan, M Kalim A, Salman Akhtar, and Jamal M Arif. 2018. "Development of in Silico Protocols to Predict Structural Insights into the Metabolic Activation Pathways of Xenobiotics." *Interdisciplinary Sciences: Computational Life Sciences* 10(2): 329–45.

Komakech, Richard, Youngmin Kang, Jun-Hwan Lee, and Francis Omujal. 2017. "A Review of the Potential of Phytochemicals from Prunus Africana (Hook f.) Kalkman Stem Bark for Chemoprevention and Chemotherapy of Prostate Cancer." *Evidence-Based Complementary and Alternative Medicine* 2017: 3014019.

Lehto, Jani, Anja Oasmaa, Yrjö Solantausta, Matti Kytö, and David Chiaramonti. 2014. "Review of Fuel Oil Quality and Combustion of Fast Pyrolysis Bio-Oils from Lignocellulosic Biomass." *Applied Energy* 116: 178–90.

Li, Miao, Ping Li, Mei Zhang, and Feng Ma. 2018. "Brucine Suppresses Breast Cancer Metastasis via Inhibiting Epithelial Mesenchymal Transition and Matrix Metalloproteinases Expressions." *Chinese Journal of Integrative Medicine* 24(1): 40–46.

Liu, Zhongzhe, Matthew Hughes, Yiran Tong, Jizhi Zhou, William Kreutter, Hugo Cortes Lopez, Simcha Singer, Daniel Zitomer, and Patrick McNamara. 2022. "Paper Mill Sludge Biochar to Enhance Energy Recovery from Pyrolysis: A Comprehensive Evaluation and Comparison." *Energy* 239: 121925.

Liu, Zhongzhe, Joseph M Norbeck, Arun S K Raju, Suhyun Kim, and Chan S Park. 2016. "Synthetic Natural Gas Production by Sorption Enhanced Steam Hydrogasification Based Processes for Improving CH4 Yield and Mitigating CO2 Emissions." *Energy Conversion and Management* 126: 256–65.

Lockowandt, Lara, José Pinela, Custódio Lobo Roriz, Carla Pereira, Rui M V Abreu, Ricardo C Calhelha, Maria José Alves, Lillian Barros, Michael Bredol, and Isabel C F R Ferreira. 2019. "Chemical Features and Bioactivities of Cornflower (Centaurea Cyanus L.) Capitula: The Blue Flowers and the Unexplored Non-Edible Part." *Industrial Crops and Products* 128: 496–503.

McNamara, Patrick, Daniel Zitomer, and Zhongzhe Liu. 2019. "Comment on "Pyrolysis of Dried Wastewater Biosolids Can Be Energy Positive"." *Water Environment Research: A Research Publication of the Water Environment Federation* 91(8): 813–15.

Nallasamy, Prakashkumar, Thenmozhi Ramalingam, Thajuddin Nooruddin, Rajasree Shanmuganathan, Pugazhendhi Arivalagan, and Suganthy Natarajan. 2020. "Polyherbal Drug Loaded Starch Nanoparticles as Promising Drug Delivery System: Antimicrobial, Antibiofilm and Neuroprotective Studies." *Process Biochemistry* 92: 355–64.

Perkins, Greg, Thallada Bhaskar, and Muxina Konarova. 2018. "Process Development Status of Fast Pyrolysis Technologies for the Manufacture of Renewable Transport Fuels from Biomass." *Renewable and Sustainable Energy Reviews* 90: 292–315.

Porwal, Omji, Sachin Kumar Singh, Dinesh Kumar Patel, Saurabh Gupta, Rahul Tripathi, and Shankar Katekhaye. 2020. "Cultivation, Collection and Processing Of Medicinal Plants." *Bioactive Phytochemicals: Drug Discovery to Product Development*, 14–30.

Rollag, Sean A, Jake K Lindstrom, Chad A Peterson, and Robert C Brown. 2022. "The Role of Catalytic Iron in Enhancing Volumetric Sugar Productivity during Autothermal Pyrolysis of Woody Biomass." *Chemical Engineering Journal* 427: 131882.

Saha, Ajoy, and B B Basak. 2020. "Scope of Value Addition and Utilization of Residual Biomass from Medicinal and Aromatic Plants." *Industrial Crops and Products* 145: 111979.

Salmerón-Manzano, Esther, Jose Antonio Garrido-Cardenas, and Francisco Manzano-Agugliaro. 2020. "Worldwide Research Trends on Medicinal Plants." *International Journal of Environmental Research and Public Health* 17(10). https://doi.org/10.3390/ijerph17103376.

Samanta, Ashis Kumar, Nasser Awwad, and Hamed Majdooa Algarni. 2020. *Chemistry and Technology of Natural and Synthetic Dyes and Pigments*. BoD–Books on Demand.

Samarth, Ravindra M, Meenakshi Samarth, and Yoshihisa Matsumoto. 2017. "Medicinally Important Aromatic Plants with Radioprotective Activity." *Future Science OA* 3(4): FSO247.

Saxena, Jyoti, Jyoti Rawat, and Raj Kumar. 2017. "Conversion of Biomass Waste into Biochar and the Effect on Mung Bean Crop Production." *CLEAN–Soil, Air, Water* 45(7): 1501020.

Sayyad, S S, B D Patil, and V S Dhumal. 2022. "Evaluation of Physico-Chemical Properties of Lassi Prepared with Optimized Level of Tulsi or Basil (Ocimum Sanctum L.)." *The Pharma Innovation Journal* 11(1): 285–88.

Sekar, V, S Chakraborty, S Mani, V K Sali, and H R Vasanthi. 2019. "Mangiferin from Mangifera Indica Fruits Reduces Post-Prandial Glucose Level by Inhibiting α-Glucosidase and α-Amylase Activity." *South African Journal of Botany* 120: 129–34.

Sousa MPDSE, Aum YKPG, Cavalcante LDA, Sales MLF, and Sousa JOPDSE. 2019. "CONSTRUÇÃO DE UM EXTRATOR SOLAR DE ÓLEOS ESSENCIAIS DE PLANTAS AROMÁTICAS E MEDICINAIS DA AMAZÔNIA." *Blucher Chemical Engineering Proceedings* Volume 1 (XIII Congresso Brasileiro de Engenharia Química em Iniciação Científica): 147–53. https://doi.org/10.1016/cobecic2019-EAT121.

Tariq, Sana, Parveen Akhter, Sidra Qayyum, and Fatima Khawar. 2021. "The Effect of Garlic Consumption with Prescribed Anti-Platelet Medicines on Platelet Count of Cardiovascular Patients." *BioSight* 2(1): 31–38.

Urquiza-Haas, Nayeli, and Emilie Cloatre. 2019. "Traditional Herbal Medicine and the Challenges of Pharmacovigilance." In Anita Lavorgna and Anna Di Ronco (eds.), *Medical Misinformation and Social Harm in Non-Science-Based Health Practices*, 133–46. Routledge.

Witkin, Jeffrey M, and Xia Li. 2013. "Curcumin, an Active Constituent of the Ancient Medicinal Herb Curcuma Longa L.: Some Uses and the Establishment and Biological Basis of Medical Efficacy." *CNS & Neurological Disorders-Drug Targets (Formerly Current Drug Targets-CNS & Neurological Disorders)* 12(4): 487–97.

Xia, Juntao, Haohao Sun, Xu-Xiang Zhang, Tong Zhang, Hongqiang Ren, and Lin Ye. 2019. "Aromatic Compounds Lead to Increased Abundance of Antibiotic Resistance Genes in Wastewater Treatment Bioreactors." *Water Research* 166: 115073.

Xin, Xing, Kirk M Torr, Ferran de Miguel Mercader, and Shusheng Pang. 2019. "Insights into Preventing Fluidized Bed Material Agglomeration in Fast Pyrolysis of Acid-Leached Pine Wood." *Energy & Fuels* 33(5): 4254–63.

Zaman, Chowdhury Zaira, Kaushik Pal, Wageeh A Yehye, Suresh Sagadevan, Syed Tawab Shah, Ganiyu Abimbola Adebisi, Emy Marliana, Rahman Faijur Rafique, and Rafie Bin Johan. 2017. *Pyrolysis: A Sustainable Way to Generate Energy from Waste*. Vol. 1. IntechOpen Rijeka.

13 Methods in Ethnopharmacology

Phytochemical Extraction, Isolation, and Detection Techniques

Adele Papetti, Raffaella Colombo, Daniela Vallelonga, Ilaria Frosi, and Chiara Milanese
University of Pavia

CONTENTS

13.1 INTRODUCTION: BACKGROUND AND DRIVING FORCES

Today, the chemical investigation of plants' constituents is an important key point in food and drug discovery. The analysis of a plant extract is a multistep process starting with the development of a suitable extraction method to obtain a high yield in bioactive compounds. The experimental design approach is widely used nowadays to optimize the extraction conditions, especially when many variables (i.e., temperature, time, pH, solid/liquid ratio, solvent extraction composition, and other

DOI: 10.1201/b22842-13

parameters specific for each technique) could affect the quality of the extract. After the optimization of this step, the extract is generally analyzed using both spectroscopic and chromatographic techniques. In fact, using techniques like UV–Visible spectrophotometry (UV-Vis), Fourier transform infrared (FT-IR), nuclear magnetic resonance (NMR), and mass spectroscopies, it is possible to have useful information about the extract constituents and their potential interaction, while the separation techniques led to isolate and chemically identify the chemical nature and the structure of each compound present in the extract.

In this chapter, the most recent and innovative extraction and analysis methods are compared with the conventional ones, highlighting the main advantages and disadvantages for each approach/technique.

13.2 EXTRACTION: A KEY STEP IN PHYTOCHEMICAL ANALYSIS

The extraction of bioactive compounds from plant material is the starting point in the development of analytical methods in phytochemistry. Sometimes, extraction methods are referred as "sample preparation technique" because they play a significant role on yield and bioactivity of the extracts (Castro-López et al. 2017; Azmir et al. 2013). Considering the great variability in phytochemicals and plant species, the extraction procedure should be optimized focusing on the specific plant matrix, targeted bioactives, and extraction process parameters. A recent common tool to improve the effectiveness of the extraction methods is the use of mathematical models to study the interaction between the parameters affecting the process and the extraction yield (Das and Dewanjee 2018). The ideal extraction method should be simple, reproducible, inexpensive, safe, and suitable for industrial applications (Blicharski and Oniszczuk 2017). Traditional techniques have been used for many decades, and one of the most frequently applied techniques was Solid-Liquid Extraction (SLE), which is based on the extraction power of different solvents and needs the application of heat and/or mixing (Castro-López et al. 2017; Azmir et al. 2013). SLE includes conventional approaches such as Soxhlet Extraction (SE), hot water bath extraction, percolation, and maceration (ME). However, they are quite expensive, due to the excessive consumption of time, energy, and high volume of toxic solvents. For these reasons, today they are mostly used as reference approaches to compare innovative methodologies (Azmir et al. 2013; Ameer, Shahbaz, and Kwon 2017).

Recently, there has been an increasing interest in the development of eco-friendly processes to obtain high-value extracts extremely rich in bioactives. They involve the use of GRAS (Generally Recognized As Safe) solvents, thus guaranteeing the absence of harmful substances in the obtained final products, which are characterized by a better yield and quality in a reduced time. In this section, some of the most promising techniques, such as Microwave-Assisted Extraction (MAE), Ultrasound-Assisted Extraction (UAE), Supercritical Fluid Extraction (SFE), Enzyme-Assisted Extraction (EAE), and Pressurized Liquid Extraction (PLE), are presented and discussed (Figure 13.1).

13.2.1 Microwave-Assisted Extraction (MAE)

MAE is a promising technique to recover phytochemicals, such as phenolic acids, flavonoids, anthocyanins, stilbenes, terpenoids, and essential oils from plant matrices. It is rapid and requires low amount of solvent to obtain high extraction yields with good reproducibility compared to the conventional methods (Ameer, Shahbaz, and Kwon 2017; Alara et al. 2019; Elez Garofulić et al. 2020).

Microwaves are composed of electric and magnetic fields that perpendicularly oscillate and impact on the plant material. Heat is generated by ionic conduction and dipole rotation mechanisms. Ionic conduction induces heat as the medium resists to the ion flow, which maintains the direction along the applied field, frequently changing direction. This results in collision between molecules and generates heat (Flórez, Conde, and Domínguez 2015). This mechanism induces alterations at cellular level with the formation of pores and fractures in the plant cell walls, which promote the release of phytochemicals, as demonstrated by scanning electron microscopy analysis performed

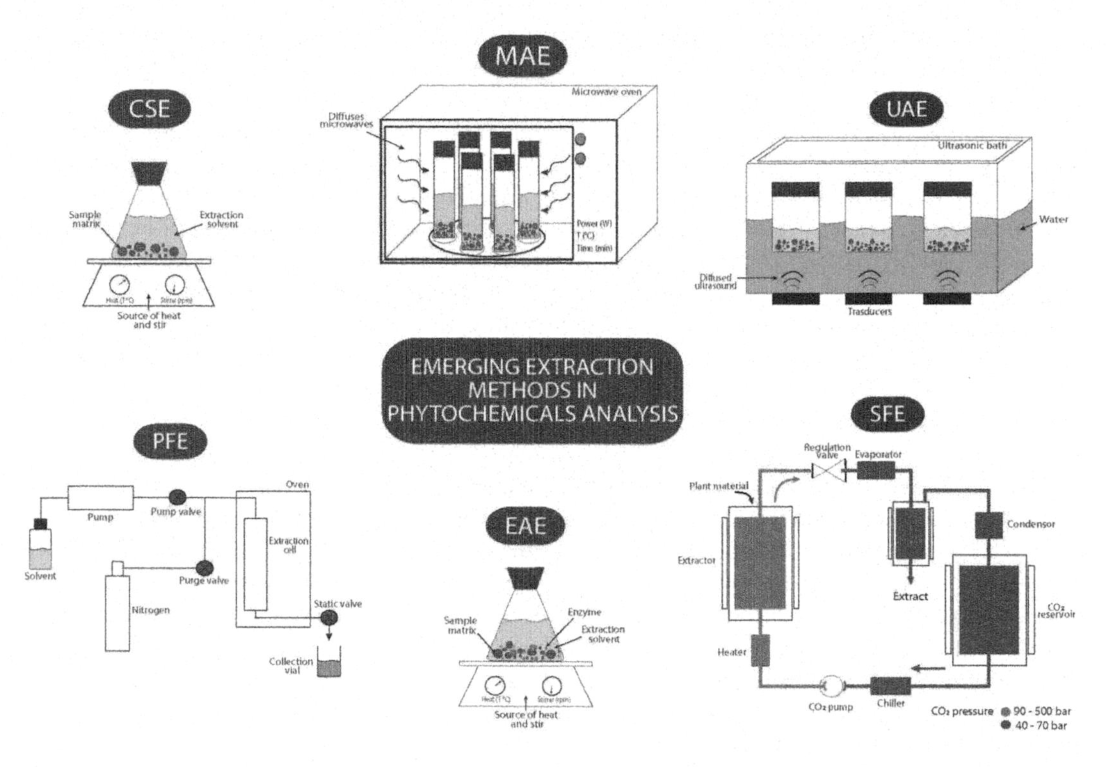

FIGURE 13.1 Schematic representation of the emerging extraction techniques equipment used in phytochemical analysis.

after microwave irradiation on Black Jamun pulp (*Syzygium cumini*) (Sharma and Dash 2021) and Bitter leaf (*Vernonia amygdalina*) (Alara et al. 2019).

The extraction mechanism involves different stages: solutes desorption from the active sites of plant matrix under high temperature and pressure conditions, solvent diffusion through the matrix, and solutes release from the matrix into the solvent (Azmir et al. 2013). The efficiency of the process is affected by different factors: microwave power, extraction time and temperature, volume and nature of the solvent, and matrix characteristics. The solvent choice is the key step to obtain an efficient extraction process. The capacity of the solvent to absorb the microwave energy is directly linked to its dielectric properties, measured through the dielectric constant and dissipation factor. High values of these parameters result in suitable solvents for MAE, such as water, ethanol, acetone, methanol, and acetonitrile. Ethanol/water mixtures are preferred as GRAS and are extremely efficient to extract bioactive compounds from plants, as demonstrated by Lovrić et al. (2017) who investigated the effect of the solvent type (ethanol and methanol), solvent volume in the mixture (50% and 70%), time (5, 15, and 25 minutes), and temperature (40, 50, and 60°C) on the extraction process of flavonoids, phenolic acids, and hydroxycinnamic acids from wild blackthorn dried flowers, using a full-factorial randomized design. Ethanol aqueous solution was more efficient than methanol to extract flavonoids, hydroxycinnamic acids, and flavonols, showing the effectiveness of the green solvent. Moreover, 70% alcohol proved to be more efficient than 50%. This is probably because a higher amount of water increases the polarity of the solvent until a certain level which is no longer suitable for the extraction. In agreement with these findings, another recent study about the recovery of phytochemicals from Roselle calyces indicated that ethanol-water mixtures have good capacity to absorb microwaves and are characterized by a faster heat up. Higher ethanol percentage leads to a decrease in dielectric constant and in dissipation factor, resulting in lower extraction yield (Pimentel-Moral et al. 2018). Therefore, an optimization of the solvent mixture in relation to the plant matrix is needed, as well as the solvent-to-material ratio. In fact, the solvent should completely soak the plant material during the microwave irradiation, and a too high solvent-to-material ratio could fail in generating an efficient mass transport, due to an uneven exposure to microwaves (Xu et al. 2017).

Temperature, microwave power, and time are other key factors in MAE procedure. Generally, an increase in temperature improves the solvent power, allowing a decrease in viscosity and surface tension and a better solvent penetration in the plant matrix. However, higher temperatures could lead to the degradation of thermolabile compounds, resulting in poor recovery. Pimientel-Moral et al. (2018) studied the impact of temperature in the range of 36–164°C, in addition to the extraction solvent composition (15–75% ethanol), and time (5–20 minutes), on the extraction yield of organic acids, phenolic compounds, flavonoids, and anthocyanins from *Hibiscus sabdariffa* calyces, using a response surface methodology. The highest yield was obtained at about 100°C, even if the highest amount of methylgallocatechin, myricetin, quercetin, and kaempferol was achieved at 164°C. On the contrary, other glycosylated flavonoids were not detected at this temperature, as thermolabile.

Time can be also set up to avoid thermal degradation. In comparison with conventional approaches, the exposure of the plant matrix to high temperature and microwave irradiation accelerates the bioactives' solubility and the extraction equilibrium can be reached in a shorter time. Therefore, the risk of thermal degradation can be limited by applying a proper extraction time. In fact, the use of MAE allows to obtain comparable or higher yield of phytochemicals in seconds or few minutes instead of longer time (from 30 minutes to hours) as for conventional approaches (Elez Garofulić et al. 2020; Alara, Abdurahman, and Ukaegbu 2018; Sharma et al. 2020). In the study of Elez Garofulić et al. (2020), a central composite design (CCD) was used to study the effect of time, temperature, and microwave power on the extraction of polyphenols from *Pistacia lentiscus* L. leaves and fruits. Time had a significant effect on total phenolic content both in leaves and in fruits, with the highest recovery obtained after 12 minutes, while longer time such as 15 minutes decreased it. By applying the optimized experimental extraction conditions, the recovery was almost like those obtained after 30 minutes using a heat reflux extraction. A similar trend was also observed by Alara et al. (2019), who investigated the extraction of polyphenols from *Vernonia amygdalina* leaves with a single factor experiments and reported an increase in total phenols content from 2 to 10 minutes, followed by a reduction for prolonged time, due to the degradation of polyphenols; also in this case, a higher yield than that obtained using Soxhlet for 4 hours confirmed the advantage of MAE in terms of extraction time and therefore energy consumption.

Microwave power is directly linked to temperature, as its increase generates a temperature increase in plant material. Considering this correlation, some authors studied only one of the two parameters. Belwal et al. (2017) used a Box-Behnken design to optimize microwave power, sample-to-solvent ratio, and solvent concentrations to extract polyphenols from *Berberis asiatica* leaves. An increase in the total flavonoids content was registered by increasing microwave power from 100 W to 300 W; for values higher than 300 W, an increase in temperature induced the degradation of thermolabile compounds.

Recently, Solvent-Free Microwave Extraction (SFME) has been developed as a green technique based on MAE principle. In this approach, the microwave energy is directly absorbed by water naturally occurred in plant material, causing the break of cell walls for increased pressure and the subsequent release of bioactives. For dried materials, samples have to be moistened with water or other solvents before extraction, for increasing the microwave energy absorption (Xu et al. 2017). This approach was mainly used in the extraction of essential oil from *Pogostemon cablin* Benth leaves (Kusuma, Altway, and Mahfud 2018), *Tymus vulgaris* L., *Melissa officinalis* L. (Khalili et al. 2018), *Cinnamonum camphora* leaves (Liu et al. 2018), and fruit peel (Liu et al. 2022).

13.2.2 Ultrasound-Assisted Extraction (UAE)

UAE is a modern green technique used to extract secondary metabolites from different plant samples, especially phenolic acids, due to its high efficiency (Alirezalu et al. 2020). Compared to capillary electrophoresis (CE), UAE is faster, allows lower extraction temperature, and reduces the solvent volume, enhancing the extract quality, and therefore, it can be considered efficient, simple, and to be of low cost, which does not require expensive instruments (Alirezalu et al. 2020; Belwal et al. 2018).

The extraction consists in mixing the sample with a solvent before the exposure to ultrasound waves under controlled temperature (Mosić et al. 2020). Ultrasound enhances the extraction mechanism, by creating cavitation bubbles, which reach an unstable point and implode creating localized high pressure and temperature zones (Lefebvre, Destandau, and Lesellier 2021; Panja 2018). For most bioactives, the frequency should be in the range of 20–200kHz to increase cell wall permeability and produce cavitation (Alara, Abdurahman, and Ukaegbu 2021). In particular, the lowest frequency (20–40kHz) generates larger cavitation bubbles in extraction solvents disrupting the plant cell walls and enhancing the tissue permeability (Dzah et al. 2020). This allows both the solvents to penetrate the inner part of the material and the extract to be washed out more easily, thus promoting the release of compounds (Daud et al. 2022).

UAE presents multiple advantages, which leads to a cost reduction: it is a green efficient way to increase mass transfer enhancing extraction yield, requires less solvent, and is compatible with any solvent (Kovačević et al. 2018; Panja 2018). A common equipment includes an ultrasonic bath and an ultrasonic probe system. Probe sonicators are in constant contact with sample, thus raising the risk of sample contamination and foam production, and the use of a bath sonicator makes difficult to control temperature and ultrasound intensity. However, with this system, a lower amount of sample can be used for obtaining better extraction yields and more than one sample can be extracted at the same time, which makes the system a lot simpler and more cost-effective (Mosić et al. 2020).

Compared to MAE and SFE, UAE is not so widely used probably because it is not so easy to scale up the process at industrial level compared to other techniques (Belwal et al. 2018). Furthermore, the exposure to intense ultrasound waves may promote deleterious effects on plant bioactive compounds generating free radicals due to oxidative pyrolysis and cavitation, which result in lower concentration and degradation of some compounds (Kovačević et al. 2018).

UAE efficiency is greatly affected by two main parameters modulating ultrasonic waves: frequency (Hz) and amplitude (MPa). Power (W) is the amplitude over time, and intensity is power over surface area (W/m). These parameters change ultrasonic waves and can differently interact with plant material (Lefebvre, Destandau, and Lesellier 2021). Hz is an important factor affecting the extraction; in fact, as ultrasound Hz increases, the production and intensity of cavitation decrease. At high Hz, the acoustic cavitation is more difficult to be induced since the cavitation bubbles need a delay to start during the rarefaction cycle and compression-rarefaction cycles can be too short to allow the increase of the cavitation bubbles. Rarefaction phase duration is indirectly correlated with ultrasonic Hz (Alara, Abdurahman, and Ukaegbu 2021; Chemat et al. 2017). The effects of UAE parameters on the extraction yield of polyphenols were analyzed by Dzah et al. (2020). Frequency below 40kHz was more effective, as the polyphenol yield increased with the increase of power up to a limit value over which the ultrasound power generated hydroxyl radicals, which could degrade polyphenols. Although the extraction yield at low ultrasound Hz seemed high, this study indicated that the optimal Hz was reactor and system specific, and therefore, an Hz optimization is needed for each extraction setup.

Other parameters that should be considered and optimized to maximize the extraction yield are solvent type and concentration, solid-to-solvent ratio, particle size, extraction time, and temperature (Christou, Stavrou, and Kapnissi-Christodoulou 2021). Typically, a temperature increase reduces solvent density and viscosity, which promote the cavitation bubble formation in the extraction medium, thus enhancing mass transfer rate and extraction yield (Arumugham et al. 2021). However, temperatures exceeding 60°C decrease the extraction yield of polyphenols due to easier hydrolyzation and oxidation reaction occurring (Dzah et al. 2020).

A similar trend could be also registered for too long extraction time: in fact, even if longer time allows plant cells to be better disrupted and, consequently, the release and diffusion of polyphenols are enhanced, when it exceeds the optimum, these compounds might be degraded due to heat generation, which leads to chemical decomposition, especially oxidation (Christou, Stavrou, and Kapnissi-Christodoulou 2021). Ultrasound efficiency was also highlighted by Alara, Abdurahman, and Ukaegbu (2021), who indicated UAE as a less expensive technique due to lower solvent volume used, higher sample volume tested in one run, and lower extraction time needed. In addition, shorter sonication time coupled with lower temperature would enhance polyphenol extraction and preservation. Similar advantages were registered by Bursać Kovačević et al. (2018), who compared CE to UAE to recover phenolic compounds from *Stevia rebaudiana* dried leaves.

Some UAE recent interesting applications are represented by the works of Lin et al. (2021) and of Palsikowski et al. (2020). Lin developed an UAE method for the optimization of the recovery of antioxidant polyphenols from India *Moringa oleifera* L. leaves. The optimal extraction conditions were 188 W, leaves/solvent ratio of 40:1 (mL/g), 52% ethanol, and 30°C for 20 minutes. These conditions led to the highest concentration of D-(+)-catechin, hyperoside, and kaempferol-3-O-rutinoside. Palsikowski optimized the polyphenols extraction from *Bauhinia forficate* leaves investigating the extraction conditions in the following ranges: temperature 40–60°C, power 20–80%, and sample-to-solvent ratio from 1:10 to 1:20 (w/v) by using a Box-Behnken experimental design. The optimal UAE condition was achieved at 80% power, 41°C, and a 1:20 sample-to-solvent ratio. Under this condition, the extraction yield was higher than that obtained by ME.

13.2.3 Supercritical Fluid Extraction (SFE)

SFE is one of the better techniques for extracting flavonoids, essential oils, carotenoids, and fatty acids from plant (Uwineza and Waśkiewicz 2020). It is one of the most technologically advanced extraction systems and finds widespread applications in environmental, chemical, food, agriculture, and pharmaceutical industries (Belwal et al. 2018). SFE is considered a green technique, as it uses green and renewable solvent, i.e., carbon dioxide (CO_2), and represents a sustainable alternative to traditional systems (Lefebvre, Destandau, and Lesellier 2021; Uwineza and Waśkiewicz 2020).

Due to its thermodynamic property, a fluid becomes supercritical when both temperature and pressure are above the critical point (Essien, Young, and Baroutian 2020). Above a critical temperature, irrespective of any high pressure, liquid phase cannot be formed from gas phase. The fluid vapor pressure at critical temperature is named critical pressure (Panja 2018), and at this state, neither liquid nor gas predominates but physicochemical properties between a gas and a liquid are exhibited. Therefore, at the critical point, fluid diffuses into the solid matrix like a gas and dissolves active materials like a liquid (Panja 2018).

CO_2 has more advantages: non-toxicity, non-flammability, good solvent power, ease to be removed, safe handling (GRAS status), low cost, and recyclability. Supercritical CO_2 (SC-CO_2) extraction uses CO_2 above critical temperature (31°C) and pressure (7.38 MPa) to extract bioactive compounds, and these conditions ensure the preservation of thermolabile phytochemicals like polyphenolic compounds (Alara, Abdurahman, and Ukaegbu 2021). SFE columns are filled with pretreated samples, and the pressurized supercritical solvents flow through the column and dissolve extractable compounds from the solid matrix out to the separator where extract and solvent are separated through pressure reduction, temperature increase, or both (Uwineza and Waśkiewicz 2020). The main advantages of SFE are the ability to alter CO_2 density, which controls solubility, by changing pressure or temperature, the production of solvent-free extracts by a simple depressurizing step (Arumugham et al. 2021; Essien, Young, and Baroutian 2020), and no alternative energy source is required, differently from MAE and UAE (Belwal et al. 2018). Low polar and small molecules are generally easily dissolved in SC-CO_2, while large and polar compounds are extracted with the addition of a co-solvent (ethanol, methanol, or water), which enhances the extraction yield by hydrogen bonds, dipole-dipole, or dipole-induced dipole interactions with the analyte (Uwineza and Waśkiewicz 2020). The typical percentage of ethanol used as a co-solvent usually ranges from 5% to 15%. Another disadvantage besides the addition of a co-solvent is the longer extraction time due to the low rate of solute diffusion from matrix to supercritical fluid (Alara, Abdurahman, and Ukaegbu 2021; Kovačević et al. 2018).

In SFE, temperature, pressure, co-solvent concentration, extraction time, and particle size are the main parameters to set up according to the target compounds. Temperature is one of the crucial factors affecting SFE process efficiency; slight temperature and pressure variations cause significant changes in SFE properties, because of their effects on compound solubility (Panja 2018; Uwineza and Waśkiewicz 2020). By increasing temperature, SC-CO_2 has low-density solvent properties and phytocompounds' volatility improves (Arumugham et al. 2021). By increasing pressure at a specific temperature, the solvent density increases, thus improving compounds' solubility (Uwineza and Waśkiewicz 2020; Essien, Young, and Baroutian 2020). For each temperature, a pressure increase is correlated with a SC-CO_2 density increase, which is directly linked to target solutes' solubility, thus improving the extraction yield (Tyśkiewicz, Konkol, and Rój 2018). Therefore, temperature should be optimized on the targeted plant matrix and, if possible, on compounds. Generally, the extraction yield increases, extending contact time between solvent and solutes till to an optimum operating time (Tyśkiewicz, Konkol, and Rój 2018), because it ensures SC-CO_2 better penetration into the matrix resulting into better dissolution and interaction of solutes with SC-CO_2 over time. Particle size should be also considered in SC-CO_2 optimization. Smaller particle size can increase the extraction rate of solutes by increasing the specific interfacial area and by reducing the solute diffusion path (Essien, Young, and Baroutian 2020). Thanks to CO_2 low polarity, SFE has been widely applied to oils and lipids (Lefebvre, Destandau, and Lesellier 2021). Hydrocarbons and other relatively low polar organic compounds with molecular weights lower than 250 Da exhibit excellent solubility in SC-CO_2, and therefore, esters, aldehydes, ethers, ketones, lactones, and epoxides are easily extractable (Uwineza and Waśkiewicz 2020).

However, in the literature, other SFE applications are reported as highly valued compounds extracted from *Stevia rebaudiana* leaves (Kovačević et al. 2018). The optimal conditions applied to extract both steviol glycosides and phenolics were 40% ethanol as a modifier, 45°C, and 225 bar. Differently, the highest steviol glycosides concentrations were achieved at 211 bar, 80°C, and 17.4% co-solvent. These results indicated that SFE is a faster method with higher yields, using lower energy and solvent in comparison with the conventional ME extraction procedure.

Another example is the extraction of curcuminoids from *Curcuma longa* and *Curcuma amada* (Nagavekar and Singhal 2019). The study on the effect of the process parameters on the extraction indicated the best conditions as follows: high pressure (35 MPa), high temperature (65°C), and 30% polar modifier (ethanol). These conditions improved both the yield and selectivity of these interesting polar compounds.

13.2.4 Enzyme-Assisted Extraction (EAE)

EAE is suitable to improve the extraction of different phytochemicals, including phenolics (Qadir et al. 2019), flavonoids (Zuorro et al. 2019), anthocyanins (Ranveer et al. 2020), but especially non-extractable polyphenols (NEPs), i.e., polymeric or low molecular weight phenolics covalently linked to plant lignocellulosic components (Domínguez-Rodríguez, Marina, and Plaza 2021; Gligor et al. 2019). In fact, as reported by Domínguez-Rodríguez, Marina, and Plaza (2021), EAE provided sweet cherry pomace (*Prunus avium L.*) extracts richer in NEPs compared to those obtained by using the most common alkaline and acid hydrolysis procedure.

EAE is also an eco-friendly alternative to conventional methods, because it can be performed under mild temperature and pressure, requires no organic solvents, and has a high substrate specificity. It is based on the use of enzymes that catalyze a hydrolytic reaction acting on the structural components of plant tissue to release phytochemicals (Qadir et al. 2019; Nadar, Rao, and Rathod 2018). The process involves few steps, including powdering and homogenization of the raw material, temperature and pH adjustment, enzyme addition and incubation, enzyme inactivation, centrifugation and/or filtration, and finally the collection of the aqueous phase of the enzymatic extract

(Belwal et al. 2020). The key factors influencing the extraction efficiency are mainly enzyme type and concentration, but also pH, incubation temperature and time, solid-to-liquid ratio, and particle size (Xu et al. 2017). Cell wall polysaccharides (mainly cellulose, hemicellulose, and pectin) usually form stable complex with bioactive compounds, either by hydrogen bonds or by hydrophobic interactions. Plants matrix acts as a barrier to solvent diffusion and the quite strong interactions before mentioned are responsible for the low extraction efficiency of phytochemicals from plant sources (Pontillo et al. 2021). Carbohydrase enzymes, such as cellulase, pectinase, hemicellulase, and protease, are extensively used in EAE to cause bioactive release. Enzymes can be used alone to perform selective hydrolysis depending on targeted compounds' localization (Habeebullah et al. 2020) or in association with other cell-wall degrading enzymes to exploit a synergistic effect, as for commercially available pre-formed enzyme blends (Saad et al. 2019).

Nguyen et al. (2021) developed a novel enzymatic approach to recover rosmarinic acid (RA) from *Rosmarinus officinalis* L. by performing a preliminary enzyme screening to identify the most efficient for the extraction. Different proteases (papain, bromelain, Protamex, Champzyme FP) and cellulase (cellulase A) were separately added to rosemary leaves macerated in water for 2 hours at 30°C. Cellulase exhibited higher extraction efficiency probably due to the selective action on the cellulosic composition of the plant cell walls, where RA is localized. A Box-Behnken design was used to optimize the extraction conditions, using different extraction times (2–6 hours), sample-to-water ratio (1:20–1:40 g/mL), enzyme amounts (1–3%), and temperature (30–50°C) (Nguyen et al. 2021). Similarly, Qadir et al. (2019) used different commercial enzyme mixtures to recover phenolics from *Morus alba* leaves. The influence of four parameters, namely, enzyme concentration, temperature, incubation time, and pH, was studied using a rotatable CCD. Zympex-014 enzyme mixture, containing different carbohydrase, amylase, and acid protease, induced the highest release of bound phenolics, thus obtaining the extract with higher antioxidant activity (Qadir et al. 2019).

Another important factor to be considered is pH, because it affects the enzymatic hydrolysis by modifying protein configuration and binding capacity to substrates, which is specific for each enzyme (Gligor et al. 2019; Nadar, Rao, and Rathod 2018). Moreover, pH can affect the dissociation of some bioactives, such as for anthocyanins from Kokum (*Garcinia indica* Choisy) rinds in acidic conditions or tannins from raspberry pomace (*Rubus ideaus L.*) in alkaline solutions (Saad et al. 2019). Extraction time and temperature should be investigated, too. Enzymes require a specific time to hydrolyze the plant material; long extraction time may increase the release of other cellular constituents, hindering the extraction of bioactives (Nguyen et al. 2021). In parallel, a temperature increase leads to an increase in the enzymatic activity, as well as a decrease in extraction solvent viscosity with high solubilization of bioactives. Nevertheless, extreme thermal values cause a gradual enzyme activity loss along with the degradation of thermolabile phytochemicals (Gligor et al. 2019).

EAE has some technical limitations that influence its industrial-scale application: high cost of enzymes and low stability of their activity, which is sensible to temperature and nutrient availability. To overcome the drawback of scale-up, EAE is increasingly used as a pretreatment in combination with conventional extraction methods or more recent UAE, MAE, and PLE (Xu et al. 2017). This matching results in enhanced extraction yields, reduced extraction times, and limited enzymes cost (Xu et al. 2017; Pontillo et al. 2021). Pontillo et al. (2021) compared a solid-liquid conventional extraction (CEM) to MAE and considered the effect of EAE as a pretreatment method for the extraction of phenolic compounds from rosemary leaves (*Rosmarinus officinalis* L.). The pretreatment with pectinolytic enzymes for 1 hour to hydrolyze cell wall polysaccharides increased by at least 30% of the phenolic yield obtained by the subsequent hydroalcoholic conventional extraction. Therefore, EAE combined with CEM led to higher-quality extracts compared to those obtained using only CEM or MAE alone. According to these findings, Zuorro et al. (2019) investigated the recovery of flavonoids from corn husks using a sustainable extraction process, consisting in an enzymatic pretreatment of the plant material with cellulase followed by heat-reflux extraction with aqueous ethanol. Enzyme and ethanol concentration, incubation time, and liquid-to-solvent ratio were optimized by a Box-Behnken design, and also in this case, the recovery of flavonoids was 30% higher than that obtained using a heat-reflux extraction performed at 80°C for 2 hours with 70% ethanol (Zuorro et al. 2019).

13.2.5 Pressurized Liquid Extraction (PLE)

PLE is an innovative technique widely used for the extraction of phenolic compounds, carotenoids, flavonoids, and anthocyanins (Kraujalienė, Pukalskas, and Venskutonis 2017; Avanza et al. 2021; Frosi et al. 2021) from different plants. The extraction method is based on the use of high pressure to maintain the extraction solvent liquid below its boiling point at working temperature, enhancing mass-transfer rate and the solubility of bioactives due to the reduction of viscosity and surface tension (Ameer, Shahbaz, and Kwon 2017). PLE has numerous advantages compared to conventional methods, such as the reduced solvent consumption, short extraction time, few operating steps, and absence of light and oxygen during the process. The "greenest" PLE uses water as extraction solvent and is commonly known as Subcritical Water Extraction (SWE) or Pressurized Hot Water Extraction (PHWE). When water is heated to 200–250°C, its dielectric constant decreases at values close to that of organic solvents, thus reaching similar polarity. For this reason, SWE can be considered an effective alternative to organic solvents in some applications (Xu et al. 2017).

PLE key parameters are the extraction solvent, temperature and pressure, time, matrix characteristics, and number of cycles (Ameer, Shahbaz, and Kwon 2017; Xu et al. 2017). Solvent must be chosen in relation to the affinity with analytes to be recovered, and PLE efficiently uses GRAS solvents, such as ethanol and water, under high temperature and pressure conditions, providing high extraction yields in a short time (Zandoná et al. 2020). Several studies on different plants agree that ethanol-water mixtures are more effective than pure organic solvents, providing extracts richest in bioactives (Zandoná et al. 2020; Avanza et al. 2021; Lasta et al. 2019). Temperature and pressure operating conditions also affect PLE. Temperature can be applied from room temperature to 200°C and pressure from 35 to 200 bar (Xu et al. 2017). The use of high temperature under reduced pressure generates thermal energy, which helps to overcome molecular bonding forces between plant matrix and solvent molecules, bringing about desorption (Ameer, Shahbaz, and Kwon 2017). High temperatures also improve extraction efficiency by reducing the solvent viscosity and surface tension, which leads to an improved penetration of the sample matrix. However, the working temperature should be carefully controlled to avoid intramolecular modification and thermal decomposition or degradation of thermolabile compounds (Nastić et al. 2020; Leyva-Jiménez et al. 2018).

Leyva-Jiménez et al. (2018) developed a PLE process to recover phytochemical compounds from *Lippia citrodora* leaves. They optimized the extraction method using a CCD to investigate the influence of temperature (40,110,180°C), extraction time (5, 12.5, 20 minutes), and ethanol percentage (15, 40, 85%) on the extraction yield. According to the results, high temperature was detrimental in increasing the response, but high temperature degraded flavonoids, decreasing their recovery. Pressure is generally increased before raising the temperature, in order to keep the solution in the liquid state (Panja 2018). A too high value of this parameter can cause air blubbing, which can lead to low solubility rates (Ameer, Shahbaz, and Kwon 2017). In the literature, different authors reported a non-effect of the pressure beyond the point at which the solvent is maintained liquid (de Aguiar et al. 2019; Lasta et al. 2019) and, for this reason, a common trend is to maintain the pressure constant at a value high enough to ensure the liquid state of the solvent during the extraction process (Avanza et al. 2021; de Aguiar et al. 2019; Fuentes et al. 2021).

Different researchers indicated an increased recovery of polyphenol by using PLE compared to other innovative methods (Colombo and Papetti 2019; Chang, Feng, and Wang 2011; Azmir et al. 2013; Milić et al. 2022). Milic et al. developed a comparative analysis of conventional solvent extraction methods and more sustainable UAE, MAE, and PLE to isolate phytochemicals from black and red currant fruits (Milić et al. 2022). MAE resulted to be the most efficient method to recover bioactives from black currants and PLE the best for red currants, in terms of both yield and antioxidant activity. PLE efficiency is also confirmed by another recent comparative study on the phytochemicals extraction from comfrey roots in which it was more efficient than SFE and a conventional solvent extraction for a wide range of phytochemicals with different polarities (Nastić et al. 2020).

13.3 ANALYTICAL SPECTROSCOPIC TECHNIQUES

Due to the interesting pharmacological activities of the compounds present in phytochemical extracts, their qualitative identification and their quantification is a fundamental step. As a first step, their detection in the plant extracts can be realized in wet chemistry by characteristic specific tests (Sharma and Dash 2021), based on organic reactions of the functional groups present in the compounds themselves, that lead to the formation of precipitates or to change in the color of the solution. For example, the presence of carotenoids can be identified by Carr-Price reaction (Jagessar et al. 2017), using a saturated solution of antimony trichloride in chloroform and leading to a change in the color of the solution from blue-green to red. Flavonoids can be recognized by the yellow color of the solution or of the precipitates obtained after alkaline reagents or lead acetate tests, respectively (Somkuwar and Kamble 2013), while for tannins, the Braymer's test can be used (Morsy 2014), consisting in the addition of ferric chloride solution (10%) to the distilled water boiled and filtered plant extract and leading, when positive, to the obtainment of a blue-green colored solution. The alkaloids detection (De, Dey, and Ghosh 2010) is based on the dissolution of the extracts in dilute hydrochloric acid, the filtration, and the subsequent treatments with potassium bismuth or potassium mercury iodides, obtaining the formation of reddish-brown or creamy-white precipitates, respectively; a saturated picric acid solution can also be used, with yellow precipitate formation.

Subsequently, analytical methods must be applied to the extracts as obtained or after suitable ad hoc reactions to structurally describe and to quantify the different characteristics of compounds present in the extract matrix. Concerning the structural determination, different absorption spectroscopic techniques, alone or in combination, have been used in the literature, thanks to the characteristic response of each different functional group in a molecule to the irradiation with light of different wavelength in a certain spectral range. In particular, the spectroscopic techniques frequently used for the analysis of herbal extracts are UV-Vis spectroscopy, IR spectroscopy, NMR spectroscopy, and mass spectroscopy (Altemimi et al. 2017).

13.3.1 UV-Visible Spectroscopy

UV-Vis spectroscopy can be utilized for the identification of bioactive compounds in both isolated and mixture form, or on solution obtained after selective reactions of some components in the extracts, and it is widely applied as a primary tool thanks to its low cost, wide availability, ease of use, relatively fast measurements, and applicability for both qualitative and quantitative purposes. Preferentially, it can be used for quantitative analysis because aromatic molecules are powerful chromophores in the UV range. One of the limits is the scarce selectivity and the possibility to determine the total content of molecules of a specific compounds class but rarely the specific compound.

Phenolic compounds including anthocyanins, tannins, polymer dyes, and phenols form complexes with iron, which have been detected by this spectroscopy: depending on their structural features, the wavelength of their maximum absorption shifts in this region of the electromagnetic spectra, from 280 nm for total phenolic extracts to 320 nm for flavones, 360 nm for phenolic acids, and 520 nm for total anthocyanins (Altemimi et al. 2017). Strategies for the simultaneous detection of different compounds can be developed: for example, in 2020, Mohamadi et al. (2020) developed a method to detect trigonelline, diosgenin, and nicotinic acid in a formulation from fenugreek extract thanks to the different λ_{max} for the three molecules (232.65 nm, 296.23 nm, and 262.60 nm, respectively).

All diarylheptanoids exhibit absorption in the range of 250–290 nm. Curcumin presents a further broad absorption band in the visible region with a maximum in the 410–430 nm range due to the extended conjugated system of the aromatic rings, the unsaturated bonds in the C7-chain, and the intramolecular H-bonding of the enol function. UV-Vis spectrophotometry is commonly used for the quantification of total curcuminoid content of medicinal plant extracts (Methods, n.d.; Green et al. 2008). The European Pharmacopoeia 9th edition publishes two official herbal drug monographs, turmeric rhizome and Javanese turmeric (*Curcumae longae rhizoma* and *Curcumae zantorrhizae rhizoma*), which require the quantitative determination of total dicinnamoyl methane derivatives (expressed as curcumin) using a spectrophotometric method for the quality control of the drugs. Absorption of ethanol extracts prepared by a reflux method is measured at 425 nm (EDQM 2017; Strasbourg 2017). These spectrophotometric methods are, however, not suitable to quantify the individual curcuminoids.

Ahmed et al. (2012) elucidated the encapsulation of curcumin preparing different lipid-based nanoemulsions and conventional emulsions. Additionally, the bioaccessibility of curcumin was investigated, when released in an *in vitro* model simulating small intestine digestion conditions. The concentration of curcumin in lipid phases was determined by spectrophotometry (Ahmed et al. 2012). Specific reactions can be used for the determination of the total content in phenolic compounds and anthocyanins, whose products can be detected by UV-Vis spectroscopy. The same applies for the total antioxidant capacity. Ozcan et al. (2021) by a modified Folin-Ciocalteu procedure (Spanos and Wrolstad 1990) determined at 765 nm wavelength the phenolic compounds content in sumac (*Rhus coriaria* L.) genotypes from southern part of Turkey, while measurements at 510 nm and 700 nm allowed to determine the total anthocyanin content thanks to a method developed by Wrolstad (1976). Finally, the procedure described by Re et al. (1999) was used for the total antioxidant capacity determination.

13.3.2 Infrared Spectroscopy

Fourier Transform InfraRed spectroscopy (FTIR) is a high-resolution analytical tool to identify the chemical constituents of a mixture and to elucidate the structural compounds and constitutes a rapid and non-destructive investigation method to fingerprint herbal extracts in liquid or in solid forms (Altemimi et al. 2017; Rasul 2018). It allows to recognize the functional groups present in the plant extract, which give absorption at different wavelength as a function of the mass of the involved atoms, the force of the bond, and the chemical surrounding. In particular, single, double, and triple C–C bonds, single and double C–O bonds, N–H and O–H bonds absorption bands are of interest for phytocomplexes' chemical analysis.

So, for example, monosaccharide types, glucosidic bonds and functional groups, and the pyranoside rings can be recognized using FT-IR spectroscopy (Nazeam et al. 2017). Moreover, the vibration bands of the bonds of some inorganic groups such as silica and silicates, phosphates, and carbonates can be detected, giving indications on the inorganic components of the samples. Anyway, also this technique allows the identification of the compound's chemical class: for the identification of a single specific compound; it must be used coupled with other tolls such as NMR analysis.

The attenuated total reflection mode, quantifying the changes that happen to an internally reflected infrared beam once it comes into contact with the chosen sample, allows to analyze small number of materials (few mg) in a solid-state form without altering it, and is attracting increasing attention for the analysis of extracts and lyophilized powders. Recently, it was employed to construct multivariate regression models for the quality control of different essential oils (Farag et al. 2018).

Chang, Feng, and Wang (2011) and Zhang (1994) collected an interesting set of data of different polysaccharides previously isolated from *Aloe arborescens* Miller constituents (Chang, Feng, and Wang 2011) assigning the IR absorption bands to the different sugar constituents present in different parts of the plant. Alkaloids were recognized as the most important constituents in several parts of *Strychnos spinosa* (spiny monkey orange) plant, and as one of the most represented class of phytochemicals in the genus *Strychnos*, recognizing that they may serve as signature molecules for members of the genus itself and explaining their broad spectrum of biological effects thanks to several different chemical structures of the present alkaloids (Adamski et al. 2020; Aremu and Moyo 2022). The determination of curcumin, demethoxycurcumin, and bisdemethoxy-curcumin in turmeric samples (Tanaka et al. 2008), and of the total curcuminoid content in rhizomes of *C. longa* (Kasemsumran et al. 2010), has been also reported. UV-Vis and IR spectroscopies were very recently used to elucidate the structure of new glycocerebrosides isolated for the first time from *Tabernaemontana contorta* Stapf. *(Apocynaceae)* and with an interesting antibacterial activity (Ebede et al. 2022).

FTIR and Raman spectra measurements have been reported for the analysis of fenugreek seeds in different forms (powder, ash, and oil), showing that fenugreek seeds contain high amounts of proteins, contrary to fats and starch present in small quantities in the seeds. Moreover, the ash of fenugreek was found significantly rich in phosphate compounds, the main constituents of fenugreek inorganic part (El-Bahy 2005).

13.3.3 Nuclear Magnetic Resonance Spectroscopy (NMR)

NMR is primarily related to the magnetic properties of certain atomic nuclei; in particular, ^{1}H-NMR and ^{13}C NMR are used to find out the hydrogen and carbon atoms present in the compounds, their chemical nature, and their bonds inside the molecular structure. ^{15}N solid-state NMR is also very useful for biological compounds studies, as it is considered a reliable probe of the nitrogen protonation state as well as of hydrogen bond interactions present at nitrogen atoms (Braga et al. 2010). NMR spectroscopy has enabled many researchers to study molecules by recording the differences among the various magnetic nuclei, thereby giving a clear picture of what the positions of these nuclei are in the molecule and of the nature and numbers of atoms in the neighboring groups. The application of NMR to phytocomplex fingerprinting is gaining popularity thanks to its efficacy, sensitivity, and non-destructive characteristics. Moreover, it allows an easy comparison of different batches of extracts collected from the same plant source, which are subject to great qualitative-quantitative variations according to a plethora of factors (Pauli et al. 2012). One-dimensional technique is routinely used, but more complicated structures of the molecules can be solved through two-dimensional NMR techniques. Considering both 1D and 2D ^{1}H NMR experiments, quantitative analyses (qHNMR) have proved to be highly suitable for the simultaneous selective recognition and quantitative determination of analytes also in complex matrixes (Pauli et al. 2012).

For example, 2D NMR spectroscopy was utilized by Ramesh et al., to investigate fenugreek galactomannan without chemical fragmentation. Sequential extraction fetched two fractions with a Gal/Man ratio of 1:1.04 and 1:1.12, respectively (Ramesh et al. 2001). Trigonelline was assessed using 1H NMR and a quantitative derived method (qHNMR) in the peel and pulp of *Balanites aegyptiaca* bioactive metabolites (Farag, Porzel, and Wessjohann 2015) and in the leaves of six species of the *Annona genus* (Machado et al. 2013), respectively: the pulp of the fruit and *Annona laurifolia* species were reported to have the highest amount of this compound, comparable to the fenugreek content. In the glycolipid (sphingolipids), the nature of the extracts was identified thanks to proton and carbon NMR data (Ebede et al. 2022). Recently, Leyva et al. (2021) developed a NMR metabolomic platform to profile the leaves of different *Vanilla* species, with the aim to determine metabolic differences among them. Thirty-six metabolites were identified and 25 quantified; malic and homocitric acids, together with two vanillin precursors, were recognized as relevant metabolic markers for species differentiation.

13.3.4 Mass Spectrometry for Chemical Compounds Identification

Mass spectrometry (MS) is a powerful analytical technique for the identification of unknown compounds, quantification of known components in a mixture, and elucidation of the structure and chemical properties of molecules also in complex matrices, and it is particularly useful in the area of natural products chemistry (Rasul 2018; Hocart 2010; Ingle et al. 2017). Among the advantages of the technique, its speed, sensitivity, selectivity, and its versatility in analyzing solids, liquids, and gases must to be quoted. In MS, organic molecules are bombarded with either electrons or lasers and converted to highly energetic charged ions: a mass spectrum is a plot of the relative abundance of a fragmented ion against the ratio of mass m /charge z of the ion itself. Using MS, relative molecular mass (molecular weight) can be determined with high accuracy, and an exact molecular formula can be determined knowing the sites where the molecule has been fragmented.

For complex matrices, tandem MS, namely, a two-step technique to analyze a sample by using either two or more mass spectrometers connected to each other (MS/MS technique) or a single mass spectrometer coupled with several analyzers arranged one after another is very useful and sensitive (Uwineza and Waśkiewicz 2020). In the first stage of MS/MS, a predetermined set of m/z ions are isolated from the rest of the ions coming from the ion source and fragmented by a chemical reaction. In the second stage, mass spectra are produced for the fragments. These methods, comprising linear ion trap (LIT) MS (which provides single-stage mass analysis that allows to obtain molecular mass information), tandem mass analysis (MS/MS), and multistaged tandem mass spectrometry (MSn), providing structural information, have been demonstrated as reliable tools for the structural elucidation of unknown compounds in complex mixtures, such as the total plant extracts.

The combination of HPLC and MS facilitates a rapid and accurate identification of chemical compounds in medicinal herbs, especially when a pure standard is unavailable. High resolution and the ability to provide precise and accurate qualitative and quantitative data established gas-chromatography (GC) coupled with MS, i.e., GC-MS analyses, as a valuable means for taxonomic research of plants. Recently, both LC-MS and GC-MS have been extensively used for the analysis of phenolic compounds, with electrospray ionization (ESI) as the preferred source due to its high ionization efficiency and using different detectors, such as the colorimetric, electrochemical, and photodiode arrays, as well as UV-Vis spectrophotometry (Büyükköroğlu et al. 2018).

Liquid chromatography followed by tandem mass spectrometry with an electrospray ionization source (LC-ESI/MS/MS) was used to investigate the phytochemical profile of *Aloe arborescens* and *Aloe barbadensis*. The results revealed that 10-H-anthracen-9-one derivatives, aloin, aloe-barbendol, and aloesaponarin II were present in *A. arborescens species*, together with naringenin and higher amounts of hydroxycinnamic acids, and confirmed the strong potential of suitable portion of Aloe leaves as an effective antioxidant (Lucini et al. 2015; Nazeam et al. 2017).

Flavonoids in oregano extracts were analyzed using high-pressure liquid chromatography-diode array detection-electrospray ionization/mass spectrometry (HPLC-DAD-ESI/MS), GC-MS, and GC-FID, obtaining the qualitative and quantitative composition and identifying 46 compounds in the leaves oil, representing more than 98% of the total composition, being carvacrol the predominant one (Stanojević et al. 2016; Knez Hrnčič et al. 2020). LC-MS/MS analysis of various fractions from *Xeroderris stuhlmannii* (Taub.) Mendonca and E.P. Sousa bark extracts confirmed the presence of 28 phytochemicals (Selemani et al. 2021), of which 12 were identified using GC-MS. Both techniques confirmed the presence of ursolic acid, roburic acid, reticuline, rotenone, and P-coumaric acid glucoside in hexane and methanol extracts. Some of these compounds in the bark extract possess antioxidant, antiviral, antitumor, antibacterial, and anti-inflammatory properties.

Tandem mass spectrometry in Selected Ion Monitoring (SIM) or Multiple-Reaction Monitoring (MRM) modes is frequently applied for the quantification of diarylheptanoids: a validated LC-ESI-MS/MS method was developed using the MRM mode to determine hirsutenone and oregonin contents of *Corylus avellana* and *C. maxima* leaf and bark samples (Alberti, Riethmüller, and Béni 2018). The most used mass analyzers for the analyses of these compounds are time-of-flight (TOF) MS, which enables accurate mass analysis, triple-quadrupole MS that provides additional structural information by the observation of the fragmentation patterns of the target analytes, quadrupole-time-of-flight (Q-TOF) that allows the exact mass measurement of both precursor and fragment ions.

13.3.5 Solid-State Techniques for the Analysis of the Separated Components

After the separation of the different components (see later for the separation techniques), some common analyses at the solid state can be applied, together with the already-discussed FT-IR and NMR techniques, for the characterization of the compounds and the evaluation of their purity. Among them, differential scanning calorimetry (DSC) for the determination of the thermal stability and X-ray powder diffraction (XRPD) for the identification of the crystallographic features of the compounds are frequently used. The combination of the results obtained by the different techniques, together with *in vitro* or *in vivo* biological assays, can be useful to detail the bioavailability of the components and to explain the pharmacological action mechanisms of the compounds.

For example, Hasa et al. (2013) characterized a standardized *Vinca minor* L. leaf dry extract by HPLC-MS, XRPD, DSC, and 13C and 15N solid-state NMR (SSNMR) using pure vincamine as a matter of comparison. Moreover, they evaluated the *in vitro* dissolution performances of the two products and their *in vivo* bioavailability in rats. The sevenfold improvement in oral bioavailability of the dry extract with respect to the pure vincamine was ascribed to interactions between the indole alkaloid and the corollary of ingredients of the dry extract, giving rise to the protonation of the alkaloid vincamine, thus enhancing its dissolution in physiological fluids.

Kaushik et al. (2021), in order to understand the effect for the treatment of dengue infection, prepared *Leucas cephalotes* plant extracts by SFE and isolated oleanolic acid by preparatory thin-layer chromatography. The compound was identified and characterized by UV-vis, FT-IR, and high-performance thin-layer chromatography (HPTLC). The structure of the compound was elucidated by ^{1}H and ^{13}C NMR. The compound was found by DSC to have a 98.27% purity and a melting point of 311.16°C.

13.4 ANALYTICAL SEPARATIVE TECHNIQUES

The choice of a separative technique in the study of phytochemicals is greatly influenced by the complexity of matrices, which can be vegetal (fruit and vegetable extracts), vegetal-derived (food and beverages), or biological (mainly plasma, bile, and urine). The phytochemical complexity requires a precise quality control to assess the qualitative-quantitative composition of different constituents, and also their potential metabolites. This represents a great challenge for the development of advanced techniques with high resolution and sensitivity (La Barbera et al. 2017; Oh and Lee 2014; Zhu et al. 2018; Stylos et al. 2017). In addition, as matrices are very different, it is evident the need of versatile and rapid procedures applied to different plants in routine and high-throughput analysis (Alarcón-Flores et al. 2013; A. Zhang et al. 2017a).

For these reasons, the most promising innovative separative tools are LC and GC coupled to high-resolution mass spectrometry (HRMS), well known as hyphenated techniques and often integrated with high-throughput platforms essential in the screening and identification of bioactive herbal constituents made of hundreds/thousands components (Zhu et al. 2018; Stylos et al. 2017; La Barbera et al. 2017; Alarcón-Flores et al. 2013; A. Zhang et al. 2017b; Dong et al. 2016). In fact, as already underlined, MS detection solves many issues in phytochemical field from quality control to chemical and metabolic profiling, allowing component identification and quantification (Zhu et al. 2018; A. Zhang et al. 2017a); in addition, MS/MS (tandem MS) allows the characterization of unknown phytocomponents thanks to the process of ions fragmentation (Stylos et al. 2017; Alarcón-Flores et al. 2013; Zhang et al. 2019).

LC-MS/MS and GC-MS/MS are often used in combination, particularly in samples with unknown components to attempt to have a complete qualitative and quantitative characterization. For example, flavonoids can be analyzed by LC-MS, while GC-MS can be applied to hydrocarbons, mono/di/triterpenes, vitamin and carotenoid derivatives, and phytosterols (Ballesteros-Vivas et al. 2019; Tor-Roca et al. 2020; Alencar Filho et al. 2020). UV-Vis detection (detectors at one fixed wavelength and photodiode-array-PDA-detector) can be used, but the identification of components is often possible only in combination with MS because of plant extracts complexity, lack of standards used as references, and presence of unknown components (Fantoukh et al. 2021).

HPTLC deserves a special discussion as a rapid screening platform often used in preliminary analysis, but nowadays also in quality control and quantity determination (Saibaba and Shanmuga Pandyan 2016). Among other techniques, multidimensional chromatographic separations, mainly two-dimensional (2D) LC-MS or 2D GC-MS, became promising to improve separation and identification, especially in complex samples with many and/or unknown components. 2D LC and 2D GC optimization involves the set-up of the two dimensions together, and this complicates the method development. 2D LC with its two-column system allows a rapid separation of components with different polarity with interesting opportunities by using the hydrophilic interaction-reversed phase (HILIC-RP) mode. In fact, HILIC-RP could represent a very important solution for the separation of polar and medium-polar phytochemicals as a substitute for CE, which have been most often used through the 2000s (Zhu et al. 2018; Dong et al. 2016; Cheng et al. 2021; Cao et al. 2014). 2D GC is a potential technique, but nowadays its limits (few available stationary phases and long analysis time) overcome its advantages and limit its diffusion (Zhu et al. 2018; Nolvachai, Kulsing, and Marriott 2017; Cao et al. 2014).

Few examples of high-speed counter-current-chromatography (HSCCC) applications will be discussed to underline the potentiality of this technique in high-throughput preparative isolation and separation of aromatic acids, flavonoids, sesquiterpenes and their glycosides, phenylalkaloids, polysaccharides, oils, and phytosterols. HSCCC is a very rapid technique that is also able to ensure a great resolution (Wu and Liang 2010; L. Zhang et al. 2017b). This part of this chapter will present the most important separative techniques in phytochemical characterization, focusing on their principles, pros and cons, and most innovative applications.

13.4.1 Liquid Chromatography (LC)

Among LC techniques, Ultra-high-Pressure Liquid Chromatography (UPLC) represents the most advanced analytical approach, thanks to its stationary-phase particle size (≤2 µm in diameter) and consequent resolution, rapidity (few minutes), and efficiency. The reduced solvent volume used as mobile phase together with shorter analysis time saves significantly costs. In combination with UPLC, the most used MS detectors are ionization-time-of-flight (ESI-TOF) (Chong et al. 2013), quadrupole (Q)-TOF (Zhang et al. 2016; Yao et al. 2017; Meng et al. 2020), and Orbitrap (Lopez-Gutierrez et al. 2014; Zou et al. 2015). Briefly, with respect to TOF, Orbitrap is exploited for very high resolution and mass accuracy, but it is more expensive (Dong et al. 2016).

Ambient ionization techniques such as ESI and Desorption Atmospheric Pressure Chemical Ionization (DAPCI) solve the MS main disadvantage consisting in long time sample pretreatment, giving the possibility to set up direct, rapid, and high-throughput analysis (Wong, So, and Yao 2016). ESI consists of ionization under atmospheric pressure, and it is ideal for moderately polar and polar micro- and macromolecules (alkaloids, organic acids, triterpenoids, and saponins), while DAPCI is useful only for small molecules with moderate polarity and good volatility (low-polarity alkaloids, low-molecular-weight triterpenoids, non-volatile ginsenosides) (Zhu et al. 2018). Over years, ESI technology advances allowed an efficient interface with UHPLC, optimizing mobile phase flow rate and dead volume (La Barbera et al. 2017). UPLC-MS platforms advantages make them suitable for multicomponent phytochemical characterization and quantification both in plant extracts and in biological matrices with rapid screening and simultaneous quantification analysis (La Barbera et al. 2017; A. Zhang et al. 2017b; H. Zhang et al. 2019).

UPLC-QTOF-MS is particularly performing in the analysis of phytochemical metabolite profiling in biological samples thanks to the great advances in data processing techniques, such as mass defect filter (MDF) approach, which is based on the difference between exact and nominal mass, allowing high mass accuracy data (Dong et al. 2016). Its interesting application is the phytochemical characterization of known and new plant metabolites, especially for medicinal plants containing potential bioactives. Because of its versatility and sensitivity, LC-MS also represents the main tool for plant metabolomics (analysis of primary and secondary metabolites, intermediates, lipids and fatty acids, and phytohormones) (Shimizu et al. 2018). To analyze secondary metabolites, untargeted and targeted approaches are available: the first one makes a complete metabolic profile, detecting the main compounds, while the second is focused on the identification and quantification of known chemically characterized metabolites (Shimizu et al. 2018; Choudhury et al. 2020).

PDA detector is mainly used in combination with MS detector to assess compound identity or stability profiling when phytochemical matrices are made of few components and to quantify significant amounts (mainly mg) (Tor-Roca et al. 2020; Braga, Pianetti, and César 2020; Seo and Song 2021; Ba et al. 2021; Dinan et al. 2020). Finally, LC-Nuclear Magnetic Resonance (NMR)-MS could represent a very performing tool in the study of complex matrices, mainly in the analysis of biological samples, as it combines LC capacity, NMR resolution, and MS sensitivity, giving important structural information in metabolite identification and pathways. Notwithstanding, the NMR limits of solvent suppression and sensitivity have been overcome by different technologies and modes; so far, NMR-MS hyphenation is still complicated, and applications remain greatly unexplored (Stylos et al. 2017). Nowadays, it is easier to find applications of NMR and MS-based platforms separately. This combination is one of the best strategies for metabolomic studies in the search for bioactive phytocomponents (Abd Ghafar et al. 2020; Tlhapi, Ramaite, and Anokwuru 2021).

13.4.2 High-Performance Thin-Layer Chromatography (HPTLC)

Thin-layer chromatography (TLC) has been widely used in the phytochemical analysis, because of its low cost, rapidity, and versatility (directly applied to raw extract for preliminary analysis of different metabolites with the exception of volatile molecules). HPTLC is the automated version of TLC, and thanks to improved stationary phases, the use of densitometric detectors represents a very promising tool for phytochemicals, particularly to create suitable platforms for screening and stability monitoring (Alencar Filho et al. 2020; Khatoon et al. 2019). HPTLC combined with densitometric scanning allows sensitive and specific determinations with a great variety of applications from phytochemical constituents to biomarkers and secondary metabolites analysis (Khatoon et al. 2019).

HPTLC is used for qualitative profiles, but also quantitative analysis with results comparable to those obtained by HPLC (Sagi et al. 2015; Viswanathan and Mukne 2016; Gupta, Patil, and Patil 2019). Recent studies showed that optimized HPTLC methods with densitometric detection still represent actual promising tools in quality control (adulterants or contaminants detection) and metabolomic analysis (herbal formula fingerprinting) (Seo and Song 2021; Viswanathan and Mukne 2016; Mulaudzi et al. 2021; Parveen et al. 2020). In addition, the direct combination of HPTLC with MS allows an efficient separation and a sensitive mass analysis, also for low molecular weight molecules, such as flavonoids and their glycosides (Kroslakova, Pedrussio, and Wolfram 2016).

13.4.3 Two-Dimensional Liquid Chromatography (2D LC)

For years, 2D LC has been widely applied in the analysis of herbal samples, also in the case of unknown components (Fantoukh et al. 2021; Iguiniz and Heinisch 2017). Online 2D LC overcame the limits of the simplest offline 2D LC, i.e., analysis time, potential sample loss and carry-over, and poor reproducibility, and represented the most used mode for phytochemicals (Nolvachai, Kulsing, and Marriott 2017; Li et al. 2016). Online 2D LC can be set up in two different configurations: heart-cutting 2D LC (LC-LC) and comprehensive 2D LC (LC×LC); notwithstanding these modes can be applied in combination, LC×LC compared to LC-LC represents the most powerful tool to analyze unknown compounds as each sample fraction in 1D is transferred and analyzed in 2D (Nolvachai, Kulsing, and Marriott 2017; Qiao et al. 2014).

2D LC can exploit different LC combining modes, allowing versatile applications in the separation of compounds of different polarity with high throughput (Dong et al. 2016). RP-LC×RP-LC is the most used mode thanks to its versatility and selectivity and also easy combination with PDA and MS detectors, and it is particularly useful for complex matrices as phytochemical extracts, while its principal disadvantage is the limited separation principle based only on RP-LC polarity (Fantoukh et al. 2021; Qiao et al. 2015). An important example of RP-LC×RP-LC approach is described by Qiao et al. (2015). 2D LC high resolution also allowed the identification and separation of unknown compounds in *Glycyrrhiza uralensis* complex herbal sample containing more than three hundreds phenolic compounds and saponins.

2D HILIC×HILIC and 2D HILIC×RP-LC are particularly promising for polar (saponins) or medium-polar (flavonoid glycosides, phlorotannins) compounds, respectively, allowing efficient and rapid separations also of complex samples (up to hundreds components) (Zhu et al. 2018; Nolvachai, Kulsing, and Marriott 2017; Fantoukh et al. 2021; Wang et al. 2015). For example, combining HILIC and RP-LC mode, it was possible to resolve complex samples containing triterpenoid saponins, characterized by polar and apolar groups, which are generally used as aging markers of plant extract, such as *Panax ginseng* (Wang et al. 2015). 2D ion exchange chromatography (IEC)×RP-LC, exploiting two different separation principles based on charge and hydrophobic interactions, can be also widely applied for complex matrices analysis, particularly combined with APCI-MS. IEC mode allowed compound concentration with good sensitivity and resolution, but often with long run-time (Nolvachai, Kulsing, and Marriott 2017).

13.4.4 Gas Chromatography (GC)

GC is the most suitable for the volatile phytocomponents analysis and also for non-volatile compounds (amino acids, fatty acids, organic acids, glucosamine, sugars, pyrimidine, and purine) after a derivatization step with an efficiency often higher or comparable to that obtained by LC techniques. Notwithstanding, GC versatility, the main disadvantages in using this technique are longer time for sample preparation than LC and the potential instability of volatile compounds. On the other hand, the hyphenation between GC and MS led access to standard spectral-wide libraries with a higher possibility to identify the unknown compounds than LC-MS (La Barbera et al. 2017; Zhu et al. 2018; Dong et al. 2016).

The most used MS-based methods coupled with GC are GC-TOF-HRMS and GC-Matrix-Assisted Laser Desorption/Ionization (MALDI)-MS (Dong et al. 2016; Bell et al. 2016). MALDI soft ionization can be directly applied to rapidly analyze complex crude extracts/mixtures, saving time and costs and allowing chemical and metabolomic profiling useful in quality and safety control (Parveen et al. 2020). On the contrary, the matrix interference in the analysis of low molecular weight compounds, mainly in biological samples, remains the main disadvantage (Dong et al. 2016). Recent MALDI-MS advances regard the elimination of matrix interference and also strategies in sample preparation with promising applications for herbal and medicinal samples (Shimizu et al. 2018).

Today, GC-MS is widely used to identify plant bioactives, discovering also new pharmaceutical molecules, and this tool is still very promising in drug discovery (Ramya 2022).

Thermal Desorption (TD)-GC is widely applied to analyze volatile (isothiocyanates, alkanes, aliphatic alcohols, carbonyl compounds, fatty acids, esters, phenols) and semi-volatile (aldehydes, ethers, esters, phenols, organic acids, ketones, amines, amides, nitroaromatics, nitrosamines) organic compounds, particularly in environmental and food field, thanks to its rapidity and robustness. TD consists of the concentration of headspace volatiles onto TD tubes, ensuring time-saving and high sensitivity analysis. An interesting TD-GC-TOF-MS application was the definition of environmental and storage conditions of salad leaves (i.e., *Eruca sativa*), monitoring and quantifying volatile phytochemicals useful for quality and stability studies (Bell et al. 2016). Finally, a recent study of Parveen et al. (2020) regards the application of HPTLC and GC-MS to have a complete metabolomic profile of different rose samples and formulations, identifying not only phytochemicals but also the potential adulterants.

13.4.5 Two-Dimensional Gas Chromatography (2D GC)

2D GC most important parameters to be optimized are internal diameter column, temperature, and flow, which can have a great impact on column loading capacity and resolution (Mommers and van der Wal 2021). In particular, GC×GC mode can be useful to solve 1D GC limits (low capacity and poor resolution), particularly in the analysis of complex phytochemical mixtures, allowing methods with high resolution if combined with MS (Ji et al. 2018; Dong et al. 2016; Jiang et al. 2015). The two most important 2D GC applications regard metabolomic and pharmacokinetic studies. Over the years, 2D GC-MS with its high resolution allowed also to obtain more accurate structural information with respect to GC-flame ionization detector (FID), and this is very important for the detection of unknown phytocomponents (Ji et al. 2018; Dong et al. 2016; Jiang et al. 2015; Shi et al. 2015). It is well known that in the past, FID has been the most commonly used GC detector with different applications to analyze C-based organic compounds; nowadays, FID with its new gas and equipment remains yet applied, but mainly in environmental and food analysis (Fantoukh et al. 2021).

In particular, 2D GC-TOF-MS represents a rapid analytical tool to identify and quantify (mass accuracy <20 ppm) more than one hundred components in a single analysis, and it is generally applied in metabolic profiling studies, because it gives accurate and important structural information for compound identification in herbal and biological samples (Jiang et al. 2015; Shi et al. 2015). In addition, over the years, the development of GC stationary phases with different polarities makes possible versatile applications also in the analysis of complex samples. GC×GC data processing advances simultaneously and can also resolve the problem of retention-time shifts, which are typical of 2D approaches compared to 1D techniques (Jiang et al. 2015; Mommers and van der Wal 2021).

13.4.6 High-Speed Counter-Current-Chromatography (HSCCC)

Recently, HSCCC represents a performing tool for rapid and efficient separations of compounds with different polarities, exploiting the different analyte partitioning in different solvent systems (Wu and Liang 2010; L. Zhang et al. 2017b). The polarity-stepwise elution mode, which consists of the use of different mobile phases (polar and apolar solvents) with increasing polarity, is a promising approach for the separation of polyphenols (Zuo et al. 2021), while pH-zone-refining CCC (PZRCCC), which is based on the use of polar (organic and water buffer) solvents, can be used for the separation of ionizable compounds (alkaloids, terpenoids, saponins, lipids, and peptides/proteins), mainly in preparative separations (Zuo et al. 2021; Song et al. 2016; Leitão et al. 2021).

The possibility to have different CCC setup has extended its application to different compounds such as terpenoids, which suffer from some difficulties (structural similarity, weak polarity, and lack of UV absorption), and saponins, which have structural similarity, a strong polarity, and a viscous consistence. In fact, HSCCC has been widely applied to these compounds, exploiting the optimization of solvent systems (mainly two-phase) in different polarities and pH or adding electrolytes. In addition, this technique can be also successfully used for ginsenosides, which are polar/moderate polar compounds (Song et al. 2016). The work of Zuo et al. (2021) is an example of a new setup of polarity-stepwise elution mode in five steps, allowing a very efficient and rapid separation of compounds with a wide range of polarity with potential applications also to complex samples.

Among HSCCC advances, the use of concentrical coil columns increased efficiency and resolution in natural products analysis (L. Zhang et al. 2017a). In addition, HSCCC is particularly promising in purifications of small amount of phytocomponents (Song et al. 2016). An interesting purification application of HSCCC regards alkaloids, as with their solubility in water and insolubility in organic solvents, they can exploit the liquid-liquid partitioning principle typical of this technique. PZRCCC remained the most applied large-scale approach for alkaloid purification and can be used alone or in combination with other HSCCC modes (Leitão et al. 2021). Notwithstanding,

HSCCC is a rapid and high-throughput technique. Thanks to their great loading capacity, often the used columns are not so robust and this can affect run reproducibility and costs (Leitão et al. 2021).

13.4.7 Capillary Electrophoresis

CE works in free solution and is particularly useful for ionic compounds, exploiting the principle of voltage in aqueous buffers and ensuring high efficiency, short analysis times, and low consumption in terms of solvents and analytes in comparison with LC. It is yet widely applied in food and safety analysis or in the analysis of biomolecules (Papetti and Colombo 2019; Colombo and Papetti 2019; Colombo and Papetti 2020). Nowadays, a small number of applications in phytochemical analysis, generally in combination with MS detectors, are reported in the literature (Zhu et al. 2018; Dong et al. 2016; Cheng et al. 2021). Among MS approaches, CE-ESI-MS has a considerable disadvantage consisting in the possible use of buffer additives (non-volatile salts, organic modifiers, surfactants), which improve CE resolution and versatility, but prevent CE hyphenation with MS. However, over the years, the use of APCI source could resolve also this limit. In addition, notwithstanding the well-known CE problem of sensitivity with respect to LC-MS, CE-TOF-MS can represent a potential application not yet explored in the analysis of phytochemical charged metabolites, mainly in biological samples (plasma and urine) (Dong et al. 2016).

REFERENCES

Abd Ghafar, Siti Zulaikha, Ahmed Mediani, M Maulidiani, R Rudiyanto, Hasanah Mohd Ghazali, Nurul Shazini Ramli, and Faridah Abas. 2020. "Complementary NMR-and MS-Based Metabolomics Approaches Reveal the Correlations of Phytochemicals and Biological Activities in Phyllanthus Acidus Leaf Extracts." *Food Research International* 136: 109312.

Adamski, Zbigniew, Linda L Blythe, Luigi Milella, and Sabino A Bufo. 2020. *Biological Activities of Alkaloids: From Toxicology to Pharmacology*. MDPI.

Aguiar, Ana Carolina de, Ana Paula da Fonseca Machado, Célio Fernando Figueiredo Angolini, Damila Rodrigues de Morais, Andressa Mara Baseggio, Marcos Nogueira Eberlin, Mário R Maróstica Junior, and Julian Martinez. 2019. "Sequential High-Pressure Extraction to Obtain Capsinoids and Phenolic Compounds from Biquinho Pepper (Capsicum Chinense)." *The Journal of Supercritical Fluids* 150: 112–21.

Ahmed, Kashif, Yan Li, David Julian McClements, and Hang Xiao. 2012. "Nanoemulsion-and Emulsion-Based Delivery Systems for Curcumin: Encapsulation and Release Properties." *Food Chemistry* 132(2): 799–807.

Alara, Oluwaseun R, Nour H Abdurahman, and Chinonso I Ukaegbu. 2018. "Soxhlet Extraction of Phenolic Compounds from Vernonia Cinerea Leaves and Its Antioxidant Activity." *Journal of Applied Research on Medicinal and Aromatic Plants* 11: 12–17.

Alara, Oluwaseun Ruth, Nour Hamid Abdurahman, and Chinonso Ishamel Ukaegbu. 2021. "Extraction of Phenolic Compounds: A Review." *Current Research in Food Science* 4: 200–14.

Alara, Oluwaseun Ruth, Nour Hamid Abdurahman, Chinonso Ishmael Ukaegbu, and Nassereldeen Ahmed Kabbashi. 2019. "Extraction and Characterization of Bioactive Compounds in Vernonia Amygdalina Leaf Ethanolic Extract Comparing Soxhlet and Microwave-Assisted Extraction Techniques." *Journal of Taibah University for Science* 13(1): 414–22.

Alarcón-Flores, María Isabel, Roberto Romero-González, José Luis Martínez Vidal, and Antonia Garrido Frenich. 2013. "Multiclass Determination of Phytochemicals in Vegetables and Fruits by Ultra High Performance Liquid Chromatography Coupled to Tandem Mass Spectrometry." *Food Chemistry* 141(2): 1120–29.

Alberti, Ágnes, Eszter Riethmüller, and Szabolcs Béni. 2018. "Characterization of Diarylheptanoids: An Emerging Class of Bioactive Natural Products." *Journal of Pharmaceutical and Biomedical Analysis* 147: 13–34.

Alencar Filho, José M T de, Hyany A P Teixeira, Pedrita A Sampaio, Emanuella C V Pereira, Isabela A e Amariz, Pedro J Rolim Neto, Larissa A Rolim, and Edigênia C da Cruz Araújo. 2020. "Phytochemical Analysis in Alternanthera Brasiliana by LC-MS/MS and GC-MS." *Natural Product Research* 34(3): 429–33.

Alirezalu, Kazem, Mirian Pateiro, Milad Yaghoubi, Abolfazl Alirezalu, Seyed Hadi Peighambardoust, and Jose M Lorenzo. 2020. "Phytochemical Constituents, Advanced Extraction Technologies and Techno-Functional Properties of Selected Mediterranean Plants for Use in Meat Products. A Comprehensive Review." *Trends in Food Science & Technology* 100: 292–306.

Altemimi, Ammar, Naoufal Lakhssassi, Azam Baharlouei, Dennis G Watson, and David A Lightfoot. 2017. "Phytochemicals: Extraction, Isolation, and Identification of Bioactive Compounds from Plant Extracts." *Plants* 6(4): 42.

Ameer, K, H M Shahbaz, and J H Kwon. 2017. "Green Extraction Methods for Polyphenols from Plant Matrices and Their Byproducts: A Review." *Comprehensive Reviews in Food Science and Food Safety* 16: 295–315.

Aremu, Adeyemi Oladapo, and Mack Moyo. 2022. "Health Benefits and Biological Activities of Spiny Monkey Orange (Strychnos Spinosa Lam.): An African Indigenous Fruit Tree." *Journal of Ethnopharmacology* 283: 114704.

Arumugham, Thanigaivelan, K Rambabu, Shadi W Hasan, Pau Loke Show, Jörg Rinklebe, and Fawzi Banat. 2021. "Supercritical Carbon Dioxide Extraction of Plant Phytochemicals for Biological and Environmental Applications–A Review." *Chemosphere* 271: 129525.

Avanza, M Victoria, Gerardo Álvarez-Rivera, Alejandro Cifuentes, José A Mendiola, and Elena Ibáñez. 2021. "Phytochemical and Functional Characterization of Phenolic Compounds from Cowpea (Vigna Unguiculata (L.) Walp.) Obtained by Green Extraction Technologies." *Agronomy* 11(1): 162.

Azmir, Jannatul, Islam Sarker Mohamed Zaidul, Mohd M Rahman, K M Sharif, A Mohamed, F Sahena, M H A Jahurul, K Ghafoor, N A N Norulaini, and A K M Omar. 2013. "Techniques for Extraction of Bioactive Compounds from Plant Materials: A Review." *Journal of Food Engineering* 117(4): 426–36.

Ba, Yinying, Ran Xiao, Qi-Jun Chen, Li-Yuan Xie, Rong-Rong Xu, Ping Yu, Xiao-Qing Chen, and Xia Wu. 2021. "Comprehensive Quality Evaluation of Polygoni Orientalis Fructus and Its Processed Product: Chemical Fingerprinting and Simultaneous Determination of Seven Major Components Coupled with Chemometric Analyses." *Phytochemical Analysis* 32(2): 141–52.

Ballesteros-Vivas, Diego, Gerardo Álvarez-Rivera, Elena Ibáñez, Fabián Parada-Alfonso, and Alejandro Cifuentes. 2019. "A Multi-Analytical Platform Based on Pressurized-Liquid Extraction, in Vitro Assays and Liquid Chromatography/Gas Chromatography Coupled to High Resolution Mass Spectrometry for Food by-Products Valorisation. Part 2: Characterization of Bioactive Compound." *Journal of Chromatography A* 1584: 144–54.

Barbera, Giorgia La, Anna Laura Capriotti, Chiara Cavaliere, Carmela Maria Montone, Susy Piovesana, Roberto Samperi, Riccardo Zenezini Chiozzi, and Aldo Laganà. 2017. "Liquid Chromatography-High Resolution Mass Spectrometry for the Analysis of Phytochemicals in Vegetal-Derived Food and Beverages." *Food Research International* 100: 28–52.

Bell, Luke, Natasha D Spadafora, Carsten T Müller, Carol Wagstaff, and Hilary J Rogers. 2016. "Use of TD-GC–TOF-MS to Assess Volatile Composition during Post-Harvest Storage in Seven Accessions of Rocket Salad (Eruca Sativa)." *Food Chemistry* 194: 26–36.

Belwal, Tarun, Indra D Bhatt, Ranbeer S Rawal, and Veena Pande. 2017. "Microwave-Assisted Extraction (MAE) Conditions Using Polynomial Design for Improving Antioxidant Phytochemicals in Berberis Asiatica Roxb. Ex DC. Leaves." *Industrial Crops and Products* 95: 393–403.

Belwal, Tarun, Farid Chemat, Petras Rimantas Venskutonis, Giancarlo Cravotto, Durgesh Kumar Jaiswal, Indra Dutt Bhatt, Hari Prasad Devkota, and Zisheng Luo. 2020. "Recent Advances in Scaling-up of Non-Conventional Extraction Techniques: Learning from Successes and Failures." *TrAC Trends in Analytical Chemistry* 127: 115895.

Belwal, Tarun, Shahira M Ezzat, Luca Rastrelli, Indra D Bhatt, Maria Daglia, Alessandra Baldi, Hari Prasad Devkota, Ilkay Erdogan Orhan, Jayanta Kumar Patra, and Gitishree Das. 2018. "A Critical Analysis of Extraction Techniques Used for Botanicals: Trends, Priorities, Industrial Uses and Optimization Strategies." *TrAC Trends in Analytical Chemistry* 100: 82–102.

Blicharski, Tomasz, and Anna Oniszczuk. 2017. "Extraction Methods for the Isolation of Isofiavonoids from Plant Material." *Open Chem* 15: 34–45.

Braga, Dario, Elena Dichiarante, Giuseppe Palladino, Fabrizia Grepioni, Michele R Chierotti, Roberto Gobetto, and Luca Pellegrino. 2010. "Remarkable Reversal of Melting Point Alternation by Co-Crystallization." *CrystEngComm* 12(11): 3534–36.

Braga, Vanessa Cristina de Carvalho, Gérson Antônio Pianetti, and Isabela Costa César. 2020. "Comparative Stability of Arbutin in Arctostaphylos Uva-ursi by a New Comprehensive Stability-indicating HPLC Method." *Phytochemical Analysis* 31(6): 884–91.

Büyükköroğlu, Gülay, Devrim Demir Dora, Filiz Özdemir, and Candan Hızel. 2018. "Techniques for Protein Analysis." In Debmalya Barh and Vasco Azevedo (eds.), *Omics Technologies and Bio-Engineering*, 317–51. Elsevier.

Cao, Ji-Liang, Jin-Chao Wei, Mei-Wan Chen, Huan-Xing Su, Jian-Bo Wan, Yi-Tao Wang, and Peng Li. 2014. "Application of Two-Dimensional Chromatography in the Analysis of Chinese Herbal Medicines." *Journal of Chromatography A* 1371: 1–14.

Castro-López, Cecilia, Janeth M Ventura-Sobrevilla, María D González-Hernández, Romeo Rojas, Juan A Ascacio-Valdés, Cristóbal N Aguilar, and Guillermo C G Martínez-Ávila. 2017. "Impact of Extraction Techniques on Antioxidant Capacities and Phytochemical Composition of Polyphenol-Rich Extracts." *Food Chemistry* 237: 1139–48.

Chang, X L, Y M Feng, and W H Wang. 2011. "Comparison of the Polysaccharides Isolated from Skin Juice, Gel Juice and Flower of Aloe Arborescens Tissues." *Journal of the Taiwan Institute of Chemical Engineers* 42(1): 13–19.

Chemat, Farid, Natacha Rombaut, Anne-Gaëlle Sicaire, Alice Meullemiestre, Anne-Sylvie Fabiano-Tixier, and Maryline Abert-Vian. 2017. "Ultrasound Assisted Extraction of Food and Natural Products. Mechanisms, Techniques, Combinations, Protocols and Applications. A Review." *Ultrasonics Sonochemistry* 34(January): 540–60. https://doi.org/10.1016/j.ultsonch.2016.06.035.

Cheng, Mengzhen, Jianqing Zhang, Lin Yang, Shijie Shen, Ping Li, Shuai Yao, Hua Qu, Jiayuan Li, Changliang Yao, and Wenlong Wei. 2021. "Recent Advances in Chemical Analysis of Licorice (Gan-Cao)." *Fitoterapia* 149: 104803.

Chong, Esther Swee Lan, Tony K McGhie, Julian A Heyes, and Kathryn M Stowell. 2013. "Metabolite Profiling and Quantification of Phytochemicals in Potato Extracts Using Ultra-high-performance Liquid Chromatography–Mass Spectrometry." *Journal of the Science of Food and Agriculture* 93(15): 3801–8.

Choudhury, Paramita, Krishna N Dutta, Akanksha Singh, Dipankar Malakar, Manoj Pillai, Narayan C Talukdar, Suman Kumar Samanta, and Rajlakshmi Devi. 2020. "Assessment of Nutritional Value and Quantitative Analysis of Bioactive Phytochemicals through Targeted LC-MS/MS Method in Selected Scented and Pigmented Rice Varietals." *Journal of Food Science* 85(6): 1781–92.

Christou, Atalanti, Ioannis J Stavrou, and Constantina P Kapnissi-Christodoulou. 2021. "Continuous and Pulsed Ultrasound-Assisted Extraction of Carob's Antioxidants: Processing Parameters Optimization and Identification of Polyphenolic Composition." *Ultrasonics Sonochemistry* 76: 105630.

Colombo, R, and Papetti, A. 2019. *Food Aroma Compounds by Capillary Electrophoresis in Food Aroma Evolution: Effects from Food Processing, Cooking and Aging*. Edited by M Bordiga and L.M.L. Nollet. Taylor & Francis.

Colombo, Raffaella, and Adele Papetti. 2020. "Pre-Concentration and Analysis of Mycotoxins in Food Samples by Capillary Electrophoresis." *Molecules* 25(15): 3441.

Das, Anup K, and Saikat Dewanjee. 2018. "Optimization of Extraction Using Mathematical Models and Computation." In Satyajit D. Sarker and Lutfun Nahar (eds.), *Computational Phytochemistry*, 75–106. Elsevier.

Daud, Nurizzati Mohd, Nicky Rahmana Putra, Roslina Jamaludin, Nur Salsabila Md Norodin, Nurul Syaza Sarkawi, Muhammad Hamiz Syukri Hamzah, Hasmida Mohd Nasir, Dayang Norulfairuz Abang Zaidel, Mohd Azizi Che Yunus, and Liza Md Salleh. 2022. "Valorisation of Plant Seed as Natural Bioactive Compounds by Various Extraction Methods: A Review." *Trends in Food Science & Technology* 119: 201–14.

De, S, Y N Dey, and A K Ghosh. 2010. "Phytochemical Investigation and Chromatographic Evaluation of the Different Extracts of Tuber of Amorphaphallus Paeoniifolius (Araceae)." *International Journal of Pharmaceutical and Biological Research* 1(5): 150–57.

Dinan, Laurence, Christine Balducci, Louis Guibout, and René Lafont. 2020. "Small-scale Analysis of Phytoecdysteroids in Seeds by HPLC-DAD-MS for the Identification and Quantification of Specific Analogues, Dereplication and Chemotaxonomy." *Phytochemical Analysis* 31(5): 643–61.

Domínguez-Rodríguez, Gloria, María Luisa Marina, and Merichel Plaza. 2021. "Enzyme-Assisted Extraction of Bioactive Non-Extractable Polyphenols from Sweet Cherry (Prunus Avium L.) Pomace." *Food Chemistry* 339: 128086.

Dong, Xin, Rui Wang, Xu Zhou, Ping Li, and Hua Yang. 2016. "Current Mass Spectrometry Approaches and Challenges for the Bioanalysis of Traditional Chinese Medicines." *Journal of Chromatography B* 1026: 15–26.

Dzah, Courage Sedem, Yuqing Duan, Haihui Zhang, Chaoting Wen, Jixian Zhang, Guangying Chen, and Haile Ma. 2020. "The Effects of Ultrasound Assisted Extraction on Yield, Antioxidant, Anticancer and Antimicrobial Activity of Polyphenol Extracts: A Review." *Food Bioscience* 35: 100547.

Ebede, Guy Roland, Joséphine Ngo Mbing, Alexis Bienvenue Nama, Nuzhat Shehla, Atta-ur Rahman, Dieudonné Emmanuel Pegnyemb, Joseph Thierry Ndongo, and Muhammad Iqbal Choudhary. 2022. "New Glycocerebrosides from the Trunk of Tabernaemontana Contorta Stapf.(Apocynaceae) and Their Antibacterial Activity." *Biochemical Systematics and Ecology* 101: 104396.

(EDQM), European Directorate for the Quality of Medicines, ed. 2017. *Javanese Turmeric Monograph.* European Pharmacopoeia (9th ed.): Strasbourg, France.

El-Bahy, G M S. 2005. "FTIR and Raman Spectroscopic Study of Fenugreek (Trigonella Foenum Graecum L.) Seeds." *Journal of Applied Spectroscopy* 72(1): 111–16.

Elez Garofulić, Ivona, Valentina Kruk, Ana Martić, Ivan Martić, Zoran Zorić, Sandra Pedisić, Sanja Dragović, and Verica Dragović-Uzelac. 2020. "Evaluation of Polyphenolic Profile and Antioxidant Activity of Pistacia Lentiscus L. Leaves and Fruit Extract Obtained by Optimized Microwave-Assisted Extraction." *Foods* 9(11): 1556.

Essien, Sinemobong O, Brent Young, and Saeid Baroutian. 2020. "Recent Advances in Subcritical Water and Supercritical Carbon Dioxide Extraction of Bioactive Compounds from Plant Materials." *Trends in Food Science & Technology* 97: 156–69.

Fantoukh, Omer I, Yan-Hong Wang, Abidah Parveen, Mohammed F Hawwal, Gadah A Al-Hamoud, Zulfiqar Ali, Amar G Chittiboyina, and Ikhlas A Khan. 2021. "Profiling and Quantification of the Key Phytochemicals from the Drumstick Tree (Moringa Oleifera) and Dietary Supplements by UHPLC-PDA-MS." *Planta Medica* 87(05): 417–27.

Farag, Mohamed A, Andrea Porzel, and Ludger A Wessjohann. 2015. "Unraveling the Active Hypoglycemic Agent Trigonelline in Balanites Aegyptiaca Date Fruit Using Metabolite Fingerprinting by NMR." *Journal of Pharmaceutical and Biomedical Analysis* 115: 383–87.

Farag, Nermeen F, Sherweit H El-Ahmady, Enas H Abdelrahman, Annette Naumann, Hartwig Schulz, Shadia M Azzam, and El-Sayeda A El-Kashoury. 2018. "Characterization of Essential Oils from Myrtaceae Species Using ATR-IR Vibrational Spectroscopy Coupled to Chemometrics." *Industrial Crops and Products* 124: 870–77.

Flórez, Noelia, Enma Conde, and Herminia Domínguez. 2015. "Microwave Assisted Water Extraction of Plant Compounds." *Journal of Chemical Technology & Biotechnology* 90(4): 590–607.

Frosi, Ilaria, Irene Montagna, Raffaella Colombo, Chiara Milanese, and Adele Papetti. 2021. "Recovery of Chlorogenic Acids from Agri-Food Wastes: Updates on Green Extraction Techniques." *Molecules* 26(15): 4515.

Fuentes, Jhunior Abrahan Marcía, Lucía López-Salas, Isabel Borrás-Linares, Miguel Navarro-Alarcón, Antonio Segura-Carretero, and Jesús Lozano-Sánchez. 2021. "Development of an Innovative Pressurized Liquid Extraction Procedure by Response Surface Methodology to Recover Bioactive Compounds from Carao Tree Seeds." *Foods* 10(2): 398.

Gligor, Octavia, Andrei Mocan, Cadmiel Moldovan, Marcello Locatelli, Gianina Crişan, and Isabel C F R Ferreira. 2019. "Enzyme-Assisted Extractions of Polyphenols–A Comprehensive Review." *Trends in Food Science & Technology* 88: 302–15.

Green, Cheryl E, Sheridan L Hibbert, Yvonne A Bailey-Shaw, Lawrence A D Williams, Sylvia Mitchell, and Eric Garraway. 2008. "Extraction, Processing, and Storage Effects on Curcuminoids and Oleoresin Yields from Curcuma Longa L. Grown in Jamaica." *Journal of Agricultural and Food Chemistry* 56(10): 3664–70.

Gupta, Priyanka, Darshana Patil, and Avinash Patil. 2019. "Qualitative HPTLC Phytochemical Profiling of Careya Arborea Roxb. Bark, Leaves and Seeds." *3 Biotech* 9(8): 1–8.

Hasa, Dritan, Beatrice Perissutti, Stefano Dall'Acqua, Michele R Chierotti, Roberto Gobetto, Iztok Grabnar, Cinzia Cepek, and Dario Voinovich. 2013. "Rationale of Using Vinca Minor Linne Dry Extract Phytocomplex as a Vincamine's Oral Bioavailability Enhancer." *European Journal of Pharmaceutics and Biopharmaceutics* 84(1): 138–44.

Hocart, Charles H. 2010. "9.10 Mass Spectrometry: An Essential Tool for Trace Identification and Quantification." *Comprehensive Natural Products II*, 327–88.

Iguiniz, Marion, and Sabine Heinisch. 2017. "Two-Dimensional Liquid Chromatography in Pharmaceutical Analysis. Instrumental Aspects, Trends and Applications." *Journal of Pharmaceutical and Biomedical Analysis* 145: 482–503.

Ingle, Krishnananda P, Amit G Deshmukh, Dipika A Padole, Mahendra S Dudhare, Mangesh P Moharil, and Vaibhav C Khelurkar. 2017. "Phytochemicals: Extraction Methods, Identification and Detection of Bioactive Compounds from Plant Extracts." *Journal of Pharmacognosy and Phytochemistry* 6(1): 32–36.

Jagessar, R, T Campus, G Georgetown, and S America. 2017. "Phytochemical Screening and TLC Profile of Montricardia Arborescens." *American Journal of Research Communication* 5(1): 129–42.

Ji, Shuai, Shuang Wang, Haishan Xu, Zhenyu Su, Daoquan Tang, Xue Qiao, and Min Ye. 2018. "The Application of On-Line Two-Dimensional Liquid Chromatography (2DLC) in the Chemical Analysis of Herbal Medicines." *Journal of Pharmaceutical and Biomedical Analysis* 160: 301–13.

Jiang, Ming, Chadin Kulsing, Yada Nolvachai, and Philip J Marriott. 2015. "Two-Dimensional Retention Indices Improve Component Identification in Comprehensive Two-Dimensional Gas Chromatography of Saffron." *Analytical Chemistry* 87(11): 5753–61.

Habeebullah, K, Sabeena Farvin, Surendraraj Alagarsamy, Zainab Sattari, Sakinah Al-Haddad, Saja Fakhraldeen, Aws Al-Ghunaim, and Faiza Al-Yamani. 2020. "Enzyme-Assisted Extraction of Bioactive Compounds from Brown Seaweeds and Characterization." *Journal of Applied Phycology* 32(1): 615–29.

Kasemsumran, Sumaporn, Vichien Keeratinijakal, Warunee Thanapase, and Yukihiro Ozaki. 2010. "Near Infrared Quantitative Analysis of Total Curcuminoids in Rhizomes of Curcuma Longa by Moving Window Partial Least Squares Regression." *Journal of Near Infrared Spectroscopy* 18(4): 263–69.

Kaushik, Sulochana, Lalit Dar, Samander Kaushik, and Jaya Parkash Yadav. 2021. "Anti-Dengue Activity of Super Critical Extract and Isolated Oleanolic Acid of Leucas Cephalotes Using in Vitro and in Silico Approach." *BMC Complementary Medicine and Therapies* 21(1): 1–15.

Khalili, Golchehreh, Ali Mazloomifar, Kambiz Larijani, Mohammad Saber Tehrani, and Parviz Aberoomand Azar. 2018. "Solvent-Free Microwave Extraction of Essential Oils from Thymus Vulgaris L. and Melissa Officinalis L." *Industrial Crops and Products* 119: 214–17.

Khatoon, Sayyada, Saba Irshad, Madan Mohan Pandey, Subha Rastogi, and Ajay Kumar Singh Rawat. 2019. "A Validated HPTLC Densitometric Method for Determination of Lupeol, β-Sitosterol and Rotenone in Tephrosia Purpurea: A Seasonal Study." *Journal of Chromatographic Science* 57(8): 688–96.

Knez Hrnčič, Maša, Darija Cör, Jana Simonovska, Željko Knez, Zoran Kavrakovski, and Vesna Rafajlovska. 2020. "Extraction Techniques and Analytical Methods for Characterization of Active Compounds in Origanum Species." *Molecules* 25(20): 4735.

Kovačević, Danijela Bursać, Marta Maras, Francisco J Barba, Daniel Granato, Shahin Roohinejad, Kumar Mallikarjunan, Domenico Montesano, Jose M Lorenzo, and Predrag Putnik. 2018. "Innovative Technologies for the Recovery of Phytochemicals from Stevia Rebaudiana Bertoni Leaves: A Review." *Food Chemistry* 268: 513–21.

Kraujalienė, V, A Pukalskas, and P R Venskutonis. 2017. "Biorefining of Goldenrod (Solidago Virgaurea L.) Leaves by Supercritical Fluid and Pressurized Liquid Extraction and Evaluation of Antioxidant Properties and Main Phytochemicals in the Fractions and Plant Material." *Journal of Functional Foods* 37: 200–208.

Kroslakova, Ivana, Simona Pedrussio, and Evelyn Wolfram. 2016. "Direct Coupling of HPTLC with MALDI-TOF MS for Qualitative Detection of Flavonoids on Phytochemical Fingerprints." *Phytochemical Analysis* 27(3–4): 222–28.

Kusuma, Heri Septya, Ali Altway, and Mahfud Mahfud. 2018. "Solvent-Free Microwave Extraction of Essential Oil from Dried Patchouli (Pogostemon Cablin Benth) Leaves." *Journal of Industrial and Engineering Chemistry* 58: 343–48.

Lasta, Heloísa Fabian Battistella, Lucas Lentz, Luiz Gustavo Gonçalves Rodrigues, Natália Mezzomo, Luciano Vitali, and Sandra Regina Salvador Ferreira. 2019. "Pressurized Liquid Extraction Applied for the Recovery of Phenolic Compounds from Beetroot Waste." *Biocatalysis and Agricultural Biotechnology* 21: 101353.

Lefebvre, Thibault, Emilie Destandau, and Eric Lesellier. 2021. "Selective Extraction of Bioactive Compounds from Plants Using Recent Extraction Techniques: A Review." *Journal of Chromatography A* 1635: 461770.

Leitão, Gilda Guimarães, Carla Monteiro Leal, Simony Carvalho Mendonça, and Rogelio Pereda-Miranda. 2021. "Purification of Alkaloids by Countercurrent Chromatography." *Revista Brasileira de Farmacognosia* 31(5): 1–23.

Leyva, Vanessa E, Juan M Lopez, Alvaro Zevallos-Ventura, Rodrigo Cabrera, Cristhian Cañari-Chumpitaz, David Toubiana, and Helena Maruenda. 2021. "NMR-Based Leaf Metabolic Profiling of V. Planifolia and Three Endemic Vanilla Species from the Peruvian Amazon." *Food Chemistry* 358: 129365.

Leyva-Jiménez, Francisco Javier, Jesús Lozano-Sánchez, Isabel Borrás-Linares, David Arráez-Román, and Antonio Segura-Carretero. 2018. "Comparative Study of Conventional and Pressurized Liquid Extraction for Recovering Bioactive Compounds from Lippia Citriodora Leaves." *Food Research International* 109: 213–22.

Li, Zheng, Kai Chen, Meng-zhe Guo, and Dao-quan Tang. 2016. "Two-dimensional Liquid Chromatography and Its Application in Traditional Chinese Medicine Analysis and Metabonomic Investigation." *Journal of Separation Science* 39(1): 21–37.

Lin, Xue, Lingfeng Wu, Xiong Wang, Linling Yao, and Lu Wang. 2021. "Ultrasonic-Assisted Extraction for Flavonoid Compounds Content and Antioxidant Activities of India Moringa Oleifera L. Leaves: Simultaneous Optimization, HPLC Characterization and Comparison with Other Methods." *Journal of Applied Research on Medicinal and Aromatic Plants* 20: 100284.

Liu, Zaizhi, Baoqin Deng, Shuailan Li, and Zhengrong Zou. 2018. "Optimization of Solvent-Free Microwave Assisted Extraction of Essential Oil from Cinnamomum Camphora Leaves." *Industrial Crops and Products* 124: 353–62.

Liu, Zaizhi, Hualan Li, Zheng Zhu, Dai Huang, Yanlong Qi, Chunhui Ma, Zhengrong Zou, and Hiyan Ni. 2022. "Cinnamomum Camphora Fruit Peel as a Source of Essential Oil Extracted Using the Solvent-Free Microwave-Assisted Method Compared with Conventional Hydrodistillation." *LWT* 153: 112549.

Lopez-Gutierrez, Noelia, Roberto Romero-González, Antonia Garrido Frenich, and José Luis Martínez Vidal. 2014. "Identification and Quantification of the Main Isoflavones and Other Phytochemicals in Soy Based Nutraceutical Products by Liquid Chromatography–Orbitrap High Resolution Mass Spectrometry." *Journal of Chromatography A* 1348: 125–36.

Lovrić, Vanja, Predrag Putnik, Danijela Bursać Kovačević, Marijana Jukić, and Verica Dragović-Uzelac. 2017. "Effect of Microwave-Assisted Extraction on the Phenolic Compounds and Antioxidant Capacity of Blackthorn Flowers." *Food Technology and Biotechnology* 55(2): 243–50.

Lucini, Luigi, Marco Pellizzoni, Roberto Pellegrino, Gian Pietro Molinari, and Giuseppe Colla. 2015. "Phytochemical Constituents and in Vitro Radical Scavenging Activity of Different Aloe Species." *Food Chemistry* 170: 501–7.

Machado, Alan Rodrigues Teixeira, Gisele Avelar Lage, Felipe da Silva Medeiros, José Dias de Souza Filho, and Lúcia Pinheiro Santos Pimenta. 2013. "Quantitative Analysis of Trigonelline in Some Annona Species by Proton NMR Spectroscopy." *Natural Products and Bioprospecting* 3(4): 158–60.

Meng, Ying, Lin Ding, Yuan Wang, Qi-ting Nie, Yang-yang Xing, and Qiang Ren. 2020. "Phytochemical Identification of Lithocarpus Polystachyus Extracts by Ultra-high-performance Liquid Chromatography–Quadrupole Time-of-flight–MS and Their Protein Tyrosine Phosphatase 1B and A-glucosidase Activities." *Biomedical Chromatography* 34(1): e4705.

Methods, ASTA Analytical. n.d. "Curcumin Content of Turmeric Spice and Oleoresins, Method 18.0."

Milić, Anita, Tatjana Daničić, Aleksandra Tepić Horecki, Zdravko Šumić, Nemanja Teslić, Danijela Bursać Kovačević, Predrag Putnik, and Branimir Pavlić. 2022. "Sustainable Extractions for Maximizing Content of Antioxidant Phytochemicals from Black and Red Currants." *Foods* 11(3): 325.

Mohamadi, Neda, Mostafa Pournamdari, Fariba Sharififar, and Mehdi Ansari. 2020. "Simultaneous Spectrophotometric Determination of Trigonelline, Diosgenin and Nicotinic Acid in Dosage Forms Prepared from Fenugreek Seed Extract." *Iranian Journal of Pharmaceutical Research: IJPR* 19(2): 153.

Mommers, John, and Sjoerd van der Wal. 2021. "Column Selection and Optimization for Comprehensive Two-Dimensional Gas Chromatography: A Review." *Critical Reviews in Analytical Chemistry* 51(2): 183–202.

Morsy, Nagy. 2014. "Phytochemical Analysis of Biologically Active Constituents of Medicinal Plants." *Main Group Chemistry* 13(1): 7–21.

Mosić, Mirjana, Aleksandra Dramićanin, Petar Ristivojević, and Dušanka Milojković-Opsenica. 2020. "Extraction as a Critical Step in Phytochemical Analysis." *Journal of AOAC International* 103(2): 365–72.

Mulaudzi, Nduvho, Chinedu P Anokwuru, Sidonie Y Tankeu, Sandra Combrinck, Weiyang Chen, Ilze Vermaak, and Alvaro M Viljoen. 2021. "Phytochemical Profiling and Quality Control of Terminalia Sericea Burch. Ex DC. Using HPTLC Metabolomics." *Molecules* 26(2): 432.

Nadar, Shamraja S, Priyanka Rao, and Virendra K Rathod. 2018. "Enzyme Assisted Extraction of Biomolecules as an Approach to Novel Extraction Technology: A Review." *Food Research International* 108: 309–30.

Nagavekar, Nupur, and Rekha S Singhal. 2019. "Supercritical Fluid Extraction of Curcuma Longa and Curcuma Amada Oleoresin: Optimization of Extraction Conditions, Extract Profiling, and Comparison of Bioactivities." *Industrial Crops and Products* 134: 134–45.

Nastić, Nataša, Isabel Borrás-Linares, Jesús Lozano-Sánchez, Jaroslava Švarc-Gajić, and Antonio Segura-Carretero. 2020. "Comparative Assessment of Phytochemical Profiles of Comfrey (Symphytum Officinale L.) Root Extracts Obtained by Different Extraction Techniques." *Molecules* 25(4): 837.

Nazeam, Jilan A, Haidy A Gad, Hala M El-Hefnawy, and Abdel-Naser B Singab. 2017. "Chromatographic Separation and Detection Methods of Aloe Arborescens Miller Constituents: A Systematic Review." *Journal of Chromatography B* 1058: 57–67.

Nguyen, Hoang Chinh, Huynh Ngoc Truc Nguyen, Meng-Yuan Huang, Kuan-Hung Lin, Dinh-Chuong Pham, Yen Binh Tran, and Chia-Hung Su. 2021. "Optimization of Aqueous Enzyme-assisted Extraction of Rosmarinic Acid from Rosemary (Rosmarinus Officinalis L.) Leaves and the Antioxidant Activity of the Extract." *Journal of Food Processing and Preservation* 45(3): e15221.

Nolvachai, Yada, Chadin Kulsing, and Philip J Marriott. 2017. "Multidimensional Gas Chromatography in Food Analysis." *TrAC Trends in Analytical Chemistry* 96: 124–37.

Oh, Ju-Hee, and Young-Joo Lee. 2014. "Sample Preparation for Liquid Chromatographic Analysis of Phytochemicals in Biological Fluids." *Phytochemical Analysis* 25(4): 314–30.

Ozcan, Akide, Zahide Susluoglu, Gozde Nogay, Muharrem Ergun, and Mehmet Sutyemez. 2021. "Phytochemical Characterization of Some Sumac (Rhus Coriaria L.) Genotypes from Southern Part of Turkey." *Food Chemistry* 358: 129779.

Palsikowski, Paula A, Letícia M Besen, Elissandro J Klein, Camila da Silva, and Edson A da Silva. 2020. "Optimization of Ultrasound-assisted Extraction of Bioactive Compounds from B. Forficata Subsp. Pruinosa." *The Canadian Journal of Chemical Engineering* 98(10): 2214–26.

Panja, Palash. 2018. "Green Extraction Methods of Food Polyphenols from Vegetable Materials." *Current Opinion in Food Science* 23: 173–82.

Papetti, Adele, and Raffaella Colombo. 2019. "High-Performance Capillary Electrophoresis for Food Quality Evaluation." In J. Zhong and X. Wang (eds.), *Evaluation Technologies for Food Quality*, 301–77. Elsevier.

Parveen, Rabea, Sultan Zahiruddin, Akshay Charegaonkar, Abhijeet Khale, and Saikat Mallick. 2020. "Chromatographic Profiling of Rose Petals in Unani Formulations (Gulkand, Arq-e-Gulab, and Rose Sharbat) Using HPTLC and GC–MS." *Journal of AOAC International* 103(3): 684–91.

Pauli, Guido F, Tanja Godecke, Birgit U Jaki, and David C Lankin. 2012. "Quantitative 1H NMR. Development and Potential of an Analytical Method: An Update." *Journal of Natural Products* 75(4): 834–51.

Pimentel-Moral, Sandra, Isabel Borrás-Linares, Jesús Lozano-Sánchez, David Arráez-Román, Antonio Martínez-Férez, and Antonio Segura-Carretero. 2018. "Microwave-Assisted Extraction for Hibiscus Sabdariffa Bioactive Compounds." *Journal of Pharmaceutical and Biomedical Analysis* 156: 313–22.

Pontillo, Antonella Rozaria Nefeli, Lydia Papakosta-Tsigkri, Theopisti Lymperopoulou, Diomi Mamma, Dimitris Kekos, and Anastasia Detsi. 2021. "Conventional and Enzyme-Assisted Extraction of Rosemary Leaves (Rosmarinus Officinalis L.): Toward a Greener Approach to High Added-Value Extracts." *Applied Sciences* 11(8): 3724.

Qadir, Rahman, Farooq Anwar, Mazhar Amjad Gilani, Sadaf Zahoor, Muhammad Misbah ur Rehman, and Muhammad Mustaqeem. 2019. "RSM/ANN Based Optimized Recovery of Phenolics from Mulberry Leaves by Enzyme-Assisted Extraction." *Czech Journal of Food Sciences* 37(2): 99–105.

Qiao, Xue, Wei Song, Shuai Ji, Yan-jiao Li, Yuan Wang, Ru Li, Rong An, De-an Guo, and Min Ye. 2014. "Separation and Detection of Minor Constituents in Herbal Medicines Using a Combination of Heart-Cutting and Comprehensive Two-Dimensional Liquid Chromatography." *Journal of Chromatography A* 1362: 157–67.

Qiao, Xue, Wei Song, Shuai Ji, Qi Wang, De-an Guo, and Min Ye. 2015. "Separation and Characterization of Phenolic Compounds and Triterpenoid Saponins in Licorice (Glycyrrhiza Uralensis) Using Mobile Phase-Dependent Reversed-Phase× Reversed-Phase Comprehensive Two-Dimensional Liquid Chromatography Coupled with Mass Spectrome." *Journal of Chromatography A* 1402: 36–45.

Ramesh, H P, Kohji Yamaki, Hiroshi Ono, and Tojiro Tsushida. 2001. "Two-Dimensional NMR Spectroscopic Studies of Fenugreek (Trigonella Foenum-Graecum L.) Galactomannan without Chemical Fragmentation." *Carbohydrate Polymers* 45(1): 69–77.

Ramya, R. 2022. "GC-MS Analysis of Bioactive Compounds in Ethanolic Leaf Extract of Hellenia Speciosa (J. Koenig) SR Dutta." *Applied Biochemistry and Biotechnology* 194(1): 176–86.

Ranveer, R C, A S Nanadane, P M Ganorkar, and A K Sahoo. 2020. "Enzyme-Assisted Extraction of Anthocyanin from Kokum (Garcinia Indica Choisy) Rinds." *European Journal of Nutrition & Food Safety* 12(10): 125–33.

Rasul, Mohammed Golam. 2018. "Extraction, Isolation and Characterization of Natural Products from Medicinal Plants." *International Journal of Basic Sciences and Applied Computing* 2(6): 1–6.

Re, Roberta, Nicoletta Pellegrini, Anna Proteggente, Ananth Pannala, Min Yang, and Catherine Rice-Evans. 1999. "Antioxidant Activity Applying an Improved ABTS Radical Cation Decolorization Assay." *Free Radical Biology and Medicine* 26(9–10): 1231–37.

Saad, Naima, François Louvet, Stéphane Tarrade, Emmanuelle Meudec, Karine Grenier, Cornelia Landolt, Tan-Sothea Ouk, and Philippe Bressollier. 2019. "Enzyme-assisted Extraction of Bioactive Compounds from Raspberry (Rubus Idaeus L.) Pomace." *Journal of Food Science* 84(6): 1371–81.

Sagi, S, B Avula, Y H Wang, and I A Khan. 2015. "Application of HPTLC in Fingerprint Analysis and Quality Control of Botanicals." *Planta Medica* 81(05): PA21.

Saibaba S V and Shanmuga Pandyan, P. 2016. "High Performance Thin Layer Chromatography: A Mini Review." *Research in Pharmacy Health Science* 2: 219–26.

Selemani, Major A, Luckmore F Kazingizi, Emily Manzombe, Lorraine Y Bishi, Cleopas Mureya, Tichaziwa T Gwata, and Freeborn Rwere. 2021. "Phytochemical Characterization and in Vitro Antibacterial Activity of Xeroderris Stuhlmannii (Taub.) Mendonca & EP Sousa Bark Extracts." *South African Journal of Botany* 142: 344–51.

Seo, Chang-Seob, and Kwang-Hoon Song. 2021. "Phytochemical Characterization for Quality Control of Phyllostachys Pubescens Leaves Using High-Performance Liquid Chromatography Coupled with Diode Array Detector and Tandem Mass Detector." *Plants* 11(1): 50.

Sharma, Basista R, Vikas Kumar, Satish Kumar, and Parmjit S Panesar. 2020. "Microwave Assisted Extraction of Phytochemicals from Ficus Racemosa." *Current Research in Green and Sustainable Chemistry* 3: 100020.

Sharma, Maanas, and Kshirod K Dash. 2021. "Deep Eutectic Solvent-based Microwave-assisted Extraction of Phytochemical Compounds from Black Jamun Pulp." *Journal of Food Process Engineering* 44(8): e13750.

Shi, Xue, Xiaoli Wei, Xinmin Yin, Yuhua Wang, Min Zhang, Cuiqing Zhao, Haiyang Zhao, Craig J McClain, Wenke Feng, and Xiang Zhang. 2015. "Hepatic and Fecal Metabolomic Analysis of the Effects of Lactobacillus Rhamnosus GG on Alcoholic Fatty Liver Disease in Mice." *Journal of Proteome Research* 14(2): 1174–82.

Shimizu, Takafumi, Mutsumi Watanabe, Alisdair R Fernie, and Takayuki Tohge. 2018. "Targeted LC-MS Analysis for Plant Secondary Metabolites." In C. Anto´nio (ed.), *Plant Metabolomics*, 171–81. Springer.

Somkuwar, Dipali O, and Vilas A Kamble. 2013. "Phytochemical Screening of Ethanolic Extracts of Stem, Leaves, Flower and Seed Kernel of Mangifera Indica L." *International Journal of Pharma and Bio Sciences* 4(2): 383–89.

Song, Hua, Jianhong Lin, Xuan Zhu, and Qing Chen. 2016. "Developments in High-speed Countercurrent Chromatography and Its Applications in the Separation of Terpenoids and Saponins." *Journal of Separation Science* 39(8): 1574–91.

Spanos, George A, and Ronald E Wrolstad. 1990. "Influence of Processing and Storage on the Phenolic Composition of Thompson Seedless Grape Juice." *Journal of Agricultural and Food Chemistry* 38(7): 1565–71.

Stanojević, LP, Jelena S Stanojević, Dragan J Cvetković, and Dušica P Ilić. 2016. "Antioxidant Activity of Oregano Essential Oil (Origanum Vulgare L.)." *Biologica Nyssana* 7(2): 131–9.

Strasbourg, France, ed. 2017. *Turmeric Rhizome Monograph*. European Pharmacopoeia (9th ed.): European Directorate for the Quality of Medicines (EDQM).

Stylos, Evgenios, Maria V Chatziathanasiadou, Aggeliki Syriopoulou, and Andreas G Tzakos. 2017. "Liquid Chromatography Coupled with Tandem Mass Spectrometry (LC–MS/MS) Based Bioavailability Determination of the Major Classes of Phytochemicals." *Journal of Chromatography B* 1047: 15–38.

Tanaka, Ken, Yosiaki Kuba, Tetsuro Sasaki, Fumiko Hiwatashi, and Katsuko Komatsu. 2008. "Quantitation of Curcuminoids in Curcuma Rhizome by Near-Infrared Spectroscopic Analysis." *Journal of Agricultural and Food Chemistry* 56(19): 8787–92.

Tlhapi, Dorcas B, Isaiah D I Ramaite, and Chinedu P Anokwuru. 2021. "Metabolomic Profiling and Antioxidant Activities of Breonadia Salicina Using 1H-NMR and UPLC-QTOF-MS Analysis." *Molecules* 26(21): 6707.

Tor-Roca, Alba, Mar Garcia-Aloy, Fulvio Mattivi, Rafael Llorach, Cristina Andres-Lacueva, and Mireia Urpi-Sarda. 2020. "Phytochemicals in Legumes: A Qualitative Reviewed Analysis." *Journal of Agricultural and Food Chemistry* 68(47): 13486–96.

Tyśkiewicz, Katarzyna, Marcin Konkol, and Edward Rój. 2018. "The Application of Supercritical Fluid Extraction in Phenolic Compounds Isolation from Natural Plant Materials." *Molecules* 23(10): 2625.

Uwineza, Pascaline Aimee, and Agnieszka Waśkiewicz. 2020. "Recent Advances in Supercritical Fluid Extraction of Natural Bioactive Compounds from Natural Plant Materials." *Molecules* 25(17): 3847. https://doi.org/10.3390/molecules25173847.

Viswanathan, Vivek, and Alka P Mukne. 2016. "Development and Validation of HPLC and HPTLC Methods for Estimation of Glabridin in Extracts of Glycyrrhiza Glabra." *Journal of AOAC International* 99(2): 374–79.

Wang, Shuangyuan, Lizhen Qiao, Xianzhe Shi, Chunxiu Hu, Hongwei Kong, and Guowang Xu. 2015. "On-Line Stop-Flow Two-Dimensional Liquid Chromatography–Mass Spectrometry Method for the Separation and Identification of Triterpenoid Saponins from Ginseng Extract." *Analytical and Bioanalytical Chemistry* 407(1): 331–41.

Wong, Melody Yee-Man, Pui-Kin So, and Zhong-Ping Yao. 2016. "Direct Analysis of Traditional Chinese Medicines by Mass Spectrometry." *Journal of Chromatography B* 1026: 2–14.

Wrolstad, R E. 1976. "Color and Pigment Analysis in Fruit Products (Bulletin 624)." *Corvallis, Oregon: Oregon Agricultural Experimental Station.*

Wu, Shihua, and Junling Liang. 2010. "Counter-Current Chromatography for High Throughput Analysis of Natural Products." *Combinatorial Chemistry & High Throughput Screening* 13(10): 932–42.

Xu, D P, Y Li, X Meng, T Zhou, Y Zhou, and J Zheng. 2017. "Natural Antioxidants in Foods and Medicinal Plants: Extraction, Assessment and Resources." *International Journal of Molecular Sciences* 18(1): 96.

Yao, Zhi-Hong, Zi-Fei Qin, Liang-Liang He, Xin-Luan Wang, Yi Dai, Ling Qin, Frank J Gonzalez, Wen-Cai Ye, and Xin-Sheng Yao. 2017. "Identification, Bioactivity Evaluation and Pharmacokinetics of Multiple Components in Rat Serum after Oral Administration of Xian-Ling-Gu-Bao Capsule by Ultra Performance Liquid Chromatography Coupled with Quadrupole Time-of-Flight Tandem Mass Spectrometr." *Journal of Chromatography B* 1041: 104–12.

Zandoná, Giovana Paula, Lucíola Bagatini, Natália Woloszyn, Juliane de Souza Cardoso, Jessica Fernanda Hoffmann, Liziane Schittler Moroni, Francieli Moro Stefanello, Alexander Junges, and Cesar Valmor Rombaldi. 2020. "Extraction and Characterization of Phytochemical Compounds from Araçazeiro (Psidium Cattleianum) Leaf: Putative Antioxidant and Antimicrobial Properties." *Food Research International* 137: 109573.

Zhang, Aihua, Hui Sun, Guangli Yan, and Xijun Wang. 2017a. "Recent Developments and Emerging Trends of Mass Spectrometry for Herbal Ingredients Analysis." *TrAC Trends in Analytical Chemistry* 94: 70–76.

Zhang, Hongye, Lu Wang, Bin Lu, Wen Qi, Fuying Jiao, Hong Zhang, and Dan Yuan. 2019. "Metabolite Profiling and Quantification of Phytochemicals of Tianma–Gouteng Granule in Human and Rat Urine Using Ultra High Performance Liquid Chromatography Coupled with Tandem Mass Spectrometry." *Journal of Separation Science* 42(17): 2762–70.

Zhang, Lihong, Yanyan Wang, Xiuyun Guo, and Shihua Wu. 2017b. "Concentrical Coils Counter-Current Chromatography for Natural Products Isolation: Salvia Miltiorrhiza Bunge as Example." *Journal of Chromatography A* 1491: 108–16.

Zhang, Qun-Qun, DONG Xin, L I U Xin-Guang, G A O Wen, L I Ping, and YANG Hua. 2016. "Rapid Separation and Identification of Multiple Constituents in Danhong Injection by Ultra-High Performance Liquid Chromatography Coupled to Electrospray Ionization Quadrupole Time-of-Flight Tandem Mass Spectrometry." *Chinese Journal of Natural Medicines* 14(2): 147–60.

Zhang, Wei-jie. 1994. "Biochemical Technology of Carbohydrate Complexes." *Hangzhou, Zhejiang.*

Zhu, Ming-Zhi, Gui-Lin Chen, Jian-Lin Wu, Na Li, Zhong-Hua Liu, and Ming-Quan Guo. 2018. "Recent Development in Mass Spectrometry and Its Hyphenated Techniques for the Analysis of Medicinal Plants." *Phytochemical Analysis* 29(4): 365–74.

Zou, Dixin, Jinfeng Wang, Bo Zhang, Suhua Xie, Qing Wang, Kexin Xu, and Ruichao Lin. 2015. "Analysis of Chemical Constituents in Wuzi-Yanzong-Wan by UPLC-ESI-LTQ-Orbitrap-MS." *Molecules* 20(12): 21373–404.

Zuo, Guang-Lei, Hyun Yong Kim, Yanymee N Guillen Quispe, Zhi-Qiang Wang, Seung Hwan Hwang, Kyong-Oh Shin, and Soon Sung Lim. 2021. "Efficient Separation of Phytochemicals from Muehlenbeckia Volcanica (Benth.) Endl. by Polarity-Stepwise Elution Counter-Current Chromatography and Their Antioxidant, Antiglycation, and Aldose Reductase Inhibition Potentials." *Molecules* 26(1): 224.

Zuorro, Antonio, Roberto Lavecchia, Ángel Darío González-Delgado, Janet Bibiana García-Martinez, and Pasqua L'Abbate. 2019. "Optimization of Enzyme-Assisted Extraction of Flavonoids from Corn Husks." *Processes* 7(11): 804.

14 Chromatographic Techniques in Phytochemistry and Analytical Techniques in Elemental Profiling

Md. Mushtaque
L.N. Mithila University

Md. Ataur Rahman
New York University Abu Dhabi

Imran Khan
Sultan Qaboos University

Ashanul Haque
University of Ha'il

CONTENTS

14.1 INTRODUCTION

Phytochemicals constitute different classes of compounds that originate from plant and plant-based products. Phytochemicals are non-nutrient compounds linked to various health benefits besides fundamental nutritional values (Thakur, Singh, and Khedkar 2020). Bioactive phytochemicals, alone or in combination, have been used by humans for medicinal purposes (Guan et al. 2021). For example, Unani, Ayurvedic, and Chinese systems of medicine heavily rely on bioactive phytochemicals. In nature, a large pool/reservoir of phytochemicals exists, which could be classified as carbohydrates, lipids, polyphenols, lignans, terpenoids, polyphenols, phytoestrogens, alkaloids, carotenoids, and others (Figure 14.1a) (Xiao and Bai 2019; Lu, Li, and Yin 2016). These large varieties of chemicals play roles such as growth and physiological processes. There is extensive literature on the bioactivity of these phytoconstituents. The use of phytochemicals in modern medicine is widespread, also as numerous drugs currently in use are either natural product or based on it. Some well-known drugs that are obtained from plants include vincristine, vinblastine, paclitaxel, and quinine, which are used worldwide.

Among phytochemicals, compounds with antioxidant properties are highly demanding. This is indeed due to their ability to prevent the deteriorating effects of free radicals produced in the body.

DOI: 10.1201/b22842-14

Free radicals such as reactive oxygen species (ROS) and reactive nitrogen species (RNS) cause oxidative stresses (imbalance between antioxidant systems and the production of oxidants) in the human body, which are associated with chronic diseases including cancer, diabetes, renal failure, and heart-related diseases. It is believed that every day, a normal human cell is encountered by millions of ROS/RNS. The primary function of antioxidants is to stabilize free radicals (ROS, RNS, hydroxyl radical) by either providing a hydrogen radical (H-atom transfer mechanism) or converting the radical into a cation (one electron transfer mechanism). Phytochemicals such as phenolic acids, tannins, flavones, carotenoids, anthocyanins, and monoterpenes are responsible for the antioxidant properties of plant-based materials (Shen et al. 2022). The antioxidant potential of these classes of compounds is linked to their ability to stabilize free radicals via multiple mechanisms. It is to be noted that the type, concentration, and nature of phytochemicals, including antioxidants, are influenced by plant types, environmental factors, and growth conditions (Cavaiuolo, Cocetta, and Ferrante 2013; Huang et al. 2016). Any change in any one of the factors has direct impact on the phytochemicals.

An extract obtained from natural products such as plants contains numerous compounds in varying amounts/levels of chemicals. Chemical fingerprinting is important to obtain an in-depth insight into the structure, activity, and application of a particular phytochemical and to identify new drug candidates. A whole range of instrumentation, including chromatographic and spectroscopic techniques, is employed for chemical fingerprints. After the phytochemicals extraction using a suitable technique (ultrasonication, homogenization, Soxhlet, etc.) and solvent (aqueous or organic), the concentrated mixture is then separated using a suitable chromatographic technique. Among chromatographic techniques, the use of thin-layer chromatography (TLC), high-performance TLC (HPTLC), high-performance liquid chromatography (HPLC), ultra-performance liquid chromatography (UPLC), gas chromatography, etc. including their variants is very common in phytochemical research (Figure 14.1b), which have been highlighted in traditional and exhaustive reviews in the past (Marston 2007). In addition, some developing chromatographic techniques, such as pH-zone-refining countercurrent chromatography (Bakri et al. 2015) and high-speed countercurrent chromatography (Baumann, Adler, and Hamburger 2001), gradually found their application in the separation of phytochemical constituents of herbal medicines. It should be noted that these methods differ in the mechanism of separation (adsorption, partition, affinity, etc.) and has both pros and cons. This chapter highlights the chromatographic techniques used to obtain information on antioxidants found in fruits, flowers, grains, herbs, etc. The main aim of this chapter is to provide readers with an update on this area of research using a selection of recent references.

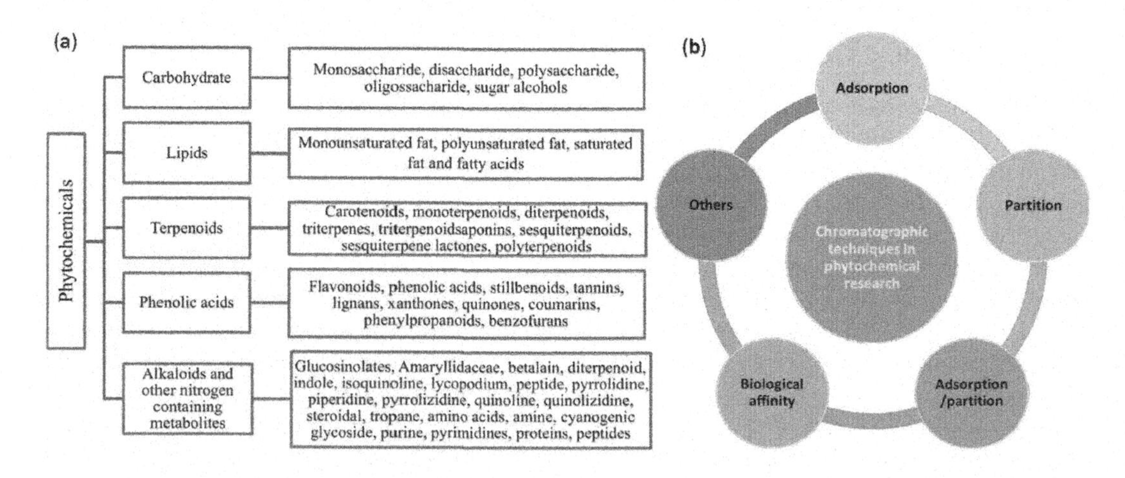

FIGURE 14.1 (a) Some common classes and examples of bioactive phytochemicals found in plants, (b) types of chromatographic techniques (based on the mechanism) used in phytochemical research. Reproduced/adapted with permission from ref. Huang et al. (2016), Marston (2007).

14.2 ANTIOXIDANT SCREENING USING HYPHENATED CHROMATOGRAPHIC TECHNIQUES

Among planar chromatography techniques, TLC and its advanced versions, such as HPTLC, rotation planar chromatography, high-speed TLC, and over-pressured liquid chromatography, are the most widely used types of planar chromatography in the modern time. Although paper chromatography has been used for the analysis of flavonoids (Wender and Gage 1949), amino acids (Thompson, Morris, and Gering 1959), auxins and growth inhibitors (Luckwill 1952), etc., in the past, its use is now very limited (mainly in undergraduate laboratories). Currently, TLC is one of the most effective and widely used traditional qualitative and quantitative techniques in chemical sciences. The advantages of TLC include low maintenance cost, cheap and widely available materials, easy handling, and fast analysis time. In addition to its routine usage in reaction monitoring and sample identification, this technique has shown potential application in pharmaceutical chemistry and natural product research (Komsta, Waksmundzka-Hajnos, and Sherma 2013; Šegan, Opsenica, and Milojković-Opsenica 2019). Both one- and two-dimensional TLC methods with normal- and reverse-phase TLCs have been developed and used for the analysis of a range of compounds. A range of TLC applications has been demonstrated, such as analysis of spices (Danciu, Hosu, and Cimpoiu 2018), synthetic and natural dyes (Danciu, Hosu, and Cimpoiu 2018; Sherma 2019), botanicals (Kowalska and Sajewicz 2022), cannabis (Tang et al. 2021), pesticides (Tang et al. 2021; Sherma 2017), and drug counterfeiting (Sherma and Rabel 2019). Despite these benefits, TLC suffers the drawbacks such as low efficacy, low sensitivity, and poor precision, spots overlapping, which often limit its applicability (Cebolla et al. 2021; Sherma and Rabel 2020; Bernard-Savary and Poole 2015; Häbe and Morlock 2016; Sherma 2018). To overcome these issues and improve the selectivity and efficiency, several modifications have been made. For example, TLC methods combined with mass spectrometry (Cheng, Huang, and Shiea 2011), Raman spectroscopy (István, Keresztury, and Szép 2003), fluorescence spectrometry, and others (Móricz et al. 2020; Rejsek et al. 2016) have been developed.

In these hyphenated methods, incorporating an additional interface enables the extraction of compounds directly from TLC plates into an appropriate source for further analysis, which is used to access more information about the analytes. Very recently, TLC-based technology has been extended for the detection of viruses (Derayea et al. 2020). When TLC is combined with a biological detection method, it is known as TLC bioautography (Marston 2011). To determine biological activity extracts, a TLC-based bioautographic assay has been developed to predict the biological activity of substances present. A slit-scanning densitometry system or imaging system is often used for the qualification and quantification analyses of TLC. Recently, there is an upsurge in the utilization of TLC-image analysis (TLC-IA) technology for pattern recognition and fingerprint analysis of plants such as saffron (Tang et al. 2021; Sereshti et al. 2018; Ristivojević et al. 2020). Although chromatographic techniques have been in use for more than a century, advances in analytical techniques have given a new dimension to them. Combined with chemometric tools, several new hyphenated chromatographic techniques have been reported to separate and identify new phytochemicals and identify medicine-related metabolites *in vivo*.

14.3 ANTIOXIDANTS FOUND IN FLOWERS, SEEDS, AND PEELS

Flowers used for beautification are also excellent sources of phytonutrients with several potential health effects (Skrajda-Brdak, Dąbrowski, and Konopka 2020). Throughout the world, flowers of different plants are used as food (salad, soup) as well as for the preparation of potions, decoction, teas and other beverages, and herbal medicines (Takahashi et al. 2020). It has been demonstrated that flowers such as roses, sunflowers, and dahlia contain a good amount of protein, carbohydrate, lipid, total dietary fiber content, and other phytochemicals (Takahashi et al. 2020). As mentioned earlier, various chromatographic techniques have been used to assess phytoconstituents present in plant materials, including the flower (Table 14.1). For example, Singh and coworkers (Dwivedi et al.

TABLE 14.1
Summary of Some Antioxidants Phytoconstituents Screening Using Hyphenated Chromatographic Techniques

		Separation technique and condition			
N°	**Plant (part)**	**Technique**	**Chromatographic conditions (column, dimension, particle size, solvents)**	**Antioxidant(s) constituents reported**	**References**
1.	*T. aestivum L* (whole wheat flour)	HPLC–ESI–TOF-MS	C_{18} (150 mm×4.6 mm, 1.8 μm); gradient elution (solvent A: 0.5% AcOH and solvent B: ACN)	Syringic acid isomers, syringic acid isomers, vanillic acid, p-coumaric acid, pinosylvin (double glycosylation), ferulic acid, dihydroferulic acid, procyanidin b-3, hinokinin, pinoresinol, etc.	(Leoncini et al. 2012; Dinelli et al. 2011)
2.	*L. macranthoides* (Buds)	HPLC–DAD–QTOF-MS/MS	C_{18} (250 mm×4.6 mm, 5 μm); gradient elution (solvent A: 0.4% AcOH in H_2O and solvent B: 0.4% AcOH in MeOH).	Cs caffeoylquinic acids, hydroxybenzoic acids, protocatechuic acid, caffeic acid, lonicerin, quercetin, chrysoeriol, etc.	(Hu et al. 2016)
3.	*A. grossedentata* (-)	HPLC–DAD–QTOF-MS/MS	C_{18} (150 mm×3.9 mm, 5); gradient elution (solvent A: 0.4% HCOOH in H_2O and solvent B: 0.1% HCOOH in MeOH)	Dihydromyricetin, myricitrin, phloridzin, hesperitin, quercetin-3-o-β-d-xyloside, phloretin, kaempferol-3-o-α-l-rhamnoside, etc.	(Gao et al. 2017)
4.	*P. granatum* L. (flowers)	HPLC–MS/MS	C_{18} column (150 mm×4.6 mm, 5 μm); gradient elution (solvent A: MeOH and solvent B: 1.0 % HCOOH in H_2O)	Pelargonidin 3,5-diglucoside and pelargonidin 3-glucoside	(Zhang, Fu, and Zhang 2011)
5.	*P. granatum* L. (flowers)	HPLC-Q-TOF-MS-MS	C_{18} column (100 mm×2.1 mm, 3.0 μm), gradient elution (solvent A: 0.1% HCOOH in H_2O and solvent B: 0.1% HCOOH in ACN)	Delphinidin-3-O-glucoside isoquercetin, ethyl brevifolincarboxylate, ethyl gallate, brevifolin, apigenin-o-diglucoside, ethyl brevifolincarboxylate, isoquercetin, ellagitannin, and others	(Smrke and Vovk 2013)
6.	*P. rockii* (Leaves and flowers)	UHPLC-ESI-HRMS	C_{18} (100 mm×2.1 mm, 1.7 μm) gradient elution (solvent A: 0.1% HCOOH in H_2O and solvent B: ACN)	Astragalin, apigenin, rhamnoglucoside, gallic acid, hydroxybenzoic acid, digalloyl glucose, and benzoyloxypaeoniflorin	(Bao et al. 2018)

(Continued)

TABLE 14.1 (*Continued*)
Summary of Some Antioxidants Phytoconstituents Screening Using Hyphenated Chromatographic Techniques

N°	Plant (part)	Separation technique and condition		Antioxidant(s) constituents reported	References
		Technique	Chromatographic conditions (column, dimension, particle size, solvents)		
7.	*P. granatum* L (flowers, leaves, and peels)	UHPLC/ QTOF-MS	C_{18} (50×2.1 mm, 1.8 µm), gradient elution (solvent A: H_2O and solvent B: MeOH)	Cyanidin 3-O-(6''-acetyl-glucoside), eriocitrin, chrysoeriol 7-o-(6''-malonyl-glucoside), dimethylmatairesinol, ferulic acid 4-O-glucoside, etc.	(Fellah et al. 2018)
8.	*C. mas* L. (stones)	UPLC/Q-TOF-MS	C_{18} (100×2.1 mm, 1.7 µm); gradient elution (solvent A: 2.0% aq. HCOOH and solvent B: can	Gallotannins and ellagitannins	(Przybylska et al. 2020)
9.	*T. turgidum* L., *T. aestivum* L (sprouted and sound wheat)	UPLC	C_{18} (30×2.1 mm, 1.8 µm). gradient elution (solvent A: 0.1% TFA and solvent B: ACN containing 0.1% TFA.	4-Hydroxybenzoic, vanillic, caffeic, syringic, p-coumaric, ferulic and sinapic	(Van Hung, Hatcher, and Barker 2011)
10.	*C. chinense* (leaves)	UPLC-PDA-ESI-MS/MS	C_{18} (100×2.1 mm, 1.7 µm); gradient elution (solvent A: 0.1% HCOOH in H_2O and solvent B: 0.1% HCOOH in ACN)	N-Caffeoyl putrescine, 5-caffeoylquinic acid, coumaroylquinic acid, apigenin-7-O-(2-O-apiosyl) hexoside, luteolin, etc.	(Herrera-Pool et al. 2021)
11.	*Triticum* sp. (sprouted and sound wheat)	Aluminum oxide 60 F_{254}	Chloroform/methanol/ acetone/ammonia 25% (10:22:53:0.2, v/v/v/v)	Anthocyanidin, delphinidin, etc.	(Sytar et al. 2018)
12.	*C. papaya* (flowers)	Silica gel 60 F_{254}	NR	NR	(Dwivedi et al. 2020)
13.	*B. variegata* L., *G. arborea* Roxb., *S. roxburghii* G. Don, and *V. inopinatum* Craib (flowers)	Silica gel 60 F_{254}	Ethyl acetate:formic acid:water (15:1:1)	Gallic acid, caffeic acid, tannic acid, luteolin, etc.	(Suksathan et al. 2021)

HPLC–DAD–QTOF-MS/MS: high-performance liquid chromatography–diode array detector–quadrupole time-of-flight tandem mass spectrometry; UHPLC/QTOF-MS: ultra-high-pressure liquid chromatography coupled to quadrupole/time-of-flight mass spectrometer; UPLC/Q-TOF-MS: ultra-performance liquid chromatography (UPLC) system, coupled with a quadrupole-time of flight (Q-TOF) MS instrument (UPLC/Synapt Q-TOF MS.
NR: not reported.

2020) studied aqueous and organic extracts of *Carica papaya* flowers using TLC. However, except for R_f values, no other information, such as the solvent system selected and spots/constituents, was provided. In contrary to this, Suksathan et al. (2021) utilized HPTLC and TLC/CMS techniques coupled with chemometric analysis to determine the phenolic content of wildflowers of Thailand (*B. variegata* L., *G. arborea* Roxb., *S. roxburghii* G. Don, and *V. inopinatum* Craib). Using HPTLC, several phytochemicals belonging to flavonoids, phenolic aldehydes, phenolic acids, tannin, anthocyanins, etc., in aqueous and ethanolic extracts were identified, and TLC/CMS further confirmed the structures (Figure 14.2). It was noted that cyanidin chloride (a flavonoid) was present in all

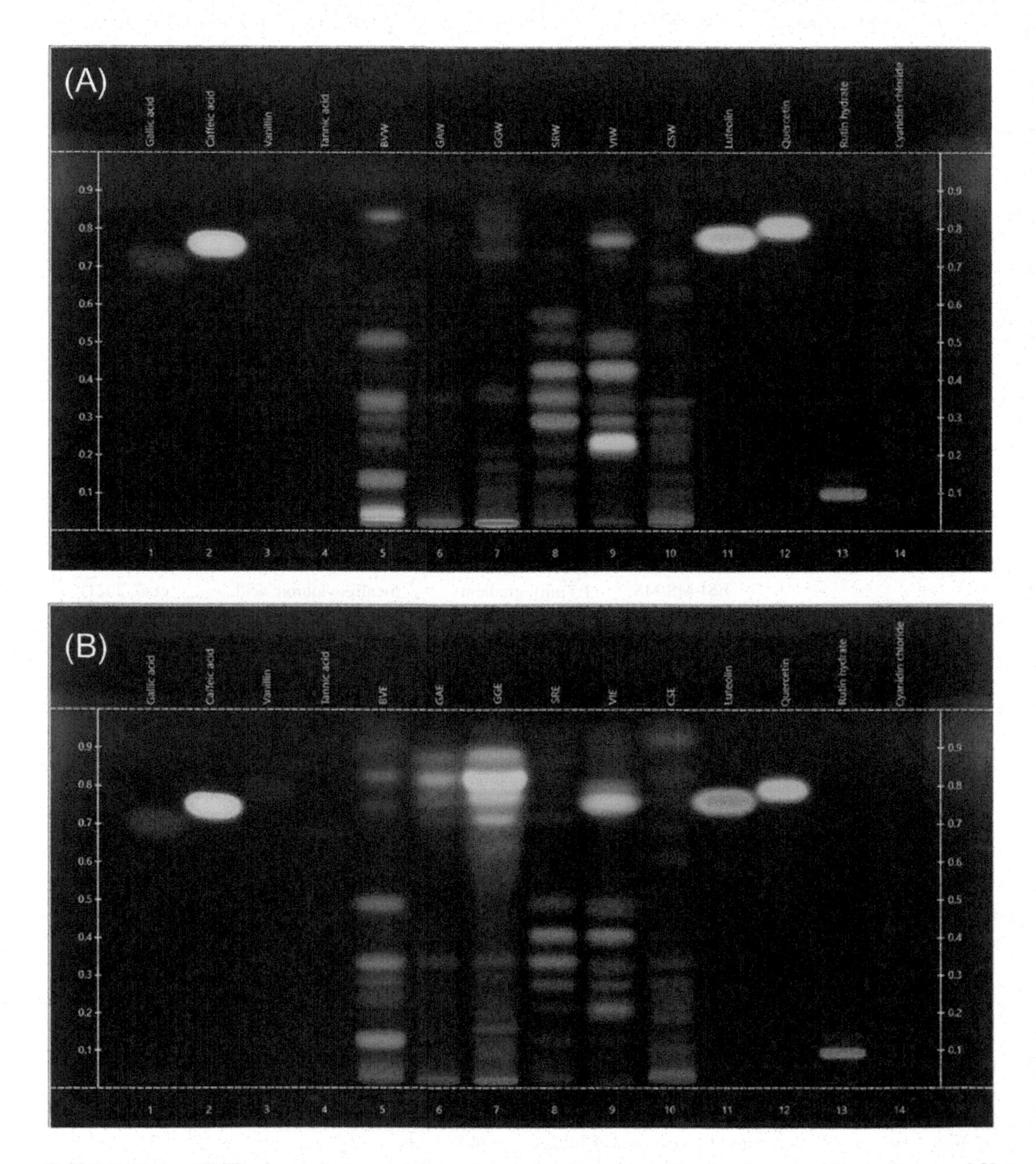

FIGURE 14.2 HPTL chromatogram of *B. variegata* L., *G. arborea* Roxb., *S. roxburghii* G. Don, and *V. inopinatum* Craib and polyphenol standards dissolved in water (A) and ethanol (B). For details, see original reference. Reproduced with permission from Suksathan et al. (2021).

the samples, irrespective of the solvent selected. In addition to the above-discussed methods, several TLC-based bioautographic assays have been reported to determine biological activity of *S. mahagoni* seed (Sahgal et al. 2009), *E. hirta* L leaves (Basma et al. 2011), and *P. atlantica* leaf (Benamar, Marouf, and Bennaceur 2018) extracts and others (Wu and Becker 2012).

Numerous traditional Chinese medicine (TCM) plants are used to prepare beverages and known for their rich antioxidant potential. For example, it was found that plants of Vitaceae (Gao et al. 2017), Caprifoliaceae (Hu et al. 2016), Paeoniaceae (Bao et al. 2018), and Leguminosae (Zhang et al. 2012) families exhibit excellent antioxidant activities. Detailed chromatographic analysis coupled with bioautographic assay revealed that several components in the plant extract are responsible for the activity. For example, Zhang and coworkers (Hu et al. 2016) reported a total of 31 antioxidants in the buds of *Lonicerae macranthoides*. Among those, they identified 20 new compounds responsible for antioxidant activity. The amount of phenolics (phenolic acids and flavonoids) varies with that of the extraction solvent selected. For example, it was found that ethyl acetate (EtOAc) was the best solvent to extract phenolics, followed by butanol (BuOH), water (H_2O), and petroleum ether (PE). They found that the antioxidant activity depends on the method selected, indicating the presence of different components.

Pomegranate (*Punica granatum* L) is a deciduous shrub with high value in ethnopharmacy due to its antioxidant, antidiabetic, and other medicinal properties. Especially, pomegranate fruits are one of the most popular fruits in Asian countries. A plethora of research has been conducted in the past using hyphenated TLC (Amir et al. 2013; Smrke and Vovk 2013), HPLC (Zhang, Fu, and Zhang 2011), and UHPLC (Fellah et al. 2018) to identify bioactive components in seeds, fruits, peels, juice, and leaf extracts. In a recent study, Huang and coworkers (Yisimayili et al. 2019) carried the phytochemical profiling of pomegranate flower extracts and their metabolites in biological samples using the HPLC-Q-TOF-MS-MS technique. A total of 67 compounds (tannins, phenolic acids, anthocyanins, and flavonoids, Figure 14.3) were identified in a sensitive and rapid manner. Of these, compounds delphinidin-3-O-glucoside, ethyl gallate, brevifolin, ellagitannin, apigenin-O-diglucoside, ethyl brevifolincarboxylate, and isoquercetin (peaks 11, 22, 29, 33, 39, 41, and 47, respectively, Figure 14.3) were identified for the first time. *In vivo* studies revealed that these compounds were

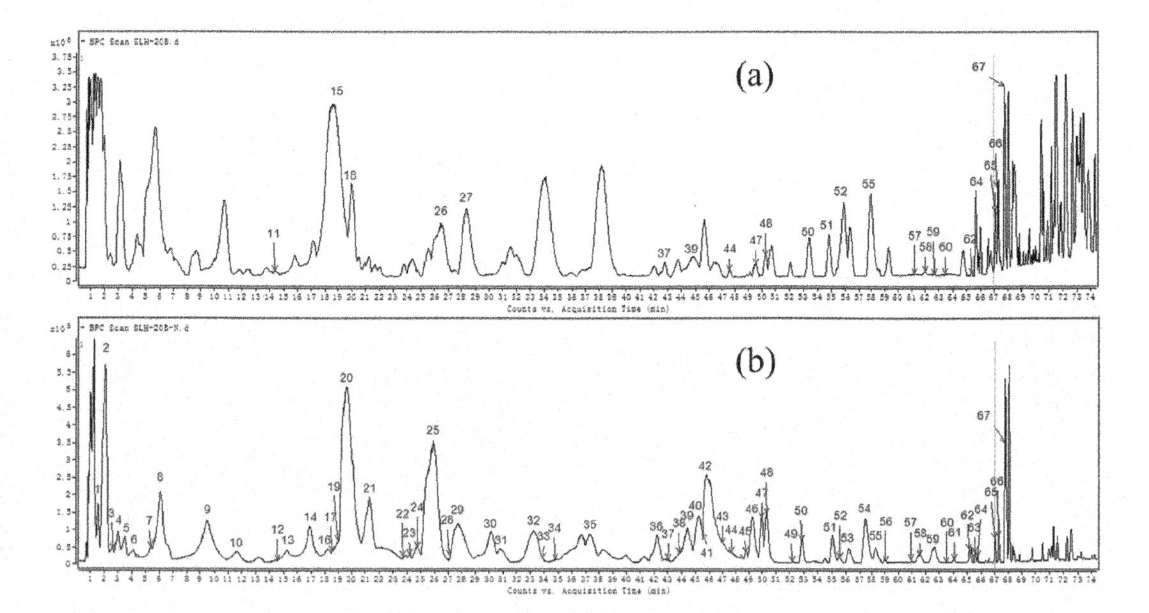

FIGURE 14.3 Base peak chromatogram (BPC) of pomegranate flower extract: (a) positive ion mode, (b) negative ion mode. For numbering, see the original paper. Reproduced with permission from Yisimayili et al. (2019) under a Creative Commons Attribution 4.0 International License.

present in the biological samples, indicating that the plasma did not absorb all compounds present in the extract. It has been demonstrated that due to the difference in the structure and concentration of the individual constituents, the antioxidant potential of the plant varies with the part used. For example, a comparative study on flowers, peels, and leaves of different varieties of Tunisian *Punica granatum* L revealed that phenolics were higher in flowers, while flavonoids and flavonols were highest in the peels (Fellah et al. 2018). The UHPLC-ESI-HRMS study on the flowers and leaves from *Paeonia rockii* indicated that leaf extracts exert more significant antioxidant activity than flower counterparts (Bao et al. 2018). Although there were >25 compounds present, only 18 compounds can be identified. Furthermore, the LC-ESI-QqQ-MS analysis of *P. rockii* fruits revealed the presence of 29 phytochemicals during its different developmental stages (Bai et al. 2021). Trans-resveratrol, benzoic acid, luteolin, and methyl gallate were predominant. An LC-MS analysis of *S. officinalis* indicated the presence of 130 polyphenols in different parts of the plants (Lachowicz et al. 2020). It was found that the amount of flavanols, mainly responsible for the antioxidant activities, was highest in leaves than in flowers, stalks, and roots. Overall, these data strongly suggest that the overall quality and benefits of a flower varies across countries (origin); therefore, the quality and benefits may vary from sample to sample.

Among several grains consumed by humans worldwide, wheat grains is at the top in terms of uses and benefits (Slavin 2004). It is now well established that whole grains are an excellent source of proteins, antioxidants, lignans, phytosterols, fatty acids, and others (Tebben, Shen, and Li 2018; Liu, Yu, and Wu 2020). Wheat grains have been reported to be laced with high amounts of phenolics and exhibit antioxidant activity (Zhou, Su, and Yu 2004). As stated earlier, the phytochemical composition not only depends on the location of the sample collected but also on the sample type. For example, Barker and coworkers (Van Hung, Hatcher, and Barker 2011) developed the very first UPLC method to study and quantify antioxidants in a short running time, with high sensitivity and resolution (Figure 14.4a-c). A comparison with the standard indicated the presence of syringic acid (peak 7) along with caffeic acid (peak 6), ferulic acid (peak 9), and sinapic acid (peak 10) in the

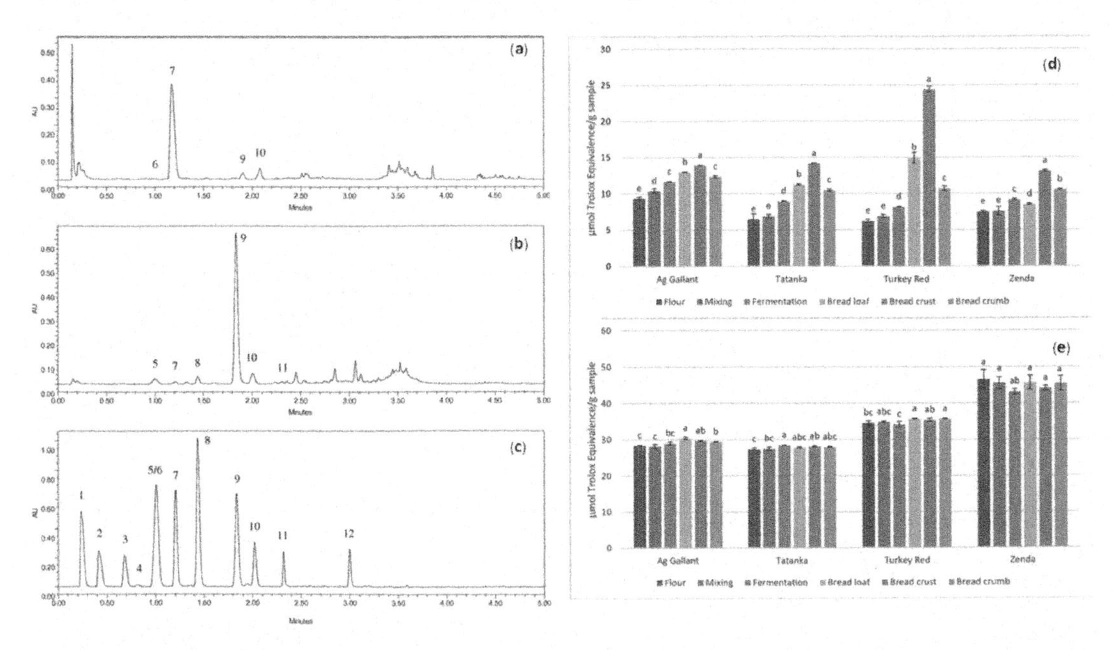

FIGURE 14.4 UPLC chromatogram of (a & b) phenolic compounds extracted under different conditions, (c) mixed standard gallic acid (1), protocatechuic acid (2), 4-hydroxybezoic acid (3), caffeic acid (4), vanillic acid (5), caffeic acid (6), syringic acid (7), p-coumaric acid (8), ferulic acid (9), sinapic acid (10), rutin (11), quercetin (12). ABTS radical scavenging activity of (d) soluble and (e) insoluble phenolic compound extracts. Reproduced with permission from Van Hung, Hatcher, and Barker (2011), Tian et al. (2021).

alcoholic extract. On the other hand, when extracted under alkaline conditions, ferulic acid (peak 9) was the major component along with vanillic acid (peak 5), syringic acid (peak 7), sinapic acid (peak 10), and rutin (peak 11). It was also found that sprouted wheat exhibits better nutritional properties than un-germinated wheat. It was noted that the amount of syringic acids, one of the seven phenolic acids found in wheat flour, increases with germination time, leading to the increased antioxidant property. A recent study conducted using a UPLC-DAD-ESI-Q-TOF-MS/MS system found that wheat flour's total phenolic and flavonoid contents also vary with bread-making, i.e., mixing, fermentation, and baking processes (Tian et al. 2021). It was found that the fermentation process generally increases the soluble phenolic content, flavonoid content, antioxidant activities, and soluble ferulic acid of whole wheat products (14 d,e). Both fermentation and baking processes generally had a positive effect on the antioxidant activity of soluble fractions. Another study on different varieties of wheat flour and wheat grass revealed the presence of anthocyanins such as delphinidin-3-o-galactoside chloride, delphinidin-3-o-glucoside chloride, and cyanindin-3-o-glucoside in varying amounts (Sharma et al. 2020). Before this study, the HPTLC analysis of phytoconstituents present in different wheat grain varieties has also been reported (Sytar et al. 2018). It was shown that the amount of phytoconstituents (such as cyanidin) and bioactivity depend on the genotype and varies in grain and sprout. These results are exciting as cyanidin and delphinidin exhibit potential bioactivity (antioxidant and anticancer) (Sharma et al. 2020).

14.4 ELEMENTAL PROFILING

Elemental profiling of plant-based materials is conducted to determine the type and amount of macro- and microminerals (metal ions) present in the plant. Minerals play a critical role in cellular signaling plants and are responsible for several biochemical reactions in humans (Cheng and Wu 2013; Soetan, Olaiya, and Oyewole 2010). The fact that elements have different roles (some are essential, while some are toxic), knowing their amount is in the product is of great interest and importance. The elemental profiling is carried out using scanning electron microscopy–energy-dispersive X-ray spectroscopy (SEM-EDX), inductively coupled plasma atomic emission spectrometry (ICP-AES), inductively coupled plasma mass spectrometry (ICP-MS), inductively coupled plasma-optical emission spectrometry (ICP-OES), flame photometer, proton-induced X-ray emission spectrometry (PIXE), etc., are used. Wu and Becker have well reviewed the science behind the use of these techniques, limitations, and applications (Cheng and Wu 2013). Herein, we discuss some examples of mineral profiling of natural products known for antioxidant properties. Likewise, for other phytochemicals, the amount and type of macro- and micronutrients (metal ions) depend on multiple factors, such as location, cultivars, part of the fruit, and method of analysis. For example, to determine the elemental composition of different cultivars of pomegranate, Lenucci and coworkers (Montefusco et al. 2021) studied different fruit parts (juices, peels, and kernels) using ICP-AES and found a total of 25 different minerals in addition to several others. Among macronutrients, the three significant elements were K, Mg, and Na in Italian cultivars. On the other hand, Moroccan cultivars studied using ICP-MS showed the presence of iodine also (Loukhmas et al. 2020). Calcium and iron were also abundant in *G. arborea* Roxb., *S. roxburghii* G. Don, and *V. inopinatum* Craib flowers (Suksathan et al. 2021). ICP-AES analysis of *C. gunnii*, *C. taii*, and *C. jiangxiensis* revealed that these plants also contain some toxic metals like As and Hg (Xiao et al. 2009). In addition to the above-discussed example, there are several more examples in which a plant-based product has been assessed for elemental composition (Table 14.2).

14.5 CONCLUSION

Humans have used natural products for nutritional and medicinal purposes for ages. Every day we use a variety of phytochemicals in the form of beverages and foods, which have a direct impact on health. Therefore, it is essential to know phytochemicals' nature, amount, and effect. Antioxidants,

TABLE 14.2
Mineral Present in Various Plant-based Products and Their Profiling Techniques

N°	Sample	Elements/minerals	Profiling technique	References
1.	*P. granatum* L	Ag, Al, As, B, Ba, Bi, Cd Co, Cr, Fe, Cu, I, In, K, Li Mg, Mn, Mo, Na, Ni, P, Pb, S, Sr, Te, Ti, Tl, V, and Zn	AAS, flame photometer, ICP-AES, ICP-MS, and ICP-OES	(Montefusco et al. 2021; Loukhmas et al. 2020; Mirdehghan and Rahemi 2007; Loukhmas et al. 2022; Al-Maiman and Ahmad 2002; Peng 2019)
2.	*B. variegata* L., *G. arborea* Roxb., *S. roxburghii* G. Don, and *V. inopinatum* Craib	Ca, Fe, P, and Na	AAS	(Suksathan et al. 2021)
3.	*C. gunnii, C. taii,* and *C. jiangxiensis*	Ca, Mg, Fe, Zn, Se, Cu, Mo, Cr, Co, Mn, Si, V, Ni, Al, Sn, Pb, Cd, As, and Hg	ICP-AES	(Xiao et al. 2009)
4.	*P. dactylifera* L	K, Ca, P, Mg, Na, Fe, Cu, Zn, and Mn	AAS	(Chaira, Mrabet, and Ferchichi 2009)
5.	*C. sinensis* L.	Na, K, Mg, Ca, S, Fe, Mn, Se, Zn, and Sr	ICP-OES	(da Silva Haas et al. 2022)
6.	*E. cicutarium*	Mg, Ca, Na, K, P, S, Al, B, Ba, Cd, Co, Cr, Cu, Fe, Li, Mn, Ni, Pb, Sr, and Zn	ICP-AES	(Bilić et al. 2020)
7.	*C. chinense*, *C. frutescens*, and *C. annuum*	K, P, Mg, Ca, S, Cu, Fe, Zn, Mo, Mn, B, Ni, Na, Al, Li, Cr, Pb, Co, Se, Cd, As, and Hg	ICP-MS, ICP-OES	(Sarpras et al. 2019; Ahmad et al. 2021; Lučić et al. 2022)
8.	*S. melongena* L.	B, Na, Mg, K, Ca, Cu, As, Rb, Sr, Pd, Cd, Sn, Cs, Ba, and Tl	ICP-MS	(Mi et al. 2020)
9.	Sesame seeds	Na, Pd, Rb, Au, Ag, B, Ba, Mo, Mn, Se, K, Pt, P, and Co	ICP-MS	(Mi et al. 2022)

which are essential to human bodies, are present in a variety of plant-based products. In this chapter, using selected examples, we highlighted the applications of different chromatographic and elemental profiling techniques in phytochemical research. Using various hyphenated techniques, the amount and type of different antioxidants can be estimated accurately. Methods such as TLC bioautography offer a rapid way to ascertain a particular component's bioactive nature. UPLC and HPLC, when hyphenated with a suitable artifact, offer a quick way to separate and identify complex mixtures. Similarly, various inorganic minerals are also present in plants, which can be determined using state-of-the-art technologies such as SEM-EDX, ICP-AES, ICP-MS, ICP-OES, PIXE, and other conventional techniques. Determination of these elements is necessary for food quality control. The discussion clearly shows that the future of hyphenated techniques in natural product research is bright.

ICP-AES, inductively coupled plasma atomic emission spectrometry; ICP-MS, inductively coupled plasma mass spectrometry; ICP-OES, inductively coupled plasma-optical emission spectrometry.

REFERENCES

Ahmad, Ilyas, Abdul Rawoof, Meenakshi Dubey, and Nirala Ramchiary. 2021. "ICP-MS Based Analysis of Mineral Elements Composition during Fruit Development in Capsicum Germplasm." *Journal of Food Composition and Analysis* 101: 103977.

Al-Maiman, Salah A, and Dilshad Ahmad. 2002. "Changes in Physical and Chemical Properties during Pomegranate (Punica Granatum L.) Fruit Maturation." *Food Chemistry* 76 (4): 437–41.

Amir, Mohd, Mohd Mujeeb, Sayeed Ahmad, Mohd Akhtar, and Kamran Ashraf. 2013. "Design Expert-Supported Development and Validation of HPTLC Method: An Application in Simultaneous Estimation of Quercetin and Rutin in Punica Granatum, Tamarindus Indica and Prunus Domestica." *Pharmaceutical Methods* 4 (2): 62–67.

Bai, Zhang-Zhen, Jing Ni, Jun-Man Tang, Dao-Yang Sun, Zhen-Guo Yan, Jing Zhang, Li-Xin Niu, and Yan-Long Zhang. 2021. "Bioactive Components, Antioxidant and Antimicrobial Activities of Paeonia Rockii Fruit during Development." *Food Chemistry* 343: 128444.

Bakri, Mahinur, Qibin Chen, Qingling Ma, Yi Yang, Abdumijit Abdukadir, and Haji Akber Aisa. 2015. "Separation and Purification of Two New and Two Known Alkaloids from Leaves of Nitraria Sibirica by PH-Zone-Refining Counter-Current Chromatography." *Journal of Chromatography B* 1006: 138–45.

Bao, Yating, Yan Qu, Jinhua Li, Yanfang Li, Xiaodong Ren, Katherine G Maffucci, Ruiping Li, Zhanguo Wang, and Rui Zeng. 2018. "In Vitro and in Vivo Antioxidant Activities of the Flowers and Leaves from Paeonia Rockii and Identification of Their Antioxidant Constituents by UHPLC-ESI-HRMSn via Pre-Column DPPH Reaction." *Molecules* 23 (2): 392.

Basma, Abu Arra, Zuraini Zakaria, Lacimanan Yoga Latha, and Sreenivasan Sasidharan. 2011. "Antioxidant Activity and Phytochemical Screening of the Methanol Extracts of Euphorbia Hirta L." *Asian Pacific Journal of Tropical Medicine* 4 (5): 386–90.

Baumann, Dietmar, Sven Adler, and Matthias Hamburger. 2001. "A Simple Isolation Method for the Major Catechins in Green Tea Using High-Speed Countercurrent Chromatography." *Journal of Natural Products* 64 (3): 353–55. https://pubs.acs.org/doi/10.1021/np0004395.

Benamar, Houari, Abderrazak Marouf, and Malika Bennaceur. 2018. "Phytochemical Composition, Antioxidant and Acetylcholinesterase Inhibitory Activities of Aqueous Extract and Fractions of Pistacia Atlantica Subsp. Atlantica from Algeria." *Journal of Herbs, Spices & Medicinal Plants* 24 (3): 229–44.

Bernard-Savary, Pierre, and Colin F Poole. 2015. "Instrument Platforms for Thin-Layer Chromatography." *Journal of Chromatography A* 1421: 184–202.

Bilić, Vanja Ljoljić, Uroš Gašić, Dušanka Milojković-Opsenica, Ivan Nemet, Sanda Rončević, Ivan Kosalec, and Jadranka Vuković Rodriguez. 2020. "First Extensive Polyphenolic Profile of Erodium Cicutarium with Novel Insights to Elemental Composition and Antioxidant Activity." *Chemistry & Biodiversity* 17 (9): e2000280.

Cavaiuolo, Marina, Giacomo Cocetta, and Antonio Ferrante. 2013. "The Antioxidants Changes in Ornamental Flowers during Development and Senescence." *Antioxidants* 2 (3): 132–55. https://mdpi-res.com/d_attachment/antioxidants/antioxidants-02-00132/article_deploy/antioxidants-02-00132.pdf?version=1375779630.

Cebolla, Vicente L, Carmen Jarne, Jesús Vela, Rosa Garriga, Luis Membrado, and Javier Galbán. 2021. "Scanning Densitometry and Mass Spectrometry for HPTLC Analysis of Lipids: The Last 10 Years." *Journal of Liquid Chromatography & Related Technologies* 44 (3–4): 148–70.

Chaira, Nizar, Abdessalem Mrabet, and A L I Ferchichi. 2009. "Evaluation of Antioxidant Activity, Phenolics, Sugar and Mineral Contents in Date Palm Fruits." *Journal of Food Biochemistry* 33 (3): 390–403.

Cheng, Sy-Chyi, Min-Zong Huang, and Jentaie Shiea. 2011. "Thin Layer Chromatography/Mass Spectrometry." *Journal of Chromatography A* 1218 (19): 2700–2711.

Danciu, Virgil, Anamaria Hosu, and Claudia Cimpoiu. 2018. "Thin-Layer Chromatography in Spices Analysis." *Journal of Liquid Chromatography & Related Technologies* 41 (6): 282–300.

Derayea, Sayed M, Mohamed A Abdel-Lateef, Mahmoud A Omar, and Ramadan Ali. 2020. "Thin-layer Chromatography/Fluorescence Detection Approach for Sensitive and Selective Determination of Hepatitis C Virus Antiviral (Velpatasvir): Application to Human Plasma." *Luminescence* 35 (7): 1048–55.

Dinelli, Giovanni, Antonio Segura-Carretero, Raffaella Di Silvestro, Ilaria Marotti, David Arráez-Román, Stefano Benedettelli, Lisetta Ghiselli, and Alberto Fernadez-Gutierrez. 2011. "Profiles of Phenolic Compounds in Modern and Old Common Wheat Varieties Determined by Liquid Chromatography Coupled with Time-of-Flight Mass Spectrometry." *Journal of Chromatography A* 1218 (42): 7670–81.

Dwivedi, Manish Kumar, Shruti Sonter, Shringika Mishra, Digvesh Kumar Patel, and Prashant Kumar Singh. 2020. "Antioxidant, Antibacterial Activity, and Phytochemical Characterization of Carica Papaya Flowers." *Beni-Suef University Journal of Basic and Applied Sciences* 9 (1): 1–11.

Fellah, Boutheina, Marwa Bannour, Gabriele Rocchetti, Luigi Lucini, and Ali Ferchichi. 2018. "Phenolic Profiling and Antioxidant Capacity in Flowers, Leaves and Peels of Tunisian Cultivars of Punica Granatum L." *Journal of Food Science and Technology* 55 (9): 3606–15. https://www.ncbi.nlm.nih.gov/pmc/articles/PMC6098781/pdf/13197_2018_Article_3286.pdf.

Gao, Qingping, Ruyi Ma, Lin Chen, Shuyun Shi, Ping Cai, Shuihan Zhang, and Haiyan Xiang. 2017. "Antioxidant Profiling of Vine Tea (Ampelopsis Grossedentata): Off-Line Coupling Heart-Cutting HSCCC with HPLC–DAD–QTOF-MS/MS." *Food Chemistry* 225: 55–61.

Guan, Ruirui, Quyet Van Le, Han Yang, Dangquan Zhang, Haiping Gu, Yafeng Yang, Christian Sonne, Su Shiung Lam, Jiateng Zhong, and Zhu Jianguang. 2021. "A Review of Dietary Phytochemicals and Their Relation to Oxidative Stress and Human Diseases." *Chemosphere* 271: 129499.

Häbe, Tim T, and Gertrud E Morlock. 2016. "Miniaturization of Instrumental Planar Chromatography with Focus on Mass Spectrometry." *Chromatographia* 79 (13): 797–810.

Herrera-Pool, Emanuel, Ana Luisa Ramos-Díaz, Manuel Alejandro Lizardi-Jiménez, Soledad Pech-Cohuo, Teresa Ayora-Talavera, Juan C Cuevas-Bernardino, Ulises García-Cruz, and Neith Pacheco. 2021. "Effect of Solvent Polarity on the Ultrasound Assisted Extraction and Antioxidant Activity of Phenolic Compounds from Habanero Pepper Leaves (Capsicum Chinense) and Its Identification by UPLC-PDA-ESI-MS/MS." *Ultrasonics Sonochemistry* 76: 105658. https://www.ncbi.nlm.nih.gov/pmc/articles/PMC8273200/pdf/main.pdf.

Hu, Xin, Lin Chen, Shuyun Shi, Ping Cai, Xuejuan Liang, and Shuihan Zhang. 2016. "Antioxidant Capacity and Phenolic Compounds of Lonicerae Macranthoides by HPLC–DAD–QTOF-MS/MS." *Journal of Pharmaceutical and Biomedical Analysis* 124: 254–60.

Huang, Yancui, Di Xiao, Britt M Burton-Freeman, and Indika Edirisinghe. 2016. "Chemical Changes of Bioactive Phytochemicals during Thermal Processing." *Reference Module in Food Science*, Elsevier. doi: 10.1016/B978-0-08-100596-5.03055-9.

Hung, Pham Van, David W Hatcher, and Wendy Barker. 2011. "Phenolic Acid Composition of Sprouted Wheats by Ultra-Performance Liquid Chromatography (UPLC) and Their Antioxidant Activities." *Food Chemistry* 126 (4): 1896–901.

István, Krisztina, Gábor Keresztury, and Andrea Szép. 2003. "Normal Raman and Surface Enhanced Raman Spectroscopic Experiments with Thin Layer Chromatography Spots of Essential Amino Acids Using Different Laser Excitation Sources." *Spectrochimica Acta Part A: Molecular and Biomolecular Spectroscopy* 59 (8): 1709–23.

Komsta, Lukasz, Monika Waksmundzka-Hajnos, and Joseph Sherma. 2013. *Thin Layer Chromatography in Drug Analysis*. CRC Press.

Kowalska, Teresa, and Mieczysław Sajewicz. 2022. "Thin-Layer Chromatography (TLC) in the Screening of Botanicals–Its Versatile Potential and Selected Applications." *Molecules* 27 (19): 6607. https://mdpi-res.com/d_attachment/molecules/molecules-27-06607/article_deploy/molecules-27-06607-v3.pdf?version=1665536920.

Lachowicz, Sabina, Jan Oszmiański, Andrzej Rapak, and Ireneusz Ochmian. 2020. "Profile and Content of Phenolic Compounds in Leaves, Flowers, Roots, and Stalks of Sanguisorba Officinalis L. Determined with the LC-DAD-ESI-QTOF-MS/MS Analysis and Their in Vitro Antioxidant, Antidiabetic, Antiproliferative Potency." *Pharmaceuticals* 13 (8): 191.

Leoncini, Emanuela, Cecilia Prata, Marco Malaguti, Ilaria Marotti, Antonio Segura-Carretero, Pietro Catizone, Giovanni Dinelli, and Silvana Hrelia. 2012. "Phytochemical Profile and Nutraceutical Value of Old and Modern Common Wheat Cultivars." PLoS ONE 7(9): e45997. doi: 10.1371/journal.pone.0045997.

Liu, Jie, Liangli Lucy Yu, and Yanbei Wu. 2020. "Bioactive Components and Health Beneficial Properties of Whole Wheat Foods." *Journal of Agricultural and Food Chemistry* 68 (46): 12904–15. https://pubs.acs.org/doi/10.1021/acs.jafc.0c00705.

Loukhmas, Sarah, Ebrahim Kerak, Meriem Outaki, Majdouline Belaqziz, and Hasnaâ Harrak. 2020. "Assessment of Minerals, Bioactive Compounds, and Antioxidant Activity of Ten Moroccan Pomegranate Cultivars." *Journal of Food Quality* 2020: 1–10.

Loukhmas, Sarah, Ebrahim Kerak, Hamza Zine, and Hasnaâ Harrak. 2022. "Assessment of Physical and Physicochemical Characteristics of Fruit Mesocarp and Peel of Ten Moroccan Pomegranate Cultivars." *Materials Today: Proceedings* 72(part 7): 3229–3942.

Lu, Baiyi, Maiquan Li, and Ran Yin. 2016. "Phytochemical Content, Health Benefits, and Toxicology of Common Edible Flowers: A Review (2000–2015)." *Critical Reviews in Food Science and Nutrition* 56 (sup1): S130–48.

Lučić, Milica, Andrijana Miletić, Aleksandra Savić, Steva Lević, Ivana Sredović Ignjatović, and Antonije Onjia. 2022. "Dietary Intake and Health Risk Assessment of Essential and Toxic Elements in Pepper (Capsicum Annuum)." *Journal of Food Composition and Analysis* 111: 104598.

Luckwill, Li C. 1952. "Application of Paper Chromatography to the Separation and Identification of Auxins and Growth-Inhibitors." *Nature* 169 (4296): 375. https://www.nature.com/articles/169375a0.pdf.

Marston, A. 2011. "Thin-Layer Chromatography with Biological Detection in Phytochemistry." *Journal of Chromatography A* 1218 (19): 2676–83.

Marston, Andrew. 2007. "Role of Advances in Chromatographic Techniques in Phytochemistry." *Phytochemistry* 68 (22–24): 2786–98.

Mi, Si, Yuhang Wang, Xiangnan Zhang, Yaxin Sang, and Xianghong Wang. 2022. "Authentication of the Geographical Origin of Sesame Seeds Based on Proximate Composition, Multi-Element and Volatile Fingerprinting Combined with Chemometrics." *Food Chemistry* 397: 133779.

Mi, Si, Wenlong Yu, Jian Li, Minxuan Liu, Yaxin Sang, and Xianghong Wang. 2020. "Characterization and Discrimination of Chilli Peppers Based on Multi-Element and Non-Targeted Metabolomics Analysis." *LWT* 131: 109742.

Mirdehghan, Seyed Hossein, and Majid Rahemi. 2007. "Seasonal Changes of Mineral Nutrients and Phenolics in Pomegranate (Punica Granatum L.) Fruit." *Scientia Horticulturae* 111 (2): 120–27.

Montefusco, Anna, Miriana Durante, Danilo Migoni, Monica De Caroli, Riadh Ilahy, Zoltán Pék, Lajos Helyes, Francesco Paolo Fanizzi, Giovanni Mita, and Gabriella Piro. 2021. "Analysis of the Phytochemical Composition of Pomegranate Fruit Juices, Peels and Kernels: A Comparative Study on Four Cultivars Grown in Southern Italy." *Plants* 10 (11): 2521. https://mdpi-res.com/d_attachment/plants/plants-10-02521/article_deploy/plants-10-02521.pdf?version=1637336735.

Móricz, Ágnes M, Virág Lapat, Gertrud E Morlock, and Péter G Ott. 2020. "High-Performance Thin-Layer Chromatography Hyphenated to High-Performance Liquid Chromatography-Diode Array Detection-Mass Spectrometry for Characterization of Coeluting Isomers." *Talanta* 219: 121306.

Peng, Yingshu. 2019. "Comparative Analysis of the Biological Components of Pomegranate Seed from Different Cultivars." *International Journal of Food Properties* 22 (1): 784–94.

Przybylska, Dominika, Alicja Z Kucharska, Iwona Cybulska, Tomasz Sozański, Narcyz Piórecki, and Izabela Fecka. 2020. "Cornus Mas L. Stones: A Valuable by-Product as an Ellagitannin Source with High Antioxidant Potential." *Molecules* 25 (20): 4646.

Rejsek, Jan, Vladimír Vrkoslav, Anu Vaikkinen, Markus Haapala, Tiina J Kauppila, Risto Kostiainen, and Josef Cvacka. 2016. "Thin-Layer Chromatography/Desorption Atmospheric Pressure Photoionization Orbitrap Mass Spectrometry of Lipids." *Analytical Chemistry* 88 (24): 12279–86.

Ristivojević, Petar, Jelena Trifković, Filip Andrić, and Dušanka Milojković-Opsenica. 2020. "Recent Trends in Image Evaluation of HPTLC Chromatograms." *Journal of Liquid Chromatography & Related Technologies* 43 (9–10): 291–99.

Sahgal, Geethaa, Surash Ramanathan, Sreenivasan Sasidharan, Mohd Nizam Mordi, Sabariah Ismail, and Sharif Mahsufi Mansor. 2009. "In Vitro Antioxidant and Xanthine Oxidase Inhibitory Activities of Methanolic Swietenia Mahagoni Seed Extracts." *Molecules* 14 (11): 4476–85. https://mdpi-res.com/d_attachment/molecules/molecules-14-04476/article_deploy/molecules-14-04476.pdf?version=1403112633.

Sarpras, M, Ilyas Ahmad, Abdul Rawoof, and Nirala Ramchiary. 2019. "Comparative Analysis of Developmental Changes of Fruit Metabolites, Antioxidant Activities and Mineral Elements Content in Bhut Jolokia and Other Capsicum Species." *LWT* 105: 363–70.

Šegan, Sandra, Dejan Opsenica, and Dušanka Milojković-Opsenica. 2019. "Thin-Layer Chromatography in Medicinal Chemistry." *Journal of Liquid Chromatography & Related Technologies* 42 (9–10): 238–48.

Sereshti, Hassan, Zahra Poursorkh, Ghazaleh Aliakbarzadeh, Shahin Zarre, and Sahar Ataolahi. 2018. "An Image Analysis of TLC Patterns for Quality Control of Saffron Based on Soil Salinity Effect: A Strategy for Data (Pre)-Processing." *Food Chemistry* 239: 831–39.

Sharma, Natasha, Vandita Tiwari, Shreya Vats, Anita Kumari, Venkatesh Chunduri, Satveer Kaur, Payal Kapoor, and Monika Garg. 2020. "Evaluation of Anthocyanin Content, Antioxidant Potential and Antimicrobial Activity of Black, Purple and Blue Colored Wheat Flour and Wheat-Grass Juice against Common Human Pathogens." *Molecules* 25 (24): 5785.

Shen, Nan, Tongfei Wang, Quan Gan, Sian Liu, Li Wang, and Biao Jin. 2022. "Plant Flavonoids: Classification, Distribution, Biosynthesis, and Antioxidant Activity." *Food Chemistry* 383: 132531.

Sherma, Joseph. 2017. "Review of Thin-Layer Chromatography in Pesticide Analysis: 2014–2016." *Journal of Liquid Chromatography & Related Technologies* 40 (5–6): 226–38.

Sherma, Joseph. 2018. "Biennial Review of Planar Chromatography: 2015–2017." *Journal of AOAC International* 101 (4): 905–13.

Sherma, Joseph. 2019. "Thin-Layer Chromatography in the Determination of Synthetic and Natural Colorants in Foods." In Nelu Grinberg and Peter W. Carr (eds.), *Advances in Chromatography*, 109–35. CRC Press.

Sherma, Joseph, and Fred Rabel. 2019. "Advances in the Thin Layer Chromatographic Analysis of Counterfeit Pharmaceutical Products: 2008–2019." *Journal of Liquid Chromatography & Related Technologies* 42 (11–12): 367–79.

Sherma, Joseph, and Fred Rabel. 2020. "Review of Advances in Planar Chromatography-Mass Spectrometry Published in the Period 2015–2019." *Journal of Liquid Chromatography & Related Technologies* 43 (11–12): 394–412.

Silva Haas, Isabel Cristina da, Juliana Santos de Espindola, Gabriela Rodrigues de Liz, Aderval S Luna, Marilde T Bordignon-Luiz, Elane Schwinden Prudêncio, Jefferson Santos de Gois, and Isabela Maia Toaldo Fedrigo. 2022. "Gravitational Assisted Three-Stage Block Freeze Concentration Process for Producing Enriched Concentrated Orange Juice (Citrus Sinensis L.): Multi-Elemental Profiling and Polyphenolic Bioactives." *Journal of Food Engineering* 315: 110802.

Skrajda-Brdak, Marta, Grzegorz Dąbrowski, and Iwona Konopka. 2020. "Edible Flowers, a Source of Valuable Phytonutrients and Their pro-Healthy Effects–A Review." *Trends in Food Science & Technology* 103: 179–99.

Slavin,Joanne.2004."WholeGrainsandHumanHealth."*NutritionResearchReviews*17(1):99–110.https://www.cambridge.org/core/services/aop-cambridge-core/content/view/D992CF3AF3244A6C85193EBF4557FD2A/S0954422404000095a.pdf/div-class-title-whole-grains-and-human-health-div.pdf.

Smrke, Samo, and Irena Vovk. 2013. "Comprehensive Thin-Layer Chromatography Mass Spectrometry of Flavanols from Juniperus Communis L. and Punica Granatum L." *Journal of Chromatography A* 1289: 119–26.

Soetan, KO, CO Olaiya, and OE Oyewole. 2010. "The Importance of Mineral Elements for Humans, Domestic Animals and Plants-A Review." *African Journal of Food Science* 4 (5): 200–222.

Suksathan, Ratchuporn, Apinya Rachkeeree, Ratchadawan Puangpradab, Kuttiga Kantadoung, and Sarana Rose Sommano. 2021. "Phytochemical and Nutritional Compositions and Antioxidants Properties of Wild Edible Flowers as Sources of New Tea Formulations." *NFS Journal* 24: 15–25.

Sytar, Oksana, Paulina Bośko, Marek Živčák, Marian Brestic, and Iryna Smetanska. 2018. "Bioactive Phytochemicals and Antioxidant Properties of the Grains and Sprouts of Colored Wheat Genotypes." *Molecules* 23 (9): 2282.

Takahashi, Jacqueline Aparecida, Flávia Augusta Guilherme Gonçalves Rezende, Marília Aparecida Fidelis Moura, Laura Ciribelli Borges Dominguete, and Denise Sande. 2020. "Edible Flowers: Bioactive Profile and Its Potential to Be Used in Food Development." *Food Research International* 129: 108868.

Tang, Tie-Xin, Hui Liu, Li-He Deng, Xin-Hua Qiu, and Jie-Fei Liang. 2021. "A Pattern Recognition Method on Smartphones for Planar Chromatography and Verification on Chromatograms of Four Herbal Medicines from Citrus Fruits." *Journal of Liquid Chromatography & Related Technologies* 44 (9–10): 484–89.

Tebben, Lauren, Yanting Shen, and Yonghui Li. 2018. "Improvers and Functional Ingredients in Whole Wheat Bread: A Review of Their Effects on Dough Properties and Bread Quality." *Trends in Food Science & Technology* 81: 10–24.

Thakur, Monika, Karuna Singh, and Renu Khedkar. 2020. "Phytochemicals: Extraction Process, Safety Assessment, Toxicological Evaluations, and Regulatory Issues." In Bhanu Prakash (ed.), *Functional and Preservative Properties of Phytochemicals*, 341–61. Elsevier.

Thompson, J F, C J Morris, and ROSE K Gering. 1959. "Purification of Plant Amino Acid for Paper Chromatography." *Analytical Chemistry* 31 (6): 1028–31.

Tian, Wenfei, Gengjun Chen, Michael Tilley, and Yonghui Li. 2021. "Changes in Phenolic Profiles and Antioxidant Activities during the Whole Wheat Bread-Making Process." *Food Chemistry* 345: 128851.

Wender, Simon H, and Thomas B Gage. 1949. "Paper Chromatography of Flavonoid Pigments." *Science* 109 (2829): 287–89.

Wu, Bei, and J Sabine Becker. 2012. "Imaging techniques for elements and element species in plant science". *Metallomics*, 4(5), 403–16.

Xiao, Jian-Hui, Dai-Min Xiao, Zhong-Hua Sun, Qing Xiong, Zong-Qi Liang, and Jian-Jiang Zhong. 2009. "Chemical Compositions and Antimicrobial Property of Three Edible and Medicinal Cordyceps Species." *Journal of Food, Agriculture and Environment* 7 (3&4): 91–100.

Xiao, Jianbo, and Weibin Bai. 2019. "Bioactive Phytochemicals." *Critical Reviews in Food Science and Nutrition* 59 (6): 827–29.

Yisimayili, Zainaipuguli, Rahima Abdulla, Qiang Tian, Yangyang Wang, Mingcang Chen, Zhaolin Sun, Zhixiong Li, Fang Liu, Haji Akber Aisa, and Chenggang Huang. 2019. "A Comprehensive Study of Pomegranate Flowers Polyphenols and Metabolites in Rat Biological Samples by High-Performance Liquid Chromatography Quadrupole Time-of-Flight Mass Spectrometry." *Journal of Chromatography A* 1604: 460472.

Zhang, Lihua, Quanjuan Fu, and Yuanhu Zhang. 2011. "Composition of Anthocyanins in Pomegranate Flowers and Their Antioxidant Activity." *Food Chemistry* 127 (4): 1444–49.

Zhang, Yu-Ping, Shu-Yun Shi, Xiang Xiong, Xiao-Qing Chen, and Mi-Jun Peng. 2012. "Comparative Evaluation of Three Methods Based on High-Performance Liquid Chromatography Analysis Combined with a 2, 2′-Diphenyl-1-Picrylhydrazyl Assay for the Rapid Screening of Antioxidants from Pueraria Lobata Flowers." *Analytical and Bioanalytical Chemistry* 402 (9): 2965–76. https://link.springer.com/article/10.1007/s00216-012-5722-3.

Zhou, Kequan, Lan Su, and Liangli Yu. 2004. "Phytochemicals and Antioxidant Properties in Wheat Bran." *Journal of Agricultural and Food Chemistry* 52 (20): 6108–14.

15 NMR-based Metabolomics and Hyphenated NMR Techniques

Md. Mushtaque
L.N. Mithila University

Syed Imran Hassan
Sultan Qaboos University

Syed Amir Ashraf and Ashanul Haque
University of Ha'il

CONTENTS

15.1 INTRODUCTION: BACKGROUND AND DRIVING FORCES

For a very long time, human beings have been using natural products (NPs) for nutritional and medicinal benefits. Since ancient times, NPs have been used as a mixture (containing several components in varying amounts with different physicochemical properties) to treat various ailments (Newman and Cragg 2012; Wachtel-Galor and Benzie 2011). However, to harness fully the power of NPs, it is essential to underpin/determine the component responsible for the bioactivity/toxicity. The isolation of chemical compounds present and their structural determination is a daunting task and requires a substantial amount of time, work, and cost. However, with the advent of modern bioanalytical techniques, isolation and identification of bioactive compounds present in an NP have become easier. One such technique is metabolomics, a data-driven technique that is used for chemical fingerprinting of a complex mixture employing multivariate data analysis (MVDA). Metabolomics deals with the study of identification and quantification of metabolites present in a given biological (cells, bio-fluids, and tissues), animal, and plant specimen (Giraudeau 2020). One main goal of metabolomics is to identify all the primary and secondary metabolites qualitatively and quantitatively. As metabolome chemical and physical properties are highly diverse, one or more analytical techniques (such as separation and detection) are coupled to identify metabolites present (De Castro et al. 2019).

Among others, multi-nuclear magnetic resonance (NMR) spectroscopy is a robust and well-established technique for structural analysis (Huang et al. 2020). In the last few decades, NMR has emerged as one of the most important analytical techniques in metabolomics. It is the preferred platform for long-term or large-scale metabolomics studies, due to the easy sample preparation

DOI: 10.1201/b22842-15

and non-destructive nature (Emwas et al. 2019). Besides, the intensity of the resonances in one-dimensional ^{1}H-NMR serves as a quantitative measure of all ^{1}H-containing molecules in the sample. NMR is highly automatable and exceptionally reproducible, making high-throughput (Guennec, Giraudeau, and Caldarelli 2014) large-scale metabolomics studies much more feasible than other techniques (e.g. LC-MS or GC-MS). Applications of NMR-based metabolomics methods have witnessed rapid growth (Nagana Gowda and Raftery 2017). With the help of global profiling and multivariate statistical data analysis, NMR-based metabolomics is employed to systematically identify the drug and endogenous metabolites (Emwas et al. 2019; Azmi et al. 2005).

A typical workflow in NMR-based methodologies includes sample preparation, sample loading, spectral acquisition, and data analysis (Figure 15.1). To obtain metabolomics data, the researcher collects NMR spectra of different nuclei either separately or simultaneously, followed by the data preprocessing phase and subsequent data analysis. Then the metabolomics study requires interpreting the results using MVDA (De Castro et al. 2019). A large number of the dataset generated is then handled by adequate supervised and unsupervised chemometric methods. In most cases, MVDA such as principal component analysis (PCA) and partial least squares-discriminant analysis (PLS-DA) are routinely employed to extract information from large datasets. Each study uses different sampling strategies, data sets, and computational algorithms. PCA is the most commonly used unsupervised multivariate technique that reduces large complex datasets of metabolic profiles to a few principal components. It can effectively identify patterns of variation between samples. Using these steps, protocols for metabolomic profiling using NMR have been developed for biological samples (Beckonert et al. 2007), marine organisms (Bayona, de Voogd, and Choi 2022), plants (Kruger, Troncoso-Ponce, and Ratcliffe 2008; Kim, Choi, and Verpoorte 2010) and others (Nanda et al. 2021; Villalón-López et al. 2018). Protocols have also been developed to distinguish different origins and identify primary and secondary metabolites in medicinal plants (Kim, Choi, and

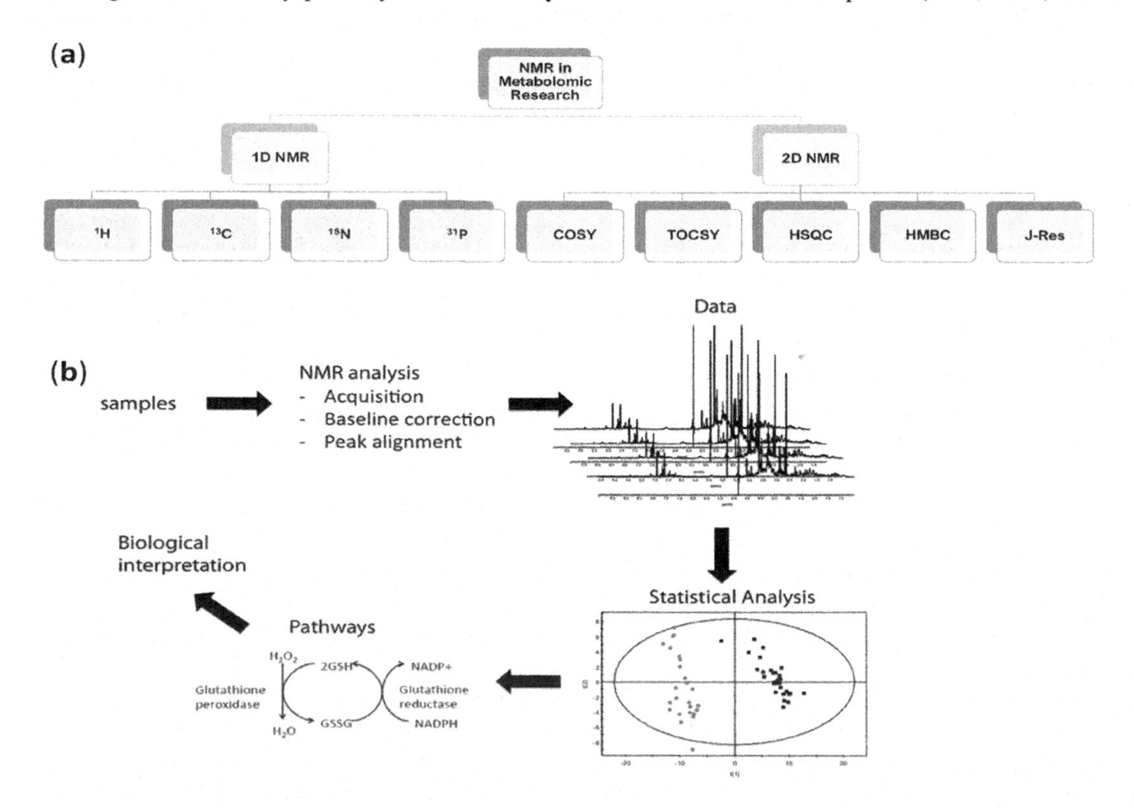

FIGURE 15.1 (a) 1D- and 2D-NMR spectroscopic techniques in metabolomic research, (b) an overview of the steps involved in an NMR-based metabolomics study. Reproduced with permission from Brennan (2014).

Verpoorte 2010). In the following subsections, we highlight NMR-based metabolomics and hyphenated NMR techniques for isolating and identifying bioactive compounds from natural sources.

15.2 APPLICATIONS OF ^{1}H NMR-BASED METABOLOMIC TECHNIQUES

15.2.1 Metabolic Profiling/Monitoring

The metabolic content of plants, seeds, and fruits varies considerably with genetic and environmental factors. The overall quality of a plant is affected by the variation in the metabolite profile over its growth periods. Therefore, it is imperative to determine the metabolomes of interest at different stages (Mediani et al. 2012). Besides, abiotic stresses such as drought, water salinity, soil, and temperatures can also affect the metabolome significantly. The NMR-based technique evolved as a useful and rapid method to monitor changes in the metabolome (Scognamiglio et al. 2019). For instance, it has been used to identify genetic variation in pea seeds (Ellis et al. 2018), phenolic content variation in olive drupes (Esposito et al. 2021), and celery (Lau, Laserna, and Li 2020) grown under different environmental conditions. Raletsena et al. (2022) studied metabolic changes in potato (*Solanum tuberosum* L.) irrigated with or without fly ash-treated acid mine drainage (AMD). However, some common metabolites were found in AMD-treated and untreated samples; cultivars treated with different levels of AMD showed changes in the ^{1}H NMR spectra, indicating AMD-induced stress. Using ^{1}HNMR-based metabolomics combined with MVA, Shaari and coworkers (Tajidin et al. 2019) performed metabolic profiling of *Andrographis paniculata* (Burm. f.) Nees. They selected leaves of different harvest ages (14, 16, 18, 20 and 22 weeks post sowing) to monitor constituents like andrographolide (ANDRO), neoandrographolide (NAG), and 14-deoxyandrographolide (DAG) along with sugars and amino acids (Chart 15.1). It was noted that the majority of the signals (aliphatic and aromatic) were relatively more intense in the spectra of young leaves than those of mature leaves. The levels of ANDRO and DAG were found to vary between the pre-flowering and flowering stages (Figure 15.2a, b), as the young leaves contain more ANDRO and DAG than the mature ones. PCA revealed that young leaves have 36–40% higher levels of diterpene lactones, while mature leaves have higher levels (20–28%) of carbohydrates (Figure 15.2c, d).

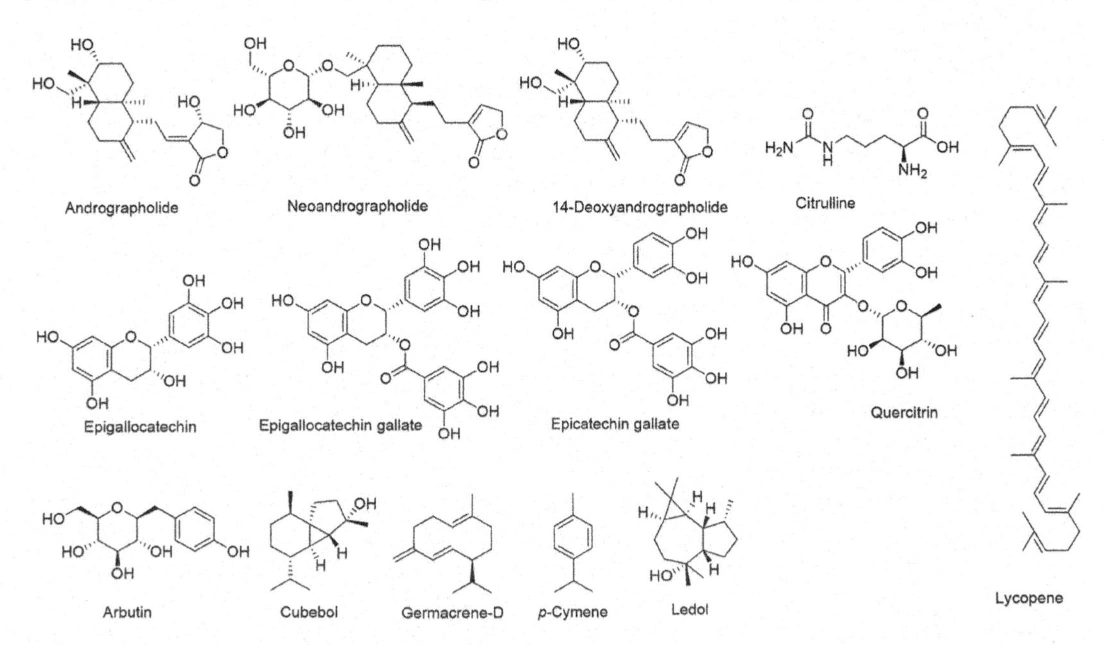

CHART 15.1 Chemical structures of some common metabolites studied using NMR-based metabolomics.

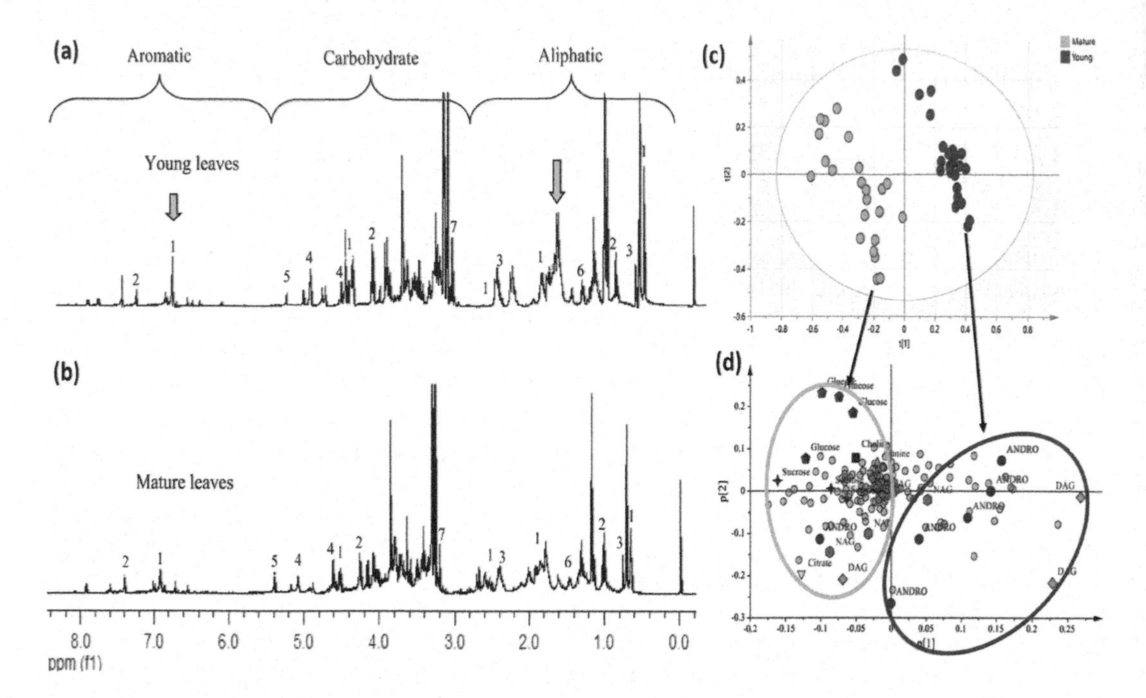

FIGURE 15.2 ^{1}H NMR spectra of young (a) and mature (b) leaves of *A. paniculata*. Andrographolide (1), neoandrographolide (2), 14-deoxyandrographolide (3), glucose (4), sucrose (5), alanine (6), and choline (7). PCA score plot for young versus mature leaves, along PC1 and PC2 (c), and PCA loading plot showing the differentiating metabolites between young and mature leaves of *A. paniculata* (d). Reproduced with permission from Tajidin et al. (2019) under a Creative Commons Attribution 4.0 International License https://www.nature.com/articles/s41598-019-52905-z#rightslink.

Results from metabolomics and regression analysis revealed higher content of bioactive compounds in young leaves, especially during pre-flowering stage.

To underpin the metabolomic differences and discriminate between red and yellow watermelon cultivars (*Citrullus lanatus* Thunb.), Ismail and coworkers (Sulaiman et al. 2020) performed metabolite profiling and MVDA of polar and non-polar extracts. The red fruits containing lycopene and a high concentration of citrulline was observed. However, the main disadvantage of this method included the absence of several components such as amino acids (methionine and lysine) and carotenoids (phytoene, phytofluene, etc.), which have been reported in water in the study by Bang et al. (2010). In a similar study, metabolomics variation with plant growth in different organs (young shoot vs flower) of tea [*Camellia sinensis* (L.) O. Ktze.] was determined (Ye et al. 2022). PCA data revealed a significant difference in principle components/metabolites such as catechins, sugars, organic acids, and amino acids in both the tea shoot and flowers varies significantly between stages (Figure 15.3). For instance, sugars (e.g., sucrose, fructose, and α- and β-glucose), polyphenols (epigallocatechin, epigallocatechin gallate, and epicatechin gallate), amino acids (glycine, alanine, tyrosine, leucine, and valine) and other chemicals were the differential metabolites between tea flowers and young shoots.

In addition to the wide applicability of ^{1}H-NMR metabolomics in active metabolite recognition, this method is also useful for identifying contaminants/adulterants found in complex mixtures of food items (solid and liquid) and differentiating between the spices/fruits grown in different regions. Suzuki et al. (2022) applied ^{1}H-NMR metabolomics coupled with PCA, hierarchical cluster analysis, and nucleotide sequence analysis to differentiate between cinnamon bark of different species, production areas, and utility applications. They found that the level of coumarin varies with the species and could demarcate between different species (Figure 15.4). Similarly, pomegranate fruit

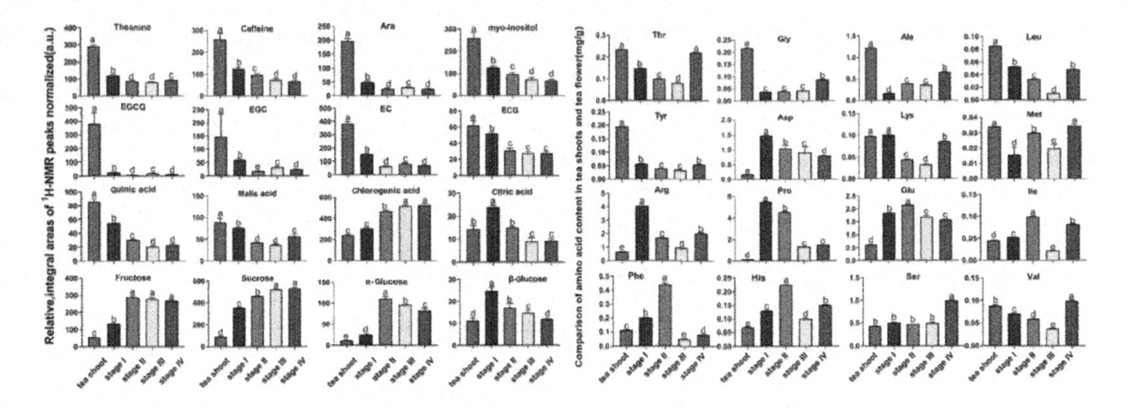

FIGURE 15.3 Relative changes or variations of individual metabolites in young tea shoots and tea flowers at four stages. Lowercase letters in the histogram indicate significant differences between samples. Reproduced with permission from Ye et al. (2022) under a Creative Commons Attribution 4.0 International License https://link.springer.com/article/10.1007/s00706-022-02928-6#rightslink.

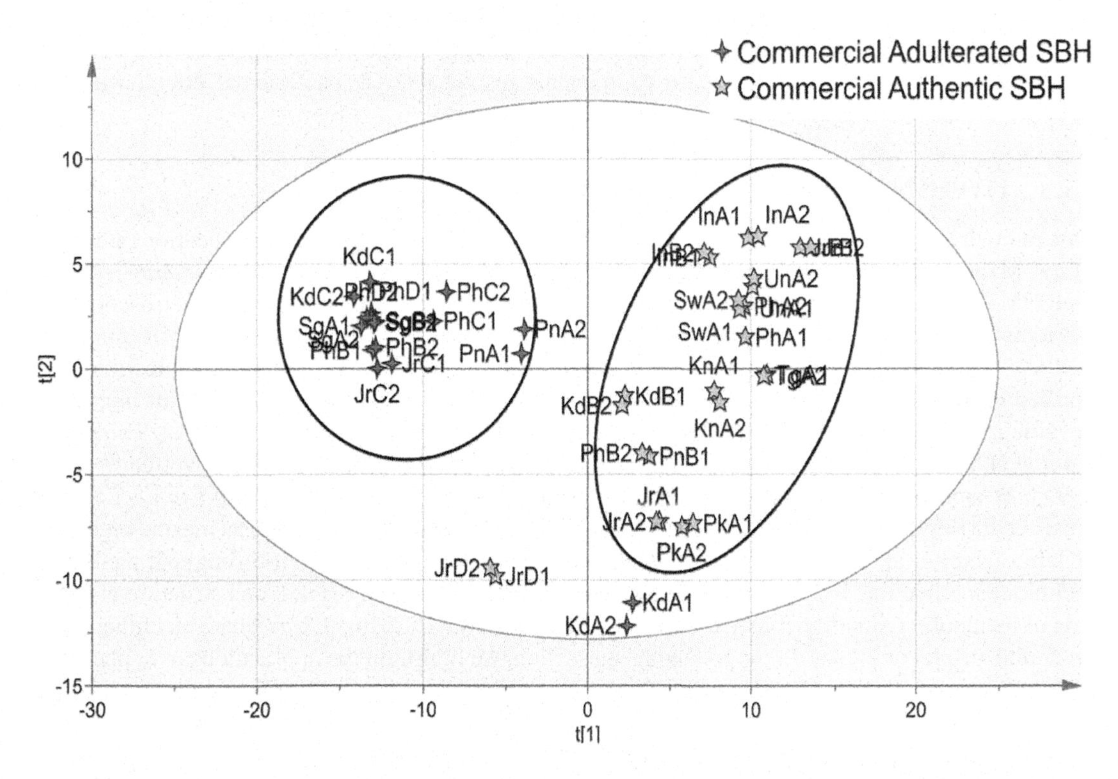

FIGURE 15.4 PCA score plot of commercial authentic SBH and commercial adulterated SBH using all the ^{1}H NMR metabolic fingerprint regions of SBH. Reproduced with permission from Yong et al. (2022).

and its juice (PJ) are well known for their medicinal value, including antidiabetic, antioxidant, anti-inflammatory, and cytoprotective effects (Les et al. 2018). PJ is composed of complex mixtures of sugars, organic acids, amino acids, and polyphenols, and its composition varies with the variety and the impurities. This makes simultaneous detection of PJ challenging using well-established analytical techniques. Besides, the currently available NMR methods are time-consuming due to different evolution times and relaxation delays. To resolve this issue, Tang and Hatzakis (2020) developed a proof-of-concept 1D- and 2D-NMR method for simultaneous detection and quantification of the above-mentioned components in PJ. In this study, the authors employed a novel 2D-NOAH method (nested NMR by ordered acquisition using ^{1}H-detection) method, which has never been used to analyze such complex mixtures. To identify adulteration in stingless bee honey (SBH), Yong et al. (2022) proposed an untargeted ^{1}H NMR coupled with PCA method. Interestingly, the developed method can be used to differentiate between authentic and impregnated samples even in the presence of 1% contaminants (Figure 15.4). The reported method is particularly useful when the current AOAC 998.12 method failed to detect such a small level of contamination.

15.2.1.1 Identification of Bioactive Metabolites

Lysiloma latisiliquum is a commonly consumed forage by animals and is known for its anthelmintic activity. To identify the anthelmintic metabolites responsible for its activity, Luis Manuel Peña-Rodríguez and coworkers (Hernández-Bolio et al. 2018) employed ^{1}H-NMR metabolomics coupled with multivariate analysis (PCA and orthogonal projections to latent structures discriminant analysis to perform bioassay-guided fractionation and metabolic profiling of *L. latisiliquum*. Statistical analysis results indicated that hydrophilic solvents are the best for extracting highly polar-glycosylated anthelmintics metabolites (quercetin and arbutin). In a similar study, Ismail and coworkers (Raja Mazlan et al. 2018) conducted studies on the extraction and identification of active metabolites present in dried berries of *Piper cubeba* L, which is known for its antibacterial activity. However, the exact entity responsible for bioactivity remains unclear; they suggested that cubebol, D-germacrene, and ledol are the main components responsible for bactericidal effect, while p-cymene, cubebol, and ledol induce anti-inflammatory activity.

15.3 HYPHENATED NMR TECHNIQUES

One main drawback of these studies is their inability to provide the absolute identification of metabolites. Many metabolites could not be detected and identified accurately, thus necessitating hyphenated NMR techniques such as 2D heteronuclear single-quantum correlation NMR spectroscopy and 2D total correlation spectroscopy. Hyphenated techniques combine spectroscopic (NMR, mass IR, and UV-visible) and chromatographic (HPLC, GC, and LC) techniques. The coupled technique is applied for quantification and structure elucidation of the metabolites present in NPs or biological specimens. For instance, they have been used to identify metabolomes in plants such as *Smirnowia iranica* H. (Lambert et al. 2005), *Monotes engleri* (Garo et al. 1998), *G. ottonis*, (Wolfender et al. 1997), *Senecio vulgaris*, Senecio mariettae Muschl. (Ndjoko et al. 1999), *Cordia linnaei* Stearn (Ioset et al. 1999), *Cordia alliodora* (Fouseki et al. 2016; Xu et al. 2012), as well as marine organism (Kleinwächter et al. 2000). Hyphenated techniques are a sophisticated high-throughput analytical technique method that is applied to metabolomics resulting in the separation and structure elucidation of metabolites simultaneously at a time (Scognamiglio et al. 2019). A hyphenated technique can be double, triple, or higher. In the following subsection, we highlight the applications of hyphenated NMR techniques (Rip 2022; "What Are Triple Hyphenated Techniques?" 2022; Gomes et al. 2018).

15.3.1 Application of Hyphenated ^{1}H NMR Techniques

15.3.1.1 Ethnopharmacological

Generally, aqueous or organic NP extract contains several metabolites with one or more pharmacological activities. Under such circumstances, identifying the fraction/component responsible for bioactivity and its chemical structure is daunting and time-consuming. High-resolution bioassay combined with hyphenated NMR techniques is one of the recent methods showing its potential ability to identify new drugs based on NPs. In this technique, fractions obtained from liquid chromatography (LC) is assessed for its activity (single, dual, triple, or quadruple high-resolution profiling), and then the fraction of interest is subjected to further isolation and structural characterization of the bioactive constituent (Liang et al. 2021; Zhao et al. 2018). Hyphenated NMR such as HPLC-NMR,

HPLC-SPE-NMR, HPLC-HRMS-SPE-NMR, and HPLC-PDA-HRMS-SPE-NMR has proven effective for the structure elucidation of individual constituents in crude extracts (Schmidt, Nyberg, and Staerk 2014). For example, HPLC-SPE-NMR has been used to identify various ethnopharmaceuticals, including enzyme inhibitors, antioxidants, and anticancer agents (Table 15.1) (Tetali

TABLE 15.1
Examples of Plant Species Analyzed through Bio-chromatogram Combined with Hyphenated-NMR Techniques

Sample	Hyphenation Mode*	Profiling	Bioactive Compounds	References
Miconia albicans	(I)	Protein tyrosine phosphatase 1B (PTP1B) inhibition	Maslinic acid derivatives	(de Cássia Lemos Lima et al. 2018)
Myrcia palustris DC	(I)	α-Glucosidase inhibition	Ellagic acid rhamnopyranosides, Myricetin, Quercetin, Casuarinin Myricetin 3-O-β-d-(6″-galloyl) galacto pyranoside, and Kaempferol 3-O-β-d-galacto pyranoside	(Wubshet, Moresco, et al. 2015)
Eremanthus crotonoides	(I)	α-Glucosidase inhibition	Quercetin, trans-tiliroside, luteolin, Quercetin-3-methyl ether, 3,5-di-O-caffeoylquinic acid n-butyl ester, 4,5-di-O-caffeoylquinic acid n-butyl ester, and others.	(Silva et al. 2016)
Lawsonia inermis L	(I)	Antileishmanial	2,4,6-Trihydroxyacetophenone-2-O-β-D-glucopyranoside, Lalioside, Luteolin-4′-O-β-D-glucopyranoside, Apigenin-4′-O-β-D-glucopyrano side, Luteolin, and Apigenin.	(Iqbal et al. 2017)
Solanum americanum	(I)	α-Glucosidase inhibitors Radical scavenging inhibitors	4-Hydroxybenzoic acid, 3-indolecarboxylic acid, *N*-trans-p-coumaroyloctopamine, *N*-trans-p-feruloyloctopamine, *N*-trans-p-coumaroyltyramine and *N*-trans-p-feruloyltyramine, along with others.	(Silva et al. 2017)
Myrcia rubella	(I)	PTP1B and α-glucosidase inhibition	4,5-Dicaffeoylquinic acid, Isoquercitrin, Quercetin-3-O-β-d-glucuronide, Kaempferol-3-O-(6″-galloyl)-β-d-glucopyrano side, Quercetin-3-O-(6″-malonyl)-glucopyranoside, Quer cetin-3-O-(6″-(E)-feruloyl)-β-d-glucopyranoside, Quercetin-3-O-(2″-(E)-sinapoyl)-glucopyranoside, Kaempferol-3-O-α-l-rhamnoside, Astragalin, and Arjunolic acid.	(Lima et al. 2018)

(*Continued*)

TABLE 15.1 (*Continued*)
Examples of Plant Species Analyzed through Bio-chromatogram Combined with Hyphenated-NMR Techniques

Sample	Hyphenation Mode*	Profiling	Bioactive Compounds	References
Machilus litseifolia	**(II)**	α-Glucosidase inhibitors	Kaempferol 3-O-(2″-E-p-coumaroyl,4″-Z-p-coumaroyl)-α-l-rhamnopyranoside, Kaempferol 3-O-(2″-E-p-coumaroyl,4″-Z-p-coumaroyl)-α-l-rhamnopyranoside, 4′-O-methyl-(2″-E-p-coumaroyl-4″-Z-p-coumaroyl) afzelin, 4′-O-methyl-(2″-Z-p-coumaroyl-4″-E-p-coumaroyl)afzelin, and several others	(Li, Kongstad, and Staerk 2019)
Ficus racemosa and others	**(I)**	PTP1B inhibition	Isoderron, Derrone, Alpinumisoflavone, and Mucusisoflavone B	(Trinh, Jäger, and Staerk 2017)
Antidesma madagascariense and others	**(I)**	Dipeptidyl peptidase IV (DPP-IV) activity	Amentoflavone	(Beidokhti et al. 2018)
Clausena excavate, Androsace umbellata, and *Oxalis corniculata*	**(I)**	Hyaluronidase inhibition	Lansiumamide B, Myricetin 3-O-β-d-glucopyranoside, Vitexin, and 4′,7-dihydroxy-5-Methoxyflavone-8-C-β-d-glucopyranoside	(Liu et al. 2015)
Dioscorea bulbifera L., *Boehmeria nivea, Tinospora sagittata,* and *Persicaria bistorta*	**(III)**	α-Glucosidase, α-amylase, and PTP1B inhibition	[1,1′-Biphenanthren]-2,2′,3,3′,6,6′,7,7′-octaol, Cassigarol D, Hederagenin, Pomolic acid, along with several others.	(Zhao et al. 2019)
Piper nigrum	**(IV)**	Monoamine oxidase A (MAO-A) inhibitors	Clorgyline and piperine	(Grosso, Jäger, and Staerk 2013)
Haplocoelum zoliolosum, Bullock and Sauvagesia erecta L.	**(I)**	Fungal plasma membrane (PM) H+-ATPase inhibition	Chebulagic acid (1) and Tellimagrandin II	(Kongstad et al. 2014)

*Hyphenation mode: HPLC-HRMS-SPE-NMR (**I**); HPLC-PDA-HRMS-SPE-NMR (**II**); HPLC-HRMS-NMR (**III**); and HPLC-SPE-ttNMR (**IV**).

et al. 2021). As mentioned above, hyphenated techniques have been employed to identify primary and secondary metabolites in complex mixtures. Staerk and coworkers (Schmidt, Nyberg, and Staerk 2014) performed a detailed metabolomic profiling of different *Allium* species/cultivars for α-glucosidase inhibitors. MVDA, α-glucosidase assay, and HPLC-SPE-NMR revealed that quercetin, N-p-coumaroyloctopamine, and N-p-coumaroyltyramine are mainly responsible for bioactivity of the *Allium* extract. Besides, the distribution of these three compounds was also mapped in 35 *Allium* species/cultivars. Staerk and coworkers (Tahtah et al. 2015) reported that *Radix Scutellariae* extract

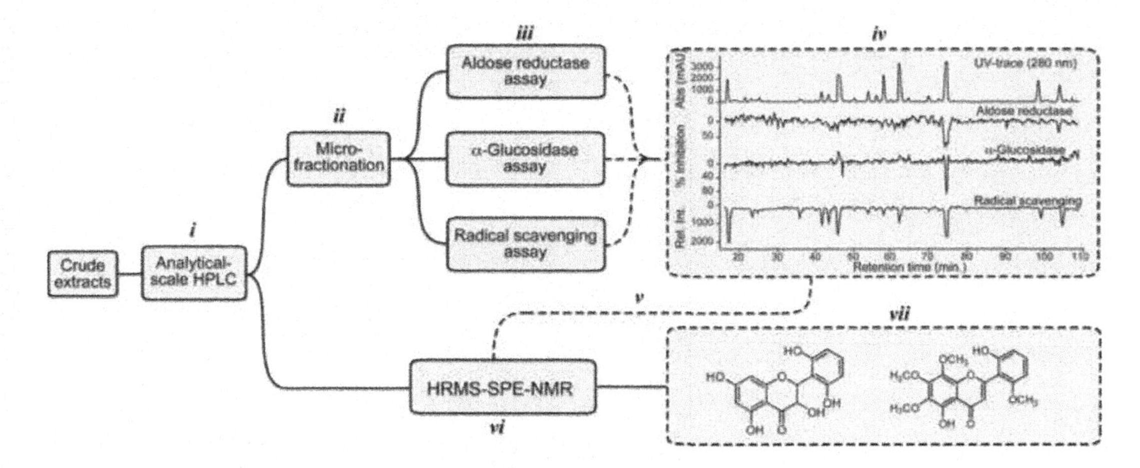

FIGURE 15.5 Schematic representation of triple high-resolution aldose reductase/α-glucosidase/radical scavenging profiling followed by HPLC-HRMS-SPE-NMR analysis of *Radix Scutellariae*. Reproduced with permission from Tahtah et al. (2015).

serves polypharmacological herbal drug, as it contains α-glucosidase and aldose reductase inhibitors as well as radical scavengers (Figure 15.5). The chemical structures of these triple-bioactive components were determined using hyphenated HPLC-HRMS-SPE-NMR technique. However, HRMS was employed to identify chromatographic peaks, and 1D- and 2D-NMR were used to confirm the stereochemistry and glycosidic bonding. Using a dual high-resolution bioactivity profiling and HPLC-HRMS-SPE-NMR, they identified protein tyrosine phosphatase-1B (PTP1B) and α-glucosidase inhibitors found in *Eremophila* spp. (Wubshet et al. 2016).

The same group identified several new branched-chain fatty acids in the leaf extract of *Eremophila* spp. (Pedersen et al. 2020). High-resolution bioactivity profiling of *E. oppositifolia* revealed moderate-to-high protein tyrosine phosphatase 1B (PTP1B) inhibition ability of most of the apolar (acetonitrile) fractions (***ii*** (R=OH), ***ii*** (R=OAc), ***iii***, **iv** (R=Ac), ***v*** (R_1=H, R_2=Ac), ***v*** (R_1=Ac, R_2=H), ***vi*** (R_1=H, R_2=Ac), ***vi*** (R_1=Ac, R_2=H), ***v*** (R_1=R_2=Ac), Figure 15.6) (Pedersen et al. 2020). The chemical structure of the metabolites (Figure 15.5) was confirmed by hyphenated HPLC-PDA-HRMS-SPE-NMR. Similar techniques have been used to identify α-glucosidase and PTP1B inhibitors in *Machilus litseifolia* (Li, Kongstad, and Staerk 2019), *Rhododendron capitatum* (Liang et al. 2021), *Myrtus communis* (Liang et al. 2021), *Solanum Americanum* (Silva et al. 2017), *Eremanthus crotonoides* (Silva et al. 2016).

Snakebite is one of the most neglected public health issues, responsible for 81,000–138,000 deaths worldwide annually ("Snakebite Envenoming" 2022). This number of other snakebite-related disabilities is much higher. In Asian countries like India and China, traditional medicinal plants have been used to treat snakebites for a long time. To identify new antibacterial to treat snakebites, Liu et al. (2014) assessed aqueous and ethanolic extract of various traditional Chinese plants. Out of 180 extracts from the 88 plant species investigated, they found that three compounds (piceid **vii**, resveratrol **viii**, and emodin **ix**, Chart 15.2) in *Polygonum cuspidatum* exhibit the antibacterial effect. The ethyl acetate extract of *P. cuspidatum* also shows α-glucosidase inhibitory activity due to the presence of procyanidin B2 (**x**, Chart 15.2), epicatechin gallate (Chart 15.1), and 3,3″-O-di-gallate. At the same time, PTP1B inhibition was attributed to the presence of (trans)-emodin-physcion bianthrone and (cis)-emodin-physcion bianthrone (Zhao et al. 2017).

In addition to the high-resolution bioassay-coupled hyphenated NMR technique, the ligand fishing bioanalytical technique is also a fast and sensitive method to identify bioactive constituents in a complex mixture. In this extraction technique, the solid support (e.g., magnetic bead) is coated with

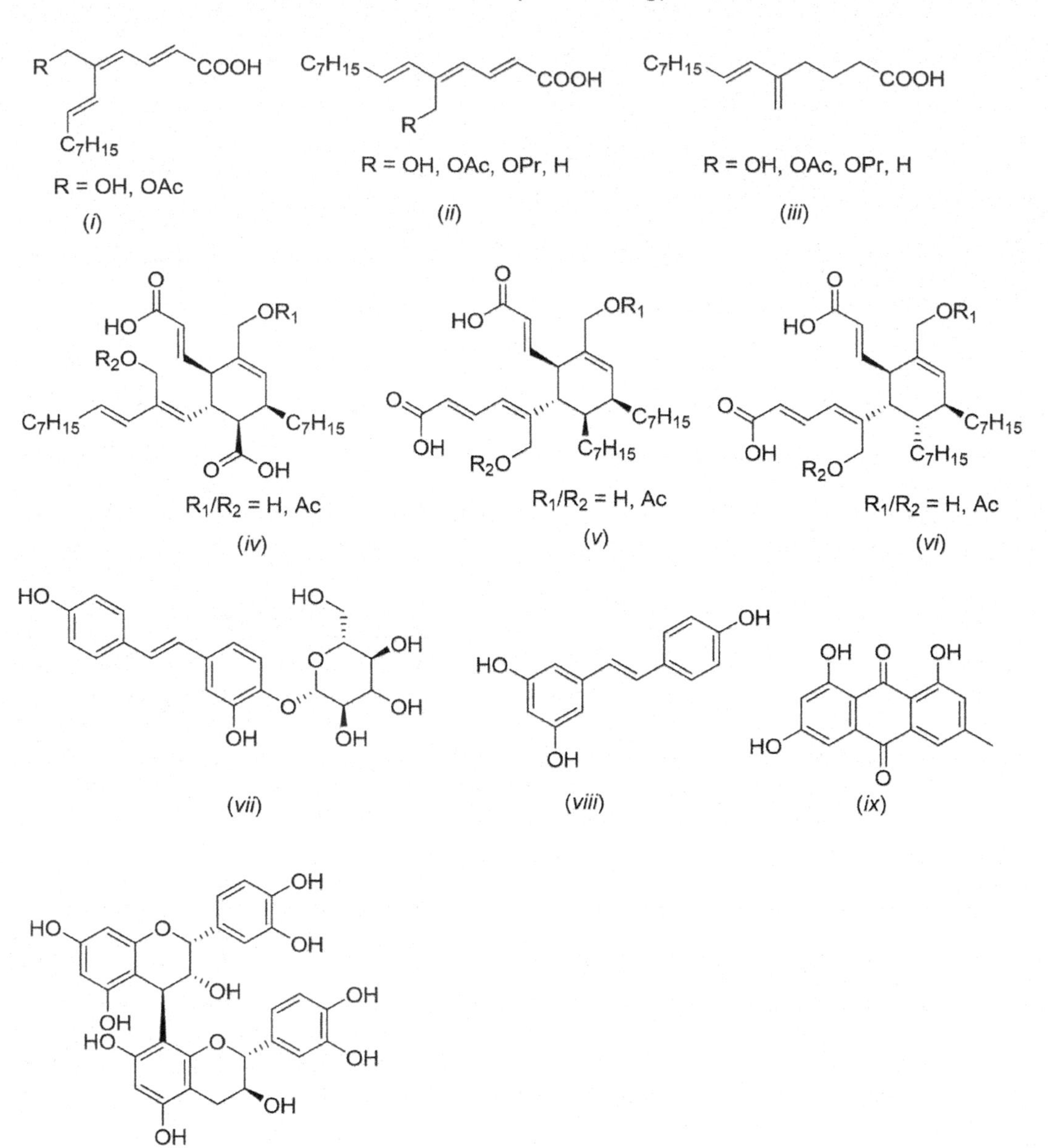

CHART 15.2 Chemical structures of monomeric branched fatty acids (i–iii), dimeric branched fatty acids type A (iv), B1 (v), and B2 (vi) isolated from *E. oppositifolia* leaf extract and antibacterial compounds (vi–x) from *Polygonum cuspidatum*.

a target (e.g., protein receptor), which is then used for fishing out the active constituent from a given mixture (Vanzolini et al. 2013). This method involves loading, washing, and elution steps and could be used for metabolite fingerprinting of plants. Figure 15.7 shows a schematic diagram of ligand fishing from *Eugenia catharinae* extract using α-glucosidase enzyme immobilized magnetic beads to identify the active constituent (Wubshet, Brighente, et al. 2015).

This method, which utilized HPLC-HRMS-SPE-NMR hyphenation, identified several anti-hyperglycemic metabolites, including myricetin 3-o-α-l-rhamnopyranoside, myricetin, quercetin,

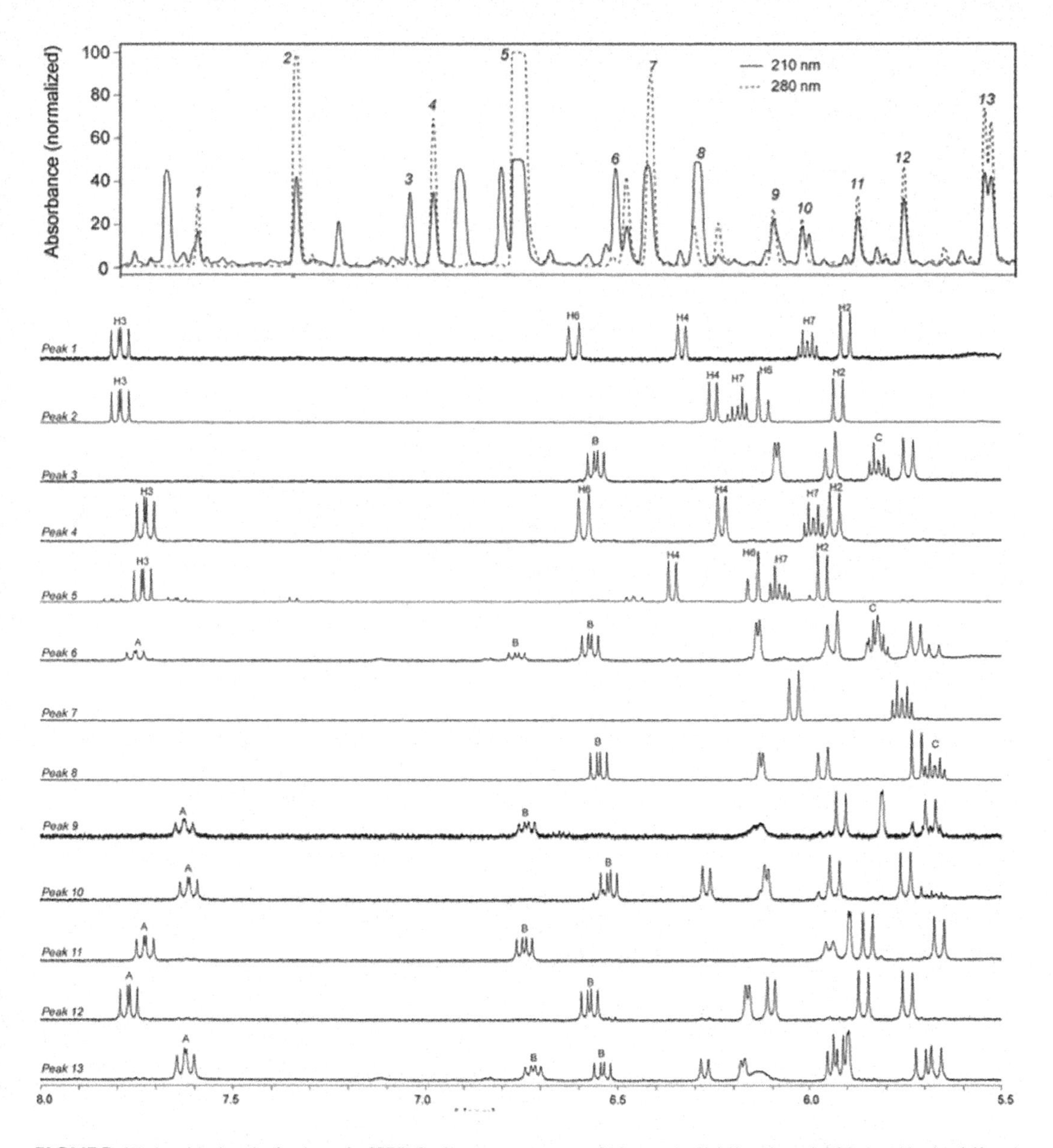

FIGURE 15.6 (a) Analytical-scale HPLC chromatogram at 210 nm (solid lines) and 280 nm (dashed lines) of acetonitrile SPE-fraction of *E. oppositifolia* leaf extract. (b) ^{1}H NMR spectra of the region δ 5.5–8.0 from HPLC-PDA-HRMS-SPE-NMR analysis of the acetonitrile SPE fraction of *E. oppositifolia* leaf extract. For numbering, see the original paper. Reprinted with permission from H.A. Pedersen, C. Ndi, S.J. Semple, B. Buirchell, B.L. Møller, D. Staerk, PTP1B-inhibiting branched-chain fatty acid dimers from *Eremophila oppositifolia* subsp. *angustifolia* identified by high-resolution PTP1B inhibition profiling and HPLC-PDA-HRMS-SPE-NMR analysis, *J. Nat. Prod.* 83(5) (2020) 1598–1610. Copyright 2020 American Chemical Society (Pedersen et al. 2020).

kaempferol, and alkylresorcinol glycosides. Using the same fishing out technique coupled with HPLC-PDA-HRMS-SPE-NMR, α-amylase inhibitors were found in the crude ethyl acetate extract of *Ginkgo biloba* leaves. Interestingly, the developed method found to be efficient even after two and half months (enzymatic activity = 75% retained) and could be applied to identify ligands from natural and artificial mixtures (Petersen et al. 2019).

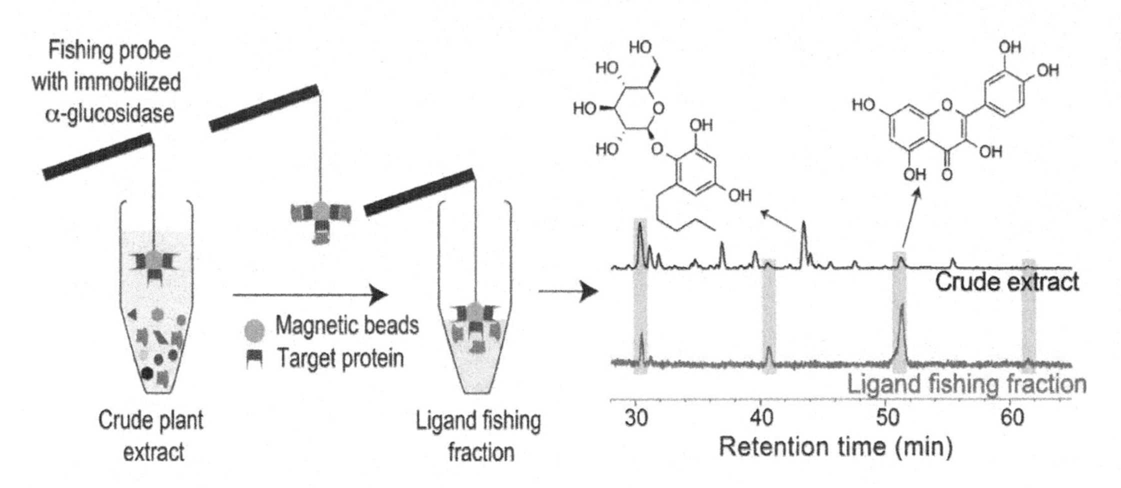

FIGURE 15.7 Fishing out approach for identifying α-glucosidase inhibitors found in *Eugenia catharinae* extract. Reprinted with permission from S.G. Wubshet, I.M. Brighente, R. Moaddel, D. Staerk, Magnetic ligand fishing as a targeting tool for HPLC-HRMS-SPE-NMR: α-glucosidase inhibitory ligands and alkylresorcinol glycosides from *E. catharinae*, *J. Nat. Prod.* 78(11) (2015) 2657–2665. Copyright 2015 American Chemical Society (Wubshet, Moresco, et al. 2015).

15.4 CONCLUSION

NMR-based metabolomics and hyphenated NMR techniques have gained a vital position in modern science, especially NP-based research. They are quite an important tool for identifying bioactive compounds from different sources. Using various combinations (e.g., LC-NMR, HPLC-NMR, and HPLC-SPE-NMR), primary and secondary metabolites found in NPs and biological samples could be determined rapidly and accurately. When combined with a high-resolution bioassay, the hyphenated NMR technique becomes a powerful tool for the real-time monitoring and identification of target-specific molecules, thereby speeding up the drug discovery process. Overall, these techniques have the potential to provide fast and reliable analyses and a deeper understanding of complex mixtures, which is otherwise hard to study.

REFERENCES

Azmi, J, J L Griffin, R F Shore, E Holmes, and J K Nicholson. 2005. "Chemometric Analysis of Biofluids Following Toxicant Induced Hepatotoxicity: A Metabonomic Approach to Distinguish the Effects of 1-Naphthylisothiocyanate from Its Products." *Xenobiotica* 35(8): 839–52.

Bang, Haejeen, Angela R Davis, Sunggil Kim, Daniel I Leskovar, and Stephen R King. 2010. "Flesh Color Inheritance and Gene Interactions among Canary Yellow, Pale Yellow, and Red Watermelon." *Journal of the American Society for Horticultural Science* 135(4): 362–68.

Bayona, Lina M, Nicole J de Voogd, and Young Hae Choi. 2022. "Metabolomics on the Study of Marine Organisms." *Metabolomics* 18(3): 1–24.

Beckonert, Olaf, Hector C Keun, Timothy M D Ebbels, Jacob Bundy, Elaine Holmes, John C Lindon, and Jeremy K Nicholson. 2007. "Metabolic Profiling, Metabolomic and Metabonomic Procedures for NMR Spectroscopy of Urine, Plasma, Serum and Tissue Extracts." *Nature Protocols* 2(11): 2692. https://www.nature.com/articles/nprot.2007.376.

Beidokhti, M N, E S Lobbens, P Rasoavaivo, D Staerk, and A K Jäger. 2018. "Investigation of Medicinal Plants from Madagascar against DPP-IV Linked to Type 2 Diabetes." *South African Journal of Botany* 115: 113–19.

Brennan, Lorraine. 2014. "NMR-Based Metabolomics: From Sample Preparation to Applications in Nutrition Research." *Progress in Nuclear Magnetic Resonance Spectroscopy* 83: 42–49.

Cássia Lemos Lima, Rita de, Kenneth T. Kongstad, Lucília Kato, Marcos José das Silva, Henrik Franzyk, and Dan Staerk. 2018. "High-Resolution PTP1B Inhibition Profiling Combined with HPLC-HRMS-SPE-NMR for Identification of PTP1B Inhibitors from Miconia Albicans." *Molecules* 23(7): 1755.

Castro, Federica De, Michele Benedetti, Laura Del Coco, and Francesco Paolo Fanizzi. 2019. "NMR-Based Metabolomics in Metal-Based Drug Research." *Molecules* 24(12): 2240.

Ellis, Noel, Chie Hattori, Jitender Cheema, James Donarski, Adrian Charlton, Michael Dickinson, Giampaolo Venditti, Péter Kaló, Zoltán Szabó, and György B Kiss. 2018. "NMR Metabolomics Defining Genetic Variation in Pea Seed Metabolites." *Frontiers in Plant Science* 9: 1022. https://www.ncbi.nlm.nih.gov/pmc/articles/PMC6056766/pdf/fpls-09-01022.pdf.

Emwas, A H, R Roy, R T McKay, L Tenori, E Saccenti, G A N Gowda, D Raftery, F Alahmari, L Jaremko, and M Jaremko. 2019. "NMR Spectroscopy for Metabolomics Research." *Metabolites* 9(7): 123.

Esposito, Assunta, Pietro Filippo De Luca, Vittoria Graziani, Brigida D'Abrosca, Antonio Fiorentino, and Monica Scognamiglio. 2021. "Phytochemical Characterization of Olea Europaea L. Cultivars of Cilento National Park (South Italy) through NMR-Based Metabolomics." *Molecules* 26(13): 3845. https://mdpi-res.com/d_attachment/molecules/molecules-26-03845/article_deploy/molecules-26-03845-v2.pdf?version=1624711226.

Fouseki, Myrto Maria, Harilaos Damianakos, George Albert Karikas, Christos Roussakis, Mahabir P Gupta, and Ioanna Chinou. 2016. "Chemical Constituents from Cordia Alliodora and C. Colloccoca (Boraginaceae) and Their Biological Activities." *Fitoterapia* 115: 9–14.

Garo, Eliane, Jean-Luc Wolfender, Kurt Hostettmann, Wolf Hiller, SÁNdor Antus, and Mavi. 1998. "Prenylated Flavanones from Monotes Engleri: On-line Structure Elucidation by LC/UV/NMR." *Helvetica Chimica Acta* 81(3-4): 754–63.

Giraudeau, Patrick. 2020. "NMR-Based Metabolomics and Fluxomics: Developments and Future Prospects." *Analyst* 145(7): 2457–72.

Gomes, Nelson G M, David M Pereira, Patricia Valentao, and Paula B Andrade. 2018. "Hybrid MS/NMR Methods on the Prioritization of Natural Products: Applications in Drug Discovery." *Journal of Pharmaceutical and Biomedical Analysis* 147: 234–49.

Grosso, Clara, Anna K Jäger, and Dan Staerk. 2013. "Coupling of a High-resolution Monoamine Oxidase-A Inhibitor Assay and HPLC–SPE–NMR for Advanced Bioactivity Profiling of Plant Extracts." *Phytochemical Analysis* 24(2): 141–47.

Guennec, Adrien Le, Patrick Giraudeau, and Stefano Caldarelli. 2014. "Evaluation of Fast 2D NMR for Metabolomics." *Analytical Chemistry* 86(12): 5946–54.

Hernández-Bolio, Gloria Ivonne, Erika Kutzner, Wolfgang Eisenreich, Juan Felipe de Jesús Torres-Acosta, and Luis Manuel Peña-Rodríguez. 2018. "The Use of 1H-NMR Metabolomics to Optimise the Extraction and Preliminary Identification of Anthelmintic Products from the Leaves of Lysiloma Latisiliquum." *Phytochemical Analysis* 29(4): 413–20.

Huang, Yanjie, Xian Li, Xingrong Peng, Adelakun Tiwalade Adegoke, Jianchao Chen, Haiguo Su, Guilin Hu, Gang Wei, and Minghua Qiu. 2020. "NMR-Based Structural Classification, Identification, and Quantification of Triterpenoids from Edible Mushroom Ganoderma Resinaceum." *Journal of Agricultural and Food Chemistry* 68(9): 2816–25.

Ioset, Jean-Robert, Jean-Luc Wolfender, Andrew Marston, Mahabir P Gupta, and Kurt Hostettmann. 1999. "Identification of Two Isomeric Meroterpenoid Naphthoquinones from Cordia Linnaei by Liquid Chromatography–Mass Spectrometry and Liquid Chromatography–Nuclear Magnetic Resonance Spectroscopy." *Phytochemical Analysis* 10(3): 137–42.

Iqbal, Kashif, Javeid Iqbal, Dan Staerk, and Kenneth T Kongstad. 2017. "Characterization of Antileishmanial Compounds from Lawsonia Inermis L. Leaves Using Semi-High Resolution Antileishmanial Profiling Combined with HPLC-HRMS-SPE-NMR." *Frontiers in Pharmacology* 8: 337. https://www.ncbi.nlm.nih.gov/pmc/articles/PMC5449460/pdf/fphar-08-00337.pdf.

Kim, Hye Kyong, Young Hae Choi, and Robert Verpoorte. 2010. "NMR-Based Metabolomic Analysis of Plants." *Nature Protocols* 5(3): 536–49. https://doi.org/10.1038/nprot.2009.237.

Kleinwächter, Peter, Karin Martin, Ingrid Groth, and Klausjürgen Dornberger. 2000. "Use of Coupled HPLC/1H NMR and HPLC/ESI-MS for the Detection and Identification of (2E, 4Z)-decadienoic Acid from a New Agromyces Species." *Journal of High Resolution Chromatography* 23(10): 609–12.

Kongstad, Kenneth T, Sileshi G Wubshet, Ane Johannesen, Lasse Kjellerup, Anne-Marie Lund Winther, Anna Katharina Jager, and Dan Staerk. 2014. "High-Resolution Screening Combined with HPLC-HRMS-SPE-NMR for Identification of Fungal Plasma Membrane H+-ATPase Inhibitors from Plants." *Journal of Agricultural and Food Chemistry* 62(24): 5595–602.

Kruger, Nicholas J, M Adrian Troncoso-Ponce, and R George Ratcliffe. 2008. "1H NMR Metabolite Fingerprinting and Metabolomic Analysis of Perchloric Acid Extracts from Plant Tissues." *Nature Protocols* 3(6): 1001–12. https://www.nature.com/articles/nprot.2008.64.

Lambert, Maja, Dan Stærk, Steen Honoré Hansen, Majid Sairafianpour, and Jerzy W Jaroszewski. 2005. "Rapid Extract Dereplication Using HPLC-SPE-NMR: Analysis of Isoflavonoids from Smirnowia iranica." *Journal of Natural Products* 68(10): 1500–509.

Lau, Hazel, Anna Karen Carrasco Laserna, and Sam Fong Yau Li. 2020. "1H NMR-Based Metabolomics for the Discrimination of Celery (Apium Graveolens L. Var. Dulce) from Different Geographical Origins." *Food Chemistry* 332: 127424.

Les, Francisco, José Miguel Arbonés-Mainar, Marta Sofía Valero, and Víctor López. 2018. "Pomegranate Polyphenols and Urolithin A Inhibit α-Glucosidase, Dipeptidyl Peptidase-4, Lipase, Triglyceride Accumulation and Adipogenesis Related Genes in 3T3-L1 Adipocyte-like Cells." *Journal of Ethnopharmacology* 220: 67–74.

Li, Tuo, Kenneth T Kongstad, and Dan Staerk. 2019. "Identification of α-Glucosidase Inhibitors in Machilus Litseifolia by Combined Use of High-Resolution α-Glucosidase Inhibition Profiling and HPLC-PDA-HRMS-SPE-NMR." *Journal of Natural Products* 82(2): 249–58. https://pubs.acs.org/doi/10.1021/acs.jnatprod.8b00609.

Liang, Chao, Louise Kjaerulff, Paul Robert Hansen, Kenneth T Kongstad, and Dan Staerk. 2021. "Dual High-Resolution α-Glucosidase and PTP1B Inhibition Profiling Combined with HPLC-PDA-HRMS-SPE-NMR Analysis for the Identification of Potentially Antidiabetic Chromene Meroterpenoids from Rhododendron Capitatum." *Journal of Natural Products* 84(9): 2454–67. https://pubs.acs.org/doi/10.1021/acs.jnatprod.1c00454.

Lima, Rita de Cássia Lemos, Lucília Kato, Kenneth Thermann Kongstad, and Dan Staerk. 2018. "Brazilian Insulin Plant as a Bifunctional Food: Dual High-Resolution PTP1B and α-Glucosidase Inhibition Profiling Combined with HPLC-HRMS-SPE-NMR for Identification of Antidiabetic Compounds in Myrcia Rubella Cambess." *Journal of Functional Foods* 45: 444–51.

Liu, Yueqiu, Mia Nielsen, Dan Staerk, and Anna K Jäger. 2014. "High-Resolution Bacterial Growth Inhibition Profiling Combined with HPLC–HRMS–SPE–NMR for Identification of Antibacterial Constituents in Chinese Plants Used to Treat Snakebites." *Journal of Ethnopharmacology* 155(2): 1276–83.

Liu, Yueqiu, Dan Staerk, Mia N Nielsen, Nils Nyberg, and Anna K Jäger. 2015. "High-Resolution Hyaluronidase Inhibition Profiling Combined with HPLC–HRMS–SPE–NMR for Identification of Anti-Necrosis Constituents in Chinese Plants Used to Treat Snakebite." *Phytochemistry* 119: 62–69.

Mazlan, Raja, Raja Nur Asila, Yaya Rukayadi, M Maulidiani, and Intan Safinar Ismail. 2018. "Solvent Extraction and Identification of Active Anticariogenic Metabolites in Piper Cubeba L. through 1H-NMR-Based Metabolomics Approach." *Molecules* 23(7): 1730.

Mediani, Ahmed, Faridah Abas, Tan Chin Ping, Alfi Khatib, and Nordin H Lajis. 2012. "Influence of Growth Stage and Season on the Antioxidant Constituents of Cosmos Caudatus." *Plant Foods for Human Nutrition* 67(4): 344–50. https://link.springer.com/article/10.1007/s11130-012-0317-x.

Nagana Gowda, G A, and Daniel Raftery. 2017. "Recent Advances in NMR-Based Metabolomics." *Analytical Chemistry* 89(1): 490–510.

Nanda, Manisha, Vinod Kumar, Neha Arora, Mikhail S Vlaskin, and Manoj K Tripathi. 2021. "1 H NMR-Based Metabolomics and Lipidomics of Microalgae." *Trends in Plant Science* 26(9): 984–85. https://www.cell.com/trends/plant-science/fulltext/S1360-1385(21)00146-1?_returnURL=https%3A%2F%2Flinkinghub.elsevier.com%2Fretrieve%2Fpii%2FS1360138521001461%3Fshowall%3Dtrue.

Ndjoko, Karine, Jean-Luc Wolfender, Erhard Röder, and Kurt Hostettmann. 1999. "Determination of Pyrrolizidine Alkaloids in Senecio Species by Liquid Chromatography/Thermospray-Mass Spectrometry and Liquid Chromatography/Nuclear Magnetic Resonance Spectroscopy." *Planta Medica* 65(06): 562–66.

Newman, David J, and Gordon M Cragg. 2012. "Natural Products as Sources of New Drugs over the 30 Years from 1981 to 2010." *Journal of Natural Products* 75(3): 311–35.

Pedersen, Hans Albert, Chi Ndi, Susan J Semple, Bevan Buirchell, Birger Lindberg Møller, and Dan Staerk. 2020. "PTP1B-Inhibiting Branched-Chain Fatty Acid Dimers from Eremophila Oppositifolia Subsp. Angustifolia Identified by High-Resolution PTP1B Inhibition Profiling and HPLC-PDA-HRMS-SPE-NMR Analysis." *Journal of Natural Products* 83(5): 1598–1610. https://pubs.acs.org/doi/10.1021/acs.jnatprod.0c00070.

Petersen, Malene J, Rita de Cássia Lemos Lima, Louise Kjaerulff, and Dan Staerk. 2019. "Immobilized α-Amylase Magnetic Beads for Ligand Fishing: Proof of Concept and Identification of α-Amylase Inhibitors in Ginkgo Biloba." *Phytochemistry* 164: 94–101.

Raletsena, Maropeng V, Samukelisiwe Mdlalose, Olusola S Bodede, Hailemariam A Assress, Adugna A Woldesemayat, and David M Modise. 2022. "1H-NMR and LC-MS Based Metabolomics Analysis of Potato (Solanum Tuberosum L.) Cultivars Irrigated with Fly Ash Treated Acid Mine Drainage." *Molecules* 27(4): 1187.

Rip, Waliul. 2022. "Hyphenated NMR Techniques." 2022. https://www.nanalysis.com/nmready-blog/2013/10/17/hyphenated-nmr-techniques.

Schmidt, Jeppe S, Nils T Nyberg, and Dan Staerk. 2014. "Assessment of Constituents in Allium by Multivariate Data Analysis, High-Resolution α-Glucosidase Inhibition Assay and HPLC-SPE-NMR." *Food Chemistry* 161: 192–98.

Scognamiglio, Monica, Vittoria Graziani, Nikolaos Tsafantakis, Assunta Esposito, Antonio Fiorentino, and Brigida D'Abrosca. 2019. "NMR-based Metabolomics and Bioassays to Study Phytotoxic Extracts and Putative Phytotoxins from Mediterranean Plant Species." *Phytochemical Analysis* 30(5): 512–23.

Silva, Eder L, Rita C Almeida-Lafetá, Ricardo M Borges, and Dan Staerk. 2017. "Dual High-Resolution Inhibition Profiling and HPLC-HRMS-SPE-NMR Analysis for Identification of α-Glucosidase and Radical Scavenging Inhibitors in Solanum Americanum Mill." *Fitoterapia* 118: 42–48. https://www.sciencedirect.com/science/article/pii/S0367326X17300242?via%3Dihub.

Silva, Eder Lana e, Jonathas Felipe Revoredo Lobo, Joachim Møllesøe Vinther, Ricardo Moreira Borges, and Dan Staerk. 2016. "High-Resolution α-Glucosidase Inhibition Profiling Combined with HPLC-HRMS-SPE-NMR for Identification of Antidiabetic Compounds in Eremanthus Crotonoides (Asteraceae)." *Molecules* 21(6): 782.

"Snakebite Envenoming." 2022. 2022. https://www.who.int/news-room/fact-sheets/detail/snakebite-envenoming#:~:text=Though the exact number of,are caused by snakebites annually.

Sulaiman, Fadzil, Amalina Ahmad Azam, Muhammad Safwan Ahamad Bustamam, Sharida Fakurazi, Faridah Abas, Yee Xuan Lee, Atira Adriana Ismail, Siti Munirah Mohd Faudzi, and Intan Safinar Ismail. 2020. "Metabolite Profiles of Red and Yellow Watermelon (Citrullus Lanatus) Cultivars Using a 1H-NMR Metabolomics Approach." *Molecules* 25(14): 3235.

Suzuki, Ryuichiro, Yuki Kasuya, Aiko Sano, Junki Tomita, Takuro Maruyama, and Masashi Kitamura. 2022. "Comparison of Various Commercially Available Cinnamon Barks Using NMR Metabolomics and the Quantification of Coumarin by Quantitative NMR Methods." *Journal of Natural Medicines* 76(1): 87–93. https://link.springer.com/article/10.1007/s11418-021-01554-6.

Tahtah, Yousof, Kenneth T Kongstad, Sileshi G Wubshet, Nils T Nyberg, Louise H Jønsson, Anna K Jäger, Sun Qinglei, and Dan Staerk. 2015. "Triple Aldose Reductase/α-Glucosidase/Radical Scavenging High-Resolution Profiling Combined with High-Performance Liquid Chromatography–High-Resolution Mass Spectrometry–Solid-Phase Extraction–Nuclear Magnetic Resonance Spectroscopy for Identification Of." *Journal of Chromatography A* 1408: 125–32.

Tajidin, Nor Elliza, Khozirah Shaari, Maulidiani Maulidiani, Nor Shariah Salleh, Bunga Raya Ketaren, and Munirah Mohamad. 2019. "Metabolite Profiling of Andrographis Paniculata (Burm. f.) Nees. Young and Mature Leaves at Different Harvest Ages Using 1H NMR-Based Metabolomics Approach." *Scientific Reports* 9(1): 1–10.

Tang, Fenfen, and Emmanuel Hatzakis. 2020. "NMR-Based Analysis of Pomegranate Juice Using Untargeted Metabolomics Coupled with Nested and Quantitative Approaches." *Analytical Chemistry* 92(16): 11177–85.

Tetali, Sarada D, Satyabrata Acharya, Aditya B Ankari, Vadthyavath Nanakram, and Agepati S Raghavendra. 2021. "Metabolomics of Withania Somnifera (L.) Dunal: Advances and Applications." *Journal of Ethnopharmacology* 267: 113469.

Trinh, Binh Thi Dieu, Anna K Jäger, and Dan Staerk. 2017. "High-Resolution Inhibition Profiling Combined with HPLC-HRMS-SPE-NMR for Identification of PTP1B Inhibitors from Vietnamese Plants." *Molecules* 22(7): 1228.

Vanzolini, Kenia Lourenço, Zhengjin Jiang, Xiaoqi Zhang, Lucas Campos Curcino Vieira, Arlene Gonçalvez Corrêa, Carmen Lucia Cardoso, Quezia Bezerra Cass, and Ruin Moaddel. 2013. "Acetylcholinesterase Immobilized Capillary Reactors Coupled to Protein Coated Magnetic Beads: A New Tool for Plant Extract Ligand Screening." *Talanta* 116: 647–52.

Villalón-López, Nayelli, José I Serrano-Contreras, Darío I Téllez-Medina, and L Gerardo Zepeda. 2018. "An 1H NMR-Based Metabolomic Approach to Compare the Chemical Profiling of Retail Samples of Ground Roasted and Instant Coffees." *Food Research International* 106: 263–70.

Wachtel-Galor, Sissi, and Iris F F Benzie. 2011. "Herbal Medicine." *Lester Packer, Ph. D.*, 1.

"What Are Triple Hyphenated Techniques?" 2022. 2022. https://www.chromatographytoday.com/news/sample-prep/67/breaking-news/what-are-triple-hyphenated-techniques/57555.

Wolfender, Jean-Luc, Sylvain Rodriguez, Kurt Hostettmann, and Wolf Hiller. 1997. "Liquid Chromatography/Ultra Violet/Mass Spectrometric and Liquid Chromatography/Nuclear Magnetic Resonance Spectroscopic Analysis of Crude Extracts of Gentianaceae Species." *Phytochemical Analysis: An International Journal of Plant Chemical and Biochemical Techniques* 8(3): 97–104.

Wubshet, Sileshi G, Inês M C Brighente, Ruin Moaddel, and Dan Staerk. 2015. "Magnetic Ligand Fishing as a Targeting Tool for HPLC-HRMS-SPE-NMR: α-Glucosidase Inhibitory Ligands and Alkylresorcinol Glycosides from Eugenia Catharinae." *Journal of Natural Products* 78(11): 2657–65. https://www.ncbi.nlm.nih.gov/pmc/articles/PMC5036580/pdf/nihms-812450.pdf.

Wubshet, Sileshi G, Henrique H Moresco, Yousof Tahtah, Inês M C Brighente, and Dan Staerk. 2015. "High-Resolution Bioactivity Profiling Combined with HPLC–HRMS–SPE–NMR: α-Glucosidase Inhibitors and Acetylated Ellagic Acid Rhamnosides from Myrcia Palustris DC.(Myrtaceae)." *Phytochemistry* 116: 246–52.

Wubshet, Sileshi G, Yousof Tahtah, Allison M Heskes, Kenneth T Kongstad, Irini Pateraki, Björn Hamberger, Birger L Møller, and Dan Staerk. 2016. "Identification of PTP1B and α-Glucosidase Inhibitory Serrulatanes from Eremophila Spp. by Combined Use of Dual High-Resolution PTP1B and α-Glucosidase Inhibition Profiling and HPLC-HRMS-SPE-NMR." *Journal of Natural Products* 79(4): 1063–72. https://pubs.acs.org/doi/10.1021/acs.jnatprod.5b01128.

Xu, Yong-Jiang, Kenn Foubert, Liene Dhooghe, Filip Lemière, Sheila Maregesi, Christina M Coleman, Yike Zou, Daneel Ferreira, Sandra Apers, and Luc Pieters. 2012. "Rapid Isolation and Identification of Minor Natural Products by LC–MS, LC–SPE–NMR and ECD: Isoflavanones, Biflavanones and Bisdihydrocoumarins from Ormocarpum Kirkii." *Phytochemistry* 79: 121–28.

Ye, Hong, Jingwei Hu, Su Peng, Wenming Zong, Shuang Zhang, Lin Tong, Chen Cao, Zenghui Liu, and Zhongwen Xie. 2022. "Determination of the Chemical Compounds of Shuchazao Tea Flowers at Different Developmental Stages and in Young Shoots Using 1H NMR-Based Metabolomics." *Monatshefte Für Chemie-Chemical Monthly* 153: 409–17.

Yong, Chin-Hong, Syahidah Akmal Muhammad, Fatin Ilyani Nasir, Mohd Zulkifli Mustafa, Baharudin Ibrahim, Simon D Kelly, Andrew Cannavan, and Eng-Keng Seow. 2022. "Detecting Adulteration of Stingless Bee Honey Using Untargeted 1H NMR Metabolomics with Chemometrics." *Food Chemistry* 368: 130808.

Zhao, Yong, Martin Xiaoyong Chen, Kenneth Thermann Kongstad, Anna Katharina Jäger, and Dan Staerk. 2017. "Potential of Polygonum Cuspidatum Root as an Antidiabetic Food: Dual High-Resolution α-Glucosidase and PTP1B Inhibition Profiling Combined with HPLC-HRMS and NMR for Identification of Antidiabetic Constituents." *Journal of Agricultural and Food Chemistry* 65(22): 4421–27. https://pubs.acs.org/doi/10.1021/acs.jafc.7b01353.

Zhao, Yong, Kenneth Thermann Kongstad, Anna Katharina Jäger, John Nielsen, and Dan Staerk. 2018. "Quadruple High-Resolution α-Glucosidase/α-Amylase/PTP1B/Radical Scavenging Profiling Combined with High-Performance Liquid Chromatography–High-Resolution Mass Spectrometry–Solid-Phase Extraction–Nuclear Magnetic Resonance Spectroscopy for Identification O." *Journal of Chromatography A* 1556: 55–63.

Zhao, Yong, Kenneth Thermann Kongstad, Yueqiu Liu, Chenghua He, and Dan Staerk. 2019. "Unraveling the Complexity of Complex Mixtures by Combining High-Resolution Pharmacological, Analytical and Spectroscopic Techniques: Antidiabetic Constituents in Chinese Medicinal Plants." *Faraday Discussions* 218: 202–18. https://pubs.rsc.org/en/content/articlelanding/2019/FD/C8FD00223A.

16 Animal Models in Phytopharmacology and Toxicological Testing of Plant Products

Ana I. Faustino-Rocha
University of Évora

Beatriz Medeiros-Fonseca, Helena Vala, Maria J. Pires, Cármen Vasconcelos-Nóbrega, and Paula A. Oliveira
Inov4Agro

CONTENTS

16.1 ANIMAL MODELS: HISTORICAL PERSPECTIVE

Comparative medicine has been practiced for many years, as it is based on the concept that species share behavioral and physiological characteristics with humans. For about 2,400 years, researchers have been aware that we can learn a lot more about ourselves by studying animals (Barré-Sinoussi and Montagutelli 2015). The similarities of anatomy and physiology between humans and animals, especially among mammals, led investigators to proceed with a mechanistic approach and to evaluate new therapies in animals before translating them to humans (Ericsson, Crim, and Franklin 2013; Franco 2013). Thus, animal experimentation began in ancient Greece, its main objective being the use of animal models for the elaboration of anatomical and physiological human models (Ericsson, Crim, and Franklin 2013; Barré-Sinoussi and Montagutelli 2015). At that time, animal vivisections were also performed to acquire knowledge about the mechanisms and functions of living organisms. In the early 20th century, animal models use increased dramatically and, therefore, the ethics of their use began to be questioned, particularly in rodents (Ericsson, Crim, and Franklin 2013). Castle (1903) began breeding mice for genetic studies to obtain strains with less genetic variability. Rudolf (1976) developed the first transgenic mouse. In 1987, Capecchi, Evans, and Smithies developed the first knockout mouse (Thomas and Capecchi 1987). In 1997, Wilmut and Campbell created the first cloned animal, Dolly the sheep (Wilmut et al. 1997). In 2002 and 2004, efforts were made for the complete sequencing of the mice and rats' genome, respectively (Waterston et al. 2002; Gibbs et al. 2004).

DOI: 10.1201/b22842-16

16.1.1 The Mouse

The mouse, also known as the house mouse, is a small mammal of the order Rodentia. Scientifically known as *Mus musculus*, it is the most used species for biomedical research. For studies that require genetically identical animals, we have several strains available, like BALB/c, FVB/n, C3H, C57BL/6, CBA, DBA/2, C57BL/10, AKR, A, 129, and SJL (Hickman et al. 2017). The most portrayed in the bibliography are BALB/c strain, through which were developed genetics and molecular studies (Fisher, Hunt, and Hood 1985; Roderick, Langley, and Leiter 1985); equine herpesvirus 9 infection investigations (El-Nahass et al. 2014); circovirus-like virus P1 studies (Wen et al. 2021), among several other investigations. Through the C57BL/6J strain can be studied type II diabetes (Surwit Cochrane et al. 1988); performed genetic studies (Mekada et al. 2009); behavioral studies (Yoshizaki, Asai, and Hara 2020); obesity studies (Siersbæk et al. 2020); brain lesions (Tarrant et al. 2020); and evaluated the use of gelatin vehicles (Martins, Matos, et al. 2022). Several studies use the FVB/n strain as an animal model, with the aim of testing plant extracts such as *Castanea sativa* (Rodrigues et al. 2020), *Brassica oleracea* L. var. *Italica* (Martins, Oliveira, et al. 2022); evaluate the use of gelatin vehicles (Martins, Matos, et al. 2022); perform obesity studies (Nascimento-Sales et al. 2017); behavioral studies (Girard et al. 2016; Murph et al. 2020); immunological studies (Knott et al. 2009); age studies (Mahler et al. 1996; Raafat et al. 2012); study of lung adenocarcinoma (Mervai et al. 2018; Rothenberger and Stabile 2020; Shafarenko et al. 1997); skin squamous cell carcinoma (Hennings et al. 1993); mammary hyperplasia (Nieto et al. 2003); cardiac hypertrophy (Yang et al. 2014); and prostate epithelium (Latonen, Kujala, and Visakorpi 2016). Research that requires outbred mice should use CD-1, ICR, and Swiss Webster. In studies requiring disease processes associated with spontaneous mutation, we have the example of the athymic nude, non-obese diabetic strains (NOD; Hickman et al. 2017). Other studies were carried out with the CD-1 strain, like CD-1 mouse fertility (Crocker et al. 2018) and hematology (Petterino et al. 2021). With the ICR strain, were carried out studies on obesity (Li et al. 2020), cloning (Tanabe et al. 2017), and toxicity (Zeng et al. 2021). Finally, we have genetically modified mice or "knock-in"/"knock-out", where a particular gene/genes are turned on or off for a certain purpose, depending on ongoing investigation. The last one is the transgenic mice (Hickman et al. 2017), in which a gene from an unrelated species was introduced into the genome. We have the example of transgenic animals for HPV16, known as K14HPV16, in which the human cytokeratin 14 promoter was inserted into its genome, which allows guiding the expression of the HPV16 viral oncogenes to the basal layer of stratified squamous epithelium, mimicking lesions identical to those caused by this virus in humans (Arbeit et al. 1994).

Numerous studies have been developed using this transgenic model, such as cervical cancer (Sepkovic et al. 2011, 2012), skin carcinogenesis (Masset et al. 2011), penile cancer (Medeiros-Fonseca et al. 2020), oral cancer (Mestre et al. 2020), cancer-associated cachexia (Peixoto da Silva et al. 2020), study of hormone receptors (de Oliveira Neto et al. 2021); test of plant extracts to assess the evolution of HPV16 lesions, such as Curcumin and Rutin (Moutinho et al. 2018), *Laurus nobilis* (Medeiros-Fonseca et al. 2018), *Tilia platyphyllos* Scop. (Ferreira et al. 2021), *Pteridium* spp. (da Costa et al. 2020), extracts of red seaweed *Porphyra umbilicalis* (S. Santos et al. 2019), *Grateloupia turuturu* (Almeida et al. 2021); test of drugs such as celecoxib (Santos et al. 2016) and parecoxib (Ferreira et al. 2019); and molecular and immunological studies with dimethylaminoparthenolide (Santos et al. 2019), microRNA-155 (Paiva et al. 2015), and microRNA-146a (Araújo et al. 2018).

16.1.2 The Rat

Rattus norvegicus are also small mammals that share a long history with humans. They were once considered as food and associated with disease and destruction. However, there are also those who treat them as pets. It is currently an animal widely used in the laboratory, due to its ease of access and similarities with humans (Hickman et al. 2017). The use of rat as a research model was recorded in 1828 (Hedrich 2000). Since then, the rats have been broadly used for research. It is described in the bibliography that the Wistar rat was used to study leptospirosis (Lindenbaum and Eylan 1982); obesity studies (Novelli et al. 2007); humane endpoints in urinary bladder carcinogenesis (Oliveira et al. 2017); chronic kidney disease (Nogueira et al. 2017); urothelial carcinoma (Padrão et al. 2018); and lung cancer (Abdel-Moneim et al. 2021). Sprague–Dawley rats were used to study liver schwannoma (Teixeira-Guedes et al. 2014), mammary carcinogenesis (Alvarado et al. 2017; Faustino-Rocha et al. 2017; Faustino-Rocha et al. 2017), and mammary cancer humane endpoints in Faustino-Rocha et al. (2019).

16.1.3 The Rabbit

Rabbits are used for scientific purposes since the 19th century. Early work on the species focused on the study of rabbit anatomy and its comparison with other species, such as the frog, with a particular emphasis in the exclusive characteristics of the rabbit's heart and circulatory system (Champneys 1874; Roy 1879). It is also portrayed in the bibliography that Louis Pasteur used rabbits in several investigations that led to the development of the first rabies vaccine in the world (Rappuoli 2014). Rabbits are useful as a model of human pregnancy (Fisher, Hunt, and Hood 1985; Ito, Ando, and Handa 2011) and for the production of polyclonal antibodies for immunological research (Hanly, Artwohl, and Bennett 1995). They are also the preferred model for toxicological studies.

16.1.4 Zebrafish

The zebrafish, scientifically known as *Danio rerio* of the family Cyprinidae, is a small fish with dark blue and yellow stripes. The zebrafish's size allows it to keep large numbers in an aquarium, it has frequent spawning and large broods, which makes it a very persuasive model. It has emerged as an important vertebrate model for studying hematopoiesis (Gore et al. 2018), and it is widely used to carry out toxicity experiments (Bambino and Chu 2017; Horzmann and Freeman 2018). The short time needed for analyses, embryos' transparency, short life cycle, high fertility, and genetic similarity make the zebrafish a preferential model for embryotoxicity studies. The toxicological studies include evaluation of toxicity of bioactive compounds or crude extracts from plants and determination of the optimal process (Chahardehi, Arsad, and Lim 2020). It is also used as a regeneration model (Lebedeva et al. 2020).

16.2 PLANT PRODUCTS: WHAT IS THEIR ORIGIN AND IMPORTANCE?

Medicinal plants have been an important font of medicinal compounds for millennia, with many of the medicines used currently originated from plants (Kinghorn and Seo 1996; Newman and Cragg 2012). A medicinal plant is any plant with compounds that can be used for therapeutic applications or as precursors for the synthesis of medicines. It is important to differentiate between medicinal plants that were tested and their curative properties were recognized scientifically, and plants that are broadly used as medicinal, but were not under a scientific validation (Evans 2009; Sofowora 1993). The plants used for curative purposes are known as traditional medicine, that is, "the sum total of the knowledge, skills and practices based on the theories, beliefs and experiences indigenous to different cultures, whether explicable or not, used in the maintenance of health, as well as in the prevention, diagnosis, improvement or treatment of physical and mental illnesses" (World Health Organization 2000).

Due to their use for the treatment of several illnesses since ancestral times, plant products are considered part of history and culture (Veeresham 2012; Salmerón-Manzano and Manzano-Agugliaro 2020). One way or another, all cultures practiced this form of medicine using the plants of their specific habit (Gurib-Fakim 2006; Houghton 1995). Traditionally, natural products afforded an unlimited font of medicinal compounds, with hundreds of plants being cultivated over the world for medicinal purposes (Kinghorn and Seo 1996; Jones, Chin, and Kinghorn 2006). At the beginning, the medicinal plants' use was instinctive, as is the animals' case (Stojanoski 1999), and the trial-and-error method was applied to treat illness and differentiate between useful and not useful plants (Folashade, Omoregie, and Ochogu 2012).

Fossil evidence dated the use of plants as medicines to the Middle Paleolithic approximately 60,000 years ago (Fabricant and Farnsworth 2001). The oldest written record of the use of medicinal plants for the preparation of drugs was found on a Sumerian clay slab from Nagpur, approximately 5,000 years old. This record included 12 recipes for the preparation of drugs and mentioned more than 250 plants. The Chinese book on roots and grasses, titled "Pen T'Sao", written by Emperor Shen Nung in 2500 BC, described 365 dried parts of plants used as medicines, many of which are still used today, like camphor, cinnamon bark, ginseng, and jimson weed (Wiart 2006). The "Ebers Papyrus" (1550 BC) written in Egypt presents a collection of 800 proscriptions referring to 700 plants or drugs, including aloe, coriander, castor oil plant, fig, garlic, onion, and pomegranate, used for therapy (Borchardt 2002; Sneader 2005; Newman and Cragg 2012). The Indian book "Vedas" (1200–1000 BC) referred the treatment using plants like clove, nutmeg, and pepper very common in that country (Petrovska 2012). In "The Iliad" and "The Odyssey" (800 BC), Homer mentioned 63 plant species from the Minoan, Mycenaean, and Egyptian pharmacotherapy. Herodotus (500 BC) referred castor oil plant, Orpheus mentioned the fragrant hellebore and garlic, and Pythagoras the cabbage, mustard, and sea onion. The works of Hippocrates (459–370 BC) contained 300 medicinal plants categorized according to their effects on body physiology: common centaury and wormwood were used against fever; garlic was used to treat intestine parasites; deadly nightshade, henbane, mandrake, and opium were used as narcotics; fragrant hellebore and haselwort as emetics; asparagus, celery, garlic, parsley, and sea onion as diuretics; and oak and pomegranate as astringents (Gorunovic and Lukic 2001). Some years later, Theophrast (371–287 BC) founded botanical science with his books "De Causis Plantarium" and "De Historia Plantarium", where he classified more than 500 medicinal plants, including cardamom, cinnamon, false hellebore, forth, fragrant hellebore, iris rhizome, mint, monkshood, and pomegranate. Due to these books, Theophrast was considered "the father of botany". Celsus (25 BC–50 AD) referred approximately 250 medicinal plants in his book "De Re Medica", such as aloe, cardamon, cinnamon, flax, henbane, pepper, and poppy (Gorunovic and Lukic 2001).

Despite all these writers, the most notable person on plant drugs in ancient history was Dioscorides, who was considered "the father of pharmacognosy". He was a military physician and pharmacognosist of Nero's Army, studying medicinal plants in those places, where he traveled with the Roman Army. Dioscorides was the author of the work "De Materia Medica" (77 AD) that provided information on the medicinal plants used until the Middle Ages and Renaissance, like their appearance, location, mode of collection, medicinal preparation, and therapeutic effects. Pliny the Elder, a Dioscorides contemporary, travelled through Germany and Spain, and described almost 1,000 medicinal plants in his work "Historia Naturalis" (Toplak Galle 2005). Galen (131–200 AD), largely known by his contributes to Anatomy and Physiology, provided the first list of drugs with similar activities in his work "De Succedanus" and added several plants to the list previously elaborated by Dioscorides and Pliny. In the 7th century, Slavic people used *Aconitum napellus* as a poison in hunting, *Allium sativum* as a remedy, *Iris germanica*, *Ocimum basilicum*, *Mentha viridis* and *Rosmarinus officinalis* in cosmetics, and *Achillea millefolium*, *Artemisia maritime* L., *Cucumis sativus*, *Lavandula officinalis*, *Sambuci flos*, *Urtica dioica*, and *Veratrum album* to treat several injurious insects, including fleas, louses, mosquitos, moths, and spiders. Charles the Great (742–814 AD) founded the medical school of Salerno and encouraged the cultivation of approximately 100 different plants on the state-owned lands in his "Capitularies" (Petrovska 2012). The Arabs added several new plants to pharmacotherapy, most of which are from India, such as aloe, cinnamon, coffee, curcuma, deadly nightshade, ginger, henbane, pepper, and saffron.

In the Middle Ages, plants used as medicines and for the preparation of drugs was mainly cultivated by the monks in the monasteries. Based on the books "Canon Medicinae" written by Avicenna

(980–1037) and "Liber Magnae Collectionis Simplicum Alimentorum Et Medicamentorum" from Ibn Baitar (1197–1248), the European physicians described more than 1,000 plants with medicinal properties. The journeys of Marco Polo (1254–1324) through the Asia, China, and Persia; the finding of America (1492); and the journey of Vasco da Gama to India (1498) led to the introduction of several plants into the Europe and the development of botanical gardens. Paracelsus (1493–1541) believed that the plants collection should be determined astrologically and was one of the supporters of chemical preparation of drugs using plants and minerals. Paracelsus sustained the "Signatura Doctrinae", according to which God labelled his own sign on the healing substances, indicating their application to treat certain disease. For example, the haselwort was a reminiscent of the liver and should be used to treat liver disorders; and the leaves of the St John's wort *Hypericum perforatum* L. appeared to be stung and then should be used to treat wounds and stings. Although the ancient people used plants with medicinal properties as simple pharmaceutical forms, like decoctions, infusions, macerations, in the 16th and 17th centuries, there was an increased demand for compound drugs, comprising products of both plant and animal origin. The drug complexity and the origin of products determined its price and search. In the 18th century, Linnaeus (1707–1788) afforded a classification and description of the plant species described until then in a manuscript titled "Species Plantarium" (1753). The polynomial system used until then for plants' naming was replaced by the binominal system—the plant name consisted of two words, the genus and species name (Jancic and Lakušic 2002).

For thousands of years, the human pharmacopeia was dominated by plant-derived products (Raskin and Ripoll 2004). Until the 18th century, the therapeutic abilities of several plants, their effects on humans, and their posology were recognized, but the active compounds responsible for the effects remained unidentified (Faridi et al. 2010). The use of plants was refined overtime, and the origin of modern science, with the development of chemical analysis and development of microscope allowed the isolation of the active principles of medical plants (Reeds 1976). So, the use of the plants alone and the characterization of their compounds to discover and develop modern drugs started in the 19th century (Veeresham 2012; Atanasov et al. 2015). Since then, the active principles of the plants started to be synthesized in laboratory to produce medicines (Atanasov et al. 2015). Morphine was the first pure natural product used for therapeutic purposes and was commercialized by Merck in 1826. In 1897, Arthur Eichengrün and Felix Hoffmann were working at Friedrich Bayer and were responsible for the creation of the first synthetic drug—aspirin. Aspirin was synthesized from salicylic acid, isolated from *Salix alba*, and then marketed by Bayer. The commercialization of these products accelerated the isolation of several compounds and development of drugs that were then marketed and are still under use, like artemisinin isolated from *Artemisia annua* to treat malaria, cannabidiol obtained from *Cannabis sativa*, and capsaicin from *Capsicum annuum* to relieve the pain, galantamine obtained from *Galanthus nivalis* for Alzheimer's disease treatment, paclitaxel isolated from *Taxus brevifolia* for breast, lung, and ovarian cancer treatment, silymarin extracted from the seeds of *Silybum marianum* to treat liver disorders, and tiotropium derived from *Atropa belladonna* to treat chronic obstructive pulmonary disease (Veeresham 2012).

It is broadly accepted that this knowledge was the origin of modern medicine and pharmacology (Kinghorn and Seo 1996; Jones, Chin, and Kinghorn 2006). Although the modern pharmaceutical industry arose from botanical medicine, synthetic approaches have become a standard for drug discovery (Schmidt et al. 2008). Nowadays, industry produces medicines based on the active principles of plants that are used as raw materials (Arceusz, Radecka, and Wesolowski 2010). It is worth to note that the domestication of plants led to the alteration of their phytochemical composition over time, with increase or decrease in some bioactive compounds. For example, modern agriculture created uniform cultivars of foxglove (*Digitalis purpurea*) with higher digoxin content (Ruiz et al. 2005; Antonious et al. 2006; Schmidt et al. 2008). It is also important to note that a better understanding of molecular biology gave new applications for old molecules (Grover, Yadav, and Vats 2002). For example, the alkaloid forskolin isolated from *Coleus forskohlii* and phytochemicals derived from *Stephania glabra* are now being rediscovered as adenylate cyclase and nitric oxide activators that may be used to prevent obesity and atherosclerosis (Veeresham 2012). The recent finding of the powerful antimalarial drug, Artemisinin, derived from *Artemisia annua* L. (sweet wormwood), a shrub of the Chinese medicine, constituted a paradigm shift. With this discovery, Tu Youyou was awarded with the Nobel Prize in Medicine/Physiology in 2015. Nowadays, the Artemisinin-based combination therapies have been recommended by the World Health Organization (WHO) to treat lethal malaria (Shen et al. 2018).

Plants remained a valuable source of therapeutic compounds due to their complex composition, consisting of several compounds with multiple actions (Schmidt et al. 2008). Today, it is wholly accepted that plants are a source of health-promoting ingredients, such as betalains, carotenoids, fiber, glucosinolates, minerals, polyphenols, and vitamins. In this way, several epidemiological studies (e.g., Aziz and Jalil 2019) have linked the plants consumption with health and concluded that a higher plants intake may reduce the incidence of metabolic or chronic diseases, like obesity or cancer. The rising interest in the use of medicinal plants is demonstrated by a high number of systematic reviews and surveys concerning to the use herbal medicines in the last 15 years (McLay et al. 2016). The use of herbal medicines both wild-collected and purchased from herbalists, pharmacies, and supermarkets is re-emerging in Europe, intimately linked with the adoption of a healthier lifestyle and the concept that the natural products are not harmful to health (Ekor 2014; Thomson et al. 2014; McLay et al. 2016). It is worth to note that the natural products may interact with other compounds/drugs and provoke adverse reactions (Shugaba et al. 2014; Ekor 2014). In contrast to the developed countries, approximately 80% of the people living in developing countries do not have access to the medicines of synthetic origin, and the traditional medicine based on the direct use of medicinal plants is still frequently used due their low cost (Devesa Jordà et al. 2004; Veeresham 2012; Parveen et al. 2015; Salmerón-Manzano and Manzano-Agugliaro 2020). The use of plants without sanitary control and without the control of possible damaging effects for human health constitute the major problem of this traditional medicine (Chan 2003). Despite several plants do not have harmful effects, like chamomile, mint or rosemary, others may contain active principles dangerous for health. For instance, the green seeds (Arceusz, Radecka, and Wesolowski 2010) of bitter melon (*Momordica charantia* L.), frequently used to cure fever and malaria, are very toxic and may cause a sharp drop in blood sugar and induce a hypoglycemic coma (Li, Zhang, and Tan 2004; Adoum 2009).

Some therapeutical approaches were passed from generation to generation with plant-based medicine persisting today (Dias, Urban, and Roessner 2012). The medicinal plants' use tends to be underreported (Walji et al. 2009), leading to an uncertainty concerning their therapeutic effectiveness, and possible acute or chronic effects (Lee et al. 2004; Welz, Emberger-Klein, and Menrad 2018). Some reviews suggest a strong correlation between use of herbal medicine and factors like age, sex, education, socioeconomic status, and failed success of conventional therapies. The plant-based products' use for improving human health evolved independently in different regions of the world, with the production, use, attitude, and regulatory aspects varying globally (Schmidt et al. 2008). Almost all pharmacopeias in the world proscribe plant drugs with medicinal value. Some countries, such as the United Kingdom and Germany, have separate herbal pharmacopoeias (Petrovska 2012). The WHO strategy, from 2014 to 2023, aimed to strengthen the role of traditional medicine, emphasizing the importance of promoting and including the use of medicinal plants in the health systems (Grover, Yadav, and Vats 2002; Veeresham 2012).

16.3 ANIMAL MODELS IN PHYTOPHARMACOLOGY AND TOXICOLOGICAL TESTING

Plant-derived drugs have contributed significantly to the development of new drugs, and it is certain that humanity would be much poorer without these natural plant-derived drugs as artemisinin, etoposide, morphine, paclitaxel, quinine, teniposide, vinblastine, vincristine, and the camptothecin derivatives topotecan and irinotecan (Kingston 2011).

While numerous studies have validated the traditional use of medicinal plants by investigating the phytochemicals present in active extracts (Van Wyk and Wink South Africa 2004; Palombo 2006), several plants have not yet been scientifically studied, lacking reports on their phytochemical constituents and their potential medicinal applications (Egamberdieva et al. 2017). Animal models provide the needed platform that will help to reveal the physiological functions of phytochemical compounds.

In Table 16.1, there is a summary of several studies related to *in vivo* phytopharmacology and toxicological testing, in the last five years (2017–2022). The information was compiled via an electronic search of the following scientific databases: Science Direct, PubMed, Web of Science,

TABLE 16.1
Animal Models Used in Phytopharmacology and for Toxicological Testing, from 2017 to 2022

Animal model	Compound	Title	Authors
db/db mouse	*Scutellaria baicalensis*	Therapeutic properties of *Scutellaria baicalensis* in db/db mice evaluated using Connectivity Map and network pharmacology	Kim et al., 2017
AG 129 mouse	*Cissampelos pareira*, *Cocculus hirsutus*, Dengue virus	*Cocculus hirsutus*-Derived Phytopharmaceutical Drug has Potent Anti-dengue Activity	Shukla et al., 2021
Male BALB/c mice. Ectopic Subcutaneous Syngeneic CT26. WT tumor model	*Mangifera indica*	Antiproliferative, Antiangiogenic, and Antimetastatic Therapy Response by Mangiferin in a Syngeneic Immunocompetent Colorectal Cancer Mouse Model Involves Changes in Mitochondrial Energy Metabolism	Rodríguez-Gonzales et al., 2021
Heterozygous Pb-PRL male mice	*Sabal serrulata* (WS R 1473) and an aqueous ethanolic extract from roots of *Urtica dioica*	Combined Sabal and Urtica Extracts (WS® 1541) Exert Anti-proliferative and Anti-inflammatory Effects in a Mouse Model of Benign Prostate Hyperplasia	Pigat et al., 2019
Mouse model of chronic dextran sulfate sodium (DSS) colitis	Sage and bitter apple (SBA)	Therapeutic efficacy of a combined sage and bitter apple phytopharmaceutical in chronic DSS-induced colitis	Hoffmann et al., 2017
Inbred male Swiss albino mice	*Camellia sinensis*, *Cucurbita maxima*, *Citrus limon* and *Camellia reticulata* peels	Citrus peels prevent cancer	Nair et al., 2018
Balb/c mice	*Tamarix stricta*	*In vitro* and *in vivo* antidiabetic activity of *Tamarix stricta* Boiss.: Role of autophagy	Bahramsoltani et al., 2020

(*Continued*)

TABLE 16.1 (*Continued*)
Animal Models Used in Phytopharmacology and for Toxicological Testing, from 2017 to 2022

Animal model	Compound	Title	Authors
Mice (ulcerative colitis induced)	*Combretum fragrans* (Combretin A and Combretin B)	In Vitro Antioxidant, Anti-inflammatory, and In Vivo Anticolitis Effects of Combretin A and Combretin B on Dextran Sodium Sulfate-Induced Ulcerative Colitis in Mice	Mbiantcha et al., 2020
FVB/N male mice	*Brassica oleracea* L. Var. *Italica*	Effect of a Sub-Chronic Oral Exposure of Broccoli (*Brassica oleracea* L. Var. *Italica*) By-Products Flour on the Physiological Parameters of FVB/N Mice: A Pilot Study	Martins et al., 2022
K14-HPV16 mice	*Pteridium* spp.	Ptaquiloside from bracken (*Pteridium* spp.) promotes oral carcinogenesis initiated by HPV16 in transgenic mice	Gil da Costa et al., 2020
Sprague Dawley Rats	Curcumin	Impact of curcumin on the pharmacokinetics of rosuvastatin in rats and dogs based on the conjugated metabolites	Zhou et al., 2017
Female Wistar rats	*Dissotis thollonii*	In Vitro Anti-Inflammatory and In Vivo Antiarthritic Activities of Aqueous and Ethanolic Extracts of *Dissotis thollonii* Cogn. (Melastomataceae) in Rats	Nguemnang et al., 2019
Wistar rats and *Mus musculus* mice	*Pentaclethra macrophylla*	Hepatoprotective effects of extracts, fractions and compounds from the stem bark of *Pentaclethra macrophylla* Benth: Evidence from in vitro and in vivo studies	Bomgning et al., 2021
Wistar rats	*Paullinia pinnata*	Aqueous and methanol extracts of *Paullinia pinnata* L. (Sapindaceae) improve inflammation, pain and histological features in CFA-induced mono-arthritis: Evidence from *in vivo* and *in vitro* studies	Tseuguem et al., 2019
Wistar rats	*Garcinia gummi-gutta*	Garcinia gummi-gutta seeds: a novel source of edible oils	Pryia Rani et al., 2022

(*Continued*)

TABLE 16.1 (*Continued*)
Animal Models Used in Phytopharmacology and for Toxicological Testing, from 2017 to 2022

Animal model	Compound	Title	Authors
Male Wistar rats and Swiss albino mice	*Ficus krishnae*	Antidiabetes constituents, cycloartenol and 24-methylenecycloartanol, from *Ficus krishnae*	Nair et al., 2020
Male Wistar rats	*Pterorhachis zenkeri*	Anti-androgenic, anti-oxidant and anti-apoptotic effects of the aqueous and methanol extracts of *Pterorhachis zenkeri*	
(Meliaceae): Evidence from in vivo and in vitro studies	Fozin Bonsou et al., 2020		
Male albino Wistar rats	*Crinum zeylanicum*	Antihypertensive Effects of the Methanol Extract and the Ethyl Acetate Fraction from *Crinum zeylanicum* (Amaryllidaceae) Leaves in L-NAME-Treated Rat	Ndjenda et al., 2021
Male Wistar rats	*Distemonanthus benthamianus*	Methanolic Extract of *Distemonanthus benthamianus* (Caesalpiniaceae) Stem Bark Suppresses Ethanol/Indomethacin Induced Chronic Gastric Injury in Rats	Marte et al., 2020
Wistar rats	*Alstonia boonei*	Effects of Aqueous and Methanolic Extracts of Stem Bark of *Alstonia boonei* De Wild. (Apocynaceae) on Dextran Sodium Sulfate-Induced Ulcerative Colitis in Wistar Rats	Adjouzem et al., 2020
Female Wistar rats	*Persea americana*	The Ethanol Extract of Avocado (*Persea americana* Mill. (Lauraceae)) Seeds Successfully Induces Implant Regression and Restores Ovarian Dynamic in a Rat Model of Endometriosis	Essono et al., 2020
Male and female Wistar rats	*Ceiba pentandra*	Hypoglycemic Properties of the Aqueous Extract from the Stem Bark of *Ceiba pentandra* in Dexamethasone-Induced Insulin Resistant Rats	Fofié et al., 2018

(*Continued*)

TABLE 16.1 (*Continued*)
Animal Models Used in Phytopharmacology and for Toxicological Testing, from 2017 to 2022

Animal model	Compound	Title	Authors
Male Wistar rats	*Paullinia pinnata*	Aqueous and Methanol Extracts of *Paullinia pinnata* (Sapindaceae) Improve Monosodium Urate-Induced Gouty Arthritis in Rat: Analgesic, Anti-Inflammatory, and Antioxidant Effects	Tseuguem et al., 2019
Wistar Rats	*Castanea sativa* Mill. Flowers	The influence of *Castanea sativa* Mill. Flower extract on hormonally and chemically induced prostate cancer in a rat model	Nascimento-Gonçalves et al., 2021
New Zealand white male rabbits	*Garcinia gummi-gutta*	Garcinia gummi-gutta seeds: a novel source of edible oils	Pryia Rani et al., 2022
Zebrafish	Capsaicin, carnosic acid, cinnamaldehyde, curcumin, diallyl trisulfide, eugenol, 6-gingerol, isoeugenol, 6-(Methylsulfinyl)hexyl isothiocyanate (6-MSITC) and quercetin	Evaluation of Antioxidant Activity of Spice-Derived Phytochemicals Using Zebrafish	Endo et al., 2020
Zebrafish	*Curcuma longa*	Phytochemical Evaluation, Embryotoxicity, and Teratogenic Effects of *Curcuma longa* Extract on Zebrafish (*Danio rerio*)	Alafiatayo et al., 2019
Zebrafish	*Allium flavum* and *Allium carinatum*	Wild edible onions—*Allium flavum* and *Allium carinatum*—successfully prevent adverse effects of chemotherapeutic drug doxorubicin	Aleksandar et al., 2019
Zebrafish	*Azadirachta indica* A. Juss ethanolic extract of the neem fruit	Antinociceptive activity of ethanolic extract of *Azadirachta indica* A. Juss (Neem, Meliaceae) fruit through opioid, glutamatergic and acid-sensitive ion pathways in adult zebrafish (Danio rerio)	Batista et al., 2018
Zebrafish and Female Sprague Dawley rats	*Zingiber ottensii*	Phytochemical and Safety Evaluations of *Zingiber ottensii* Valeton Essential Oil in Zebrafish Embryos and Rats	Thitinarongwate et al., 2021

(*Continued*)

TABLE 16.1 (*Continued*)
Animal Models Used in Phytopharmacology and for Toxicological Testing, from 2017 to 2022

Animal model	Compound	Title	Authors
Mongrel Dogs	*Libidibia ferrea*	Efficacy of Phytopharmaceuticals From the Amazonian Plant *Libidibia ferrea* for Wound Healing in Dogs	Américo et al., 2020
Beagle dogs	Curcumin	Impact of curcumin on the pharmacokinetics of rosuvastatin in rats and dogs based on the conjugated metabolites	Zhou et al., 2017

and Scopus. The search terms were as follows: "animal models of phytochemicals" alone or in combination with "*in vivo* models", "pharmacology", "plant products", and "toxicological testing".

It is evident from the search results that several plant products have been tested as potential therapy for several diseases, such as arthritis, cancer, colitis, dengue, diabetes, endometriosis, gastric injury, and hypertension. It is also clear that the rodent models, namely mice and rats, were the animal models most frequently used for phytopharmacology and toxicological testing, similarly with other areas of research. It is worth to note that *in vivo* studies are the ultimate way to determine the clinical importance of plant products. Choosing the correct animal model is essential to the success of new drugs discovery, and the study purpose, the species available, the most advantageous species concerning technical questions and animals' well-being, practical aspects (animals' availability, housing, manipulation, equipment required), and ethical and scientific considerations should also be considered. The mice and rats are commonly used because their physiology and genetics are well known, they are easy to maintain and manipulate, they are relatively cheap, and most importantly they are mammals with several similarities with humans, in term of their anatomy, physiology, genetics, and biochemistry.

16.4 CONCLUSIONS AND FUTURE PERSPECTIVES

Natural products have historically served as the most significant source of new leads for pharmaceutical development and inspiration for a large fraction of the current pharmacopoeia. Before the 20th century, 80% of all medicines were extracted from roots, barks, and leaves, but with the evolution of the "synthetic era", the advancement in the industrial sector and industrial medicine has led to the use of herbs and phytopharmaceuticals to a secondary plan for a long period of time. Phytopharmacology and toxicological testing includes a lot of plants and plant-based compounds and continues to expand rapidly across the world with many people now resorting to these products for the treatment of various health challenges. In 2013, it was estimated that up to 4 billion people (representing 80% of the world's population) living in the developing world rely on herbal medicinal products as a primary source of healthcare, and traditional medical practice that involves the use of herbs was viewed as an integral part of the culture in those communities. Currently, a new approach to the traditional methods is being accepted worldwide. The purpose of animal studies is to test hypothesis on how the specific factor affects the specific species under certain circumstances. Their value is speeding up the process of drug discovery by generating ideas or suggesting hypotheses that might be relevant to humans. The differences in biological characteristics (e.g., anatomy, physiology, and metabolic rate) between animals and humans should be kept in mind, while studying qualitative questions devoted to the identification of the diet component responsible for a specific function in the body, or quantitative questions concerning a dose of a diet component necessary to cause the specific reaction in the body. A comparative biology continues to be a useful tool for choosing an appropriate animal model. An impressive number of plant species and plant products are traditionally used, however, although presumed safe, some of them have been associated with acute or chronic intoxications. Most of those plants were not sufficiently studied to guarantee a safe use; thereby, new investigations are necessary to discover new phytochemicals, to give new uses for known phytochemicals, to ensure their safe use, and to guarantee a better therapeutic approach for several human diseases, thus improving the humans' lifetime and quality of life.

Funding: This work was supported by National Funds by FCT—Portuguese Foundation for Science and Technology (project ID: UIDB/04033/2020 and PhD; grant no. 2020.07675.BD).

REFERENCES

Abdel-Moneim, Adel, Osama M Ahmed, Abd El-Twab, M Sanaa, Mohamed Y Zaky, and Lamiaa N Bakry. 2021. "Prophylactic Effects of Cynara Scolymus L. Leaf and Flower Hydroethanolic Extracts against Diethylnitrosamine/Acetylaminoflourene-Induced Lung Cancer in Wistar Rats." *Environmental Science andPollution Research* 28 (32): 43515–27.

Adoum, Oumar Al-Moubarak. 2009. "Determination of Toxicity Levels of Some Savannah Plants Using Brine Shrimp Test (BST)." *Bayero Journal of Pure Sciences* 2 (1): 135–38.

Almeida, José, Tiago Ferreira, Susana Santos, Maria J Pires, Rui M Gil da Costa, Rui Medeiros, Margarida M S M Bastos, Maria J Neuparth, Ana I Faustino-Rocha, and Helena Abreu. 2021. "The Red Seaweed Grateloupia Turuturu Prevents Epidermal Dysplasia in HPV16-Transgenic Mice." *Nutrients* 13 (12): 4529.

Alvarado, Antonieta, Rui M Gil da Costa, Ana I Faustino-Rocha, Rita Ferreira, Carlos Lopes, Paula A Oliveira, and Bruno Colaço. 2017. "Effects of Exercise Training on Breast Cancer Metastasis in a Rat Model." *International Journal of Experimental Pathology* 98 (1): 40–46.

Antonious, G F, T S Kochhar, R L Jarret, and J C Snyder. 2006. "Antioxidants in Hot Pepper: Variation among Accessions." *Journal of Environmental Science and Health, Part B* 41 (7): 1237–43. https://doi.org/10.1080/03601230600857114.

Araújo, R, J M O Santos, M Fernandes, F Dias, H Sousa, J Ribeiro, Mmsm Bastos, et al. 2018. "Expression Profile of MicroRNA-146a along HPV-Induced Multistep Carcinogenesis: A Study in HPV16 Transgenic Mice." *Journal of Cancer Research Clinical Oncology* 144 (2): 241–48. https://doi.org/10.1007/s00432-017-2549-5.

Arbeit, J M, K Münger, P M Howley, and D Hanahan. 1994. "Progressive Squamous Epithelial Neoplasia in K14-Human Papillomavirus Type 16 Transgenic Mice." *Journal of Virology* 68 (7): 4358–68. https://doi.org/10.1128/jvi.68.7.4358-4368.1994.

Arceusz, Agnieszka, Iwona Radecka, and Marek Wesolowski. 2010. "Identification of Diversity in Elements Content in Medicinal Plants Belonging to Different Plant Families." *Food Chemistry* 120 (1): 52–58.

Atanasov, Atanas G, Birgit Waltenberger, Eva-Maria Pferschy-Wenzig, Thomas Linder, Christoph Wawrosch, Pavel Uhrin, Veronika Temml, Limei Wang, Stefan Schwaiger, and Elke H Heiss. 2015. "Discovery and Resupply of Pharmacologically Active Plant-Derived Natural Products: A Review." *Biotechnology Advances* 33 (8): 1582–1614.

Aziz, N A A, and A M M Jalil. 2019. "Bioactive Compounds, Nutritional Value, and Potential Health Benefits of Indigenous Durian (Durio Zibethinus Murr.): A Review." *Foods (Basel, Switzerland)* 8 (3): 96.

Bambino, K, and J Chu. 2017. "Zebrafish in Toxicology and Environmental Health." *Current Topics in Development Biology* 124: 331–67. https://doi.org/10.1016/bs.ctdb.2016.10.007.

Barré-Sinoussi, F, and X Montagutelli. 2015. "Animal Models Are Essential to Biological Research: Issues and Perspectives." *Future Science OA* 1 (4): Fso63. https://doi.org/10.4155/fso.15.63.

Borchardt, John K. 2002. "The Beginnings of Drug Therapy: Ancient Mesopotamian Medicine." *Drug News* 15 (3): 187–92.

Castle, W E. 1903. "Mendel's Law of Heredity." *Science* 18 (456): 396–406. https://doi.org/10.1126/science.18.456.396.

Chahardehi, A M, H Arsad, and V Lim. 2020. "Zebrafish as a Successful Animal Model for Screening Toxicity of Medicinal Plants." *Plants (Basel)* 9 (10). https://doi.org/10.3390/plants9101345.

Champneys, F. 1874. "The Septum Atriorum of the Frog and the Rabbit." *J Anat Physiol* 8 (Pt 2): 340–52.

Chan, K. 2003. "Some Aspects of Toxic Contaminants in Herbal Medicines." *Chemosphere* 52 (9): 1361–71. https://doi.org/10.1016/s0045-6535(03)00471-5.
Costa, Rui M Gil da, Tiago Neto, Diogo Estêvão, Magda Moutinho, Ana Félix, Rui Medeiros, Carlos Lopes, Margarida M S M Bastos, and Paula A Oliveira. 2020. "Ptaquiloside from Bracken (Pteridium Spp.) Promotes Oral Carcinogenesis Initiated by HPV16 in Transgenic Mice." *Food and Function* 11 (4): 3298–305.
Crocker, K, M D Calder, N A Edwards, D H Betts, and A J Watson. 2018. "CD-1 Mouse Fertility Rapidly Declines and Is Accompanied with Early Pregnancy Loss under Conventional Housing Conditions." *Theriogenology* 108: 245–54. https://doi.org/10.1016/j.theriogenology.2017.12.018.
Devesa Jordà, F, J Pellicer Bataller, J Ferrando Ginestar, A Borghol Hariri, M Bustamante Balén, J Ortuño Cortés, I Ferrando Marrades, et al. 2004. "Consumption of Medicinal Herbs in Patients Attending a Gastroenterology Outpatient Clinic." *Gastroenterol Hepatol* 27 (4): 244–49. https://doi.org/-10.1016/s0210-5705(03)70453-1.
Dias, Daniel A, Sylvia Urban, and Ute Roessner. 2012. "A Historical Overview of Natural Products in Drug Discovery." *Metabolites* 2 (2): 303–36.
Egamberdieva, Dilfuza, Nazim Mamedov, Elisa Ovidi, Antonio Tiezzi, and Lyle Craker. 2017. "Phytochemical and Pharmacological Properties of Medicinal Plants from Uzbekistan: A Review." *Journal of Medicinally Active Plants* 5 (2): 59–75.
Ekor, Martins. 2014. "The Growing Use of Herbal Medicines: Issues Relating to Adverse Reactions and Challenges in Monitoring Safety." *Frontiers in Pharmacology* 4: 177.
El-Nahass, E, K M El-Dakhly, N El-Habashi, S I Anwar, H Sakai, A Hirata, A Okada, R Abo-Sakaya, H Fukushi, and T Yanai. 2014. "Susceptibility of BALB/c-Nu/Nu Mice and BALB/c Mice to Equine Herpesvirus 9 Infection." *Veterinary Pathology* 51 (3): 581–90. https://doi.org/10.1177/0300985813493932.
Ericsson, A C, M J Crim, and C L Franklin. 2013. "A Brief History of Animal Modeling." *Missouri Medicine* 110 (3): 201–5.
Evans, William Charles. 2009. *Trease and Evans' Pharmacognosy*. Elsevier Health Sciences.
Fabricant, Daniel S, and Norman R Farnsworth. 2001. "The Value of Plants Used in Traditional Medicine for Drug Discovery." *Environmental Health Perspectives* 109 (suppl 1): 69–75.
Faridi, Pouya, Mohammad M Zarshenas, Zohreh Abolhassanzadeh, and Abdolali Mohagheghzadeh. 2010. "Collection and Storage of Medicinal Plants in The Canon of Medicine." *Pharmacognosy Journal* 2 (8): 216–18.
Faustino-Rocha, Ana I, Adelina Gama, Paula A Oliveira, Katrien Vanderperren, Jimmy H Saunders, Maria J Pires, Rita Ferreira, and Mario Ginja. 2017. "Modulation of Mammary Tumor Vascularization by Mast Cells: Ultrasonographic and Histopathological Approaches." *Life Sciences* 176: 35–41.
Faustino-Rocha, Ana I, Mário Ginja, Rita Ferreira, and Paula A Oliveira. 2019. "Studying Humane Endpoints in a Rat Model of Mammary Carcinogenesis." *Iranian Journal of Basic Medical Sciences* 22 (6): 643.
Faustino-Rocha, Ana I, Adelina Gama, Paula A Oliveira, Katrien Vanderperren, Jimmy H Saunders, Maria J Pires, Rita Ferreira, and Mario Ginja. 2017. "A Contrast-Enhanced Ultrasonographic Study About the Impact of Long-term Exercise Training on Mammary Tumor Vascularization." *Journal of Ultrasound in Medicine* 36 (12): 2459–66.
Ferreira, Tiago, Sandra Campos, Mónica G Silva, Rita Ribeiro, Susana Santos, José Almeida, Maria João Pires, Rui Miguel Gil da Costa, Cláudia Córdova, and António Nogueira. 2019. "The Cyclooxigenase-2 Inhibitor Parecoxib Prevents Epidermal Dysplasia in HPV16-Transgenic Mice: Efficacy and Safety Observations." *International Journal of Molecular Sciences* 20 (16): 3902.
Ferreira, Tiago, Elisabete Nascimento-Gonçalves, Sara Macedo, Inês Borges, Adelina Gama, Rui M Gil da Costa, Maria J Neuparth, Germano Lanzarin, Carlos Venâncio, and Luís Félix. 2021. "Toxicological and Anti-Tumor Effects of a Linden Extract (Tilia Platyphyllos Scop.) in a HPV16-Transgenic Mouse Model." *Food* 12 (9): 4005–14.
Fisher, Douglas A, S W Hunt 3rd, and Leroy Hood. 1985. "Structure of a Gene Encoding a Murine Thymus Leukemia Antigen, and Organization of Tla Genes in the BALB/c Mouse." *The Journal of Experimental Medicine* 162 (2): 528–45.
Folashade, Oluyemisi, Henry Omoregie, and Peter Ochogu. 2012. "Standardization of Herbal Medicines-A Review." *International Journal of Biodiversity* 4 (3): 101–12.
Franco, Nuno Henrique. 2013. "Animal Experiments in Biomedical Research: A Historical Perspective." *Animals* 3 (1): 238–73.
Gibbs, R A, G M Weinstock, M L Metzker, D M Muzny, E J Sodergren, S Scherer, G Scott, et al. 2004. "Genome Sequence of the Brown Norway Rat Yields Insights into Mammalian Evolution." *Nature* 428: 493–521.

Girard, Stéphane D, Guy Escoffier, Michel Khrestchatisky, and François S Roman. 2016. "The FVB/N Mice: A Well Suited Strain to Study Learning and Memory Processes Using Olfactory Cues." *Behavioural Brain Research* 296: 254–59.

Gore, Aniket V, Laura M Pillay, Marina Venero Galanternik, and Brant M Weinstein. 2018. "The Zebrafish: A Fintastic Model for Hematopoietic Development and Disease." *Wiley Interdisciplinary Reviews: Developmental Biology* 7 (3): e312.

Gorunovic, M, and Lukic, P. 2001. *Pharmacognosy*. Beograd.

Grover, J K, S Yadav, and V Vats. 2002. "Medicinal Plants of India with Anti-Diabetic Potential." *Journal of Ethnopharmacology* 81 (1): 81–100. https://doi.org/10.1016/s0378-8741(02)00059-4.

Gurib-Fakim, Ameenah. 2006. "Medicinal Plants: Traditions of Yesterday and Drugs of Tomorrow." *Molecular Aspects of Medicine* 27 (1): 1–93. https://doi.org/10.1016/j.mam.2005.07.008.

Hanly, W C, J E Artwohl, and B T Bennett. 1995. "Review of Polyclonal Antibody Production Procedures in Mammals and Poultry." *ILAR Journal* 37 (3): 93–118. https://doi.org/10.1093/ilar.37.3.93.

Hedrich, Hans J. 2000. "History, Strains and Models." In G J Krinke (ed.), *The Laboratory Rat*, 3–16. Academic Press.

Hennings, Henry, Adam B Glick, David T Lowry, Ljubicka S Krsmanovic, Linda M Sly, and Stuart H Yuspa. 1993. "FVB/N Mice: An Inbred Strain Sensitive to the Chemical Induction of Squamous Cell Carcinomas in the Skin." *Carcinogenesis* 14 (11): 2353–58.

Hickman, D L, J Johnson, T H Vemulapalli, J R Crisler, and R Shepherd. 2017. "Commonly Used Animal Models." *Principles of Animal Research for Graduate and Undergraduate Students*. 117–75. doi: 10.1016/B978-0-12-802151-4.00007-4.

Horzmann, Katharine A, and Jennifer L Freeman. 2018. "Making Waves: New Developments in Toxicology with the Zebrafish." *Toxicological Sciences* 163 (1): 5–12.

Houghton, Peter J. 1995. "The Role of Plants in Traditional Medicine and Current Therapy." *The Journal of Alternative Medicine, Complementary* 1 (2): 131–43.

Ito, Takumi, Hideki Ando, and Hiroshi Handa. 2011. "Teratogenic Effects of Thalidomide: Molecular Mechanisms." *Cellular andMolecular Life Sciences* 68 (9): 1569–79.

Jaenisch, Rudolf. 1976. "Germ Line Integration and Mendelian Transmission of the Exogenous Moloney Leukemia Virus." *Proceedings of the National Academy of Sciences* 73 (4): 1260–64.

Jancic, R, and B Lakušic. 2002. "Botanika Farmaceutika." *Farmaceutski Fakultet, Zavod Za Botaniku, Beograd.*

Jones, William P, Young-Won Chin, and A Douglas Kinghorn. 2006. "The Role of Pharmacognosy in Modern Medicine and Pharmacy." *Current Drug Targets* 7 (3): 247–64.

Kinghorn, A Douglas, and Eun-Kyoung Seo. 1996. "Plants as Sources of Drugs." *Agricultural Materials as Renewable* Resources (Chapter 12, pp. 179–193). ACS *Symposium* Series. Vol. 647. ACS Publications.

Kingston, D G. 2011. "Modern Natural Products Drug Discovery and Its Relevance to Biodiversity Conservation." *Journal of Natural Products* 74 (3): 496–511. https://doi.org/10.1021/np100550t.

Knott, M L, S P Hogan, H Wang, K I Matthaei, and L A Dent. 2009. "FVB/N Mice Are Highly Resistant to Primary Infection with Nippostrongylus Brasiliensis." *Parasitology* 136 (1): 93–106. https://doi.org/10.1017/s0031182008005192.

Latonen, L, P Kujala, and T Visakorpi. 2016. "Incidence of Mucinous Metaplasia in the Prostate of FVB/N Mice (Mus Musculus)." *Comparative Medicine* 66 (4): 286–89.

Lebedeva, L, B Zhumabayeva, T Gebauer, I Kisselev, and Z Aitasheva. 2020. "Zebrafish (Danio Rerio) as a Model for Understanding the Process of Caudal Fin Regeneration." *Zebrafish* 17 (6): 359–72. https://doi.org/10.1089/zeb.2020.1926.

Lee, G B W, T C Charn, Z H Chew, and T P Ng. 2004. "Complementary and Alternative Medicine Use in Patients with Chronic Diseases in Primary Care Is Associated with Perceived Quality of Care and Cultural Beliefs." *Family Practice* 21 (6): 654–60.

Li, Jinglei, Haishan Wu, Yuting Liu, and Liu Yang. 2020. "High Fat Diet Induced Obesity Model Using Four Strains of Mice: Kunming, C57BL/6, BALB/c and ICR." *Experimental Animals* 69(3): 19–148.

Li, S Y, B L Zhang, and F P Tan. 2004. "Glycopenia Coma Caused by Taking Both Dimethylbiguanide and Momordica Charantia Buccal Tablet in One Patient." *Chinese Journal of New Drugs Andclinical Remedies* 23 (3): 189–90.

Lindenbaum, I, and E Eylan. 1982. "Leptospirosis in Rattus Norvegicus and Rattus Rattus in Israel." *Israel Journal of Medical Sciences* 18 (2): 271–75.

Mahler, J F, W Stokes, P C Mann, M Takaoka, and R R Maronpot. 1996. "Spontaneous Lesions in Aging FVB/N Mice." *Toxicol Pathol* 24 (6): 710–16. https://doi.org/10.1177/019262339602400606.

Martins, T, A F Matos, J Soares, R Leite, M J Pires, T Ferreira, B Medeiros-Fonseca, E Rosa, P A Oliveira, and L M Antunes. 2022. "Comparison of Gelatin Flavors for Oral Dosing of C57BL/6J and FVB/N Mice." *Journal of American Association for Laboratory Animal Science* 61 (1): 89–95. https://doi.org/10.30802/-aalas-jaalas-21-000045.

Martins, T, P A Oliveira, M J Pires, M J Neuparth, G Lanzarin, L Félix, C Venâncio, et al. 2022. "Effect of a Sub-Chronic Oral Exposure of Broccoli (Brassica Oleracea L. Var. Italica) By-Products Flour on the Physiological Parameters of FVB/N Mice: A Pilot Study." *Foods* 11 (1). https://doi.org/10.3390/foods11010120.

Masset, A, C Maillard, N E Sounni, N Jacobs, F Bruyére, P Delvenne, M Tacke, et al. 2011. "Unimpeded Skin Carcinogenesis in K14-HPV16 Transgenic Mice Deficient for Plasminogen Activator Inhibitor." *International Journal of Cancer* 128 (2): 283–93. https://doi.org/10.1002/ijc.25326.

McLay, James S, Abdul R Pallivalappila, Ashalatha Shetty, Binita Pande, Moza Al Hail, and Derek Stewart. 2016. "'Asking the Right Question'. A Comparison of Two Approaches to Gathering Data on'Herbals' Use in Survey Based Studies." *PloS One* 11 (2): e0150140.

Medeiros-Fonseca, B, V F Mestre, Bruno Colaço, Maria João Pires, Tânia Martins, R M Gil da Costa, Maria João Neuparth, Rui Medeiros, Magda S S Moutinho, and Maria Inês Dias. 2018. "Laurus Nobilis (Laurel) Aqueous Leaf Extract's Toxicological and Anti-Tumor Activities in HPV16-Transgenic Mice." *Food* 9 (8): 4419–28.

Medeiros-Fonseca, B, V F Mestre, D Estêvão, D F Sánchez, S Cañete-Portillo, M J Fernández-Nestosa, F Casaca, et al. 2020. "HPV16 Induces Penile Intraepithelial Neoplasia and Squamous Cell Carcinoma in Transgenic Mice: First Mouse Model for HPV-Related Penile Cancer." *Journal of Pathology* 251 (4): 411–19. https://doi.org/10.1002/path.5475.

Mekada, K, K Abe, A Murakami, S Nakamura, H Nakata, K Moriwaki, Y Obata, and A Yoshiki. 2009. "Genetic Differences among C57BL/6 Substrains." *Experimental Animals* 58 (2): 141–49. https://doi.org/10.1538/expanim.58.141.

Mervai, Z, K Egedi, I Kovalszky, and K Baghy. 2018. "Diethylnitrosamine Induces Lung Adenocarcinoma in FVB/N Mouse." *BMC Cancer* 18 (1): 157. https://doi.org/10.1186/s12885-018-4068–4.

Mestre, V F, B Medeiros-Fonseca, D Estêvão, F Casaca, S Silva, A Félix, F Silva, et al. 2020. "HPV16 Is Sufficient to Induce Squamous Cell Carcinoma Specifically in the Tongue Base in Transgenic Mice." *Journal of Pathology* 251 (1): 4–11. https://doi.org/10.1002/path.5387.

Moutinho, M S S, S Aragão, D Carmo, F Casaca, S Silva, J Ribeiro, H Sousa, et al. 2018. "Curcumin and Rutin Down-Regulate Cyclooxygenase-2 and Reduce Tumor-Associated Inflammation in HPV16-Transgenic Mice." *Anticancer Research* 38 (3): 1461–66. https://doi.org/10.21873/anticanres.12371.

Murph, M M, S Liu, W Jia, H Nguyen, M A MacFarlane, S S Smyth, S S Kuppa, and K K Dobbin. 2020. "Diet-Regulated Behavior: FVB/N Mice Fed a Lean Diet Exhibit Increased Nocturnal Bouts of Aggression between Littermates." *Lab Animal* 54 (2): 159–70. https://doi.org/10.1177/0023677219834582.

Nascimento-Sales, M, I Fredo-da-Costa, A C B Borges Mendes, S Melo, T T Ravache, T G B Gomez, F Gaisler-Silva, et al. 2017. "Is the FVB/N Mouse Strain Truly Resistant to Diet-Induced Obesity?" *Physiological Reports* 5 (9). https://doi.org/10.14814/phy2.13271.

Newman, David J, and Gordon M Cragg. 2012. "Natural Products as Sources of New Drugs over the 30 Years from 1981 to 2010." *Journal of Natural Products* 75 (3): 311–35.

Nieto, A I, G Shyamala, J J Galvez, G Thordarson, L M Wakefield, and R D Cardiff. 2003. "Persistent Mammary Hyperplasia in FVB/N Mice." *Comparative Medicine* 53 (4): 433–38.

Nogueira, A, H Vala, C Vasconcelos-Nóbrega, A I Faustino-Rocha, C A Pires, A Colaço, P A Oliveira, and M J Pires. 2017. "Long-Term Treatment with Chaethomellic Acid A Reduces Glomerulosclerosis and Arteriolosclerosis in a Rat Model of Chronic Kidney Disease." *Biomedicine and Pharmacotherapy* 96: 489–96. https://doi.org/10.1016/j.biopha.2017.09.137.

Novelli, E L, Y S Diniz, C M Galhardi, G M Ebaid, H G Rodrigues, F Mani, A A Fernandes, A C Cicogna, and J L Novelli Filho. 2007. "Anthropometrical Parameters and Markers of Obesity in Rats." *Lab Animal* 41 (1): 111–19. https://doi.org/10.1258/002367707779399518.

Oliveira, M, E Nascimento-Gonçalves, J Silva, P A Oliveira, R Ferreira, L Antunes, R Arantes-Rodrigues, and A I Faustino-Rocha. 2017. "Implementation of Humane Endpoints in a Urinary Bladder Carcinogenesis Study in Rats." *In Vivo* 31 (6): 1073–80. https://doi.org/10.21873/invivo.11172.

Oliveira Neto, C P de, B Medeiros-Fonseca, D Estêvão, V F Mestre, N R Costa, F E de Andrade, P A Oliveira, et al. 2021. "Differential Incidence of Tongue Base Cancer in Male and Female HPV16-Transgenic Mice: Role of Female Sex Hormone Receptors." *Pathogens* 10 (10). https://doi.org/10.3390/pathogens10101224.

Padrão, A I, R Nogueira-Ferreira, R Vitorino, D Carvalho, C Correia, M J Neuparth, M J Pires, et al. 2018. "Exercise Training Protects against Cancer-Induced Cardiac Remodeling in an Animal Model of Urothelial Carcinoma." *Archives of Biochemical Biophysics* 645: 12–18. https://doi.org/10.1016/j.abb.2018.03.013.

Paiva, I, R M Gil da Costa, J Ribeiro, H Sousa, M Bastos, A Faustino-Rocha, C Lopes, P A Oliveira, and R Medeiros. 2015. "A Role for MicroRNA-155 Expression in Microenvironment Associated to HPV-Induced Carcinogenesis in K14-HPV16 Transgenic Mice." *PLoS One* 10 (1): e0116868. https://doi.org/-10.1371/journal.pone.0116868.

Palombo, E A. 2006. "Phytochemicals from Traditional Medicinal Plants Used in the Treatment of Diarrhoea: Modes of Action and Effects on Intestinal Function." *Phytotherapy Research* 20 (9): 717–24. https://doi.org/10.1002/ptr.1907.

Parveen, Abida, Bushra Parveen, Rabea Parveen, and Sayeed Ahmad. 2015. "Challenges and Guidelines for Clinical Trial of Herbal Drugs." *Journal of Pharmacy Sciences* 7 (4): 329.

Peixoto da Silva, J M O Santos, V F Mestre, B Medeiros-Fonseca, P A Oliveira, M S M Bastos M, R M Gil da Costa, and R Medeiros. 2020. "Human Papillomavirus 16-Transgenic Mice as a Model to Study Cancer-Associated Cachexia." *International Journal of Molecular Sciences* 21 (14). https://doi.org/10.3390/ijms21145020.

Petrovska, B B. 2012. "Historical Review of Medicinal Plants' Usage." *Pharmacognosy Review*, 6 (11), 1–5.

Petterino, C, A L Caffull, D B Chuchu, and M E Hartness. 2021. "What Is Your Diagnosis? A Flag for Platelet Clumping in the Peripheral Blood of a Female Crl:CD-1 Mouse on an ADVIA Hematology Analyzer." *Veterinary Clinical Pathology* 50 (4): 611–14. https://doi.org/10.1111/vcp.12986.

Raafat, A, L Strizzi, K Lashin, E Ginsburg, D McCurdy, D Salomon, G H Smith, D Medina, and R Callahan. 2012. "Effects of Age and Parity on Mammary Gland Lesions and Progenitor Cells in the FVB/N-RC Mice." *PLoS One* 7 (8): e43624. https://doi.org/10.1371/journal.pone.0043624.

Rappuoli, R. 2014. "Inner Workings: 1885, the First Rabies Vaccination in Humans." *Proceedings of National Academy of Sciences of the United States of America* 111 (34): 12273. https://doi.org/10.1073/pnas.1414226111.

Raskin, Ilya, and Christophe Ripoll. 2004. "Can an Apple a Day Keep the Doctor Away?" *Current Pharmaceutical Design* 10 (27): 3419–29.

Reeds, Karen Meier. 1976. "Renaissance Humanism and Botany." *Annals of Science* 33 (6): 519–42.

Roderick, T H, S H Langley, and E H Leiter. 1985. "Some Unusual Genetic Characteristics of BALB/c and Evidence for Genetic Variation among BALB/c Substrains." *Current Topics in Microbiology and Immunology* 122: 9–18. https://doi.org/10.1007/978-3-642-70740-7_2.

Rodrigues, P, T Ferreira, E Nascimento-Gonçalves, F Seixas, R M Gil da Costa, T Martins, M J Neuparth, et al. 2020. "Dietary Supplementation with Chestnut (Castanea Sativa) Reduces Abdominal Adiposity in FVB/n Mice: A Preliminary Study." *Biomedicines* 8 (4). https://doi.org/10.3390/biomedicines8040075.

Rothenberger, N J, and L P Stabile. 2020. "Induction of Lung Tumors and Mutational Analysis in FVB/N Mice Treated with the Tobacco Carcinogen 4-(Methylnitrosamino)-1-(3-Pyridyl)-1-Butanone." *Methods in Molecular Biology* 2102: 149–60. https://doi.org/10.1007/978-1-0716-0223-2_7.

Roy, C S. 1879. "The Form of the Pulse-Wave: As Studied in the Carotid of the Rabbit." *Journal of Physiology* 2 (1): 66–90.11. https://doi.org/10.1113/jphysiol.1879.sp000046.

Ruiz, David, José Egea, Francisco A Tomás-Barberán, and María I Gil. 2005. "Carotenoids from New Apricot (Prunus Armeniaca L.) Varieties and Their Relationship with Flesh and Skin Color." *Journal of Agricultural Chemistry* 53 (16): 6368–74.

Salmerón-Manzano, Esther, and Francisco Manzano-Agugliaro. 2020. "Worldwide Research on Low Cost Technologies through Bibliometric Analysis." *Inventions* 5 (1): 9.

Santos, C, T Neto, P Ferreirinha, H Sousa, J Ribeiro, Mmsm Bastos, A I Faustino-Rocha, et al. 2016. "Celecoxib Promotes Degranulation of CD8(+) T Cells in HPV-Induced Lesions of K14-HPV16 Transgenic Mice." *Life Sci* 157: 67–73. https://doi.org/10.1016/j.lfs.2016.05.040.

Santos, J M O, A Moreira-Pais, T Neto, S Peixoto da Silva, P A Oliveira, R Ferreira, J Mendes, et al. 2019. "Dimethylaminoparthenolide Reduces the Incidence of Dysplasia and Ameliorates a Wasting Syndrome in HPV16-Transgenic Mice." *Drug Development Research* 80 (6): 824–30. https://doi.org/10.1002/ddr.21565.

Santos, S, T Ferreira, J Almeida, M J Pires, A Colaço, S Lemos, R M Gil da Costa, et al. 2019. "Dietary Supplementation with the Red Seaweed Porphyra Umbilicalis Protects against DNA Damage and Pre-Malignant Dysplastic Skin Lesions in HPV-Transgenic Mice." *Marine Drugs* 17 (11). https://doi.org/10.3390/md17110615.

Schmidt, Barbara, David M Ribnicky, Alexander Poulev, Sithes Logendra, William T Cefalu, and Ilya Raskin. 2008. "A Natural History of Botanical Therapeutics." *Metabolism* 57: S3–9.

Sepkovic, D W, L Raucci, J Stein, A D Carlisle, K Auborn, H B Ksieski, T Nyirenda, and H L Bradlow. 2012. "3,3'-Diindolylmethane Increases Serum Interferon-γ Levels in the K14-HPV16 Transgenic Mouse Model for Cervical Cancer." *In Vivo* 26 (2): 207–11.

Sepkovic, D W, J Stein, A D Carlisle, H B Ksieski, K Auborn, L Raucci, T Nyirenda, and H L Bradlow. 2011. "Results from a Dose-Response Study Using 3,3'-Diindolylmethane in the K14-HPV16 Transgenic Mouse Model: Cervical Histology." *Cancer Prevention Research (Phila)* 4 (6): 890–96. https://doi.org/-10.1158/1940-6207.Capr-10-0369.

Shafarenko, M, J Mahler, C Cochran, A Kisielewski, E Golding, R Wiseman, and T Goodrow. 1997. "Similar Incidence of K-Ras Mutations in Lung Carcinomas of FVB/N Mice and FVB/N Mice Carrying a Mutant P53 Transgene." *Carcinogenesis* 18 (7): 1423–26. https://doi.org/10.1093/carcin/18.7.1423.

Shen, Qian, Lida Zhang, Zhihua Liao, Shengyue Wang, Tingxiang Yan, P U Shi, Meng Liu, Xueqing Fu, Qifang Pan, and Yuliang Wang. 2018. "The Genome of Artemisia Annua Provides Insight into the Evolution of Asteraceae Family and Artemisinin Biosynthesis." *Molecular Plant* 11 (6): 776–88.

Shugaba, Ali Ishaq, Musa Baba Tanko Umar, Chioma Uzokwe, Gana Joseph Umaru, Muhammed Bello Muhammad, Francis Shinku, Ahmed Muhammed Rabiu, and Rene Mathew. 2014. "The Effect of Yoyo Cleanser Bitters on the Cerebellum of Adult Male Wistar Rat." *Sky Journal of Medicine and Medical Sciences* 2(5): pp. 21–30.

Siersbæk, M S, N Ditzel, E K Hejbøl, S M Præstholm, L K Markussen, F Avolio, L Li, et al. 2020. "C57BL/6J Substrain Differences in Response to High-Fat Diet Intervention." *Science Reports* 10 (1): 14052. https://-doi.org/10.1038/s41598-020-70765-w.

Sneader, Walter. 2005. *Drug Discovery: A History*. John Wiley & Sons.

Sofowora, A. 1993. *Medicinal Plant and Traditional Medicine in Africa. Ibadan-Owerri-Kaduna-Lagos.* Spectrum Book Ltd, 158.

Stojanoski, N. 1999. "Development of Health Culture in Veles and Its Region from the Past to the End of the 20th Century." *Veles: Society of Science* 13: 34.

Surwit Cochrane C, McCubbin JA, Feinglos MN, R S Kuhn C M. 1988. "Diet-Induced Type II Diabetes in C57BL/6J Mice." *Diabetes* 37: 1163–67.

Tanabe, Yoshiaki, Hiroki Kuwayama, Sayaka Wakayama, Hiroaki Nagatomo, Masatoshi Ooga, Satoshi Kamimura, Satoshi Kishigami, and Teruhiko Wakayama. 2017. "Production of Cloned Mice Using Oocytes Derived from ICR-Outbred Strain." *Reproduction* 154 (6): 859–66.

Tarrant, James C, Patrick Savickas, Lorna Omodho, Marco Spinazzi, and Enrico Radaelli. 2020. "Spontaneous Incidental Brain Lesions in C57BL/6J Mice." *Veterinary Pathology* 57 (1): 172–82.

Teixeira-Guedes, Catarina Isabel, Ana Isabel Faustino-Rocha, Daniela Talhada, José Alberto Duarte, Rita Ferreira, Fernanda Seixas, and Paula Alexandra Oliveira. 2014. "A Liver Schwannoma Observed in a Female Sprague-Dawley Rat Treated with MNU." *Experimental Pathology* 66 (2–3): 125–28.

Thomas, Kirk R, and Mario R Capecchi. 1987. "Site-Directed Mutagenesis by Gene Targeting in Mouse Embryo-Derived Stem Cells." *Cell* 51 (3): 503–12.

Thomson, Patricia, Jenny Jones, Matthew Browne, and Stephen J Leslie. 2014. "Psychosocial Factors That Predict Why People Use Complementary and Alternative Medicine and Continue with Its Use: A Population Based Study." *Complementary Therapies in Clinical Practice* 20 (4): 302–10.

Toplak Galle, K. 2005. "Domestic Medicinal Plants." *Zagreb: Mozaic Book*, 60–61.

Veeresham, Ciddi. 2012. "Natural Products Derived from Plants as a Source of Drugs." *Journal of Advanced Pharmaceutical Technology* 3 (4): 200.

Walji, Rishma, Heather Boon, Joanne Barnes, Zubin Austin, G Ross Baker, and Sandy Welsh. 2009. "Adverse Event Reporting for Herbal Medicines: A Result of Market Forces." *Healthcare Policy* 4 (4): 77.

Waterston RH, Lindblad-Toh K, Birney E, Rogers J, Abril JF, Agarwal P, Agarwala R, et al. 2002. "Mouse Genome Sequencing Consortium: Initial Sequencing and Comparative Analysis of the Mouse Genome." *Nature* 420: 520–62.

Welz, Alexandra N, Agnes Emberger-Klein, and Klaus Menrad. 2018. "Why People Use Herbal Medicine: Insights from a Focus-Group Study in Germany." *BMC Complementary Medicine* 18 (1): 1–9.

Wen, Libin, Xiaojing Gao, Shaoyang Sheng, Qi Xiao, Wei Wang, and Kongwang He. 2021. "Characterization of Porcine Circovirus-like Virus P1 Replication and Lesions in BALB/c Mice." *Virology* 556: 33–38.

Wiart, C. 2006. *Ethnopharmacology of Medicinal Plants: Asia and the Pacific (pp. 1–50).* Humana Press. 10: 971–78.

Wilmut, I, A E Schnieke, J McWhir, A J Kind, and K H Campbell. 1997. "Viable Offspring Derived from Fetal and Adult Mammalian Cells." *Nature* 385 (6619): 810–13. https://doi.org/10.1038/385810a0.

World Health Organization. 2000. *General Guidelines for Methodologies on Research and Evaluation of Traditional Medicine*. World Health Organization.

Wyk, B Van, and M Wink 2004. *Medicinal Plants of the World*. Briza Publications.

Yang, C Z, A J Tian, Z H Meng, J M Wu, Y Y Zhang, L J Guo, and Z J Li. 2014. "[Establishment of a FVB/N mouse model of cardiac hypertrophy by isoprenaline]." *Beijing Da Xue Xue Bao Yi Xue Ban* 46 (6): 906–10.

Yoshizaki, K, M Asai, and T Hara. 2020. "High-Fat Diet Enhances Working Memory in the Y-Maze Test in Male C57BL/6J Mice with Less Anxiety in the Elevated Plus Maze Test." *Nutrients* 12 (7). https://doi.org/10.3390/nu12072036.

Zeng, T, W Guo, L Jiang, Q Luo, Z Shi, B Lei, J Zhang, and Z Cai. 2021. "Integration of Omics Analysis and Atmospheric Pressure MALDI Mass Spectrometry Imaging Reveals the Cadmium Toxicity on Female ICR Mouse." *Science of the Total Environmental* 801: 149803. https://doi.org/10.1016/j.scitotenv.2021.149803.

17 Computational Phytochemistry in Drug Discovery

Databases and Tools

Ilma Shakeel, Taj Mohammad, and Md. Imtaiyaz Hassan
Jamia Millia Islamia

CONTENTS

17.1 INTRODUCTION

Due to the drastic increase in the information on biological micro- and macromolecules, the application of *in-silico* drug discovery has been improved drastically. It can be broadly applied to every stage of drug discovery and also in workflow development, which includes identifying and validating the targets, lead discovery, preclinical test and optimization. Computational drug discovery is a productive move towards economizing and accelerating the developmental process and discovery of drugs. A plethora of computational software, tools, and databases have been created to help medicinal chemists carry out contemporary drug discovery programmes more successfully.

In recent decades, there has been a remarkable increase in the incorporation of artificial intelligence, computational techniques and mathematical modelling in phytochemical research, specifically in screening natural plant material, plant metabolomics, chemical taxonomy, chemical fingerprinting, phylogenetics and biosynthetic studies, prediction of toxicological and pharmacological properties (*in silico* studies or virtual screening), and automated determination of the

DOI: 10.1201/b22842-17

structure of phytochemicals based on spectroscopic data, was observed (Sarker and Nahar 2017). Phytochemical informatics was initially developed for some aspects, which deal with data related to phytochemicals or/and their respective sources (Ehrman, Barlow, and Hylands 2010). However, the starting of the new era in phytochemical research is now termed 'Computational Phytochemistry'.

Computational phytochemistry combines the digital age, statistics, mathematical algorithms, and large databases to combine modelling and theories with experimental observations. The two basic building blocks of computational phytochemistry are the creation of simulations and models of physical processes associated with phytochemical protocols and applying data analysis techniques and statistics to find useful information from vast databases. (Sarker and Nahar 2018).

The upswing of well-designed and user-friendly computational tools which enables cheminformatics (CI)-based drug design and structure-based drug design (SBDD) has provided the medicinal chemist with the armament of databases, applications, and tools that efficiently supplements the entire drug design process (Figure 17.1) and hence boosts the efficiency and speed

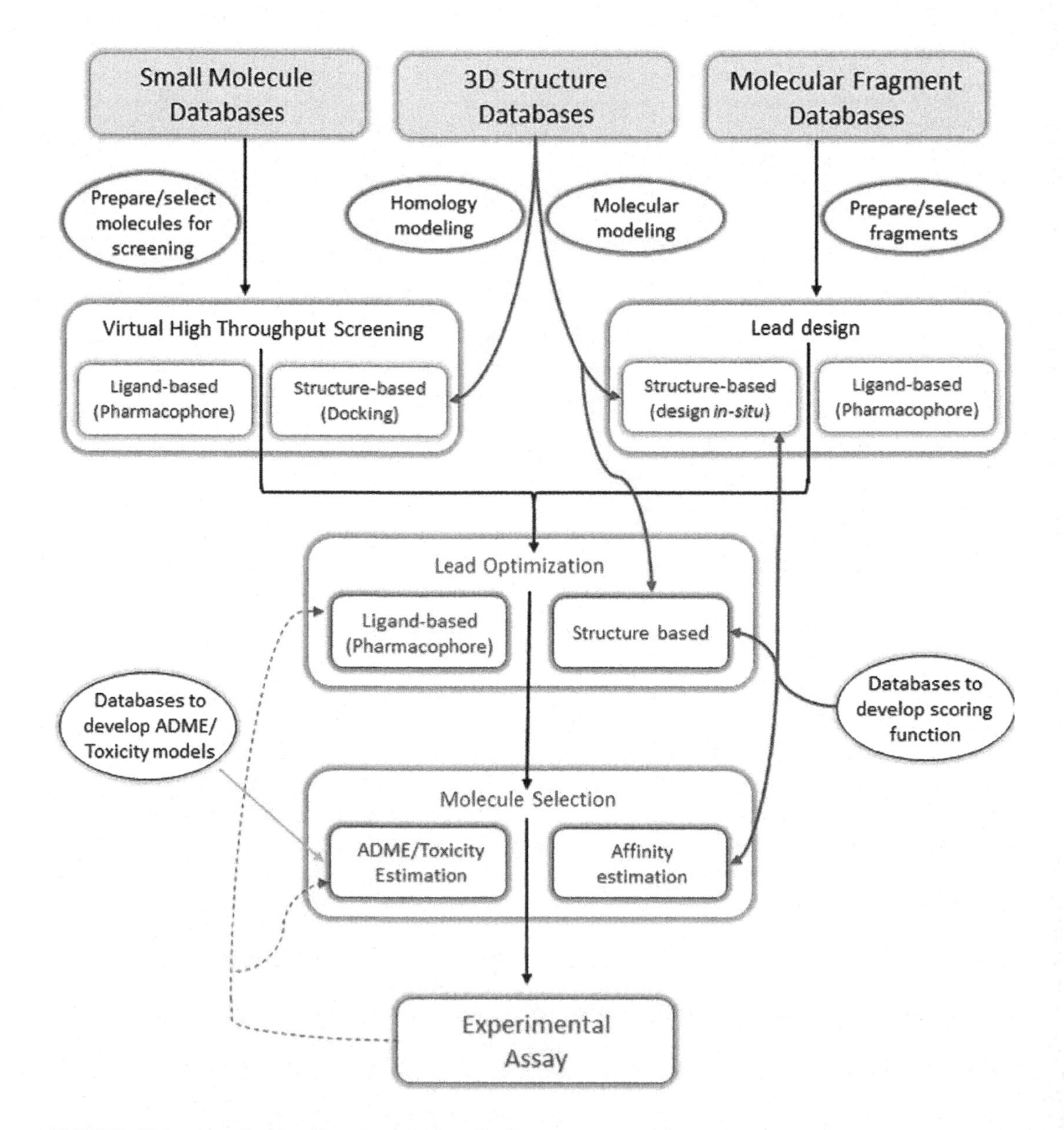

FIGURE 17.1 A typical in silico drug design pipeline.

of the design and makes the test analysis cycle. Recent computational tools and applications go beyond all the areas of drug discovery; hence, most pharmacists can employ the information effectively in innumerable drug discoveries. Here, we exhibit an overview of all the essential computational platforms and methods successfully applied in the field of computational phytochemistry.

17.2 DATABASES

Several remarkable platforms and methodologies have been constructed and developed whose main aim is to focus on computational drug discovery and development. In this section, various platforms and methods have been discussed, which involve the identification of the target, conformational sampling, docking-based virtual screening, calculation of molecular similarities, scoring function, sequence-based drug design and virtual library design.

17.2.1 Protein Data Bank (PDB)

RCSB PDB (Research Collaboratory for Structural Bioinformatics Protein Data Bank) is a US-based data centre for the global protein data bank with chronicles of 3D structures of macro-biological molecules, including proteins, RNA and DNA. This data is very important in the area of education and research in health, fundamental biology, energy and biotechnology. PDB was accepted as the first open access and leading global digital platform which provides the resource of all medicines, biology, and experimental data which can be used for scientific discovery.

PDB has stored data of 3D structure, which supports a huge advancement in our knowledge and understanding of protein architecture, resulting in the recent breakthrough discovery in the prediction of protein structure by artificial intelligence and machine learning methods. The vision of RCSB PDB is to provide free knowledge of 3D structure, function and evolution of biological macromolecules, thereby expanding the frontier of fundamental biology, biotechnology and biomedicines.

17.2.2 UniProt

UniProt is an open access database for protein sequences and their functional information. It comes under the UniProt consortium, which includes various European bioinformatics organizations and a US foundation (UniProt Consortium 2015). It holds enormous information on protein's biological function derived from various research literature. The vision of uniport is to equip the scientific community with an encyclopaedia of high-quality and freely available data on protein sequences and also about functional information (O'Donovan et al. 2002).

UniProt is made up of four components [UniProtKB (with Swiss-Prot and TrEMBL as subparts), UniParc, and UniRef], where each component is optimized for different uses:

UniProtKB (UniProt Knowledgebase): It is a pivotal access point for extensively curated data for protein information which includes classification, function and cross reference (UniProt Consortium 2010). UniProtKB further comprises two sections:
UniProtKB/Swiss-Prot: It manually annotates and reviews.
UniProtKB/TrEMBL: It annotates automatically and is not reviewed.

UniRef (UniProt Reference Clusters): This database provides a set of clustered sequences from UniProtKB and provides selective UniProt archive records to fully cover the sequence space at various resolutions while masking unnecessary sequences (Suzek et al. 2007).

UniParc (UniProt Archive): It is a comprehensive repository that keeps a record of sequences and their identifiers (Leinonen et al. 2004).

17.2.3 TarFisDock

It is a web tool for automatizing the method of searching for protein small molecule interactions from a repository of protein structure. It provides a potential drug target database (PDTD), a reverse protein-ligand docking programme and a target database having 698 protein structures which cover 15 therapeutic areas (Li, Guo, et al. 2006). In comparison to conventional protein-ligand docking, reverse protein-ligand interaction docking is intended to seek probable protein targets (Li, Guo, et al. 2006). Then, this web server searches the potential binding protein against our required small molecule using a docking approach. For ranking the proteins, the protein-ligand interaction energy terms from the programme DOCK are used. TarFisDock may play a pivotal role in the identification of the target, the mechanism of study of ancient drugs and discovery of probes from natural products. TarFisDock is accessible at the website: http://www.dddc.ac.cn/tarfisdock/.

17.2.4 Potential Drug Target Database

A potential drug target database (PDTD) is a web-based and easily available protein database for drug target identification. It contains more than 1,100 protein entries with their 3D structures presented in PDB (Gao et al. 2008). The data for PDTD were extracted from several online databases and literature, such as Drug Bank, Therapeutic Target Database and Thomson Pharma. This database screens very diverse information about 830 potential and known drug targets, which includes active site structure and protein in both mol2 and PDB formats, biological function and related diseases and also covers regulating pathways (Gao et al. 2008). Biochemical functions and nosology characterize every target. PDTD supports the search function using keywords such as by searching target name, PDB ID, as well as disease name. PDTD-generated data set could be visualized by plugging in a molecular visualization tool, and the data can be downloaded freely. Similar to TarFisDock, PDTD could be utilized in identifying binding proteins for small molecules (Gao et al. 2008). The probable application of PDTD includes the identification of in silico drug targets, virtual screening, and discovering the secondary effect of an already known drug or an existing target. PDTD is accessible from the website: http://www.dddc.ac.cn/pdtd/.

17.2.4.1 PharmMapper

PharmMapper is an openly accessible web server designed to identify potential targets for the available small molecule probe, including natural products, drugs or novel synthesized compounds with their binding target not identified, using the pharmacophore mapping approach (Liu, Ouyang, et al. 2010). Pharmacophore is an alternative method apart from docking, where a spatial arrangement of the features is made for a small molecule to interact with its specific target receptors. PharmMapper has very high efficiency and is a robust method for mapping. It also possesses high throughput abilities, which help identify the potential drug targets from the available databases within a few hours (Liu, Ouyang, et al. 2010).

PharmMapper contains a large in-house repository of pharmacophore databases taken from all the available targets in the Drug Bank, PDTD, Target Bank and BindingDB. A little over 7,000 receptor-based pharmacophore models (which cover 1,627 drug targets, among which 459 are human protein targets) are accessed and stored by PharmMapper. The recent version of 2017 contains a database of about 23,236 proteins, of which about 16,159 pharmacophore models are predicted as druggable binding sites and about 52,431 pharmacophore models have a higher than 6.0 pKd value (Wang et al. 2017).

17.2.4.2 SuperPred

SuperPred is a prediction-based web server for predicting the targets of a compound and for anatomical therapeutic chemical (ATC) code (Nickel et al. 2014). ATC is the drug classification system published by WHO. Its classification of drugs is based on chemical and therapeutic characteristics. SuperPred web server's target prediction and ATC prediction are based on the model of machine learning by utilizing Morgan fingerprints of length 2,048 and logistic regression (Nickel et al. 2014). The drug classification for the compound could be performed at the drug classification site, whereas the target prediction for the input compound could be performed at the site of the target prediction.

17.2.4.3 PharmTargetDB

PharmTargetDB is a tool that gathers all the public information available on particular target(s) (linked to 3D structures, disease, safety, novelty, ligandability, etc.) and rearranges it in a user-friendly format for the researcher to analyze (De Cesco, Davis, and Brennan 2020). Furthermore, it contains a target scoring system based on the prudent attributes of good potential therapeutic targets and machine learning to classify novel targets with challenging or promising tractability (De Cesco, Davis and Brennan 2020). The databases of TargetDB come from a vast variety of data sources, including the processed or pre-aggregated data from TCRD or UniProt. At the same time, other data comes directly from source APIs, including OpenTargets for disease association and Human Protein Atlas for protein expression levels.

17.2.4.4 DrugBank

DrugBank is a highly curated resource that collaborates detailed drug data with information on drug action and comprehensive drug targets. DrugBank has been broadly used to promote in silico drug target discovery, drug screening or docking, drug design, prediction of drug interaction and drug metabolism and general education of pharmaceuticals (Wishart et al. 2008). The updated version of DrugBank contains approximately 4,900 drug entries and about 60% more FDA-approved biotech drugs and small molecules, including 10% more experimental drugs. Each entry of DrugCard has more than 100 data fields where half information includes the data for chemicals/drugs and the other half includes pharmacogenomics, pharmacological and molecular biology data (Wishart et al. 2008). Numerous new fields have been added to DrugBank, such as drug–drug interaction, food–drug interaction and experimental absorption, distribution, metabolism and excretion data. Textual and structural queries have been significantly improvized to simplify and power the searches in DrugBank databases. DrugBank is available at the website: http://www.drugbank.ca.

17.2.4.5 BindingDB

BindingDB is a freely accessible database which contains about 20,000 experimentally determined binding affinity of ligand–protein complexes, about 11,000 small molecule ligands and about 110 protein targets which include mutational variants and isoforms (Liu et al. 2007). BindingDB supports a wide variety of queries, such as search by substructure, chemical structure and similarity, protein and ligand names, protein sequences, molecular weights and affinity ranges (Liu et al. 2007). Data generated by the queries in BindingDB could be downloaded as curated files for further analysis. BindingDB data is linked to the structural data in PDB through PDB ID and sequence and chemical searches and to the literature in PubMed through PubMed ID. BindingDB is available at the website: http://www.bindingdb.org.

17.2.4.6 PDBBind

PDBBind database is an extensive pool of experimentally determined data of binding affinities, such as IC_{50}, *K*d and *K*i, for ligand–protein complexes deposited in the PDB (Wang et al. 2005). PDBBind provides a network between the structural information and energetics of ligand–protein complexes, which is of great deal to several studies on molecular recognition in biological systems. The data is updated annually in PDBBind databases to cope with growth in the depository of PDB.

The latest version of PDBBind provides two basic types of information (Liu et al. 2015):

i. Basic structural information of about 13,000 complexes between protein–protein, protein–ligand, protein–nucleic acid and nucleic acid–ligand (small molecule). Users can open and examine these complexes online and search via the chemical structure of ligands/small molecules included in these complex structures.
ii. Structural information and data for binding affinities (IC_{50}, *K*d and *K*i) of about 12,995 biomolecular complexes, which includes ligand–protein (10,656), ligand–nucleic acid (87), nucleic acid–protein (660) and protein–protein (1,592) complexes, which is the largest of one of such collection so far.

The team of PDBBind collected all these data from about 28,000 original references and also cross-checked them to ensure accuracy, and they should match the complex structures from PDB. Moreover, a high-quality data set of ligand–protein interaction is produced for scoring studies and docking as a 'Core Set' and 'Refined Set'.

17.2.4.7 Techniques

Several remarkable algorithms and methodologies have been developed whose main aim is to focus on computational drug discovery and development. In this section, several techniques or algorithms have been discussed which are used for molecular docking and virtual screening, molecular dynamic simulation and pharmacophore mapping.

17.2.4.7.1 Resources in Molecular Docking and Virtual Screening

Docking is the finding of low-energy binding mode of the ligand (small molecule) with the active site of protein (macromolecule) or receptors of known structures. To solve the problem of docking computationally, one requires an accurate representation of efficient algorithms and molecular energetics to search for potent binding modes. This section includes various platforms and methodologies involved in docking-based virtual screening.

17.2.4.7.1.1 GAsDock GAsDock is a flexible and fast docking programme based on the improvized multi-population genetic algorithm. Accuracy and speed of docking and screening efficiency of GAsDock were evaluated from the docking results of HIV-1 reverse transcriptase (RT) and thymidine kinase (TK) enzymes, with their 10 known and widely available inhibitors and 990 random selected ligands (Li et al. 2004). In the case of TK, 9 of the 10 known inhibitors were accurately docked in the active site of a protein, and a root-mean-square deviation (RMSD) value between X-ray crystal structure and docking is less than 1.7Å. In the case of HIV-1 RT, the binding poses (orientation and conformation) of 9 of the 10 known inhibitors were reproduced by GAsDock with the RMSD value less than 2.0Å. Docking time is proportional to the number of available rotatable bonds in ligands, with an approximate time taken by GAsDock to finish a docking simulation of under 60s for a ligand (small molecule) with not more than 20 rotatable bonds (Li et al. 2004).

17.2.4.7.1.2 AutoDock AutoDock is an automated docking tool designed to analyze how ligands (small molecules) bind to the receptors of known 3D structures of proteins (Morris et al. 2001). New modifications and improvizations have been added timely over the years to equip AutoDock with new functionalities and multiple search engines. AutoDock has multiple applications in protein–protein docking, SBDD, virtual screening (HTS), X-ray crystallography, combinatorial library design, chemical mechanism studies and lead optimization. Two generations of AutoDock software have been distributed, namely AutoDock 4 and AutoDock Vina (Huey, Morris, and Forli 2012).

AutoDock 4: It is made up of two main programmes: AutoDock (it performs docking of the ligand to the set of grids specific for the target protein) and Autogrid (it pre-calculates the grids).

AutoDock Vina: It is an instant virtual programme where pre-calculations of grid maps and choosing of atom types are not required. Instead, it calculates the grid map internally for the atom necessary types.

AutoDockTools: A user-friendly graphical interface also helps set the bonds that will be rotatable in a small molecule ligand and analyzes docking.

17.2.4.7.1.3 InstaDock InstaDock uses QuickVina-W (a modified version of AutoDock Vina) and is a single-click GUI for docking calculations (Mohammad, Mathur, and Hassan 2021). InstaDock is specifically designed for the convenience of non-bioinformatic background people who don't have expertise in computational programmes and algorithms. InstaDock aids in the onboard analysis of docking and visualization of the results in just a single click (Mohammad, Mathur, and Hassan 2021). InstaDock is one of the most interactive and straightforward interfaces in the history of GUIs for molecular docking and virtual screening. InstaDock is freely available through https://hassanlab.org/instadock.

17.2.4.7.1.4 GOLD Genetic optimization for ligand docking (GOLD) is a computerized ligand docking programme which uses genetic algorithms to scrutinize the complete range of conformational flexibility of ligands with protein's partial flexibility and comply with the fundamental requirements that a ligand should displace loosely bound water while binding (Nurisso et al. 2012). Various modifications and enhancements were performed timely in GOLD to increase the applicability and reliability of the algorithm substantially. This advanced algorithm was tested on the data set of 100 complexes from the databases of PDB. GOLD displays a 71% success rate in determining the experimental binding mode when docking the ligand back to the binding sites.

17.2.4.7.1.5 FlexX FlexX is an automated docking programme used to dock ligands into protein's binding sites. Hence, this is used in the designing process of specific protein–ligand combinations. FlexX combines a model appropriate for the physicochemical properties of the docking ligands with the efficient method for sampling the ligand's conformational space (Jones et al. 1997). A flexible ligand can adopt a vast variety of different conformations. FlexX method of docking includes the ligand conformational space based on a discrete model and uses a tree-searching technique for putting the ligand incrementally into the protein's active site (Jones et al. 1997). Hashing technique (adapted from computer vision) is used for placing the first fragment of a small molecule (ligand) into the protein.

17.2.4.7.1.6 DOCK The algorithm of DOCK addresses the docking of a rigid body using the geometric matching algorithm to superimpose the ligands onto the negative image of binding pockets (Jones et al. 1997). Some important features were added to the DOCK programme over the years that improvized the ability of the algorithm to find the binding mode with the lowest energy, including on-the-fly optimization, force-field-based scoring, algorithm(s) for the docking of flexible ligands and an improvized matching algorithm for rigid docking body (Jones et al. 1997). DOCK can be used in multiple applications, such as to predict the binding mode of protein–ligand complexes, to search the databases of ligand for compounds that either binds a particular protein and nucleic acid targets or inhibits the enzymatic activity, to the exam in the possibility of the binding orientation of protein–DNA and protein–protein complexes and much more.

17.2.4.7.1.7 Surflex Surflex is a fully automated flexible molecular docking approach that combines the Hammerhead docking system's scoring features with search engines that use the surface-based molecular similarity method to produce potential poses for molecular fragments quickly (Jain 2003). Results are given by comparing 81 protein–ligand complexes with significant structural diversity to X-ray crystallographic investigations to assess the precision and dependability of docking. Over 90% of complexes have one of the top-rank poses within 2.5Å RMSD, with the best score by Surflex docking position being under 2.5Å RMSD in about 80% of complexes (Jain 2003). The success rate of Surflex was remarkably better, with a true positive rate of >80% and a false-positive rate of <1%. The amount of time needed for docking increased roughly linearly with the number of rotatable bonds, taking a few seconds per stiff molecule and an additional 10 seconds per rotatable link.

17.2.4.7.1.8 Glide Glide describes a thorough systematic search of a docked ligand's orientation, conformation and position space within a rigid 3D structure of a known protein receptor. A few hundred (or perhaps even more) surviving candidate poses are subjected to torsional flexible energy optimizations on an OPLS-AA non-bonded potential grid following an initial rough scoring and placement phase that significantly reduces the search space (s) (Friesner et al. 2004). In some circumstances, utilizing Monte Carlo sampling of conformational poses to narrow down the best options is essential to getting an appropriate docking posture. A model energy function that includes force-field-based and empirical variables is used to choose the best-docked pose. By re-docking ligands from 282 co-crystallized PDB complexes beginning from conformationally optimized ligand geometry that bears no memory of presently docked postures, the accuracy of docking in Glide is examined (Friesner et al. 2004). Errors in the geometry for top-rank poses are <1A in roughly half of the cases and >2A in only about one-third of the cases. The comparison of the published RMSD data shows that for ligands with up to 20 rotatable bonds.

17.2.4.7.2 For Molecular Dynamics Simulations

Molecular dynamics (MD) simulations predict how an atom in the molecular system (or in protein) moves over time based on a standard model of physics (governing interatomic interactions) (Hollingsworth and Dror 2018). MD simulation captures the behavioural patterns of proteins or different biomolecules in complete atomic details at a fine temporal resolution level. In this section, various platforms and methodologies involve in molecular dynamics simulations.

17.2.4.7.2.1 GROMACS GROMACS is a widely used, openly assessable and freely available software code in chemistry, used mainly for molecular dynamics simulations of biomolecules (proteins, nucleic acids and lipids). GROMACS is the most popular and fastest software available, which can run on CPUs (central processing units) and GPUs (graphics processing units) (Abraham et al. 2015; Kutzner et al. 2007). It provides a rich set of calculation type, analysis and preparation tools. GROMACS also supports various advanced techniques for the calculation of free energy. Version 5 of GROMACS reached the heights of performance through various enhanced and new parallelization algorithms. These work on every level, such as multi-threading, SIMD register inside core, state-of-the-art 3D domain decomposition, heterogeneous CPU-GPU acceleration and ensemble level parallelization via built-in replica exchange and separate Copernicus frameworks (Abraham et al. 2015).

17.2.4.7.2.2 CHARMM (CHARMM-GUI CHARMMing.org) CHARMM-GUI is an ideal web-based programme to build complex systems interactively and prepare their input with reproducible and well-established simulation protocol for the state-of-the-art molecular dynamic simulation by utilizing widely used simulation packages, including NAMD, CHARMM, AMBER, GROMACS, LAMMPS, GENESIS, Desmond, Tinker and OpenMM (Jo et al. 2008). CHARMM-GUI mainly focusses on helping the users to attain the desired task, such as solvating a protein or building a membrane system, by providing a streamlined interface (Jo et al. 2008). CHARMM-GUI has a user-friendly interface that makes it accessible to all users, whether new users with little experience in modelling tools or professionals, especially for the batch generation of the system. It is a well-recognized tool used by many researchers to carry out novel and innovative molecular modelling and simulation communities.

17.2.4.7.2.3 WebGro The ultimate aim of developing WebGro is to make a fully automated and user-friendly online computational biology tool for performing molecular dynamics simulations of protein (macromolecule) alone or in complex with small molecules (ligands) (Paz, Batchelor, and Pedersen 2004). WebGro uses the GROMACS simulation package to perform fully solvated molecular dynamics simulations. Users have to submit their protein file only in the.pdb extension format, and WebGro automatically performs simulations and trajectory analysis on the input data (Paz, Batchelor, and Pedersen 2004). Parameters at the input page are set to default based on the published data related to the GROMACS simulation. The simulation results are stored in the WebGro server, allowing users to view their results in a different session. WebGro is available for research purposes on the website: http://webgro.ae.iastate.edu.

17.2.4.7.2.4 Amber Assisted model building with energy refinement (Amber) evolved from a programme in the late 1970s constructed to perform assisted model building with the refinement of energy (Case et al. 2005). Amber now contains several programmes that embody powerful modern computational chemistry tools, focussing on calculating free power and molecular dynamics of nucleic acids, proteins and carbohydrates. The term Amber refers to two things (i) set of molecular simulation programmes, including demos and source codes and (ii) set of molecular mechanical force fields for simulating biomolecules (Case et al. 2005).

17.2.4.7.2.5 SwissParam SwissParam presents a fast-force field generation tool which can generate parameters and topologies for arbitrary small organic molecules (Zoete et al. 2011). The output files of SwissParam can be used with GROMACS and CHARMM. The parameters and topologies generated by the SwissParam are used by several docking software, such as EADock DSS and EADock2, to describe the ligand (small molecule). In contrast, the protein (rigid molecule) is described by CHARMM force field (Zoete et al. 2011). In toto, SwissParam parameters and topologies to describe small organic molecules (ligands), together with the CHARMM22/27 description of the targeted protein, help in computer-aided drug design discovery. SwissParam is available freely for academic and research purposes on www.swissparam.ch.

17.2.4.7.2.6 YASARA Dynamics YASARA is a molecular modelling, molecular simulation and graphics programme developed in 1993 for Windows, MacOS, Linux and Android. YASARA has an intuitive user interface, shutter glasses, photorealistic graphics, autostereoscopic display, support for affordable virtual reality headsets and input devices which makes YASARA a new level of artificial reality interaction programme, which allows one to focus on the goal and forget about the programme details (Land and Humble 2018). YASARA is powered by portable vector language, a new developmental framework that performs remarkably well above other traditional software (Land and Humble 2018). Users can push or pull a molecule while working with the dynamic models on YASARA instead of static pictures. The initial stages of YASARA view are free, while the higher stages, such as YASARA Dynamics, YASARA Model and YASARA structure, require a paid licence.

17.2.4.7.2.7 Pharmacophore Mapping Pharmacophore is an alternative method apart from docking, where the spatial arrangement of the features is made for a small molecule to interact with its specific target receptor (macromolecule) (Jain and Chincholikar 2004). Pharmacophore searches find molecules with different chemistry, but they have the correct geometry of functional groups. Pharmacophore mapping involves using a pharmacophore map concerning the application of molecular modellings, such as database searching, *de novo* compound synthesis and 3D-QSAR (Jain and Chincholikar 2004). In this section, various platforms and methodologies have been discussed which are involved in pharmacophore mapping.

17.2.4.7.2.7.1 PharmMapper Server PharmMapper is an open access and freely available web server designed to find the potential target molecule for the available small molecules (drugs, newly designed inhibitors and natural products with unidentified binding targets) using the approach of pharmacophore mapping (Liu, Xie, et al. 2010). PharmMapper benefits from highly efficient and very robust triangle hashing mapping methods that impart a high throughput ability that only takes an average of about 1 hour to screen the whole PharmTargetDB. PharmMapper is available for research and academic purpose at http://59.78.96.61/pharmmapper.

17.2.4.7.2.7.2 ChemMapper Server ChemMapper is a freely accessible web server to predict poly-pharmacology effects and modes of action for small molecules based on 3D similarities. It serves for computational drug discovery based on the concept that small molecules with 3D similarities may have comparatively similar target association profiles (Gong et al. 2013). ChemMapper contains about 3,50,000 chemical structures with bioactivity and associated target annotations and about 30,00,000 non-annotated compounds from public and commercial chemical catalogues for virtual screening (Gong et al. 2013). ChemMapper uses the in-house SHAFTS method, which combines the strength of chemical feature matching and superimposition of molecular shapes to perform ranking, 3D similarity searching and superimposition (Gong et al. 2013). It is used in various chemogenomics, polypharmacology, novel bioactive compound identification, drug repurposing and virtual screening, and it is available for research and academic purposes at http://lilab.ecust.edu.cn/chemmapper/.

17.2.4.8 Miscellaneous

Several programmes that implement various kinds of molecular docking include HADDOCK (De Vries, Van Dijk, and Bonvin 2010), DOCK Blaster (Li et al. 2014), SwissDock (Grosdidier, Zoete, and Michielin 2011) or CB-Dock Blind Docking Server (Liu et al. 2020). Web servers like 3-D QSAR (Ragno 2019), E-Dragon (Tetko et al. 2005) and MOLFEAT (Nahar et al. 2022) are used to carry out QSAR-related tasks such as molecular descriptor calculation and QSAR modelling. Another impressive method in the area of CADD that has made the switch from command line programmes to the web is MD. The Gromacs server (Noel et al. 2010), Vienna-PTM (Margreitter, Petrov, and Zagrovic 2013), MoSGrid (Krüger et al. 2014), WebGro (https://simlab.uams.edu/) and MDWeb (Hospital et al. 2012) are a few applications of MD services available online.

17.2.4.9 Notable Findings

RhoA, is one of the well-studied members of Rho GTPase family. It is crucial for many cellular functions, including cytoskeletal reorganization, gene expression, differentiation, proliferation and apoptosis (Ridley 2015). Treating cardiovascular disorders may benefit from targeting this protein (Nossaman and Kadowitz 2009). Several novel small molecule RhoA inhibitors were found in the SPECS database using an integrative approach (Deng et al. 2011; Ou-Yang et al. 2012). Here, eight of the substances demonstrated high RhoA inhibition activities, and two showed considerable inhibitory effects on PE-induced contraction of the thoracic aorta rings (Ou-Yang et al. 2012).

Insulin-like growth factor-1 receptor (IGF-1R) is essential for signalling pathways that control cell development, proliferation and cell death (Sadagurski et al. 2006). IGF-1R has been demonstrated to be overexpressed in various human malignancies, raising the possibility that it could be a useful target for cancer treatment (Fan et al. 2013). IGF-1R inhibitors were found using a hierarchical combination of molecular docking and pharmacophore-based virtual screening (Liu, Xie, et al. 2010). A pharmacophore model and crystal structures of IGF-1R and its inhibitor, pyridine-2-one, were generated (Liu, Xie, et al. 2010; Kumar and Zhang 2015). This approach was used to screen the SPECS database. The top hits were then docked with Glide29, 64 to the ATP-binding site. Thanks to this approach, many new thiazolidine-2,4-dione analogues were found to be promising IGF-1R inhibitors; the compounds show good inhibitory activity against IGF-1R (Ou-Yang et al. 2012).

Finding new inhibitors for the p90 ribosomal S6 protein kinase 2 is a potential use for the LBDD programme SHAFTS (RSK2) (Ou-Yang et al. 2012). Numerous human illnesses, including breast cancer, prostate cancer, and squamous cell carcinoma, have been associated with RSK2 overexpression and abnormal activation (Dar et al. 2018). Chemotype switching guided by the SHAFTS computation was used to find 16 compounds with IC_{50} lower than 20 mol/L that would be missed by traditional 2D approaches using the putative 3D conformations of two weakly binding, moderately active RSK2 inhibitors as the query templates (Ou-Yang et al. 2012; Jiang et al. 2021). The most effective compounds specifically exhibit low micromolar inhibitory actions for RSK2, and one substance also displays strong anti-migration activity in MDA-MB-231 tumour cells (Ou-Yang et al. 2012).

Another work used a *de novo* drug design methodology to identify several novel inhibitors of cyclophilin A (CypA) (Ni et al. 2009). Numerous biological processes depend on CypA, such as the speed at which proteins fold and unfold, the inhibition of calcineurin's serine/threonine phosphatase activity, the facilitation of viral replication and infection, and the induction of neuroprotective and neurotrophic effects (Guichou et al. 2006). Additionally, it has been noted that CypA has only overexpressed in cancer cells, especially solid tumours, indicating that CypA is a key regulator of carcinogenesis (Choi et al. 2007; Li, Zhai, et al. 2006). The LD1.0 tool created a targeted combinatorial library of 255 compounds utilizing the fragment structures of previously identified CypA inhibitors (Ou-Yang et al. 2012; Li, Zhang, et al. 2006). Sixteen compounds were chosen for synthesis and bioassay by using virtual screening that targets CypA (Ou-Yang et al. 2012). The experiment outcomes revealed that these substances had significant CypA inhibitory effects. The strongest substance among the discovered novel CypA inhibitors has a binding affinity and inhibitory activity that are roughly ten times stronger than the strongest inhibitor previously discovered (Ou-Yang et al. 2012).

REFERENCES

Abraham, Mark James, Teemu Murtola, Roland Schulz, Szilárd Páll, Jeremy C Smith, Berk Hess, and Erik Lindahl. 2015. "GROMACS: High Performance Molecular Simulations through Multi-Level Parallelism from Laptops to Supercomputers." *SoftwareX* 1: 19–25.

Case, David A, Thomas E Cheatham III, Tom Darden, Holger Gohlke, Ray Luo, Kenneth M Merz Jr, Alexey Onufriev, Carlos Simmerling, Bing Wang, and Robert J Woods. 2005. "The Amber Biomolecular Simulation Programs." *Journal of Computational Chemistry* 26(16): 1668–88.

Cesco, Stephane De, John B Davis, and Paul E Brennan. 2020. "TargetDB: A Target Information Aggregation Tool and Tractability Predictor." *PloS One* 15(9): e0232644.

Choi, Kyu Jin, Yu Ji Piao, Min Jin Lim, Jin Hwan Kim, Joohun Ha, Wonchae Choe, and Sung Soo Kim. 2007. "Overexpressed Cyclophilin A in Cancer Cells Renders Resistance to Hypoxia-and Cisplatin-Induced Cell Death." *Cancer Research* 67(8): 3654–62.

Dar, Khalid B, Aashiq Hussain Bhat, Shajrul Amin, Rabia Hamid, Suhail Anees, Syed Anjum, Bilal Ahmad Reshi, Mohammad Afzal Zargar, Akbar Masood, and Showkat Ahmad Ganie. 2018. "Modern Computational Strategies for Designing Drugs to Curb Human Diseases: A Prospect." *Current Topics in Medicinal Chemistry* 18(31): 2702–19.

Deng, Jing, Enguang Feng, Sheng Ma, Yan Zhang, Xiaofeng Liu, Honglin Li, Huang Huang, Jin Zhu, Weiliang Zhu, and Xu Shen. 2011. "Design and Synthesis of Small Molecule RhoA Inhibitors: A New Promising Therapy for Cardiovascular Diseases?" *Journal of Medicinal Chemistry* 54(13): 4508–22.

Ehrman, T M, D J Barlow, and P J Hylands. 2010. "Phytochemical Informatics and Virtual Screening of Herbs Used in Chinese Medicine." *Current Pharmaceutical Design* 16(15): 1785–98.

Fan, Gaofeng, Guang Lin, Robert Lucito, and Nicholas K Tonks. 2013. "Protein-Tyrosine Phosphatase 1B Antagonized Signaling by Insulin-like Growth Factor-1 Receptor and Kinase BRK/PTK6 in Ovarian Cancer Cells." *Journal of Biological Chemistry* 288(34): 24923–34.

Friesner, Richard A, Jay L Banks, Robert B Murphy, Thomas A Halgren, Jasna J Klicic, Daniel T Mainz, Matthew P Repasky, Eric H Knoll, Mee Shelley, and Jason K Perry. 2004. "Glide: A New Approach for Rapid, Accurate Docking and Scoring. 1. Method and Assessment of Docking Accuracy." *Journal of Medicinal Chemistry* 47(7): 1739–49.

Gao, Zhenting, Honglin Li, Hailei Zhang, Xiaofeng Liu, Ling Kang, Xiaomin Luo, Weiliang Zhu, Kaixian Chen, Xicheng Wang, and Hualiang Jiang. 2008. "PDTD: A Web-Accessible Protein Database for Drug Target Identification." *BMC Bioinformatics* 9(1): 1–7.

Gong, Jiayu, Chaoqian Cai, Xiaofeng Liu, Xin Ku, Hualiang Jiang, Daqi Gao, and Honglin Li. 2013. "ChemMapper: A Versatile Web Server for Exploring Pharmacology and Chemical Structure Association Based on Molecular 3D Similarity Method." *Bioinformatics* 29(14): 1827–29.

Grosdidier, Aurelien, Vincent Zoete, and Olivier Michielin. 2011. "SwissDock, a Protein-Small Molecule Docking Web Service Based on EADock DSS." *Nucleic Acids Research* 39(suppl_2): W270–77.

Guichou, Jean-François, Julien Viaud, Clément Mettling, Guy Subra, Yea-Lih Lin, and Alain Chavanieu. 2006. "Structure-Based Design, Synthesis, and Biological Evaluation of Novel Inhibitors of Human Cyclophilin A." *Journal of Medicinal Chemistry* 49(3): 900–10.

Hollingsworth, Scott A, and Ron O Dror. 2018. "Molecular Dynamics Simulation for All." *Neuron* 99(6): 1129–43.

Hospital, Adam, Pau Andrio, Carles Fenollosa, Damjan Cicin-Sain, Modesto Orozco, and Josep Lluís Gelpí. 2012. "MDWeb and MDMoby: An Integrated Web-Based Platform for Molecular Dynamics Simulations." *Bioinformatics* 28(9): 1278–79.

Huey, Ruth, Garrett M Morris, and Stefano Forli. 2012. "Using AutoDock 4 and AutoDock Vina with AutoDockTools: A Tutorial." *The Scripps Research Institute Molecular Graphics Laboratory* 10550: 92037.

Jain, Ajay N. 2003. "Surflex: Fully Automatic Flexible Molecular Docking Using a Molecular Similarity-Based Search Engine." *Journal of Medicinal Chemistry* 46(4): 499–511.

Jain, Sanmati K, and A Chincholikar. 2004. "Pharmacophore Mapping and Drug Design." *Indian Journal of Pharmaceutical Sciences* 66(1): 11.

Jiang, Zhenla, Jianrong Xu, Aixia Yan, and Ling Wang. 2021. "A Comprehensive Comparative Assessment of 3D Molecular Similarity Tools in Ligand-Based Virtual Screening." *Briefings in Bioinformatics* 22(6): bbab231.

Jo, Sunhwan, Taehoon Kim, Vidyashankara G Iyer, and Wonpil Im. 2008. "CHARMM-GUI: A Web-based Graphical User Interface for CHARMM." *Journal of Computational Chemistry* 29(11): 1859–65.

Jones, Gareth, Peter Willett, Robert C Glen, Andrew R Leach, and Robin Taylor. 1997. "Development and Validation of a Genetic Algorithm for Flexible Docking." *Journal of Molecular Biology* 267(3): 727–48.

Krüger, Jens, Richard Grunzke, Sandra Gesing, Sebastian Breuers, André Brinkmann, Luis de la Garza, Oliver Kohlbacher, Martin Kruse, Wolfgang E Nagel, and Lars Packschies. 2014. "The MoSGrid Science Gateway–a Complete Solution for Molecular Simulations." *Journal of Chemical Theory and Computation* 10(6): 2232–45.

Kumar, Ashutosh, and Kam Y J Zhang. 2015. "Hierarchical Virtual Screening Approaches in Small Molecule Drug Discovery." *Methods* 71: 26–37.

Kutzner, Carsten, David Van Der Spoel, Martin Fechner, Erik Lindahl, Udo W Schmitt, Bert L De Groot, and Helmut Grubmüller. 2007. "Speeding up Parallel GROMACS on High-latency Networks." *Journal of Computational Chemistry* 28(12): 2075–84.

Land, Henrik, and Maria Svedendahl Humble. 2018. "YASARA: A Tool to Obtain Structural Guidance in Biocatalytic Investigations." Methods in Molecular Biology 1685: 43–67.

Leinonen, Rasko, Federico Garcia Diez, David Binns, Wolfgang Fleischmann, Rodrigo Lopez, and Rolf Apweiler. 2004. "UniProt Archive." *Bioinformatics* 20(17): 3236–37.

Li, Honglin, Zhenting Gao, Ling Kang, Hailei Zhang, Kun Yang, Kunqian Yu, Xiaomin Luo, Weiliang Zhu, Kaixian Chen, and Jianhua Shen. 2006. "TarFisDock: A Web Server for Identifying Drug Targets with Docking Approach." *Nucleic Acids Research* 34(suppl_2): W219–24.

Li, Hongjian, Kwong-Sak Leung, Pedro J Ballester, and Man-Hon Wong. 2014. "Istar: A Web Platform for Large-Scale Protein-Ligand Docking." *PloS One* 9(1): e85678.

Li, Honglin, Chunlian Li, Chunshan Gui, Xiaomin Luo, Kaixian Chen, Jianhua Shen, Xicheng Wang, and Hualiang Jiang. 2004. "GAsDock: A New Approach for Rapid Flexible Docking Based on an Improved Multi-Population Genetic Algorithm." *Bioorganic & Medicinal Chemistry Letters* 14(18): 4671–76.

Li, Jian, Jian Zhang, Jing Chen, Xiaomin Luo, Weiliang Zhu, Jianhua Shen, Hong Liu, Xu Shen, and Hualiang Jiang. 2006. "Strategy for Discovering Chemical Inhibitors of Human Cyclophilin A: Focused Library Design, Virtual Screening, Chemical Synthesis and Bioassay." *Journal of Combinatorial Chemistry* 8(3): 326–37.

Li, Min, Qihui Zhai, Uddalak Bharadwaj, Hao Wang, Fei Li, William E Fisher, Changyi Chen, and Qizhi Yao. 2006. "Cyclophilin A Is Overexpressed in Human Pancreatic Cancer Cells and Stimulates Cell Proliferation through CD147." *Cancer* 106(10): 2284–94.

Liu, Tiqing, Yuhmei Lin, Xin Wen, Robert N Jorissen, and Michael K Gilson. 2007. "BindingDB: A Web-Accessible Database of Experimentally Determined Protein–Ligand Binding Affinities." *Nucleic Acids Research* 35(suppl_1): D198–201.

Liu, Xiaofeng, Sisheng Ouyang, Biao Yu, Yabo Liu, Kai Huang, Jiayu Gong, Siyuan Zheng, Zhihua Li, Honglin Li, and Hualiang Jiang. 2010. "PharmMapper Server: A Web Server for Potential Drug Target Identification Using Pharmacophore Mapping Approach." *Nucleic Acids Research* 38(suppl_2): W609–14.

Liu, Xiaofeng, Hua Xie, Cheng Luo, Linjiang Tong, Yi Wang, Ting Peng, Jian Ding, Hualiang Jiang, and Honglin Li. 2010. "Discovery and SAR of Thiazolidine-2, 4-Dione Analogues as Insulin-like Growth Factor-1 Receptor (IGF-1R) Inhibitors via Hierarchical Virtual Screening." *Journal of Medicinal Chemistry* 53(6): 2661–65.

Liu, Yang, Maximilian Grimm, Wen-tao Dai, Mu-chun Hou, Zhi-Xiong Xiao, and Yang Cao. 2020. "CB-Dock: A Web Server for Cavity Detection-Guided Protein–Ligand Blind Docking." *Acta Pharmacologica Sinica* 41(1): 138–44.

Liu, Zhihai, Yan Li, Li Han, Jie Li, Jie Liu, Zhixiong Zhao, Wei Nie, Yuchen Liu, and Renxiao Wang. 2015. "PDB-Wide Collection of Binding Data: Current Status of the PDBbind Database." *Bioinformatics* 31(-3): 405–12.

Margreitter, Christian, Drazen Petrov, and Bojan Zagrovic. 2013. "Vienna-PTM Web Server: A Toolkit for MD Simulations of Protein Post-Translational Modifications." *Nucleic Acids Research* 41(W1): W422–26.

Mohammad, Taj, Yash Mathur, and Md Imtaiyaz Hassan. 2021. "InstaDock: A Single-Click Graphical User Interface for Molecular Docking-Based Virtual High-Throughput Screening." *Briefings in Bioinformatics* 22(4): bbaa279.

Morris, Garrett M, David S Goodsell, Ruth Huey, William E Hart, Scott Halliday, Rik Belew, and Arthur J Olson. 2001. AutoDock: Automated Dockingof Flexible Ligands to Receptors. http://autodock.scripps.edu/faqs-help/manual/autodock-3-user-sguide/AutoDock3.0.5_UserGuide.pdf.

Nahar, Rufiat, Seigo Iwata, Daiki Morita, Yuhei Tahara, Yasunobu Sugimoto, Makoto Miyata, and Shinsaku Maruta. 2022. "Multimerization of Small G-Protein H-Ras Induced by Chemical Modification at Hyper Variable Region with Caged Compound." *The Journal of Biochemistry* 171(2): 215–25.

Ni, Shuaishuai, Yaxia Yuan, Jin Huang, Xiaona Mao, Maosheng Lv, Jin Zhu, Xu Shen, Jianfeng Pei, Luhua Lai, and Hualiang Jiang. 2009. "Discovering Potent Small Molecule Inhibitors of Cyclophilin A Using de Novo Drug Design Approach." *Journal of Medicinal Chemistry* 52(17): 5295–98.

Nickel, Janette, Bjoern-Oliver Gohlke, Jevgeni Erehman, Priyanka Banerjee, Wen Wei Rong, Andrean Goede, Mathias Dunkel, and Robert Preissner. 2014. "SuperPred: Update on Drug Classification and Target Prediction." *Nucleic Acids Research* 42(W1): W26–31.

Noel, Jeffrey K, Paul C Whitford, Karissa Y Sanbonmatsu, and Jose N Onuchic. 2010. "SMOG@ Ctbp: Simplified Deployment of Structure-Based Models in GROMACS." *Nucleic Acids Research* 38(suppl_2): W657–61.

Nossaman, Bobby D, and Philip J Kadowitz. 2009. "The Role of the RhoA/Rho-Kinase Pathway in Pulmonary Hypertension." *Current Drug Discovery Technologies* 6(1): 59–71.

Nurisso, Alessandra, Juan Bravo, Pierre-Alain Carrupt, and Antoine Daina. 2012. "Molecular Docking Using the Molecular Lipophilicity Potential as Hydrophobic Descriptor: Impact on GOLD Docking Performance." *Journal of Chemical Information and Modeling* 52(5): 1319–27.

O'Donovan, Claire, Maria Jesus Martin, Alexandre Gattiker, Elisabeth Gasteiger, Amos Bairoch, and Rolf Apweiler. 2002. "High-Quality Protein Knowledge Resource: SWISS-PROT and TrEMBL." *Briefings in Bioinformatics* 3(3): 275–84.

Ou-Yang, Si-Sheng, Jun-Yan Lu, Xiang-Qian Kong, Zhong-Jie Liang, Cheng Luo, and Hualiang Jiang. 2012. "Computational Drug Discovery." *Acta Pharmacologica Sinica* 33(9): 1131–40.

Paz, Joel O, William D Batchelor, and Palle Pedersen. 2004. "WebGro: A Web-based Soybean Management Decision Support System." *Agronomy Journal* 96(6): 1771–79.

Ragno, Rino. 2019. "Www. 3d-Qsar. Com: A Web Portal That Brings 3-D QSAR to All Electronic Devices—the Py-CoMFA Web Application as Tool to Build Models from Pre-Aligned Datasets." *Journal of Computer-Aided Molecular Design* 33(9): 855–64.

Ridley, Anne J. 2015. "Rho GTPase Signalling in Cell Migration." *Current Opinion in Cell Biology* 36: 103–12.

Sadagurski, Marianna, Shoshana Yakar, Galina Weingarten, Martin Holzenberger, Christopher J Rhodes, Dirk Breitkreutz, Derek LeRoith, and Efrat Wertheimer. 2006. "Insulin-like Growth Factor 1 Receptor Signaling Regulates Skin Development and Inhibits Skin Keratinocyte Differentiation." *Molecular and Cellular Biology* 26(7): 2675–87.

Sarker, Satyajit D, and Lutfun Nahar. 2017. "Computer-Aided Phytochemical Research." *Trends in Phytochemical Research* 1(1): 1–2.

Sarker, S.D. and L. Nahar. 2018. "An Introduction to Computational Phytochemistry." In Satyajit Sarker and Lutfun Nahar (eds.), *Computational Phytochemistry*, 1–41. Elsevier.

Suzek, Baris E, Hongzhan Huang, Peter McGarvey, Raja Mazumder, and Cathy H Wu. 2007. "UniRef: Comprehensive and Non-Redundant UniProt Reference Clusters." *Bioinformatics* 23(10): 1282–88.

Tetko, Igor V, Johann Gasteiger, Roberto Todeschini, Andrea Mauri, David Livingstone, Peter Ertl, Vladimir A Palyulin, Eugene V Radchenko, Nikolay S Zefirov, and Alexander S Makarenko. 2005. "Virtual Computational Chemistry Laboratory–Design and Description." *Journal of Computer-Aided Molecular Design* 19(6): 453–63.

UniProt Consortium. 2010. "The Universal Protein Resource (UniProt) in 2010." *Nucleic Acids Research* 38(suppl_1): D142–48.

UniProt Consortium. 2015. "UniProt: A Hub for Protein Information." *Nucleic Acids Res* 43(D1): D204–212.

Vries, Sjoerd J De, Marc Van Dijk, and Alexandre M J J Bonvin. 2010. "The HADDOCK Web Server for Data-Driven Biomolecular Docking." *Nature Protocols* 5(5): 883–97.

Wang, Renxiao, Xueliang Fang, Yipin Lu, Chao-Yie Yang, and Shaomeng Wang. 2005. "The PDBbind Database: Methodologies and Updates." *Journal of Medicinal Chemistry* 48(12): 4111–19.

Wang, Xia, Yihang Shen, Shiwei Wang, Shiliang Li, Weilin Zhang, Xiaofeng Liu, Luhua Lai, Jianfeng Pei, and Honglin Li. 2017. "PharmMapper 2017 Update: A Web Server for Potential Drug Target Identification with a Comprehensive Target Pharmacophore Database." *Nucleic Acids Research* 45(W1): W356–60.

Wishart, David S, Craig Knox, An Chi Guo, Dean Cheng, Savita Shrivastava, Dan Tzur, Bijaya Gautam, and Murtaza Hassanali. 2008. "DrugBank: A Knowledgebase for Drugs, Drug Actions and Drug Targets." *Nucleic Acids Research* 36(suppl_1): D901–6.

Zoete, Vincent, Michel A Cuendet, Aurélien Grosdidier, and Olivier Michielin. 2011. "SwissParam: A Fast Force Field Generation Tool for Small Organic Molecules." *Journal of Computational Chemistry* 32(11): 2359–68.

18 Nanoformulations and Herbal Drug Development

Arif Jamal Siddiqui, Syed Amir Ashraf, Riadh Badraoui, and Fevzi Bardakci
University of Ha'il

Ritu Singh
Central University of Rajasthan

Sadaf Jahan
Majmaah University

Sanjeev Kumar
Central University of Jharkhand

CONTENTS

DOI: 10.1201/b22842-18

18.1 INTRODUCTION

With the advances in science, technology, genomics, and combinatorial chemistry, a huge range of new, effective, and specific therapeutics are being generated. The nano-sized particles, specifically known as nanoparticles, are a large class of materials that encloses specific matter with a size of less than 100 nanometers (Jeevanandam et al. 2017). Because of their tunable physiochemical and biological performances, nanoparticles or nanomaterials have made significant advances in various industrial sectors (Subhan, Choudhury, and Neogi 2021). In the 21st century, nanotechnology has shown revolutionary developments in a wide range of fields including the biological system treatment, monitoring, diagnosis, therapeutics, drug delivery, etc. The major disadvantages of conventional drugs include their nonspecificity, poor solubility, and incompetency to enter inside the cells, all of which provides a remarkable prospect for nanoparticles to play a substantial role. Since antiquity, herbal medicines have been widely used in different parts of the globe. Herbal drugs now hold a prominent status in the pharmaceutical industry because their effects are well known, and in addition, their side effects are minimal. Furthermore, when compared to synthetic drugs, the herbal drug has a symmetrical method of interest for fabricating nanoparticles (Khan, Saeed, and Khan 2019). Although herbal drugs have powerful pharmacological effects against a wide range of diseases, they have only a restricted effect on the human beings. In highly acidic pH of the stomach, the activity of herbal drugs gets reduced or they get metabolized by the liver before reaching the bloodstream (Sandhiya and Ubaidulla 2020). Their lower kinetic performance, incapability to cross lipid membranes, high molecular size and weight, and poor absorption lead to decreased bioavailability and efficacy (Kesarwani and Gupta 2013). Furthermore, due to the aforementioned drawbacks, some of the extracts are not used clinically. In order to overcome these limitations, carriers are utilized as an alternate method to modify and increase the kinetics and dynamics of a drug molecule on a biological system. In recent decades, herbal drugs with nanocarriers have received much attention due to its unique physicochemical properties. Nano-herbal systems have been observed to improve the activity of herbal drugs and overcome the problems associated with plant medicine, thus having a great future potential (Figure 18.1).

Nanoparticles are classified in a variety of ways, such as categorization based on the type of material, size, surface, and shape (Khan, Saeed, and Khan 2019). The classifications are also based on the type of coating material and the ligand attached to nanoparticles. The use of nanocarriers in herbal remedies increases the surface area, solubility, and bioavailability of the drug, as well as facilitates precise drug targeting, which is the process of releasing a drug molecule over a small area of the system for a long time in order to stimulate a response in the diseased tissue.

The carriers are selected on the basis of the kinetic property of the moiety and could be organic or inorganic in nature (Su 2020). Because of the common problems associated with most of the drugs like poor solubility, poor bioavailability, high toxicity, high dosage, nonspecific delivery, *in-vivo* degradation, short circulating half-lives, and poor stability, the means of drug delivery can influence the efficacy and potential for commercialization (Senapati et al. 2018). Thus, there is a corresponding need for safer and more effective methods and devices for drug delivery. Among all the above-mentioned problems, solubility is a significant property which determines the drug's concentration in systemic circulation which is desirable for pharmacological response. The poor properties of drug led to ineffective absorption and poor pharmacokinetics, ultimately resulting in high clinical failure (Savjani, Gajjar, and Savjani 2012; Patra et al. 2018).

18.2 HERBAL MEDICINES

These are also known as botanical medicine or phytomedicine. As per WHO, herbal medicines are "finished, labeled medicinal products that contains as active ingredients aerial or underground parts of plants, or other plant material, or combinations thereof, whether in the crude state or as plant preparations". Any plant-derived substances like juices, gums, fatty/essential oils, etc., are examples

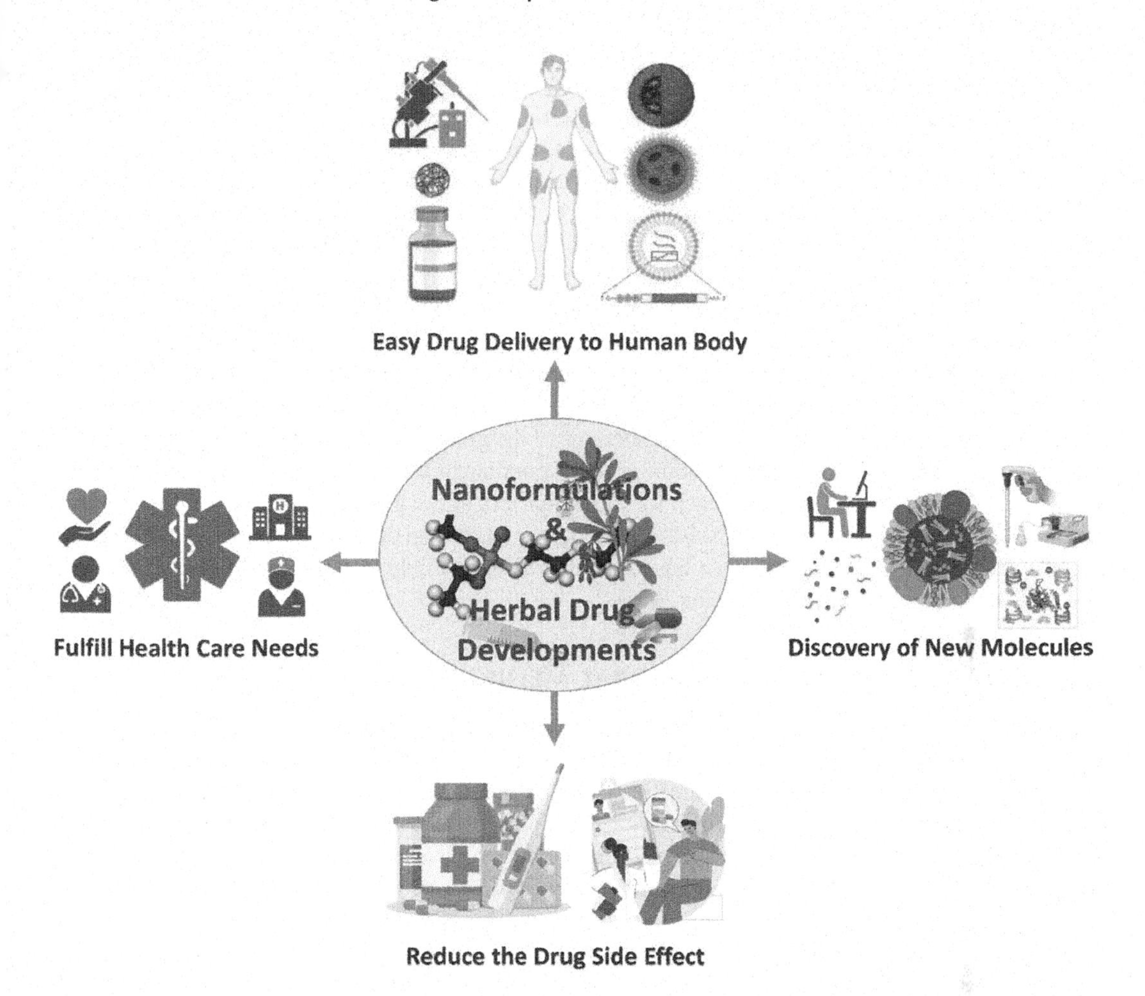

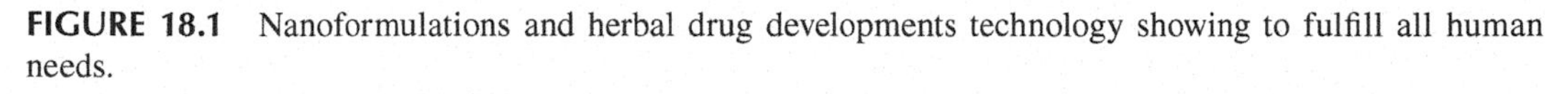
FIGURE 18.1 Nanoformulations and herbal drug developments technology showing to fulfill all human needs.

of plant materials (Sajid et al. 2019). Excipients may also be present in herbal medicines. Any herbal medicine which is combined with chemically defined active substances is not considered as herbal medicines. In few countries, phytomedicines contain organic/inorganic active ingredients that are not derived from plants. Within the field of phytotherapy, a distinction should be made between rational therapy, which is comparable to the treatment using synthetic chemical medicines, and other forms of therapy derived from the traditional use of medicinal plants and preparations made from them (Pan et al. 2014). While the effectiveness of the products used in "rational phytotherapy" has been established through appropriate pharmacological investigations and clinical studies in patients, the worth of a large number of phytopharmaceuticals has yet to be established in this way. Traditional medicines and herbalism have been used for the treatment of various ailments since ancient times. Herbalism is a type of traditional medicine that relies on the use of plants and plant extracts. Botanical medicine, medical herbalism, herbal medicine, herbology, and phytotherapy are the terms used to describe herbalism. Herbal medicine can sometimes be expanded to include fungi, bee products, minerals, shells, and animal parts (Falzon and Balabanova 2017). Many plants produce substances such as aromatic substances, phenols, tannins, etc., which are beneficial to the health of humans and animals. Many are secondary metabolites, of which at least 12,000 have been isolated, with more being investigated (Falzon and Balabanova 2017). These compounds (particularly the alkaloids) are often used by plants to protect themselves from microorganisms, insects, and herbivores. Medicinal compounds can be found in different herbs and spices used in seasonal foods. Herbs are making a comeback, and a herbal resurgence is taking place around the world as they are safe in comparison to synthetics (Wink 2015). Though the medicinal and aromatic properties of varied herbs were treasured for periods, the synthetic products outshone their significance. The blind reliance on synthetics is now ending up and the society is turning back to safer natural herbs (Pan et al. 2014). Various factors, such as population growth, insufficient drug supply, prohibitive treatment costs, side effects of several allopathic drugs, and the development of resistance to currently used drugs for infectious diseases, lead to reliance over herbal medicines. As per the estimates, 80% of the world's population can't afford the synthetic pharmaceutical drugs and must rely on traditional herbal medicines. This is well documented in the medicinal plant inventory, which lists over 20,000 species (Guan et al. 2021; Karunamoorthi et al. 2013). Despite the overwhelming influences, our reliance on modern medicines and great advances in synthetic drugs, large sections of the global population continue to prefer plant-based drugs. Plant-based drugs are becoming more popular in many developing countries because modern life-saving drugs are out of reach for three quarters of the global population, despite the fact that many of these countries spend 40–50% of their total wealth on drugs and healthcare. In developing countries, increased usage of plant-based medicines would help in reducing financial burden in future (Karunamoorthi et al. 2013; Ekor 2013).

18.3 PHARMACOKINETICS IN HERBAL MEDICINE

The most important point to remember while studying herbal pharmacokinetics is that their oral or topical administration is preferred over direct injection into bloodstream. Hence, the study of bioavailability or active constituents in plants is critical. The degree of absorption of active substances into the bloodstream after oral administration is known as bioavailability, and this is a factor of preparation which is used to supply active substance doses. Drugs intended for oral administration are typically designed to have a high bioavailability (Bhattaram et al. 2002; Zhang et al. 2017). In contrast, phytochemicals have lower bioavailability, which can be compounded to be the preferred dosage preparation. Because it is accepted in modern drug development that medicinal plants act on a chemical level in the body, understanding herbal pharmacokinetics is essential. The bioavailability of a molecule is determined by a number of factors that determine how the molecule crosses the gastrointestinal tract barrier and survives in the bloodstream (Zhang et al. 2017; Yang et al. 2017). The following parameters should be taken into account when evaluating herbal pharmacokinetics:

(1) More information on traditional and anecdotal uses of medicinal plants, as well as better data for rational dosage. (2) A better understanding of scientific data, particularly in vitro/in vivo studies in which the active compounds are given by injection. (3) A better understanding of plant safety and toxicity, as well as the potential for herb–drug interactions. (4) Supporting evidence for herbal medicine's synthetic nature. (5) Methods for increasing the bioavailability of herbal medicines, thus their efficacy. (6) Molecule size: large molecules may have around 1% or less bioavailability due to pinocytosis. (7) The molecule's lipid solubility: higher lipid solubility results in higher bioavailability. (8) A molecule's water solubility causes it to dissolve in digestive juices and then cross the lipid membrane, implying that purely water-soluble molecules will have low bioavailability. (9) Bioavailability is usually reduced when a molecule is ionized. (10) Specific factors relating to gut wall crossing, such as active transportation. Factors in the gut: food interaction, gut stability, gastric emptying, and so on. (11) First-pass metabolism in the liver and gut metabolism. (12) Patient-specific factors, including the impact of pathological factors. (13) The absorption and bioavailability of plant constituents can be affected by the presence or absence of food. Herbal pharmacokinetics is a unique field that is extremely difficult to understand for the reasons listed below. (14) The chemical complexity of plant medicines as well as the potential for constituent interactions. The differences in bioavailability between the compound's molecules. (15) Because the active components are not always known, the components in the plant that need to be studied cannot always be identified.

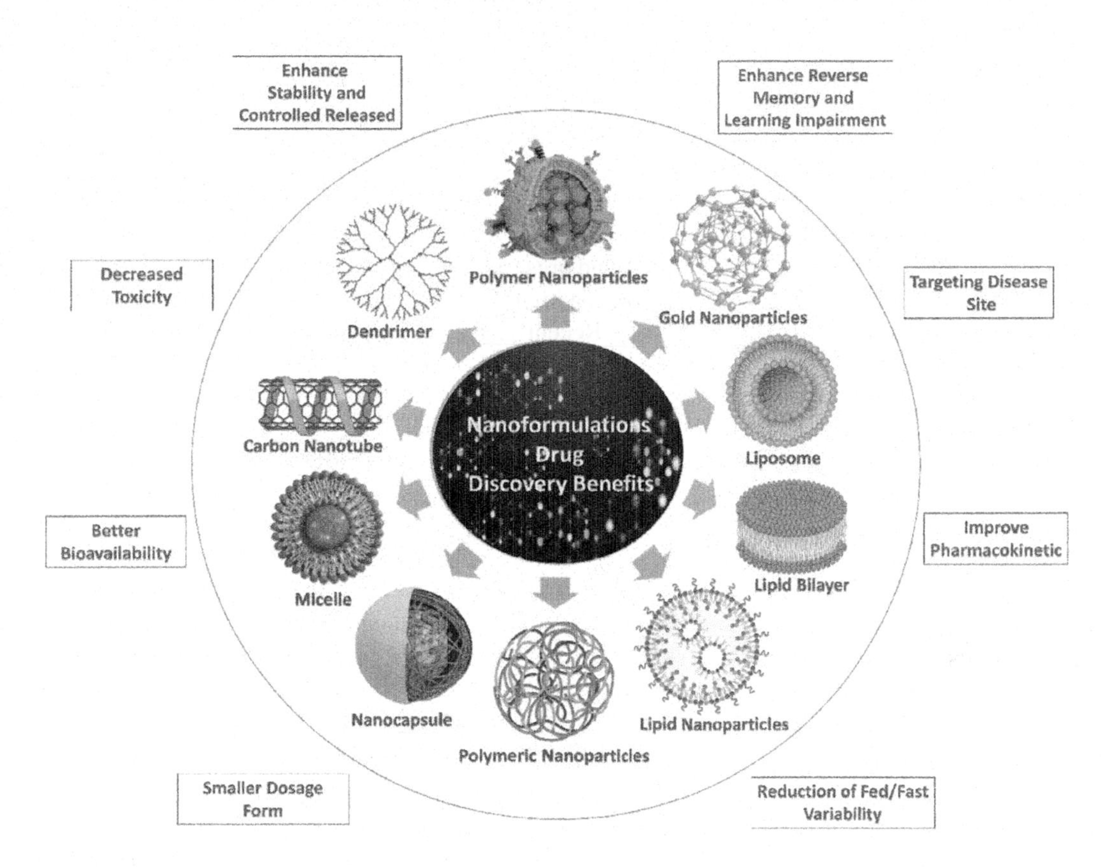

FIGURE 18.2 Types of nanoformulations in drug development and their possible role in pharmacokinetics.

(16) Herbal medicines are not designed for predictable pharmacokinetic properties, and natural compounds, in particular, are frequently metabolized in the digestive tract, acting as prodrugs. (17) Large polar molecules are frequently involved, which are likely to have low and variable bioavailability (Figure 18.2).

Pharmacokinetics of herbal medicines is required to ensure efficacy, safety, and the avoidance of side effects. Their complex composition, unknown active constituents, and low plasma concentrations of metabolites made pharmacokinetic studies much difficult. Animal tests are typically inexpensive and simple to conduct, but the parameters derived from them do not always apply to humans. As a result, clinical pharmacokinetic parameters of herbal medicines have become a valuable source of evidence for rational herbal remedies. For some of the herbal drugs, the data on clinical pharmacokinetics and clinical drug interactions are available in literature, thanks to advances in active compound knowledge and the availability of highly sensitive and selective analytical methods (Rombolà et al. 2020).

18.4 IMPORTANCE OF PHARMACOKINETICS

Understanding the interaction between herbs and pharmaceuticals requires pharmacokinetics data. There is not much information available over pharmacokinetics of herbal drugs due to the lack of research and insufficient information. The distinction between pharmacodynamics and pharmacokinetics in classical pharmacology is expressed by this dichotomy (Borse, Singh, and Nivsarkar 2019; Ekor 2013). The physiological (or that of its constituents) effects of a given dose of herbal medicine depend on the effective tissue concentration of the remedy, which in turn remains dependent on pharmacokinetic parameters. Herbal pharmacokinetics could deliver practitioners with useful facts to help them prescribe herbs safely and effectively. It may also make it possible to make useful predictions, such as about possible interactions between herbal remedies and conventional pharmaceuticals (Turfus et al. 2017; Waller, Sampson, and Hitchings 2021). Drug–drug interactions make up the majority of traditional pharmacokinetic literature, but herb–drug interactions are gaining attraction in the popular press and among physicians, who are becoming more aware of their patients' widespread and often undisclosed use of herbal medicines, as well as the potential for significant pharmacokinetic interactions between herbs and prescription pharmaceuticals. The situation for the data on herb–drug interactions is very unlike, owing to the fact that pharmaceuticals and herbs are governed by different regulatory frameworks. Recently, in a review published on pharmacokinetics, it was revealed that a very few human studies have been carried out in this field with not so large number of herbs and that too with normal or healthy volunteers. Taking into account the complexity of herbal remedies and the manifold factors that influence pharmacokinetics, this lack of data is likely to persist for some time (Turfus et al. 2017; Saghir and Ansari 2018).

18.5 NANOFORMULATIONS

Nanotherapeutics recently demonstrated a significant improvement in therapeutic efficacy by designing and developing nanoformulations. This call is about biological nanoformulation (like proteins, peptides, nucleic acids, and enzymes). The biologicals can be efficiently transported through the biological barriers to the target organs, tissues, and cells with the right formulation (Siddiqui et al. 2020; Lopalco and Denora 2018). By designing and developing nanoformulations, nanotherapeutics recently demonstrated a significant increase in therapeutic efficacy.

18.6 LIPOSOMES

A.D. Bangham and his colleague R.W. Thorne first explained the concept of liposomes (Weissig 2017). Gerald Weissman proposed a structural arrangement named as "liposome" and defined as "microscopic vesicular structure contained lipid bilayers (one or more)." Nowadays, liposomes are used in cosmetics as drug delivery systems (Senior 1990; Tenchov et al. 2021). Liposomes have the advantage of being very biocompatible, making them a good choice for the delivery of antibacterial, antiviral, antiparasite, fungicidal, transdermal transporters, diagnostic instruments, enzymes, and vaccine adjuvants (Çağdaş, Sezer, and Bucak 2014). Phospholipids are natural and biocompatible compounds. These are amphiphilic compounds that self-organize in the aqueous phase. Liposomes are being studied for intracellular drug delivery since they are similar to biological components. They have the ability to combine with cell components. Furthermore, they are commonly employed as a cell model for the penetration of various medicines. Liposomes have a higher surface area due to their lower particle size. As a result, its contact time with the cell membrane and intracellular passage increases. As a result, lipid vesicles containing therapeutic compounds of 100 nm size offer more benefits, which are difficult to achieve due to the presence of the blood-brain barrier and the brain-skin barrier (Tapeinos, Battaglini, and Ciofani 2017). Liposomes' unique ability to encapsulate pharmaceuticals in both an aqueous and a lipid layer makes them ideal for both hydrophilic and hydrophobic medications. Lipophilic medications are generally encased in liposomes' lipid bilayers, while hydrophilic drugs stay in the aqueous core. The poorly water-soluble substance had issues with entrapment and storage stability (Tapeinos, Battaglini, and Ciofani 2017; Yingchoncharoen, Kalinowski, and Richardson 2016).

18.7 LIPID NANOPARTICLES

Lipid nanoparticles include lipid drug conjugates, solid lipid nanoparticles (SLN), and nanostructured lipid carriers (NLC) (Hou et al. 2021; Scioli Montoto, Muraca, and Ruiz 2020). Nanoparticles have gained considerable attention in the delivery of drugs to the skin. Broadly classified into polymeric and lipid-based systems, these nanoparticles have made progress in various aspects of drug-delivery technologies. Polymeric nanoparticles are exploited in the delivery of drugs via the transdermal pathway resulting in systemic concentration, whereas lipid-based systems are mainly used to retain and treat the dermatological conditions (Stefanov and Andonova 2021). Lipid-based systems have a unique ability to mimic the existing environment of the skin, thereby being well-tolerated and nonirritant. Lipids can encapsulate and solubilize a large number of active ingredients and is suitable for several skin types. Lipids and oils are known to provide nourishment to the skin and help in maintaining the moisture barrier at the interface. Lipids are also able to be formulated into aesthetically pleasing forms and enhance the patient experience (Yingchoncharoen, Kalinowski, and Richardson 2016). Lipid-based systems range from nanoemulsions, vesicular carriers, and nanoparticulates. Of the several types of nanoparticles, nanostructured lipid carriers (NLCs) have emerged as a favorable and well-studied carrier for the dermal delivery of the active molecules. NLCs-based carrier systems are the second generation of solid lipid nanoparticles (SLNs) and consist of a well-blended composition of one or more solid and liquid lipids with the addition of ionic or nonionic surfactants. SLNs have been reported to be exhibiting an erratic drug release and drug expulsion upon storage. SLNs have limited stability, and due to the presence of solid lipids alone, the solubility and entrapment of drugs are limited (Yingchoncharoen, Kalinowski, and Richardson 2016; Naseri, Valizadeh, and Zakeri-Milani 2015). In the context of dermal drug delivery, NLCs offer competitive advantages to the formulator conferring extended stability, the formation of occlusive layer upon application and the ease of incorporation into the cream or gel bases. NLCs are also highly regarded for their unique ability to confer extended stability, including the unavoidable exposure to light, and humid conditions. The third generation of SLNs, called the smartLipids®, are known today, which consists of more than ten different lipids and oils, making it a challenging task to manufacture and optimize (Müller et al. 2007). Moreover, any possible interactions and instabilities during the formulation of smartLipids® go unrecognized and may result in the formation of an inferior product.

18.8 NANOSTRUCTURED LIPID CARRIERS (NLCs)

NLCs are made from a mixture of solid and liquid lipids along with two or more surfactants. The unique feature of NLCs lies in the liquid lipids. Drugs tend to show a large solubility within the liquid lipids when compared to the solid counterparts. NLCs contain irregularly crystallized matrix owing to the presence of liquid lipids allowing higher retention of the drug with negligible drug expulsion (Beloqui et al. 2016; Naseri, Valizadeh, and Zakeri-Milani 2015). With a possibility of obtaining particles sizes of less than 100 nm, NLCs favor a generation of an occlusive layer at the skin-air interface that prevent water loss from the skin tissues. NLCs are uniquely shown to be prepared from ingredients that have a generally-regarded-as-safe (GRAS) status indicating their biocompatibility and biodegradability. Being made from lipids and oils, they are solid at room temperatures and thus can be easily incorporated into the cream and gel bases. Several cosmetic products containing NLCs are approved for human use and are available in the market (Khosa, Reddi, and Saha 2018; Stefanov and Andonova 2021).

18.9 PHYTOSOMES

Phytosomes are defined as "phyto" means plants and "some" means cell-like. Phytosome is a newly introduced patented technology developed by Indena that incorporates standardized plant extract into phospholipids, mainly phophatidylcholine, to give lipid compatible molecular complexes to improve absorption and bioavailability (Sharma and Roy 2010; Alharbi et al. 2021). The phospholipid subatomic structure incorporated a water-dissolvable heads and two fat soluble tails; as a result of their double dissolvability, phospholipid acts as an effective emulsifier. An emulsifier is a material that can consolidates two fluids that typically will not combine well. By consolidating the emulsifying activity of the phospholipid, with standardized plant extract, the phytosome gives drastically upgraded bioavailability and convey quicker and enhanced retention in the intestinal tract, and in light of the fact that not every plant property is as bioavailable as others, going along with them to phospholipids creates a powerful medium for increased absorption of active constituents of herb. These are biocompatible and biodegradable delivery system formed of plant's active constituent and phospholipid formulations, i.e., liposomes having plant secondary metabolite (Lu et al. 2019; Gnananath, Nataraj, and Rao 2017). The flavonoids and terpenoids component of these plant extracts enables them to directly bind with phosphatidylcholine (PC), leading to the formation of little microspheres, cell, or vesicles. In case of phytosomes, a hybrid bond forms between PC and polyphenol, producing a lipid-soluble complex which has lower polarity and higher capability to cross the biological membrane, thereby enhancing the bioavailability (Yang et al. 2020; Anjana et al. 2017). Accordingly, the phytosome is another leap-forward innovation for the advancement in herbal innovation. The phytosome procedure heightens the herbal products by enhancing ingestion, expanding bioavailability, and upgrading delivery to the tissue. Some natural compounds are not extremely bioavailable; however, binding them to phosphatidylcholine produces an exceptionally bioavailable form of the herbal compound. A few investigations have demonstrated that the body utilizes phytosomal formulation more efficiently than that of nonphytosome atoms.

18.10 NANOEMULSIONS

Nanotechnology contains technological development based on the nanometer scale, typically 0.1–100 nm. In pharmaceuticals, utilization of nanoinnovation and medicine has developed in course of most recent couple of years. Nanoemulsions are mostly of three types based on their content (Gupta et al. 2016):

1. Oil in water nanoemulsion, where oil as drops is invisible in the continuous aqua phase.
2. Water contains in oil nanoemulsion: in this formulation, water drop is invisible in the regular oil phase.
3. Bicontinuous nanoemulsion contains microdomains of oil and water—interdispersed within the system.

These nanoemulsions are also alternatively known as submicron emulsions, ultrafine emulsions, and mini emulsions. Phase behavior studies revealed that the size of the droplets largely depends on the surfactant phase structure at the inversion point induced by either temperature or the composition. Studies on nanoemulsion formation by the phase inversion temperature method have shown a relationship between minimum droplet size and complete solubilization of the oil in a microemulsion bicontinuous phase independently of whether the initial phase equilibrium is single or multiphase (Gupta et al. 2016; Sheth et al. 2020). Nanoemulsions gain strength against sedimentation or creaming with Ostwald ripening due to their narrow drop shape, and this is the main mechanism of nanoemulsion breakdown. The main use of nanoemulsions is the production of nanoparticles by adopting a polymerizable monomer as the disperse phase where nanoemulsion drops act as nanoreactors (Lovelyn and Attama 2011; Fryd and Mason 2012).

18.10.1 Advantages of Nanoemulsion

Various advantages of nanoemulsion are described below: (1) It creates the plasma concentration profiles; (2) enhances the bioavailability of drugs; (3) eives ultralow interfacial tension and high oil in water interfacial areas; (4) provides an advantage on existing self-emulsifying system in terms of rapid onset of action; (5) provides reduced intersubject variability in terms of gastrointestinal fluid volume; (6) has high kinetic strength and optical transparency resembling to microemulsions; (7) the design of nanoemulsion is shorter than visible wavelength, thus remains optically transparent, even at huge loading; (8) nanoemulsions are prepared in many application forms such as creams, liquids, sprays, and foams; (9) nanoemulsion is nontoxic and nonirritant, so it can be simply utilized to skin and mucous membranes.

18.10.2 Transferosomes

Gregor Cevc coined the term "transferosome" and the concept behind it in 1991. A transferosome is a complex aggregate that is highly adaptable and stress-responsive in its broadest sense. An ultradeformable vesicle surrounded by an aqueous core and a complex lipid bilayer, it is the most preferable form of transferosome. Herein, the vesicle remains self-regulating and self-optimizing, enabling the transfersome to efficiently cross various transport barriers and then act as a drug carrier for noninvasive targeted drug delivery and therapeutic agent sustained release (Fernández-García et al. 2020). In this regard, transdermal delivery is an intriguing option because it is convenient as well as safe. This has a number of benefits over conventional options such as the avoidance of first-pass metabolism, predictable and extended duration of activity, minimizing undesirable side effects, utility of short half-life drugs, improved physiological and pharmacological response, avoiding drug-level fluctuations, inter- and intrapatient variations, and patient convenience (Opatha, Titapiwatanakun, and Chutoprapat 2020). In order to improve the efficacy of material transfer across intact skin, penetration enhancers, iontophoresis, sonophoresis, and colloidal carriers such as lipid vesicles (liposomes and proliposomes) and nonionic surfactant vesicles (niosomes and proniosomes) have been used (Opatha, Titapiwatanakun, and Chutoprapat 2020; Reddy et al. 2015; Sapkota and Dash 2021).

18.10.3 Features and Limitation of Transferosomes

Because transferosomes have a structure that combines hydrophobic and hydrophilic moieties, they can accommodate drug molecules with variable solubilities. Transferosomes can deform and pass through constriction up to ten times smaller than their own diameter without any measurable loss. This high deformability allows intact vesicles to penetrate more easily. Low and high molecular weight drugs, such as analgesics, anesthetics, corticosteroids, sex hormone, insulin, gap junction protein, and albumin, can be carried by them. Because they are made from natural phospholipids, they are biocompatible and biodegradable, similar to liposomes (Sapkota and Dash 2021). They have high entrapment efficiency, approaching 90% in the case of lipophilic drugs. They prevent the encapsulated drug from being degraded by the body's metabolism. They serve as a depot, gradually releasing their contents. They can be used for both systemic and topical drug delivery. As the procedure is straightforward, they are easy to scale up and do not necessitate the use of pharmaceutically unacceptable additives (Reddy et al. 2015; Rajan et al. 2011; Solanki et al. 2016).

18.10.4 Niosomes

Niosomes are liposomes made of nonionic surfactants. These are microscopic lamellar structures ranging in size from 10 to 1,000 nm. As compared to liposomes, they are more flexible due to their nonionic surfactant bilayer. The majority of them are made with cholesterol as an excipient. Other excipients can be used as well. Niosomes have a greater ability to penetrate than liposomes. They have a bilayer structure similar to liposomes, but the materials used to make niosomes make them more stable, and thus, niosomes offer many more benefits than liposomes (Marianecci et al. 2014; Sadeghi-Ghadi et al. 2021).

18.10.5 Ethosomes

For many years, the importance of vesicles in cellular communication and particle transportation has been recognized. Several studies have been carried out to study the properties of vesicle structures in order to improve the drug delivery within their cavities and tagging the vesicle for cell specificity. Vesicles would also enable drug release rate control over a longer period of time, shielding the drug from immune response or other removal systems, and releasing just the right amount of drug and maintaining that concentration for longer periods of time (Paiva-Santos et al. 2021; Ainbinder et al. 2010). The main advantages of ethosome technology are (1) enhanced permeation, (2) passive and noninvasive, (3) suitable for a wide range of drugs with various chemical properties, (4) suitable for a wide range of applications, (5) made from safe and approved materials, (6) protected by international patents, (7) proven feasibility—scientifically and commercially, and (8) well established and published technology.

18.10.6 Cyclodextrins

Cyclodextrins (CD) are used extensively in the pharmaceutical industry. Because of their bioadaptability and versatility, cyclodextrins can be used to alleviate the undesirable properties of drug molecules in a variety of nanotechnology applications by forming inclusion complexes (Loftsson 2021). In fact, several new and promising CD derivatives are constantly being tested for their ability to increase complexity in order to obtain higher drug solubility as well as stabilizing or solubilizing agents. The use of CD in nano- and microcarrier systems can improve the ability to encapsulate the guest molecule, improve drug stability, and effectively regulate drug release rate, among other things (Loftsson 2021; Jacob and Nair 2018).

18.10.7 Structure of Cyclodextrins

CDs are interesting molecular hosts in and of themselves; their structural modifications provide a lot of their utility. CDs are truncated cone-shaped toroidal molecules with lipophilic inner cavities and a hydrophilic outer surface. CDs are attractive not only because of their inclusion complexation with guest molecules but also because of the large number of hydroxyl groups in the glucose units, which are 18, 21, and 24 for, and -CD, respectively (Jansook, Ogawa, and Loftsson 2018). The ability of molecular recognition gets affected if the hydroxyl groups of CDs are changed. The secondary hydroxyl groups are on the wider side of the ring in cyclodextrin's structure, while the primary hydroxyl groups are on the narrower side of the torus (Jansook, Ogawa, and Loftsson 2018; Jambhekar and Breen 2016). The hydroxyl groups are oriented toward the exterior of the cone, and the CH groups with the H-1, H-2, and H-4 protons are mainly located on the exterior side of the molecule, making the external portion of CDs hydrophilic in nature. Two rings of CH groups (H-3 and H-5) and a ring of glycosidic ether oxygen are present in the interior of the torus, with H-6 located near the cavity. The host–guest complex is formed when a low-polarity central void can encapsulate a wide variety of guest molecules resulting in a stable association without the formation of covalent bonds (Crini et al. 2018; Sá Couto Salústio and Cabral-Marques 2014). The use of multiple CDs and/or pharmaceutical additives in combination will result in a more balanced oral bioavailability with longer therapeutic effects. Furthermore, the optimal combination of CDs and other pharmaceutical excipients or carriers, such as nanoparticles, liposomes, and others, promotes the advancement of advanced dosage forms (Crini et al. 2018; Sá Couto Salústio and Cabral-Marques 2014).

18.10.8 Polymeric Nanomicelles

Amphiphilic polymers with wonderful hydrophobic and hydrophilic segments are used to make polymeric nanomicelles. In aqueous solution, the polymer self-assembles to form micelles, with the water-insoluble section forming the core and the hydrophilic section forming the corona. Sometimes, self-assembly does not occur spontaneously, and micelle formation is aided by external factors such as temperature (Zheng et al. 2021). Above the CMC, there is a self-meeting. The solubilization of the entire supramolecular structure is facilitated by the hydrophilic segments that make up the corona. Polymeric micelles are distinguished in solution by their low CMC and excellent kinetic and thermodynamic stability. Polymers used to make nanomicelles should ideally be biodegradable and/or biocompatible (Zong et al. 2021). The most widely studied core-forming polymers include poly (lactide), poly (propylene oxide) (PPO), poly (glycolide), poly (lactide-co-glycolide), poly (-caprolactone) (PCL), and poly (ethylene glycol) (PEG).

18.10.9 Surfactant Nanomicelle

Surfactants are amphiphilic molecules with a hydrophilic head and a hydrophobic tail that have a hydrophilic head and a hydrophobic tail. Surfactant molecules' hydrophilic heads can be dipolar/zwitterionic, charged or anionic/cationic, neutral/nonionic, or neutral/nonionic (Sohrabi, Mansouri, and Karimi 2022). Sodium dodecyl sulfate (SDS, anionic surfactant) and dodecyltrimethylammonium bromide (DTAB, cationic surfactant) are two commonly used surfactants in the nanomicellar method. A hydrophobic tail usually consists of an extended chain hydrocarbon, with a halogenated/oxygenated hydrocarbon/siloxane chain appearing only rarely. Micelles are formed when surfactants are dissolved in water at a concentration higher than the CMC. For nanomicellar formulation, intermolecular forces such as Van der Waal interactions, hydrogen bonding, hydrophobic, stearic, and electrostatic interactions must be stable. The noncovalent aggregation of surfactant monomers is what determines the shape of nanomicelles. The shape and size of nanomicelles may be determined by changes in the surfactant chemical structure and environmental factors such as surfactant concentration, pH, temperature, and ionic strength (Sohrabi, Mansouri, and Karimi 2022; Vaishya et al. 2014).

18.10.10 Polyionic Nanomicelle

Micelles of the polyion complex (PIC) have been studied extensively as a nanocarrier system for gene and antigen oligonucleotide delivery. PIC micelles, in particular, have been extensively investigated for the delivery of ionic hydrophilic therapeutics. Electrostatic interactions between polyion copolymers (composed of neutral and ionic segments) and the oppositely charged ionic class form this superior class of micelles. The water-soluble block copolymer has very narrow poly dispersity. The block copolymer's neutral segment is usually PEG, while the ionic segment is counterbalanced by oppositely charged classes to form a hydrophobic core (Nayak et al. 2021; Bose et al. 2021).

18.10.11 Safety Considerations

While evaluating the novel topical herbal nanoformulations, safety assessment is a much-needed task. The nanoparticles, just like the herbal drugs, can enter the systemic circulation and may have unexpected interactions with immune system cells or generate free radicals. Nanoparticles have been reported to be toxic not only to keratinocytes and fibroblasts in the skin but also to the cells in the immune system and other organs (Siddiqui et al. 2020; de la Harpe et al. 2019). Given that nanocomponents could easily get entry into the soil and water via leaching or run off, potential environmental impact of nanocomponents on plants, aquatic species, and humans should be investigated (Nehate et al. 2018). It is difficult to formulate uniform guidelines because each nanoformulation contains different nanocomponents, and each formulation must be evaluated individually based on the nanocomponents added. It is critical to develop in vitro methods for assessing the safety of each nanocomponent.

18.10.12 Herbal Drug Loading

Herbal drugs are gaining popularity for their use in treating a wide array of diseases with fewer side effects and better therapeutic results (Sandhiya and Ubaidulla 2020). On the other hand, herbal extracts have a few limitations, such as being unstable in highly acidic pH, having a high first-pass metabolism, and so on, which can lead to drug levels in the blood being below the therapeutic concentration, resulting in less or no therapeutic effect. To overcome these limitations, herbal drugs are loaded into novel carriers. The carriers help in reducing drug degradation and severe side effects caused by drug accumulation in nontargeted areas (Sandhiya and Ubaidulla 2020). There are different techniques of nanoparticle loading which are listed below:

18.10.13 High-Pressure Homogenization Techniques

High-pressure homogenization is a well-known and widely used method for producing parenteral emulsions (Shah et al. 2022). The use of high-pressure homogenization techniques to prepare lipid nanoparticles has been extensively developed and practiced. For the preparation of lipid nanoparticles, two different temperature-based approaches, cold and hot homogenization techniques, can be used. Prior to high-pressure homogenization, the active compound is dissolved or dispersed in the melted lipid in both processes (Vinchhi, Patel, and Patel 2021). High-pressure homogenizers force a liquid through a narrow gap at high pressure (100–2,000 bar). The forming particles are disrupted down to the submicron range by extremely high shear stress and cavitation forces. This technique allows for large-scale production of lipid nanoparticles with regulatory approval, as production lines for parenteral lipid emulsions have been in use for a long time. This is the main issue with the other techniques that are currently available (Vinchhi, Patel, and Patel 2021; Jiang et al. 2021). High-energy conditions of temperature and pressure, on the other hand, cast doubt on its applicability in certain situations.

a. **Hot high-pressure homogenization technique:** A pre-emulsion of the drug-loaded lipid melt and the emulsifier solution is prepared with a high-shear mixing device for hot homogenization (such as UltraTurrax) (Vinchhi, Patel, and Patel 2021). The pre-emulsion is then homogenized under high pressure at temperatures above the lipid's melting point. Lipid nanoparticles are formed by cooling the sample to room temperature or temperatures below the melting point of the lipid. The melted lipid and drug are dispersed in a hot aqueous surfactant solution at the same temperature using high-speed stirring; the resulting hot pre-emulsion is homogenized several times to yield nanoemulsion. This hot nanoemulsion cools to form an aqueous dispersion of lipid nanoparticles. It can be used to entrap lipophilic and nonlipophilic drugs in a lipid matrix. Because the exposure time to high temperatures is relatively short, temperature-sensitive compounds can be processed by hot HPH (Kasongo, Müller, and Walker 2012). This procedure, however, is not the most appropriate for hydrophilic drugs. The drug will partition to the water phase during homogenization of the melted lipid phase, resulting in a low-encapsulation rate.
b. **Cold high-pressure homogenization technique:** In this method, solid lipid-containing drugs are milled to microparticles and then dispersed in an emulsifier solution, as opposed to hot homogenization. The presuspension is then homogenized under high pressure at or below the room temperature (Kasongo, Müller, and Walker 2012; Souto and Müller 2006). The lipid is melted above its melting point in the cold high-pressure homogenization (HPH) technique, and the drug is dispersed or dissolved in it. The system is cooled with dry ice or liquid nitrogen. After solidification, the lipid mass is ground with a ball or mortar mill to produce lipid microparticles with diameters ranging from 50 to 100 micrometers (Duong, Nguyen, and Maeng 2020). Microemulsions are created by stirring these microparticles into a cold surfactant solution. The microparticles are broken down to nanoparticles by passing it through a high-pressure homogenizer at/or below room temperature. Due to the melting of the lipid in the first step of the process, the cold HPH technique reduces the sample's thermal exposure but does not completely eliminate it. As a result, temperature-sensitive compounds could benefit from this technique. This method can also be used to

incorporate hydrophilic compounds that may partition from the liquid lipid phase to the water phase during the hot HPH (Duong, Nguyen, and Maeng 2020). The PI and mean particle size of lipid particles prepared as per this technique are slightly higher than those obtained using the hot HPH technique (Vinchhi, Patel, and Patel 2021). To further reduce the particle size and minimize polydispersity, homogenization cycles can be increased.

c. **Melt emulsification ultrasound homogenization technique:** Among the techniques used for the preparation of lipid carriers, the most common is hot homogenization. Nonetheless, it necessitates the use of specific devices that are rarely found in research labs. To make lipid nanoparticles, ultrasonication was used instead of high-pressure homogenization (Yuan et al. 2007; Agrawal et al. 2021). The extreme conditions created within the collapsing cavitational bubbles of the inner phase, which lead to size reduction, are the basis for this technology. This method uses the same procedure as hot high-pressure homogenization, but instead of a homogenizer, an ultrasonication device is used. If the operating parameters are standardized, ultrasonic processing is quick and repeatable. The operating temperature, ultrasonication time, and power must all be optimized. Ultrasound probes are practically self-cleaning, have minimal sample loss, and can be used in large-scale manufacturing. However, when using high-shear homogenizers and ultrasonicators, inhomogeneous power distributions are more likely to take place than when using high-pressure homogenizers, which have a homogeneous power distribution due to the small size of the homogenizing gap (Agrawal et al. 2021).
d. **Solvent displacement or injection technique:** The technique was first described for the synthesis of preformed polymers into liposomes and polymeric nanoparticles. This method has recently been used to make lipid nanoparticles (Sala et al. 2017; Schubert and Müller-Goymann 2003). It is based on lipid precipitation in solution from a dissolved lipid. A lipid solution in a water-miscible solvent or a water-miscible solvent mixture is injected rapidly into an aqueous phase with or without surfactant. An o/w emulsion is created by injecting the organic phase into the aqueous phase while gently stirring with a magnetic stirrer. The oil phase is made up of a semipolar water-miscible solvent such as methanol, acetone, or ethanol, into which the lipid component is dissolved, and then the active compound is dissolved or dispersed (Sala et al. 2017). Surfactant is present in the aqueous phase. After evaporation of the solvent, distillation can be used to displace the solvent, and lipid nanoparticles are formed. The size of the particles is determined by the preparation conditions, such as lipid concentration, injected amount, emulsifier concentration, and solvent used (Salatin et al. 2017). The benefits include the use of pharmaceutically acceptable organic solvents, the elimination of the need for high-pressure homogenization, ease of handling, and a quick production process that does not require technologically sophisticated equipment. The use of organic solvents, however, has a number of drawbacks (Duong, Nguyen, and Maeng 2020).
e. **Emulsification-solvent diffusion technique:** This method is commonly used to form polymeric nanoparticles. Partially water-miscible solvents (e.g., benzyl alcohol, ethyl formate, tetrahydrofuran) are used in this technique (Chaudhary et al. 2021; Bouchemal et al. 2004). Mutual saturation with water ensures the initial thermodynamic equilibrium of both liquids. The lipid is then added to the water-saturated solvent and emulsified at high temperatures with a solvent-saturated aqueous surfactant solution. Due to diffusion of the organic solvent from the emulsion droplets to the continuous phase, lipid nanoparticles precipitate after the addition of excess water (typical ratio: 1:5–1:10). An o/w emulsion is created using this procedure, which consists of the organic phase of a partially water-soluble solvent that has been previously saturated with water to ensure the two liquids' initial thermodynamic equilibrium (water and solvent). In the saturated solvent, the lipid dissolves, followed by the drug in the organic phase. This organic phase is then mechanically emulsified in an aqueous solution containing surfactant to form an o/w emulsion (Bouchemal et al. 2004;

Nagavarma et al. 2012). When too much water is added to the system, the solvent diffuses into the external phase, causing the lipid to precipitate. Ultrafiltration or distillation can be used to remove the solvent. After the organic solvent has been removed completely, an aqueous dispersion of lipid nanoparticles is obtained. The dispersion is relatively dilute, similar to the production of lipid nanoparticles via microemulsions, and must be concentrated via ultrafiltration or freeze drying. This method can achieve particle sizes of around 100 nm and very narrow particle size distributions (Nagavarma et al. 2012).

f. **Microemulsion technique:** Various research groups have adapted this method since the early days of lipid nanoparticle formulation (Shah et al. 2014; Zhang et al. 2018). When an excess amount of outer phase is added to a hot microemulsion under cooling conditions, the system breaks down and converts to nanoemulsion, which recrystallizes the internal oil or lipid phase, forming particles. Briefly, a microemulsion is formed by mixing melted lipid-containing drug with surfactant and cosurfactant-containing aqueous phase prepared at the same temperature as the lipid in a specific ratio (Zhang et al. 2018). When a hot microemulsion is diluted in excess cold water, the microemulsion breaks down, converting it into an ultrafine nanoemulsion, which then recrystallizes the internal lipid phase, forming lipid particles. The dilution with water and temperature drop results in breaking of microemulsion (Tartaro et al. 2020). The process parameters that could affect the lipid particle size and structure include microemulsion composition, temperature, dispersing device for adding the microemulsion to low temperature water, and lyophilization. It has the advantages of requiring no specialized equipment, requiring no energy for production, and allowing for scale-up production of lipid nanoparticles (Chauhan et al. 2020). The disadvantage is the dilution of the particle suspension with water, which creates problem in removing excess water (Malik, Wani, and Hashim 2012). Furthermore, high concentrations of surfactants and cosurfactants in the formulation may raise regulatory concerns. Surfactant removal can, however, be accomplished using a suitable procedure such as dialysis, ultrafiltration, or ultracentrifugation, but this will add a step to the procedure.

g. **Emulsification-solvent evaporation technique:** This method is similar to the solvent evaporation method for production of polymeric nanoparticles and microparticles in o/w emulsions (Paswan and Saini 2017; Gundloori, Singam, and Killi 2019). Herein, the lipid is dissolved in a water-immiscible organic solvent (e.g., toluene, chloroform), which is then emulsified in an aqueous phase before the solvent is evaporated under reduced pressure. The lipid precipitates and forms nanoparticles after the solvent evaporates. To begin, the lipid is dissolved in a water-immiscible organic solvent followed by dispersion/dissolution of the drug (Gundloori, Singam, and Killi 2019). Mechanical stirring emulsifies the organic phase in an o/w surfactant-containing aqueous phase. Nanoparticle dispersion is achieved by the precipitation of the lipid in the aqueous medium after quick removal of the solvent by evaporation from the obtained o/w emulsion under mechanical stirring or reduced pressure. The solvent evaporation step must be completed quickly to avoid particle aggregation (Vinothini and Rajan 2019; Staff, Landfester, and Crespy 2013). Because this method avoids the use of heat during the formulation, it can also be used to incorporate highly thermolabile drugs. Solvent residues in the final dispersion may cause issues, which are a regulatory concern. These dispersions are typically quite dilute because of limited solubility of lipid, thus it must be concentrated by lyophilization, ultrafiltration, or evaporation (Staff, Landfester, and Crespy 2013).

h. **Multiple emulsion techniques:** This is a w/o/w double emulsion-based modified solvent emulsification-evaporation method (Figure 18.3). For the preparation of hydrophilic drug encapsulated SLN, it uses emulsification followed by solvent evaporation. In the external water phase of a w/o/w double emulsion, a stabilizer is added to prevent drug partitioning to the external water phase during solvent evaporation (Khan, Saeed, and Khan 2019). It has the benefits and drawbacks of the previously described emulsification solvent evaporation

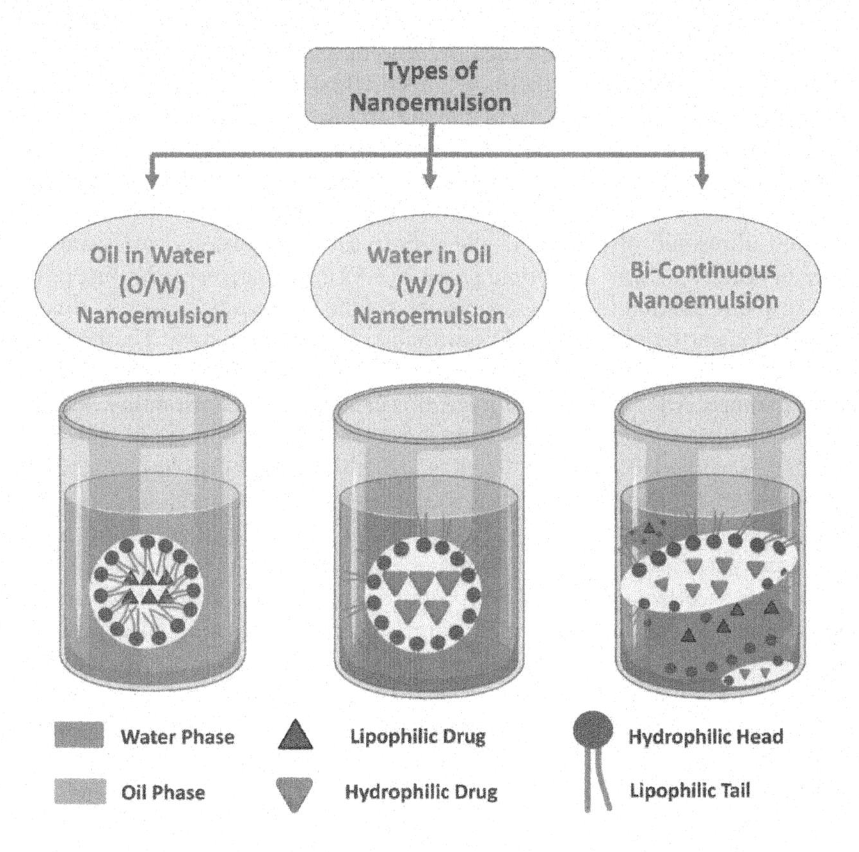

FIGURE 18.3 Different types of nanoemulsion techniques.

technique, but it can be used to incorporate hydrophilic molecules like peptides and proteins (L. Chen et al. 2020).

i. **Membrane-contactor technique:** This is a simple procedure wherein lipid is passed through membrane pores while maintaining a pressure that keeps the system above the lipid's melting temperature, resulting in the formation of minute droplets (Luis 2018). The formed droplets are swept away by the circulating aqueous phase inside the membrane module. The formation of SLN occurs when the preparation is cooled to room temperature. Aqueous phase and lipid phase temperatures, speed of circulating aqueous phase, lipid phase pressure, and membrane pore size are all process variables that affect SLN size and lipid flux (Luis 2018; Lammari et al. 2021). This method has the advantages of a simple procedure, SLN size control via appropriate process parameter selection, and scaling-up capabilities.
j. **Phase-inversion technique:** This method basically consists of two steps. First, the formulation components (lipid, surfactant, and water) are added in the predetermined ratio and mixed with magnetic stirrer followed by heating and cooling cycles, and later on, the dilution of the mixture at low temperatures (Figoli, Marino, and Galiano 2016). At a rate of 40 °C/min, three cycles of heating and cooling from room temperature to 850 °C and then back to 600 °C are done. The emulsion will invert as a result of this thermal treatment (Figoli, Marino, and Galiano 2016). In the next step, the mixture is rapidly cooled and diluted with cold water that led to the formation of nanometer-sized lipid particles. The formed particles were then subjected to slow magnetic stirring for prevention of their aggregation (Gohil and Choudhury 2019; Kausar 2017). This method has the advantage of

allowing thermolabile drugs to be used because thermal degradation is unlikely to occur after a short heating period. Furthermore, this method does not require the use of organic solvents. The size parameters of the obtained particles are influenced by variations in formulation content proportion, so they must be optimized (Kausar 2017).

k. **Film ultrasonication dispersion technique:** High-speed stirring or sonication can also be used to make lipid nanoparticles (Mahbubul et al. 2015). After the evaporation of the solvent and ultrasonic dispersion in the presence of an aqueous surfactant solution at higher temperature, a thin film of lipid phase was formed; successive cooling of the system resulted in the formation of lipid nanoparticles. One of the advantages of this method is that the equipment required is readily available in any laboratory. The most significant disadvantage of this method is the wider particle size distribution, which can reach into the micrometer range. After long-term storage, this causes physical instability, such as particle growth. Long ultrasonication cycles can lead to metal contamination, which is a major issue with this method (Kaboorani, Riedl, and Blanchet 2013; Sandhya et al. 2021).

l. **Supercritical PGSS technique:** This is a relatively new SLN production technique which utilizes supercritical carbon dioxide ($scCO_2$). It is a one-step method for encapsulating drugs in lipid particles that are free of organic solvents. For this method, carbon dioxide (99.99%) is an excellent solvent. The procedure involves dissolving $scCO_2$ in the bulk of a melted lipid with dissolved/dispersed drugs, then allowing its rapid expansion through a nozzle, causing the melt to atomize, the gas to completely evaporate, and the SLN to precipitate (Tutek et al. 2021; Soh and Lee 2019). The advantages of this method include (a) one-step procedure, (b) the absence of an organic solvent, and (c) low processing temperatures (Soh and Lee 2019). The main disadvantage of this method is frequent blockage of the nozzle.

18.10.14 Application of Nanoformulations

Researchers have studied nanomaterials extensively for use in the diagnosis, monitoring, treatment, and prevention of disease in the field of nanomedicine. The properties of nanomaterial such as high surface reactivity, large surface area-to-volume ratio, and the dominance of quantum effects improve their strength, electrical properties, optical properties, magnetic properties, and in vivo behavior. Some of the drug molecules do not get adsorbed efficiently owing to their low bioavailability, thus are not able to elicit enough pharmacological effects. Nanoparticulate drug delivery systems have been used to change the pharmacokinetic properties of the drug molecules to impact its pharmacodynamics performance used in therapeutic application (Scioli Montoto, Muraca, and Ruiz 2020). In comparison to conventional formulations, nanoformulations have several advantages such as improved solubility and bioavailability, targeted drug delivery, sustained drug release, lower doses, and reduced side effects. A variety of polymer-based and lipid-based nanoparticulate systems have been introduced to facilitate the delivery of therapeutic and imaging agents for various medical applications (Scioli Montoto, Muraca, and Ruiz 2020). The applications of nanomedicine range from the nanomaterials and biological devices to nanoelectronic biosensors and from drug delivery to futuristic applications. However, the toxicity of nanomaterials and their ultimate fate in the environment is a major concern of the present time. Distinguished with superior dissolution enhancements and high-solubility efficiencies, nanoformulations are known to expedite highly sensitive design of drug-polymer composites (Scioli Montoto, Muraca, and Ruiz 2020). Nanomaterials could be used well in the development of diagnostic devices, contrast agents, analytical tools, drug delivery vehicles, etc., as the size of these nanomaterials is much similar to that of biological molecules. The overall drug consumption as well as side effects can be considerably curtailed by the deposition of the active agent in the morbid region at suitable doses. Each year, more than $65 billion is squandered due to poor bioavailability. Smaller devices are less invasive and can potentially be implanted inside the body, and biochemical reaction times are much faster when using the "nanoscale regime" for medical technologies. These devices tend to be more sensitive and faster as compared to traditional drug delivery systems (Y. Chen et al. 2020). The efficacy of drug delivery via nanomedicine depends on a number of factors including (a) efficient drug encapsulation, (b) successful drug delivery to the targeted region of the body, and (c) successful drug release. The drug molecules could be used more efficiently through a triggered response. In this case, the drugs are implanted in the body, and they get activate only when they come into contact with a specific signal. Drug-delivery systems may also be able to prevent tissue damage through regulated drug release; reduce drug clearance rates; or lower the volume of distribution and reduce the effect on nontarget tissue (Y. Chen et al. 2020). Besides dissolution, the manufacturing cost and scaling-up expenses are also considered as an important aspect for the development of an ideal drug formulation (Khiev et al. 2021). Toward this end, the technology of "nanoising" the poorly soluble drugs in the presence of carrier nanoparticles has been identified as the most promising and exciting among the existing drug formulation technologies (Khiev et al. 2021). The role of the carrier matrix in various drug formulations is the crucial part in the design and development of the nanoformulations which is the important subject of the present investigations in arriving at the right nanoformulation for some of the NSAID drugs (Jeevanandam et al. 2017).

REFERENCES

Agrawal, Mukta, Shailendra Saraf, Madhulika Pradhan, Ravish J Patel, Gautam Singhvi, and Amit Alexander. 2021. "Design and Optimization of Curcumin Loaded Nano Lipid Carrier System Using Box-Behnken Design." *Biomedicine & Pharmacotherapy* 141: 111919.

Ainbinder, D, D Paolino, M Fresta, and E Touitou. 2010. "Drug Delivery Applications with Ethosomes." *Journal of Biomedical Nanotechnology* 6(5): 558–68.

Alharbi, Waleed S, Fahad A Almughem, Alshaimaa M Almehmady, Somayah J Jarallah, Wijdan K Alsharif, Nouf M Alzahrani, and Abdullah A Alshehri. 2021. "Phytosomes as an Emerging Nanotechnology Platform for the Topical Delivery of Bioactive Phytochemicals." *Pharmaceutics* 13(9): 1475.

Anjana, Rani, S Kumar, H Sharma, and R Khar. 2017. "Phytosome Drug Delivery of Natural Products: A Promising Technique for Enhancing Bioavailability." *International Journal of Drug Delivery Technology* 7(03): 157–65.

Beloqui, Ana, María Ángeles Solinís, Alicia Rodríguez-Gascón, António J Almeida, and Véronique Préat. 2016. "Nanostructured Lipid Carriers: Promising Drug Delivery Systems for Future Clinics." *Nanomedicine: Nanotechnology, Biology and Medicine* 12(1): 143–61.

Bhattaram, Venkatesh Atul, Ulrike Graefe, Claudia Kohlert, Markus Veit, and Hartmut Derendorf. 2002. "Pharmacokinetics and Bioavailability of Herbal Medicinal Products." *Phytomedicine* 9: 1–33.

Borse, Swapnil P, Devendra P Singh, and Manish Nivsarkar. 2019. "Understanding the Relevance of Herb–Drug Interaction Studies with Special Focus on Interplays: A Prerequisite for Integrative Medicine." *Porto Biomedical Journal* 4(2): e15.

Bose, Anamika, Debanwita Roy Burman, Bismayan Sikdar, and Prasun Patra. 2021. "Nanomicelles: Types, Properties and Applications in Drug Delivery." *IET Nanobiotechnology* 15(1): 19–27.

Bouchemal, Kawthar, S Briançon, E Perrier, and H Fessi. 2004. "Nano-Emulsion Formulation Using Spontaneous Emulsification: Solvent, Oil and Surfactant Optimisation." *International Journal of Pharmaceutics* 280(1–2): 241–51.

Çağdaş, Melis, Ali Demir Sezer, and Seyda Bucak. 2014. "Liposomes as Potential Drug Carrier Systems for Drug Delivery." *Application of Nanotechnology in Drug Delivery* 1: 1–50.

Chaudhary, Sunita A, Dasharath M Patel, Jayvadan K Patel, and Deepa H Patel. 2021. "Solvent Emulsification Evaporation and Solvent Emulsification Diffusion Techniques for Nanoparticles." In Jayvadan K. Patel and Yashwant V. Pathak (eds.), *Emerging Technologies for Nanoparticle Manufacturing*, 287–300. Springer.

Chauhan, Iti, Mohd Yasir, Madhu Verma, and Alok Pratap Singh. 2020. "Nanostructured Lipid Carriers: A Groundbreaking Approach for Transdermal Drug Delivery." *Advanced Pharmaceutical Bulletin* 10(2): 150.

Chen, Lijuan, Fen Ao, Xuemei Ge, and Wen Shen. 2020. "Food-Grade Pickering Emulsions: Preparation, Stabilization and Applications." *Molecules* 25(14): 3202.

Chen, Yan, Yao Lu, Robert J Lee, and Guangya Xiang. 2020. "Nano Encapsulated Curcumin: And Its Potential for Biomedical Applications." *International Journal of Nanomedicine* 15: 3099.

Crini, Grégorio, Sophie Fourmentin, Éva Fenyvesi, Giangiacomo Torri, Marc Fourmentin, and Nadia Morin-Crini. 2018. "Cyclodextrins, from Molecules to Applications." *Environmental Chemistry Letters* 16(4): 1361–75.

Duong, Van-An, Thi-Thao-Linh Nguyen, and Han-Joo Maeng. 2020. "Preparation of Solid Lipid Nanoparticles and Nanostructured Lipid Carriers for Drug Delivery and the Effects of Preparation Parameters of Solvent Injection Method." *Molecules* 25(20): 4781.

Ekor, M. 2013. "The Growing Use of Herbal Medicines: Issues Relating to Adverse Reactions and Challenges in Monitoring Safety." *Frontiers in Neurology* 4, 1–10.

Falzon, Charles C, and Anna Balabanova. 2017. "Phytotherapy: An Introduction to Herbal Medicine." *Primary Care: Clinics in Office Practice* 44(2): 217–27.

Fernández-García, Raquel, Aikaterini Lalatsa, Larry Statts, Francisco Bolás-Fernández, M Paloma Ballesteros, and Dolores R Serrano. 2020. "Transferosomes as Nanocarriers for Drugs across the Skin: Quality by Design from Lab to Industrial Scale." *International Journal of Pharmaceutics* 573: 118817.

Figoli, A, T Marino, and F Galiano. 2016. "Polymeric Membranes in Biorefinery." *Membrane Technologies for Biorefining*, 29–59.

Fryd, Michael M, and Thomas G Mason. 2012. "Advanced Nanoemulsions." *Annual Review of Physical Chemistry* 63: 493–518.

Gnananath, Kattamanchi, Kalakonda Sri Nataraj, and Battu Ganga Rao. 2017. "Phospholipid Complex Technique for Superior Bioavailability of Phytoconstituents." *Advanced Pharmaceutical Bulletin* 7(1): 35.

Gohil, Jaydevsinh M, and Rikarani R Choudhury. 2019. "Introduction to Nanostructured and Nano-Enhanced Polymeric Membranes: Preparation, Function, and Application for Water Purification." In Sabu Thomas, Daniel Pasquini, Shao-Yuan Leu, and Deepu A. Gopakumar (eds.), *Nanoscale Materials in Water Purification*, 25–57. Elsevier.

Guan, Juan, Wei Chen, Min Yang, Ercan Wu, Jun Qian, and Changyou Zhan. 2021. "Regulation of in Vivo Delivery of Nanomedicines by Herbal Medicines." *Advanced Drug Delivery Reviews* 174: 210–28.

Gundloori, Rathna V N, Amarnath Singam, and Naresh Killi. 2019. "Nanobased Intravenous and Transdermal Drug Delivery Systems." In Shyam S. Mohapatra, Shivendu Ranjan, Nandita Dasgupta, Raghvendra Kumar Mishra, Sabu Thomas (eds.), *Applications of Targeted Nano Drugs and Delivery Systems*, 551–94. Elsevier.

Gupta, Ankur, H Burak Eral, T Alan Hatton, and Patrick S Doyle. 2016. "Nanoemulsions: Formation, Properties and Applications." *Soft Matter* 12(11): 2826–41.

Hou, Xucheng, Tal Zaks, Robert Langer, and Yizhou Dong. 2021. "Lipid Nanoparticles for MRNA Delivery." *Nature Reviews Materials* 6(12): 1078–94.

Jacob, Shery, and Anroop B Nair. 2018. "Cyclodextrin Complexes: Perspective from Drug Delivery and Formulation." *Drug Development Research* 79(5): 201–17.

Jambhekar, Sunil S, and Philip Breen. 2016. "Cyclodextrins in Pharmaceutical Formulations I: Structure and Physicochemical Properties, Formation of Complexes, and Types of Complex." *Drug Discovery Today* 21(2): 356–62.

Jansook, Phatsawee, Noriko Ogawa, and Thorsteinn Loftsson. 2018. "Cyclodextrins: Structure, Physicochemical Properties and Pharmaceutical Applications." *International Journal of Pharmaceutics* 535(1–2): 272–84.

Jeevanandam, Jaison, Yii S Aing, Yen S Chan, Sharadwata Pan, and Michael K Danquah. 2017. "Nanoformulation and Application of Phytochemicals as Antimicrobial Agents." In Alexandru Mihai Grumezescu (ed.) *Antimicrobial Nanoarchitectonics*, 61–82. Elsevier.

Jiang, Yu Shan, Shao Bing Zhang, Shu Yan Zhang, and Yun Xuan Peng. 2021. "Comparative Study of High-intensity Ultrasound and High-pressure Homogenization on Physicochemical Properties of Peanut Protein-stabilized Emulsions and Emulsion Gels." *Journal of Food Process Engineering* 44(6): e13710.

Kaboorani, Alireza, Bernard Riedl, and Pierre Blanchet. 2013. "Ultrasonication Technique: A Method for Dispersing Nanoclay in Wood Adhesives." *Journal of Nanomaterials* 2013: 1–19.

Karunamoorthi, Kaliyaperumal, Kaliyaperumal Jegajeevanram, Jegajeevanram Vijayalakshmi, and Embialle Mengistie. 2013. "Traditional Medicinal Plants: A Source of Phytotherapeutic Modality in Resource-Constrained Health Care Settings." *Journal of Evidence-Based Complementary and Alternative Medicine* 18(1): 67–74. https://doi.org/10.1177/2156587212460241.

Kasongo, Kasongo Wa, Rainer H Müller, and Roderick B Walker. 2012. "The Use of Hot and Cold High Pressure Homogenization to Enhance the Loading Capacity and Encapsulation Efficiency of Nanostructured Lipid Carriers for the Hydrophilic Antiretroviral Drug, Didanosine for Potential Administration to Paediatric Patients." *Pharmaceutical Development and Technology* 17(3): 353–62.

Kausar, Ayesha. 2017. "Phase Inversion Technique-Based Polyamide Films and Their Applications: A Comprehensive Review." *Polymer-Plastics Technology and Engineering* 56(13): 1421–37.

Kesarwani, Kritika, and Rajiv Gupta. 2013. "Bioavailability Enhancers of Herbal Origin: An Overview." *Asian Pacific Journal of Tropical Biomedicine* 3(4): 253–66.

Khan, Ibrahim, Khalid Saeed, and Idrees Khan. 2019. "Nanoparticles: Properties, Applications and Toxicities." *Arabian Journal of Chemistry* 12(7): 908–31.

Khiev, Dawin, Zeinab A Mohamed, Riddhi Vichare, Ryan Paulson, Sofia Bhatia, Subhra Mohapatra, Glenn P Lobo, Mallika Valapala, Nagaraj Kerur, and Christopher L Passaglia. 2021. "Emerging Nano-Formulations and Nanomedicines Applications for Ocular Drug Delivery." *Nanomaterials* 11(1): 173.

Khosa, Archana, Satish Reddi, and Ranendra N Saha. 2018. "Nanostructured Lipid Carriers for Site-Specific Drug Delivery." *Biomedicine & Pharmacotherapy* 103: 598–613.

la Harpe, Kara M de, Pierre P D Kondiah, Yahya E Choonara, Thashree Marimuthu, Lisa C du Toit, and Viness Pillay. 2019. "The Hemocompatibility of Nanoparticles: A Review of Cell–Nanoparticle Interactions and Hemostasis." *Cells* 8(10): 1209.

Lammari, Narimane, Mohamad Tarhini, Karim Miladi, Ouahida Louaer, Abdeslam Hassen Meniai, Souad Sfar, Hatem Fessi, and Abdelhamid Elaïssari. 2021. "Encapsulation Methods of Active Molecules for Drug Delivery." In Eric Chappel (ed.), *Drug Delivery Devices and Therapeutic Systems*, 289–306. Elsevier.

Loftsson, Thorsteinn. 2021. "Cyclodextrins in Parenteral Formulations." *Journal of Pharmaceutical Sciences* 110(2): 654–64.

Lopalco, Antonio, and Nunzio Denora. 2018. "Nanoformulations for Drug Delivery: Safety, Toxicity, and Efficacy." *Methods in Molecular Biology* 1800: 347–65.

Lovelyn, Charles, and Anthony A Attama. 2011. "Current State of Nanoemulsions in Drug Delivery." *Journal of Biomaterials and Nanobiotechnology* 2(05): 626.

Lu, Mei, Qiujun Qiu, Xiang Luo, Xinrong Liu, Jing Sun, Cunyang Wang, Xiangyun Lin, Yihui Deng, and Yanzhi Song. 2019. "Phyto-Phospholipid Complexes (Phytosomes): A Novel Strategy to Improve the Bioavailability of Active Constituents." *Asian Journal of Pharmaceutical Sciences* 14(3): 265–74.

Luis, Patricia. 2018. "Membrane Contactors." In Patricia Luis (ed.), *Fundamental Modelling of Membrane Systems*, 153–208. Elsevier.

Mahbubul, I M, Rahman Saidur, M A Amalina, E B Elcioglu, and T Okutucu-Ozyurt. 2015. "Effective Ultrasonication Process for Better Colloidal Dispersion of Nanofluid." *Ultrasonics Sonochemistry* 26: 361–69.

Malik, M A, M Y Wani, and M A Hashim. 2012. "Microemulsion Method: A Novel Route to Synthesize Organic and Inorganic This Article Is Licensed under a Creative Commons Attribution-NonCommercial 3.0 Unported Licence. Nanomaterials: 1st Nano Update, Arabian J." *Arabian Journal of Chemistry* 5: 397–417.

Marianecci, Carlotta, Luisa Di Marzio, Federica Rinaldi, Christian Celia, Donatella Paolino, Franco Alhaique, Sara Esposito, and Maria Carafa. 2014. "Niosomes from 80s to Present: The State of the Art." *Advances in Colloid and Interface Science* 205: 187–206.

Müller, R H, R D Petersen, A Hommoss, and J Pardeike. 2007. "Nanostructured Lipid Carriers (NLC) in Cosmetic Dermal Products." *Advanced Drug Delivery Reviews* 59(6): 522–30.

Nagavarma, B V N, Hemant K S Yadav, AVLS Ayaz, L S Vasudha, and H G Shivakumar. 2012. "Different Techniques for Preparation of Polymeric Nanoparticles-a Review." *Asian Journal of Pharmaceutical and Clinical Research* 5(3): 16–23.

Naseri, Neda, Hadi Valizadeh, and Parvin Zakeri-Milani. 2015. "Solid Lipid Nanoparticles and Nanostructured Lipid Carriers: Structure, Preparation and Application." *Advanced Pharmaceutical Bulletin* 5(3): 305.

Nayak, Kritika, Manisha Vinayak Choudhari, Swati Bagul, Tejas Avinash Chavan, and Manju Misra. 2021. "Ocular Drug Delivery Systems." In Eric Chappel (ed.), *Drug Delivery Devices and Therapeutic Systems*, 515–66. Elsevier.

Nehate, Chetan, Aji Alex Moothedathu Raynold, V Haridas, and Veena Koul. 2018. "Comparative Assessment of Active Targeted Redox Sensitive Polymersomes Based on PPEGMA-SS-PLA Diblock Copolymer with Marketed Nanoformulation." *Biomacromolecules* 19(7): 2549–66.

Opatha, Shakthi Apsara Thejani, Varin Titapiwatanakun, and Romchat Chutoprapat. 2020. "Transfersomes: A Promising Nanoencapsulation Technique for Transdermal Drug Delivery." *Pharmaceutics* 12(9): 855.

Paiva-Santos, Ana Cláudia, Ana Luísa Silva, Catarina Guerra, Diana Peixoto, Miguel Pereira-Silva, Mahdi Zeinali, Filipa Mascarenhas-Melo, Ricardo Castro, and Francisco Veiga. 2021. "Ethosomes as Nanocarriers for the Development of Skin Delivery Formulations." *Pharmaceutical Research* 38(6): 947–70.

Pan, Si-Yuan, Gerhard Litscher, Si-Hua Gao, Shu-Feng Zhou, Zhi-Ling Yu, Hou-Qi Chen, Shuo-Feng Zhang, Min-Ke Tang, Jian-Ning Sun, and Kam-Ming Ko. 2014. "Historical Perspective of Traditional Indigenous Medical Practices: The Current Renaissance and Conservation of Herbal Resources." *Evidence-Based Complementary and Alternative Medicine* 2014: 525340.

Paswan, Suresh K, and T R Saini. 2017. "Purification of Drug Loaded PLGA Nanoparticles Prepared by Emulsification Solvent Evaporation Using Stirred Cell Ultrafiltration Technique." *Pharmaceutical Research* 34(12): 2779–86.

Patra, Jayanta Kumar, Gitishree Das, Leonardo Fernandes Fraceto, Estefania Vangelie Ramos Campos, Maria del Pilar Rodriguez-Torres, Laura Susana Acosta-Torres, Luis Armando Diaz-Torres, Renato Grillo, Mallappa Kumara Swamy, and Shivesh Sharma. 2018. "Nano Based Drug Delivery Systems: Recent Developments and Future Prospects." *Journal of Nanobiotechnology* 16(1): 1–33.

Rajan, Reshmy, Shoma Jose, V P Biju Mukund, and Deepa T Vasudevan. 2011. "Transferosomes-A Vesicular Transdermal Delivery System for Enhanced Drug Permeation." *Journal of Advanced Pharmaceutical Technology & Research* 2(3): 138.

Reddy, Y Dastagiri, A B Sravani, V Ravisankar, P Ravi Prakash, Y Siva Rami Reddy, and N Vijaya Bhaskar. 2015. "Transferosomes a Novel Vesicular Carrier for Transdermal Drug Delivery System." *Journal of Innovations in Pharmaceutical and Biological Science* 2(2): 193–208.

Rombolà, Laura, Damiana Scuteri, Straface Marilisa, Chizuko Watanabe, Luigi Antonio Morrone, Giacinto Bagetta, and Maria Tiziana Corasaniti. 2020. "Pharmacokinetic Interactions between Herbal Medicines and Drugs: Their Mechanisms and Clinical Relevance." *Life* 10(7): 106.

Sá Couto Salústio, P., & Cabral-Marques, H, A. 2014. "CyclodextrinsCyclodextrins (CDs)." *Polysaccharides*, 1–36. https://doi.org/10.1007/978-3-319-03751-6_22-1.

Sadeghi-Ghadi, Zaynab, Pedram Ebrahimnejad, Fereshteh Talebpour Amiri, and Ali Nokhodchi. 2021. "Improved Oral Delivery of Quercetin with Hyaluronic Acid Containing Niosomes as a Promising Formulation." *Journal of Drug Targeting* 29(2): 225–34.

Saghir, S A, and R A Ansari. 2018. *Pharmacokinetics. Reference Module in Biomedical Sciences.* Elsevier. https://doi. org/10.1016/B978-0-12-801238-3.62154-2.

Sajid, Mohammad, Swaranjit Singh Cameotra, Mohd Sajjad Ahmad Khan, and Iqbal Ahmad. 2019. "Nanoparticle-Based Delivery of Phytomedicines: Challenges and Opportunities." *New Look to Phytomedicine*, 597–623.

Sala, M, K Miladi, G Agusti, A Elaissari, and H Fessi. 2017. "Preparation of Liposomes: A Comparative Study between the Double Solvent Displacement and the Conventional Ethanol Injection—From Laboratory Scale to Large Scale." *Colloids and Surfaces A: Physicochemical and Engineering Aspects* 524: 71–78.

Salatin, Sara, Jaleh Barar, Mohammad Barzegar-Jalali, Khosro Adibkia, Farhad Kiafar, and Mitra Jelvehgari. 2017. "Development of a Nanoprecipitation Method for the Entrapment of a Very Water Soluble Drug into Eudragit RL Nanoparticles." *Research in Pharmaceutical Sciences* 12(1): 1.

Sandhiya, V, and U Ubaidulla. 2020. "A Review on Herbal Drug Loaded into Pharmaceutical Carrier Techniques and Its Evaluation Process." *Future Journal of Pharmaceutical Sciences* 6(1): 1–16.

Sandhya, Madderla, D Ramasamy, K Sudhakar, K Kadirgama, and W S W Harun. 2021. "Ultrasonication an Intensifying Tool for Preparation of Stable Nanofluids and Study the Time Influence on Distinct Properties of Graphene Nanofluids–A Systematic Overview." *Ultrasonics Sonochemistry* 73: 105479.

Sapkota, Rachana, and Alekha K Dash. 2021. "Liposomes and Transferosomes: A Breakthrough in Topical and Transdermal Delivery." *Therapeutic Delivery* 12(2): 145–58.

Savjani, Ketan T, Anuradha K Gajjar, and Jignasa K Savjani. 2012. "Drug Solubility: Importance and Enhancement Techniques." *International Scholarly Research Notices* 2012: 195727.

Schubert, M A, and C C Müller-Goymann. 2003. "Solvent Injection as a New Approach for Manufacturing Lipid Nanoparticles–Evaluation of the Method and Process Parameters." *European Journal of Pharmaceutics and Biopharmaceutics* 55(1): 125–31.

Scioli Montoto, Sebastián, Giuliana Muraca, and María Esperanza Ruiz. 2020. "Solid Lipid Nanoparticles for Drug Delivery: Pharmacological and Biopharmaceutical Aspects." *Frontiers in Molecular Biosciences* 7: 587997.

Senapati, Sudipta, Arun Kumar Mahanta, Sunil Kumar, and Pralay Maiti. 2018. "Controlled Drug Delivery Vehicles for Cancer Treatment and Their Performance." *Signal Transduction and Targeted Therapy* 3(1): 1–19.

Senior, Judith H. 1990. "Liposomes in Vivo: Prospects for Liposome-Based Pharmaceuticals in the 1990s." *Biotechnology and Genetic Engineering Reviews* 8(1): 279–318.

Shah, Ayaz Ali, Tahir Hussain Seehar, Kamaldeep Sharma, and Saqib Sohail Toor. 2022. "Biomass Pretreatment Technologies." In Sunil K. Maity, Kalyan Gayen and Tridib Kumar Bhowmick (eds.), *Hydrocarbon Biorefinery*, 203–28. Elsevier.

Shah, Rohan M, François Malherbe, Daniel Eldridge, Enzo A Palombo, and Ian H Harding. 2014. "Physicochemical Characterization of Solid Lipid Nanoparticles (SLNs) Prepared by a Novel Microemulsion Technique." *Journal of Colloid and Interface Science* 428: 286–94.

Sharma, Shalini, and Ram Kumar Roy. 2010. "Phytosomes: An Emerging Technology." *International Journal of Pharmaceutical Research and Development* 2(5): 1–7.
Sheth, Tanvi, Serena Seshadri, Tamás Prileszky, and Matthew E Helgeson. 2020. "Multiple Nanoemulsions." *Nature Reviews Materials* 5(3): 214–28.
Siddiqui, Lubna, Harshita Mishra, Sushama Talegaonkar, and Mahendra Rai. 2020. "Nanoformulations: Opportunities and Challenges." In S. Talegaonkar and M. Rai (eds.), *Nanoformulations in Human Health*. Springer. https://doi.org/10.1007/978-3-030-41858-8_1.
Soh, Soon Hong, and Lai Yeng Lee. 2019. "Microencapsulation and Nanoencapsulation Using Supercritical Fluid (SCF) Techniques." *Pharmaceutics* 11(1): 21.
Sohrabi, Beheshteh, Fereshteh Mansouri, and Shokooh Karimi. 2022. "The Natural Non-Ionic Magnetic Surfactants: Nanomicellar and Interfacial Properties." *Journal of Nanostructure in Chemistry* 12(5): 889–902.
Solanki, Dharmendra, Lalit Kushwah, Mohit Motiwale, and Vicky Chouhan. 2016. "Transferosomes-a Review." *World Journal of Pharmacy and Pharmaceutical Sciences* 5(10): 435–49.
Souto, E B, and R H Müller. 2006. "Investigation of the Factors Influencing the Incorporation of Clotrimazole in SLN and NLC Prepared by Hot High-Pressure Homogenization." *Journal of Microencapsulation* 23(4): 377–88.
Staff, Roland H, Katharina Landfester, and Daniel Crespy. 2013. "Recent Advances in the Emulsion Solvent Evaporation Technique for the Preparation of Nanoparticles and Nanocapsules." *Hierarchical Macromolecular Structures: 60 Years after the Staudinger Nobel Prize II, Advances in Polymer Science* 262: 329–44.
Stefanov, Stefan R, and Velichka Y Andonova. 2021. "Lipid Nanoparticulate Drug Delivery Systems: Recent Advances in the Treatment of Skin Disorders." *Pharmaceuticals* 14(11): 1083.
Su, S. 2020. "M. Kang P." *Recent Advances in Nanocarrier-Assisted Therapeutics Delivery Systems. Pharmaceutics* 12: 837.
Subhan, Md Abdus, Kristi Priya Choudhury, and Newton Neogi. 2021. "Advances with Molecular Nanomaterials in Industrial Manufacturing Applications." *Nanomanufacturing* 1(2): 75–97.
Tapeinos, Christos, Matteo Battaglini, and Gianni Ciofani. 2017. "Advances in the Design of Solid Lipid Nanoparticles and Nanostructured Lipid Carriers for Targeting Brain Diseases." *Journal of Controlled Release* 264: 306–32.
Tartaro, Giuseppe, Helena Mateos, Davide Schirone, Ruggero Angelico, and Gerardo Palazzo. 2020. "Microemulsion Microstructure (s): A Tutorial Review." *Nanomaterials* 10(9): 1657.
Tenchov, Rumiana, Robert Bird, Allison E Curtze, and Qiongqiong Zhou. 2021. "Lipid Nanoparticles— from Liposomes to MRNA Vaccine Delivery, a Landscape of Research Diversity and Advancement." *ACS Nano* 15(11): 16982–15.
Turfus, S, R Delgoda, D Picking, and B J Gurley. 2017. *Pharmacokinetics Pharmacognosy*. Academic.
Tutek, Karol, Anna Masek, Anna Kosmalska, and Stefan Cichosz. 2021. "Application of Fluids in Supercritical Conditions in the Polymer Industry." *Polymers* 13(5): 729.
Vaishya, R D, V Khurana, S Patel, and A K Mitra. 2014. "Controlled Ocular Drug Delivery with Nanomicelles." *Wiley Interdisciplinary Reviews: Nanomedicine and Nanobiotechnology* 6: 422–37.
Vinchhi, Preksha, Jayvadan K Patel, and Mayur M Patel. 2021. "High-Pressure Homogenization Techniques for Nanoparticles." In Jayvadan K. Patel and Yashwant V. Pathak (eds.), *Emerging Technologies for Nanoparticle Manufacturing*, 263–85. Springer.
Vinothini, Kandasamy, and Mariappan Rajan. 2019. "Mechanism for the Nano-Based Drug Delivery System." In Shyam S. Mohapatra, Shivendu Ranjan, Nandita Dasgupta, Raghvendra Kumar Mishra, and Sabu Thomas (eds.), *Characterization and Biology of Nanomaterials for Drug Delivery*, 219–63. Elsevier.
Waller, Derek G, Anthony Sampson, and Andrew Hitchings. 2021. *Medical Pharmacology and Therapeutics E-Book*. Elsevier Health Sciences.
Weissig, Volkmar. 2017. "Liposomes Came First: The Early History of Liposomology." *Liposomes*, 1–15.
Wink, Michael. 2015. "Modes of Action of Herbal Medicines and Plant Secondary Metabolites." *Medicines* 2(3): 251–86.
Yang, Bingyan, Yixin Dong, Fei Wang, and Yu Zhang. 2020. "Nanoformulations to Enhance the Bioavailability and Physiological Functions of Polyphenols." *Molecules* 25(20): 4613.
Yang, Rong, Tuo Wei, Hannah Goldberg, Weiping Wang, Kathleen Cullion, and Daniel S Kohane. 2017. "Getting Drugs across Biological Barriers." *Advanced Materials* 29(37): 1606596.
Yingchoncharoen, Phatsapong, Danuta S Kalinowski, and Des R Richardson. 2016. "Lipid-Based Drug Delivery Systems in Cancer Therapy: What Is Available and What Is yet to Come." *Pharmacological Reviews* 68(3): 701–87.

Yuan, Hong, Lei-Lei Wang, Yong-Zhong Du, Jian You, Fu-Qiang Hu, and Su Zeng. 2007. "Preparation and Characteristics of Nanostructured Lipid Carriers for Control-Releasing Progesterone by Melt-Emulsification." *Colloids and Surfaces B: Biointerfaces* 60(2): 174–79.

Zhang, Kunming, Guangli Yan, Aihua Zhang, Hui Sun, and Xijun Wang. 2017. "Recent Advances in Pharmacokinetics Approach for Herbal Medicine." *RSC Advances* 7(46): 28876–88.

Zhang, Yong-Tai, Zhi Wang, Li-Na Shen, Yan-Yan Li, Ze-Hui He, Qing Xia, and Nian-Ping Feng. 2018. "A Novel Microemulsion-Based Isotonic Perfusate Modulated by Ringer's Solution for Improved Microdialysis Recovery of Liposoluble Substances." *Journal of Nanobiotechnology* 16(1): 1–11.

Zheng, Xue, Jizhen Xie, Xing Zhang, Weiting Sun, Heyang Zhao, Yantuan Li, and Cheng Wang. 2021. "An Overview of Polymeric Nanomicelles in Clinical Trials and on the Market." *Chinese Chemical Letters* 32(1): 243–57.

Zong, Lanlan, Haiyan Wang, Xianqiao Hou, Like Fu, Peirong Wang, Hongliang Xu, Wenjie Yu, Yuxin Dai, Yonghui Qiao, and Xuefeng Wang. 2021. "A Novel GSH-Triggered Polymeric Nanomicelles for Reversing MDR and Enhancing Antitumor Efficiency of Hydroxycamptothecin." *International Journal of Pharmaceutics* 600: 120528.

19 Novel Phytochemicals Targeting the Signaling Pathways of Anticancer Stem Cell

A Novel Approach Against Cancer

Urvashi Bhardwaj, Shouvik Mukherjee, and Shaheen Ali
Jamia Hamdard

Arif Jamal Siddiqui
University of Ha'il

Danish Iqbal, Sami G. Almalki, Suliman A. Alsagaby, and Sadaf Jahan
Majmaah University

Uzair Ahmad Ansari
CSIR-Indian Institute of Toxicology Research (CSIR-IITR)
CSIR-Human Resource Development Centre (CSIR-HRDC) Campus,
Academy of Scientific and Innovative Research (ACSIR)

CONTENTS

DOI: 10.1201/b22842-19

19.1 INTRODUCTION

Cancer is a disease that results from excessive cell proliferation in a specific region of the body, and it can be benign or malignant. Despite progress in disease biology and drug development, it is usually fatal; thus, global efforts to design effective medications and treatment techniques for illness management are underway. The most prevalent reasons for cancer treatment failure are metastasis, recurrence, heterogeneity, resistance to chemotherapy and radiotherapy, and a lack of immunological surveillance leading to the development of more aggressive tumors which spread rapidly. This therapy resistance is owned by the cancer stem cells (CSCs); they retain self-renewal properties as the normal stem cells do, hence, described as immortal tumor cells. Because CSCs have unique and yet-to-be-explored properties, it necessitates the creation of novel, diversified, and multitargeted cancer therapeutic techniques. With the growing side effects of radiotherapy and chemotherapy on normal stem cells, the natural phytochemicals obtained from natural sources comprising therapeutic capabilities become evident. They have been proven to have the ability to target diverse populations of cancer cells and CSCs in cancer. Furthermore, they can target cancer critical signaling pathways, while leaving normal cells unharmed or with low toxicity. We have discussed some major phytochemicals targeting signaling pathways, and they could be good modulators of CSCs. The cancer stem cells are a small subpopulation of cells found in tumors; these cells have the capability of self-renewal, differentiation, and tumorigenicity (Yu et al. 2012). The history of the CSCs can be dated back for around 150 years ago. Julius Cohnheim in 1877 postulated a theory that tells that the existence in the body of "embryonic rests" which remained underutilized in the course of development is said to be the origin of tumor formation (Cooper 2009). The concept was not novel, as John Muller in 1838 had already classified tumors as aberrant continuations of embryonic cell growth based on physical similarities. Cohnheim went beyond morphological similarity by understanding the origin of tumors, which depends on the existence of embryonic cells that are still present or persistent in the body. Rudolf Virchow at the time emphasized a connection between embryonic and tumor development, stating that both involve cell division and multiplication. He stated that because of their embryonic origins, suppose these cells had the required blood supply, then they would multiply uncontrolled. They then develop malignancies, which are thought to be the result of "errors" during the process of development (Cooper 2009; Maehle 2011). Embryonic cells even can also be the source for a normal cell growth as seen throughout puberty or pregnancy, according to Cohnheim (Cooper 2009). Max Askanzy was able to demonstrate the validity of this theory using rat tumors that resembled teratomas (Maehle 2011).

In the late 1800s, Theodor Boveri came up with his theory named chromosomal theory of cancer; he believed that the juvenile characteristics of such cancer cells are most often side consequences of the aberrant chromosomal distribution, mainly the gain or loss of chromosomes during embryonic asymmetrical segregation. In 1953, Stevens came up with a theory of "pluripotent embryonic stem cell" (Cooper 2009). In 1964, G. Barry Pierce, it is interesting that starting in the late 1960s, Pierce used these findings to build a cancer hypothesis in which tumors are produced by nongenetic "induction" or disturbance of the "developmental field," rather than by mutation caused in a single cell that then multiplies anarchically (Pierce 1967). It was in 1994, when John E Dick and his colleagues from their laboratory showed some interesting features of leukemia, as they showed that just a few uncommon mouse acute myeloid lymphoma (AML) cells have the capability to cause leukemia in other mice. Furthermore, they demonstrated that these cells had a strong ability for self-renewal and that this is a key trait of stem cells, through a series of successive transplants. CSCs produce cells that are comparable to the ones which were transplanted (through self-renewal) as well as cells which have lost their possible tumorigenic potential due to differentiation (Bonnet and Dick 1997; Lapidot et al. 1994). CSCs have now been discovered in a different type of malignancies including solid tumors; this was proved in 2000 in breast cancer when CSC was able to induce cancer, and cells having alternate phenotypes failed to form any tumor (Hermann et al. 2010; Al-Hajj et al. 2003). Thus, the idea of the CSC was refined; further, changes came to the concept in the 2010s, and now it means that these CSCs can self-renew, also differentiate into defined progenies, as well as initiate and also retain tumor growth (in vivo) (Zhang et al. 2015).

19.2 ORIGIN, KEY CHARACTERISTICS OF CSCS, AND THEIR SIMILARITIES AND DIFFERENCES WITH NORMAL STEM CELLS

19.2.1 Origin of CSCs

Cancer stem cells may lead to tumors, but tumors may generate from multiple such stem cell clones. Treatment-resistant cells may or may not have any distinguishing (stem cell) traits other than the fact that they are the exception to the norm or that they are the ones who are protected by their microenvironment (Ghaffari 2011). The first theory states that if a typical SC transforms into a CSC, it means it already contains a full stem-cell package, including differentiation into multiple progenitor cells, immunological surveillance protection, and emigration to far-flung locations. When a stem cell "goes bad," cancer begins, and the key stem-cell features of a cancer cell are usually inherited (Nguyen et al. 2012). The second theory states that CSCs are reinforced with a partial stem-cell package, when they emerge from any cell that takes on stem-cell properties. And then it progresses to malignant behaviors (Pattabiraman and Weinberg 2014). The third theory states that when cancer emcompasses of dedifferentiated normal SCs, it is provided with a distinct package of good and bad SCs. As a result of the bad cell's interaction with a good SC, cancer develops. CSCs are attracted to normal stem cells in this phenomenon and play a crucial role in the establishment and progress of the malignancy. Thus, one of the most often asked questions concerning the stem-cell origins of cancer idea is whether cancer is caused by stem cells or by cells with stem-cell-like features (Takahashi et al. 2007). According to one theory, only a small percentage of cancerous cells have stem cell characteristics and they multiply themselves to prolong the disease, similar to how good stem cells regenerate and nourish our organs and tissues (Yoo and Hatfield 2008). According to this view, cancer cells that do not have stem cells cannot withstand drugs for lengthy periods. However, a cell must endure a considerable number of essential changes in the DNA sequences that regulate the cell to become malignant (Mayer, Klotz, and Venkateswaran 2015). Although there are different theories about stem cells, they do have distinct characteristics features that make them different from normal stem cells; the next section will deal with key characteristics of CSCs.

19.2.2 Key Characteristics of a CSC

CSC does have the ability for self-renewal; they can produce quiescent CSC and even a dedicated progenitor that may arise in the mass of cells, causing tumors. They use pathways like notch, RAS/MAPK, Hedgehog, PI3K/AKT, and BMI-1 (epigenetic level) to perform self-renewability. This leads to the ability to arrange a hierarchy of cellular derivatives, including progenitors and differentiating cells. Since cancer cells have an extended life span, they show extended telomerase activity and telomeres. To grow, they secrete growth hormones and cytokines, which not only lead to angiogenesis by secreting angiopoietic factors but also to diverse karyotypes and phenotypically diverse progeny (Bapat et al. 2010). However, some characteristics of CSC are similar to that of normal stem cells, which we shall discuss in the next section.

19.2.3 Similarities of Cancer and Normal Stem Cells

CSCs do share some resemblance to normal stem cells like they have the ability to self-renewal, stay at the quiescent stage, and develop further progenies. They express alike surface receptors such as CXCR4, CD133, a6 integrin, c-kit, c-met, and LIF-R. They do self-renewal through similar pathways. Perform angiogenesis and extended telomerase activity, as well as secrete growth hormones and cytokines. However, the fate of stem cells is that they die and are not carcinogenic (Rossi et al. 2020); this leads us to discuss the differences between them in the next section.

19.2.4 Differences between Normal Stem Cells and CSCs

CSCs unlike normal stem cells can grow uncontrollably (Rossi et al. 2020; Rahman et al. 2016). There are also other differences which we shall discuss in Table 19.1 represented below.

19.3 SIGNALING PATHWAYS OF CSCS/SELF-RENEWAL PATHWAYS

The signaling pathways which are known to regulate the function of a cell such as survival, proliferation, differentiation, and cell death are sometimes unusually functioning or repressing in CSCs. Signaling pathways in CSCs may be constitutively active or inadequately modulated as a consequence of epigenetic and/or genetic alterations, resulting in uncontrollable proliferation (Yang et al. 2020; Beachy 2004). In another aspect, cancer-derived extracellular vesicles also transform normal cells to tumorigenic cells via modulating the pathways (Jahan et al. 2022). We have outlined some of the most essential self-renewal pathways such as Notch, Wnt, Hedgehog, Hippo, as well as other signaling pathways in this section, as well as an account of the phytochemicals that can influence some of these processes.

TABLE 19.1
Representing Differences between Cancer and Normal Stem Cell

Sl. No.	Cancer Stem Cell	Normal Stem Cell	References
1	Ability to self-renewal but in an extensive and indefinite way	Ability to self-renewal but having extensive and not in an indefinite way	(Topcul, Topcul, and Cetin 2013)
2	They use their ability to endure tumor	They use their ability for tissue regeneration	(Rossi et al. 2020)
3	They can evade the immune response	Normal stem cells like MSCs (mesenchymal stem cells) do immunomodulation	(Rossi et al. 2020)
4	They have the ability for immune suppression	Somatic stem cells (SSCs) can regulate the immune response during inflammation	(Rossi et al. 2020)
5	Tumorigenic capacity	Organogenic capacity	(Topcul, Topcul, and Cetin 2013)
6	Abnormal karyotype or individual collection of chromosomes	Normal karyotype	(Topcul, Topcul, and Cetin 2013)
7	They are rare and not frequent or rare in tumors.	They are normally rare in adult tissues	(Topcul, Topcul, and Cetin 2013)
8	They are less mitotically active	They are quiescent most of the time	(Topcul, Topcul, and Cetin 2013)
9	They can be detected through various markers	They are similar types of cells on the surface of the tissue	(Topcul, Topcul, and Cetin 2013)
10	They may lead to heterozygous populations since they lead to phenotypically diverse populations	They have a homozygous population of cells, as the progenies of normal stem cells has the same genotypic and phenotypic characteristics as that of their parental cells	(Rahman et al. 2016)
11	They have a high renewal capacity but the tight junction is lost	They have renewal capacity, but however tight junctions are not lost	(Rahman et al. 2016)
12	For self-renewal and tumor induction, they use the Wnt/β-catenin, Jak/Stat, TGF-β, Notch, and Sonic Hedgehog pathways.	Embryonic stem cells (ESCs) retain the property to differentiate for a long time. However, SSCs (somatic stem cells) differentiate on number of factors like environmental and internal signals, such as genetic and epigenetic processes, and external signals like cytokines and growth factors. Thus, the self-renewal process is highly regulated in normal stem cells.	(O'Rourke 2001; Medema 2013; Friedmann-Morvinski and Verma 2014; Takebe et al. 2015; Rossi et al. 2020)

19.3.1 Notch

Notch signaling is a molecular cell-signaling route that is extremely conserved and affects multiple processes in CSCs such as cell morphology, including pluripotent progenitor cell activation, metastasis, cell renewal, apoptosis inhibition, cell proliferation, and cell survival, cell differentiation, and cell fate specification (Tsakonas, Rand, and Lake 1999; Venkatesh et al. 2018). Since the Notch signaling pathway is so crucial in the link between angiogenesis and CSC self-renewal, it is becoming an increasingly popular target for CSC removal (Zhu et al. 2020). Notch has been identified as an oncogene as well as a suppressor gene, as there were different Notch ligands and receptors that are expressed for different cancers, including glioblastoma, leukemia, breast, gastrointestinal, pancreatic, colon, and lung cancers (Androutsellis-Theotokis et al. 2006; Bolós, Grego-Bessa, and

La Pompa de 2007; Parr, Watkins, and Jiang 2004; van Es and Clevers 2005; Sikandar et al. 2010). It uses membrane-bound ligands and receptors. Notch is a big membrane protein that acts as a receptor, as it is produced in signal-receiving cells. Generally, there are four kinds of Notch receptors (known as Notch1–4) in mammals, each with a complex dimeric structure made up of a transmembrane monomer connected to an extracellular monomer via noncovalent interactions (Bolós et al. 2009), and Jagged 1, Jagged 2, delta-like ligand (DLL-1, DLL-3, and DLL-4) are Notch ligands. The DSL (Delta/Serrate/Lag2) amino-terminal domain, which is followed by a number of epidermal growth factor-like domains which are varied, is a common characteristic of these ligands (Bray 2006). The binding of Notch ligand to notch receptors activates the transcription of targeted genes, the metalloprotease. S2 Cleavage occurs when Tumor Necrosis Factor-converting enzyme (TACE)/ADAM cleaves the receptor near the transmembrane domain on the face. The signal-sending cell eventually endocytosis the extracellular portion released, because the ligand is attached to it. The enzyme γ-secretase then performs a second cleavage in the transmembrane domain, releasing a Notch intracellular domain (NICD) into nucleus. After translocation into the nucleus, NICD binds to transcription factors and coactivators followed by transcription of the HES and HEY notch genes and HERP family genes, the proto-oncogenes myelocytomatosis oncogene (MYC) (Schweisguth 2004; Venkatesh et al. 2018; Gómez-del Arco et al. 2010; Andersson, Sandberg, and Lendahl 2011).

This signaling pathway is essential for the differentiation of stem cells and their maintenance because it is engaged in cellular proliferation and differentiation. Notch signaling plays a significant role in the self-renewal, as well as survival of the CSCs. As notch is triggered by γ-secretase, it may be targeted and blocked with Gamma Secretase Inhibitor (GSI), which effectively represses CSC and prevents cancer recurrence by slowing angiogenesis and inducing tumor cell apoptosis (Venkatesh et al. 2018). Along with GSI, ligands expression, ubiquitination, and endocytosis, Notch receptors expression, Notch activation, heterodimer dissociation, assembly of the coactivator complex along with Notch and CBF1, suppressor of hairless, Lag-1 (CSL), Notch post-translational changes, heterodimerization of Notch transcriptional complexes, as well as expression of Notch targets could be used as a targeted therapeutic tool in the removal of CSCs (Espinoza and Miele 2013).

19.3.2 Wnt/β-catenin

The Wnt pathway, which controls stem cell pluripotency via transcription factors and adhesion molecules, is an evolutionarily conserved mechanism that controls cell self-renewal, differentiation, cell proliferation, movement, and polarity, and its aberrant expression or activation can lead to tumors (Radtke 2005; Reya and Clevers 2005; Liu, Dontu, and Wicha 2005; Saito-Diaz et al. 2013; Katoh 2017). The three major pathways in the Wnt signaling cascade include the noncanonical Wnt pathway (β-catenin independent), planar cell polarity pathway, canonical Wnt pathway, and noncanonical Wnt–calcium pathway. The Wnt pathway includes Fizzled receptors (FZD) and Wnt ligands. In the canonical pathway and noncanonical pathway, in the absence of Wnt ligands, glycogen synthase kinase 3 (GSK3) phosphorylates β-catenin, resulting in β-catenin breakdown via β-transducin repeat-containing protein (also known as β-TrCP) ubiquination and inhibiting β-catenin translocation from the cytoplasm to the nucleus. Wnt ligand attaches to receptors such as Frizzled (FZD) and lipoprotein receptor-related protein (LRP)-5/-6 receptors, which were phosphorylated by GSK3 and CK1 allowing AXIN2 to translocate to the membrane, and attaches to the cytoplasmic tail of LRP5/6. Following the recruitment of the Dishevelled (DVL) protein to FZD, the degradation complex is inactivated, resulting in catenin buildup in the cytosol. β-catenin can then translocate to the nucleus, where it plays a role to activate the T-cell factor/lymphoid enhancer factor (also known as TCF–LEF) transcription complex, causing the production of a variety of genes with different functions via interaction with coactivators (Niehrs 2012; Holland et al. 2013; Kim et al. 2017; Katoh 2017; Yang et al. 2020). Wnt signaling abnormal activity is meant to increase CSC growth and malignant transformation leading to the deterioration of tumor. As a result, CSC inhibition can be mediated through this pathway. When β-catenin is mutated, it can play a role in cancer-related WNT signaling abnormalities. Targeting the signaling system has resulted in the development of a variety of medications that have the ability to obstruct or interfere with crucial aspects of carcinogenesis, invasiveness of tumors, as well as metastasis (Tran and Zheng 2017; Zhan, Rindtorff, and Boutros 2017).

19.3.3 Hedgehog

The Hh-signaling pathway is critical for stem cells, progenitor cells, and self-renewal capability, as well as influencing proliferation, motility, adhesion, tissue polarity, and cell fate (Habib and O'Shaughnessy 2016; Yang et al. 2010). The Hh signaling network includes extracellular ligands such as Sonic Hedgehog (SHh), Indian Hedgehog (IHh), and Desert Hedgehog (DHh), the transmembrane protein receptors PTCH (Patched 1 (PTCH1) and PTCH2), the transmembrane protein smoothened (SMO), intermediate transduction, and downstream molecules glioma-associated protein 1/2 GLI (Merchant and Matsui 2010). Hh ligands bind to PTCH receptors on target cell membranes in the presence of ligand signals, triggering a series of intracellular signal transduction cascades. When there exists absence of ligand signal, the PTCH receptor binds to SMO and suppresses its function, preventing signaling (Lee, Sun, and Veltmaat 2013; Dandawate et al. 2013; Zhou and Kalderon 2011; Briscoe and Thérond 2013; Robbins, Fei, and Riobo 2012). Furthermore, Hh signaling has been linked to cancer growth and maintenance, as well as chemotherapy resistance, leading to a more aggressive phenotype (Chen et al. 2011).

19.3.4 Hippo

The early tumor is influenced by the Hippo pathway, which involves in the regulation for the proliferation and maintenance of stem cells through participating in cellular reprogramming and tissue size regulation (Mo, Park, and Guan 2014). Two types of kinase complexes make up the Hippo pathway, one contains MST1 as well as MST2 (MST1/2), while the other contains LATS1 and LATS2 (LATS1/2), as well as the adaptor proteins such as SAV1 and MOB1. As the hippo signaling pathway is activated, LATS1/2 is being phosphorylated and activated by MST1/2. YAP/TAZ is then inactivated by LATS1/2; after this, it is translocated to the cytoplasm, blocking TEAD (TEA domain family member)-mediated gene expression and therefore CSC proliferation (Ma et al. 2019; Yu, Zhao, and Guan 2015). In terms of regulation and function, the Hippo and Wnt pathways are very similar (Fu, Plouffe, and Guan 2017). Increased YAP1 and/or TAZ activity is caused by alterations in the Hippo pathway, which leads to the proliferation of CSC populations, which contributes to the development of solid tumors (Lamar, Motilal Nehru, and Weinberg 2018).

19.3.5 Other Signaling Pathways

In addition to the key signaling pathways described above, JAK-STAT, PI3-K/Akt, MAPK/ERK, NF-κB, PRAR and TGF/SMAD have been linked or implicated in the survival or maintenance of CSCs.

19.3.5.1 PI3K/Akt

Phosphatidylinositol3-kinase (PI3K) is a phosphatidylinositol kinase that is found inside cells. The regulatory and the catalytic subunits p85 and p110 have serine/threonine kinase as well as phosphatidylinositol kinase activity, respectively. Akt is actually a serine/threonine kinase that is expressed as Akt1,2,3. Furthermore, the mTOR (mammalian target of rapamycin) complex, a conserved kind of serine/threonine kinase, is one of AKT's primary downstream target genes which will further form the mTOR complexes, namely, mTOR 1 and 2 (LoPiccolo et al. 2008; Carracedo and Pandolfi 2008; Dey, De, and Leyland-Jones 2017; Wang, Chen, and Hay 2017). PI3K influences or affects both ovarian cancer as well as epithelial to mesenchymal transition (Deng et al. 2019). Therefore, this signaling pathway is important or essential for survival as well as proliferation of cells.

19.3.5.2 MAPK/ERK

The RAS-Mitogen-Activated Protein Kinase (MAPK) pathway, which is also known as an extracellular signal-regulated kinase (ERK), is critical for increasing cell adhesion, proliferation, migration, and survival by transducing cytokine and growth factor signals to Receptor Tyrosine Kinases (RTK) (Katz, Amit, and Yarden 2007; Ding et al. 2009). MAPKK-kinases phosphorylate and activate Ser/Thr protein kinases, which are further activated by interaction with small GTPase family proteins that connect MAPK components to cell surface receptors. The SOS-Ras-Raf-MAPK signaling cascade is deregulated in a wide range of human malignancies, which makes sense given its essential role in cell proliferation. Further, transcription factors such as Jun and Fos are activated when ERK1/2 translocate to the nucleus. MAPK activation has been observed to be suppressed by curcumin, resveratrol, silibinin, and indole-3 carbinol in both inflammation and cancer (Parekh et al. 2011; Sarkar and Li 2009).

19.3.5.3 JAK-STAT

The signal transducer and activator of transcription (STAT)/Janus kinase (JAK) pathway plays a crucial role in influencing cell fates such as death, differentiation, proliferation, and immune response (Kiger et al. 2001). JAK activation occurs when granulocyte/macrophage colony stimulating factor, interleukin-2–7, interferon, or growth hormone like PDGF, EGF bind to the receptor. In the nucleus, first, the activated STAT forms a dimer and then binds to target genes in order to regulate transcription (Quintás-Cardama and Verstovsek 2013; Liu et al. 2011). Impaired STAT signaling may contribute to tumor development indirectly by weakening tumor immune surveillance (Wang, Fahrmann, et al. 2018).

19.3.5.4 NF-κB

Nuclear Factor-κB or NF-κB performs various functions, such as mediating immune response as well as cell development, proliferation and survival, invasion as well as metastasis. It is a fast-inducible transcription factor made up of five distinct proteins, namely, RelB, p65, NF-N1, c-Rel, and NF-B2. Toll-like receptors or TLR, TNF receptors (Tumor Necrosis Factor Receptors), IL-1 receptors (Interleukin-1 receptor), as well as antigen receptors are all activated after contacting with ligands like IL-1, TNF, or lipopolysaccharides (Oeckinghaus and Ghosh 2009; Perkins and Gilmore 2006). The activation and phosphorylation of IκB kinase or IκK proteins occur when these receptors are stimulated, resulting in the phosphorylation of IκB proteins (Vazquez-Santillan et al. 2015; Hayden and Ghosh 2008). The possible expressions of important angiogenesis factors as well as adhesion molecules including IL-8, VEGF, and growth-regulated oncogene (GRO) are also increased when NF-B is activated (Gonzalez-Torres et al. 2017).

19.3.5.5 TGF-β/SMAD

Transforming growth factor (TGF) is involved in a variety of processes, such as cell differentiation, proliferation, apoptosis, homeostasis, and also embryo development. The type II receptor is recruited and phosphorylated by the TGF-ligand, which binds to the receptor. The SMAD protein (R-SMAD) is then phosphorylated by the type-I receptor, and R-SMAD and co-SMAD form complexes, which are then translocated to the nucleus, where it activates multiple target genes' transcription. TGF/SMAD signaling is activated in human malignancies, according to numerous studies. Dkk-3, a secreted protein, prevents prostate cancer migration and invasion by inhibiting

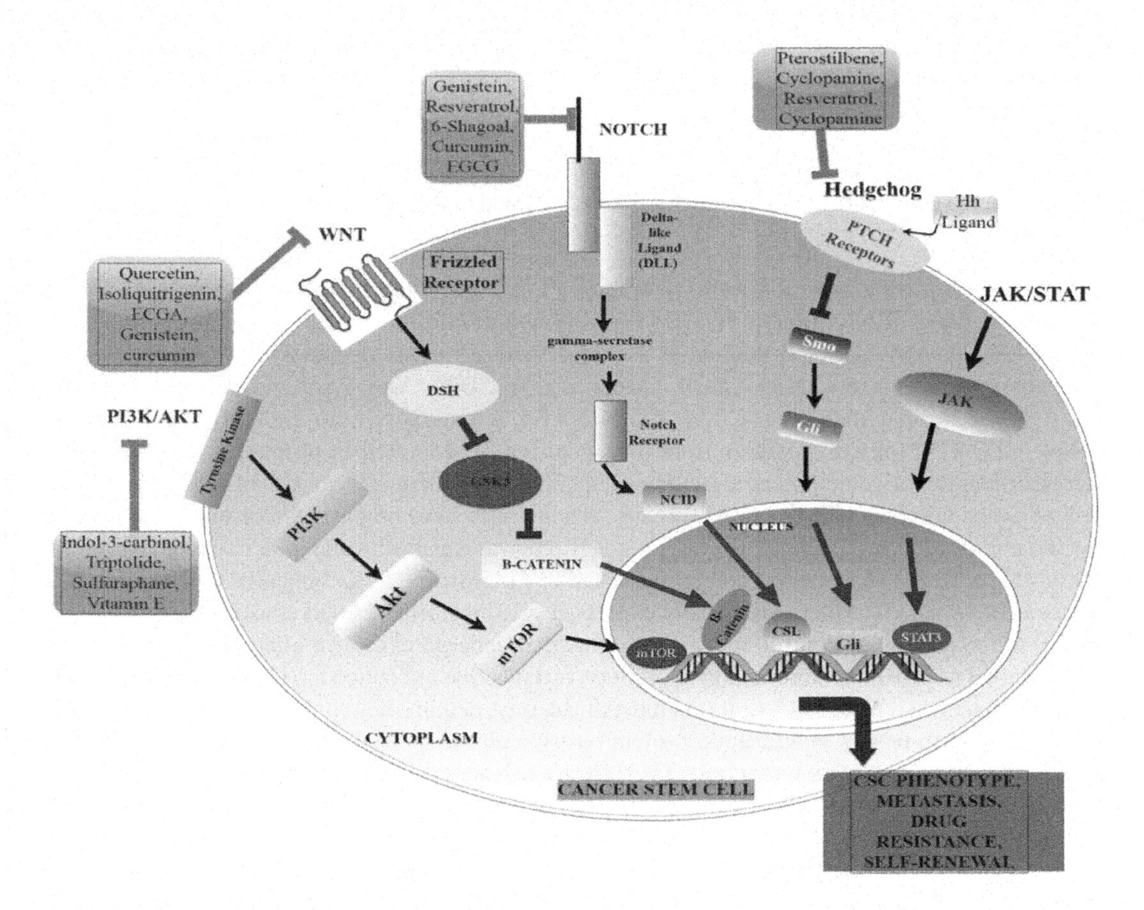

FIGURE 19.1 Different signaling pathways in CSCs along with phytochemicals suppression in various pathways.

TGF-induced production of matrix metallopeptidase 9 (MMP9) and MMP13 (Romero et al. 2016; Weiss and Attisano 2013; Liu et al. 2012; Liang et al. 2016; Kim et al. 2018).

19.3.5.6 PPAR

Peroxisomes proliferator-activated receptors, also known as PPARs, are nuclear transcription factors, and they are activated by ligands. There are three varieties of PPARs: PPARα, PPARβ, and PPARγ. Natural ligands of PPAR include eicosane acids, unsaturated fatty acids, very low-density lipoprotein (VLDL), oxidized low-density lipoprotein (oxLDL), and derivatives of linoleic acids.

In breast CSCs, as a tumor suppressor, the PPAR binds and thus activates a canonical response element in miR-15a gene, therefore reducing the $CD49^{high}/CD24^{+}$ MSC population and inhibiting angiogenesis (Kramer, Wu, and Crowe 2016; Yousefnia et al. 2018) (Figure 19.1).

19.4 MARKERS INVOLVED IN CSCs

The most frequently studied CSC surface markers are CD44, CD133, CD24, CD29, CD90, epithelial-specific antigen (ESA), and aldehyde dehydrogenase 1 (ALDH1), which is used to identify and isolate these cells (Yu et al. 2012). However, there are also other markers such as CD10, CD13, CD26, CD117, CD274, EpCAM, CCL-1, TIM-3, Galactin-9, BMI-, Nanog, Oct-3/4, Sox-2, OV6(+),

LGR5, TLR-4^+, α2δ1, K19, and DCLK1 which have been recently reported (Koike 2018; Kim and Ryu 2017; Darwish et al. 2016; Shafiei et al. 2019; Walcher et al. 2020; Yamashita et al. 2013). Below we shall study CSC markers in brief.

19.4.1 Cluster of Differentiation as Surface Markers for CSCs

19.4.1.1 CD44

CD44 (cluster of differentiation 44) is a multifunctional and multistructural glycoprotein and is encoded by the greatly conservative CD44 gene found in chromosome 11 in humans (Goldstein et al. 1989). Notably, CD44 is marked as a unique surface marker discretely or in a combination with other markers like CD133, CD24, C-Met, and/or CD34 (Yan, Zuo, and Wei 2015). In the breast cancer, subpopulations of $CD44^+$ $CD24^{-/low}$ lineage markers were found in CSCs. Isolated $CD44^+$ $CD24^{-/low}$ $lineage^-$ markers from breast cancer incorporated in immunosuppressive mice demonstrated tumorigenic potential (Al-Hajj et al. 2003). In gastric cancer, CD44 cells were identified as tumor initiators. $CD44^+$ cells were also displayed to have properties of a stem cell, since it has the ability to form differentiated progeny, self-renewal capability, and give rise to $CD44^-$ cells (Su et al. 2011). CD44 is a good cell surface marker for gastric CSCs. Typically, a CD44 cell has 20 exons of which 10 of them are usually expressed in all isoforms (called constant exons) and the rest 10 are expressed as central or variable exons that undergo extensive alternative splicing in a region called as proximal stem region to produce variants that are spliced. CD44s isoform lacks all variant exons, whereas CD44v isoforms most of the time acquire new functions mostly as a coreceptor for growth factors, nonreceptor protein tyrosine kinase, and so on (Wang, Zuo, et al. 2018). A recent study shows that a variant of CD44, CD44v6, functions probably as a CSC marker in the colon cancer (Todaro et al. 2014). Variants of CD44, CD44(v8–10), acts as a unique CSC marker for gastric cancer (Lau et al. 2014). However, CD44v3 acts as a good marker for head and neck cancer (Bourguignon, Wong, and Earle 2014). All isoforms of $CD44^+$ that are found in different cancers such as cancer of salivary gland, laryngeal and nasopharyngeal, head and neck, stomach, colon, lung, ovarian, prostate, and even glioma and leukemia are considered to be CSC markers. CD44s (standard isoform) are considered a good CSC marker in pancreatic cancer, $CD44v9^+$ acts as a good CSC marker for stomach cancer, $CD44v2^+$, $v6^+$, $v9^+$ act as a good CSC marker for colon cancer. CD44v6+ is also found in prostate cancer and is proven to be a good CSC marker (Yan, Zuo, and Wei 2015).

19.4.1.2 CD24 and CD29

CD24 (cluster of differentiation 24) is a kind of protein encoded by a gene called CD24 gene; it is a glycosyl phosphatidylinositol (GPI)-linked glycoprotein (Hough et al. 1994). Lee et al. found CD24 as a functional marker as CSC in hepatocellular carcinoma. NANOG (it is a homeobox protein and a transcription factor that helps stem cells maintain their pluripotency) was found as a downstream effector of CD24 signaling, which functioned through signal transducer and activator of transcription 3 or STAT 3 activation, and has shown to be substantially linked to CD24 in the cell lines of HCC and also in clinical data (Lee et al. 2011). Wang et al. demonstrated that $CD24^+$, $CD44^+$, and $EpCAM^{high}$ isolated from three extrahepatic cholangiocarcinoma grafts exhibited higher tumorigenicity than $CD24^-$, $CD44^-$, and $EpCAM^{low/-}$. They reported that these tumorigenic cells revealed several properties of stem cells, such as self-renewal as well as the ability to give rise to heterogeneous progeny. Treatment with anti-miR-21 lowered tumor development with a subsequent reduction in $CD24^+$. This somehow proves that CD24 is a good CSC marker for diagnostic applications (Wang et al. 2011). Some reports recently stated that both CD44 and CD24 act as CSCs in ovarian and colorectal cancers (Yeung et al. 2010). Because they have an epithelial phenotype, ovarian cancer cell lines have a high expression of CD24 (Jaggupilli and Elkord 2012). CD24 has been associated or linked with ovarian cancer metastasis in many studies, and this might be exploited as a potential marker for CSCs (Wei et al. 2010; Gao et al. 2009, 2010; Su et al. 2009). Hurt et al. isolated cells of CD44$^+$/CD24$^-$ having prostate cancer and discovered that this phenotype had tumor-initiating capabilities along with clonogenic and differentiation potential (Hurt et al. 2008). In a study belonging to cells of pancreatic cancer, it was found that in 0.5–1% of the pancreatic cancer cells, CD44$^+$/CD24$^+$/ESA$^+$ cells were observed and indicated to be CSCs (Lee, Dosch, and Simeone 2008). However, more research is required in order to understand the role of the CD24 cells as a CSC marker (Jaggupilli and Elkord 2012). CD29 (cluster of differentiation 29), also known as integrin beta-1, is a cell surface receptor present in humans and is encoded by the ITGB1 gene (Goodfellow 1989). Recently reported, CD29 is considered to be a good CSC marker, since it promoted epithelial-to-mesenchymal transition in the squamous cell carcinoma (Geng et al. 2013).

19.4.1.3 CD133

CD133 (cluster of differentiation 133) is a glycosylated protein, discovered or identified as a potential target of AC133 monoclonal antibody (MAbs) specific for the CD34$^+$ population in hematopoietic stem cells (Grosse-Gehling et al. 2013; Yin et al. 1997). In the lung, pancreatic, breast, brain, liver, prostate, colon, ovary, and head and neck malignancies, the CD133 marker has identified CSC populations, and CD133$^+$ populations unambiguously form tumors in immunocompromised mice more effectively than CD133$^-$ populations (Grosse-Gehling et al. 2013). CD133 is mostly found on the proliferating cells' surface. However, it has also been discovered on the surface of differentiated epithelial cells in a variety of tissues (Kim and Ryu 2017). *In vitro*, the CD133 (+) cell subpopulation from human brain tumors has stem cell characteristics. When transplanted into an immune-deficient mouse brain, they had the capability for self-renewal and perfect replication of the original tumor (Ai et al. 2011; Bourseau-Guilmain et al. 2012). In prostate cancer, CD133 coexpressed with CXCR4 (receptor of CXCL12) expressed in metastasizing cancer cells (Stratford et al. 2011). In pancreatic and head carcinoma-collected specimens from 80 patients showed higher levels of CD133+ specimens alongside vascular endothelial growth factor (VEGF)-C, CXCR4, Ki67, CD34, as well as cytokeratin expressions. CD133 expression was linked to lymphatic invasion, histological type, VEGF-C expression, and lymph node metastasis, suggesting that CD133 is a prognostic factor in its way. Therefore, it shows the relevance of CD133 as a CSC biomarker (Kim and Ryu 2017; Wu and Wu 2009).

19.4.1.4 CD90

CD90 or cluster of differentiation 90, originally found as a thymocyte antigen, which is strongly N-glycosylated, glycophosphatidylinositol (GPI) anchored conserved cell surface protein, has a single V-like immunoglobulin domain (Wikipedia contributors 2021). CD90+ cells from hepatocellular carcinoma cell lines may form tumor nodules in mice having immune deficiency, and CD90 is also found to be a good marker for CSCs in the brain and insulinoma (Ksander et al. 2014; Schatton et al. 2008; Miyauchi et al. 1987). Yamashita et al. found that CD90+ HCC cells displayed vascular endothelial cell characteristics and had high metastatic potential. *In vitro*, CD90+ HCC cells expressed a lot of c-KIT and were chemosensitive to imatinib mesylate (Yamashita et al. 2013). This shows the importance of CD90 as a CSC biomarker (Yamashita et al. 2013; Koike 2018).

19.4.1.5 Effect of Other CD CSC Markers

The "classical" CSC markers have been reported as CD123 and CD33 for AML. CD123 has been linked to enhanced proliferation and differentiation, but CD33 has been the most widely utilized marker for AML stem cells in the past. CD33 is abundantly expressed on blasts and at increased densities in CML (Walcher et al. 2020). In an experiment, Tamai et al. discovered that CD274 has a new role in human CC (cholangiocarcinoma) as a negative regulator of CSC-related phenotypes. According to *in vitro* investigations, CD274^{low} cells had various CSC-related traits, including ALDH activity, ROS species generation, and a dormant state in the cell cycle (Tamai et al. 2014). CD10 is a unique marker for therapeutic resistance as well as for CSCs in the head and neck squamous cell

(HNSCC) and breast carcinomas, according to the recent research (Fukusumi et al. 2014; Maguer-Satta et al. 2011). CD117 (also known as c-Kit) is a stem cell factor receptor, usually expressed at very low levels in cells of a normal tissue (Kim and Ryu 2017). It is also involved in the survival, as well as self-renewal signaling in a variety of cells (Miettinen and Lasota 2005). For example, in the human epithelial ovarian cancer, CD44+ and CD117+ cells do have cancer stem cell-like features, demonstrating enhanced chemoresistance (Chen et al. 2013). CD26 (also known as dipeptidyl peptidase-4 or DPP4) is basically a serine DPP4 found upon the surface of several cell types (Kim and Ryu 2017). CD26 has been identified as a good CSC marker for both the leukemic as well as for colorectal CSCs in studies (Herrmann et al. 2014; Pang et al. 2010).

19.4.2 Markers Other Than CDs

19.4.2.1 EpCAM

Epithelial cell adhesion molecule or EpCAM is a type of glycoprotein, originally marked as a marker for carcinoma (Trzpis et al. 2007). Overexpression of EpCAM is related to cell life proliferation in renal cell carcinoma (Seligson et al. 2004). It is reported that complexes of CD44v4-v7-tetraspanin-EpCAM-claudin-7 found in tumor cell lines can influence cell–cell adhesion, cell–matrix adhesion, and also pose apoptotic resistance; additionally, they are also involved to promote metastasis (Schmidt et al. 2004). The CD44v4-v7-tetraspanin-EpCAM-claudin-7 complex is also found in pancreatic and colorectal carcinomas and is reported to be a poor prognosis in such carcinomas for such complexes of EpCAM (Trzpis et al. 2007). Xie et al. reported that EpCAM is distributed differently in normal and different malignant carcinomas such as signet-ring cell carcinoma, squamous cell carcinoma, as well as adenocarcinomas (Xie et al. 2005).

19.4.2.2 CLL-1

C-type lectin-like molecule-1 is a newly discovered myeloid lineage restricted cell surface marker that is highly expressed in over 90% of AML patient myeloid blasts and leukemia stem cells (Lu et al. 2014). In AML, increased CLL-1 expression is linked to a poor prognosis and a quicker relapse (Darwish et al. 2016; Van Rhenen et al. 2007). Therefore, CLL-1 can be a good CSC marker; however, post chemotherapy, its role as a biomarker is unclear and further research is required to understand its role (Walcher et al. 2020).

19.4.2.3 TIM3, Galactin-9, and BMI-1

A nonclassical LSC (Leukemia Stem Cells) markers are T-cell immunoglobulin and mucin 3 (Walcher et al. 2020). LSCs are the primary reason for acute myeloid lymphoma (AML) relapses and drug resistance. In promoting cellular proliferation and survival, LSCs depend on the B-cell-specific Moloney murine leukemia virus integration site (also called BMI-1). When deprived of nutrients, LSCs upregulate the signaling pathway in cancer progression. When LSCs were analyzed with quantitative RT-PCR, they showed higher expressions of CLL-1, TIM-3, and BMI-1 and correlate or correspond to bad prognosis in the patients having AML (Darwish et al. 2016). In an experiment using xenograft mice, the stem cell capabilities of TIM-3+ were also validated (Kikushige et al. 2010). According to the studies, there is also a strong link between TIM-3 and galactin-9, which is highly expressed in LSCs. The binding of TIM-3 and galactin-9 activates NF-κB, as well as β-catenin signaling that effect changes in the gene expression which also includes upregulation of MCL-1 (Myeloid leukemia 1) that aids LSC survival (Kikushige et al. 2015). It is also found that TIM-3/galactin-9 signaling is important since deprivation of galactin-9 accelerates apoptosis in leukemia. Galactin-9 generated by LSCs binds and then stimulates TIM-3 in AML as well as LSCs, inducing the autocrine effect (Kikushige and Miyamoto 2013). TIM-3 is usually expressed on $CD4^+$ Th1 cells' surface. It is involved in Th1 cell immunity modulation and tolerance (Sabatos et al.

2003). On the other side, many studies suggest the potential role of Polycomb-group (PcG) proteins, like BMI-1 in breast cancer. BMI-1 is also involved in lymphomagenesis. Markers such as BMI-1 and EZH2 showed strong coexpression in varieties of (B-cell lymphoma) non-Hodgkin lymphomas' cells (van Kemenade et al. 2001). A study suggests that BMI-1 surpasses cell-cycle arrest leading to proliferation using Hedgehog signaling pathway (Liu et al. 2006). BMI-1 overexpression also protects LSCs and blast cells against apoptosis (Chandler et al. 2014; Darwish et al. 2016).

19.4.3 Intracellular Markers

19.4.3.1 ALDH

Aldehyde dehydrogenase (ALDH) is a type of enzyme which has the capability to catalyze aldehydes (Marchitti et al. 2008). ALDH cells have stem-like characteristics including clonogenic development, self-renewal, tumor-initiating potential, and drug resistance (Walcher et al. 2020; Vassalli 2019). In an experiment of nonsmall cell lung cancer (NSCLC) revealed two CSC subpopulations which were described as higher expression of CD133 and ALDH (Walcher et al. 2020). In tumors such as breast, liver, colon, lung, prostate, pancreas, ovary, esophagus, stomach, brain, bone skin, and bone marrow, ALDH^{br} have been identified as having tumor-initiating capability. ALDH^{br} populations are abundant in biopsies of the patients having (metastatic) melanoma. ALDH^{br} and ALDH^{dim} cells, on contrary, have similar aggressive behavior and resistance to antimelanoma drugs (Vassalli 2019). Cytotoxic drugs like doxorubicin, cyclophosphamide, hydroperoxycyclophosphamide (4-HC), arabinofuranosyl cytidine (Ara-C), cisplatin, as well as dacarbazine face tumor resistance due to the activity of ALDH (Sládek et al. 2002; Honoki et al. 2010; Hilton 1984). However, more research will open new theories and understanding about ALDH as a CSC biomarker (Vassalli 2019).

19.4.3.2 NANOG

Nanog is a transcription factor with a homeobox domain which is commonly observed in human cancers (Jeter et al. 2015). In colorectal cancer, its expression in CD133 cells was highly expressed. Nanog is also linked to liver and lymph node metastasis in a linear fashion (Cui et al. 2015). Nanog and Oct-3/4 expression in breast cancer patients is linked to the epithelial-to-mesenchymal transition as well as bad prognosis (Yang, Zhang, and Yang 2018; Wang, Lu, et al. 2014). Nanog expression is linked to self-renewal, invasion, metastasis, and drug resistance in hepatocarcinoma cell lines (Xiang et al. 2016).

19.4.3.3 Oct-3/4

Octamer-binding transcription factor 4 or Oct3/4 (known as POUF51) belongs to the POU homeobox gene family (Gene ID: 5460) (Takeda, Seino, and Bell 1992). It is a regulator of pluripotency of stem cells in mammals (Walcher et al. 2020). In many cancers, Oct3/4 is upregulated, and it supports neoplastic transformation and therapy resistance; in addition, Oct-4 participation in TIC functions is associated with self-renewal, epithelial-to-mesenchymal transformation, and metastasis (Wang and Herlyn 2015). Increased cell mobility and liver metastasis are often caused by Oct-3/4 in colorectal cancer (Fujino and Miyoshi 2019; Dai et al. 2013). Another study on lung cancer found that the expression of Oct-3/4 is linked to the ability of self-renewal as well as metastasis (Prabavathy, Swarnalatha, and Ramadoss 2018). Oct-3/4 in hepatocellular carcinoma is related to recurrence and cell size of tumors (Zhao et al. 2016). However, much more research is needed to understand its role as a potential CSC biomarker (Li et al. 2016).

19.4.3.4 SOX2

SOX2 is a transcription factor which is engaged in phenotypic maintenance and comes from the SRY-related HMG-box (SOX) family (Lundberg et al. 2016). As revealed in colorectal cancers, its abnormal expression leads to higher treatment resistance and asymmetric divisions (Takeda et al. 2018). SOX2 expression is linked to state of the stem cell as well as reduced expression of the CDX2 or caudal-related homeobox transcription factor 2, suggesting that it might be used as a prognostic marker for bad prognosis (Lundberg et al. 2016) (Takeda et al. 2018). SOX2 expression in gastric cancers is linked to tumor stage and even bad prognosis (Basati, Mohammadpour, and Emami Razavi 2020; Zhang et al. 2016). The transcription factors, such as the Nanog, SOX2, and even Oct-3/4, as well as the overexpression of CD44 and CD133, all correlate with the development of tumor spheroids in vitro (Zhang et al. 2016). Furthermore, further study is needed in order to understand and prove SOX2 as a good CSC biomarker (Walcher et al. 2020).

19.4.3.5 LGR5

Leucin-rich repeat-containing G protein-coupled receptor 5 or LGR5 gene encodes a composition of the Wnt receptor complex that works as a receptor for the R-spondin family of the Wnt signal agonists (Tomita and Hara 2020). Interestingly, LGR5 is discovered to be upregulated in hepatocellular carcinomas, basal carcinomas, and ovarian and colorectal tumors (Tanese et al. 2008; McClanahan et al. 2006). LGR5 influences the Wnt/β-catenin signaling and thus stimulates proliferation, as well as the renewal of the cancer stem cell (Yang et al. 2015; Kemper et al. 2012). Indeed, LGR5 was demonstrated as a potential functional biomarker of CSCs, contributes to CSC proliferation, as well as self-renewal (Xu et al. 2019). Wnt signaling was shown to be enhanced in the CSCs because of increased localization of β-catenin inside the region of the nucleus. In breast cancer, cells are shown to improve tumor formation and cancer cell mobility and also epithelial-to-mesenchymal transition via Wnt/β-catenin signaling (Yang et al. 2015) regulates or controls the malignant phenotype in a subset of glioblastoma (CSCs) stem cells generated from the patients, suggesting that it might be used as a possible glioblastoma prognostic marker (Xie et al. 2019). LGR5 has been marked as a unique biomarker of the CSCs in a variety of human malignancies, which includes adenocarcinoma, glioblastoma, colorectal, and breast cancers, in multiple investigations employing genetic lineage tracing analyses or detection through antibodies against the LGR5 (von Rahden et al. 2011; Liu et al. 2006; Kleist et al. 2011; Shimokawa et al. 2017). LGR5 is highly expressed in colorectal adenomas, cancers, hepatocellular carcinoma, basal cell carcinoma, and neuroblastoma (Fan et al. 2010; Becker, Huang, and Mashimo 2008; Uchida et al. 2010; Tanese et al. 2008; Yamamoto et al. 2003; Vieira et al. 2015). It is also a novel CSC biomarker for cancers in stomach, pancreas, liver, colon, and ovary (Jiang et al. 2013; Ong et al. 2012; Wang, Yuen, et al. 2014).

19.5 PHYTOCHEMICALS AS CSCs MODULATORS

Plants not only furnish basic nutrients for life but also bioactive phytochemicals that aid in disease avoidance and health enhancement. Phytochemicals have lately appeared as modulators of major cellular signaling pathways; nevertheless, macro- and micronutrients in plants have long been regarded to be one of the fundamental components of human well-being (Yoo et al. 2018). Phytochemicals, also known as phytonutrients, are naturally occurring bioactive compounds found in foods, for example, fruits, vegetables, whole grain products, nuts and seeds, legumes, tea, and dark chocolate. Despite the fact that there are thousands of phytochemicals, only a small number have been isolated, characterized, or otherwise located in plants (Xiao and Bai 2019). Phytochemicals have been shown to target various pathways in cancer cells, therefore regarded as potential options for anticancer drug development, and it is also known to have pro-apoptotic, antioxidant, as well as antiproliferative effects on a number of malignancies where the CSCs were detected (Mukherjee et al. 2014; Taylor and Jabbarzadeh 2017; Dandawate et al. 2013). This section will shed light on the efficacy of a few well-known phytochemicals in the treatment of CSCs.

19.5.1 Curcumin

Curcumin (1,7-bis(4-hydroxy-3-methoxyphenyl)-1,6-heptadiene-3,5-dione), commonly referred as diferuloylmethane, is a significant natural polyphenol present in the rhizome of *Curcuma longa* (also called turmeric), as well as in other Curcuma species. Because of its antioxidant, anti-inflammatory, antimicrobial, antimutagenic, also having anticancer characteristics, *Curcuma longa* has long been employed as a medicinal herb (Hewlings and Kalman 2017; Lestari and Indrayanto 2014; Teow et al. 2016; Mansouri et al. 2020). Curcumin, a polyphenol, has been demonstrated to target several signaling components, while also displaying cellular activity, supporting its numerous health advantages (Gupta, Patchva, and Aggarwal 2013). Curcumin has been demonstrated in several studies to have anticancer qualities in various types of cancers by affecting cell growth as well as metastasis while also encouraging cell death. Curcumin also helps to prevent cancer from developing (Giordano and Tommonaro 2019).

Curcumin has the ability to target particular molecules, as well as pathways while avoiding resistance and toxicity, despite several "targeted" chemotherapeutic drugs that incur toxicity and resistance (Zhou, Beevers, and Huang 2011). Several studies have revealed that combining curcumin with chemotherapeutic drugs can improve efficacy, leading to the speculation that adding two or even more targeted compounds to attack cancer cells' resistance mechanisms, such as eradication of resistant lung cancer cell lines (Chanvorachote et al. 2009). Curcumin improved cisplatin-induced metastatic suppression as well as apoptosis of the highly migratory CSC subpopulation (CD166/EpCAM+) in nonsmall cell lung cancer (known as NSCLC) cell lines, suggesting that it might be possibly used as a supplement to the standard chemotherapy to mitigate tumor progression as well as to control metastasis (Baharuddin et al. 2016). Curcumin, alone or along with cisplatin, suppressed properties of CSC in the follow-up study of Satar et al., suggesting that it could be a useful treatment strategy for preventing the surfacing of chemoresistance in NSCLC by eliminating CSCs (Satar, Ismail, and Yahaya 2021).

Dasatinib (a Src inhibitor) and curcumin combined therapy suppressed cellular proliferation, invasion, as well as colonosphere development in the chemo-resistant colon cancer cells enriched in CSCs (lowering the CSC population), as evidenced by lower expression of CSC-specific markers CD44, CD133, CD166, and ALDH (Nautiyal et al. 2011). Zhu et al. (2017) showed curcumin's inhibitory effects on lung CSCs, they also found that the Wnt/β-catenin and Sonic Hh pathways play a key role in curcumin interference of lung CSCs.

19.5.2 Lycopene

Lycopene is an antioxidant, anti-inflammatory, and cancer-preventive red carotenoid found in ripe tomatoes, grapefruits, and red watermelon (Imran et al. 2020; Cha et al. 2017; Qi et al. 2021). Lycopene prevents DNA oxidation, scavenging free radicals, and dousing singlet oxygen. Consequently, it averts the conversion of normal cells into cancerous, as well as modulating gene functions, carcinogen-metabolizing enzymes, apoptosis, and immune function (Lu et al. 2011). Lycopene has been shown to suppress not just cancer cells along with CSCs in pancreatic cancer and other cancer types (Li et al. 2011).

Lycopene suppressed NF-κB activity, which prohibited the cell growth (Lu et al. 2011) in the prostrate, breast, as well as in endometrial cancer cells (Hung et al. 2008). Lycopene restricts cell growth by lowering the NF-kB binding affinity in LNCaP cells, which affects proteins related to the cell cycle (cell-cycle proteins) as well as apoptosis-mediating proteins such as p21, cyclinD1, p53, p27, Bas (Bcl-2 associated X-protein), or Bcl-2 (Tang et al. 2008). In a study, lycopene inhibited activation of Akt, as well as nonphosphorylated β-catenin protein levels in the cells of colon cancer, while enhancing/increasing the phosphorylated form of β-catenin, which was linked to lower cyclin D1 protein expression. As a result, lycopene may suppress Wnt/β-catenin signaling through the Akt/GSK3β/β-catenin pathway (Li et al. 2011). In prostate PC-3 (a cell line for prostate cancer), colon HCT-116, HT-29 (a cell line of human colon adenocarcinoma), as well as lung BEN cancer cells, lycopene suppresses HMG-CoA reductase expression, cell proliferation, and Ras were noticed (Palozza et al. 2010). Recently, in animal studies, Rakic et al. demonstrated that lycopene or lycopene metabolites potentially be the therapeutic agents that inhibit the development of cancers such as lung cancer and skin carcinogenesis through several molecular pathways (Mustra Rakic et al. 2019).

19.5.3 Berberine

Berberine ($C_{20}H_{18}NO_4$), a small molecule isoquinoline alkaloid derived from the rhizomes of *Coptis chinensis* and *Hydrastis canadensis*, has been used to cure bacterial diarrhea for centuries; therefore, it is also derived from *Berberis aristate*, *Berberis aetnensis* C. Pesl., *Berberis vulgaris*, *Coptis japonica*, *Phellondendron amurense*, *Coptis rhizome*, and *Tinosora cordifolia* (Habtemariam 2016; Neag et al. 2018). Additionally, berberine has antitumor effects by interfering with carcinogenesis and numerous characteristics of tumor growth and has wide biological functions such as anti-inflammation, antioxidative, and antidiabetic (Liu et al. 2019). Berberine suppresses the manufacture and expressions of TNF-α, IL-6, monocyte chemoattractant protein 1 (MCP1), as well as COX-2. Moreover, it could even alter prostaglandin E2 (PGE2) levels and reduce the expression of significant metastasis genes like MMP2 and MMP9. Therefore, Raf/MEK/ERK and NF-κB pathways may be involved in some of these effects (McCubrey et al. 2017).

Berberine is known to have reduced colorectal cancer's malignant behavior by aiming at AMPK or TGF-1/SMAD signaling pathway, arresting the cell cycle, also endorsing programmed cell death (apoptosis), and suppressing growth factor signaling pathways (Wang, Wang, et al. 2021; Zhao et al. 2021; Hallajzadeh et al. 2020). Consequently, a study of berberine delivery was shown to be harmless and beneficial in colorectal cancer therapy clinics in lowering the chance for colorectal adenoma recurrence, as well as found as a potential alternative for chemoprevention after the procedure of polypectomy (Chen et al. 2020). According to Ruan et al., berberine binds to the retinoid X receptor alpha (RXRα) in colon cancer cells and reduces β-catenin signaling by suppressing β-catenin transcription (Ruan et al. 2017). Berberine is also known to suppress the proliferation or growth of multiple myeloma cells through downregulating the expression of miR-21 (Luo et al. 2014). A study found that berberine suppressed self-renewal and properties of metastasis (*in vitro*); it also slowed tumor growth *in vivo* through downregulating miR-21 in oral cancer (Lin et al. 2017). El-Benhawy et al. showed that berberine suppressed cell proliferation unaided and in combination with ionizing radiation (IR), and it also has a radio-sensitizing effect that can possibly be activated by lowering the expression of the CSC markers OCT4 as well as SOX2 (El-Benhawy et al. 2020).

19.5.4 Quercetin

Quercetin belongs to the flavonoid group, which is an antioxidant flavonol and is commonly found as quercetin glycoside. Quercetin and its compounds are phytochemicals that occur naturally and have intriguing biological properties. Quercetin is known to possess anti-inflammatory, antidiabetic, antioxidant, antibacterial, anti-Alzheimer's, antiarthritic, cardiovascular, and wound-healing properties, and it also has anticancer effects against several cancer cell lines (Salehi et al. 2020). Quercetin, a common dietary flavonoid, has revealed to prohibit cancer cell proliferation in multitude of cell types, including liver, breast, prostate, leukemia, gastric, lung, and colon cancer, also in animal models (S. Li et al. 2018). In various malignancies, quercetin has been found to influence a range of pathways.

Quercetin eliminates tumor cells selectively without affecting healthy cells. According to research that quercetin might increase cancer cell chemosensitivity in a dose-dependent form, the effect of quercetin in reversing MDR in cancer cells by reducing YB1 activity, including the mechanism, remained unknown (S. Li et al. 2018). Quercetin was discovered to reverse MDR in breast

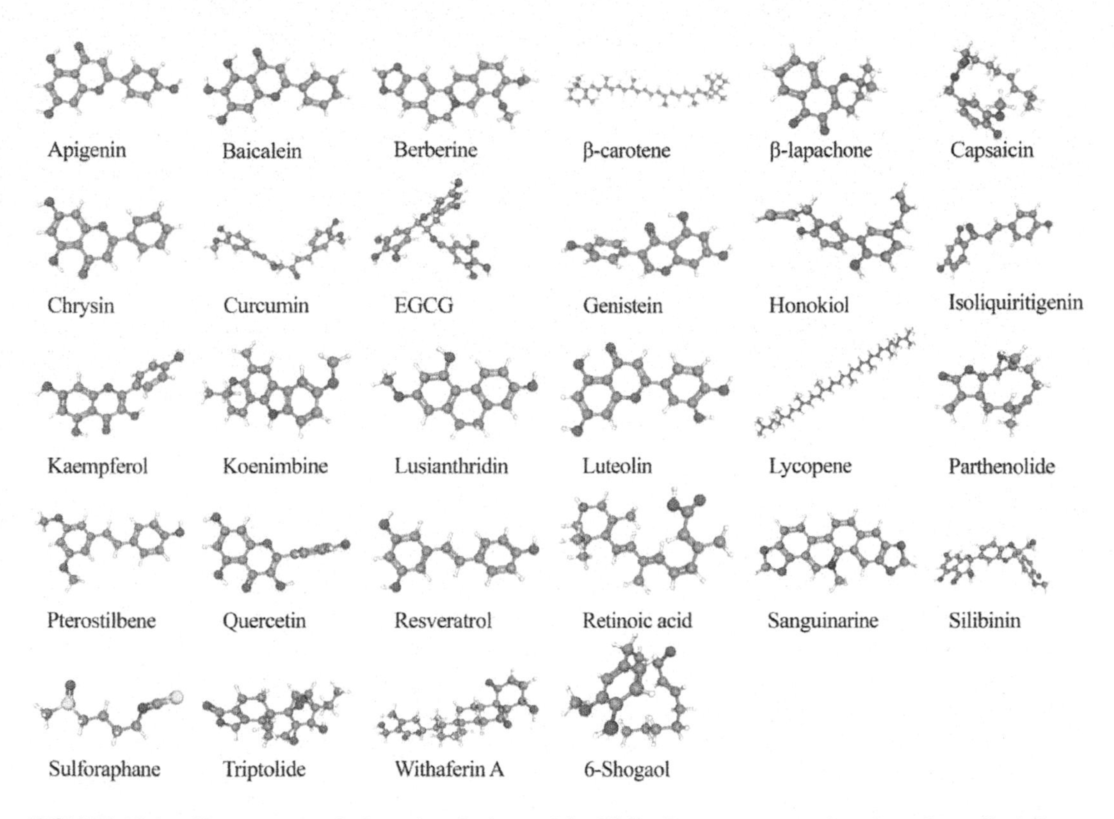

FIGURE 19.2 3D structure of phytochemicals used in CSCs. Structures are taken from https://pubchem.ncbi.nlm.nih.gov/.

cancer cells by downregulating Pgp expression and removing BCSCs via YB1 nuclear translocation. Nontoxic doses of quercetin, along with antitumor active quantities, can destroy cancer cells in combination with chemotherapy drugs (S. Li et al. 2018). Quercetin prohibits the growth of the lung as well as breast cancer cells by lowering CDK6 expression; moreover, increasing the amount of reactive oxygen species (ROS) levels causes cell death process, called ferroptosis. Additionally, quercetin results to death of cells by activating lysosomes (Wang, Ma, et al. 2021; Yousuf et al. 2020). Through lowering the production of VEGF-A, MMP-2 as well as MMP-9 in esophageal cancer cells, quercetin decreased angiogenesis and metastasis (Yin et al. 2021). In breast cancer, the PI3K/Akt/mTOR axis is accountable for CSC development, as well as for progression. As a result, quercetin inhibits the PI3K/Akt/mTOR axis in cell of breast cancer and results to arrest in the cell cycle during the G1 phase (X. Li et al. 2018) (Figure 19.2).

19.5.5 Sulforaphane

Sulforaphane [1-isothiocyanato-4(methylsulfinyl)butane] is an effective phytochemical predominantly found in cruciferous vegetables, such as broccoli, a rich source of glucosinolates. Glucosinolates are hydrolyzed by the enzyme myrosinase resulting in the production of organic aglycone moieties. Therefore, it is considered a natural antioxidant. Sulforaphane is the most efficient chemopreventive compound known to target numerous cell processes (Zhang et al. 1992). According to an early study, sulforaphane is thought to alter tumorigenesis and tumor formation by limiting the initial stage of cancer by suppressing phase 1 metabolism enzymes and increasing phase 2 metabolism enzymes (Clarke, Dashwood, and Ho 2008).

Additional research demonstrated that sulforaphane can influence a variety of biological functions, including decreasing cell proliferation, triggering apoptosis, and suppressing angiogenesis and metastasis (Zhang and Tang 2007). Sulforaphane has been demonstrated to reduce NF-κB-induced antiapoptosis in CSCs or tumor-initiating cells, or to downregulate the Wnt/β-catenin pathway in breast CSCs (Kallifatidis et al. 2008; Li et al. 2010). By encoding the most significant transcription factor, c-Myc (a proto-oncogene) is involved in tumor initiation and progression. c-Myc controls the expression of multiple genes engaged in a variety of biological processes, including cell proliferation, apoptosis, differentiation, and stem cell self-renewal (Wilson et al. 2004; Dang 1999). As a result, by upregulating miR-214, sulforaphane effectively suppressed c-MYC expression. SFN/miR-214/c-MYC signaling's functional impact on CSC characteristics and chemoresistance was also studied (Li et al. 2017). Sulforaphane therapy suppressed human pancreatic CSC-derived spheres in an in vitro model. Moreover, sulforaphane decreased Smo, Gli1, and Gli2 expression, along with Gli1 and Gli2 nuclear translocation and transcriptional activity (Rodova et al. 2012).

19.5.6 Resveratrol

Resveratrol (trans-3,4′,5-trihydroxystilbene) is a naturally occurring small polyphenolic compound that can be isolated from mulberries, berries, peanuts, and grapes. In 1940, it was discovered in the root of white hellebore (*Veratrum grandiflorum*) (Huang et al. 2014; Chen, Kuo, et al. 2012). It modulates signal transduction pathways that control cell division and proliferation, apoptosis, inflammation, angiogenesis, and metastasis, and hence is considered a promising anticancer therapeutic by some (Shukla and Singh 2011). A previous research found that resveratrol inhibited pancreatic CSCs in transgenic mice, which was linked to the inhibition of pluripotency-preserving proteins and the epithelial-mesenchymal transition (Shankar et al. 2011). The suppression of JAK2/STAT3 signaling was involved in the elimination of resveratrol-induced osteosarcoma stem cells. It is a multitargeted drug that inhibits the expression of several cytokines that trigger the JAK/STAT pathway (Ahmad et al. 2018; Peng and Jiang 2018). Resveratrol inhibited fatty acid synthase (FAS) gene expression and increased proapoptotic genes including DAPK2 and BNIP3 expression in cancer stem-like cells (CD24(–)/CD44(+)/ESA(+), which were isolated from both ER and ER-breast cancer cell lines. Inhibited cell viability and mammosphere development, as well as induced proapoptotic effects, were seen as a result of these changes (Pandey et al. 2011). Resveratrol has been shown to target ovarian CSCs through both ROS-dependent and ROS-independent mechanisms. Resveratrol, alone or in combination with the tyrosine kinase inhibitor imatinib mesylate, may obstruct melanoma CSC-associated signaling molecules, hence inhibiting melanoma melanogenesis and CSC features (Chen et al. 2013). Resveratrol's potential to decrease K562 cell proliferation may be due to its ability to modulate the SphK1 pathway via controlling S1P and ceramide levels, which in turn affects the proliferation and death of leukemia cells (Tian and Yu 2015). Through mitochondrial metabolic regulation, the resveratrol analog 4′-Bromo-resveratrol (4-BR) has been shown to significantly decrease SIRT3 activity and suppress melanoma cell proliferation (Nguyen, Gertz, and Steegborn 2013; George et al. 2019).

19.5.7 Epigallocatechin Gallate (EGCG)

Epigallocatechin-gallate is the main polyphenolic component of dried green tea leaves and plays antitumor roles in various cancers. Green tea's most abundant and biologically active catechin is EGCG (Du et al. 2012). The antiproliferative, proapoptotic, antiangiogenic, and anti-invasive properties of ECGC were observed (Gan et al. 2018). The anticancer effect of EGCG and green tea extracts is mediated by a number of mechanisms, including antioxidant activity and detoxification system activation, cell cycle alteration, suppression of the mitogen-activated protein kinase (MAPK) and receptor protein kinase (RTKs) pathways, inhibition of clonal expansion of the tumor-initiating stem cell population, and production of epigenetic changes in gene expression (Thawonsuwan et al. 2010; Lambert and Elias 2010; Shimizu et al. 2011; Lin et al. 2012). Two EGCG analogs from the library reduce cell proliferation and mammosphere formation, as well as suppress CD44+ high/CD24 low cell populations and activate the AMPK signaling pathway in MDA-MB-231 human breast cancer cells (Chen, Pamu, et al. 2012). HNSC (head and neck squamous carcinoma) CSCs' self-renewal ability and chemosensitivity are regulated by EGCG. EGCG reduced the expression of the stem cell markers OCT4, SOX2, NANOG, and CD44, in particular (Lee et al. 2013). LiCl-triggered Wnt/β-catenin activation reduced the inhibitory effects of EGCG on Wnt/β-catenin activation, tumor sphere formation, lung CSCs markers expression, cell proliferation, and apoptosis induction (Zhu et al. 2017).

19.5.8 Genistein

The primary isoflavone phytoestrogen, genistein (4′,5,7-trihydroxyisoflavone), was discovered in soy products. Genistein inhibits carcinogenesis by regulating a variety of signaling pathways, including programmed cell death, cell cycle, and angiogenesis (Banerjee et al. 2008; Li et al. 2012). Genistein reduces the ability of gastric cancer cells to self-renew. It was found to inhibit the induction of GCSC markers and could prevent ERK1/2 activity, which is consistent with its inhibitory effects (Kim et al. 2007; Huang et al. 2014). Breast cancer stem cells (BCSCs) are selectively inhibited by genistein, which is associated with a downregulation in the Hedgehog–Gli1 self-renewal pathway (Fan et al. 2013). In nasopharyngeal CSCs (NCSCs), genistein blocked the SHH pathway by lowering the expression levels of SHH, SMO, and GLI1 (Zhang et al. 2019). The ability of genistein to strongly suppress CD44 expression as well as the tumorigenicity of PCa TCs suggests that it could be useful in targeting prostate CSCs (Zhang et al. 2012). Genistein reduces the expression of CD44 marker in bladder epithelial cells by inhibiting arsenic-induced phosphorylation of HER2. This resulted in the downregulation of HER2 targets such as ERK, AKT, and STAT3 signaling pathways (Chu et al. 2022). Genistein has been shown to stimulate differentiation of breast CSCs by modulating the PI3K and MEK pathways, resulting in a decrease of CD44+/CD24-/ESA+ cells and elevation of differentiated cells markers (Liu et al. 2016). By decreasing CD163, CD44, p-STAT3, IL-8, IL-10, and upregulating IL-12 and nitric oxide, genistein decreased clonogenic and sphere-formation capacities in SKOV3 ovarian cancer cells, resulting in stemness inhibition (Ning et al. 2019). Genistein, a natural chemical with no known side effects, may offer a new, safe, and effective therapy option for breast cancer.

19.5.9 Silibinin

Silibinin ($C_{25}H_{22}O_{10}$) also known as milk thistle is a compound found in the seeds of *Silybum marianum* (L.) Gaertn. In Europe and Asia, silibinin is a hepatoprotective medication used to treat the hepatic injury caused by bile duct inflammation, cirrhosis, fatty liver, mushroom poisoning, and Hepatitis. The hepatoprotective properties of silibinin and silymarin are principally due to their potent antioxidant properties *in vitro* and *in vivo* (Mao et al. 2018). Silibinin has been shown to activate cellular checkpoints and cyclin-dependent kinase inhibitors (CDKIs), and CDK levels, and produce severe cell cycle arrest in cancer cells. Along with cell cycle arrest, silibinin targets CDK-CDKI interaction, CDK kinase activity, and Rb phosphorylation (Mateen et al. 2010). Silibinin induced apoptosis and autophagy in MCF-7 cells, and is followed by downregulation of AKT, mTOR, and ERK (Zheng et al. 2015). The downregulation of HER-2/neu expression and the generation of senescent-like growth arrest and death via p53 mediated pathway were primarily responsible for silibinin's anticancer activity (Provinciali et al. 2007). Silibinin also inhibits cancer cell migration and invasion by targeting MAPK signaling; inhibition of ERK1/2 by silibinin results in a decrease in MMP2, uPA, and invasiveness of A549 cells (Chen et al. 2005). A combination therapy with sorafenib and silibinin targets both hepatocellular carcinoma cells (HCC) and CSCs synergistically by reducing the phosphorylation of

STAT3/ERK/AKT (Mao et al. 2018). In ovarian cancer, silibinin reversed the epithelial-mesenchymal transition (EMT) process as well as decreased PI3K/AKT, Smad2/3 molecules *in vitro* and *in vivo*; silibinin effectively slowed tumor growth (Maleki et al. 2022).

19.5.10 Parthenolide

Parthenolide which is produced from the leaves of a medicinal plant *Tanacetum parthenium* is a sesquiterpene lactone. It has anti-inflammatory, redox-modulating, and epigenetic properties for cancer stem and progenitor cells (Freund et al. 2020). Parthenolide inhibits signal transducer and activator of transcription proteins (STAT proteins). Preventing STAT3 phosphorylation on Tyr705 prevents STAT3 dimerization, dependent gene expression, and nuclear translocation. It also inhibits the activity of ubiquitin-specific peptidase 7 (USP7) and WNT signaling (Sztiller-Sikorska and Czyz 2020). The PI3K/AKT pathway and the expression of cancer stem cell markers were both suppressed by parthenolide (Liu et al. 2021). By targeting AEBP1/PI3K/AKT signaling, Dimethylamino Micheliolide (ACT001), produced from parthenolide, inhibited glioblastoma stem cells proliferation and glioma sphere formation and prolonged animal lifespan (Hou et al. 2021).

TABLE 19.2
Phytochemicals targeting signaling pathways and CSCs

Sl. No.	Phytochemical	Targeted Pathway	Cancer	References
1	Luteolin	Caspase-8 (extrinsic pathway)	Breast cancer	(Yang et al. 2014)
2	Piperine	Wnt signaling	Colorectal cancer	(de Almeida et al. 2020)
3	Cyclopamine	Sonic hedgehog signaling	Gastric cancer	(Song et al. 2011)
4	Isoliquiritigenin	β-Catenin signaling	Breast cancer	(Wang, Wang, et al. 2014)
5	Retinoic acid	Protein Kinase C (PKC)	Breast cancer	(Berardi et al. 2021)
6	β-Lapachone	SIRT6/AKT/XIAP signaling	Liver cancer	(Zhao et al. 2021)
7	Apigenin	PI3K/AKT signaling MAPK/ERK signaling	Lung cancer Colorectal cancer Prostate cancer	(Yan et al. 2017)
8	Triptolide	Notch signaling	Triple negative breast cancer	(Ramamoorthy et al. 2021)
9	Chrysin	NF-kB signaling/Twist axis	Cervical cancer	(Dong et al. 2019)
10	Capsaicin	ERK 1/2, p38 MAPK, or JNK signaling	Gastric cancer	(Park et al. 2014)
11	β-Carotene	NOTCH1 and Sox2, and Wnt/β-catenin signaling	Colorectal cancer	(Kim, Kim, and Kim 2019)
12	Pterostilbene	NF-κB signaling	Breast cancer	(Leek and Harris 2002)
13	Koenimbin	Wnt/β-catenin signaling	Breast cancer	(Ahmadipour et al. 2015)
14	Lusianthridin	Src-STAT3-c-Myc	Lung cancer	(Bhummaphan et al. 2019)
15	Baicalein	Sonic Hedgehog (SH) signaling	Pancreatic cancer	(Song et al. 2018)
16	Sanguinarine	Wnt signaling	Lung cancer	(Yang et al. 2016)
17	Honokiol	Sonic Hedgehog signaling	Colon cancer	(Ponnurangam et al. 2012)
18	Withaferin A	Hedgehog signaling	Pancreatic cancer Prostate cancer Breast cancer	(Yoneyama et al. 2015)
19	6-Shogaol	Notch signaling	Breast cancer	(Ray, Vasudevan, and Sengupta 2015)

19.5.11 Kaempferol

Kaempferol (3,5,7-trihydroxy-2-(4-hydroxyphenyl)-4H-chromen-4-one) is a flavonoid antioxidant prevalent in fruits and vegetables; therefore, dietary kaempferol has been shown in numerous studies to minimize the risk of chronic diseases, particularly cancer. Kaempferol, which has anti-inflammatory effects, prevents DNA from binding to NF-B and myeloid differentiation factor 88. It also inhibits IL-6, IL-1, IL-18, and TNF production while raising mRNA and protein expression of the Nrf2 gene. TLR4 (toll-like receptor 4) is likewise blocked (Zhang et al. 2017; Tang et al. 2015; Saw et al. 2014). Kaempferol suppresses cancer cell growth by inducing apoptosis, cell cycle arrest in the G2/M phase, downregulation of signaling pathways including (PI3K)/(AKT), expression of EMT markers such as N-cadherin, E-cadherin, Snail, and Slug, and matrix metallopeptidase 2 (MMP2) (Imran et al. 2020; Marfe et al. 2009) (Table 19.2).

19.6 CONCLUSION

Artificial materials are progressively getting responsible for serious illnesses like cancer. As a result, a variety of medicines and procedures have been considered to battle the malignancy, yet resistance to them still exists. Conventional therapies, including chemotherapy and radiation therapy, have shown effectiveness against a variety of tumors, but they perform poorly against CSC-specific targets, resulting in tumor regrowth and metastasis. Naturally occurring phytochemicals have gained a lot of interest because of their ability to disrupt many signaling pathways and target CSCs. Consequently, phytochemicals provide strong evidence that abnormal regulation of CSC signaling pathways can result in dysregulation of self-renewal, apoptosis, proliferation and, most significantly, anticancer drug resistance. Importantly, there are no or few side effects related to the usage of plant-derived substances. Phytochemicals are expected to have a significant impact on CSC signaling pathways like NF-kB, MAPK/ERK, and others. In addition, more phytochemicals have yet to be uncovered, necessitating further studies to achieve positive outcomes.

REFERENCES

Abdul Satar, Nazilah, Mohd Nazri Ismail, and Badrul Hisham Yahaya. 2021. "Synergistic Roles of Curcumin in Sensitising the Cisplatin Effect on a Cancer Stem Cell-like Population Derived from Non-Small Cell Lung Cancer Cell Lines." *Molecules* 26(4): 1056.

Ahmad, Sheikh F, Mushtaq A Ansari, Ahmed Nadeem, Saleh A Bakheet, Mohammad Z Alzahrani, Musaad A Alshammari, Wael A Alanazi, Abdullah F Alasmari, and Sabry M Attia. 2018. "Resveratrol Attenuates Pro-Inflammatory Cytokines and Activation of JAK1-STAT3 in BTBR T+ Itpr3tf/J Autistic Mice." *European Journal of Pharmacology* 829: 70–78.

Ahmadipour, Fatemeh, Mohamed Ibrahim Noordin, Syam Mohan, Aditya Arya, Mohammadjavad Paydar, Chung Yeng Looi, Yeap Swee Keong, Ebrahimi Nigjeh Siyamak, Somayeh Fani, and Maryam Firoozi. 2015. "Koenimbin, a Natural Dietary Compound of Murraya Koenigii (L) Spreng: Inhibition of MCF7 Breast Cancer Cells and Targeting of Derived MCF7 Breast Cancer Stem Cells (CD44+/CD24–/Low): An in Vitro Study." *Drug Design, Development and Therapy* 9: 1193.

Ai, Zhilong, Hongtao Pan, Tao Suo, Chentao Lv, Yueqi Wang, Saixiong Tong, and Houbao Liu. 2011. "Arsenic Oxide Targets Stem Cell Marker CD133/Prominin-1 in Gallbladder Carcinoma." *Cancer Letters* 310(2): 181–87.

Al-Hajj, Muhammad, Max S Wicha, Adalberto Benito-Hernandez, Sean J Morrison, and Michael F Clarke. 2003. "Prospective Identification of Tumorigenic Breast Cancer Cells." *Proceedings of the National Academy of Sciences* 100(7): 3983–88.

Almeida, Gracielle C de, Luiz F S Oliveira, Danilo Predes, Harold H Fokoue, Ricardo M Kuster, Felipe L Oliveira, Fabio A Mendes, and Jose G Abreu. 2020. "Piperine Suppresses the Wnt/β-Catenin Pathway and Has Anti-Cancer Effects on Colorectal Cancer Cells." *Scientific Reports* 10(1): 1–12.

Andersson, Emma R, Rickard Sandberg, and Urban Lendahl. 2011. "Notch Signaling: Simplicity in Design, Versatility in Function." *Development* 138(17): 3593–612.

Androutsellis-Theotokis, Andreas, Ronen R Leker, Frank Soldner, Daniel J Hoeppner, Rea Ravin, Steve W Poser, Maria A Rueger, Soo-Kyung Bae, Raja Kittappa, and Ronald D G McKay. 2006. "Notch Signalling Regulates Stem Cell Numbers in Vitro and in Vivo." *Nature* 442(7104): 823–26.

Baharuddin, Puteri, Nazilah Satar, Kamal Shaik Fakiruddin, Norashikin Zakaria, Moon Nian Lim, Narazah Mohd Yusoff, Zubaidah Zakaria, and Badrul Hisham Yahaya. 2016. "Curcumin Improves the Efficacy of Cisplatin by Targeting Cancer Stem-like Cells through P21 and Cyclin D1-Mediated Tumour Cell Inhibition in Non-Small Cell Lung Cancer Cell Lines." *Oncology Reports* 35(1): 13–25.

Banerjee, Sanjeev, Yiwei Li, Zhiwei Wang, and Fazlul H Sarkar. 2008. "Multi-Targeted Therapy of Cancer by Genistein." *Cancer Letters* 269(2): 226–42.

Bapat, Sharmila A, Anagha Krishnan, Avinash D Ghanate, Anjali P Kusumbe, and Rajkumar S Kalra. 2010. "Gene Expression: Protein Interaction Systems Network Modeling Identifies Transformation-Associated Molecules and Pathways in Ovarian CancerSystems Network Analyses of Serous Ovarian Cancer." *Cancer Research* 70(12): 4809–19.

Basati, Gholam, Hadiseh Mohammadpour, and Amirnader Emami Razavi. 2020. "Association of High Expression Levels of SOX2, NANOG, and OCT4 in Gastric Cancer Tumor Tissues with Progression and Poor Prognosis." *Journal of Gastrointestinal Cancer* 51(1): 41–47.

Beachy, P A. 2004. "Karhadkar SS, Berman DM." *Tissue Repair and Stem Cell Renewal in Carcinogenesis. Nature* 432: 324–31.

Becker, Laren, Qin Huang, and Hiroshi Mashimo. 2008. "Immunostaining of Lgr5, an Intestinal Stem Cell Marker, in Normal and Premalignant Human Gastrointestinal Tissue." *TheScientificWorldJournal* 8: 1168–76.

Berardi, Damian Emilio, Lizeth Ariza Bareño, Natalia Amigo, Luciana Cañonero, Maria de las Nieves Pelagatti, Andrea Nora Motter, María Agustina Taruselli, María Inés Díaz Bessone, Stefano Martin Cirigliano, and Alexis Edelstein. 2021. "All-Trans Retinoic Acid and Protein Kinase C α/B1 Inhibitor Combined Treatment Targets Cancer Stem Cells and Impairs Breast Tumor Progression." *Scientific Reports* 11(1): 1–17.

Bhummaphan, Narumol, Nalinrat Petpiroon, Ornjira Prakhongcheep, Boonchoo Sritularak, and Pithi Chanvorachote. 2019. "Lusianthridin Targeting of Lung Cancer Stem Cells via Src-STAT3 Suppression." *Phytomedicine* 62: 152932.

Bolós, V, J Grego-Bessa, and JL La Pompa de. 2007. "Notch Signaling in Development and Cancer." *Endocrine Reviews* 28: 339–63.

Bolós, Victoria, Moisés Blanco, Vanessa Medina, Guadalupe Aparicio, Silvia Díaz-Prado, and Enrique Grande. 2009. "Notch Signalling in Cancer Stem Cells." *Clinical and Translational Oncology* 11(1): 11–19.

Bonnet, Dominique, and John E Dick. 1997. "Human Acute Myeloid Leukemia Is Organized as a Hierarchy That Originates from a Primitive Hematopoietic Cell." *Nature Medicine* 3(7): 730–37.

Bourguignon, Lilly Y W, Gabriel Wong, and Christine Earle. 2014. "Hyaluronan-CD44v3 Interaction with Oct4/Sox2/Nanog Promotes MiR-302 Expression Leading to Self-Renewal, Clonal Formation and Cisplatin Resistance in Cancer Stem Cells from Head and Neck Squamous Cell Carcinoma." *Cancer Research* 74(19_Supplement): 3861.

Bourseau-Guilmain, Erika, Jérôme Bejaud, Audrey Griveau, Nolwenn Lautram, François Hindré, M Weyland, Jean-Pierre Benoit, and Emmanuel Garcion. 2012. "Development and Characterization of Immuno-Nanocarriers Targeting the Cancer Stem Cell Marker AC133." *International Journal of Pharmaceutics* 423(1): 93–101.

Bray, Sarah J. 2006. "Notch Signalling: A Simple Pathway Becomes Complex." *Nature Reviews Molecular Cell Biology* 7(9): 678–89.

Briscoe, James, and Pascal P Thérond. 2013. "The Mechanisms of Hedgehog Signalling and Its Roles in Development and Disease." *Nature Reviews Molecular Cell Biology* 14(7): 416–29.

Carracedo, A1, and P P Pandolfi. 2008. "The PTEN–PI3K Pathway: Of Feedbacks and Cross-Talks." *Oncogene* 27(41): 5527–41.

Cha, Jae Hoon, Woo Kyoung Kim, Ae Wha Ha, Myung Hwan Kim, and Moon Jeong Chang. 2017. "Anti-Inflammatory Effect of Lycopene in SW480 Human Colorectal Cancer Cells." *Nutrition Research and Practice* 11(2): 90–96.

Chandler, Hollie, Harshil Patel, Richard Palermo, Sharon Brookes, Nik Matthews, and Gordon Peters. 2014. "Role of Polycomb Group Proteins in the DNA Damage Response–a Reassessment." *PLoS One* 9(7): e102968.

Chanvorachote, Pithi, Varisa Pongrakhananon, Sumalee Wannachaiyasit, Sudjit Luanpitpong, Yon Rojanasakul, and Ubonthip Nimmannit. 2009. "Curcumin Sensitizes Lung Cancer Cells to Cisplatin-Induced Apoptosis through Superoxide Anion-Mediated Bcl-2 Degradation." *Cancer Investigation* 27(6): 624–35.

Chen, Bao-Yuan, Chia-Hung Kuo, Yung-Chuan Liu, Li-Yi Ye, Jiann-Hwa Chen, and Chwen-Jen Shieh. 2012. "Ultrasonic-Assisted Extraction of the Botanical Dietary Supplement Resveratrol and Other Constituents of Polygonum Cuspidatum." *Journal of Natural Products* 75(10): 1810–13.

Chen, Di, Sreedhar Pamu, Qiuzhi Cui, Tak Hang Chan, and Q Ping Dou. 2012. "Novel Epigallocatechin Gallate (EGCG) Analogs Activate AMP-Activated Protein Kinase Pathway and Target Cancer Stem Cells." *Bioorganic & Medicinal Chemistry* 20(9): 3031–37.

Chen, Junsong, Jing Wang, Dengyu Chen, Jie Yang, Cuiping Yang, Yunxia Zhang, Hongyi Zhang, and Jun Dou. 2013. "Evaluation of Characteristics of CD44+ CD117+ Ovarian Cancer Stem Cells in Three Dimensional Basement Membrane Extract Scaffold versus Two Dimensional Monocultures." *BMC Cell Biology* 14(1): 1–11.

Chen, Pei-Ni, Yih-Shou Hsieh, Hui-Ling Chiou, and Shu-Chen Chu. 2005. "Silibinin Inhibits Cell Invasion through Inactivation of Both PI3K-Akt and MAPK Signaling Pathways." *Chemico-Biological Interactions* 156(2–3): 141–50.

Chen, Xiaoli, Shilpa Lingala, Shiva Khoobyari, Jan Nolta, Mark A Zern, and Jian Wu. 2011. "Epithelial Mesenchymal Transition and Hedgehog Signaling Activation Are Associated with Chemoresistance and Invasion of Hepatoma Subpopulations." *Journal of Hepatology* 55(4): 838–45.

Chen, Ying-Xuan, Qin-Yan Gao, Tian-Hui Zou, Bang-Mao Wang, Si-De Liu, Jian-Qiu Sheng, Jian-Lin Ren, Xiao-Ping Zou, Zhan-Ju Liu, and Yan-Yan Song. 2020. "Berberine versus Placebo for the Prevention of Recurrence of Colorectal Adenoma: A Multicentre, Double-Blinded, Randomised Controlled Study." *The Lancet Gastroenterology & Hepatology* 5(3): 267–75.

Chen, Yu-Jen, Ying-Yin Chen, Yi-Feng Lin, Hsuan-Yun Hu, and Hui-Fen Liao. 2013. "Resveratrol Inhibits Alpha-Melanocyte-Stimulating Hormone Signaling, Viability, and Invasiveness in Melanoma Cells." *Evidence-Based Complementary and Alternative Medicine* 2013.

Chu, Man, Cheng Zheng, Cheng Chen, Gendi Song, Xiaoli Hu, and Zhi-wei Wang. 2022. "Targeting Cancer Stem Cells by Nutraceuticals for Cancer Therapy." *Seminars in Cancer Biology* 85: 234–45.

Clarke, John D, Roderick H Dashwood, and Emily Ho. 2008. "Multi-Targeted Prevention of Cancer by Sulforaphane." *Cancer Letters* 269(2): 291–304.

Cooper, Melinda. 2009. "Regenerative Pathologies: Stem Cells, Teratomas and Theories of Cancer." *Medicine Studies* 1(1): 55–66.

Cui, Junwei, Peng Li, Xiaoling Liu, Hui Hu, and Wei Wei. 2015. "Abnormal Expression of the Notch and Wnt/β-Catenin Signaling Pathways in Stem-like ALDHhiCD44+ Cells Correlates Highly with Ki-67 Expression in Breast Cancer." *Oncology Letters* 9(4): 1600–6.

Dai, Xinzheng, Jing Ge, Xuehao Wang, Xiaofeng Qian, Chuanyong Zhang, and Xiangcheng Li. 2013. "OCT4 Regulates Epithelial-Mesenchymal Transition and Its Knockdown Inhibits Colorectal Cancer Cell Migration and Invasion." *Oncology Reports* 29(1): 155–60.

Dandawate, Prasad, Subhash Padhye, Aamir Ahmad, and Fazlul H Sarkar. 2013. "Novel Strategies Targeting Cancer Stem Cells through Phytochemicals and Their Analogs." *Drug Delivery and Translational Research* 3(2): 165–82.

Dang, Chi V. 1999. "C-Myc Target Genes Involved in Cell Growth, Apoptosis, and Metabolism." *Molecular and Cellular Biology* 19(1): 1–11.

Darwish, Noureldien H E, Thangirala Sudha, Kavitha Godugu, Osama Elbaz, Hasan A Abdelghaffar, Emad E A Hassan, and Shaker A Mousa. 2016. "Acute Myeloid Leukemia Stem Cell Markers in Prognosis and Targeted Therapy: Potential Impact of BMI-1, TIM-3 and CLL-1." *Oncotarget* 7(36): 57811.

Deng, Junli, Xupeng Bai, Xiaojie Feng, Jie Ni, Julia Beretov, Peter Graham, and Yong Li. 2019. "Inhibition of PI3K/Akt/MTOR Signaling Pathway Alleviates Ovarian Cancer Chemoresistance through Reversing Epithelial-Mesenchymal Transition and Decreasing Cancer Stem Cell Marker Expression." *BMC Cancer* 19(1): 1–12.

Dey, Nandini, Pradip De, and Brian Leyland-Jones. 2017. "PI3K-AKT-MTOR Inhibitors in Breast Cancers: From Tumor Cell Signaling to Clinical Trials." *Pharmacology & Therapeutics* 175: 91–106.

Ding, Wei, Marialena Mouzaki, Hanning You, Joshua C Laird, Jose Mato, Shelly C Lu, and C Bart Rountree. 2009. "CD133+ Liver Cancer Stem Cells from Methionine Adenosyl Transferase 1A–Deficient Mice Demonstrate Resistance to Transforming Growth Factor (TGF)-β–Induced Apoptosis." *Hepatology* 49 (4): 1277–86.

Dong, Weilei, A Chen, Xiaocheng Chao, Xiang Li, Y Cui, Chang Xu, Jianguo Cao, Yingxia Ning, and X Cao. 2019. "Chrysin Inhibits Proinflammatory Factor-Induced EMT Phenotype and Cancer Stem Cell-like Features in HeLa Cells by Blocking the NF-KB/Twist Axis." *Cell Physiol Biochem* 52(5): 1236–50.

Du, Guang-Jian, Zhiyu Zhang, Xiao-Dong Wen, Chunhao Yu, Tyler Calway, Chun-Su Yuan, and Chong-Zhi Wang. 2012. "Epigallocatechin Gallate (EGCG) Is the Most Effective Cancer Chemopreventive Polyphenol in Green Tea." *Nutrients* 4(11): 1679–91.

El-Benhawy, Sanaa A, Heba G El-Sheredy, Heba B Ghanem, and Amira A Abo El-Soud. 2020. "Berberine Can Amplify Cytotoxic Effect of Radiotherapy by Targeting Cancer Stem Cells." *Breast Cancer Management* 9(2): BMT41.

Es, Johan H van, and Hans Clevers. 2005. "Notch and Wnt Inhibitors as Potential New Drugs for Intestinal Neoplastic Disease." *Trends in Molecular Medicine* 11(11): 496–502.

Espinoza, Ingrid, and Lucio Miele. 2013. "Notch Inhibitors for Cancer Treatment." *Pharmacology & Therapeutics* 139(2): 95–110.

Fan, Panhong, Shujun Fan, Huan Wang, Jun Mao, Yu Shi, Mohammed M Ibrahim, Wei Ma, Xiaotang Yu, Zhenhuan Hou, and Bo Wang. 2013. "Genistein Decreases the Breast Cancer Stem-like Cell Population through Hedgehog Pathway." *Stem Cell Research & Therapy* 4(6): 1–10.

Fan, Xiang-Shan, Hong-Yan Wu, Hui-Ping Yu, Qiang Zhou, Yi-Fen Zhang, and Qin Huang. 2010. "Expression of Lgr5 in Human Colorectal Carcinogenesis and Its Potential Correlation with β-Catenin." *International Journal of Colorectal Disease* 25(5): 583–90.

Freund, Robert R A, Philipp Gobrecht, Dietmar Fischer, and Hans-Dieter Arndt. 2020. "Advances in Chemistry and Bioactivity of Parthenolide." *Natural Product Reports* 37(4): 541–65.

Friedmann-Morvinski, Dinorah, and Inder M Verma. 2014. "Dedifferentiation and Reprogramming: Origins of Cancer Stem Cells." *EMBO Reports* 15(3): 244–53.

Fu, Vivian, Steven W Plouffe, and Kun-Liang Guan. 2017. "The Hippo Pathway in Organ Development, Homeostasis, and Regeneration." *Current Opinion in Cell Biology* 49: 99–107.

Fujino, Shiki, and Norikatsu Miyoshi. 2019. "Oct4 Gene Expression in Primary Colorectal Cancer Promotes Liver Metastasis." *Stem Cells International* 2019: 7896524.

Fukusumi, T, H Ishii, M Konno, T Yasui, S Nakahara, Y Takenaka, Y Yamamoto, S Nishikawa, Y Kano, and H Ogawa. 2014. "CD10 as a Novel Marker of Therapeutic Resistance and Cancer Stem Cells in Head and Neck Squamous Cell Carcinoma." *British Journal of Cancer* 111(3): 506–14.

Gan, Ren-You, Hua-Bin Li, Zhong-Quan Sui, and Harold Corke. 2018. "Absorption, Metabolism, Anti-Cancer Effect and Molecular Targets of Epigallocatechin Gallate (EGCG): An Updated Review." *Critical Reviews in Food Science and Nutrition* 58(6): 924–41.

Gao, M Q, Y P Choi, S Kang, J H Youn, and N H Cho. 2010. "CD24+ Cells from Hierarchically Organized Ovarian Cancer Are Enriched in Cancer Stem Cells." *Oncogene* 29(18): 2672–80.

Gao, Ming-Qing, Yan-Tao Han, Li Zhu, Shou-Guo Chen, Zhen-Yu Hong, and Chun-Bo Wang. 2009. "Cytotoxicity of Natural Extract from Tegillarca Granosa on Ovarian Cancer Cells Is Mediated by Multiple Molecules." *Clinical and Investigative Medicine* 32(5): E368–75.

Geng, Songmei, Yuanyuan Guo, Qianqian Wang, Lan Li, and Jianli Wang. 2013. "Cancer Stem-like Cells Enriched with CD29 and CD44 Markers Exhibit Molecular Characteristics with Epithelial–Mesenchymal Transition in Squamous Cell Carcinoma." *Archives of Dermatological Research* 305(1): 35–47.

George, Jasmine, Minakshi Nihal, Chandra K Singh, and Nihal Ahmad. 2019. "4′-Bromo-resveratrol, a Dual Sirtuin-1 and Sirtuin-3 Inhibitor, Inhibits Melanoma Cell Growth through Mitochondrial Metabolic Reprogramming." *Molecular Carcinogenesis* 58(10): 1876–85.

Ghaffari, Saghi. 2011. "Cancer, Stem Cells and Cancer Stem Cells: Old Ideas, New Developments." *F1000 Medicine Reports* 3.

Giordano, Antonio, and Giuseppina Tommonaro. 2019. "Curcumin and Cancer." *Nutrients* 11(10): 2376.

Goldstein, Leslie A, David F H Zhou, Louis J Picker, Catherine N Minty, Robert F Bargatze, Jie F Ding, and Eugene C Butcher. 1989. "A Human Lymphocyte Homing Receptor, the Hermes Antigen, Is Related to Cartilage Proteoglycan Core and Link Proteins." *Cell* 56(6): 1063–72.

Gómez-del Arco, Pablo, Mariko Kashiwagi, Audrey F Jackson, Taku Naito, Jiangwen Zhang, Feifei Liu, Barbara Kee, Marc Vooijs, Freddy Radtke, and Juan Miguel Redondo. 2010. "Alternative Promoter Usage at the Notch1 Locus Supports Ligand-Independent Signaling in T Cell Development and Leukemogenesis." *Immunity* 33(5): 685–98.

Gonzalez-Torres, Carolina, Javier Gaytan-Cervantes, Karla Vazquez-Santillan, Edna Ayerim Mandujano-Tinoco, Gisela Ceballos-Cancino, Alfredo Garcia-Venzor, Cecilia Zampedri, Paulina Sanchez-Maldonado, Raul Mojica-Espinosa, and Luis Enrique Jimenez-Hernandez. 2017. "NF-KB Participates in the Stem Cell Phenotype of Ovarian Cancer Cells." *Archives of Medical Research* 48(4): 343–51.

Goodfellow, P J. 1989. "Gene Symbol Report." HUGO Gene Nomenclature Committee. https://www.genenames.org/data/gene-symbol-report/#!/hgnc_id/HGNC:6153.

Grosse-Gehling, Philipp, Christine A Fargeas, Claudia Dittfeld, Yvette Garbe, Malcolm R Alison, Denis Corbeil, and Leoni A Kunz-Schughart. 2013. "CD133 as a Biomarker for Putative Cancer Stem Cells in Solid Tumours: Limitations, Problems and Challenges." *The Journal of Pathology* 229(3): 355–78.

Gupta, Subash C, Sridevi Patchva, and Bharat B Aggarwal. 2013. "Therapeutic Roles of Curcumin: Lessons Learned from Clinical Trials." *The AAPS Journal* 15(1): 195–218.

Habib, Joyce G, and Joyce A O'Shaughnessy. 2016. "The Hedgehog Pathway in Triple-negative Breast Cancer." *Cancer Medicine* 5(10): 2989–3006.

Habtemariam, Solomon. 2016. "Berberine and Inflammatory Bowel Disease: A Concise Review." *Pharmacological Research* 113: 592–99.

Hallajzadeh, Jamal, Parisa Maleki Dana, Moein Mobini, Zatollah Asemi, Mohammad Ali Mansournia, Mehran Sharifi, and Bahman Yousefi. 2020. "Targeting of Oncogenic Signaling Pathways by Berberine for Treatment of Colorectal Cancer." *Medical Oncology* 37(6): 1–9.

Hayden, Matthew S, and Sankar Ghosh. 2008. "Shared Principles in NF-KB Signaling." *Cell* 132(3): 344–62.

Hermann, Patrick C, Sonu Bhaskar, Michele Cioffi, and Christopher Heeschen. 2010. "Cancer Stem Cells in Solid Tumors." *Seminars in Cancer Biology*, 20:77–84.

Herrmann, Harald, Irina Sadovnik, Sabine Cerny-Reiterer, Thomas Rülicke, Gabriele Stefanzl, Michael Willmann, Gregor Hoermann, Martin Bilban, Katharina Blatt, and Susanne Herndlhofer. 2014. "Dipeptidylpeptidase IV (CD26) Defines Leukemic Stem Cells (LSC) in Chronic Myeloid Leukemia." *Blood, The Journal of the American Society of Hematology* 123(25): 3951–62.

Hewlings, Susan J, and Douglas S Kalman 2017. "Curcumin: A Review of Its' Effects on Human Health." *Foods* 6: 10–92.

Hilton, John. 1984. "Role of Aldehyde Dehydrogenase in Cyclophosphamide-Resistant L1210 Leukemia." *Cancer Research* 44(11): 5156–60.

Holland, Jane D, Alexandra Klaus, Alistair N Garratt, and Walter Birchmeier. 2013. "Wnt Signaling in Stem and Cancer Stem Cells." *Current Opinion in Cell Biology* 25(2): 254–64.

Honoki, Kanya, Hiromasa Fujii, Atsushi Kubo, Akira Kido, Toshio Mori, Yasuhito Tanaka, and Toshifumi Tsujiuchi. 2010. "Possible Involvement of Stem-like Populations with Elevated ALDH1 in Sarcomas for Chemotherapeutic Drug Resistance." *Oncology Reports* 24(2): 501–5.

Hou, Yanli, Bowen Sun, Wenxue Liu, Bo Yu, Qiqi Shi, Fei Luo, Yongrui Bai, and Haizhong Feng. 2021. "Targeting of Glioma Stem-like Cells with a Parthenolide Derivative ACT001 through Inhibition of AEBP1/PI3K/AKT Signaling." *Theranostics* 11(2): 555.

Hough, Margaret R, Patricia M Rosten, Tracy L Sexton, Robert Kay, and R Keith Humphries. 1994. "Mapping of CD24 and Homologous Sequences to Multiple Chromosomal Loci." *Genomics* 22(1): 154–61.

Huang, Weifeng, Chunpeng Wan, Qicong Luo, Zhengjie Huang, and Qi Luo. 2014. "Genistein-Inhibited Cancer Stem Cell-like Properties and Reduced Chemoresistance of Gastric Cancer." *International Journal of Molecular Sciences* 15(3): 3432–43.

Hung, Chi-Feng, Tur-Fu Huang, Bing-Huei Chen, Jiunn-Min Shieh, Pi-Hui Wu, and Wen-Bin Wu. 2008. "Lycopene Inhibits TNF-α-Induced Endothelial ICAM-1 Expression and Monocyte-Endothelial Adhesion." *European Journal of Pharmacology* 586(1–3): 275–82.

Hurt, Elaine M, Brian T Kawasaki, George J Klarmann, Suneetha B Thomas, and William L Farrar. 2008. "CD44+ CD24– Prostate Cells Are Early Cancer Progenitor/Stem Cells That Provide a Model for Patients with Poor Prognosis." *British Journal of Cancer* 98(4): 756–65.

Imran, Muhammad, Fereshteh Ghorat, Iahtisham Ul-Haq, Habib Ur-Rehman, Farhan Aslam, Mojtaba Heydari, Mohammad Ali Shariati, Eleonora Okuskhanova, Zhanibek Yessimbekov, and Muthu Thiruvengadam. 2020. "Lycopene as a Natural Antioxidant Used to Prevent Human Health Disorders." *Antioxidants* 9(8): 706.

Jaggupilli, Appalaraju, and Eyad Elkord. 2012. "Significance of CD44 and CD24 as Cancer Stem Cell Markers: An Enduring Ambiguity." *Clinical and Developmental Immunology* 2012: 708036.

Jahan, Sadaf, Shouvik Mukherjee, Shaheen Ali, Urvashi Bhardwaj, Ranjay Kumar Choudhary, Santhanaraj Balakrishnan, Asma Naseem, Shabir Ahmad Mir, Saeed Banawas, and Mohammed Alaidarous. 2022. "Pioneer Role of Extracellular Vesicles as Modulators of Cancer Initiation in Progression, Drug Therapy, and Vaccine Prospects." *Cells* 11(3): 490.

Jeter, Collene R, Tao Yang, Junchen Wang, Hsueh-Ping Chao, and Dean G Tang. 2015. "Concise Review: NANOG in Cancer Stem Cells and Tumor Development: An Update and Outstanding Questions." *Stem Cells* 33(8): 2381–90.

Jiang, Xiaomo, Huai-Xiang Hao, Joseph D Growney, Steve Woolfenden, Cindy Bottiglio, Nicholas Ng, Bo Lu, Mindy H Hsieh, Linda Bagdasarian, and Ronald Meyer. 2013. "Inactivating Mutations of RNF43

Confer Wnt Dependency in Pancreatic Ductal Adenocarcinoma." *Proceedings of the National Academy of Sciences* 110(31): 12649–54.

Kallifatidis, Georgios, Vanessa Rausch, B Baumann, A Apel, B M Beckermann, A Groth, J Mattern, Z Li, A Kolb, and P Altevogt. 2008. "Sulforaphane Eradicates Pancreatic Cancer Stem Cells by NF-KB." *Anticancer Research*, 28: 3313.

Katoh, Masaru. 2017. "Canonical and Non-Canonical WNT Signaling in Cancer Stem Cells and Their Niches: Cellular Heterogeneity, Omics Reprogramming, Targeted Therapy and Tumor Plasticity." *International Journal of Oncology* 51(5): 1357–69.

Katz, Menachem, Ido Amit, and Yosef Yarden. 2007. "Regulation of MAPKs by Growth Factors and Receptor Tyrosine Kinases." *Biochimica et Biophysica Acta (BBA)-Molecular Cell Research* 1773(8): 1161–76.

Kemenade, Folkert J van, Frank M Raaphorst, Tjasso Blokzijl, Elly Fieret, Karien M Hamer, David P E Satijn, Arie P Otte, and Chris J L M Meijer. 2001. "Coexpression of BMI-1 and EZH2 Polycomb-Group Proteins Is Associated with Cycling Cells and Degree of Malignancy in B-Cell Non-Hodgkin Lymphoma." *Blood, The Journal of the American Society of Hematology* 97(12): 3896–901.

Kemper, Kristel, Pramudita R Prasetyanti, Wim De Lau, Hans Rodermond, Hans Clevers, and Jan Paul Medema. 2012. "Monoclonal Antibodies against Lgr5 Identify Human Colorectal Cancer Stem Cells." *Stem Cells* 30(11): 2378–86.

Kiger, Amy A, D Leanne Jones, Cordula Schulz, Madolyn B Rogers, and Margaret T Fuller. 2001. "Stem Cell Self-Renewal Specified by JAK-STAT Activation in Response to a Support Cell Cue." *Science* 294 (5551): 2542–45.

Kikushige, Yoshikane, and Toshihiro Miyamoto. 2013. "TIM-3 as a Novel Therapeutic Target for Eradicating Acute Myelogenous Leukemia Stem Cells." *International Journal of Hematology* 98(6): 627–33.

Kikushige, Yoshikane, Toshihiro Miyamoto, Junichiro Yuda, Siamak Jabbarzadeh-Tabrizi, Takahiro Shima, Shin-ichiro Takayanagi, Hiroaki Niiro, Ayano Yurino, Kohta Miyawaki, and Katsuto Takenaka. 2015. "A TIM-3/Gal-9 Autocrine Stimulatory Loop Drives Self-Renewal of Human Myeloid Leukemia Stem Cells and Leukemic Progression." *Cell Stem Cell* 17(3): 341–52.

Kikushige, Yoshikane, Takahiro Shima, Shin-ichiro Takayanagi, Shingo Urata, Toshihiro Miyamoto, Hiromi Iwasaki, Katsuto Takenaka, Takanori Teshima, Toshiyuki Tanaka, and Yoshimasa Inagaki. 2010. "TIM-3 Is a Promising Target to Selectively Kill Acute Myeloid Leukemia Stem Cells." *Cell Stem Cell* 7(6): 708–17.

Kim, Daeun, Yerin Kim, and Yuri Kim. 2019. "Effects of β-Carotene on Expression of Selected MicroRNAs, Histone Acetylation, and DNA Methylation in Colon Cancer Stem Cells." *Journal of Cancer Prevention* 24(4): 224.

Kim, Eun-Kyung, Kang-Beom Kwon, Mi-Young Song, Sang-Wan Seo, Sung-Joo Park, Sun-O Ka, Lv Na, Kyung-Ah Kim, Do-Gon Ryu, and Hong-Seob So. 2007. "Genistein Protects Pancreatic β Cells against Cytokine-Mediated Toxicity." *Molecular and Cellular Endocrinology* 278(1–2): 18–28.

Kim, Jee-Heun, So-Yeon Park, Youngsoo Jun, Ji-Young Kim, and Jeong-Seok Nam. 2017. "Roles of Wnt Target Genes in the Journey of Cancer Stem Cells." *International Journal of Molecular Sciences* 18(8): 1604.

Kim, Jong Bin, Seulki Lee, Hye Ri Kim, Seo-Young Park, Minjong Lee, Jung-Hwan Yoon, and Yoon Jun Kim. 2018. "Transforming Growth Factor-β Decreases Side Population Cells in Hepatocellular Carcinoma in Vitro." *Oncology Letters* 15(6): 8723–28.

Kim, Won-Tae, and Chun Jeih Ryu. 2017. "Cancer Stem Cell Surface Markers on Normal Stem Cells." *BMB Reports* 50(6): 285.

Kleist, Britta, Li Xu, Guojun Li, and Christian Kersten. 2011. "Expression of the Adult Intestinal Stem Cell Marker Lgr5 in the Metastatic Cascade of Colorectal Cancer." *International Journal of Clinical and Experimental Pathology* 4(4): 327.

Koike, Naoto. 2018. "The Role of Stem Cells in the Hepatobiliary System and in Cancer Development: A Surgeon's Perspective." In Yun-Wen Zheng (ed.), *Stem Cells and Cancer in Hepatology*, 211–53. Elsevier.

Kramer, K, J Wu, and D L Crowe. 2016. "Tumor Suppressor Control of the Cancer Stem Cell Niche." *Oncogene* 35(32): 4165–78.

Ksander, Bruce R, Paraskevi E Kolovou, Brian J Wilson, Karim R Saab, Qin Guo, Jie Ma, Sean P McGuire, Meredith S Gregory, William J B Vincent, and Victor L Perez. 2014. "ABCB5 Is a Limbal Stem Cell Gene Required for Corneal Development and Repair." *Nature* 511(7509): 353–57.

Lamar, John M, Vijeyaluxmy Motilal Nehru, and Guy Weinberg. 2018. "Epithelioid Hemangioendothelioma as a Model of YAP/TAZ-Driven Cancer: Insights from a Rare Fusion Sarcoma." *Cancers* 10(7): 229.

Lambert, Joshua D, and Ryan J Elias. 2010. "The Antioxidant and Pro-Oxidant Activities of Green Tea Polyphenols: A Role in Cancer Prevention." *Archives of Biochemistry and Biophysics* 501(1): 65–72.

Lapidot, Tsvee, Christian Sirard, Josef Vormoor, Barbara Murdoch, Trang Hoang, Julio Caceres-Cortes, Mark Minden, Bruce Paterson, Michael A Caligiuri, and John E Dick. 1994. "A Cell Initiating Human Acute Myeloid Leukaemia after Transplantation into SCID Mice." *Nature* 367(6464): 645–48.

Lau, Wen Min, Eileen Teng, Hui Shan Chong, Kirsten Anne Pagaduan Lopez, Amy Yuh Ling Tay, Manuel Salto-Tellez, Asim Shabbir, Jimmy Bok Yan So, and Shing Leng Chan. 2014. "CD44v8–10 Is a Cancer-Specific Marker for Gastric Cancer Stem CellsCD44v8–10 Marks Gastric Cancer Stem Cells." *Cancer Research* 74(9): 2630–41.

Lee, Cheong J, Joseph Dosch, and Diane M Simeone. 2008. "Pancreatic Cancer Stem Cells." *Journal of Clinical Oncology* 26(17): 2806–12.

Lee, May Yin, Li Sun, and Jacqueline M Veltmaat. 2013. "Hedgehog and Gli Signaling in Embryonic Mammary Gland Development." *Journal of Mammary Gland Biology and Neoplasia* 18(2): 133–38.

Lee, Sang Hyuk, Hyo Jung Nam, Hyun Jung Kang, Hye Won Kwon, and Young Chang Lim. 2013. "Epigallocatechin-3-Gallate Attenuates Head and Neck Cancer Stem Cell Traits through Suppression of Notch Pathway." *European Journal of Cancer* 49(15): 3210–18.

Lee, Terence Kin Wah, Antonia Castilho, Vincent Chi Ho Cheung, Kwan Ho Tang, Stephanie Ma, and Irene Oi Lin Ng. 2011. "CD24+ Liver Tumor-Initiating Cells Drive Self-Renewal and Tumor Initiation through STAT3-Mediated NANOG Regulation." *Cell Stem Cell* 9(1): 50–63.

Leek, Russell D, and Adrian L Harris. 2002. "Tumor-Associated Macrophages in Breast Cancer." *Journal of Mammary Gland Biology and Neoplasia* 7(2): 177–89.

Lestari, Maria L A D, and Gunawan Indrayanto. 2014. "Curcumin." *Profiles of Drug Substances, Excipients and Related Methodology* 39: 113–204.

Li, Qian-Qian, You-Ke Xie, Yue Wu, Lin-Lin Li, Ying Liu, Xiao-Bo Miao, Qiu-Zhen Liu, Kai-Tai Yao, and Guang-Hui Xiao. 2017. "Sulforaphane Inhibits Cancer Stem-like Cell Properties and Cisplatin Resistance through MiR-214-Mediated Downregulation of c-MYC in Non-Small Cell Lung Cancer." *Oncotarget* 8(7): 12067.

Li, Qing-Shan, Cui-Yun Li, Zi-Lin Li, and Hai-Liang Zhu. 2012. "Genistein and Its Synthetic Analogs as Anticancer Agents." *Anti-Cancer Agents in Medicinal Chemistry (Formerly Current Medicinal Chemistry-Anti-Cancer Agents)* 12(3): 271–81.

Li, Shizheng, Qian Zhao, Bo Wang, Song Yuan, Xiuyan Wang, and Kun Li. 2018. "Quercetin Reversed MDR in Breast Cancer Cells through Down-regulating P-gp Expression and Eliminating Cancer Stem Cells Mediated by YB-1 Nuclear Translocation." *Phytotherapy Research* 32(8): 1530–36.

Li, Shuang-Jiang, Jian Huang, Xu-Dong Zhou, Wen-Biao Zhang, Yu-Tian Lai, and Guo-Wei Che. 2016. "Clinicopathological and Prognostic Significance of Oct-4 Expression in Patients with Non-Small Cell Lung Cancer: A Systematic Review and Meta-Analysis." *Journal of Thoracic Disease* 8(7): 1587.

Li, Xiuli, Na Zhou, Jin Wang, Zhijie Liu, Xiaohui Wang, Qin Zhang, Qingyan Liu, Lifeng Gao, and Rong Wang. 2018. "Quercetin Suppresses Breast Cancer Stem Cells (CD44+/CD24–) by Inhibiting the PI3K/Akt/MTOR-Signaling Pathway." *Life Sciences* 196: 56–62.

Li, Yanyan, Max S Wicha, Steven J Schwartz, and Duxin Sun. 2011. "Implications of Cancer Stem Cell Theory for Cancer Chemoprevention by Natural Dietary Compounds." *The Journal of Nutritional Biochemistry* 22(9): 799–806.

Li, Yanyan, Tao Zhang, Hasan Korkaya, Suling Liu, Hsiu-Fang Lee, Bryan Newman, Yanke Yu, Shawn G Clouthier, Steven J Schwartz, and Max S Wicha. 2010. "Sulforaphane, a Dietary Component of Broccoli/Broccoli Sprouts, Inhibits Breast Cancer Stem CellsSulforaphane Inhibits Breast Cancer Stem Cells." *Clinical Cancer Research* 16(9): 2580–90.

Liang, Yu, Fengyu Zhu, Haojie Zhang, Demeng Chen, Xiuhong Zhang, Qian Gao, and Yang Li. 2016. "Conditional Ablation of TGF-β Signaling Inhibits Tumor Progression and Invasion in an Induced Mouse Bladder Cancer Model." *Scientific Reports* 6(1): 1–9.

Lin, Che-Yi, Pei-Ling Hsieh, Yi-Wen Liao, Chih-Yu Peng, Ming-Yi Lu, Ching-Hsuan Yang, Cheng-Chia Yu, and Chia-Ming Liu. 2017. "Berberine-Targeted MiR-21 Chemosensitizes Oral Carcinomas Stem Cells." *Oncotarget* 8(46): 80900.

Lin, Chien-Hung, Yao-An Shen, Peir-Haur Hung, Yuan-Bin Yu, and Yann-Jang Chen. 2012. "Epigallocathechin Gallate, Polyphenol Present in Green Tea, Inhibits Stem-like Characteristics and Epithelial-Mesenchymal Transition in Nasopharyngeal Cancer Cell Lines." *BMC Complementary and Alternative Medicine* 12(1): 1–12.

Liu, Dandan, Yanyan Han, Lei Liu, Xinxiu Ren, Han Zhang, Shujun Fan, Tao Qin, and Lianhong Li. 2021. "Parthenolide Inhibits the Tumor Characteristics of Renal Cell Carcinoma." *International Journal of Oncology* 58(1): 100–110.

Liu, Li, Jingyan Fan, Guihai Ai, Jie Liu, Ning Luo, Caixia Li, and Zhongping Cheng. 2019. "Berberine in Combination with Cisplatin Induces Necroptosis and Apoptosis in Ovarian Cancer Cells." *Biological Research* 52(1): 1–14.

Liu, Lucy, Sangkil Nam, Yan Tian, Fan Yang, Jun Wu, Yan Wang, Anna Scuto, Panos Polychronopoulos, Prokopios Magiatis, and Leandros Skaltsounis. 2011. "6-Bromoindirubin-3′-Oxime Inhibits JAK/STAT3 Signaling and Induces Apoptosis of Human Melanoma CellsBromoindirubin Inhibits JAK/STAT3 Signaling." *Cancer Research* 71(11): 3972–79.

Liu, Suling, Gabriela Dontu, Ilia D Mantle, Shivani Patel, Nam-shik Ahn, Kyle W Jackson, Prerna Suri, and Max S Wicha. 2006. "Hedgehog Signaling and Bmi-1 Regulate Self-Renewal of Normal and Malignant Human Mammary Stem Cells." *Cancer Research* 66(12): 6063–71.

Liu, Suling, Gabriela Dontu, and Max S Wicha. 2005. "Mammary Stem Cells, Self-Renewal Pathways, and Carcinogenesis." *Breast Cancer Research* 7(3): 1–10.

Liu, Yanchen, Tianbiao Zou, Shuhuai Wang, Hong Chen, Dongju Su, Xiaona Fu, Qingyuan Zhang, and Xinmei Kang. 2016. "Genistein-Induced Differentiation of Breast Cancer Stem/Progenitor Cells through a Paracrine Mechanism." *International Journal of Oncology* 48(3): 1063–72.

Liu, Zhao, Abhik Bandyopadhyay, Robert W Nichols, Long Wang, Andrew P Hinck, Shui Wang, and Lu-Zhe Sun. 2012. "Blockade of Autocrine TGF-β Signaling Inhibits Stem Cell Phenotype, Survival, and Metastasis of Murine Breast Cancer Cells." *Journal of Stem Cell Research & Therapy* 2(1): 1.

LoPiccolo, Jaclyn, Gideon M Blumenthal, Wendy B Bernstein, and Phillip A Dennis. 2008. "Targeting the PI3K/Akt/MTOR Pathway: Effective Combinations and Clinical Considerations." *Drug Resistance Updates* 11(1–2): 32–50.

Lu, Hua, Quan Zhou, Vishal Deshmukh, Hardeep Phull, Jennifer Ma, Virginie Tardif, Rahul R Naik, Claire Bouvard, Yong Zhang, and Seihyun Choi. 2014. "Targeting Human C-type Lectin-like Molecule-1 (CLL1) with a Bispecific Antibody for Immunotherapy of Acute Myeloid Leukemia." *Angewandte Chemie* 126(37): 9999–10003.

Lu, Rui, Hongxia Dan, Ruiqing Wu, Wenxia Meng, Na Liu, Xin Jin, Min Zhou, Xin Zeng, Gang Zhou, and Qianming Chen. 2011. "Lycopene: Features and Potential Significance in the Oral Cancer and Precancerous Lesions." *Journal of Oral Pathology & Medicine* 40(5): 361–68.

Lundberg, Ida V, Sofia Edin, Vincy Eklöf, Åke Öberg, Richard Palmqvist, and Maria L Wikberg. 2016. "SOX2 Expression Is Associated with a Cancer Stem Cell State and Down-Regulation of CDX2 in Colorectal Cancer." *BMC Cancer* 16(1): 1–11.

Luo, Xiaochuang, Jingyi Gu, Rongxuan Zhu, Maoxiao Feng, Xuejiao Zhu, Yumin Li, and Jia Fei. 2014. "Integrative Analysis of Differential MiRNA and Functional Study of MiR-21 by Seed-Targeting Inhibition in Multiple Myeloma Cells in Response to Berberine." *BMC Systems Biology* 8(1): 1–10.

Ma, Shenghong, Zhipeng Meng, Rui Chen, and Kun-Liang Guan. 2019. "The Hippo Pathway: Biology and Pathophysiology." *Annual Review of Biochemistry* 88: 577–604.

Maehle, Andreas-Holger. 2011. "Ambiguous Cells: The Emergence of the Stem Cell Concept in the Nineteenth and Twentieth Centuries." *Notes and Records of the Royal Society* 65(4): 359–78.

Maguer-Satta, Véronique, Marion Chapellier, Emmanuel Delay, and Elodie Bachelard-Cascales. 2011. "CD10: A Tool to Crack the Role of Stem Cells in Breast Cancer." *Proceedings of the National Academy of Sciences* 108(49): E1264.

Maleki, Narges, Negar Yavari, Maryam Ebrahimi, Ahmad Faisal Faiz, Roya Khosh Ravesh, Aysan Sharbati, Mohammad Panji, Keivan Lorian, Abdollah Gravand, and Mojtaba Abbasi. 2022. "Silibinin Exerts Anti-Cancer Activity on Human Ovarian Cancer Cells by Increasing Apoptosis and Inhibiting Epithelial-Mesenchymal Transition (EMT)." *Gene* 823: 146275.

Mansouri, Kamran, Shna Rasoulpoor, Alireza Daneshkhah, Soroush Abolfathi, Nader Salari, Masoud Mohammadi, Shabnam Rasoulpoor, and Shervin Shabani. 2020. "Clinical Effects of Curcumin in Enhancing Cancer Therapy: A Systematic Review." *BMC Cancer* 20(1): 1–11.

Mao, Jie, Hongbao Yang, Tingting Cui, Pan Pan, Nadia Kabir, Duo Chen, Jinyan Ma, Xingyi Chen, Yijun Chen, and Yong Yang. 2018. "Combined Treatment with Sorafenib and Silibinin Synergistically Targets Both HCC Cells and Cancer Stem Cells by Enhanced Inhibition of the Phosphorylation of STAT3/ERK/AKT." *European Journal of Pharmacology* 832: 39–49.

Marchitti, Satori A, Chad Brocker, Dimitrios Stagos, and Vasilis Vasiliou. 2008. "Non-P450 Aldehyde Oxidizing Enzymes: The Aldehyde Dehydrogenase Superfamily." *Expert Opinion on Drug Metabolism & Toxicology* 4(6): 697–720.

Marfe, Gabriella, Marco Tafani, Manuela Indelicato, Paola Sinibaldi-Salimei, Valentina Reali, Bruna Pucci, Massimo Fini, and Matteo Antonio Russo. 2009. "Kaempferol Induces Apoptosis in Two Different Cell

Lines via Akt Inactivation, Bax and SIRT3 Activation, and Mitochondrial Dysfunction." *Journal of Cellular Biochemistry* 106(4): 643–50.

Mateen, Samiha, Alpna Tyagi, Chapla Agarwal, Rana P Singh, and Rajesh Agarwal. 2010. "Erratum: Silibinin Inhibits Human Nonsmall Cell Lung Cancer Cell Growth through Cell-cycle Arrest by Modulating Expression and Function of Key Cell-cycle Regulators." *Molecular Carcinogenesis* 49(9): 849.

Mayer, M J, L H Klotz, and V Venkateswaran. 2015. "Metformin and Prostate Cancer Stem Cells: A Novel Therapeutic Target." *Prostate Cancer and Prostatic Diseases* 18(4): 303–9.

McClanahan, Terrill, Sandra Koseoglu, Kathleen Smith, Jeffrey Grein, Eric Gustafson, Stuart Black, Paul Kirschmeier, and Ahmed A Samatar. 2006. "Identification of Overexpression of Orphan G Protein-Coupled Receptor GPR49 in Human Colon and Ovarian Primary Tumors." *Cancer Biology & Therapy* 5(4): 419–26.

McCubrey, James A, Kvin Lertpiriyapong, Linda S Steelman, Steve L Abrams, Li V Yang, Ramiro M Murata, Pedro L Rosalen, Aurora Scalisi, Luca M Neri, and Lucio Cocco. 2017. "Effects of Resveratrol, Curcumin, Berberine and Other Nutraceuticals on Aging, Cancer Development, Cancer Stem Cells and MicroRNAs." *Aging (Albany NY)* 9(6): 1477.

Medema, Jan Paul. 2013. "Cancer Stem Cells: The Challenges Ahead." *Nature Cell Biology* 15(4): 338–44.

Merchant, Akil A, and William Matsui. 2010. "Targeting Hedgehog—a Cancer Stem Cell PathwayHedgehog Signaling in Cancer." *Clinical Cancer Research* 16(12): 3130–40.

Miettinen, Markku, and Jerzy Lasota. 2005. "KIT (CD117): A Review on Expression in Normal and Neoplastic Tissues, and Mutations and Their Clinicopathologic Correlation." *Applied Immunohistochemistry & Molecular Morphology* 13(3): 205–20.

Miyauchi, Jun, Colm A Kelleher, Yu-Chung Yang, Gordon G Wong, Steven C Clark, Mark D Minden, Salomon Minkin, and Ernest A McCulloch. 1987. "The Effects of Three Recombinant Growth Factors, IL-3, GM-CSF, and G-CSF, on the Blast Cells of Acute Myeloblastic Leukemia Maintained in Short-Term Suspension Culture." *Blood* 70(3): 657–63.

Mo, Jung-Soon, Hyun Woo Park, and Kun-Liang Guan. 2014. "The Hippo Signaling Pathway in Stem Cell Biology and Cancer." *EMBO Reports* 15(6): 642–56.

Mukherjee, Shravanti, Shilpi Saha, Argha Manna, Minakshi Mazumdar, Samik Chakraborty, Shrutarshi Paul, and Tanya Das. 2014. "Targeting Cancer Stem Cells by Phytochemicals: A Multimodal Approach to Colorectal Cancer." *Current Colorectal Cancer Reports* 10(4): 431–41.

Mustra Rakic, Jelena, Chun Liu, Sudipta Veeramachaneni, Dayong Wu, Ligi Paul, C-Y Oliver Chen, Lynne M Ausman, and Xiang-Dong Wang. 2019. "Lycopene Inhibits Smoke-Induced Chronic Obstructive Pulmonary Disease and Lung Carcinogenesis by Modulating Reverse Cholesterol Transport in FerretsLycopene, COPD, and Lung Carcinogenesis." *Cancer Prevention Research* 12(7): 421–32.

Nautiyal, Jyoti, Sanjeev Banerjee, Shailender S Kanwar, Yingjie Yu, Bhaumik B Patel, Fazlul H Sarkar, and Adhip P N Majumdar. 2011. "Curcumin Enhances Dasatinib-induced Inhibition of Growth and Transformation of Colon Cancer Cells." *International Journal of Cancer* 128(4): 951–61.

Neag, Maria A, Andrei Mocan, Javier Echeverría, Raluca M Pop, Corina I Bocsan, Gianina Crişan, and Anca D Buzoianu. 2018. "Berberine: Botanical Occurrence, Traditional Uses, Extraction Methods, and Relevance in Cardiovascular, Metabolic, Hepatic, and Renal Disorders." *Frontiers in Pharmacology* 9: 557.

Nguyen, Giang Thi Tuyet, Melanie Gertz, and Clemens Steegborn. 2013. "Crystal Structures of Sirt3 Complexes with 4′-Bromo-Resveratrol Reveal Binding Sites and Inhibition Mechanism." *Chemistry & Biology* 20 (11): 1375–85.

Nguyen, Long V, Robert Vanner, Peter Dirks, and Connie J Eaves. 2012. "Cancer Stem Cells: An Evolving Concept." *Nature Reviews Cancer* 12(2): 133–43.

Niehrs, Christof. 2012. "The Complex World of WNT Receptor Signalling." *Nature Reviews Molecular Cell Biology* 13(12): 767–79.

Ning, Yingxia, Weifeng Feng, Xiaocheng Cao, Kaiqun Ren, Meifang Quan, A Chen, Chang Xu, Yebei Qiu, Jianguo Cao, and Xiang Li. 2019. "Genistein Inhibits Stemness of SKOV3 Cells Induced by Macrophages Co-Cultured with Ovarian Cancer Stem-like Cells through IL-8/STAT3 Axis." *Journal of Experimental & Clinical Cancer Research* 38(1): 1–15.

O'Rourke, P Pearl. 2001. "Human Pluripotent Stem Cells: NIH Guidelines." *Molecular Aspects of Medicine* 22(3): 165–70.

Oeckinghaus, Andrea, and Sankar Ghosh. 2009. "The NF-KB Family of Transcription Factors and Its Regulation." *Cold Spring Harbor Perspectives in Biology* 1(4): a000034.

Ong, Choon Kiat, Chutima Subimerb, Chawalit Pairojkul, Sopit Wongkham, Ioana Cutcutache, Willie Yu, John R McPherson, George E Allen, Cedric Chuan Young Ng, and Bernice Huimin Wong. 2012. "Exome Sequencing of Liver Fluke–Associated Cholangiocarcinoma." *Nature Genetics* 44(6): 690–93.

Palozza, Paola, Maria Colangelo, Rossella Simone, Assunta Catalano, Alma Boninsegna, Paola Lanza, Giovanni Monego, and Franco O Ranelletti. 2010. "Lycopene Induces Cell Growth Inhibition by Altering Mevalonate Pathway and Ras Signaling in Cancer Cell Lines." *Carcinogenesis* 31(10): 1813–21.

Pandey, Puspa R, Hiroshi Okuda, Misako Watabe, Sudha K Pai, Wen Liu, Aya Kobayashi, Fei Xing, Koji Fukuda, Shigeru Hirota, and Tam otsu Sugai. 2011. "Resveratrol Suppresses Growth of Cancer Stem-like Cells by Inhibiting Fatty Acid Synthase." *Breast Cancer Research and Treatment* 130(2): 387–98.

Pang, Roberta, Wai Lun Law, Andrew C Y Chu, Jensen T Poon, Colin S C Lam, Ariel K M Chow, Lui Ng, Leonard W H Cheung, Xiao R Lan, and Hui Y Lan. 2010. "A Subpopulation of CD26+ Cancer Stem Cells with Metastatic Capacity in Human Colorectal Cancer." *Cell Stem Cell* 6(6): 603–15.

Parekh, Palak, Leena Motiwale, Nishigandha Naik, and K V K Rao. 2011. "Downregulation of Cyclin D1 Is Associated with Decreased Levels of P38 MAP Kinases, Akt/PKB and Pak1 during Chemopreventive Effects of Resveratrol in Liver Cancer Cells." *Experimental and Toxicologic Pathology* 63(1–2): 167–73.

Park, Seon-Young, Ji-Young Kim, Su-Mi Lee, Chung-Hwan Jun, Sung-Bum Cho, Chang-Hwan Park, Young-Eun Joo, Hyun-Soo Kim, Sung-Kyu Choi, and Jong-Sun Rew. 2014. "Capsaicin Induces Apoptosis and Modulates MAPK Signaling in Human Gastric Cancer Cells." *Molecular Medicine Reports* 9(2): 499–502.

Parr, C, G Watkins, and W G Jiang. 2004. "The Possible Correlation of Notch-1 and Notch-2 with Clinical Outcome and Tumour Clinicopathological Parameters in Human Breast Cancer." *International Journal of Molecular Medicine* 14(5): 779–86.

Pattabiraman, Diwakar R, and Robert A Weinberg. 2014. "Tackling the Cancer Stem Cells—What Challenges Do They Pose?" *Nature Reviews Drug Discovery* 13(7): 497–512.

Peng, Lihua, and Dianming Jiang. 2018. "Resveratrol Eliminates Cancer Stem Cells of Osteosarcoma by STAT3 Pathway Inhibition." *PLoS One* 13(10): e0205918.

Perkins, N D, and T D Gilmore. 2006. "Good Cop, Bad Cop: The Different Faces of NF-KB." *Cell Death & Differentiation* 13(5): 759–72.

Pierce, G Barry. 1967. "Teratocarcinoma: Model for a Developmental Concept of Cancer." *Current Topics in Developmental Biology* 2: 223–46.

Ponnurangam, Sivapriya, Joshua Mammen, Satish Ramalingam, Zhiyun He, Youcheng Zhang, Shahid Umar, Dharmalingam Subramaniam, and Shrikant Anant. 2012. "Honokiol in Combination with Radiation Targets Notch Signaling to Inhibit Colon Cancer Stem CellsHonokiol Radiosensitizes Colon Cancer Stem Cells." *Molecular Cancer Therapeutics* 11(4): 963–72.

Prabavathy, D, Y Swarnalatha, and Niveditha Ramadoss. 2018. "Lung Cancer Stem Cells—Origin, Characteristics and Therapy." *Stem Cell Investigation* 5: 6.

Provinciali, Mauro, Francesca Papalini, Fiorenza Orlando, Sara Pierpaoli, Alessia Donnini, Paolo Morazzoni, Antonella Riva, and Arianna Smorlesi. 2007. "Effect of the Silybin-Phosphatidylcholine Complex (IdB 1016) on the Development of Mammary Tumors in HER-2/Neu Transgenic Mice." *Cancer Research* 67(5): 2022–29.

Qi, Wang Jia, Wang Shi Sheng, Chu Peng, Ma Xiaodong, and Tang Ze Yao. 2021. "Investigating into Anti-Cancer Potential of Lycopene: Molecular Targets." *Biomedicine & Pharmacotherapy* 138: 111546.

Quintás-Cardama, Alfonso, and Srdan Verstovsek. 2013. "Molecular Pathways: JAK/STAT Pathway: Mutations, Inhibitors, and ResistanceJAK/STAT Pathway and Resistance." *Clinical Cancer Research* 19(8): 1933–40.

Radtke, F. 2005. "Clevers H." Self-Renewal and Cancer of the Gut: Two Sides of a Coin. *Science* 307: 1904–9.

Rahden, Burkhard H A von, Stefan Kircher, Maria Lazariotou, Christoph Reiber, Luisa Stuermer, Christoph Otto, Christoph T Germer, and Martin Grimm. 2011. "LgR5 Expression and Cancer Stem Cell Hypothesis: Clue to Define the True Origin of Esophageal Adenocarcinomas with and without Barrett's Esophagus?" *Journal of Experimental & Clinical Cancer Research* 30(1): 1–11.

Rahman, MdShaifur, Hossen Mohammad Jamil, Naznin Akhtar, K M T Rahman, Rashedul Islam, and S M Asaduzzaman. 2016. "Stem Cell and Cancer Stem Cell: A Tale of Two Cells." *Progress in Stem Cell* 3(02): 97–108.

Ramamoorthy, Prabhu, Prasad Dandawate, Roy A Jensen, and Shrikant Anant. 2021. "Celastrol and Triptolide Suppress Stemness in Triple Negative Breast Cancer: Notch as a Therapeutic Target for Stem Cells." *Biomedicines* 9(5): 482.

Ray, Anasuya, Smreti Vasudevan, and Suparna Sengupta. 2015. "6-Shogaol Inhibits Breast Cancer Cells and Stem Cell-like Spheroids by Modulation of Notch Signaling Pathway and Induction of Autophagic Cell Death." *PLoS One* 10(9): e0137614.

Reya, Tannishtha, and Hans Clevers. 2005. "Wnt Signalling in Stem Cells and Cancer." *Nature* 434(7035): 843–50.

Rhenen, Anna Van, Guus A M S Van Dongen, Angèle Kelder, Elwin J Rombouts, Nicole Feller, Bijan Moshaver, Marijke Stigter-van Walsum, Sonja Zweegman, Gert J Ossenkoppele, and Gerrit Jan Schuurhuis. 2007. "The Novel AML Stem Cell–Associated Antigen CLL-1 Aids in Discrimination between Normal and Leukemic Stem Cells." *Blood, The Journal of the American Society of Hematology* 110(7): 2659–66.

Robbins, David J, Dennis Liang Fei, and Natalia A Riobo. 2012. "The Hedgehog Signal Transduction Network." *Science Signaling* 5(246): re6.
Rodova, Mariana, Junsheng Fu, Dara Nall Watkins, Rakesh K Srivastava, and Sharmila Shankar. 2012. "Sonic Hedgehog Signaling Inhibition Provides Opportunities for Targeted Therapy by Sulforaphane in Regulating Pancreatic Cancer Stem Cell Self-Renewal." *PLoS One* 7(9): e46083.
Romero, Diana, Zainab Al-Shareef, Irantzu Gorroño-Etxebarria, Stephanie Atkins, Frances Turrell, Jyoti Chhetri, Nora Bengoa-Vergniory, Christoph Zenzmaier, Peter Berger, and Jonathan Waxman. 2016. "Dickkopf-3 Regulates Prostate Epithelial Cell Acinar Morphogenesis and Prostate Cancer Cell Invasion by Limiting TGF-β-Dependent Activation of Matrix Metalloproteases." *Carcinogenesis* 37(1): 18–29.
Rossi, Fiorella, Hunter Noren, Richard Jove, Vladimir Beljanski, and Karl-Henrik Grinnemo. 2020. "Differences and Similarities between Cancer and Somatic Stem Cells: Therapeutic Implications." *Stem Cell Research & Therapy* 11(1): 1–16.
Ruan, H, Y Y Zhan, J Hou, B Xu, B Chen, Y Tian, D Wu, Y Zhao, Y Zhang, and X Chen. 2017. "Berberine Binds RXRα to Suppress β-Catenin Signaling in Colon Cancer Cells." *Oncogene* 36(50): 6906–18.
Sabatos, Catherine A, Sumone Chakravarti, Eugene Cha, Anna Schubart, Alberto Sánchez-Fueyo, Xin Xiao Zheng, Anthony J Coyle, Terry B Strom, Gordon J Freeman, and Vijay K Kuchroo. 2003. "Interaction of Tim-3 and Tim-3 Ligand Regulates T Helper Type 1 Responses and Induction of Peripheral Tolerance." *Nature Immunology* 4(11): 1102–10.
Saito-Diaz, Kenyi, Tony W Chen, Xiaoxi Wang, Curtis A Thorne, Heather A Wallace, Andrea Page-McCaw, and Ethan Lee. 2013. "The Way Wnt Works: Components and Mechanism." *Growth Factors* 31(1): 1–31.
Salehi, Bahare, Laura Machin, Lianet Monzote, Javad Sharifi-Rad, Shahira M Ezzat, Mohamed A Salem, Rana M Merghany, Nihal M El Mahdy, Ceyda Sibel Kılıç, and Oksana Sytar. 2020. "Therapeutic Potential of Quercetin: New Insights and Perspectives for Human Health." *Acs Omega* 5(20): 11849–72.
Sarkar, Fazlul H, and Yiwei Li. 2009. "Harnessing the Fruits of Nature for the Development of Multi-Targeted Cancer Therapeutics." *Cancer Treatment Reviews* 35(7): 597–607.
Saw, Constance Lay Lay, Yue Guo, Anne Yuqing Yang, Ximena Paredes-Gonzalez, Christina Ramirez, Douglas Pung, and Ah-Ng Tony Kong. 2014. "The Berry Constituents Quercetin, Kaempferol, and Pterostilbene Synergistically Attenuate Reactive Oxygen Species: Involvement of the Nrf2-ARE Signaling Pathway." *Food and Chemical Toxicology* 72: 303–11.
Schatton, Tobias, George F Murphy, Natasha Y Frank, Kazuhiro Yamaura, Ana Maria Waaga-Gasser, Martin Gasser, Qian Zhan, Stefan Jordan, Lyn M Duncan, and Carsten Weishaupt. 2008. "Identification of Cells Initiating Human Melanomas." *Nature* 451(7176): 345–49.
Schmidt, Dirk-Steffen, Pamela Klingbeil, Martina Schnölzer, and Margot Zöller. 2004. "CD44 Variant Isoforms Associate with Tetraspanins and EpCAM." *Experimental Cell Research* 297(2): 329–47.
Schweisguth, François. 2004. "Regulation of Notch Signaling Activity." *Current Biology* 14(3): R129–38.
Seligson, David B, Allan J Pantuck, Xueli Liu, Yunda Huang, Steven Horvath, Matthew H T Bui, Ken-ryu Han, Adrian J L Correa, Mervi Eeva, and Sheila Tze. 2004. "Epithelial Cell Adhesion Molecule (KSA) Expression: Pathobiology and Its Role as an Independent Predictor of Survival in Renal Cell Carcinoma." *Clinical Cancer Research* 10(8): 2659–69.
Shafiei, Somayeh, Elham Kalantari, Leili Saeednejad Zanjani, Maryam Abolhasani, Mohammad Hossein Asadi Lari, and Zahra Madjd. 2019. "Increased Expression of DCLK1, a Novel Putative CSC Maker, Is Associated with Tumor Aggressiveness and Worse Disease-Specific Survival in Patients with Bladder Carcinomas." *Experimental and Molecular Pathology* 108: 164–72.
Shankar, Sharmila, Dara Nall, Su-Ni Tang, Daniel Meeker, Jenna Passarini, Jay Sharma, and Rakesh K Srivastava. 2011. "Resveratrol Inhibits Pancreatic Cancer Stem Cell Characteristics in Human and KrasG12D Transgenic Mice by Inhibiting Pluripotency Maintaining Factors and Epithelial-Mesenchymal Transition." *PLoS One* 6(1): e16530.
Shimizu, Masahito, Seiji Adachi, Muneyuki Masuda, Osamu Kozawa, and Hisataka Moriwaki. 2011. "Cancer Chemoprevention with Green Tea Catechins by Targeting Receptor Tyrosine Kinases." *Molecular Nutrition & Food Research* 55(6): 832–43.
Shimokawa, Mariko, Yuki Ohta, Shingo Nishikori, Mami Matano, Ai Takano, Masayuki Fujii, Shinya Sugimoto, Takanori Kanai, and Toshiro Sato. 2017. "Visualization and Targeting of LGR5+ Human Colon Cancer Stem Cells." *Nature* 545(7653): 187–92.
Shukla, Yogeshwer, and Richa Singh. 2011. "Resveratrol and Cellular Mechanisms of Cancer Prevention." *Annals of the New York Academy of Sciences* 1215(1): 1–8.
Sikandar, Shaheen S, Kira T Pate, Scott Anderson, Diana Dizon, Robert A Edwards, Marian L Waterman, and Steven M Lipkin. 2010. "NOTCH Signaling Is Required for Formation and Self-Renewal of

Tumor-Initiating Cells and for Repression of Secretory Cell Differentiation in Colon CancerNOTCH in Colon Cancer–Initiating Cells." *Cancer Research* 70(4): 1469–78.

Silberstein, Marc and Jean-Pascal Capp. 2012. *Nouveau Regard Sur Le Cancer.* Belin, 166.

Sládek, Norman E, Rahn Kollander, Lakshmaiah Sreerama, and David T Kiang. 2002. "Cellular Levels of Aldehyde Dehydrogenases (ALDH1A1 and ALDH3A1) as Predictors of Therapeutic Responses to Cyclophosphamide-Based Chemotherapy of Breast Cancer: A Retrospective Study." *Cancer Chemotherapy and Pharmacology* 49(4): 309–21.

Song, Libin, Xiangyuan Chen, Peng Wang, Song Gao, Chao Qu, and Luming Liu. 2018. "Effects of Baicalein on Pancreatic Cancer Stem Cells via Modulation of Sonic Hedgehog Pathway." *Acta Biochimica et Biophysica Sinica* 50(6): 586–96.

Song, Zhou, Wen Yue, Bo Wei, Ning Wang, Tao Li, Lidong Guan, Shuangshuang Shi, Quan Zeng, Xuetao Pei, and Lin Chen. 2011. "Sonic Hedgehog Pathway Is Essential for Maintenance of Cancer Stem-like Cells in Human Gastric Cancer." *PLoS One* 6(3): e17687.

Stratford, Eva W, Russell Castro, Anna Wennerstrøm, Ruth Holm, Else Munthe, Silje Lauvrak, Bodil Bjerkehagen, and Ola Myklebost. 2011. "Liposarcoma Cells with Aldefluor and CD133 Activity Have a Cancer Stem Cell Potential." *Clinical Sarcoma Research* 1(1): 1–11.

Su, Dan, HongXin Deng, Xia Zhao, Xi Zhang, LiJuan Chen, XianCheng Chen, ZhengYu Li, Yu Bai, YongSheng Wang, and Qian Zhong. 2009. "Targeting CD24 for Treatment of Ovarian Cancer by Short Hairpin RNA." *Cytotherapy* 11(5): 642–52.

Su, Ying-Jhen, Hsin-Mei Lai, Yi-Wen Chang, Guan-Ying Chen, and Jia-Lin Lee. 2011. "Direct Reprogramming of Stem Cell Properties in Colon Cancer Cells by CD44." *The EMBO Journal* 30(15): 3186–99.

Sztiller-Sikorska, Malgorzata, and Malgorzata Czyz. 2020. "Parthenolide as Cooperating Agent for Anti-Cancer Treatment of Various Malignancies." *Pharmaceuticals* 13(8): 194.

Takahashi, Kazutoshi, Koji Tanabe, Mari Ohnuki, Megumi Narita, Tomoko Ichisaka, Kiichiro Tomoda, and Shinya Yamanaka. 2007. "Induction of Pluripotent Stem Cells from Adult Human Fibroblasts by Defined Factors." *Cell* 131(5): 861–72.

Takebe, Naoko, Lucio Miele, Pamela Jo Harris, Woondong Jeong, Hideaki Bando, Michael Kahn, Sherry X Yang, and S Percy Ivy. 2015. "Targeting Notch, Hedgehog, and Wnt Pathways in Cancer Stem Cells: Clinical Update." *Nature Reviews Clinical Oncology* 12(8): 445–64.

Takeda, Jun, Susumu Seino, and Graeme I Bell. 1992. "Human Oct3 Gene Family: CDNA Sequences, Alternative Splicing, Gene Organization, Chromosomal Location, and Expression at Low Levels in Adult Tissues." *Nucleic Acids Research* 20(17): 4613–20.

Takeda, Koki, Tsunekazu Mizushima, Yuhki Yokoyama, Haruka Hirose, Xin Wu, Yamin Qian, Katsuya Ikehata, Norikatsu Miyoshi, Hidekazu Takahashi, and Naotsugu Haraguchi. 2018. "Sox2 Is Associated with Cancer Stem-like Properties in Colorectal Cancer." *Scientific Reports* 8(1): 1–9.

Tamai, Keiichi, Mao Nakamura, Masamichi Mizuma, Mai Mochizuki, Misa Yokoyama, Hiroyuki Endo, Kazunori Yamaguchi, Takayuki Nakagawa, Masaaki Shiina, and Michiaki Unno. 2014. "Suppressive Expression of CD 274 Increases Tumorigenesis and Cancer Stem Cell Phenotypes in Cholangiocarcinoma." *Cancer Science* 105(6): 667–74.

Tanese, Keiji, Mariko Fukuma, Taketo Yamada, Taisuke Mori, Tsutomu Yoshikawa, Wakako Watanabe, Akira Ishiko, Masayuki Amagai, Takeji Nishikawa, and Michiie Sakamoto. 2008. "G-Protein-Coupled Receptor GPR49 Is up-Regulated in Basal Cell Carcinoma and Promotes Cell Proliferation and Tumor Formation." *The American Journal of Pathology* 173(3): 835–43.

Tang, Feng-Yao, Chung-Jin Shih, Li-Hao Cheng, Hsin-Jung Ho, and Hung-Jiun Chen. 2008. "Lycopene Inhibits Growth of Human Colon Cancer Cells via Suppression of the Akt Signaling Pathway." *Molecular Nutrition & Food Research* 52(6): 646–54.

Tang, Xi-lan, Jian-xun Liu, Wei Dong, Peng Li, Lei Li, Jin-cai Hou, Yong-qiu Zheng, Cheng-ren Lin, and Jun-guo Ren. 2015. "Protective Effect of Kaempferol on LPS plus ATP-Induced Inflammatory Response in Cardiac Fibroblasts." *Inflammation* 38(1): 94–101.

Taylor, Wesley F, and Ehsan Jabbarzadeh. 2017. "The Use of Natural Products to Target Cancer Stem Cells." *American Journal of Cancer Research* 7(7): 1588.

Teow, Sin-Yeang, Kitson Liew, Syed A Ali, Alan Soo-Beng Khoo, and Suat-Cheng Peh. 2016. "Antibacterial Action of Curcumin against Staphylococcus Aureus: A Brief Review." *Journal of Tropical Medicine* 2016.

Thawonsuwan, J, V Kiron, S Satoh, A Panigrahi, and V Verlhac. 2010. "Epigallocatechin-3-Gallate (EGCG) Affects the Antioxidant and Immune Defense of the Rainbow Trout, Oncorhynchus Mykiss." *Fish Physiology and Biochemistry* 36(3): 687–97.

Tian, Hongying, and Zhongcui Yu. 2015. "Resveratrol Induces Apoptosis of Leukemia Cell Line K562 by Modulation of Sphingosine Kinase-1 Pathway." *International Journal of Clinical and Experimental Pathology* 8(3): 2755.

Todaro, Matilde, Miriam Gaggianesi, Veronica Catalano, Antonina Benfante, Flora Iovino, Mauro Biffoni, Tiziana Apuzzo, Isabella Sperduti, Silvia Volpe, and Gianfranco Cocorullo. 2014. "CD44v6 Is a Marker of Constitutive and Reprogrammed Cancer Stem Cells Driving Colon Cancer Metastasis." *Cell Stem Cell* 14(3): 342–56.

Tomita, Hiroyuki, and Akira Hara. 2020. "Serrated Lesions and Stem Cells on Drug Resistance and Colon Cancer." In Chi Hin Cho and Tao Hu (eds.), *Drug Resistance in Colorectal Cancer: Molecular Mechanisms and Therapeutic Strategies*, 75–82. Elsevier.

Topcul, Mehmet, Funda Topcul, and Idil Cetin. 2013. "Effects of Femara and Tamoxifen on Proliferation of FM3A Cells in Culture." *Asian Pacific Journal of Cancer Prevention* 14(5): 2819–22.

Tran, Freddi Huan, and Jie J Zheng. 2017. "Modulating the Wnt Signaling Pathway with Small Molecules." *Protein Science* 26(4): 650–61.

Trzpis, Monika, Pamela M J McLaughlin, Lou M F H de Leij, and Martin C Harmsen. 2007. "Epithelial Cell Adhesion Molecule: More than a Carcinoma Marker and Adhesion Molecule." *The American Journal of Pathology* 171(2): 386–95.

Tsakonas, S Artavanis, Matthew D Rand, and Robert J Lake. 1999. "Notch Signaling: Cell Fate Control and Signal Integration in Development." *Science* 284(5415): 770–76.

Uchida, Hiroshi, Ken Yamazaki, Mariko Fukuma, Taketo Yamada, Tetsu Hayashida, Hirotoshi Hasegawa, Masaki Kitajima, Yuko Kitagawa, and Michiie Sakamoto. 2010. "Overexpression of Leucine-rich Repeat-containing G Protein-coupled Receptor 5 in Colorectal Cancer." *Cancer Science* 101(7): 1731–37.

Vassalli, Giuseppe. 2019. "Aldehyde Dehydrogenases: Not Just Markers, but Functional Regulators of Stem Cells." *Stem Cells International* 2019: 3904645.

Vazquez-Santillan, K, J Melendez-Zajgla, L Jimenez-Hernandez, G Martinez-Ruiz, and V Maldonado. 2015. "NF-KB Signaling in Cancer Stem Cells: A Promising Therapeutic Target?" *Cellular Oncology* 38(5): 327–39.

Venkatesh, Vandana, Raghu Nataraj, Gopenath S Thangaraj, Murugesan Karthikeyan, Ashok Gnanasekaran, Shanmukhappa B Kaginelli, Gobianand Kuppanna, Chandrashekrappa Gowdru Kallappa, and Kanthesh M Basalingappa. 2018. "Targeting Notch Signalling Pathway of Cancer Stem Cells." *Stem Cell Investigation* 5: 5.

Vieira, Gabriella Cunha, S Chockalingam, Zsombor Melegh, Alexander Greenhough, Sally Malik, Marianna Szemes, Ji Hyun Park, Abderrahmane Kaidi, Li Zhou, and Daniel Catchpoole. 2015. "LGR5 Regulates Pro-Survival MEK/ERK and Proliferative Wnt/β-Catenin Signalling in Neuroblastoma." *Oncotarget* 6(37): 40053.

Walcher, Lia, Ann-Kathrin Kistenmacher, Huizhen Suo, Reni Kitte, Sarah Dluczek, Alexander Strauß, André-René Blaudszun, Tetyana Yevsa, Stephan Fricke, and Uta Kossatz-Boehlert. 2020. "Cancer Stem Cells—Origins and Biomarkers: Perspectives for Targeted Personalized Therapies." *Frontiers in Immunology* 11: 1280.

Wang, Dan, Ping Lu, Hao Zhang, Minna Luo, Xin Zhang, Xiaofei Wei, Jiyue Gao, Zuowei Zhao, and Caigang Liu. 2014. "Oct-4 and Nanog Promote the Epithelial-Mesenchymal Transition of Breast Cancer Stem Cells and Are Associated with Poor Prognosis in Breast Cancer Patients." *Oncotarget* 5(21): 10803.

Wang, Kai, Siu Tsan Yuen, Jiangchun Xu, Siu Po Lee, Helen H N Yan, Stephanie T Shi, Hoi Cheong Siu, Shibing Deng, Kent Man Chu, and Simon Law. 2014. "Whole-Genome Sequencing and Comprehensive Molecular Profiling Identify New Driver Mutations in Gastric Cancer." *Nature Genetics* 46(6): 573–82.

Wang, Liang, Xiangsheng Zuo, Keping Xie, and Daoyan Wei. 2018. "The Role of CD44 and Cancer Stem Cells." *Methods in Molecular Biology* 1692: 31–42.

Wang, Min, Juan Xiao, Min Shen, Yu Yahong, Rui Tian, Feng Zhu, Jianxin Jiang, Zhiyong Du, Jun Hu, and Wensong Liu. 2011. "Isolation and Characterization of Tumorigenic Extrahepatic Cholangiocarcinoma Cells with Stem Cell-like Properties." *International Journal of Cancer* 128(1): 72–81.

Wang, Neng, Zhiyu Wang, Cheng Peng, Jieshu You, Jiangang Shen, Shouwei Han, and Jianping Chen. 2014. "Dietary Compound Isoliquiritigenin Targets GRP78 to Chemosensitize Breast Cancer Stem Cells via β-Catenin/ABCG2 Signaling." *Carcinogenesis* 35(11): 2544–54.

Wang, Qi, Xinyu Chen, and Nissim Hay. 2017. "Akt as a Target for Cancer Therapy: More Is Not Always Better (Lessons from Studies in Mice)." *British Journal of Cancer* 117(2): 159–63.

Wang, Tianyi, Johannes Francois Fahrmann, Heehyoung Lee, Yi-Jia Li, Satyendra C Tripathi, Chanyu Yue, Chunyan Zhang, Veronica Lifshitz, Jieun Song, and Yuan Yuan. 2018. "JAK/STAT3-Regulated Fatty

Acid β-Oxidation Is Critical for Breast Cancer Stem Cell Self-Renewal and Chemoresistance." *Cell Metabolism* 27(1): 136–150. e5.

Wang, Ying-Jie, and Meenhard Herlyn. 2015. "The Emerging Roles of Oct4 in Tumor-Initiating Cells." *American Journal of Physiology-Cell Physiology* 309(11): C709–18.

Wang, Zehao, Lixia Wang, Boya Shi, Xiuli Sun, Yinrong Xie, Haonan Yang, Chengting Zi, Xuanjun Wang, and Jun Sheng. 2021. "Demethyleneberberine Promotes Apoptosis and Suppresses TGF-β/Smads Induced EMT in the Colon Cancer Cells HCT-116." *Cell Biochemistry and Function* 39(6): 763–70.

Wang, Zi-Xuan, Jing Ma, Xin-Yu Li, Yong Wu, Huan Shi, Yao Chen, Guang Lu, Han-Ming Shen, Guo-Dong Lu, and Jing Zhou. 2021. "Quercetin Induces P53-independent Cancer Cell Death through Lysosome Activation by the Transcription Factor EB and Reactive Oxygen Species-dependent Ferroptosis." *British Journal of Pharmacology* 178(5): 1133–48.

Wei, Xiaolong, David Dombkowski, Katia Meirelles, Rafael Pieretti-Vanmarcke, Paul P Szotek, Henry L Chang, Frederic I Preffer, Peter R Mueller, Jose Teixeira, and David T MacLaughlin. 2010. "Müllerian Inhibiting Substance Preferentially Inhibits Stem/Progenitors in Human Ovarian Cancer Cell Lines Compared with Chemotherapeutics." *Proceedings of the National Academy of Sciences* 107(44): 18874–79.

Weiss, Alexander, and Liliana Attisano. 2013. "The TGFbeta Superfamily Signaling Pathway." *Wiley Interdisciplinary Reviews: Developmental Biology* 2(1): 47–63.

Wikipedia contributors. 2021. "CD90." Wikipedia. 2021. https://en.wikipedia.org/wiki/CD90.

Wilson, Anne, Mark J Murphy, Thordur Oskarsson, Konstantinos Kaloulis, Michael D Bettess, Gabriela M Oser, Anne-Catherine Pasche, Christian Knabenhans, H Robson MacDonald, and Andreas Trumpp. 2004. "C-Myc Controls the Balance between Hematopoietic Stem Cell Self-Renewal and Differentiation." *Genes & Development* 18(22): 2747–63.

Wu, Yaojiong, and Philip Yuguang Wu. 2009. "CD133 as a Marker for Cancer Stem Cells: Progresses and Concerns." *Stem Cells and Development* 18(8): 1127–34.

Xiang, Yan, Ting Yang, Bing-yao Pang, Ying Zhu, and Yong-ning Liu. 2016. "The Progress and Prospects of Putative Biomarkers for Liver Cancer Stem Cells in Hepatocellular Carcinoma." *Stem Cells International* 2016: 7614971.

Xiao, Jianbo, and Weibin Bai. 2019. "Bioactive Phytochemicals." *Critical Reviews in Food Science and Nutrition* 59(6): 827–29.

Xie, Xin, Chun-Yan Wang, Yun-Xin Cao, Wei Wang, Ran Zhuang, Li-Hua Chen, Na-Na Dang, Liang Fang, and Bo-Quan Jin. 2005. "Expression Pattern of Epithelial Cell Adhesion Molecule on Normal and Malignant Colon Tissues." *World Journal of Gastroenterology: WJG* 11(3): 344.

Xie, Yuan, Anders Sundström, Naga P Maturi, E-Jean Tan, Voichita D Marinescu, Malin Jarvius, Malin Tirfing, Chuan Jin, Lei Chen, and Magnus Essand. 2019. "LGR5 Promotes Tumorigenicity and Invasion of Glioblastoma Stem-like Cells and Is a Potential Therapeutic Target for a Subset of Glioblastoma Patients." *The Journal of Pathology* 247(2): 228–40.

Xu, Liangliang, Weiping Lin, Longping Wen, and Gang Li. 2019. "Lgr5 in Cancer Biology: Functional Identification of Lgr5 in Cancer Progression and Potential Opportunities for Novel Therapy." *Stem Cell Research & Therapy* 10(1): 1–9.

Yamamoto, Yoshiya, Michiie Sakamoto, Gen Fujii, Hitomi Tsuiji, Kengo Kenetaka, Masahiro Asaka, and Setsuo Hirohashi. 2003. "Overexpression of Orphan G-Protein–Coupled Receptor, Gpr49, in Human Hepatocellular Carcinomas with β-Catenin Mutations." *Hepatology* 37(3): 528–33.

Yamashita, Taro, Masao Honda, Yasunari Nakamoto, Masayo Baba, Kouki Nio, Yasumasa Hara, Sha Sha Zeng, Takehiro Hayashi, Mitsumasa Kondo, and Hajime Takatori. 2013. "Discrete Nature of EpCAM+ and CD90+ Cancer Stem Cells in Human Hepatocellular Carcinoma." *Hepatology* 57(4): 1484–97.

Yan, Xiaohui, Miao Qi, Pengfei Li, Yihong Zhan, and Huanjie Shao. 2017. "Apigenin in Cancer Therapy: Anti-Cancer Effects and Mechanisms of Action." *Cell & Bioscience* 7(1): 1–16.

Yan, Yongmin, Xiangsheng Zuo, and Daoyan Wei. 2015. "Concise Review: Emerging Role of CD44 in Cancer Stem Cells: A Promising Biomarker and Therapeutic Target." *Stem Cells Translational Medicine* 4(9): 1033–43.

Yang, Fan, Jiaming Zhang, and Hua Yang. 2018. "OCT4, SOX2, and NANOG Positive Expression Correlates with Poor Differentiation, Advanced Disease Stages, and Worse Overall Survival in HER2+ Breast Cancer Patients." *OncoTargets and Therapy* 11: 7873.

Yang, Jia, Zhihong Fang, Jianchun Wu, Xiaoling Yin, Yuan Fang, Fanchen Zhao, Shiguo Zhu, and Yan Li. 2016. "Construction and Application of a Lung Cancer Stem Cell Model: Antitumor Drug Screening and Molecular Mechanism of the Inhibitory Effects of Sanguinarine." *Tumor Biology* 37(10): 13871–83.

Yang, L, G Xie, Q Fan, and J Xie. 2010. "Activation of the Hedgehog-Signaling Pathway in Human Cancer and the Clinical Implications." *Oncogene* 29(4): 469–81.

Yang, Liqun, Pengfei Shi, Gaichao Zhao, Jie Xu, Wen Peng, Jiayi Zhang, Guanghui Zhang, Xiaowen Wang, Zhen Dong, and Fei Chen. 2020. "Targeting Cancer Stem Cell Pathways for Cancer Therapy." *Signal Transduction and Targeted Therapy* 5(1): 1–35.

Yang, Lu, Hailin Tang, Yanan Kong, Xinhua Xie, Jianping Chen, Cailu Song, Xiaoping Liu, Feng Ye, Ning Li, and Neng Wang. 2015. "LGR5 Promotes Breast Cancer Progression and Maintains Stem-like Cells through Activation of Wnt/β-Catenin Signaling." *Stem Cells* 33(10): 2913–24.

Yang, Mon-Yuan, Chau-Jong Wang, Nai-Fang Chen, Wen-Hsin Ho, Fung-Jou Lu, and Tsui-Hwa Tseng. 2014. "Luteolin Enhances Paclitaxel-Induced Apoptosis in Human Breast Cancer MDA-MB-231 Cells by Blocking STAT3." *Chemico-Biological Interactions* 213: 60–68.

Yeung, Trevor M, Shaan C Gandhi, Jennifer L Wilding, Ruth Muschel, and Walter F Bodmer. 2010. "Cancer Stem Cells from Colorectal Cancer-Derived Cell Lines." *Proceedings of the National Academy of Sciences* 107(8): 3722–27.

Yin, Amy H, Sheri Miraglia, Esmail D Zanjani, Graca Almeida-Porada, Makio Ogawa, Anne G Leary, Johanna Olweus, John Kearney, and David W Buck. 1997. "AC133, a Novel Marker for Human Hematopoietic Stem and Progenitor Cells." *Blood, The Journal of the American Society of Hematology* 90(12): 5002–12.

Yin, Zhaofa, Juan Li, Le Kang, Xiangyang Liu, Jianguo Luo, Ling Zhang, Yuting Li, and Jiarong Cai. 2021. "Epigallocatechin-3-gallate Induces Autophagy-related Apoptosis Associated with LC3B II and Beclin Expression of Bladder Cancer Cells." *Journal of Food Biochemistry* 45(6): e13758.

Yoneyama, Tatsuro, Midori A Arai, Samir K Sadhu, Firoj Ahmed, and Masami Ishibashi. 2015. "Hedgehog Inhibitors from Withania Somnifera." *Bioorganic & Medicinal Chemistry Letters* 25(17): 3541–44.

Yoo, Min-Hyuk, and Dolph L Hatfield. 2008. "The Cancer Stem Cell Theory: Is It Correct?" *Molecules and Cells* 26(5): 514–16.

Yoo, Sunyong, Kwansoo Kim, Hojung Nam, and Doheon Lee. 2018. "Discovering Health Benefits of Phytochemicals with Integrated Analysis of the Molecular Network, Chemical Properties and Ethnopharmacological Evidence." *Nutrients* 10(8): 1042.

Yousefnia, Saghar, Sara Momenzadeh, Farzad Seyed Forootan, Kamran Ghaedi, and Mohammad Hossein Nasr Esfahani. 2018. "The Influence of Peroxisome Proliferator-Activated Receptor γ (PPARγ) Ligands on Cancer Cell Tumorigenicity." *Gene* 649: 14–22.

Yousuf, Mohd, Parvez Khan, Anas Shamsi, Mohd Shahbaaz, Gulam Mustafa Hasan, Qazi Mohd Rizwanul Haque, Alan Christoffels, Asimul Islam, and Md Imtaiyaz Hassan. 2020. "Inhibiting CDK6 Activity by Quercetin Is an Attractive Strategy for Cancer Therapy." *Acs Omega* 5(42): 27480–91.

Yu, Fa-Xing, Bin Zhao, and Kun-Liang Guan. 2015. "Hippo Pathway in Organ Size Control, Tissue Homeostasis, and Cancer." *Cell* 163(4): 811–28.

Yu, Zuoren, Timothy G Pestell, Michael P Lisanti, and Richard G Pestell. 2012. "Cancer Stem Cells." *The International Journal of Biochemistry & Cell Biology* 44(12): 2144–51.

Zhan, Tailan, Niklas Rindtorff, and Michael Boutros. 2017. "Wnt Signaling in Cancer." *Oncogene* 36(11): 1461–73.

Zhang, Linlin, Lei Li, Min Jiao, Dapeng Wu, Kaijie Wu, Xiang Li, Guodong Zhu, Lin Yang, Xinyang Wang, and Jer-Tsong Hsieh. 2012. "Genistein Inhibits the Stemness Properties of Prostate Cancer Cells through Targeting Hedgehog–Gli1 Pathway." *Cancer Letters* 323(1): 48–57.

Zhang, Qi, Wan-Shuang Cao, Xue-Qi Wang, Min Zhang, Xiao-Min Lu, Jia-Qi Chen, Yue Chen, Miao-Miao Ge, Cai-Yun Zhong, and Hong-Yu Han. 2019. "Genistein Inhibits Nasopharyngeal Cancer Stem Cells through Sonic Hedgehog Signaling." *Phytotherapy Research* 33(10): 2783–91.

Zhang, Ruihua, Xia Ai, Yongjie Duan, Man Xue, Wenxiao He, Cunlian Wang, Tong Xu, Mingju Xu, Baojian Liu, and Chunhong Li. 2017. "Kaempferol Ameliorates H9N2 Swine Influenza Virus-Induced Acute Lung Injury by Inactivation of TLR4/MyD88-Mediated NF-KB and MAPK Signaling Pathways." *Biomedicine & Pharmacotherapy* 89: 660–72.

Zhang, Shanshan, Xianling Guo, Jianrui Song, Kai Sun, Yujiao Song, and Lixin Wei. 2015. "Effect of Autophagy on Chemotherapy-Induced Apoptosis and Growth Inhibition." In M.A. Hayat (ed.), *Autophagy: Cancer, Other Pathologies, Inflammation, Immunity, Infection, and Aging*, 145–56. Elsevier.

Zhang, Xiaowei, Ruixi Hua, Xiaofeng Wang, Mingzhu Huang, Lu Gan, Zhenhua Wu, Jiejun Zhang, Hongqiang Wang, Yufan Cheng, and Jin Li. 2016. "Identification of Stem-like Cells and Clinical Significance of Candidate Stem Cell Markers in Gastric Cancer." *Oncotarget* 7(9): 9815–31.

Zhang, Yuesheng, Paul Talalay, Cheon-Gyu Cho, and Gary H Posner. 1992. "A Major Inducer of Anticarcinogenic Protective Enzymes from Broccoli: Isolation and Elucidation of Structure." *Proceedings of the National Academy of Sciences* 89(6): 2399–2403.

Zhang, Yuesheng, and Li Tang. 2007. "Discovery and Development of Sulforaphane as a Cancer Chemopreventive Phytochemical 1." *Acta Pharmacologica Sinica* 28(9): 1343–54.

Zhao, R C, J Zhou, K F Chen, J Gong, J Liu, J Y He, P Guan, B Li, and Y Qin. 2016. "The Prognostic Value of Combination of CD90 and OCT4 for Hepatocellular Carcinoma after Curative Resection." *Neoplasma* 63(2): 288–98.

Zhao, Wenxiu, Lingxiang Jiang, Ting Fang, Fei Fang, Yingchun Liu, Ye Zhao, Yuting You, Hao Zhou, Xiaolin Su, and Jiangwei Wang. 2021. "β-Lapachone Selectively Kills Hepatocellular Carcinoma Cells by Targeting NQO1 to Induce Extensive DNA Damage and PARP1 Hyperactivation." *Frontiers in Oncology* 11: 747282.

Zheng, Nan, Ping Zhang, Huai Huang, Weiwei Liu, Toshihiko Hayashi, Linghe Zang, Ye Zhang, Lu Liu, Mingyu Xia, and Shin-ichi Tashiro. 2015. "ERα Down-Regulation Plays a Key Role in Silibinin-Induced Autophagy and Apoptosis in Human Breast Cancer MCF-7 Cells." *Journal of Pharmacological Sciences* 128(3): 97–107.

Zhou, H, C S Beevers, and C Huang. 2011. "The Targets of Curcumin." *Current Drug Targets* 12: 332–47.

Zhou, Qianhe, and Daniel Kalderon. 2011. "Hedgehog Activates Fused through Phosphorylation to Elicit a Full Spectrum of Pathway Responses." *Developmental Cell* 20(6): 802–14.

Zhu, Jian-Yun, Xue Yang, Yue Chen, Y E Jiang, Shi-Jia Wang, Yuan Li, Xiao-Qian Wang, Yu Meng, Ming-Ming Zhu, and Xiao Ma. 2017. "Curcumin Suppresses Lung Cancer Stem Cells via Inhibiting Wnt/B-catenin and Sonic Hedgehog Pathways." *Phytotherapy Research* 31(4): 680–88.

Zhu, Qingyun, Yingying Shen, Xiguang Chen, Jun He, Jianghua Liu, and Xuyu Zu. 2020. "Self-Renewal Signalling Pathway Inhibitors: Perspectives on Therapeutic Approaches for Cancer Stem Cells." *OncoTargets and Therapy* 13: 525.

20 Evolving Challenges and Opportunities in Plant-based Drug Discovery and Development

Raffaella Colombo, Valeria Cavalloro, Adele Papetti, Ilaria Frosi, Daniela Rossi, Simona Collina, Emanuela Martino, and Pasquale Linciano
University of Pavia

CONTENTS

DOI: 10.1201/b22842-20

20.1 PLANT-DERIVED SMALL MOLECULES

The use of plant-derived medicines to treat many pathological conditions dates back thousands of years ago. Since the early days, mankind exploited natural resources to find remedies for the cure of different diseases. Such knowledge was passed through generations creating the invaluable cultural baggage of Traditional Medicine (Dias, Urban, and Roessner 2012). Only during the 19th century, the isolation of morphine from opium as a pure pharmacological active compound by Sertürner (Krishnamurti and Rao 2016; Schmitz 1985) posed the first milestone of what will become the Nature-Aided Drug Discovery (NADD). From that moment ahead, the investigation of natural sources, and particularly of plants, represents a systematic and scientific approach in finding drugs for fighting human diseases. Conversely, plant-derived vaccines represent a new and exciting frontier of medicine. Plant cells synthesize and assemble complex human proteins by using mainly similar mechanisms (expression and some post-translational modifications-PTMs) to those of animal cells protein. Wild-type or genetically modified plants may produce vaccine antigens, plantibodies (mAbs), and antibody fragments which can be extracted and purified by conventional methods. These biological products are particularly effective, safe, and stable and can be produced in efficient way in terms of amounts, rapidity, and costs (Peeters et al. 2001; Fischer 2003; Floss, Falkenburg, and Conrad 2007; Chung et al. 2021). The purpose of this chapter is to focus on the up-to-date approaches and breakthrough technologies in the discovery of new drugs, vaccine antigens, and vaccines from plant sources.

20.1.1 Decline and Rise, Pitfalls, and Chance of Plant-based Drug Discovery

Plant-based Drug Discovery (PBDD) exploits the identification, isolation, and pharmacological study of plant secondary metabolites (SMs) for their potential in treatment of human diseases. Plants SMs are not directly involved in the essential functions of the vegetal cells, but they are produced in response to stress factors and to ward off attacks from animals, environmental insults, and other vexillary purposes. Many important life-saving drugs used in modern medicine are plant-derived drugs, i.e., morphine and codeine (from *Papaverum somniferum*) (Krishnamurti and Rao 2016; Schmitz 1985), salicylic acid (from *Salix* spp.), digoxin and digitoxin (from *Digitalis* spp.), quinine (from *Cinchona spp.*), artemisinin (from *Artemisia annua*) (Martino, Tarantino, et al. 2019), vinblastine and vincristine (from *Catharanthus roseus*) (Martino et al. 2018), paclitaxel, also known as taxol (from *Taxus brevifolia)* (Gabetta et al. 1995), and many others. Once the pharmacological properties have been associated with a natural compound, it can be produced by extraction from the plant or obtained by semi- or total synthesis, depending on production costs or on the availability of plant sources. An emblematic example is taxol, discovered in 1962. Owing to the difficulties harvesting paclitaxel from the bark of *Taxus brevifolia* and to the complexity of its total synthesis, the production of paclitaxel for clinical use became possible about 15 years later, when the precursor 10-deacetyl-baccatin III was extracted from the common yew *Taxus baccata* and then converted by chemical modification in taxol (Gabetta et al. 1995). Therefore, the term "natural drugs" is referring to drugs of natural origin, although they are produced by total synthesis or semi-synthesis or through biotechnological approaches.

SMs have some points of strength. First of all, they are characterized by peculiar and unique chemical scaffolds which cover a distinct region of the chemical space occupied by synthetic small molecules (Lachance et al. 2012; Chen et al. 2018). A chemoinformatic analysis revealed that the 83% of the chemical scaffolds found in nature are unique, since synthons for their synthesis are commonly not commercially available, or the synthetic procedure for their total synthesis is prohibitive (Lachance et al. 2012). From a structural standpoint, SMs are characterized by a high number of stereogenic centers, a great structure complexity and molecular rigidity, a high number of hydrogen bond acceptor and donor functional groups containing oxygen atoms rather than nitrogen and halogens. Although these characteristics might confer to SMs properties that exceed the Lipinski's rule of five (RO5), they are overall suitable for the development, as testified by the number of natural-inspired oral drugs approved in the past 20 years (DeGoey et al. 2018). More specifically, around two-third of the new drugs introduced in the market between 1981 and 2019 and approximately 15% of the new chemical entities (NCE) in clinical trials are of natural origin or inspired to NP (Newman and Cragg 2020). The opportunity for the discovery of new potential drugs is still wide, since only 6% of plants species existing on Earth have been investigated for their biological properties, and most traditional plant-based medicines are still unexplored from a scientific standpoint.

It is well recognized that the structural diversity and complementarity to synthetic compounds, and the abundance of SMs make plants a valuable source of new drug precursors, prototypes, or new chemical entities of therapeutic value in these field. Despite this, starting from 1990s, the majority of pharmaceutical companies have reduced natural product Research and Development (R&D) expenditures, since several drawbacks and limitations are associated with PBDD, as briefly explained here below. A PBDD programme usually starts with the screening of extracts towards a panel of biological targets or cell-lines of interest. Raw natural extracts are often not compatible as such with a target-based or a phenotypic screening, since they do not satisfy the requirement for performing a high-throughput screening (HTS), due to the characteristics of the crude extracts or of SM. The high viscosity of most plant extracts might not guarantee a trustworthy control during the dispensation, especially using robotized liquid handling system; in addition some compounds might aggregate or precipitate when they enter in contact with the aqueous environment of the assay. Sometimes, SM might non-specifically bind with the proteins, or they could possess an intrinsic fluorescence or fluorescent quenching activity, or they could be coloured thus interfering with the HTS endpoint measurement in fluorometric or colorimetric-based assays, respectively (e.g., tannins, chlorophylls). This can result in misleading assay outcomes. Furthermore, the presence in the extract of cytotoxic compounds might interfere with cell-based assays masking the biological effect desired or leading to positive false (e.g., saponins). For these reasons, removing such compounds from the primary extract before testing is essential and particularly challenging. De-replication is another drawback, because it is necessary to ascertain that the biological activity is not mediated by already known bioactive compounds. Once compounds possessing the biological activity of interest have been identified, it is necessary to deciphering the structure of the molecule/s responsible/s for the observed biological activity. Moreover, efforts are required for deepening their mechanism of actions. An additional limitation occurs whether the optimization of pharmacodynamic and pharmacokinetic properties is necessary. The modification of the chemical structure-of SMs for structure activity relationship studies may in fact represent a challenge, especially when the SMs are not synthetically accessible with the chemical knowledge available to date or when the SM chemical manipulation is not allowed owing to the liability of SM structure. Lastly, a relevant drawback for pharmaceutical companies is related to intellectual properties issues. It is not always possible to gain intellectual property right (IPR) on unmodified natural product but solely on semi-synthetic derivatives. The IPR of SM strictly depends on the regulatory laws of the single states. As an example, whilst in most European countries, it is possible to patent newly isolated SMs; this is not possible in USA. However, the patent application requires the detailed description of the procedure for SM isolation and strong evidence of the biological effect. Conversely, for already known natural compounds, it is occasionally possible to gain IPR by claiming new uses and formulations. Accordingly, IPR stuffs represent for pharmaceutical companies a barrier to marketing the new products and to recover the cost of the R&D process. Furthermore, the production of adequate quantities of the active compound needed for drug profiling might results a limitation. Lastly, the production of natural drugs is often labour-intensive and time-consuming. The supply of raw plant material is another critical issue, especially in case of indigenous plants. Since 1992, several countries prohibited the collection and export of indigenous plant materials. Therefore, companies need to obtain by these countries the permission to collect plant materials and to negotiate the sharing of the profit with them.

All the criticisms described above (difficulties in elucidating the biological mechanism, in isolating pure chemical compounds, in acquiring IPR and performing clinical trials, as well as in establishing standardized procedures) posed challenges to the development of natural products as drugs. Facing these challenges requires a huge effort and economical investment for the pharmaceutical companies to bring in the market new natural-derived drugs, compared to synthetic small molecules. Therefore, a 30-years long break in PBDD (from 80s to 2010s) occurred. In the last decade, a reversal trend was observed. Pharmaceutical companies and academia renewed their interest in natural products research by successfully and rapidly integrating rational approaches for isolation and identification of bioactive SMs using hyphenated technologies, computational studies, and other new techniques (Li and Vederas 2009).

20.2 SELECTION OF BOTANICAL SOURCES

The selection of plant candidates for the identification of potential bioactive SMs is the first and crucial step of a successful PBDD programme. Experimental methodologies are widely applied, and recently, they may be complemented by in silico methodologies and by "omics" approach. A brief description of the possible approaches is presented here below.

20.2.1 Experimental Approaches

The knowledge-based ethnopharmacological approach poses its rationale on the empirical usage of plants in traditional medicine. The identification of bioactive compounds in such plant species requires an interdisciplinary connection between scientific disciplines such as ethnobotany, phytochemistry, biochemistry, pharmacology, medicinal chemistry, and humanistic disciplines such as anthropology, archaeology, and history. Similarly, the phytochemical approach allowed to the selection of the plant source, based on its already known SMs content. As an example, *Ginko biloba* extracts, known to be enriched in bioactive SMs against metalloproteins, have been investigated for the treatment of Alzheimer's disease and other neurodegenerative disorders. Both ethnopharmacological and phytochemical approaches are safe and are associated to a high success rate. Conversely, random screening approach requires the investigation of plants without any prior information regarding their composition, use, or biological activity. The plants species is selected from local/national areas, and their extracts are systematically screened over a panel of biological targets or cells. This approach is a high win-high risk approach. It has the advantage to reduce the bias derived from the ethnomedical data and, most importantly, can lead to the identification of completely brand-new bioactive compounds that alternatively could not be expected by the existing knowledge. A successful example of the drug discovered following this approach is camptothecin (Martino et al. 2017), a potent anti-cancer drug identified by the National Cancer Institute (NCI) in the United States. *Per contra*, the probability of finding bioactive compounds is very low. The taxonomic approach is based on the assumption that plant species, belonging to the same family, may produce analogue secondary metabolite chemotypes that can be thus exploited for the same medical need. The chemotaxonomic approach combines taxonomic and phytochemical approaches by selecting plant species based on their belonging to the same family and having a similar phytochemical content.

For instance, certain plant families such as Ranunculaceae, Berberidaceae, and Papaveraceae are known to have high concentrations of active alkaloids. Therefore, plants belonging to these families have the potential to be valuable sources of analogous active metabolites. Lastly, in ecological approach, the plant is chosen for its ability to protect itself from external damages. In recent years, this approach has led to select extremophile plants because of the ability to withstand to extreme climatic conditions and to quench toxic Reactive Oxigen Species (ROS) (Das and Roychoudhury 2014).

20.2.2 In Silico Approaches

The recent advent of chemoinformatic approaches gave a significant contribution to PBDD, as they allow a focused selection of promising SMs. These approaches have become possible nowadays, thanks to the introduction of processors and servers with high computing power, thus allowing the development of computational approaches based on artificial intelligence and machine learning (Maia et al. 2020). Compared to the costs of an experimental investigation, the expenditures for *in silico* approaches are negligible, and a growing number of valuable open-sources and web-based tools have been becoming available besides the expensive commercial software. However, the advent of big data cannot overlook the importance of having databases that provide easy and comprehensive access to chemical, biological, ethnobotanical, and phytochemical information about small molecules (SMs), as well as facilitate the management and processing of this large volume of data. Encyclopaedic and general Natural Product (NP) databases, databases loaded with SMs used in traditional medicines, and specialist databases focusing on certain habitats, geographical locations, creatures, biological activities, or even specific NP classes are all examples of NP databases. There are several excellent and extensive reviews of NP databases, published over the past years (Chen, de Bruyn Kops, and Kirchmair 2017; Yongye, Waddell, and Medina-Franco 2012; Füllbeck et al. 2006; Sorokina and Steinbeck 2020; Medina-Franco 2020; Chen and Kirchmair 2020). Virtual screening approaches ranging from basic, fast methods based on 2D molecular fingerprint similarity to more complex 3D methods based on molecular shape similarity, pharmacophore models, molecular interaction fields, or docking may be employed, alone or in combination (Santana et al. 2021; Linciano, Cavalloro, et al. 2020; Maia et al. 2020). Recently, machine learning techniques have been recognized as a useful tool for virtual screening of bioactive SMs (Kirchweger and Rollinger 2018).

20.2.3 'Omics' Approaches

The advances in computational methodologies and in analytical techniques open the door to the use of 'omics' approaches consisting in drawing the metabolomic profile of a plant extract (Zhao, Zhang, and Li 2018). The metabolome is the qualitative and the quantitative collection of all the metabolites present in a biological sample (Schripsema and Dagnino 2014). Metabolomics can offer reliable information on the metabolite composition of NP extracts, assisting in the prioritization of SMs for separation, accelerating de-replication (Hubert, Nuzillard, and Renault 2017), and annotating undiscovered analogues and novel scaffolds. Profiling the metabolome can be done using either nuclear magnetic resonance (NMR) or mass spectrometry (MS) coupled with separation techniques, including gas chromatography (GC-MS), liquid chromatography (LC-MS), and ultraperformance liquid chromatography (UPLC-MS) (Zhang, Sun, et al. 2012; Wolfender et al. 2015). NMR is the most suitable technique for the metabolomic approach because it offers several advantages. Firstly, it is a non-destructive technique. Secondly, it allows the detection of different classes of plant metabolites simultaneously. Additionally, NMR signals are proportional to the molar concentration, making possible the comparison among the metabolites. Lastly, NMR facilitates structure elucidation, and recognition of some classes of compounds present in the extract (Markley et al. 2017). Conversely, NMR is less sensitive compared to other analytical techniques (e.g., MS) so that larger amount of sample is required, and the signals overlapping may complicate the analysis of the spectra, although multidimensional NMR spectroscopy can help to bypass this problem (Kim, Choi, and Verpoorte 2010). NMR-based metabolomics has been successfully applied in PBDD (Khoo et al. 2015; Maulidiani et al. 2015; Kuhnen et al. 2010). The binning procedure serves to remedy the inaccuracies in chemical shifts, for the digitalization of NMR data in numeric values and for reducing the complexity of the data set. The binning leads to the conversion of NMR information into numeric data. Once the metabolomic fingerprint is obtained, the data are interpreted to generate a hypothesis through multivariate analysis. Data from metabolomics may be combined with data from other omics approaches like transcriptomics, proteomics, and genomic (Zhang et al. 2021; Martínez-Esteso et al. 2015). There have been a few successful attempts where this sequence-based approach has successfully guided the search for new natural products (Bumpus et al. 2009).

20.3 PLANT EXTRACTS PREPARATION

Along the years, several procedures have been set up to extract SMs from vegetal matrices and raw materials. Once SMs of interest have been identified, a proper methodology must be set up to ensure an exhaustive recovery. Solvent extraction (SE) is based on the capability of the appropriate solvent to penetrate within the vegetal matrix and to dissolve the compounds herein present. This way, the solutes diffuse out from the plant drugs and can be collected. Hence, the efficiency of the extraction is affected by the diffusivity and properties of the solvent used, the solubility of the SMs of interest in the solvent, the particle size of the raw material, solvent-solid ratio, and the extraction temperature and duration. For these reasons, the selection of a proper solvent is critical. Over the years, different protocols and techniques have been developed for the solvent-mediated extraction of metabolites from natural matrices, named solid-liquid extractions (SLEs). There is not a single method that could be considered as the golden standard for the extraction of bioactive compounds from plants, since qualitative and quantitative extraction yield from plants depend mostly on choosing the proper method of extraction. SLEs are classified in conventional (or traditional) and non-conventional (or innovative) methodologies. Conventional extraction methods are maceration, decoctions, percolation, digestion, reflux extraction, and hydro-distillation (Picot-Allain et al. 2021). These techniques require large volumes of organic solvents, often volatile, flammable and toxic solvents, long extraction times, and high temperatures strictly related to thermal decomposition of thermolabile compounds. Non-conventional methodologies exploit technologies able to facilitate the SMs extraction and recovery. According to the source of energy and extraction mechanism, they can be distinguished in microwave-assisted extraction (MAE), ultrasound-assisted extraction (UAE), pressurized solvent extraction (PLE), supercritical fluid extraction (SFE), enzyme-assisted extraction (EAE), and micellar extraction (Picot-Allain et al. 2021). In comparison with conventional techniques, they are faster, need less solvents, reduce possible breakdown of natural products, and are more ecosustainable (Picot-Allain et al. 2021).

20.3.1 Microwave-assisted Extraction

MAE exploits microwave irradiation to heat solvents containing samples and to cause a more effective disruption of the cellular structures (cell walls and cellular membranes) of the vegetal matrices improving the porosity of the biological matrix, thus favouring the release of the cellular content into the solvent and speeding up the extraction process (Figure 20.1A) (Cavalloro et al. 2021). Hence, MAE is more advanced than the traditional SE method, since it rapidly heats the matrix internally and externally without a thermal gradient, so that the SMs can be extracted efficiently and protectively using less energy and solvent volume. Across the years, MAE has been declined in different ways to improve the extraction yields of active compounds easily affected by thermal degradation and oxidation. Nitrogen-protected microwave-assisted extraction (NPMAE) exploits nitrogen (or other inert gases such as argon) to pressurize the extraction vessel, thus preventing the oxidation of the active compounds during the extraction process. Similarly, it has been possible to improve the

A. Microwave-Assisted Extraction (MAE)

Plant source
Microwave oven
Plant cell
mw
Cell wall rupture
Release of cell content

Dynamic microwave-assisted extraction (DMAE)

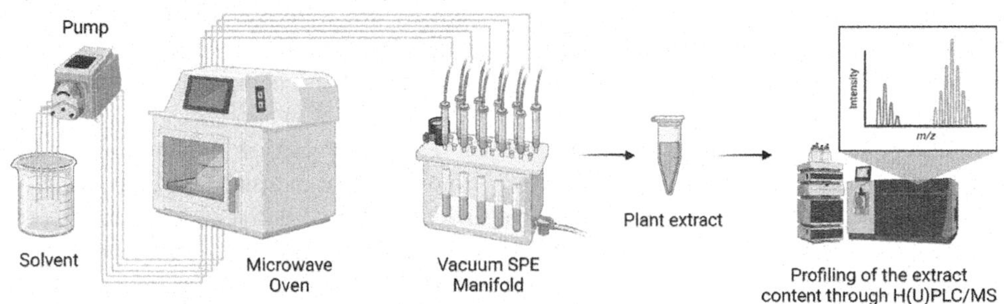

B. Ultrasound-Assisted Extraction (UAE)

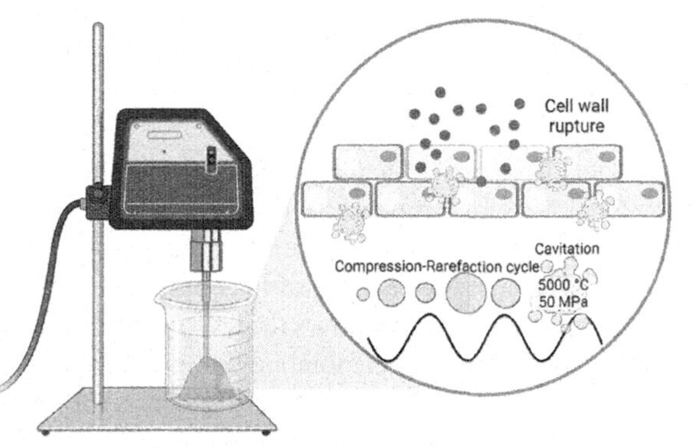

C. Supercritical Fluid Extraction

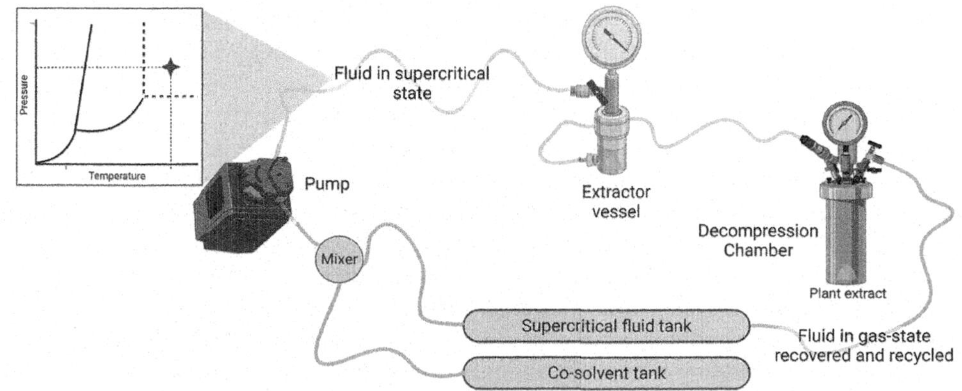

FIGURE 20.1 (A) Effect of mw-induced heating on plant cells and schematic representation of a multi-sample dynamic microwave-assisted extraction-solid-phase extraction (DMAE-SPE) system. (B) Effects of acoustic-induced cavitation phenomenon on plant cells in a typical UAE. (C) Schematic representation of an apparatus for supercritical fluid extraction.

extraction yield of oxygen and thermal sensitive compounds by irradiation of the sample at low pressure. In vacuum microwave-assisted extraction (VMAE), the vacuum pressure enhances the mass transfer of compounds into the solvent and reduces the thermal degradation, thanks to the lowering of the solvent boiling point (Pasquet et al. 2011; Kapoore et al. 2018; Xiao et al. 2009). The microwave energy may be applied also for the extraction in solvent-free conditions (solvent-free microwave extraction—SFME) (Li et al. 2013). In this case, the plant material is directly placed into the microwave reactor, without the addition of any solvent. The heating of the water contained into plant material distends the natural matrices and causes the rupture of the cells releasing their content. This methodology can be successfully applied for the extraction of the essential oil. The efficiency of MAE depends on several factors, such as solvent, solvent-drug ratio, temperature, time, pressure, microwave power, water content, and characteristic of the matrix. Each of these parameters should not be considered alone, but they are all linked together, and the comprehension of the effects and influences of these factors are pivotal for MAE efficiency (Cavalloro et al. 2021). However, the availability of safe instrumentations suitable for scale-up process and industrial applications has been represented a technical barrier for a long time. The first industrial-scale ovens have only recently appeared, so that in the near future, pharmaceutical, and nutraceutical industries would benefit from the MAE technology. Lastly, MAE was coupled in line with analytical techniques, giving rise to dynamic microwave-assisted extraction (DMAE) (Ericsson and Colmsjö 2000). This hyphenated technique assures an overall reduction of the extraction time, avoid the necessity to perform extraction cycle since the process is continuous, and lessens the risks of analyte loss and contamination as the system runs continuously in a closed and automated manner (Figure 20.1A).

20.3.2 Ultrasound-assisted Extraction

UAE is a solid/liquid extraction, based on the acoustic-induced cavitation phenomenon (Ashokkumar 2011). This causes a very local raise in temperature that enhances the solubility of the analytes (as in the case of lipids), together with the local pressure increase that facilitates the penetration of the solvent into the sample matrix and eases the release of the compounds (Figure 20.1B). Moreover, the implosion of cavitation bubbles can hit the surface of the solid matrix and disintegrate the cells. As a result, UAE is suitable for the extraction of thermolabile compounds, and it is environment-friendly, because it requires short extraction time and minimum quantity of solvents. Such as for MAE, the reliability and reproducibility of results are influenced by process variables, such as ultrasonic power, frequency, temperature, reactor design, solvents, solvent–sample ratio, or particle size and structure (Chemat et al. 2017). Ultrasound technology is easily adaptable to conventional extractive methodologies (Khadhraoui et al. 2021). UAE has been successfully combined with instantaneous controlled pressure drop process (DIC, Détente Instantanée Contrôlée). DIC extraction exploits the injection of steam into the material followed by a drastic drop of pressure by applying vacuum that causes the auto-vaporization of the moisture inside the material with the release of the compounds within. DIC could thus induce particle fragmentation and exudation; meanwhile, the ultrasound cavitation could enhance the erosion of solid particles and the mass transfer from the inside to the outside of the treated matrix. Combining DIC with UAE in extraction from plant matrix resulted in yield improvement and time reduction in comparison with hydro-distillation and SE. UAE poses less limitation in the hardware for scale-up and industrial application (Kiss et al. 2018). The ultrasonic bath is the cheapest and commonly used ultrasonic device; although for extraction application, high-power ultrasonic probes are preferred because of the ultrasonic intensity produced and for the capability of directly immersing the probe into the reactor resulting in a direct delivery of ultrasound in the extraction media with minimal ultrasonic energy loss. The probes differ in lengths, diameters, and tip geometries, and they might be properly selected according to the application and to the sample volume to be sonicated in order to maximize the extraction yield, and are easily employed for industrial scale-up (Kiss et al. 2018). UAE is nowadays applied for the industrial extraction of vindoline, catharanthine, and vinblastine from *Vinca* spp. (Yang, Wang, Zu, et al. 2011; Martino et al. 2018). Sono-Soxhlet methods combine the advances of ultrasound extraction, by applying ultrasound in the extraction chamber to enhance the extraction and migration of metabolites from solid matrix to fresh solvent repeatedly provided by Soxhlet (Djenni et al. 2013; Luque-García and Luque de Castro 2004). The combination of ultrasound cavitation with Clevenger distillation at atmospheric or reduced pressure resulted in the development of Sono-Clevenger. It has found great applicability in the flavour and fragrance industry for the extraction of essential oils from plant materials, resulting in reduced extraction times and enhanced yields compared to the traditional hydro-distillation. Similarly, UAE may be combined with MAE, giving rise to the so-called ultrasound microwave-assisted extraction (UMAE), in order to benefit from the double simultaneous irradiation with mw and ultrasounds (Cravotto et al. 2008). The ultrasound cavitation promotes the release of the compounds from the vegetal matrix; meanwhile, the fast heating induced by mw favours the migration of the dissolved molecules. This hybrid technique resulted in high efficiency and in the dramatically reduction of extraction time due to the additive or even synergistic effects of the two sources of energy in the extraction process. Although this technique is still pioneering, several applications of UMAE for plant extraction are reported in literature highlighting its great potential in both academic and industrial research activities.

20.3.3 Pressurized Liquid Extraction

Although the PLE was developed about 60 years ago, it remains an excellent alternative to the traditional extraction methods. PLE exploits high pressure to increase the boiling point of the solvents. In this harsh condition, the liquid enhances its solvent and mass transfer properties and its capability to penetrate cells and tissues of plant drugs. As a result, the extraction yield increases, and in the meantime, time and solvent consumption decrease. More in detail, the PLE equipment provides protection from oxygen and light for sensitive compounds, whilst this methodology is not particularly suitable for thermolabile compounds unless a thorough optimization of the extraction parameters is performed (Alvarez-Rivera et al. 2020).

20.3.4 Super-critical Fluid Extraction

SFE is an advanced technique that exploits super-critical fluids as solvent (Uwineza and Waśkiewicz 2020). Super-critical fluids are usually gases or liquids compressed and heated over their critical point. In these conditions, fluids acquire both gas and liquid phase properties such as diffusion, viscosity, and surface tension of a gas and density, and solvation power of a liquid. The low viscosity and high diffusivity enhance the transport properties of the fluid that can easily diffuse through the matrix, resulting in faster extraction rates (Figure 20.1C). Moreover, the solvation power of a super-critical fluid is strictly dependent from its density that can be regulated by modifying its pressure and temperature. Carbon dioxide (CO_2) is the ideal and most employed super-critical fluid in SFE because of a series of advantages such as chemical inactivity, accessibility, higher selectivity, diffusivity, and ecology. Additionally, it is recognized as safe and employable in food treatment such as for the extraction of caffeine from coffee. Thanks to its low polarity, hydrophobic compounds and small molecules are easily dissolved in supercritical CO_2; the extraction of more polar compounds might be enhanced with the addition of co-solvents such as ethanol, methanol, or water. Moreover, the low critical temperature and pressure (31°C and 74 bars) make super-critical CO_2 suitable for the extraction of heat-sensitive compounds. Ethane, propane, and dimethyl ether have been used as super-critical solvents for the extraction of bioactive compounds from plants. These solvents have higher polarizability than CO_2, resulting in a stronger eluant power for more polar compounds and co-solvents. Despite these advantages, super-critical fluid-assisted techniques still present some limitations, such as the need of expensive equipment, high energy costs mainly related to the fluid compression, and a careful cost/effective set-up. Preparative systems (pilot- or industrial-scale) are used to deal with greater quantities of samples, and their configuration is determined by the degree of automation. When operating on a pilot size, these devices are used to extract grams of compounds, whereas on an industrial scale, they are utilized to extract kilograms. Additionally, this equipment may have one or more separators that collect the extract and depressurize the solvent. They are often outfitted with independent temperature and pressure control, allowing for the fractionation of extracted substances by progressive depressurization. Depending on their differential solubility in the super-critical fluid, various chemicals are produced inside each separator. The most volatile substances can be recovered if the separator is coupled with a cooling mechanism. Therefore, nowadays, their application is limited to high value-added products. However, there still is room for improvement of the SFE technology, and its application in pharmaceutical and nutraceutical industries will open a new scenario for a more sustainable, cleaner, and environmental-friendly extraction process in the research of bioactive compounds. Similar to other procedures, SFE offers the possibility of direct coupling with analytical chromatography such as GC or super-critical fluid chromatography (SFC) exploiting the super-critical fluid as carrier. SFE offers numerous benefits, including fast processing time, suitability for extracting volatile and thermolabile compounds, higher productivity in terms of increased yields, reduced solvent use, and protection of environment using safe solvents.

20.3.5 Enzyme-assisted Extraction and Ionic Liquid-Based Extraction

EAE is another promising technique for extracting bioactive compounds from plant material by exploiting the hydrolytic action of certain enzymes such as glucose oxidase, amyloglucosidase, hemicellulose, amylase, cellulase, pectinase, etc., depending on the material and the purpose of the extraction (Streimikyte, Viskelis, and Viskelis 2022; Puri, Sharma, and Barrow 2012). These enzymes cause the breakdown of phytochemicals and cellular structures and facilitate the extraction process, so that it can be successfully used as a pre-treatment step (Sowbhagya and Chitra 2010). Enzymes can improve extraction efficiency by catalyzing reactions with high specificity and under moderate circumstances, increasing extraction yields and improving product quality (Puri, Sharma, and Barrow 2012). However, there are significant constraints in terms of the cost of enzymes required to digest huge volumes of raw material (Pinelo, Arnous, and Meyer 2006; Puri, Sharma, and Barrow 2012). Furthermore, scaling up the process may be problematic since enzymes might respond differently even with slight variations in the process environment. If these issues are addressed, EAE has the potential to be a highly promising approach for processing plant matrices. Several recent applications have shown the usefulness of ionic liquids (ILs) for secondary metabolite extraction from the plant material (ionic liquid-based extraction) (Xiao, Chen, and Li 2018). The use of IL or IL-based materials in conjunction with LL, UAE, MAE, high-performance liquid chromatography (HPLC), solid-phase extraction (SPE), and other techniques has resulted in the effective extraction or separation of bioactive chemicals from plants. To modify the adsorption/affinity capability of silica and polymers, ILs were used as co-surfactants, electrolytes, or adjuvants, or as supporting materials. In general, it was shown that when appropriately selected, IL-based solvents and materials may provide greater extraction yields and purification factors than standard solvents and materials. Furthermore, tailor-made ILs may be created by carefully selecting anions and cations, resulting in great selectivity throughout the extraction process. In contrast, as with organic solvents, simple evaporation of the ILs is not an option. Other separation methods must be developed to make ILs appealing as an alternative solvent on a large scale. The extraction of artemisinin from *Artemisia annua* is the most established example (Lapkin 2015).

20.3.6 Micellar Extraction

Micellar extraction exploits the interesting behaviour of surfactant in water. Surfactants are amphiphilic substances with hydrophilic and lipophilic components. They agglomerate in water at a particular concentration, forming an oil-soluble phase that allows hydrophobic chemicals to be dissolved. Non-ionic surfactants have a distinct character because, at a certain temperature, the micelles become unstable and degrade. Two phases are created at this so-called cloud point. Depending on the extracted compound's composition, it can be found in either the surfactant-rich or water-rich phase. The initial phase in cloud-point extraction (CPE) is the micellar extraction of the plant material below the cloud temperature. After the hydrophobic molecules have been removed, a temperature increase is applied to separate the surfactant-rich phase from the matrix. The literature has several examples of CPE of biomolecules including proteins, polyphenols, and triterpenes. Plant material extraction employing ionic surfactants is uncommon in the literature due to the complexities of extract removal and solvent recovery. Apart from the difficulties in recovering the target chemicals from the extraction medium, micellar extraction has other advantages, such as a high capacity to concentrate analytes with almost quantifiable yields. Furthermore, several ecological-friendly and low-cost surfactants are accessible. In comparison to the necessity for organic solvents in plant extraction, only trace quantities of surfactants are required. Furthermore, because of the comparatively moderate extracting conditions, the extraction of thermally sensitive chemicals is possible. Design of Experiment (DoE), response surface methodology, and other statistical approaches are of great support in determining the best conditions to achieve the highest yield of the metabolite of interest from the natural source (Aydar 2018; Malacrida et al. 2019).

20.3.7 Analytical Fingerprint of Crude Extracts

The analysis of crude plant extracts is another important milestone of PBDD process. Particularly, crude extract analysis is essential for drawing the extract fingerprint, for determining their composition, and for guiding the isolation of bioactive SMs (Martino, Della Volpe, et al. 2019; Amri et al. 2017). Crude extract analysis is initially performed by chromatographic techniques, such as thin layer chromatography (TLC), high-performance thin-layer chromatography (HPTLC), GC, and HPLC (Salam, Lyles, and Quave 2019). Briefly, TLC is commonly used for routine analysis both for qualitative and semi-quantitative purposes and also to monitor the fractionation process, since it is simple, has a low cost, and great versatility. Accordingly, TLC and HPLTC are useful techniques for drawing the fingerprint of a given extract (Wagner and Bladt 1996). Of note, nowadays, many manuals are available, which can be exploited as a reference for analysing natural extracts, identifying unknown drugs in natural matrices, or monitoring the purity or constituents of a given extract; TLC is also a useful technique to screen the biological activity of the SMs and to monitor bio-guided assay fractionations, thanks to the procedure known as bio-autography (Urbain and Simões-Pires 2020; Dewanjee et al. 2015). An example of bio-autography consists of exposing the TLC plates to microbial suspensions and visualizing zones with no microbial growth to identify metabolites with anti-microbial activity (Jesionek et al. 2015). Another well-known application consists in exploiting DPPH (2,2-diphenyl-1-picryl-hydrazyl-hydrate) solution as a stain to identify anti-oxidant metabolites. In detail, the yellow-coloured spots are the ones responsible for the metabolites with anti-oxidant activity (Cieśla et al. 2012). For drawing extract fingerprints, GC may be used as a powerful technique, since it is cheap, fast, requires low sample amounts, giving in the meantime a high resolution of complex mixtures. Nevertheless, it is suitable only to analyze essential oil and volatile compounds, since sample injection requires analyte vaporization. Conversely, HPLC is extensively used both in industry as well as in academia, and can be applied both for analytical and preparative purposes. The analysis of plant extracts by HPLC is versatile, although it requires an accurate set-up of the method, by selecting a proper combination of stationary and mobile phases together with an appropriate detector (i.e., photodiode array-PDA, circular dichroism, refraction index, and mass spectrometer-MS-detectors). Moreover, HPLC/PDA/MS analysis provides additional information for identification and structural elucidation of unknown metabolites. Very recently, the new concept of optimum performance laminar chromatography (OPLC) has been developed. OPLC combines the advantages of the user-friendly interface and resolution of HPLC with the capacity of flash chromatography and multi-dimensionality of TLC. Such as in HPLC, in OPLC, the mobile phase is pumped at higher pressure (up to 50 bars) and at constant linear velocity through a stationary phase which is packed into a flat planar column. Both HPLC and OPLC are analytical and preparative tools, suitable for research and quality control laboratories (Wagner and Bladt 1996).

To sum up, the hyphenation of HPLC or GC separation with MS and recently with NMR may allow the identification of all constituents present in crude extracts, even with complex composition, or in pre-fractionated bioactive fractions. The most abundant compounds or the main compounds present in the extracts can be identified before proceeding with fractionation, thus helping in the choice of the most suitable separation methodology, thus accelerating the entire process.

20.4 IDENTIFICATION OF SECONDARY METABOLITES ENDOWED WITH BIOLOGICAL ACTIVITY

Plant extracts are complex mixtures of hundreds of chemical compounds. As a result, the evaluation and identification of the compound responsible for the bioactivity are extremely difficult. The development of reliable methods for screening bioactive SMs in plant extract is demanding (Wang et al. 2011).

20.4.1 Bio-guided Fractionation

Bio-guided fractionation approach consists in performing fractionation and purification of crude extracts guided by biological assays. Briefly, after a first fractionation, the fractions are screened for biological activity. The fractions that show an interesting biological profile are further fractionated and tested in an iterative circle until the bioactive ingredient is identified (Malacrida et al. 2019; Pellavio et al. 2017; Gaggeri et al. 2013). Thus, the extraction and the biological screening proceed side-by-side. One of the main drawbacks of this approach is the progressively loss of activity during the fractionation process, and most importantly, a high probability to isolate already known compounds. Thus, de-replication, namely the identification of known compounds responsible for the activity of an extract before bioassay-guided fractionation, is mandatory (Reynolds 2017). A good de-replication method allows rapid and reliable identification of already known metabolites and of ubiquitous interfering compounds such as tannins, fatty acids, and saponins, thus avoiding the time-consuming and useless isolation and characterization processes. Progress in MS, molecular networking, combined with computational MS tools and genome mining methods, as well as advances in NMR spectroscopy has contributed to advance de-replication (Gaudêncio and Pereira 2015). The spectroscopic data collected for all the constituent of the extract are compared by means of elaborate algorithms with the spectral patterns in the de-replication database for the identification of already known compounds (Zani and Carroll 2017). The success of a de-replication method strongly depends on the availability of databases providing experimental analytical data for known SMs. Unfortunately, there is no universal database (Sorokina and Steinbeck 2020), although the Global Natural Social Molecular Networking (GNPS) aimed to develop a compressive platform for de-replication involving all the researcher worldwide who want to contribute to this project by supplying spectral data for known SMs (Wang, Carver, et al. 2016). The main disadvantage of bio-guided fractionation approach is represented by the difficulty to identify synergic effect of SMs. To overcome this issue and to evaluate the synergic effect of SMs, combinations of different fractions may be assayed. It is conceivable for potentiators to wind up in inactive fractions during the isolation process and hence be neglected when using bioactivity guided fractionation (Junio et al. 2011; Dettweiler et al. 2020; Caesar and Cech 2019; García, Torres, and Macías 2015; Britton et al. 2018).

20.4.2 Direct Phytochemical Isolation

The direct phytochemical isolation strategy focuses on the whole chemical characterization of a plant extract. Special attention is placed on the isolation of new compounds utilizing de-replication technologies (Appendino et al. 2009; Citti et al. 2019; Linciano, Citti, et al. 2020). Biological testing is performed on the isolated compounds. Screening pure compounds provides numerous advantages, including improved compatibility with HTS and avoiding many of the challenges associated with extract testing, as discussed above. However, time and effort must be committed in advance for the separation of natural chemicals, with no certainty that the individual compounds will eventually exhibit significant bioactivity.

20.4.3 Cell-based Screening

Cell-based screening assays have been extensively used for the identification of SMs from plant extracts with successful outcomes (Qiu et al. 2015; Hong et al. 2011; Sun et al. 2015; Yang et al. 2021; Zhang et al. 2009; Feng et al. 2022; Kang et al. 2021; Sklirou et al. 2021). This method requires the treatment of cell cultures with the extract; after washing out the cells to remove the unbounded compounds, the cells are digested and the bound compounds are detected and analyzed using chromatographic methods coupled with UV or MS detectors. One variant of this approach exploits hollow fibres. The cells are injected into the fiber lumen, and the charged hollow fibers are bent into a U-shape and soaked into the plant extracts to fish out the active compounds (Figure 20.2A) (Hu et al. 2019). This approach is relatively simple, fast, and allows high enrichment and low solvent consumption and low cost. A main drawback results in a concentration gradient between the inside and outside of the fibre that could lead to positive false, and it requires a long incubation time, and it is not translatable for HTS.

20.4.4 Equilibrium Dialysis, Microdialysis, and Ultrafiltration

Equilibrium dialysis, micro-dialysis, and ultrafiltration were also successfully used to fish-out from plant extract bioactive SMs (Wang, Liu, et al. 2016). In equilibrium dialysis and micro-dialysis, both extract and cells or targets are mixed in a dialysis bag. After the proper incubation time, any unbound compounds diffuse from the membrane, while the complex SM-target is retained into the bag. Once the equilibrium is reached, the dialysate is collected and analyzed through HPLC or UPLC-UV/MS (Figure 20.2B). Dialysis technique is simple, and the only external factor that could affect the process is the dialysis bag or tube. Moreover, micro-dialysis has the additional advantage to be coupled to "on-line" or "off-line" systems for detection. Unfortunately, as for cell-based screening, dialysis suffers from long incubation time, and the screening cannot be performed in HTS. Van Breemen (van Breemen et al. 1997) is the pioneer for ultrafiltration in screening for SMs (Yan et al. 2014; Nikolic et al. 2000; Liu et al. 2007; Choi et al. 2011; Tang, Si, and Liu 2015; Chen and Guo 2017; Chen et al. 2016; Ghani 2015). In ultrafiltration, the mixture of target and plant extract is filtered through a semi-permeable membrane. Unbound compounds are filtered off, while in the donor well, the compounds bounded to the target are retained (Figure 20.1C). Compared to dialysis, ultrafiltration allows to cut down the incubation time, making this approach simplest and most reliable. However, a disadvantage of the ultrafiltration is that the unbound compounds cannot be completely removed unless repeated manual washing is performed.

20.4.5 Chromatographic-based Approaches

Chromatographic-based methods are useful tools for the screening of bioactive compounds. The most exploited chromatographic methods are based on *bio-affinity chromatography* [cellular membrane affinity chromatography (CMAC) (Stephen, El Omri, and Ciesla 2018), cellular membrane chromatography (CMC) (Ma et al. 2021), immobilized liposome chromatography (ILC) (Zhang, Li, et al. 2012)] that uses carrier properly modified with the biological target as a stationary phase. So that, once the plant extract is flowed through the column, solely the potential active compounds are selectively retained by the stationary phase, whereas the other components are eluted out. After releasing the bounded compound from the stationary phase, the eluted compound is collected and analyzed. In CMAC, transmembrane target proteins are immobilized onto the stationary phase

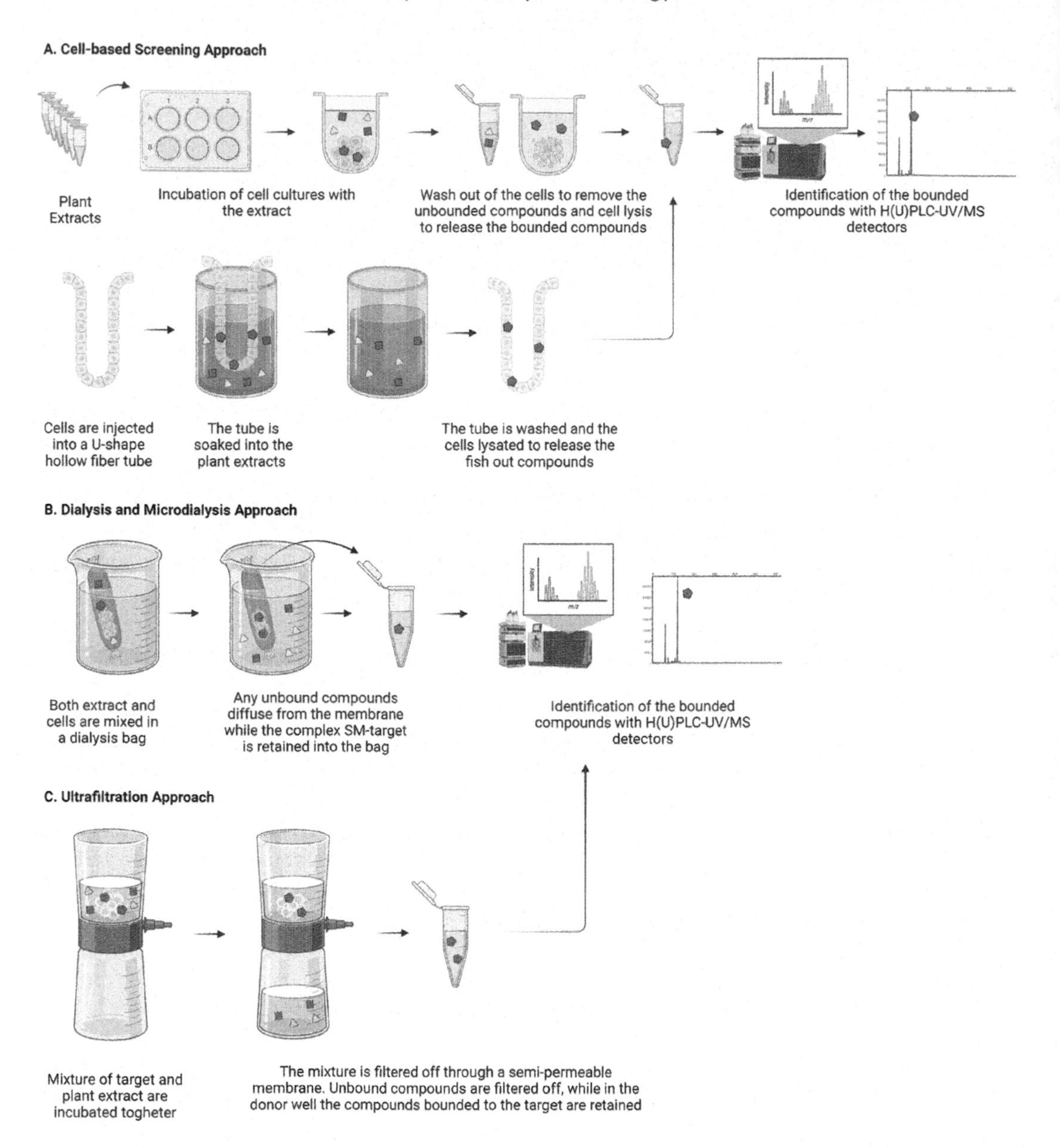

FIGURE 20.2 Visual representation of the main phases for the identification of bioactive secondary metabolites in plant extracts through a cell-based screening (A), dialysis and microdialysis (B), and ultrafiltration approaches (C).

(Figure 20.3A). Several important works that use this approach for screening of SMs are reported (Stephen, El Omri, and Ciesla 2018). CMC, instead, uses cell membrane fragments adsorbed on silica gel as a stationary phase (Figure 20.3B). Although uses for many years, initially this technique suffered of low specificity, sensitivity, and selectivity, but the recent use of cell membranes with increased expression levels of the target of interest allowed to improve the results of this approach. Moreover, passing from a one-dimension column to 2D-CMC-HPLC in an online mode resulted in a higher column efficiency, peak capacity, and structural identification. 2D-CMC-HPLC approaches such as CMC-offline-GC (HPLC)/MS (Yang, Wang, Zhang, et al. 2011; Hou et al. 2009) and CMC-online-HPLC/MS (Wang, Sun, et al. 2010; Wang, Ren, et al. 2010) were used to screen and analyze SM-active compounds. However, CMC assay is complex and expensive, and it has been partially replaced by ILC. ILC has been extensively used for screening and analysis of potential cell permeable components of SMs with higher efficiency than other method (Figure 20.3C) (Hou et al. 2020; Sheng et al. 2005). Overall, these methods are complicated and show poor reproducibility, which may restrict the extensive applications of bioaffinity chromatography. To overcome these limitations, the immobilization of target molecules for ligand fishing has been developed and applied in PBDD. The pioneering work used Abs immobilized on magnetic beads (MB) to capture a specific ligand from complex mixture. From this moment ahead, magnetic nanoparticles (made of magnetite) were extensively used for protein immobilization and for fishing bioactive compounds from plant extract (Figure 20.3D). MBs offer suspension stability, high surface area, ease surface modification, and an easy-to-handle liquid-solid separation. The target-derivatized MBs are suspended into the solution of the plant extract, and after the appropriate incubation time, the MBs are easily removed from the solution, thanks to their magnetism. After being washed to remove unbounded compounds, the target–ligand complex is dissociated and the solution containing the fished-out compounds are analyzed using (U)HPLC/MS methods. One main limitation of ferrite MBs is their tendency to aggregate, which has been solved by coating MBs with stabilizing agents such as silica or polymers. MBs were also used as a support for cell membranes with noticeable advantage over CMC. Besides MBs, other materials were used for fishing system, such as agarose beads bound to DNA, HF, and halloysite nanotube. The development of this approach recently resulted in an online platform based on capillary enzyme reactors (ICERs), which was applied for the screening of acetylcholinesterase (AChE) inhibitors in the extracts of *Corydalis yanhusuo* (Wang et al. 2018). All these methods result to be very effective, and a number of successful examples of their applications for the analysis and identification of bioactive SMs from plant extracts are reported in literature. However, each one has its own advantages/disadvantages, in terms of equipment, applicability, complexity, solvent consumption and compatibility, time, efficiency, etc. Moreover, due to the complexity of the extract compositions, the risk to result in false positives, false negatives, and missed binders is high. Besides the intrinsic limitations, certain methods have been more successfully applied in PBDD programs with ligand fishing being the newest and most promising approach in the identification of active compounds from the plant matrices.

20.4.6 Structural Elucidation of Bioactive Secondary Metabolites

Once the compounds responsible for the biological activity have been identified, their structural elucidation is mandatory. The structure of isolated SMs may be deciphered by combining several spectroscopic techniques such as FT-IR, NMR, and mass spectrometry analysis, as it is usually performed for identifying organic compounds (Citti et al. 2019; Linciano, Citti, et al. 2020). The advent of increasingly high-resolution mass spectrometers and high-field NMR (1-1.2 GHz) allows to solve complex structure-related issues using very low amounts of samples, sometimes less than 1 mg. Moreover, as stated in paragraph Section 20.1.1, SMs are usually endowed with both architectural and stereochemical complexity. Chirality is a key feature of natural products, and stereochemistry is crucial for biological activity. Accordingly, focusing on the evaluation of enantiomeric composition (usually performed via enantioselective chromatography, for deepening this aspect, see review (Ward and Ward 2012; Fernandes 2019; Gumustas, Ozkan, and Chankvetadze 2018), the absolute configuration (AC) assignment is mandatory. AC can be assigned using several methods, including X-ray crystallography, NMR methods, chemical correlation method, and stereo-controlled total organic synthesis. Particularly, X-ray crystallography is used in the first assignment of the AC of a chiral compound, and it is still considered the most reliable technique. Conversely, NMR spectroscopy can be used only if a chiral auxiliary is added during the experiment. NMR auxiliaries may include, as an example, chiral derivatizing agents, chiral solvation agents, and chiral hosting compounds (Seco, Quiñoá, and Riguera 2004; Mishra and Suryaprakash 2017). Lastly, stereo-controlled organic synthesis is time-consuming, expensive, and highly dependent on the correct AC of both

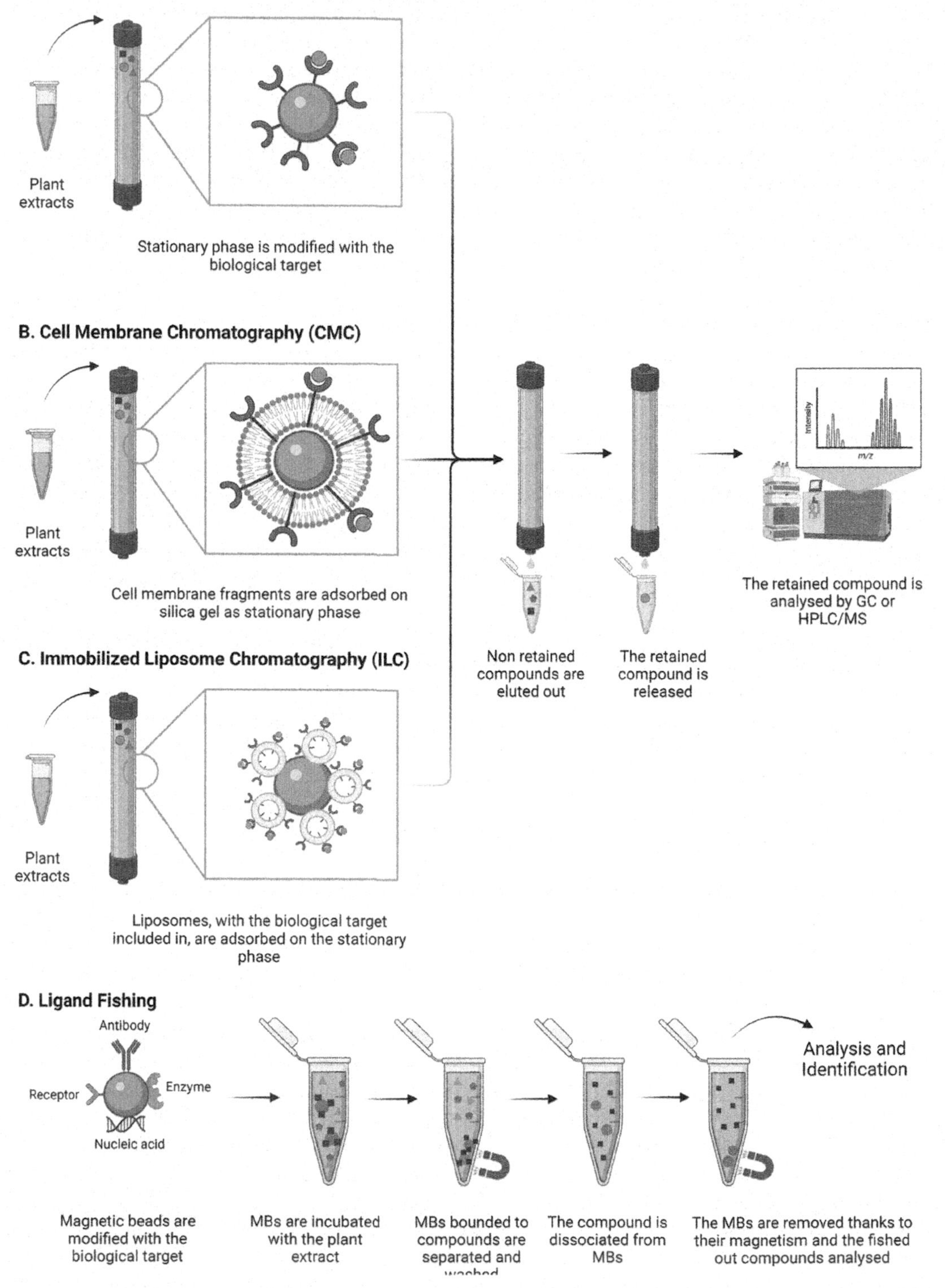

FIGURE 20.3 Schematic workflow for bio-affinity chromatography-based [A. cellular membrane affinity chromatography (CMAC) (Stephen, El Omri, and Ciesla 2018) (Stephen, El Omri, and Ciesla 2018), B. cellular membrane chromatography (CMC), C. immobilized liposome chromatography (ILC)] and ligand-fishing methods (D) for the screening of bioactive SM from plant extracts.

starting materials and products and not always applicable. As an alternative, the AC assignment study may be performed by a full set of chiroptical spectroscopies in comparison with calculated spectra obtained by ab initio methods (Density functional theory - DFT) (Crawford 2006). More in details, optical rotatory dispersion (ORD), electronic circular dichroism (ECD), and vibrational circular dichroism (VCD) spectra may be recorded and compared to the calculated ones. The combined use of the three techniques is particularly advised for the AC assigned of natural products (Linciano et al. 2022; Rossi et al. 2016; Linciano, Citti, et al. 2020; Citti et al. 2019). These techniques are non-destructive and can be measured directly in solution, without the need for crystallization (Ward and Ward 2012; Gumustas, Ozkan, and Chankvetadze 2018).

20.5 PLANT-DERIVED RECOMBINANT PROTEINS

Over the last decades, the development of protein-based drugs, such as vaccines, monoclonal antibodies (mAbs), enzymes, cytokines, and growth factors, grew very quickly (Lagassé et al. 2017; Gaughan 2016).

In particular, since the late 1990s, the production of plant-derived vaccines, mAbs (named also plantibodies) and antibody fragments, began to emerge with many important issues, such as the balance of benefit and risk and relation between delivery methods and immunity but with the well-known advantages of efficient production in terms of amounts, rapidity, and low costs, conversely. The possibility to be produced locally and in bulk represents a great advantage to ensure rapid and economic production processes (Daniell, Streatfield, and Wycoff 2001; Sohrab et al. 2017; Peeters et al. 2001; Fischer 2003; Chung et al. 2021). For years, vaccine and antibody production in plants have been very versatile as many forms or derivatives can be expressed and used in many applications for human and veterinary scopes, such as therapy, diagnosis, immunomodulation, and intracellular immunization. In addition, these biological products are particularly effective, safe, and stable. In fact, plant cells synthesize and assemble complex human proteins by using mainly similar mechanisms (expression and some PTMs) to those of animal cells protein (Peeters et al. 2001; Fischer 2003; Floss, Falkenburg, and Conrad 2007; Chung et al. 2021). Although plant-based production platforms have many advantages, as cited, the main problem of plant-derived protein products can regard the different PTMs (for example, glycosylation) between plants and humans and the consequent necessity to find new expression strategies (Ko 2014; Chung et al. 2021). In addition, the limiting steps to overcome often still remain purification and characterization approaches, and so it is mandatory to have efficient extraction procedures and suitable analytical techniques (Temporini et al. 2020; Daniell, Streatfield, and Wycoff 2001).

Plant-derived human or animal vaccines have been studied for different administration routes (mucosal, subcutaneous, intramuscular, intravenous, topical), and thanks also to plants' advantage to give stable products, they seem to be particularly promising for mucosal (oral or nasal) immunization. In fact, plant-derived mucosal vaccines have been widely studied for many human (Norovirus, Hepatitis B Virus, Rabies Virus, Enteric pathogens) and animal (Transmissible Gastroenteritis Virus, *Actinobacillus Pleuropneumoniae*) viruses (Mason and Herbst-Kralovetz 2012; Tokuhara 2018). A lot of vaccines and antibodies have shown promising results in efficacy and safety in pre-clinical and clinical trials (Sohrab et al. 2017). Nowadays, only a plant-based vaccine against Newcastle disease virus in poultry (authorization by the United States Department of Agriculture) and a plant-made mAb fragment, which is a single-chain fragment variable (scFv) mAb used in the production of a recombinant Hepatitis B virus (HBV) vaccine (authorization in Cuba), have been licensed. In addition, in 2012, US Food and Drug Administration (FDA) has approved only one human therapeutic plant-based vaccine produced in carrot cells, named as Elelyso® (taliglucerase alfa), which is effective against Gaucher's disease, a mitochondrial enzyme deficit pathology. This problem is particularly due to the fact that vaccines fall within the classification of genetically modified (GM) crops (Dhama et al. 2020; Chung et al. 2021).

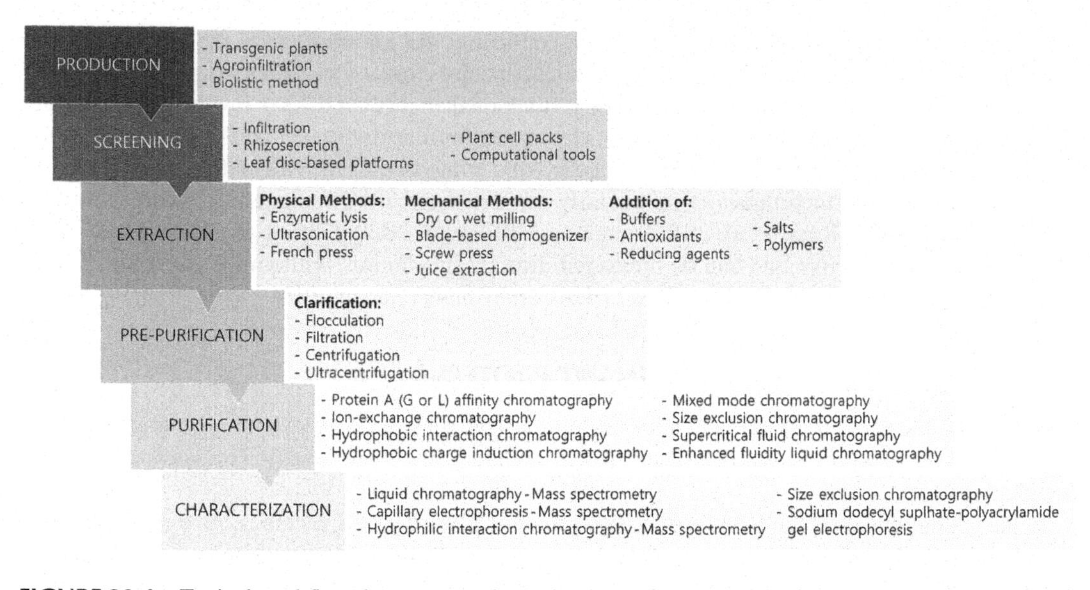

FIGURE 20.4 Typical workflow that resumes the main phases for the preparation of plant-derived recombinant proteins, focusing on the innovative approaches that will be discussed in Section 20.5.

For years, the benefits and future possibilities of these products have been discussed and well-known, and plants represent potential important sources of vaccines against global human pathologies such as acquired immunodeficiency syndrome (AIDS), seasonal influenza, or Ebola virus infection (Tremouillaux-Guiller et al. 2020; Ward et al. 2020; Tripathy et al. 2021). In addition, the search for plant-based vaccines is widely open also against recent diseases, such as coronavirus disease 2019 (COVID-19) or niche and orphan diseases (Dhama et al. 2020; Chung et al. 2021).

To obtain high quantity of stable and highly purified plant-based proteins, the procedures from extraction to clarification and purification must be carefully set up with a robust optimization, taking into consideration that instability with not desired modifications, such as deamidation, isomerization, and truncation, can be frequent during these processes and storage (Łojewska et al. 2016). In the characterization step, the analytical approaches must answer to many critical quality (product- and process-related) and safety attributes to meet the standards for approval. So they must be sensitive, giving accurate and precise information on primary sequence, stability, glycosylation profiling, and aggregation state (Temporini et al. 2020). Figure 20.4 depicts the typical workflow for the preparation (from expression to analytical characterization) of plant-derived recombinant proteins, together with the novel and most promising cutting-edge approaches that will be discussed in the following paragraphs.

20.6 PRODUCTION AND SCREENING APPROACHES

In general, the production of plant-based vaccines/mAbs occurs based on transgenic plants or in alternative agroinfiltration. The first common process relies on a stable transformation system (integration of a transgene—nuclear or plasmid—in the plant genome) via genetically modified plant virus by using *Agrobacterium tumefaciens* and has a long-time development phase (about two years) to obtain a routine method (direct gene transfer method). The second transformation process is named agrobacterium-mediated gene transfer and is a transient expression method in which *Agrobacterium* species (mainly *tumefaciens* and *rhizogenes*) work as vectors to infect plants (indirect gene transfer method). This is rapid (days–weeks) and can produce large amounts of products (mg), but its scale-up remains expensive (Laere et al. 2016; Fischer 2003; Chung et al.

2021). Other production methods are promising but remain still little applied and consist of biolistic (named also as gene gun or micro-projectile bombardment method) or polyethylene glycol (PEG) treatments. Biolistic method is used as an alternative to agrobacterium-mediated transformation and consists of a direct and stable introduction of a DNA coating into the plant genome without using vectors but micro-carriers, such as gold or tungsten. On the contrary, PEG can be used as a stimulant, which promotes DNA uptake, also in association with other gene delivery methods; it is economic but more complex to realize (Dubey et al. 2018; Laere et al. 2016; Chung et al. 2021).

For years, phage-antibody display represents an efficient strategy to produce libraries of mAbs with high affinity and specificity in short period of time (almost two weeks). An antibody gene is fused to the gene on its surface, and antibody or antibody fragments produced remain on the surface, giving origin to a wide range of different linkages and by consequence large antibody libraries (Griffiths and Duncan 1998; Kandari and Bhatnagar 2021). Over time, different strategies have been widely applied in mAb production in plants, mainly for the humanization of mAb carbohydrate profile, from the elimination of mAb immunogenicity to the introduction of human genes or regulation of protein glycosylation (Sheshukova, Komarova, and Dorokhov 2016).

In addition, recent advances in computational tools could represent promising strategies also in vaccines and plantibodies design. In fact, methods such as molecular dynamics simulations, structure activity relationships, and virtual screening can be very useful in the crucial issue of epitope prediction, and some bioinformatic methods have contributed to an optimization of epitope processing. The epitope prediction is particularly important for vaccine design; nowadays, many servers are available to predict different epitopes (T cell or B cell epitopes), and a lot of databases allow to identify potential epitopes fundamental in the binding with specific proteins, as for example HIV proteins involved in AIDS (Soria-Guerra et al. 2015; Dubey et al. 2018).

For screening, infiltration procedure of plants or plants tissues has been widely used, but unfortunately, it lacks reproducibility and scalability. On the contrary, the production from plant cell cultures can represent partially a solution, giving origin to a continuous and controlled process. The only limit of this approach is the production transfer from plant cells to whole plants (Fischer et al. 2004).

In large-scale production of proteins and mAbs, the rhizosecretion could represent an interesting solution, also to simplify the following extraction and purification steps. It consists of a continuous production method from roots, obtaining high yields of homogeneous products directly during the plant life cycle. This system works in hydroponic medium, in which it is possible to control growth media composition, ensuring low cost and efficient productions (Madeira et al. 2016). Leaf disc-based platforms are a very useful approach for the high-throughput screening of transient gene expression in plants. In fact, leaf age and position influence the product accumulation in terms of quantity and uniformity. They consist of a leaf disc infiltration approach, which exploits leaf discs obtained from leaves of adult plants, which are infiltrated and then selected, according to their infiltration capacity, allowing a higher accumulation and stability of protein products (Piotrzkowski, Schillberg, and Rasche 2012). In large-scale production of proteins and mAbs, rhizosecretion could represent an interesting solution, also to simplify the following extraction and purification steps. It consists of a continuous production method from roots, obtaining high yields of homogeneous products directly during the plant life cycle. This system works in hydroponic medium, in which it is possible to control the growth media composition, ensuring low cost and efficient production (Madeira et al. 2016).

Among the new platforms for recombinant protein expression, nowadays, plant cell packs (PCPs), which are named also "cookies", ensure scalable and high-throughput screening methods. The PCP method is a versatile expression system, which exploits a rapid infusion with bacterial suspension into plant cell aggregates (Rademacher et al. 2019; Gengenbach, Opdensteinen, and Buyel 2020; Poborilova et al. 2020). In addition, the recent automation of PCP allowed the possibility to have complete platforms, named as "robot cookies", which work from cultivation to infiltration and extraction automatically. This ensures high-throughput and cost-effective methods (Gengenbach, Opdensteinen, and Buyel 2020).

20.7 EXTRACTION TECHNIQUES

Regarding plant-derived proteins, the set-up of the extraction process is mainly correlated to product expression way/strategy and type of plant tissues (whole-plant and *in vitro* cell/tissue cultures). In the case of secreted products, which are directly present in solutions or media, for example, in cell culture supernatant or hydroponic fluids, no specific extraction operation is needed. When an extraction is required, techniques can be chosen based on physical or mechanical principles and different parameters such as temperature, solvent/buffer nature, and addition of denaturants can be set up in relation to the desired protein products, potential plant tissues contaminants (chlorophyll, alkaloids or DNA and RNA), and small- or large-scale processes (Buyel 2015; Buyel, Twyman, and Fischer 2015).

Tissue homogenization is the basic process to extract biopharmaceuticals and can occur exploiting nonmechanical methods with enzymatic lysis, ultrasonication, or French press in presence of cell culture systems (Blanco-Pascual et al. 2014; Bi, Tian, and Row 2011; Wang and Zhang 2012; Buyel, Twyman, and Fischer 2015) and by using mechanical procedures, such as dry or wet milling for seeds, or blade-based homogenizers and/or screw presses for leafy biomass (Kim, Yoo, and Lamsal 2013; Buyel and Fischer 2014b). Mild extraction conditions of non-mechanical approaches are useful only in the case of small-scale investigation, while mechanical processes are necessary for large-scale production, ensuring low-cost platforms in contrast to physical-type extractions.

The addition of aqueous buffer (citrate, phosphate, and carbonate) can be useful during the extraction process to stabilize salt content and pH, facilitating protein solubility and recovery. Furthermore, the presence of additives as antioxidants (ascorbic acid or sodium sulphite/bisulfite) can improve product stability, while reducing agents (β-mercaptoethanol or dithiothreitol) can prevent product aggregation (Buyel, Twyman, and Fischer 2015).

Juice extraction is a recent interesting extraction approach comparable to homogenization and filtration to obtain protein products from biomass with a low consumption of solvent volumes and by consequence low costs. In the past, this method was used for the extraction of small molecules from biomass, but the combination of extraction and solid-liquid separation represents a performing integrated method also to extract proteins and mAbs. A juicer or screw press under an increased pressure induces a continuous homogenization, allowing a process which is able to ensure concentrated products notwithstanding the non-optimal recovery (≤65%) (Buyel and Fischer 2015).

Aqueous two-phase separation (ATPS) is an interesting approach, which combines extraction and purification (see Purification technologies paragraph) based on the addition of salts (citrate and phosphate) and/or polymers (PEG) above critical concentrations. The type and concentration of polymers and composition and pH of buffers can be optimized to obtain the two-phase system, which allows a selective separation of proteins. Polymers are expensive, but the combination of different downstreaming process (DSP) steps, i.e., extraction and purification, allows rapid times, which balance the total costs (Buyel, Twyman, and Fischer 2015).

Regarding proteins production, the extraction process can be complicated and causes a loss of material because of the degradation problems of plant proteases released during the extraction process, but to overcome this disadvantage, the extraction process can be performed on ice or into liquid nitrogen or adding protease inhibitors (Pillay et al. 2014; Clemente et al. 2019). In addition, immediately after the extraction procedure and before the purification step, Western blot technique can be carried out to identify the potential sequences prone to degradation directly in plants and to optimize recovery and costs (Hehle et al. 2015). This is particularly recommended in the presence of new recombinant protein products, as it allows to optimize extraction procedures or with truncated and degraded products without epitope, which cannot be recognized by the antibodies of a classical enzyme-linked immunosorbent assay (ELISA) (Cao et al. 2017).

20.8 PRE-PURIFICATION PROCEDURES

20.8.1 Clarification

Next to the extraction and before the purification step, the early stages of DSP include also the clarification procedure, which consists of an additional pre-treatment before the purification step useful to separate the desired product from insoluble aggregates. In fact, during extraction, contaminants (DNA, RNA, alkaloids, chlorophyll, and soluble proteins) can be released by plant tissues and they must be removed to prevent the presence of interferents, which can precipitate and compromise recovery and costs during the purification process.

Filtration and centrifugation are the two clarification procedures, often used in combination, and among their parameters, the process rate must be carefully set-up as it influences the stability of samples and the potential formation of degradation products due to oxidation, deamidation, and proteolysis (Liu et al. 2010; Buyel 2015; Buyel, Twyman, and Fischer 2015). Micro-filtration, bag, or depth filtration are different models or configurations for filtration with various materials, size rating, and capacity loads but with the same potential scalability. Depth filters are very versatile, ensuring single-use devices able to remove large undesirable particulates or products, such as host cell proteins (HCPs), DNA, and pigments. Centrifugation and ultracentrifugation, which differ for the processing speed and costs, ensure a continuous clarification process, but their scale-up is very difficult, because of the fundamental necessity of cleaning procedures to avoid cross-product/process contaminations (Buyel and Fischer 2014a; Buyel, Twyman, and Fischer 2015).

To improve the efficiency of the clarification step, flocculation can be used before filtration or centrifugation. It consists of an aggregation process carried out by the addition of synthetic charged polymers or natural electrolytes (mainly chitosan and polyphosphate), often used also in combination, which can increase the particle diameter of aggregates, facilitating the precipitation and solid-liquid separation by filters or centrifuges (Buyel and Fischer 2015). Flocculation procedure can be optimized mainly in terms of pH, and polymer type or concentration and can be easily combined with extraction procedures, such as juice extraction or particular filtration systems (depth filters), used during clarification, allowing an increased filtration capacity. It takes about 30 minutes, but low costs remain consistent because of the inexpensive polymers added as flocculants. A disadvantage is the necessity to set up a procedure to remove the added polymers (Kang et al. 2013; Buyel and Fischer 2014a, 2014b; Buyel, Twyman, and Fischer 2015).

20.8.2 Other Procedures

Viscosity of extracts is a key parameter, which can compromise the purification step efficiency. This parameter can be carefully optimized after the clarification process, adding a dilution or concentration step by using membranes or precipitation reagents (ammonium sulphate, PEG, caprylic acid), depending on the extract concentration. For example, during the purification step, in all the chromatographic approaches, a dilution procedure can be necessary to prevent the formation of aggregates and their precipitation, ensuring the presence of soluble products, while by using protein affinity chromatography (AFC) (See Purification technologies paragraph), a concentration step can be added to ensure a higher binding efficiency (Ayyar et al. 2012).

20.9 PURIFICATION TECHNOLOGIES

The purification step in plant-derived pharmaceuticals is particularly complex because of the low concentration of products in plants. Chromatography remains the most efficient approach, but often the presence of HCPs must be resolved before the use of this technique. To eliminate HCPs, precipitation with polymers (PEG) or heat precipitation, which is possible when the products are stable at high temperatures, is suggested in case of small-scale or large-scale processes, respectively. In addition, the use of genetic fusion tags can make more efficient the purification step; they consist in peptides or proteins (as for example oleosins) that can induce a phase separation between HCPs and water-soluble proteins, which are then separated by filtration or centrifugation. Nowadays, this process is well standardized and scalable (Buyel, Twyman, and Fischer 2015).

20.9.1 Chromatographic Methods

The liquid chromatographic (LC) approach remains the most used for the purification of therapeutic plant-based proteins (Buyel, Twyman, and Fischer 2015). Protein A Affinity chromatography is one of the most used techniques, thanks to its versatility, selectivity, and applicability also to complex samples, and notwithstanding the high costs to make it scalable, it is nowadays well standardized. It exploits an adsorption principle based on reversible interactions between a ligand on a solid support and a protein, and this must be carefully optimized. The supports are commercially available in different materials (natural, inorganic, or organic), and formats and ligands must be chosen based on their stability and specificity; the most used ligand to purify mAbs is Protein A, which is a bacteria-derived product, mainly immobilized on agarose (Arora, Saxena, and Ayyar 2017; Liu et al. 2010; Buyel, Twyman, and Fischer 2015). Another important parameter to set up is the pH of the buffer system, which is important to maximize the binding and then elution. Also Protein G, which differs from Protein A in molecular weight, binding sites, and operative pH, can be used for mAbs or their fragments (Hnasko and McGarvey 2015).

For purification of those compounds, as for example secretory immunoglobulin (SIgA) complexes made of polymeric IgA and secretory component, which do not bind Protein A- or Protein G-based matrices, L-sepharose columns with Protein L can represent a robust alternative support (Paul et al. 2014).

Ion-exchange chromatography (IEX) represents an alternative chromatographic approach, which ensures stable products as Protein A AFC, and includes two modes: cation-exchange chromatography (CEX) or anion-exchange chromatography (AEX), in which negative or positive resins are used. MAbs (with neutral/basic pI values) and process-related impurities (with negative charges) are purified by CEX and AEX, respectively. The resins have a capacity which is correlated to sample, and loading flow speed and a careful buffer optimization (pH, conductivity, gradient) can influence the product quantity and purity (Müller-Späth and Morbidelli 2014; Maria et al. 2015).

Hydrophobic interaction chromatography (HIC) can be used orthogonally in combination with Protein A AFC or IEX techniques, exploiting the formation of hydrophobic interactions between proteins and resins with ligands. It is widely applied to separate aggregates (as HCPs and DNA) with IEX. It is based on the use of buffers, in which a regulation of ionic force promotes the interaction and then elution (Liu et al. 2010). In hydrophobic charge induction chromatography (HCIC), the hydrophobic interactions are maximized by using heterocycle ligands and not buffers at high concentrations. One limit is the easy formation of strong non-specific interactions, but its use in combination with IEX can overcome this problem (Liu et al. 2010). Mixed-mode chromatography (MMC), named also multimodal chromatography, can be used in combination or in alternative with Protein A AFC. Resins with different mechanisms (hydrogen, hydrophilic, and hydrophobic interactions) increase the purification step selectivity, and interesting applications exploit MMC resins to separate mAbs and aggregates, simultaneously (Maria et al. 2015). Size exclusion chromatography (SEC) remains the least used chromatographic approach in the purification step due to a difficult scalability related with its low loading capacity. It can separate aggregates with different molecular weights, and it is mainly useful during clarification (Brusotti et al. 2018; Buyel, Twyman, and Fischer 2015; Arora, Saxena, and Ayyar 2017). Recent alternatives to LC in the purification step of biomolecules (peptides and proteins) not yet well explored could be SFC or enhanced fluidity liquid chromatography (EFLC), in which the mobile phase consists mainly of pressurized carbon dioxide or hydro-organic/purely organic solvents and low percentage of pressurized carbon dioxide, respectively. SFC is a green approach, in which pressurized carbon dioxide properties allow often rapid and efficient processes, obtaining intact products, and also a low-cost scalability, thanks to the low solvent consumption in comparison to LC (Schiavone et al. 2019; Govender et al. 2020; Molineau, Hideux, and West 2021).

20.10 ANALYTICAL CHARACTERIZATION PROCEDURES

In the presence of plant-based proteins, characterization is particularly important to ensure their quality, which is strictly correlated to the consistency of protein synthesis (primary sequence and glycosylation profiling), proteolysis degree, stability, and aggregation state. Product- and process-related impurities can compromise the quality of protein products, so specific characterization assays are fundamental during each process development phase to assess their identity, quantity, purity, impurities, potency, and stability. For protein biopharmaceuticals, chromatographic and/or mass spectrometric analytical approaches (LC-electrospray ionization (ESI)-MS/mS and LC-MALDI-MS/MS) are still the election techniques also in the characterization step, thanks to their technological advancements in many parameters (stationary phase morphology, column type, and dimensions, instrumentation performances) (Fekete et al. 2016; Sandra, Vandenheede, and Sandra 2014).

20.10.1 Primary Sequence and Stability

The characterization of protein products is fundamental to assess the identity both at the primary sequence and intact protein levels. To identify the primary sequence, the mainly used approach is peptide mapping by LC-MS/MS, which requires accurate digestion processes with trypsin or other proteases or by using chemical reagents (for example Edman degradation). This allows to localize site-specific proteolytic fragmentations and individuate proteoforms different for number of amino acids and folding (Mouchahoir and Schiel 2018; Hehle et al. 2015). Therefore, the assessment of primary sequence gives important information on protein stability/degradation and consequent protein activity. It allows also to set up some strategic approaches (silencing of genes of the proteases, directly in plant tissues; addition of specific protease inhibitors; mutation of the amino acids prone to the undesired proteolysis), which can protect proteins during their production (Niemer et al. 2014; Mandal et al. 2016; Pillay et al. 2014). In addition, to assess degradation and separate protein fragments, sodium dodecyl sulfate-polyacrylamide gel electrophoresis (SDS-PAGE) under non-reducing conditions, and ELISA can be applied also at the end of the purification process prior to the characterization step (Loos et al. 2014; Hehle et al. 2015). The characterization of protein products is fundamental to assess identity both at the primary sequence and intact protein levels. To identify the primary sequence, the mainly used approach is peptide mapping by LC-MS/MS, which requires accurate digestion processes with trypsin or other proteases or by using chemical reagents (e.g., Edman degradation). This allows to localize site-specific proteolytic fragmentations and individuate proteoforms different for number of amino acids and folding (Mouchahoir and Schiel 2018; Hehle et al. 2015).

20.10.2 Glycosylation Profiling

PTMs as glycosylation in transgenic plants and plant cell cultures are different in comparison to human process. Notwithstanding the existence of bioengineered plants, in which for example, glycosylation can modified to be similar to humans; this problem makes the availability of suitable techniques able to study glycosylation pattern mandatory also to avoid glyco-engineering process. PTMs influence therapeutic efficacy, serum half-life, and potential side effects of protein products, and glycosylation represents the most critical quality attribute of plant-derived products. The first analytical approaches in the analysis of glycosylation profiling date back to 1990s, and over the years, high-resolution mass spectrometry (HRMS), mainly, ESI-Time-of-flight

(TOF) or ESI-Q-TOF, coupled to Reverse Phase LC (RPLC), Capillary Electrophoresis (CE), or more recently, to Hydrophilic Interaction Liquid Chromatography (HILIC) are proved to be the ideal approach to obtain glycosylation profile at the intact protein level, identifying different glycoforms in terms of composition and abundance with accurate mass measurements (Webster and Thomas 2012; Montero-Morales and Steinkellner 2018). Glycosylation position can be evaluated at the glycopeptide level after digestion with trypsin or other proteases. In addition, glycan structure can be characterized by using mainly enzymatic (Peptide N-glycosidase-PNGase F and A) or hydrazinolysis treatments to release N-glycans or O-glycans, respectively, then both can be treated with exoglycosidases, ensuring the possibility to reach the monosaccharide units. Glycans are then analyzed mainly by RPLC- and HILIC-MS or MS/MS mode, and this combination allows the identification (structure) and quantification (accurate mass and relative abundance) of all the released glycans (Tekoah et al. 2004; Yang and Bartlett 2019; Yu et al. 2018; Shubhakar et al. 2015).

Different glycosylation profiles allow to deeply characterize protein products variants and to correlate glycosylation and the potential assembly of the variants. This helps to interfere in the protein assembly to modulate it and its consequent efficacy (Lai et al. 2014; Teh et al. 2014). With these studies, it has been also demonstrated that the degree of glycosylation is not always directly correlated to an uncorrected protein assembly and function (Vamvaka et al. 2016).

20.10.3 Aggregation

The evaluation of aggregation state of plant-based protein products is part of characterization studies, as also the aggregation process can influence protein function. For example, IgM variants are particularly prone, not only to glycosylation but also to aggregation, and plant-derived IgM are characterized by a higher number of hexamers in comparison to mammalian IgM. A series of analytical techniques as SDS-PAGE and SEC can be useful to study aggregation and verify the correct plant-derived oligomerization (Loos et al. 2014).

20.11 CONCLUSION

Drug discovery has an ancient origin, which relies in nature and particularly in plant kingdom. Plants are a rich source of pharmaceutical ingredients and of novel compounds endowed with pharmacological properties *per se*, or after chemical modification. The perceived failure of current drug discovery paradigms to adequately tap into the vast reservoir latent in natural products calls for more efficient and impactful approaches to be devised. Both previous published research as well as insights from industry stakeholders reaffirm this latent potential. Accordingly, in the last decade, the world assisted to a resurgence of scientific interest and research trends in the plant-derived natural product drug discovery. The revived interest of medicinal chemists in natural product drug development is an engine for the developing of innovative techniques useful to reduce the technical limitations associated with natural product development and to solve the main drawbacks associated with PBDD. At the same time, in recent years, the importance of plant-based production systems for recombinant proteins has increased significantly. Nowadays, in fact, many successful platforms for plant-based protein production provided high quality, stability, and yields. For good outcomes, an interdisciplinary approach integrating traditional and ethnopharmacological knowledge, phytochemistry, botany, analytical chemistry, appropriate biological screening methodologies, and current drug development technologies is essential. In conclusion, plants remain a promising pool for the discovery of various bioactive SMs that can be directly developed or used as starting points for optimization into novel drugs, together with the production of plant-based proteins. Thus, we believe that the scientific and technological advances discussed in this chapter provide a strong basis for PDBB to continue making major contributions to human health.

REFERENCES

Alvarez-Rivera, Gerardo, Monica Bueno, Diego Ballesteros-Vivas, Jose A. Mendiola, and Elena Ibañez. 2020. "Pressurized Liquid Extraction." In Colin Poole (ed.), *Liquid-Phase Extraction*, 375–98. Elsevier. https://doi.org/10.1016/B978-0-12-816911-7.00013iX.

Amri, Bédis, Emanuela Martino, Francesca Vitulo, Federica Corana, Leila Bettaieb-Ben Kaâb, Marta Rui, Daniela Rossi, Michela Mori, Silvia Rossi, and Simona Collina. 2017. "Marrubium Vulgare L. Leave Extract: Phytochemical Composition, Antioxidant and Wound Healing Properties." *Molecules* 22 (11): 1851. https://doi.org/10.3390/molecules22111851.

Appendino, Giovanni, Orazio Taglialatela-Scafati, Adriana Romano, Federica Pollastro, Cristina Avonto, and Patrizia Rubiolo. 2009. "Genepolide, a Sesterpene γ-Lactone with a Novel Carbon Skeleton from Mountain Wormwood (Artemisia Umbelliformis)." *Journal of Natural Products* 72 (3): 340–44. https://doi.org/10.1021/np800468m.

Arora, Sushrut, Vikas Saxena, and B. Vijayalakshmi Ayyar. 2017. "Affinity Chromatography: A Versatile Technique for Antibody Purification." *Methods* 116 (March): 84–94. https://doi.org/10.1016/j.ymeth.2016.12.010.

Ashokkumar, Muthupandian. 2011. "The Characterization of Acoustic Cavitation Bubbles – An Overview." *Ultrasonics Sonochemistry* 18 (4): 864–72. https://doi.org/10.1016/j.ultsonch.2010.11.016.

Aydar, Alev Yüksel. 2018. "Utilization of Response Surface Methodology in Optimization of Extraction of Plant Materials." InValter Silva (ed.), *Statistical Approaches with Emphasis on Design of Experiments Applied to Chemical Processes*. InTech. https://doi.org/10.5772/intechopen.73690.

Ayyar, B. Vijayalakshmi, Sushrut Arora, Caroline Murphy, and Richard O'Kennedy. 2012. "Affinity Chromatography as a Tool for Antibody Purification." *Methods* 56 (2): 116–29. https://doi.org/10.1016/j.ymeth.2011.10.007.

Bi, Wentao, Minglei Tian, and Kyung Ho Row. 2011. "Ultrasonication-Assisted Extraction and Preconcentration of Medicinal Products from Herb by Ionic Liquids." *Talanta* 85 (1): 701–6. https://doi.org/10.1016/j.talanta.2011.04.054.

Blanco-Pascual, Nuria, Ailén Alemán, Maria Del Carmine Gómez-Guillén, and Montero P. Montero. 2014. "Enzyme-Assisted Extraction of κ/ι-Hybrid Carrageenan from Mastocarpus Stellatus for Obtaining Bioactive Ingredients and Their Application for Edible Active Film Development." *Food and Function* 5 (2): 319–29. https://doi.org/10.1039/C3FO60310E.

Breemen, Richard B. van, Chao-Ran Huang, Dejan Nikolic, Charles P. Woodbury, Yong-Zhong Zhao, and Duane L. Venton. 1997. "Pulsed Ultrafiltration Mass Spectrometry: A New Method for Screening Combinatorial Libraries." *Analytical Chemistry* 69 (11): 2159–64. https://doi.org/10.1021/ac970132j.

Britton, Emily R., Joshua J. Kellogg, Olav M. Kvalheim, and Nadja B. Cech. 2018. "Biochemometrics to Identify Synergists and Additives from Botanical Medicines: A Case Study with Hydrastis Canadensis (Goldenseal)." *Journal of Natural Products* 81 (3): 484–93. https://doi.org/10.1021/acs.jnatprod.7b00654.

Brusotti, Gloria, Enrica Calleri, Raffaella Colombo, Gabriella Massolini, Francesca Rinaldi, and Caterina Temporini. 2018. "Advances on Size Exclusion Chromatography and Applications on the Analysis of Protein Biopharmaceuticals and Protein Aggregates: A Mini Review." *Chromatographia* 81 (1): 3–23. https://doi.org/10.1007/s10337-017-3380-5.

Bumpus, Stefanie B, Bradley S Evans, Paul M Thomas, Ioanna Ntai, and Neil L Kelleher. 2009. "A Proteomics Approach to Discovering Natural Products and Their Biosynthetic Pathways." *Nature Biotechnology* 27 (10): 951–56. https://doi.org/10.1038/nbt.1565.

Buyel, J. F., and R. Fischer. 2015. "A Juice Extractor Can Simplify the Downstream Processing of Plant-Derived Biopharmaceutical Proteins Compared to Blade-Based Homogenizers." *Process Biochemistry* 50 (5): 859–66. https://doi.org/10.1016/j.procbio.2015.02.017.

Buyel, J. F., R. M. Twyman, and R. Fischer. 2015. "Extraction and Downstream Processing of Plant-Derived Recombinant Proteins." *Biotechnology Advances* 33 (6): 902–13. https://doi.org/10.1016/j.biotechadv.2015.04.010.

Buyel, Johannes. 2015. "Process Development Strategies in Plant Molecular Farming." *Current Pharmaceutical Biotechnology* 16 (11): 966–82. https://doi.org/10.2174/138920101611150902115413.

Buyel, Johannes F., and Rainer Fischer. 2014a. "Flocculation Increases the Efficacy of Depth Filtration during the Downstream Processing of Recombinant Pharmaceutical Proteins Produced in Tobacco." *Plant Biotechnology Journal* 12 (2): 240–52. https://doi.org/10.1111/pbi.12132.

———. 2014b. "Scale-down Models to Optimize a Filter Train for the Downstream Purification of Recombinant Pharmaceutical Proteins Produced in Tobacco Leaves." *Biotechnology Journal* 9 (3): 415–25. https://doi.org/10.1002/biot.201300369.

———. 2015. "Synthetic Polymers Are More Effective than Natural Flocculants for the Clarification of Tobacco Leaf Extracts." *Journal of Biotechnology* 195 (February): 37–42. https://doi.org/10.1016/j.jbiotec.2014.12.018.

Buyel, Johannes Felix, and Rainer Fischer. 2014. "Downstream Processing of Biopharmaceutical Proteins Produced in Plants." *Bioengineered* 5 (2): 138–42. https://doi.org/10.4161/bioe.28061.

Caesar, Lindsay K., and Nadja B. Cech. 2019. "Synergy and Antagonism in Natural Product Extracts: When 1 + 1 Does Not Equal 2." *Natural Product Reports* 36 (6): 869–88. https://doi.org/10.1039/C9NP00011A.

Cao, Zhen, Wei Zhang, Xiangxue Ning, Baomin Wang, Yunjun Liu, and Qing X. Li. 2017. "Development of Monoclonal Antibodies Recognizing Linear Epitope: Illustration by Three Bacillus Thuringiensis Crystal Proteins of Genetically Modified Cotton, Maize, and Tobacco." *Journal of Agricultural and Food Chemistry* 65 (46): 10115–22. https://doi.org/10.1021/acs.jafc.7b03426.

Cavalloro, Valeria, Emanuela Martino, Pasquale Linciano, and Simona Collina. 2021. "Microwave-Assisted Solid Extraction from Natural Matrices." In Gennadiy I. Churyumov (ed.), *Microwave Heating - Electromagnetic Fields Causing Thermal and Non-Thermal Effects*. IntechOpen. https://doi.org/10.5772/intechopen.95440.

Chemat, Farid, Natacha Rombaut, Anne-Gaëlle Sicaire, Alice Meullemiestre, Anne-Sylvie Fabiano-Tixier, and Maryline Abert-Vian. 2017. "Ultrasound Assisted Extraction of Food and Natural Products. Mechanisms, Techniques, Combinations, Protocols and Applications. A Review." *Ultrasonics Sonochemistry* 34 (January): 540–60. https://doi.org/10.1016/j.ultsonch.2016.06.035.

Chen, Gui-Lin, Yong-Qiang Tian, Jian-Lin Wu, Na Li, and Ming-Quan Guo. 2016. "Antiproliferative Activities of Amaryllidaceae Alkaloids from Lycoris Radiata Targeting DNA Topoisomerase I." *Scientific Reports* 6 (1): 38284. https://doi.org/10.1038/srep38284.

Chen, Guilin, and Mingquan Guo. 2017. "Screening for Natural Inhibitors of Topoisomerases I from Rhamnus Davurica by Affinity Ultrafiltration and High-Performance Liquid Chromatography–Mass Spectrometry." *Frontiers in Plant Science* 8 (September). https://doi.org/10.3389/fpls.2017.01521.

Chen, Ya, Christina de Bruyn Kops, and Johannes Kirchmair. 2017. "Data Resources for the Computer-Guided Discovery of Bioactive Natural Products." *Journal of Chemical Information and Modeling* 57 (9): 2099–111. https://doi.org/10.1021/acs.jcim.7b00341.

Chen, Ya, Marina Garcia de Lomana, Nils-Ole Friedrich, and Johannes Kirchmair. 2018. "Characterization of the Chemical Space of Known and Readily Obtainable Natural Products." *Journal of Chemical Information and Modeling* 58 (8): 1518–32. https://doi.org/10.1021/acs.jcim.8b00302.

Chen, Ya, and Johannes Kirchmair. 2020. "Cheminformatics in Natural Product-based Drug Discovery." *Molecular Informatics* 39 (12): 2000171. https://doi.org/10.1002/minf.202000171.

Choi, Yongsoo, Katherine Jermihov, Sang-Jip Nam, Megan Sturdy, Katherine Maloney, Xi Qiu, Lucas R. Chadwick, et al. 2011. "Screening Natural Products for Inhibitors of Quinone Reductase-2 Using Ultrafiltration LC–MS." *Analytical Chemistry* 83 (3): 1048–52. https://doi.org/10.1021/ac1028424.

Chung, Young Hun, Derek Church, Edward C. Koellhoffer, Elizabeth Osota, Sourabh Shukla, Edward P. Rybicki, Jonathan K. Pokorski, and Nicole F. Steinmetz. 2021. "Integrating Plant Molecular Farming and Materials Research for Next-Generation Vaccines." *Nature Reviews Materials*, December. https://doi.org/10.1038/s41578-021-00399-5.

Cieśla, Łukasz, Jakub Kryszeń, Anna Stochmal, Wiesław Oleszek, and Monika Waksmundzka-Hajnos. 2012. "Approach to Develop a Standardized TLC-DPPH Test for Assessing Free Radical Scavenging Properties of Selected Phenolic Compounds." *Journal of Pharmaceutical and Biomedical Analysis* 70 (November): 126–35. https://doi.org/10.1016/j.jpba.2012.06.007.

Citti, Cinzia, Pasquale Linciano, Fabiana Russo, Livio Luongo, Monica Iannotta, Sabatino Maione, Aldo Laganà, et al. 2019. "A Novel Phytocannabinoid Isolated from Cannabis Sativa L. with an in Vivo Cannabimimetic Activity Higher than Δ9-Tetrahydrocannabinol: Δ9-Tetrahydrocannabiphorol." *Scientific Reports* 9 (1): 1–13. https://doi.org/10.1038/s41598-019-56785-1.

Clemente, Marina, Mariana Corigliano, Sebastián Pariani, Edwin Sánchez-López, Valeria Sander, and Víctor Ramos-Duarte. 2019. "Plant Serine Protease Inhibitors: Biotechnology Application in Agriculture and Molecular Farming." *International Journal of Molecular Sciences* 20 (6): 1345. https://doi.org/10.3390/ijms20061345.

Cravotto, Giancarlo, Luisa Boffa, Stefano Mantegna, Patrizia Perego, Milvio Avogadro, and Pedro Cintas. 2008. "Improved Extraction of Vegetable Oils under High-Intensity Ultrasound and/or Microwaves." *Ultrasonics Sonochemistry* 15 (5): 898–902. https://doi.org/10.1016/j.ultsonch.2007.10.009.

Crawford, T. Daniel. 2006. "Ab Initio Calculation of Molecular Chiroptical Properties." *Theoretical Chemistry Accounts* 115 (4): 227–45. https://doi.org/10.1007/s00214-005-0001-4.

Daniell, Henry, Stephen J Streatfield, and Keith Wycoff. 2001. "Medical Molecular Farming: Production of Antibodies, Biopharmaceuticals and Edible Vaccines in Plants." *Trends in Plant Science* 6 (5): 219–26. https://doi.org/10.1016/S1360-1385(01)01922-7.

Das, Kaushik, and Aryadeep Roychoudhury. 2014. "Reactive Oxygen Species (ROS) and Response of Antioxidants as ROS-Scavengers during Environmental Stress in Plants." *Frontiers in Environmental Science* 2 (December). https://doi.org/10.3389/fenvs.2014.00053.

DeGoey, David A., Hui-Ju Chen, Philip B. Cox, and Michael D. Wendt. 2018. "Beyond the Rule of 5: Lessons Learned from AbbVie's Drugs and Compound Collection." *Journal of Medicinal Chemistry* 61 (7): 2636–51. https://doi.org/10.1021/acs.jmedchem.7b00717.

Dettweiler, Micah, Lewis Marquez, Max Bao, and Cassandra L. Quave. 2020. "Quantifying Synergy in the Bioassay-Guided Fractionation of Natural Product Extracts." Edited by Branislav T. Šiler. *PLOS ONE* 15 (8): e0235723. https://doi.org/10.1371/journal.pone.0235723.

Dewanjee, Saikat, Moumita Gangopadhyay, Niloy Bhattacharya, Ritu Khanra, and Tarun K. Dua. 2015. "Bioautography and Its Scope in the Field of Natural Product Chemistry." *Journal of Pharmaceutical Analysis* 5 (2): 75–84. https://doi.org/10.1016/j.jpha.2014.06.002.

Dhama, Kuldeep, Senthilkumar Natesan, Mohd. Iqbal Yatoo, Shailesh Kumar Patel, Ruchi Tiwari, Shailendra K Saxena, and Harapan Harapan. 2020. "Plant-Based Vaccines and Antibodies to Combat COVID-19: Current Status and Prospects." *Human Vaccines & Immunotherapeutics* 16 (12): 2913–20. https://doi.org/10.1080/21645515.2020.1842034.

Dias, Daniel A, Sylvia Urban, and Ute Roessner. 2012. "A Historical Overview of Natural Products in Drug Discovery." *Metabolites* 2 (2): 303–36.

Djenni, Zoubida, Daniella Pingret, Timothy J. Mason, and Farid Chemat. 2013. "Sono–Soxhlet: In Situ Ultrasound-Assisted Extraction of Food Products." *Food Analytical Methods* 6 (4): 1229–33. https://doi.org/10.1007/s12161-012-9531-2.

Dubey, Kashyap Kumar, Garry A Luke, Caroline Knox, Punit Kumar, Brett I Pletschke, Puneet Kumar Singh, and Pratyoosh Shukla. 2018. "Vaccine and Antibody Production in Plants: Developments and Computational Tools." *Briefings in Functional Genomics* 17 (5): 295–307. https://doi.org/10.1093/bfgp/ely020.

Ericsson, Magnus, and Anders Colmsjö. 2000. "Dynamic Microwave-Assisted Extraction." *Journal of Chromatography A* 877 (1–2): 141–51. https://doi.org/10.1016/S0021-9673(00)00246-6.

Fekete, Szabolcs, Davy Guillarme, Pat Sandra, and Koen Sandra. 2016. "Chromatographic, Electrophoretic, and Mass Spectrometric Methods for the Analytical Characterization of Protein Biopharmaceuticals." *Analytical Chemistry* 88 (1): 480–507. https://doi.org/10.1021/acs.analchem.5b04561.

Feng, Juan, Yu-Peng Li, Youtian Hu, Yueyang Zhou, Hua Zhang, and Fang Wu. 2022. "Novel Quinic Acid Glycerates from Tussilago Farfara Inhibit Polypeptide GalNAc-Transferase." *ChemBioChem* 23 (3). https://doi.org/10.1002/cbic.202100539.

Fernandes, Carla. 2019. "Chiral Stationary Phases for Liquid Chromatography : Recent Developments." *Molecules (Basel, Switzerland)* 24 (5): 865. https://doi.org/10.3390/molecules24050865.

Fischer, R. 2003. "Production of Antibodies in Plants and Their Use for Global Health." *Vaccine* 21 (7–8): 820–25. https://doi.org/10.1016/S0264-410X(02)00607-2.

Fischer, Rainer, Eva Stoger, Stefan Schillberg, Paul Christou, and Richard M Twyman. 2004. "Plant-Based Production of Biopharmaceuticals." *Current Opinion in Plant Biology* 7 (2): 152–58. https://doi.org/10.1016/j.pbi.2004.01.007.

Floss, Doreen Manuela, Dieter Falkenburg, and Udo Conrad. 2007. "Production of Vaccines and Therapeutic Antibodies for Veterinary Applications in Transgenic Plants: An Overview." *Transgenic Research* 16 (3): 315–32. https://doi.org/10.1007/s11248-007-9095-x.

Füllbeck, Melanie, Elke Michalsky, Mathias Dunkel, and Robert Preissner. 2006. "Natural Products: Sources and Databases." *Natural Product Reports* 23 (3): 347–56. https://doi.org/10.1039/B513504B.

Gabetta, Bruno, Paolo de Bellis, Roberto Pace, Giovanni Appendino, Luciano Barboni, Elisabetta Torregiani, Pierluigi Gariboldi, and Davide Viterbo. 1995. "10-Deacetylbaccatin III Analogues from Taxus Baccata." *Journal of Natural Products* 58 (10): 1508–14. https://doi.org/10.1021/np50124a005.

Gaggeri, Raffaella, Daniela Rossi, Maria Daglia, Flavio Leoni, Maria Antonia Avanzini, Melissa Mantelli, Markus Juza, and Simona Collina. 2013. "An Eco-Friendly Enantioselective Access to (R)-Naringenin as Inhibitor of Proinflammatory Cytokine Release." *Chemistry & Biodiversity* 10 (8): 1531–38. https://doi.org/10.1002/cbdv.201200227.

García, Benito, Ascensión Torres, and Francisco Macías. 2015. "Synergy and Other Interactions between Polymethoxyflavones from Citrus Byproducts." *Molecules* 20 (11): 20079–106. https://doi.org/10.3390/molecules201119677.

Gaudêncio, Susana P., and Florbela Pereira. 2015. "Dereplication: Racing to Speed up the Natural Products Discovery Process." *Natural Product Reports* 32 (6): 779–810. https://doi.org/10.1039/C4NP00134F.

Gaughan, Christopher L. 2016. "The Present State of the Art in Expression, Production and Characterization of Monoclonal Antibodies." *Molecular Diversity* 20 (1): 255–70. https://doi.org/10.1007/s11030-015-9625-z.

Gengenbach, Benjamin Bruno, Patrick Opdensteinen, and Johannes Felix Buyel. 2020. "Robot Cookies – Plant Cell Packs as an Automated High-Throughput Screening Platform Based on Transient Expression." *Frontiers in Bioengineering and Biotechnology* 8 (May). https://doi.org/10.3389/fbioe.2020.00393.

Ghani, Usman. 2015. "Re-Exploring Promising α-Glucosidase Inhibitors for Potential Development into Oral Anti-Diabetic Drugs: Finding Needle in the Haystack." *European Journal of Medicinal Chemistry* 103 (October): 133–62. https://doi.org/10.1016/j.ejmech.2015.08.043.

Govender, Kamini, Tricia Naicker, Sooraj Baijnath, Anil Amichund Chuturgoon, Naeem Sheik Abdul, Taskeen Docrat, Hendrik Gerhardus Kruger, and Thavendran Govender. 2020. "Sub/Supercritical Fluid Chromatography Employing Water-Rich Modifier Enables the Purification of Biosynthesized Human Insulin." *Journal of Chromatography B* 1155 (October): 122126. https://doi.org/10.1016/j.jchromb.2020.122126.

Griffiths, Andrew D, and Alexander R Duncan. 1998. "Strategies for Selection of Antibodies by Phage Display." *Current Opinion in Biotechnology* 9 (1): 102–8. https://doi.org/10.1016/S0958-1669(98)80092-X.

Gumustas, Mehmet, Sibel A. Ozkan, and Bezhan Chankvetadze. 2018. "Analytical and Preparative Scale Separation of Enantiomers of Chiral Drugs by Chromatography and Related Methods." *Current Medicinal Chemistry* 25 (33): 4152–88. https://doi.org/10.2174/0929867325666180129094955.

Hehle, Verena K., Raffaele Lombardi, Craig J. van Dolleweerd, Mathew J. Paul, Patrizio Di Micco, Veronica Morea, Eugenio Benvenuto, Marcello Donini, and Julian K-C. Ma. 2015. "Site-Specific Proteolytic Degradation of IgG Monoclonal Antibodies Expressed in Tobacco Plants." *Plant Biotechnology Journal* 13 (2): 235–45. https://doi.org/10.1111/pbi.12266.

Hnasko, Robert M., and Jeffery A. McGarvey. 2015. "Affinity Purification of Antibodies." In, 29–41. https://doi.org/10.1007/978-1-4939-2742-5_3.

Hong, Min, Xin-Zhi Wang, Liang Wang, Yong-Qing Hua, Hong-Mei Wen, and Jin-Ao Duan. 2011. "Screening of Immunomodulatory Components in Yu-Ping-Feng-San Using Splenocyte Binding and HPLC." *Journal of Pharmaceutical and Biomedical Analysis* 54 (1): 87–93. https://doi.org/10.1016/j.jpba.2010.08.016.

Hou, Xiaofang, Mingzhe Zhou, Qiao Jiang, Sicen Wang, and Langchong He. 2009. "A Vascular Smooth Muscle/Cell Membrane Chromatography–Offline-Gas Chromatography/Mass Spectrometry Method for Recognition, Separation and Identification of Active Components from Traditional Chinese Medicines." *Journal of Chromatography A* 1216 (42): 7081–87. https://doi.org/10.1016/j.chroma.2009.08.062.

Hou, Xiaorong, Xiaoyi Lou, Qian Guo, Lan Tang, and Weiguang Shan. 2020. "Development of an Immobilized Liposome Chromatography Method for Screening and Characterizing α-Glucosidase-Binding Compounds." *Journal of Chromatography B* 1148 (July): 122097. https://doi.org/10.1016/j.jchromb.2020.122097.

Hu, Shuang, Xuan Chen, Run-Qin Wang, Li Yang, and Xiao-Hong Bai. 2019. "Natural Product Applications of Liquid-Phase Microextraction." *TrAC Trends in Analytical Chemistry* 113 (April): 340–50. https://doi.org/10.1016/j.trac.2018.11.006.

Hubert, Jane, Jean-Marc Nuzillard, and Jean-Hugues Renault. 2017. "Dereplication Strategies in Natural Product Research: How Many Tools and Methodologies behind the Same Concept?" *Phytochemistry Reviews* 16 (1): 55–95. https://doi.org/10.1007/s11101-015-9448-7.

Jesionek, Wioleta, Ágnes M Móricz, Ágnes Alberti, Péter G Ott, Béla Kocsis, Györgyi Horváth, and Irena M Choma. 2015. "TLC-Direct Bioautography as a Bioassay Guided Method for Investigation of Antibacterial Compounds in Hypericum Perforatum L." *Journal of AOAC International* 98 (4): 1013–20. https://doi.org/10.5740/jaoacint.14-233.

Junio, Hiyas A., Arlene A. Sy-Cordero, Keivan A. Ettefagh, Johnna T. Burns, Kathryn T. Micko, Tyler N. Graf, Scott J. Richter, Robert E. Cannon, Nicholas H. Oberlies, and Nadja B. Cech. 2011. "Synergy-Directed Fractionation of Botanical Medicines: A Case Study with Goldenseal (Hydrastis Canadensis)." *Journal of Natural Products* 74 (7): 1621–29. https://doi.org/10.1021/np200336g.

Kandari, Divya, and Rakesh Bhatnagar. 2021. "Antibody Engineering and Its Therapeutic Applications." *International Reviews of Immunology*, August, 1–28. https://doi.org/10.1080/08830185.2021.1960986.

Kang, Chungwon, Soyoun Kim, Euiyeon Lee, Jeahee Ryu, Minhyeong Lee, and Youngeun Kwon. 2021. "Genetically Encoded Sensor Cells for the Screening of Glucocorticoid Receptor (GR) Effectors in Herbal Extracts." *Biosensors* 11 (9): 341. https://doi.org/10.3390/bios11090341.

Kang, Yun (Kenneth), James Hamzik, Michael Felo, Bo Qi, Julia Lee, Stanley Ng, Gregory Liebisch, et al. 2013. "Development of a Novel and Efficient Cell Culture Flocculation Process Using a Stimulus Responsive

Polymer to Streamline Antibody Purification Processes." *Biotechnology and Bioengineering* 110 (11): 2928–37. https://doi.org/10.1002/bit.24969.

Kapoore, Rahul, Thomas Butler, Jagroop Pandhal, and Seetharaman Vaidyanathan. 2018. "Microwave-Assisted Extraction for Microalgae: From Biofuels to Biorefinery." *Biology* 7 (1): 18. https://doi.org/10.3390/biology7010018.

Khadhraoui, B., V. Ummat, B. K. Tiwari, A. S. Fabiano-Tixier, and F. Chemat. 2021. "Review of Ultrasound Combinations with Hybrid and Innovative Techniques for Extraction and Processing of Food and Natural Products." *Ultrasonics Sonochemistry* 76 (August): 105625. https://doi.org/10.1016/j.ultsonch.2021.105625.

Khoo, Leng Wei, Ahmed Mediani, Nur Khaleeda Zulaikha Zolkeflee, Sze Wei Leong, Intan Safinar Ismail, Alfi Khatib, Khozirah Shaari, and Faridah Abas. 2015. "Phytochemical Diversity of Clinacanthus Nutans Extracts and Their Bioactivity Correlations Elucidated by NMR Based Metabolomics." *Phytochemistry Letters* 14 (December): 123–33. https://doi.org/10.1016/j.phytol.2015.09.015.

Kim, Hye Kyong, Young Hae Choi, and Robert Verpoorte. 2010. "NMR-Based Metabolomic Analysis of Plants." *Nature Protocols* 5 (3): 536–49. https://doi.org/10.1038/nprot.2009.237.

Kim, Tae Hyun, Chang Geun Yoo, and B. P. Lamsal. 2013. "Front-End Recovery of Protein from Lignocellulosic Biomass and Its Effects on Chemical Pretreatment and Enzymatic Saccharification." *Bioprocess and Biosystems Engineering* 36 (6): 687–94. https://doi.org/10.1007/s00449-013-0892-8.

Kirchweger, Benjamin, and Judith M. Rollinger. 2018. "Virtual Screening for the Discovery of Active Principles from Natural Products." In Valdir Cechinel Filho (ed.), *Natural Products as Source of Molecules with Therapeutic Potential*, 333–64. Springer International Publishing. https://doi.org/10.1007/978-3-030-00545-0_9.

Kiss, Anton A, Rob Geertman, Matthias Wierschem, Mirko Skiborowski, Bjorn Gielen, Jeroen Jordens, Jinu J John, and Tom Van Gerven. 2018. "Ultrasound-Assisted Emerging Technologies for Chemical Processes." *Journal of Chemical Technology & Biotechnology* 93 (5): 1219–27. https://doi.org/10.1002/jctb.5555.

Ko, Kisung. 2014. "Expression of Recombinant Vaccines and Antibodies in Plants." *Monoclonal Antibodies in Immunodiagnosis and Immunotherapy* 33 (3): 192–98. https://doi.org/10.1089/mab.2014.0049.

Krishnamurti, Chandrasekhar, and SSCChakra Rao. 2016. "The Isolation of Morphine by Serturner." *Indian Journal of Anaesthesia* 60 (11): 861. https://doi.org/10.4103/0019-5049.193696.

Kuhnen, Shirley, Juliana Bernardi Ogliari, Paulo Fernando Dias, Maiara da Silva Santos, Antônio Gilberto Ferreira, Connie C. Bonham, Karl Vernon Wood, and Marcelo Maraschin. 2010. "Metabolic Fingerprint of Brazilian Maize Landraces Silk (Stigma/Styles) Using NMR Spectroscopy and Chemometric Methods." *Journal of Agricultural and Food Chemistry* 58 (4): 2194–200. https://doi.org/10.1021/jf9037776.

Lachance, Hugo, Stefan Wetzel, Kamal Kumar, and Herbert Waldmann. 2012. "Charting, Navigating, and Populating Natural Product Chemical Space for Drug Discovery." *Journal of Medicinal Chemistry* 55 (13): 5989–6001. https://doi.org/10.1021/jm300288g.

Laere, Erna, Anna Pick Kiong Ling, Ying Pei Wong, Rhun Yian Koh, Mohd Azmi Mohd Lila, and Sobri Hussein. 2016. "Plant-Based Vaccines: Production and Challenges." *Journal of Botany* 2016 (April): 1–11. https://doi.org/10.1155/2016/4928637.

Lagassé, H. A. Daniel, Aikaterini Alexaki, Vijaya L. Simhadri, Nobuko H. Katagiri, Wojciech Jankowski, Zuben E. Sauna, and Chava Kimchi-Sarfaty. 2017. "Recent Advances in (Therapeutic Protein) Drug Development." *F1000Research* 6 (February): 113. https://doi.org/10.12688/f1000research.9970.1.

Lai, Huafang, Junyun He, Jonathan Hurtado, Jake Stahnke, Anja Fuchs, Erin Mehlhop, Sergey Gorlatov, Andreas Loos, Michael S. Diamond, and Qiang Chen. 2014. "Structural and Functional Characterization of an Anti-West Nile Virus Monoclonal Antibody and Its Single-Chain Variant Produced in Glycoengineered Plants." *Plant Biotechnology Journal* 12 (8): 1098–1107. https://doi.org/10.1111/pbi.12217.

Lapkin, Alexei A. 2015. "Green Extraction of Artemisinin from Artemisia Annua L." In Farid Chemat and Jochen Strube (eds.), *Green Extraction of Natural Products*, 333–56. Wiley-VCH Verlag GmbH & Co. KGaA. https://doi.org/10.1002/9783527676828.ch10.

Li, Jesse W.-H., and John C. Vederas. 2009. "Drug Discovery and Natural Products: End of an Era or an Endless Frontier?" *Science* 325 (5937): 161–65. https://doi.org/10.1126/science.1168243.

Li, Ying, Anne Sylvie Fabiano-Tixier, Maryline Abert Vian, and Farid Chemat. 2013. "Solvent-Free Microwave Extractionof Bioactive Compounds Providesa Tool for Green Analytical Chemistry." *Trends in Analytical Chemistry* 47: 1–11. https://doi.org/10.1016/j.trac.2013.02.007.

Linciano, Pasquale, Valeria Cavalloro, Emanuela Martino, Johannes Kirchmair, Roberta Listro, Daniela Rossi, and Simona Collina. 2020. "Tackling Antimicrobial Resistance with Small Molecules Targeting LsrK: Challenges and Opportunities." *Journal of Medicinal Chemistry* 63 (24): 15243–57. https://doi.org/10.1021/acs.jmedchem.0c01282.

Linciano, Pasquale, Cinzia Citti, Livio Luongo, Carmela Belardo, Sabatino Maione, Maria Angela Vandelli, Flavio Forni, et al. 2020. "Isolation of a High-Affinity Cannabinoid for the Human CB1 Receptor from a Medicinal Cannabis Sativa Variety: Δ9-Tetrahydrocannabutol, the Butyl Homologue of Δ9-Tetrahydrocannabinol." *Journal of Natural Products* 83 (1): 88–98. https://doi.org/10.1021/acs.jnatprod.9b00876.

Linciano, Pasquale, Rita Nasti, Roberta Listro, Marialaura Amadio, Alessia Pascale, Donatella Potenza, Francesca Vasile, et al. 2022. "Chiral 2-phenyl-3-hydroxypropyl Esters as PKC-alpha Modulators: HPLC Enantioseparation, NMR Absolute Configuration Assignment, and Molecular Docking Studies." *Chirality* 34 (3): 498–513. https://doi.org/10.1002/chir.23406.

Liu, Dongting, Jian Guo, Yan Luo, David J. Broderick, Michael I. Schimerlik, John M. Pezzuto, and Richard B. van Breemen. 2007. "Screening for Ligands of Human Retinoid X Receptor-α Using Ultrafiltration Mass Spectrometry." *Analytical Chemistry* 79 (24): 9398–402. https://doi.org/10.1021/ac701701k.

Liu, Hui F., Junfen Ma, Charles Winter, and Robert Bayer. 2010. "Recovery and Purification Process Development for Monoclonal Antibody Production." *MAbs* 2 (5): 480–99. https://doi.org/10.4161/mabs.2.5.12645.

Łojewska, Ewelina, Tomasz Kowalczyk, Szymon Olejniczak, and Tomasz Sakowicz. 2016. "Extraction and Purification Methods in Downstream Processing of Plant-Based Recombinant Proteins." *Protein Expression and Purification* 120 (April): 110–17. https://doi.org/10.1016/j.pep.2015.12.018.

Loos, Andreas, Clemens Gruber, Friedrich Altmann, Ulrich Mehofer, Frank Hensel, Melanie Grandits, Chris Oostenbrink, Gerhard Stadlmayr, Paul G. Furtmüller, and Herta Steinkellner. 2014. "Expression and Glycoengineering of Functionally Active Heteromultimeric IgM in Plants." *Proceedings of the National Academy of Sciences* 111 (17): 6263–68. https://doi.org/10.1073/pnas.1320544111.

Luque-García, J.L, and M.D Luque de Castro. 2004. "Ultrasound-Assisted Soxhlet Extraction: An Expeditive Approach for Solid Sample Treatment." *Journal of Chromatography A* 1034 (1–2): 237–42. https://doi.org/10.1016/j.chroma.2004.02.020.

Ma, Weina, Cheng Wang, Rui Liu, Nan Wang, Yanni Lv, Bingling Dai, and Langchong He. 2021. "Advances in Cell Membrane Chromatography." *Journal of Chromatography A* 1639 (February): 461916. https://doi.org/10.1016/j.chroma.2021.461916.

Madeira, Luisa M., Tim H. Szeto, Maurice Henquet, Nicole Raven, John Runions, Jon Huddleston, Ian Garrard, Pascal M. W. Drake, and Julian K-C. Ma. 2016. "High-Yield Production of a Human Monoclonal IgG by Rhizosecretion in Hydroponic Tobacco Cultures." *Plant Biotechnology Journal* 14 (2): 615–24. https://doi.org/10.1111/pbi.12407.

Maia, Eduardo Habib Bechelane, Letícia Cristina Assis, Tiago Alves de Oliveira, Alisson Marques da Silva, and Alex Gutterres Taranto. 2020. "Structure-Based Virtual Screening: From Classical to Artificial Intelligence." *Frontiers in Chemistry* 8 (April). https://doi.org/10.3389/fchem.2020.00343.

Malacrida, Alessio, Valeria Cavalloro, Emanuela Martino, Arianna Cassetti, Gabriella Nicolini, Roberta Rigolio, Guido Cavaletti, Barbara Mannucci, Francesca Vasile, Marcello Di Giacomo, et al. 2019. "Anti-Multiple Myeloma Potential of Secondary Metabolites from Hibiscus Sabdariffa." *Molecules* 24 (13): 2500. https://doi.org/10.3390/molecules24132500.

Mandal, Manoj K., Houtan Ahvari, Stefan Schillberg, and Andreas Schiermeyer. 2016. "Tackling Unwanted Proteolysis in Plant Production Hosts Used for Molecular Farming." *Frontiers in Plant Science* 7 (March). https://doi.org/10.3389/fpls.2016.00267.

Maria, Sophie, Gilles Joucla, Bertrand Garbay, Wilfrid Dieryck, Anne-Marie Lomenech, Xavier Santarelli, and Charlotte Cabanne. 2015. "Purification Process of Recombinant Monoclonal Antibodies with Mixed Mode Chromatography." *Journal of Chromatography A* 1393 (May): 57–64. https://doi.org/10.1016/j.chroma.2015.03.018.

Markley, John L, Rafael Brüschweiler, Arthur S Edison, Hamid R Eghbalnia, Robert Powers, Daniel Raftery, and David S Wishart. 2017. "The Future of NMR-Based Metabolomics." *Current Opinion in Biotechnology* 43 (February): 34–40. https://doi.org/10.1016/j.copbio.2016.08.001.

Martínez-Esteso, María J., Ascensión Martínez-Márquez, Susana Sellés-Marchart, Jaime A. Morante-Carriel, and Roque Bru-Martínez. 2015. "The Role of Proteomics in Progressing Insights into Plant Secondary Metabolism." *Frontiers in Plant Science* 6 (July). https://doi.org/10.3389/fpls.2015.00504.

Martino, Emanuela, Giuseppe Casamassima, Sonia Castiglione, Edoardo Cellupica, Serena Pantalone, Francesca Papagni, Marta Rui, Angela Marika Siciliano, and Simona Collina. 2018. "Vinca Alkaloids and Analogues as Anti-Cancer Agents: Looking Back, Peering Ahead." *Bioorganic & Medicinal Chemistry Letters* 28 (17): 2816–26. https://doi.org/10.1016/j.bmcl.2018.06.044.

Martino, Emanuela, Marilù Tarantino, Maddalena Bergamini, Veronica Castelluccio, Adriana Coricello, Marta Falcicchio, Eleonora Lorusso, and Simona Collina. 2019. "Artemisinin and Its Derivatives; Ancient Tradition Inspiring the Latest Therapeutic Approaches against Malaria." *Future Medicinal Chemistry* 11 (12): 1443–59. https://doi.org/10.4155/fmc-2018-0337.

Martino, Emanuela, Serena Della Volpe, Valeria Cavalloro, Bedis Amri, Leila Bettaeib Been Kaab, Giorgio Marrubini, Daniela Rossi, and Simona Collina. 2019. "The Use of a Microwave-assisted Solvent Extraction Coupled with HPLC-UV/PAD to Assess the Quality of *Marrubium Vulgare L.* (White Horehound) Herbal Raw Material." *Phytochemical Analysis* 30 (4): 377–84. https://doi.org/10.1002/pca.2820.

Martino, Emanuela, Serena Della Volpe, Elisa Terribile, Emanuele Benetti, Mirena Sakaj, Adriana Centamore, Andrea Sala, and Simona Collina. 2017. "The Long Story of Camptothecin: From Traditional Medicine to Drugs." *Bioorganic & Medicinal Chemistry Letters* 27 (4): 701–7. https://doi.org/10.1016/j.bmcl.2016.12.085.

Mason, H. S., and M. M. Herbst-Kralovetz. 2012. "Plant-Derived Antigens as Mucosal Vaccines." Current Topics in Microbiology and Immunology 354: 101–20. https://doi.org/10.1007/82_2011_158.

Maulidiani, M., Bassem Y. Sheikh, Ahmed Mediani, Leong Sze Wei, Intan Safinar Ismail, Faridah Abas, and Nordin H. Lajis. 2015. "Differentiation of Nigella Sativa Seeds from Four Different Origins and Their Bioactivity Correlations Based on NMR-Metabolomics Approach." *Phytochemistry Letters* 13 (September): 308–18. https://doi.org/10.1016/j.phytol.2015.07.012.

Medina-Franco, José L. 2020. "Towards a Unified Latin American Natural Products Database: LANaPD." *Future Science OA* 6 (8). https://doi.org/10.2144/fsoa-2020-0068.

Mishra, Sandeep Kumar, and Nagarajarao Suryaprakash. 2017. "Some New Protocols for the Assignment of Absolute Configuration by NMR Spectroscopy Using Chiral Solvating Agents and CDAs." *Tetrahedron: Asymmetry* 28 (10): 1220–32. https://doi.org/10.1016/j.tetasy.2017.09.017.

Molineau, Jérémy, Maria Hideux, and Caroline West. 2021. "Chromatographic Analysis of Biomolecules with Pressurized Carbon Dioxide Mobile Phases – A Review." *Journal of Pharmaceutical and Biomedical Analysis* 193 (January): 113736. https://doi.org/10.1016/j.jpba.2020.113736.

Montero-Morales, Laura, and Herta Steinkellner. 2018. "Advanced Plant-Based Glycan Engineering." *Frontiers in Bioengineering and Biotechnology* 6 (June). https://doi.org/10.3389/fbioe.2018.00081.

Mouchahoir, Trina, and John E. Schiel. 2018. "Development of an LC-MS/MS Peptide Mapping Protocol for the NISTmAb." *Analytical and Bioanalytical Chemistry* 410 (8): 2111–26. https://doi.org/10.1007/s00216-018-0848-6.

Müller-Späth, Thomas, and Massimo Morbidelli. 2014. "Purification of Human Monoclonal Antibodies and Their Fragments." *Methods in Molecular Biology* 1060: 331–51. https://doi.org/10.1007/978-1-62703-586-6_17.

Newman, David J., and Gordon M. Cragg. 2020. "Natural Products as Sources of New Drugs over the Nearly Four Decades from 01/1981 to 09/2019." *Journal of Natural Products* 83 (3): 770–803. https://doi.org/10.1021/acs.jnatprod.9b01285.

Niemer, Melanie, Ulrich Mehofer, Juan Antonio Torres Acosta, Maria Verdianz, Theresa Henkel, Andreas Loos, Richard Strasser, et al. 2014. "The Human Anti-HIV Antibodies 2F5, 2G12, and PG9 Differ in Their Susceptibility to Proteolytic Degradation: Down-regulation of Endogenous Serine and Cysteine Proteinase Activities Could Improve Antibody Production in Plant-based Expression Platforms." *Biotechnology Journal* 9 (4): 493–500. https://doi.org/10.1002/biot.201300207.

Nikolic, Dejan, Sohrab Habibi-Goudarzi, David G. Corley, Stefan Gafner, John M. Pezzuto, and Richard B. van Breemen. 2000. "Evaluation of Cyclooxygenase-2 Inhibitors Using Pulsed Ultrafiltration Mass Spectrometry." *Analytical Chemistry* 72 (16): 3853–59. https://doi.org/10.1021/ac0000980.

Pasquet, Virginie, Jean-René Chérouvrier, Firas Farhat, Valérie Thiéry, Jean-Marie Piot, Jean-Baptiste Bérard, Raymond Kaas, et al. 2011. "Study on the Microalgal Pigments Extraction Process: Performance of Microwave Assisted Extraction." *Process Biochemistry* 46 (1): 59–67. https://doi.org/10.1016/j.procbio.2010.07.009.

Paul, Matthew, Rajko Reljic, Katja Klein, Pascal MW Drake, Craig van Dolleweerd, Martin Pabst, Markus Windwarder, et al. 2014. "Characterization of a Plant-Produced Recombinant Human Secretory IgA with Broad Neutralizing Activity against HIV." *MABs* 6 (6): 1585–97. https://doi.org/10.4161/mabs.36336.

Peeters, Koen, Chris De Wilde, Geert De Jaeger, Geert Angenon, and Ann Depicker. 2001. "Production of Antibodies and Antibody Fragments in Plants." *Vaccine* 19 (17–19): 2756–61. https://doi.org/10.1016/S0264-410X(00)00514-4.

Pellavio, Giorgia, Marta Rui, Laura Caliogna, Emanuela Martino, Giulia Gastaldi, Simona Collina, and Umberto Laforenza. 2017. "Regulation of Aquaporin Functional Properties Mediated by the Antioxidant Effects of Natural Compounds." *International Journal of Molecular Sciences* 18 (12). https://doi.org/10.3390/ijms18122665.

Picot-Allain, Carene, Mohamad Fawzi Mahomoodally, Gunes AK, and Gokhan Zengin. 2021. "Conventional versus Green Extraction Techniques — a Comparative Perspective." *Current Opinion in Food Science* 40 (August): 144–56. https://doi.org/10.1016/j.cofs.2021.02.009.

Pillay, Priyen, Urte Schlüter, Stefan van Wyk, Karl Josef Kunert, and Barend Juan Vorster. 2014. "Proteolysis of Recombinant Proteins in Bioengineered Plant Cells." *Bioengineered* 5 (1): 15–20. https://doi.org/10.4161/bioe.25158.

Pinelo, Manuel, Anis Arnous, and Anne S. Meyer. 2006. "Upgrading of Grape Skins: Significance of Plant Cell-Wall Structural Components and Extraction Techniques for Phenol Release." *Trends in Food Science & Technology* 17 (11): 579–90. https://doi.org/10.1016/j.tifs.2006.05.003.

Piotrzkowski, Natalia, Stefan Schillberg, and Stefan Rasche. 2012. "Tackling Heterogeneity: A Leaf Disc-Based Assay for the High-Throughput Screening of Transient Gene Expression in Tobacco." Edited by Peter Meyer. *PLoS ONE* 7 (9): e45803. https://doi.org/10.1371/journal.pone.0045803.

Poborilova, Zuzana, Helena Plchova, Noemi Cerovska, Cornelius J. Gunter, Inga I. Hitzeroth, Edward P. Rybicki, and Tomas Moravec. 2020. "Transient Protein Expression in Tobacco BY-2 Plant Cell Packs Using Single and Multi-Cassette Replicating Vectors." *Plant Cell Reports* 39 (9): 1115–27. https://doi.org/10.1007/s00299-020-02544-w.

Puri, Munish, Deepika Sharma, and Colin J. Barrow. 2012. "Enzyme-Assisted Extraction of Bioactives from Plants." *Trends in Biotechnology* 30 (1): 37–44. https://doi.org/10.1016/j.tibtech.2011.06.014.

Qiu, Jing-ying, Xu Chen, Xiao-xiao Zheng, Xiang-lan Jiang, Dong-zhi Yang, Yan-yan Yu, Qian Du, Dao-quan Tang, and Xiao-xing Yin. 2015. "Target Cell Extraction Coupled with LC-MS/MS Analysis for Screening Potential Bioactive Components in Ginkgo Biloba Extract with Preventive Effect against Diabetic Nephropathy." *Biomedical Chromatography* 29 (2): 226–32. https://doi.org/10.1002/bmc.3264.

Rademacher, Thomas, Markus Sack, Daniel Blessing, Rainer Fischer, Tanja Holland, and Johannes Buyel. 2019. "Plant Cell Packs: A Scalable Platform for Recombinant Protein Production and Metabolic Engineering." *Plant Biotechnology Journal* 17 (8): 1560–66. https://doi.org/10.1111/pbi.13081.

Reynolds, W. F. 2017. "Natural Product Structure Elucidation by NMR Spectroscopy." In Simone Badal and Rupika Delgoda (eds.), *Pharmacognosy: Fundamentals, Applications and Strategies*, 567–96. Elsevier. https://doi.org/10.1016/B978-0-12-802104-0.00029-9.

Rossi, Daniela, Rita Nasti, Annamaria Marra, Silvia Meneghini, Giuseppe Mazzeo, Giovanna Longhi, Maurizio Memo, et al. 2016. "Enantiomeric 4-Acylamino-6-Alkyloxy-2 Alkylthiopyrimidines As Potential A 3 Adenosine Receptor Antagonists: HPLC Chiral Resolution and Absolute Configuration Assignment by a Full Set of Chiroptical Spectroscopy." *Chirality* 28 (5): 434–40. https://doi.org/10.1002/chir.22599.

Salam, Akram M., James T. Lyles, and Cassandra L. Quave. 2019. "Methods in the Extraction and Chemical Analysis of Medicinal Plants." *Methods and Techniques in Ethnobiology and Ethnoecology*. Springer Protocols Handbooks. New York, NY: Humana Press, 257–83. https://doi.org/10.1007/978-1-4939-8919-5_17.

Sandra, Koen, Isabel Vandenheede, and Pat Sandra. 2014. "Modern Chromatographic and Mass Spectrometric Techniques for Protein Biopharmaceutical Characterization." *Journal of Chromatography A* 1335 (March): 81–103. https://doi.org/10.1016/j.chroma.2013.11.057.

Santana, Kauê, Lidiane Diniz do Nascimento, Anderson Lima e Lima, Vinícius Damasceno, Claudio Nahum, Rodolpho C. Braga, and Jerônimo Lameira. 2021. "Applications of Virtual Screening in Bioprospecting: Facts, Shifts, and Perspectives to Explore the Chemo-Structural Diversity of Natural Products." *Frontiers in Chemistry* 9 (April). https://doi.org/10.3389/fchem.2021.662688.

Schiavone, Nicole M., Raffeal Bennett, Michael B. Hicks, Gregory F. Pirrone, Erik L. Regalado, Ian Mangion, and Alexey A. Makarov. 2019. "Evaluation of Global Conformational Changes in Peptides and Proteins Following Purification by Supercritical Fluid Chromatography." *Journal of Chromatography B* 1110–1111 (March): 94–100. https://doi.org/10.1016/j.jchromb.2019.02.012.

Schmitz, R. 1985. "Friedrich Wilhelm Sertürner and the Discovery of Morphine." *Pharmacy in History* 27 (2): 61–74.

Schripsema, Jan, and Denise Dagnino. 2014. "Metabolomics: Experimental Design, Methodology and Data Analysis." In Robert A. Meyers (ed.), *Encyclopedia of Analytical Chemistry*, 1–17. John Wiley & Sons, Ltd. https://doi.org/10.1002/9780470027318.a9939.

Seco, José Manuel, Emilio Quiñoá, and Ricardo Riguera. 2004. "The Assignment of Absolute Configuration by NMR." *Chemical Reviews* 104 (1): 17–118. https://doi.org/10.1021/cr000665j.

Sheng, Liang-Hong, Song-Lin Li, Liang Kong, Xue-Guo Chen, Xi-Qin Mao, Xing-Ye Su, Han-Fa Zou, and Ping Li. 2005. "Separation of Compounds Interacting with Liposome Membrane in Combined Prescription of Traditional Chinese Medicines with Immobilized Liposome Chromatography." *Journal of Pharmaceutical and Biomedical Analysis* 38 (2): 216–24. https://doi.org/10.1016/j.jpba.2005.01.008.

Sheshukova, E. V., T. V. Komarova, and Y. L. Dorokhov. 2016. "Plant Factories for the Production of Monoclonal Antibodies." *Biochemistry (Moscow)* 81 (10): 1118–35. https://doi.org/10.1134/S0006297916100102.

Shubhakar, Archana, Karli R. Reiding, Richard A. Gardner, Daniel I. R. Spencer, Daryl L. Fernandes, and Manfred Wuhrer. 2015. "High-Throughput Analysis and Automation for Glycomics Studies." *Chromatographia* 78 (5–6): 321–33. https://doi.org/10.1007/s10337-014-2803-9.

Sklirou, Aimilia D., Maria T. Angelopoulou, Aikaterini Argyropoulou, Eliza Chaita, Vasiliki Ioanna Boka, Christina Cheimonidi, Katerina Niforou, et al. 2021. "Phytochemical Study and In Vitro Screening Focusing on the Anti-Aging Features of Various Plants of the Greek Flora." *Antioxidants* 10 (8): 1206. https://doi.org/10.3390/antiox10081206.

Sohrab, Sayed Sartaj, Mohd. Suhail, Mohammad A. Kamal, Azamal Husen, and Esam I. Azhar. 2017. "Recent Development and Future Prospects of Plant-Based Vaccines." *Current Drug Metabolism* 18 (9). https://doi.org/10.2174/1389200218666170711121810.

Soria-Guerra, Ruth E., Ricardo Nieto-Gomez, Dania O. Govea-Alonso, and Sergio Rosales-Mendoza. 2015. "An Overview of Bioinformatics Tools for Epitope Prediction: Implications on Vaccine Development." *Journal of Biomedical Informatics* 53 (February): 405–14. https://doi.org/10.1016/j.jbi.2014.11.003.

Sorokina, Maria, and Christoph Steinbeck. 2020. "Review on Natural Products Databases: Where to Find Data in 2020." *Journal of Cheminformatics* 12 (1): 20. https://doi.org/10.1186/s13321-020-00424-9.

Sowbhagya, H. B., and V. N. Chitra. 2010. "Enzyme-Assisted Extraction of Flavorings and Colorants from Plant Materials." *Critical Reviews in Food Science and Nutrition* 50 (2): 146–61. https://doi.org/10.1080/10408390802248775.

Stephen, Cayman, Abdelfatteh El Omri, and Lukasz M. Ciesla. 2018. "Cellular Membrane Affinity Chromatography (CMAC) in Drug Discovery from Complex Natural Matrices." *ADMET and DMPK* 6 (3): 200–214. https://doi.org/10.5599/admet.535.

Streimikyte, Paulina, Pranas Viskelis, and Jonas Viskelis. 2022. "Enzymes-Assisted Extraction of Plants for Sustainable and Functional Applications." *International Journal of Molecular Sciences* 23 (4): 2359. https://doi.org/10.3390/ijms23042359.

Sun, Min, Limei Huang, Jianliang Zhu, Wenjie Bu, Jing Sun, and Zhaohui Fang. 2015. "Screening Nephroprotective Compounds from Cortex Moutan by Mesangial Cell Extraction and UPLC." *Archives of Pharmacal Research* 38 (6): 1044–53. https://doi.org/10.1007/s12272-014-0469-3.

Tang, Ping, Shihui Si, and Liangliang Liu. 2015. "Analysis of Bovine Serum Albumin Ligands from Puerariae Flos Using Ultrafiltration Combined with HPLC-MS." *Journal of Chemistry* 2015: 1–6. https://doi.org/10.1155/2015/648361.

Teh, Audrey Y.-H., Daniel Maresch, Katja Klein, and Julian K.-C. Ma. 2014. "Characterization of VRC 01, a Potent and Broadly Neutralizing Anti- HIV m A b, Produced in Transiently and Stably Transformed Tobacco." *Plant Biotechnology Journal* 12 (3): 300–11. https://doi.org/10.1111/pbi.12137.

Tekoah, Yoram, Kisung Ko, Hilary Koprowski, David J Harvey, Mark R Wormald, Raymond A Dwek, and Pauline M Rudd. 2004. "Controlled Glycosylation of Therapeutic Antibodies in Plants." *Archives of Biochemistry and Biophysics* 426 (2): 266–78. https://doi.org/10.1016/j.abb.2004.02.034.

Temporini, Caterina, Raffaella Colombo, Enrica Calleri, Sara Tengattini, Francesca Rinaldi, and Gabriella Massolini. 2020. "Chromatographic Tools for Plant Derived Recombinant Antibodies Purification and Characterization." *Journal of Pharmaceutical and Biomedical Analysis* 179 (February): 112920. https://doi.org/10.1016/j.jpba.2019.112920.

Tokuhara, Daisuke. 2018. "Challenges in Developing Mucosal Vaccines and Antibodies against Infectious Diarrhea in Children." *Pediatrics International* 60 (3): 214–23. https://doi.org/10.1111/ped.13497.

Tremouillaux-Guiller, Jocelyne, Khaled Moustafa, Kathleen Hefferon, Goabaone Gaobotse, and Abdullah Makhzoum. 2020. "Plant-Made HIV Vaccines and Potential Candidates." *Current Opinion in Biotechnology* 61 (February): 209–16. https://doi.org/10.1016/j.copbio.2020.01.004.

Tripathy, Satyajit, Barsha Dassarma, Manojit Bhattacharya, and Motlalepula Gilbert Matsabisa. 2021. "Plant-Based Vaccine Research Development against Viral Diseases with Emphasis on Ebola Virus Disease: A Review Study." *Current Opinion in Pharmacology* 60 (October): 261–67. https://doi.org/10.1016/j.coph.2021.08.001.

Urbain, Aurélie, and Claudia Avello Simões-Pires. 2020. "Thin-Layer Chromatography for the Detection and Analysis of Bioactive Natural Products." In Robert A. Meyers (ed.) *Encyclopedia of Analytical Chemistry*, 1–29. Wiley. https://doi.org/10.1002/9780470027318.a9907.pub2.

Uwineza, Pascaline Aimee, and Agnieszka Waśkiewicz. 2020. "Recent Advances in Supercritical Fluid Extraction of Natural Bioactive Compounds from Natural Plant Materials." *Molecules* 25 (17): 3847. https://doi.org/10.3390/molecules25173847.

Vamvaka, Evangelia, Richard M. Twyman, Andre Melro Murad, Stanislav Melnik, Audrey Yi-Hui Teh, Elsa Arcalis, Friedrich Altmann, et al. 2016. "Rice Endosperm Produces an Underglycosylated and Potent

Form of the HIV-Neutralizing Monoclonal Antibody 2G12." *Plant Biotechnology Journal* 14 (1): 97–108. https://doi.org/10.1111/pbi.12360.

Wagner, Hildebert, and Sabine Bladt. 1996. *Plant Drug Analysis*. Springer Berlin Heidelberg. https://doi.org/10.1007/978-3-642-00574-9.

Wang, Bochu, Jia Deng, Yimeng Gao, Liancan Zhu, Rui He, and Yingqian Xu. 2011. "The Screening Toolbox of Bioactive Substances from Natural Products: A Review." *Fitoterapia* 82 (8): 1141–51. https://doi.org/10.1016/j.fitote.2011.08.007.

Wang, Lan, Jing Ren, Meng Sun, and Sicen Wang. 2010. "A Combined Cell Membrane Chromatography and Online HPLC/MS Method for Screening Compounds from Radix Caulophylli Acting on the Human A1A-Adrenoceptor." *Journal of Pharmaceutical and Biomedical Analysis* 51 (5): 1032–36. https://doi.org/10.1016/j.jpba.2009.11.007.

Wang, Lu, Shu Liu, Junpeng Xing, Zhiqiang Liu, and Fengrui Song. 2016. "Characterization of Interaction Property of Multi-Components in Gardenia Jasminoides with Aldose Reductase by Microdialysis Combined with Liquid Chromatography Coupled to Mass Spectrometry." *Rapid Communications in Mass Spectrometry* 30 (August): 87–94. https://doi.org/10.1002/rcm.7620.

Wang, Lvhuan, Yumei Zhao, Yanyan Zhang, Tingting Zhang, Jeroen Kool, Govert W. Somsen, Qiqin Wang, and Zhengjin Jiang. 2018. "Online Screening of Acetylcholinesterase Inhibitors in Natural Products Using Monolith-Based Immobilized Capillary Enzyme Reactors Combined with Liquid Chromatography-Mass Spectrometry." *Journal of Chromatography A* 1563 (August): 135–43. https://doi.org/10.1016/j.chroma.2018.05.069.

Wang, Mingxun, Jeremy J Carver, Vanessa V Phelan, Laura M Sanchez, Neha Garg, Yao Peng, Don Duy Nguyen, et al. 2016. "Sharing and Community Curation of Mass Spectrometry Data with Global Natural Products Social Molecular Networking." *Nature Biotechnology* 34 (8): 828–37. https://doi.org/10.1038/nbt.3597.

Wang, Sicen, Meng Sun, Yanmin Zhang, Hui Du, and Langchong He. 2010. "A New A431/Cell Membrane Chromatography and Online High Performance Liquid Chromatography/Mass Spectrometry Method for Screening Epidermal Growth Factor Receptor Antagonists from Radix Sophorae Flavescentis." *Journal of Chromatography A* 1217 (32): 5246–52. https://doi.org/10.1016/j.chroma.2010.06.037.

Wang, Xiaoqin, and Xuewu Zhang. 2012. "Optimal Extraction and Hydrolysis of Chlorella Pyrenoidosa Proteins." *Bioresource Technology* 126 (December): 307–13. https://doi.org/10.1016/j.biortech.2012.09.059.

Ward, Brian J, Alexander Makarkov, Annie Séguin, Stéphane Pillet, Sonia Trépanier, Jiwanjeet Dhaliwall, Michael D Libman, Timo Vesikari, and Nathalie Landry. 2020. "Efficacy, Immunogenicity, and Safety of a Plant-Derived, Quadrivalent, Virus-like Particle Influenza Vaccine in Adults (18–64 Years) and Older Adults (≥65 Years): Two Multicentre, Randomised Phase 3 Trials." *The Lancet* 396 (10261): 1491–1503. https://doi.org/10.1016/S0140-6736(20)32014-6.

Ward, Timothy J., and Karen D. Ward. 2012. "Chiral Separations: A Review of Current Topics and Trends." *Analytical Chemistry* 84 (2): 626–35. https://doi.org/10.1021/ac202892w.

Webster, Diane E., and Merlin C. Thomas. 2012. "Post-Translational Modification of Plant-Made Foreign Proteins; Glycosylation and Beyond." *Biotechnology Advances* 30 (2): 410–18. https://doi.org/10.1016/j.biotechadv.2011.07.015.

Wolfender, Jean-Luc, Guillaume Marti, Aurélien Thomas, and Samuel Bertrand. 2015. "Current Approaches and Challenges for the Metabolite Profiling of Complex Natural Extracts." *Journal of Chromatography A* 1382 (February): 136–64. https://doi.org/10.1016/j.chroma.2014.10.091.

Xiao, Jiao, Gang Chen, and Ning Li. 2018. "Ionic Liquid Solutions as a Green Tool for the Extraction and Isolation of Natural Products." *Molecules* 23 (7): 1765. https://doi.org/10.3390/molecules23071765.

Xiao, Xiao-Hua, Jun-Xia Wang, Gang Wang, Jia-Yue Wang, and Gong-Ke Li. 2009. "Evaluation of Vacuum Microwave-Assisted Extraction Technique for the Extraction of Antioxidants from Plant Samples." *Journal of Chromatography A* 1216 (51): 8867–73. https://doi.org/10.1016/j.chroma.2009.10.087.

Yan, Yunyan, Xuan Chen, Shuang Hu, and Xiaohong Bai. 2014. "Applications of Liquid-Phase Microextraction Techniques in Natural Product Analysis: A Review." *Journal of Chromatography A* 1368 (November): 1–17. https://doi.org/10.1016/j.chroma.2014.09.068.

Yang, Jian, Yuan Li, Jiaxing Li, Jie Yuan, Sheng Wang, Liangyun Zhou, Li Zhou, Chuanzhi Kang, and Lanping Guo. 2021. "High-throughput Screening of Secondary Metabolites by Sorbus Pohuashanensis Cells under Environmental Stress Using UHPLC-QTOF Combined with AntDAS." *Physiologia Plantarum* 173 (4): 2216–25. https://doi.org/10.1111/ppl.13572.

Yang, Lei, Han Wang, Yuan-gang Zu, Chunjian Zhao, Lin Zhang, Xiaoqiang Chen, and Zhonghua Zhang. 2011. "Ultrasound-Assisted Extraction of the Three Terpenoid Indole Alkaloids Vindoline, Catharanthine and Vinblastine from Catharanthus Roseus Using Ionic Liquid Aqueous Solutions." *Chemical Engineering Journal* 172 (2–3): 705–12. https://doi.org/10.1016/j.cej.2011.06.039.

Yang, Xiangkun, and Michael G. Bartlett. 2019. "Glycan Analysis for Protein Therapeutics." *Journal of Chromatography B* 1120 (July): 29–40. https://doi.org/10.1016/j.jchromb.2019.04.031.

Yang, Xingxin, Yanwei Wang, Xiaoxia Zhang, Ruimiao Chang, and Xiaoni Li. 2011. "Screening Vasoconstriction Inhibitors from Traditional Chinese Medicines Using a Vascular Smooth Muscle/Cell Membrane Chromatography-Offline-Liquid Chromatography-Mass Spectrometry." *Journal of Separation Science* 34 (19): 2586–93. https://doi.org/10.1002/jssc.201100366.

Yongye, Austin B., Jacob Waddell, and José L. Medina-Franco. 2012. "Molecular Scaffold Analysis of Natural Products Databases in the Public Domain." *Chemical Biology & Drug Design* 80 (5): 717–24. https://doi.org/10.1111/cbdd.12011.

Yu, Aiying, Jingfu Zhao, Wenjing Peng, Alireza Banazadeh, Seth D. Williamson, Mona Goli, Yifan Huang, and Yehia Mechref. 2018. "Advances in Mass Spectrometry-based Glycoproteomics." *Electrophoresis* 39 (24): 3104–22. https://doi.org/10.1002/elps.201800272.

Zani, Carlos L., and Anthony R. Carroll. 2017. "Database for Rapid Dereplication of Known Natural Products Using Data from MS and Fast NMR Experiments." *Journal of Natural Products* 80 (6): 1758–66. https://doi.org/10.1021/acs.jnatprod.6b01093.

Zhang, Aihua, Hui Sun, Ping Wang, Ying Han, and Xijun Wang. 2012. "Modern Analytical Techniques in Metabolomics Analysis." *The Analyst* 137 (2): 293–300. https://doi.org/10.1039/C1AN15605E.

Zhang, Cong, Jian Li, Li Xu, and Zhi-Guo Shi. 2012. "Fast Immobilized Liposome Chromatography Based on Penetrable Silica Microspheres for Screening and Analysis of Permeable Compounds." *Journal of Chromatography A* 1233 (April): 78–84. https://doi.org/10.1016/j.chroma.2012.02.013.

Zhang, Hong-Wei, Chao Lv, Li-Jun Zhang, Xin Guo, Yi-Wen Shen, Dale G. Nagle, Yu-Dong Zhou, San-Hong Liu, Wei-Dong Zhang, and Xin Luan. 2021. "Application of Omics- and Multi-Omics-Based Techniques for Natural Product Target Discovery." *Biomedicine & Pharmacotherapy* 141 (September): 111833. https://doi.org/10.1016/j.biopha.2021.111833.

Zhang, Wenping, Jin Sun, Yongjun Wang, Xiaohong Liu, Yinghua Sun, Rong Lu, and Zhonggui He. 2009. "Screening and Identification of Permeable Components of Radix et Rhizoma Rhei Extract by Use of Immobilized Artificial Membrane Chromatography." *Chromatographia* 70 (9–10): 1321–26. https://doi.org/10.1365/s10337-009-1342-2.

Zhao, Qi, Jia-Le Zhang, and Fei Li. 2018. "Application of Metabolomics in the Study of Natural Products." *Natural Products and Bioprospecting* 8 (4): 321–34. https://doi.org/10.1007/s13659-018-0175-9.

Index

Note: **Bold** page numbers refer to tables and *italic* page numbers refer to figures.